DATA CLUSTERING

Algorithms and Applications

Chapman & Hall/CRC
Data Mining and Knowledge Discovery Series

SERIES EDITOR

Vipin Kumar

University of Minnesota
Department of Computer Science and Engineering
Minneapolis, Minnesota, U.S.A.

AIMS AND SCOPE

This series aims to capture new developments and applications in data mining and knowledge discovery, while summarizing the computational tools and techniques useful in data analysis. This series encourages the integration of mathematical, statistical, and computational methods and techniques through the publication of a broad range of textbooks, reference works, and handbooks. The inclusion of concrete examples and applications is highly encouraged. The scope of the series includes, but is not limited to, titles in the areas of data mining and knowledge discovery methods and applications, modeling, algorithms, theory and foundations, data and knowledge visualization, data mining systems and tools, and privacy and security issues.

PUBLISHED TITLES

ADVANCES IN MACHINE LEARNING AND DATA MINING FOR ASTRONOMY
Michael J. Way, Jeffrey D. Scargle, Kamal M. Ali, and Ashok N. Srivastava

BIOLOGICAL DATA MINING
Jake Y. Chen and Stefano Lonardi

COMPUTATIONAL INTELLIGENT DATA ANALYSIS FOR SUSTAINABLE DEVELOPMENT
Ting Yu, Nitesh V. Chawla, and Simeon Simoff

COMPUTATIONAL METHODS OF FEATURE SELECTION
Huan Liu and Hiroshi Motoda

CONSTRAINED CLUSTERING: ADVANCES IN ALGORITHMS, THEORY, AND APPLICATIONS
Sugato Basu, Ian Davidson, and Kiri L. Wagstaff

CONTRAST DATA MINING: CONCEPTS, ALGORITHMS, AND APPLICATIONS
Guozhu Dong and James Bailey

DATA CLUSTERING: ALGORITHMS AND APPLICATIONS
Charu C. Aggarawal and Chandan K. Reddy

DATA CLUSTERING IN C++: AN OBJECT-ORIENTED APPROACH
Guojun Gan

DATA MINING FOR DESIGN AND MARKETING
Yukio Ohsawa and Katsutoshi Yada

DATA MINING WITH R: LEARNING WITH CASE STUDIES
Luís Torgo

FOUNDATIONS OF PREDICTIVE ANALYTICS
James Wu and Stephen Coggeshall

GEOGRAPHIC DATA MINING AND KNOWLEDGE DISCOVERY, SECOND EDITION
Harvey J. Miller and Jiawei Han

HANDBOOK OF EDUCATIONAL DATA MINING
Cristóbal Romero, Sebastian Ventura, Mykola Pechenizkiy, and Ryan S.J.d. Baker

INFORMATION DISCOVERY ON ELECTRONIC HEALTH RECORDS
Vagelis Hristidis

INTELLIGENT TECHNOLOGIES FOR WEB APPLICATIONS
Priti Srinivas Sajja and Rajendra Akerkar

INTRODUCTION TO PRIVACY-PRESERVING DATA PUBLISHING: CONCEPTS AND TECHNIQUES
Benjamin C. M. Fung, Ke Wang, Ada Wai-Chee Fu, and Philip S. Yu

KNOWLEDGE DISCOVERY FOR COUNTERTERRORISM AND LAW ENFORCEMENT
David Skillicorn

KNOWLEDGE DISCOVERY FROM DATA STREAMS
João Gama

MACHINE LEARNING AND KNOWLEDGE DISCOVERY FOR ENGINEERING SYSTEMS HEALTH MANAGEMENT
Ashok N. Srivastava and Jiawei Han

MINING SOFTWARE SPECIFICATIONS: METHODOLOGIES AND APPLICATIONS
David Lo, Siau-Cheng Khoo, Jiawei Han, and Chao Liu

MULTIMEDIA DATA MINING: A SYSTEMATIC INTRODUCTION TO CONCEPTS AND THEORY
Zhongfei Zhang and Ruofei Zhang

MUSIC DATA MINING
Tao Li, Mitsunori Ogihara, and George Tzanetakis

NEXT GENERATION OF DATA MINING
Hillol Kargupta, Jiawei Han, Philip S. Yu, Rajeev Motwani, and Vipin Kumar

PRACTICAL GRAPH MINING WITH R
Nagiza F. Samatova, William Hendrix, John Jenkins, Kanchana Padmanabhan, and Arpan Chakraborty

RELATIONAL DATA CLUSTERING: MODELS, ALGORITHMS, AND APPLICATIONS
Bo Long, Zhongfei Zhang, and Philip S. Yu

SERVICE-ORIENTED DISTRIBUTED KNOWLEDGE DISCOVERY
Domenico Talia and Paolo Trunfio

SPECTRAL FEATURE SELECTION FOR DATA MINING
Zheng Alan Zhao and Huan Liu

STATISTICAL DATA MINING USING SAS APPLICATIONS, SECOND EDITION
George Fernandez

SUPPORT VECTOR MACHINES: OPTIMIZATION BASED THEORY, ALGORITHMS, AND EXTENSIONS
Naiyang Deng, Yingjie Tian, and Chunhua Zhang

TEMPORAL DATA MINING
Theophano Mitsa

TEXT MINING: CLASSIFICATION, CLUSTERING, AND APPLICATIONS
Ashok N. Srivastava and Mehran Sahami

THE TOP TEN ALGORITHMS IN DATA MINING
Xindong Wu and Vipin Kumar

UNDERSTANDING COMPLEX DATASETS: DATA MINING WITH MATRIX DECOMPOSITIONS
David Skillicorn

DATA CLUSTERING

Algorithms and Applications

Edited by

Charu C. Aggarwal

Chandan K. Reddy

CRC Press
Taylor & Francis Group
Boca Raton London New York

CRC Press is an imprint of the
Taylor & Francis Group, an **informa** business
A CHAPMAN & HALL BOOK

CRC Press
Taylor & Francis Group
6000 Broken Sound Parkway NW, Suite 300
Boca Raton, FL 33487-2742

International Standard Book Number-13: 978-1-4665-5821-2 (Hardback)

Library of Congress Cataloging-in-Publication Data

Data clustering : algorithms and applications / [edited by] Charu C. Aggarwal, Chandan K. Reddy.
pages cm. -- (Chapman & Hall/CRC data mining and knowledge discovery series)
Includes bibliographical references and index.
ISBN 978-1-4665-5821-2 (hardback)
1. Document clustering. 2. Cluster analysis. 3. Data mining. 4. Machine theory. 5. File organization (Computer science) I. Aggarwal, Charu C., editor of compilation. II. Reddy, Chandan K., 1980- editor of compilation.

QA278.D294 2014
519.5'35--dc23 2013008698

Visit the Taylor & Francis Web site at
http://www.taylorandfrancis.com

and the CRC Press Web site at
http://www.crcpress.com

Contents

Preface

The problem of clustering is perhaps one of the most widely studied in the data mining and machine learning communities. This problem has been studied by researchers from several disciplines over five decades. Applications of clustering include a wide variety of problem domains such as text, multimedia, social networks, and biological data. Furthermore, the problem may be encountered in a number of different scenarios such as streaming or uncertain data. Clustering is a rather diverse topic, and the underlying algorithms depend greatly on the data domain and problem scenario.

Therefore, this book will focus on three primary aspects of data clustering. The first set of chapters will focus on the core methods for data clustering. These include methods such as probabilistic clustering, density-based clustering, grid-based clustering, and spectral clustering. The second set of chapters will focus on different problem domains and scenarios such as multimedia data, text data, biological data, categorical data, network data, data streams and uncertain data. The third set of chapters will focus on different detailed insights from the clustering process, because of the subjectivity of the clustering process, and the many different ways in which the same data set can be clustered. How do we know that a particular clustering is good or that it solves the needs of the application? There are numerous ways in which these issues can be explored. The exploration could be through interactive visualization and human interaction, external knowledge-based supervision, explicitly examining the multiple solutions in order to evaluate different possibilities, combining the multiple solutions in order to create more robust ensembles, or trying to judge the quality of different solutions with the use of different validation criteria.

The clustering problem has been addressed by a number of different communities such as pattern recognition, databases, data mining and machine learning. In some cases, the work by the different communities tends to be fragmented and has not been addressed in a unified way. This book will make a conscious effort to address the work of the different communities in a unified way. The book will start off with an overview of the basic methods in data clustering, and then discuss progressively more refined and complex methods for data clustering. Special attention will also be paid to more recent problem domains such as graphs and social networks.

The chapters in the book will be divided into three types:

- **Method Chapters:** These chapters discuss the *key techniques* which are commonly used for clustering such as feature selection, agglomerative clustering, partitional clustering, density-based clustering, probabilistic clustering, grid-based clustering, spectral clustering, and non-negative matrix factorization.

- **Domain Chapters:** These chapters discuss the specific methods used for different *domains* of data such as categorical data, text data, multimedia data, graph data, biological data, stream data, uncertain data, time series clustering, high-dimensional clustering, and big data. Many of these chapters can also be considered application chapters, because they explore the specific characteristics of the problem in a particular domain.

- **Variations and Insights:** These chapters discuss the *key variations* on the clustering process such as semi-supervised clustering, interactive clustering, multi-view clustering, cluster ensembles, and cluster validation. Such methods are typically used in order to obtain detailed insights from the clustering process, and also to explore different possibilities on the clustering process through either supervision, human intervention, or through automated generation of alternative clusters. The methods for cluster validation also provide an idea of the quality of the underlying clusters.

This book is designed to be comprehensive in its coverage of the entire area of clustering, and it is hoped that it will serve as a knowledgeable compendium to students and researchers.

Editor Biographies

Charu C. Aggarwal is a Research Scientist at the IBM T. J. Watson Research Center in Yorktown Heights, New York. He completed his B.S. from IIT Kanpur in 1993 and his Ph.D. from Massachusetts Institute of Technology in 1996. His research interest during his Ph.D. years was in combinatorial optimization (network flow algorithms), and his thesis advisor was Professor James B. Orlin. He has since worked in the field of performance analysis, databases, and data mining. He has published over 200 papers in refereed conferences and journals, and has applied for or been granted over 80 patents. He is author or editor of nine books, including this one. Because of the commercial value of the above-mentioned patents, he has received several invention achievement awards and has thrice been designated a Master Inventor at IBM. He is a recipient of an IBM Corporate Award (2003) for his work on bioterrorist threat detection in data streams, a recipient of the IBM Outstanding Innovation Award (2008) for his scientific contributions to privacy technology, and a recipient of an IBM Research Division Award (2008) for his scientific contributions to data stream research. He has served on the program committees of most major database/data mining conferences, and served as program vice-chairs of the SIAM Conference on Data Mining (2007), the IEEE ICDM Conference (2007), the WWW Conference (2009), and the IEEE ICDM Conference (2009). He served as an associate editor of the *IEEE Transactions on Knowledge and Data Engineering Journal* from 2004 to 2008. He is an associate editor of the *ACM TKDD Journal*, an action editor of the *Data Mining and Knowledge Discovery Journal*, an associate editor of *ACM SIGKDD Explorations*, and an associate editor of the *Knowledge and Information Systems Journal*. He is a fellow of the IEEE for "*contributions to knowledge discovery and data mining techniques*," and a life-member of the ACM.

Chandan K. Reddy is an Assistant Professor in the Department of Computer Science at Wayne State University. He received his Ph.D. from Cornell University and M.S. from Michigan State University. His primary research interests are in the areas of data mining and machine learning with applications to healthcare, bioinformatics, and social network analysis. His research is funded by the National Science Foundation, the National Institutes of Health, Department of Transportation, and the Susan G. Komen for the Cure Foundation. He has published over 40 peer-reviewed articles in leading conferences and journals. He received the Best Application Paper Award at the ACM SIGKDD conference in 2010 and was a finalist of the INFORMS Franz Edelman Award Competition in 2011. He is a member of IEEE, ACM, and SIAM.

Contributors

Ayan Acharya
University of Texas
Austin, Texas

Charu C. Aggarwal
IBM T. J. Watson Research Center
Yorktown Heights, New York

Amrudin Agovic
Reliancy, LLC
Saint Louis Park, Minnesota

Mohammad Al Hasan
Indiana University - Purdue University
Indianapolis, Indiana

Salem Alelyani
Arizona State University
Tempe, Arizona

David C. Anastasiu
University of Minnesota
Minneapolis, Minnesota

Bill Andreopoulos
Lawrence Berkeley National Laboratory
Berkeley, California

James Bailey
The University of Melbourne
Melbourne, Australia

Arindam Banerjee
University of Minnesota
Minneapolis, Minnesota

Sandra Batista
Duke University
Durham, North Carolina

Shiyu Chang
University of Illinois at Urbana-Champaign
Urbana, Illinois

Wei Cheng
University of North Carolina
Chapel Hill, North Carolina

Hongbo Deng
University of Illinois at Urbana-Champaign
Urbana, Illinois

Cha-charis Ding
University of Texas
Arlington, Texas

Martin Ester
Simon Fraser University
British Columbia, Canada

S M Faisal
The Ohio State University
Columbus, Ohio

Joydeep Ghosh
University of Texas
Austin, Texas

Dimitrios Gunopulos
University of Athens
Athens, Greece

Jiawei Han
University of Illinois at Urbana-Champaign
Urbana, Illinois

Alexander Hinneburg
Martin-Luther University
Halle/Saale, Germany

Thomas S. Huang
University of Illinois at Urbana-Champaign
Urbana, Illinois

U Kang
KAIST
Seoul, Korea

George Karypis
University of Minnesota
Minneapolis, Minnesota

Dimitrios Kotsakos
University of Athens
Athens, Greece

Tao Li
Florida International University
Miami, Florida

Zhongmou Li
Rutgers University
New Brunswick, New Jersey

Huan Liu
Arizona State University
Tempe, Arizona

Jialu Liu
University of Illinois at Urbana-Champaign
Urbana, Illinois

Srinivasan Parthasarathy
The Ohio State University
Columbus, Ohio

Guo-Jun Qi
University of Illinois at Urbana-Champaign
Urbana, Illinois

Chandan K. Reddy
Wayne State University
Detroit, Michigan

Andrea Tagarelli
University of Calabria
Arcavacata di Rende, Italy

Jiliang Tang
Arizona State University
Tempe, Arizona

Hanghang Tong
IBM T. J. Watson Research Center
Yorktown Heights, New York

Goce Trajcevski
Northwestern University
Evanston, Illinois

Min-Hsuan Tsai
University of Illinois at Urbana-Champaign
Urbana, Illinois

Shen-Fu Tsai
Microsoft Inc.
Redmond, Washington

Bhanukiran Vinzamuri
Wayne State University
Detroit, Michigan

Wei Wang
University of California
Los Angeles, California

Hui Xiong
Rutgers University
New Brunswick, New Jersey

Mohammed J. Zaki
Rensselaer Polytechnic Institute
Troy, New York

Arthur Zimek
University of Alberta
Edmonton, Canada

Chapter 1

An Introduction to Cluster Analysis

Charu C. Aggarwal
IBM T. J. Watson Research Center
Yorktown Heights, NY
charu@us.ibm.com

1.1 Introduction

The problem of data clustering has been widely studied in the data mining and machine learning literature because of its numerous applications to summarization, learning, segmentation, and target marketing [46, 47, 52]. In the absence of specific labeled information, clustering can be considered a concise model of the data which can be interpreted in the sense of either a summary or a generative model. The basic problem of clustering may be stated as follows:

Given a set of data points, partition them into a set of groups which are as similar as possible.

Note that this is a very rough definition, and the variations in the problem definition can be significant, depending upon the specific model used. For example, a generative model may define similarity on the basis of a probabilistic generative mechanism, whereas a distance-based approach will use a traditional distance function for quantification. Furthermore, the specific data type also has a significant impact on the problem definition.

Some common application domains in which the clustering problem arises are as follows:

- **Intermediate Step for other fundamental data mining problems:** Since a clustering can be considered a form of data summarization, it often serves as a key intermediate step for many fundamental data mining problems such as classification or outlier analysis. A compact summary of the data is often useful for different kinds of application-specific insights.

- **Collaborative Filtering:** In collaborative filtering methods, the clustering provides a summarization of like-minded users. The ratings provided by the different users for each other are used in order to perform the collaborative filtering. This can be used to provide recommendations in a variety of applications.

- **Customer Segmentation:** This application is quite similar to collaborative filtering, since it creates groups of similar customers in the data. The major difference from collaborative filtering is that instead of using rating information, arbitrary attributes about the objects may be used for clustering purposes.

- **Data Summarization:** Many clustering methods are closely related to dimensionality reduction methods. Such methods can be considered a form of data summarization. Data summarization can be helpful in creating compact data representations, which are easier to process and interpret in a wide variety of applications.

- **Dynamic Trend Detection:** Many forms of dynamic and streaming algorithms can be used to perform trend detection in a wide variety of social networking applications. In such applications, the data is dynamically clustered in a streaming fashion and can be used in order to determine important patterns of changes. Examples of such streaming data could be multidimensional data, text streams, streaming time-series data, and trajectory data. Key trends and events in the data can be discovered with the use of clustering methods.

- **Multimedia Data Analysis:** A variety of different kinds of documents such as images, audio or video, fall in the general category of multimedia data. The determination of similar segments has numerous applications, such as the determination of similar snippets of music or similar photographs. In many cases, the data may be multimodal and may contain different types. In such cases, the problem becomes even more challenging.

- **Biological Data Analysis:** Biological data has become pervasive in the last few years, because of the success of the human genome effort and the increasing ability to collect different kinds of gene expression data. Biological data is usually structured either as sequences or as networks. Clustering algorithms provide good ideas of the key trends in the data, as well as the unusual sequences.

- **Social Network Analysis:** In these applications, the structure of a social network is used in order to determine the important communities in the underlying network. Community detection has important applications in social network analysis, because it provides an important understanding of the community structure in the network. Clustering also has applications to social network summarization, which is useful in a number of applications.

The aforementioned list of applications is not exhaustive by any means; nevertheless it represents a good cross-section of the wide diversity of problems which can be addressed with clustering algorithms.

The work in the data clustering area typically falls into a number of broad categories;

- **Technique-centered:** Since clustering is a rather popular problem, it is not surprising that numerous methods, such as probabilistic techniques, distance-based techniques, spectral techniques, density-based techniques, and dimensionality-reduction based techniques, are used for the clustering process. Each of these methods has its own advantages and disadvantages, and may work well in different scenarios and problem domains. Certain kinds of data types such as high dimensional data, big data, or streaming data have their own set of challenges and often require specialized techniques.

- **Data-Type Centered:** Different applications create different kinds of data types with different properties. For example, an ECG machine will produce time series data points which are highly correlated with one another, whereas a social network will generated a mixture of document and structural data. Some of the most common examples are categorical data, time series data, discrete sequences, network data, and probabilistic data. Clearly, the nature of the data greatly impacts the choice of methodology used for the clustering process. Furthermore, some data types are more difficult than others because of the separation between different kinds of attributes such as behavior or contextual attributes.

- **Additional Insights from Clustering Variations:** A number of insights have also been designed for different kinds of clustering variations. For example, visual analysis, supervised analysis, ensemble-analysis, or multiview analysis can be used in order to gain additional insights. Furthermore, the issue of cluster validation is also important from the perspective of gaining specific insights about the performance of the clustering.

This chapter will discuss each of these issues in detail, and will also discuss how the organization of the book relates to these different areas of clustering. The chapter is organized as follows. The next section discusses the common techniques which are used in cluster analysis. Section 1.3 explores the use of different data types in the clustering process. Section 1.4 discusses the use of different variations of data clustering. Section 1.5 offers the conclusions and summary.

1.2 Common Techniques Used in Cluster Analysis

The clustering problems can be addressed using a wide variation of methods. In addition, the data preprocessing phase requires dedicated techniques of its own. A number of good books and surveys discuss these issues [14, 20, 31, 37, 46, 47, 48, 52, 65, 80, 81]. The most common techniques which are used for clustering are discussed in this section.

1.2.1 Feature Selection Methods

The feature selection phase is an important preprocessing step which is needed in order to enhance the quality of the underlying clustering. Not all features are equally relevant to finding the clusters, since some may be more noisy than other. Therefore, it is often helpful to utilize a preprocessing phase in which the noisy and irrelevant features are pruned from contention. Feature selection and dimensionality reduction are closely related. In feature selection, original subsets of the features are selected. In dimensionality reduction, linear combinations of features may be used in techniques such as principal component analysis [50] in order to further enhance the feature selection effect. The advantage of the former is greater interpretability, whereas the advantage of the latter is that a lesser number of transformed directions is required for the representation process. Chapter 2 of this book will discuss such feature selection methods in detail. A comprehensive book on feature selection may be found in [61].

It should be noted that feature selection can also be integrated directly into the clustering algorithm to gain better locality specific insights. This is particularly useful, when different features are relevant to different localities of the data. The motivating factor for high dimensional subspace clustering algorithms is the failure of global feature selection algorithms. As noted in [9]: "... *in many real data examples, some points are correlated with respect to a given set of dimensions and others are correlated with respect to different dimensions. Thus, it may not always be feasible to prune off too many dimensions without at the same time incurring a substantial loss of information*" p.61. Therefore, the use of local feature selection, by integrating the feature selection process into the algorithm, is the best way of achieving this goal. Such local feature selections can also be extended to the dimensionality reduction problem [8, 19] and are sometimes referred to as *local dimensionality reduction*. Such methods are discussed in detail in Chapter 9. Furthermore, Chapter 2 also discusses the connections of these classes of methods to the problem of feature selection.

1.2.2 Probabilistic and Generative Models

In probabilistic models, the core idea is to model the data from a *generative process*. First, a specific form of the generative model (e.g., mixture of Gaussians) is assumed, and then the parameters of this model are estimated with the use of the Expectation Maximization (EM) algorithm [27]. The available data set is used to estimate the parameters in such as way that they have a *maximum likelihood fit* to the generative model. Given this model, we then estimate the generative probabilities (or fit probabilities) of the underlying data points. Data points which fit the distribution well will have high fit probabilities, whereas anomalies will have very low fit probabilities.

The broad principle of a mixture-based generative model is to *assume* that the data were generated from a mixture of k distributions with the probability distributions $\mathcal{G}_1 \dots \mathcal{G}_k$ with the use of the following process:

- Pick a data distribution with prior probability α_i, where $i \in \{1 \dots k\}$, in order to pick one of the k distributions. Let us assume that the rth one is picked.
- Generate a data point from $\mathcal{G}_r$.

The probability distribution $\mathcal{G}_r$ is picked from a host of different possibilities. Note that this generative process requires the determination of several parameters such as the prior probabilities and the model parameters for each distribution $\mathcal{G}_r$. Models with different levels of flexibility may be designed depending upon whether the prior probabilities are specified as part of the problem setting, or whether interattribute correlations are assumed within a component of the mixture. Note that the model parameters and the probability of assignment of data points to clusters are dependent on one another in a circular way. Iterative methods are therefore desirable in order to resolve this circularity. The generative models are typically solved with the use of an EM approach, which starts off with a random or heuristic initialization and then iteratively uses two steps to resolve the circularity in computation:

- (E-Step) Determine the expected probability of assignment of data points to clusters with the use of current model parameters.
- (M-Step) Determine the optimum model parameters of each mixture by using the assignment probabilities as weights.

One nice property of EM-models is that they can be generalized relatively easily to different kinds of data, as long as the generative model for each component is properly selected for the individual mixture component $\mathcal{G}_r$. Some examples are as follows:

- For numerical data, a Gaussian mixture model may be used in order to model each component $\mathcal{G}_r$. Such a model is discussed in detail in Chapter 3.
- For categorical data, a Bernoulli model may be used for $\mathcal{G}_r$ in order to model the generation of the discrete values.
- For sequence data, a Hidden Markov Model (HMM) may be used for $\mathcal{G}_r$ in order to model the generation of a sequence. Interestingly, an HMM is itself a special kind of mixture model in which the different components of the mixture are dependent on each other though transitions. Thus, the clustering of sequence data with a mixture of HMMs can be considered a two-level mixture model.

Generative models are among the most fundamental of all clustering methods, because they try to understand the underlying *process* through which a cluster is generated. A number of interesting connections exist between other clustering methods and generative models, by considering special cases in terms of prior probabilities or mixture parameters. For example, the special case in which each prior probability is fixed to the same value and all mixture components are assumed to have the same radius along all dimensions reduces to a soft version of the k-means algorithm. These connections will be discussed in detail in Chapter 3.

1.2.3 Distance-Based Algorithms

Many special forms of generative algorithms can be shown to reduce to distance-based algorithms. This is because the mixture components in generative models often use a distance function within the probability distribution. For example, the Gaussian distribution represents data generation probabilities in terms of the euclidian distance from the mean of the mixture. As a result, a generative model with the Gaussian distribution can be shown to have a very close relationship with the k-means algorithm. In fact, many distance-based algorithms can be shown to be reductions from or simplifications of different kinds of generative models.

Distance-based methods are often desirable because of their simplicity and ease of implementation in a wide variety of scenarios. Distance-based algorithms can be generally divided into two types:

- *Flat:* In this case, the data is divided into several clusters in one shot, typically with the use of partitioning representatives. The choice of the partitioning representative and distance function is crucial and regulates the behavior of the underlying algorithm. In each iteration, the data points are assigned to their closest partitioning representatives, and then the representative is adjusted according to the data points assigned to the cluster. It is instructive to compare this with the iterative nature of the EM algorithm, in which soft assignments are performed in the E-step, and model parameters (analogous to cluster representatives) are adjusted in the M-step. Some common methods for creating the partitions are as follows:

• Flat:

- *k-Means:* In these methods, the partitioning representatives correspond to the mean of each cluster. Note that the partitioning representative is not drawn from the original data set, but is created as a function of the underlying data. The euclidian distance is used in order to compute distances. The *k*-Means method is considered one of the simplest and most classical methods for data clustering [46] and is also perhaps one of the most widely used methods in practical implementations because of its simplicity.
- *k-Medians:* In these methods, the median along each dimension, instead of the mean, is used to create the partitioning representative. As in the case of the *k*-Means approach, the partitioning representatives are not drawn from the original data set. The *k*-Medians approach is more stable to noise and outliers, because the median of a set of values is usually less sensitive to extreme values in the data. It should also be noted that the term "*k*-Medians" is sometimes overloaded in the literature, since it is sometimes also used to refer to a *k*-Medoid approach (discussed below) in which the partitioning representatives are drawn from the original data. In spite of this overloading and confusion in the research literature, it should be noted that the *k*-Medians and *k*-Medoid methods should be considered as distinct techniques. Therefore, in several chapters of this book, the *k*-Medians approach discussed is really a *k*-Medoids approach, though we have chosen to be consistent within the specific research paper which is being described. Nevertheless, it would be useful to note the overloading of this term in order to avoid confusion.
- *k-Medoids:* In these methods, the partitioning representative is sampled from the original data. Such techniques are particularly useful in cases, where the data points to be clustered are arbitrary objects, and it is often not meaningful to talk about functions of these objects. For example, for a set of network or discrete sequence objects, it may not be meaningful to talk about their mean or median. In such cases, partitioning representatives are drawn from the data, and iterative methods are used in order to improve the quality of these representatives. In each iteration, one of the representatives is replaced with a representative from the current data, in order to check if the quality of the clustering improves. Thus, this approach can be viewed as a kind of hill climbing method. These methods generally require many more iterations than *k*-Means and *k*-Medoids methods. However, unlike the previous two methods, they can be used in scenarios where it is not meaningful to talk about means or medians of data objects (eg. structural data objects).

• *Hierarchical:* In these methods, the clusters are represented hierarchically through a *dendogram*, at varying levels of granularity. Depending upon whether this hierarchical representation is created in top-down or bottom-up fashion, these representations may be considered either agglomerative or divisive.

  - *Agglomerative:* In these methods, a bottom-up approach is used, in which we start off with the individual data points and successively merge clusters in order to create a tree-like structure. A variety of choices are possible in terms of how these clusters may be merged, which provide different tradeoffs between quality and efficiency. Some examples of these choices are *single-linkage*, *all-pairs linkage*, *centroid-linkage*, and *sampled-linkage* clustering. In single-linkage clustering, the shortest distance between any pair of points in two clusters is used. In all-pairs linkage, the average over all pairs is used, whereas in sampled linkage, a sampling of the data points in the two clusters is used for calculating the average distance. In centroid-linkage, the distance between the centroids is used. Some variations of these methods have the disadvantage of *chaining*, in which larger clusters are naturally biased toward having closer distances to other points and will therefore attract a successively larger number of points. Single linkage clustering is particularly susceptible to this phenomenon. A number of data domains such as network clustering are also more susceptible to this behavior.

- *Divisive:* In these methods, a top-down approach is used in order to successively partition the data points into a tree-like structure. Any flat clustering algorithm can be used in order to perform the partitioning at each step. Divisive partitioning allows greater flexibility in terms of both the hierarchical structure of the tree and the level of balance in the different clusters. It is not necessary to have a perfectly balanced tree in terms of the depths of the different nodes or a tree in which the degree of every branch is exactly two. This allows the construction of a tree structure which allows different tradeoffs in the balancing of the node depths and node weights (number of data points in the node). For example, in a top-down method, if the different branches of the tree are unbalanced in terms of node weights, then the largest cluster can be chosen preferentially for division at each level. Such an approach [53] is used in *METIS* in order to create well balanced clusters in large social networks, in which the problem of cluster imbalance is particularly severe. While *METIS* is not a distance-based algorithm, these general principles apply to distance-based algorithms as well.

Distance-based methods are very popular in the literature, because they can be used with almost any data type, as long as an appropriate distance function is created for that data type. Thus, the problem of clustering can be reduced to the problem of finding a distance function for that data type. Therefore, distance function design has itself become an important area of research for data mining in its own right [5, 82]. Dedicated methods also have often been designed for specific data domains such as categorical or time series data [32, 42]. Of course, in many domains, such as high-dimensional data, the quality of the distance functions reduces because of many irrelevant dimensions [43] and may show both errors and concentration effects, which reduce the statistical significance of data mining results. In such cases, one may use either the redundancy in larger portions of the pairwise distance matrix to abstract out the noise in the distance computations with spectral methods [19] or projections in order to directly find the clusters in relevant subsets of attributes [9]. A discussion of many hierarchical and partitioning algorithms is provided in Chapter 4 of this book.

1.2.4 Density- and Grid-Based Methods

Density- and grid-based methods are two closely related classes, which try to explore the data *space* at high levels of granularity. The density at any particular point in the data space is defined either in terms of the number of data points in a prespecified volume of its locality or in terms of a smoother kernel density estimate [74]. Typically, the data space is explored at a reasonably high level of granularity and a postprocessing phase is used in order to "put together" the dense regions of the data space into an arbitrary shape. Grid-based methods are a specific class of density-based methods in which the individual regions of the data space which are explored are formed into a grid-like structure. Grid-like structures are often particularly convenient because of greater ease in putting together the different dense blocks in the post-processing phase. Such grid-like methods can also be used in the context of high-dimensional methods, since the lower dimensional grids define clusters on subsets of dimensions [6].

A major advantage of these methods is that since they explore the data space at a high level of granularity, they can be used to reconstruct the entire shape of the data distribution. Two classical methods for density-based methods and grid-based methods are DBSCAN [34] and STING [83], respectively. The major challenge of density-based methods is that they are naturally defined on data points in a continuous space. Therefore, they often cannot be meaningfully used in a discrete or noneuclidian space, unless an embedding approach is used. Thus, many arbitrary data types such as time-series data are not quite as easy to use with density-based methods without specialized transformations. Another issue is that density computations becomes increasingly difficult to define with greater dimensionality because of the greater number of cells in the underlying grid structure and the sparsity of the data in the underlying grid. A detailed discussion of density-based and grid-based methods is provided in Chapters 5 and 6 of this book.

1.2.5 Leveraging Dimensionality Reduction Methods

Dimensionality reduction methods are closely related to both feature selection and clustering, in that they attempt to use the closeness and correlations between dimensions to reduce the dimensionality of representation. Thus, dimensionality reduction methods can often be considered a vertical form of clustering, in which columns of the data are clustered with the use of either correlation or proximity analysis, as opposed to the rows. Therefore, a natural question arises, as to whether it is possible to perform these steps *simultaneously*, by clustering rows and columns of the data together. The idea is that simultaneous row and column clustering is likely to be more effective than performing either of these steps individually. This broader principle has led to numerous algorithms such as matrix factorization, spectral clustering, probabilistic latent semantic indexing, and co-clustering. Some of these methods such as spectral clustering are somewhat different but are nevertheless based on the same general concept. These methods are also closely related to projected clustering methods, which are commonly used for high dimensional data in the database literature. Some common models will be discussed below.

1.2.5.1 Generative Models for Dimensionality Reduction

In these models, a generative probability distribution is used to model the relationships between the data points and dimensions in an integrated way. For example, a generalized Gaussian distribution can be considered a mixture of arbitrarily correlated (oriented) clusters, whose parameters can be learned by the EM-algorithm. Of course, this is often not easy to do robustly with increasing dimensionality due to the larger number of parameters involved in the learning process. It is well known that methods such as EM are highly sensitive to overfitting, in which the number of parameters increases significantly. This is because EM methods try to retain *all* the information in terms of soft probabilities, rather than making the hard choices of point and dimension selection by nonparametric methods. Nevertheless, many special cases for different data types have been learned successfully with generative models.

A particularly common dimensionality reduction method is that of topic modeling in text data [45]. This method is also sometimes referred to as *Probabilistic Latent Semantic Indexing (PLSI)*. In topic modeling, a cluster is associated with a set of words and a set of documents simultaneously. The main parameters to be learned are the probabilities of assignments of words (dimensions) to topics (clusters) and those of the documents (data points) to topics (clusters). Thus, this naturally creates a soft clustering of the data from *both* a row and column perspective. These are then learned in an integrated way. These methods have found a lot of success in the text mining literature, and many methods, such as *Latent Dirichlet Allocation (LDA)*, which vary on this principle, with the use of more generalized priors have been proposed [22]. Many of these models are discussed briefly in Chapter 3.

1.2.5.2 Matrix Factorization and Co-Clustering

Matrix factorization and co-clustering methods are also commonly used classes of dimensionality reduction methods. These methods are usually applied to data which is represented as sparse nonnegative matrices, though it is possible in principle to generalize these methods to other kinds of matrices as well. However, the real attraction of this approach is the additional *interpretability* inherent in *nonnegative* matrix factorization methods, in which a data point can be expressed as a *nonnegative* linear combination of the concepts in the underlying data. Nonnegative matrix factorization methods are closely related to co-clustering, which clusters the rows and columns of a matrix simultaneously.

Let A be a nonnegative $n \times d$ matrix, which contains n data entries, each of which has a dimensionality of d. In most typical applications such as text data, the matrix A represents small nonnegative quantities such as word frequencies and is not only nonnegative, but also sparse. Then, the matrix A can be *approximately* factorized into two nonnegative low rank matrices U and V of sizes $n \times k$ and $k \times d$, respectively. As we will discuss later, these matrices are the representatives for the clusters on the rows and the columns, respectively, when exactly k clusters are used in order to represent both rows and columns. Therefore, we have

$$A \approx U \cdot V \tag{1.1}$$

The residual matrix R represents the noise in the underlying data:

$$R = A - U \cdot V \tag{1.2}$$

Clearly, it is desirable to determine the factorized matrices U and V, such that the sum of the squares of the residuals in R is minimized. This is equivalent to determining nonnegative matrices U and V, such that the Froebinius norm of $A - U \cdot V$ is minimized. This is a constrained optimization problem, in which the constraints correspond to the nonnegativity of the entries in U and V. Therefore, a Lagrangian method can be used to learn the parameters of this optimization problem. A detailed discussion of the iterative approach is provided in [55].

The nonnegative $n \times k$ matrix U represents the coordinates of each of the n data points into each of the k newly created dimensions. A high positive value of the entry (i, j) implies that data point i is closely related to the newly created dimension j. Therefore, a trivial way to perform the clustering would be to assign each data point to the newly created dimension for which it has the largest component in U. Alternatively, if data points are allowed to belong to multiple clusters, then for each of the k columns in V, the entries with value above a particular threshold correspond to the document clusters. Thus, the newly created set of dimensions can be made to be synonymous with the clustering of the data set. In practice, it is possible to do much better, by using k-means on the new representation. One nice characteristic of the nonnegative matrix factorization method is that the size of an entry in the matrix U tells us how much a particular data point is related to a particular concept (or newly created dimension). The price of this greater interpretability is that the newly created sets of dimensions are typically not orthogonal to one another. This brings us to a discussion of the physical interpretation of matrix V.

The $k \times d$ matrix V provides the actual representation of each of the k newly created dimensions in terms of the original d dimensions. Thus, each row in this matrix is one component of this newly created axis system, and the rows are not necessarily orthogonal to one another (unlike most other dimensionality reduction methods). A large positive entry implies that this newly created dimension is highly related to a particular dimension in the underlying data. For example, in a document clustering application, the entries with large positive values in each row represent the most common words in each cluster. A thresholding technique can be used to identify these words. Note the similarity of this approach with a projection-based technique or the softer PLSI technique. In the context of a document clustering application, these large positive entries provide the *word clusters* in the underlying data. In the context of text data, each document can therefore be approximately expressed (because of the factorization process) as a nonnegative linear combination of at most k word-cluster vectors. The specific weight of that component represents the importance of that component, which makes the decomposition highly interpretable. Note that this interpretability is highly dependent on nonnegativity. Conversely, consider the word-membership vector across the corpus. This can be expressed in terms of at most k document-cluster vectors. This provides an idea of how important each document cluster is to that word.

At this point, it should be evident that the matrices U and V *simultaneously* provide the clusters on the rows (documents) and columns (words). This general principle, when applied to sparse non-negative matrices, is also referred to as co-clustering. Of course, nonnegative matrix factorization is only one possible way to perform the co-clustering. A variety of graph-based spectral methods and other information theoretic methods can also be used in order to perform co-clustering [29, 30, 71]. Matrix factorization methods are discussed in Chapter 7. These techniques are also closely related to *spectral methods* for dimensionality reduction, as discussed below.

1.2.5.3 Spectral Methods

Spectral methods are an interesting technique for dimensionality reduction, which work with the similarity (or distance) matrix of the underlying data, instead of working with the original points and dimensions. This, of course, has its own set of advantages and disadvantages. The major advantage is that it is now possible to work with arbitrary objects for dimensionality reduction, rather than simply data points which are represented in a multi-dimensional space. In fact, spectral methods also perform the dual task of *embedding* these objects into a euclidian space, while performing the dimensionality reduction. Therefore, spectral methods are extremely popular for performing clustering on arbitrary objects such as node sets in a graph. The disadvantage of spectral methods is that since they work with an $n \times n$ similarity matrix, the time complexity for even creating the similarity matrix scales with the *square* of the number of *data points*. Furthermore, the process of determining the eigenvectors of this matrix can be extremely expensive, unless a very small number of these eigenvectors is required. Another disadvantage of spectral methods is that it is much more difficult to create lower dimensional representations for data points, unless they are part of the original sample from which the similarity matrix was created. For multidimensional data, the use of such a large similarity matrix is rather redundant, unless the data is extremely noisy and high dimensional.

Let $\mathcal{D}$ be a database containing n points. The first step is to create an $n \times n$ matrix W of weights, which represents the pairwise similarity between the different data points. This is done with the use of the *heat kernel*. For any pair of data points $\overline{X_i}$ and $\overline{X_j}$, the *heat kernel* defines a similarity matrix W of weights:

$$W_{ij} = exp(-||\overline{X_i} - \overline{X_j}||^2/t) \tag{1.3}$$

Here t is a user-defined parameter. Furthermore, the value of W_{ij} is set to 0 if the distance between $\overline{X_i}$ and $\overline{X_j}$ is greater than a given threshold. The similarity matrix may also be viewed as the adjacency matrix of a graph, in which each node corresponds to a data item, and the weight of an edge corresponds to the similarity between these data items. Therefore, spectral methods reduce the problem to that of finding optimal cuts in this graph, which correspond to partitions which are weakly connected by edges representing similarity. Hence, spectral methods can be considered a graph-based technique for clustering of any kinds of data, by converting the similarity matrix into a network structure. Many variants exist in terms of the different choices for constructing the similarity matrix W. Some simpler variants use the mutual k-nearest neighbor graph, or simply the binary graph in which the distances are less than a given threshold. The matrix W is symmetric.

It should be noted that even when distances are very noisy, the similarity matrix encodes a significant amount of information because of its exhaustive nature. It is here that spectral analysis is useful, since the noise in the similarity representation can be abstracted out with the use of eigenvector-analysis of this matrix. Thus, these methods are able to recover and sharpen the latent information in the similarity matrix, though at a rather high cost, which scales with the square of the number of data points.

First, we will discuss the problem of mapping the points onto a 1-dimensional space. The generalization to the k-dimensional case is relatively straightforward. We would like to map the data points in $\mathcal{D}$ into a set of points $y_1 \ldots y_n$ on a line, in which the similarity maps onto euclidian distances on this line. Therefore, it is undesirable for data points which are very similar to be mapped onto distant points on this line. We would like to determine values of y_i which minimize the following objective function O:

$$O = \sum_{i=1}^{n} \sum_{j=1}^{n} W_{ij} \cdot (y_i - y_j)^2 \tag{1.4}$$

The objective function O can be rewritten in terms of the Laplacian matrix L of W. The Laplacian matrix is defined as $D - W$, where D is a diagonal matrix satisfying $D_{ii} = \sum_{j=1}^{n} W_{ij}$. Let $\overline{y} = (y_1 \ldots y_n)$. The objective function O can be rewritten as follows:

$$O = 2 \cdot \overline{y}^T \cdot L \cdot \overline{y} \tag{1.5}$$

We need to incorporate a scaling constraint in order to ensure that the trivial value of $y_i = 0$ for all i is not selected by the problem. A possible scaling constraint is as follows:

$$\overline{y}^T \cdot D \cdot \overline{y} = 1 \tag{1.6}$$

Note that the use of the matrix D provides different weights to the data items, because it is assumed that nodes with greater similarity to different data items are more involved in the clustering process. This optimization formulation is in generalized eigenvector format, and therefore the value of $\overline{y}$ is optimized by selecting the smallest eigenvalue for which the generalized eigenvector relationship $L \cdot \overline{y} = \lambda \cdot D \cdot \overline{y}$ is satisfied. In practice however, the smallest generalized eigenvalue corresponds to the trivial solution, where $\overline{y}$ is the (normalized) unit vector. This trivial eigenvector is noninformative. Therefore, it can be discarded, and it is not used in the analysis. The second-smallest eigenvalue then provides an optimal solution, which is more informative.

This model can be generalized to determining all the eigenvectors in ascending order of eigenvalue. Such directions correspond to successive orthogonal directions in the data, which result in the best mapping. This results in a set of n eigenvectors $\overline{e_1}, \overline{e_2} \ldots \overline{e_n}$ (of which the first is trivial), with corresponding eigenvalues $0 = \lambda_1 \leq \lambda_2 \ldots \leq \lambda_n$. Let the corresponding vector representation of the data points along each eigenvector be denoted by $\overline{y}^1 \ldots \overline{y}^n$.

What do the small and large magnitude eigenvectors intuitively represent in the new transformed space? By using the ordering of the data items along a small magnitude eigenvector to create a cut, the weight of the edges across the cut is likely to be small. Thus, this represents a cluster in the space of data items. At the same time, if the top k *longest* eigenvectors are picked, then the vector representations $\overline{y}^n \ldots \overline{y}^{n-k+1}$ provide a $n \times k$ matrix. This provides a k-dimensional embedded representation of the entire data set of n points, which preserves the maximum amount of information. Thus, spectral methods can be used in order to simultaneously perform clustering and dimensionality reduction. In fact, spectral methods can be used to recover the entire lower dimensional (possibly nonlinear) shape of the data, though arguably at a rather high cost [78]. The specific local dimensionality reduction technique of this section is discussed in detail in [19]. An excellent survey on spectral clustering methods may be found in [35, 63]. These methods are also discussed in detail in Chapter 8 of this book.

1.2.6 The High Dimensional Scenario

The high dimensional scenario is particularly challenging for cluster analysis because of the large variations in the behavior of the data attributes over different parts of the data. This leads to numerous challenges in many data mining problems such as clustering, nearest neighbor search, and outlier analysis. It should be noted that many of the algorithms for these problems are dependent upon the use of distances as an important subroutine. However, with increasing dimensionality, the distances seem to increasingly lose their effectiveness and statistical significance because of irrelevant attributes. The premise is that a successively smaller *fraction* of the attributes often remains relevant with increasing dimensionality, which leads to the blurring of the distances and increasing concentration effects, because of the averaging behavior of the irrelevant attributes. *Concentration effects* refer to the fact that when many features are noisy and uncorrelated, their additive effects will lead to all pairwise distances between data points becoming similar. The noise and concentration effects are problematic in two ways for distance-based clustering (and many other) algorithms:

- An increasing noise from irrelevant attributes may cause errors in the distance representation, so that it no longer properly represents the *intrinsic distance* between data objects.

- The concentration effects from the irrelevant dimensions lead to a reduction in the statistical significance of the results from distance-based algorithms, if used directly with distances that have not been properly denoised.

The combination of these issues above leads to questions about whether full-dimensional distances are truly meaningful [8, 21, 43]. While the natural solution to such problems is to use feature selection, the problem is that different attributes may be relevant to different localities of the data. This problem is inherent in high-dimensional distance functions and nearest neighbor search. As stated in [43]: "*... One of the problems of the current notion of nearest neighbor search is that it tends to give equal treatment to all features (dimensions), which are however not of equal importance. Furthermore, the importance of a given dimension may not even be independent of the query point itself*" p. 506. These noise and concentration effects are therefore a problematic *symptom* of (locally) irrelevant, uncorrelated, or noisy attributes, which tend to impact the effectiveness and statistical significance of full-dimensional algorithms. This has lead to significant efforts to redesign many data mining algorithms such as clustering, which are dependent on the notion of distances [4]. In particular, it would seem odd that data mining algorithms should behave poorly with increasing dimensionality at least from a qualitative perspective when a larger number of dimensions clearly provides more information. The reason is that conventional distance measures were generally designed for many kinds of low-dimensional spatial applications which are not suitable for the high-dimensional case. While high-dimensional data contain more information, they are also more complex. Therefore, naive methods will do worse with increasing dimensionality because of the noise effects of locally irrelevant attributes. Carefully designed distance functions can leverage the greater information in these dimensions and show *improving* behavior with dimensionality [4, 11] at least for a few applications such as similarity search.

Projected clustering methods can be considered a form of *local feature selection*, or *local dimensionality reduction*, in which the feature selection or transformation is performed specific to different localities of the data. Some of the earliest methods even refer to these methods as local-dimensionality reduction [23] in order to emphasize the local feature selection effect. Thus, a projected cluster is defined as a set of clusters $C_1 \dots C_k$, along with a corresponding set of subspaces $\mathcal{E}_1 \dots \mathcal{E}_k$, such that the points in C_i cluster well in the subspace represented by $\mathcal{E}_i$. Thus, these methods are a form of local feature selection, which can be used to determine the relevant clusters in the underlying data. Note that the subspace $\mathcal{E}_i$ could represent a subset of the original attributes, or it could represent a transformed axis system in which the clusters are defined on a small set of orthogonal attributes. Some of the earliest projected clustering methods are discussed in [6, 9, 10]. The literature on this area has grown significantly since 2000, and surveys on the area may be found in [67, 54]. An overview of algorithms for high-dimensional data will be provided in Chapter 9.

It should also be pointed out that while pairwise distances are often too noisy or concentrated to be used meaningfully with off-the-shelf distance-based algorithms, larger portions of the *entire* distance matrix often retain a significant amount of latent structure due to the inherent redundancies in representing different pairwise distances. Therefore, even when there is significant noise and concentration of the distances, there will always be some level of consistent variations over the similarity matrix because of the impact of the consistent attributes. This redundancy can be leveraged in conjunction with spectral analysis [19, 78] to filter out the noise and concentration effects from full-dimensional distances and implicitly recover the local or global lower dimensional shape of the underlying data in terms of enhanced distance representations of newly embedded data in a lower dimensional space. In essence, this approach enhances the contrasts of the distances by carefully examining how the noise (due to the many irrelevant attributes) and the minute correlations (due to the smaller number of relevant attributes) relate to the different pairwise distances. The general principle of these techniques is that distances are measured differently in high-dimensional space, depending upon how the data is distributed. This in turn depends upon the relevance and relationships of different attributes in different localities. These results are strong evidence for the fact that proper distance function design is highly dependent on its ability to recognize the relevance and relationships of different attributes in different data localities.

Thus, while (naively designed) pairwise distances cannot be used meaningfully in high dimensional data with *off-the-shelf* algorithms because of noise and concentration effects, they often do retain sufficient *latent* information *collectively* when used carefully in conjunction with denoising methods such as spectral analysis. Of course, the price for this is rather high, since the size of the similarity matrix scales with the square of the number of data points. The projected clustering method is generally a more efficient way of achieving an approximately similar goal, since it works directly with the data representation, rather than building a much larger and more redundant intermediate representation such as the distance matrix.

1.2.7 Scalable Techniques for Cluster Analysis

With advances in software and hardware technology, data collection has become increasingly easy in a wide variety of scenarios. For example, in social sensing applications, users may carry mobile or wearable sensors, which may result in the continuous accumulation of data over time. This leads to numerous challenges when real-time analysis and insights are required. This is referred to as the *streaming* scenario, in which it is assumed that a single pass is allowed over the data stream, because the data are often too large to be collected within limited resource constraints. Even when the data are collected offline, this leads to numerous scalability issues, in terms of integrating with traditional database systems or in terms of using the large amounts of data in a distributed setting, with the big data framework. Thus, varying levels of challenges are possible, depending upon the nature of the underlying data. Each of these issues is discussed below.

1.2.7.1 I/O Issues in Database Management

The most fundamental issues of scalability arise when a data mining algorithm is coupled with a traditional database system. In such cases, it can be shown that the major bottleneck arises from the I/O times required for accessing the objects in the database. Therefore, algorithms which use sequential scans of the data, rather than methods which randomly access the data records, are often useful. A number of classical methods were proposed in the database literature in order to address these scalability issues.

The easiest algorithms to extend to this case are flat partitioning methods which use sequential scans over the database in order to assign data points to representatives. One of the first methods [66] in this direction was *CLARANS*, in which these representatives are determined with the use of a k-medoids approach. Note that the k-medoids approach can still be computationally quite intensive because each iteration requires trying out new partitioning representatives through an exchange process (from the current set). If a large number of iterations is required, this will increase the number of passes over the data set. Therefore, the *CLARANS* method uses sampling, by performing the iterative hill climbing over a smaller sample, in order to improve the efficiency of the approach. Another method in this direction is *BIRCH* [87], which generalizes a k-means approach to the clustering process. The *CURE* method proposed in [41] finds clusters of nonspherical shape by using more than one representative point per cluster. It combines partitioning and sampling to ensure a more efficient clustering process. A number of scalable algorithms for clustering are discussed in Chapter 11.

1.2.7.2 Streaming Algorithms

The streaming scenario is particularly challenging for clustering algorithms due to the requirements of *real-time* analysis, and the *evolution* and *concept-drift* in the underlying data. While database-centric algorithms require a *limited* number of passes over the data, streaming algorithms require exactly one pass, since the data cannot be stored at all. In *addition to* this challenge, the analysis typically needs to be performed in real time, and the changing patterns in the data need to be properly accounted for in the analysis.

In order to achieve these goals, virtually all streaming methods use a summarization technique to create *intermediate representations*, which can be used for clustering. One of the first methods in this direction uses a *microclustering approach* [7] to create and maintain the clusters from the underlying data stream. Summary statistics are maintained for the microclusters to enable an effective clustering process. This is combined with a pyramidal time frame to capture the evolving aspects of the underlying data stream. Stream clustering can also be extended to other data types such as discrete data, massive-domain data, text data, and graph data. A number of unique challenges also arises in the distributed setting. A variety of stream clustering algorithms is discussed in detail in Chapter 10.

1.2.7.3 The Big Data Framework

While streaming algorithms work under the assumption that the data are too large to be stored explicitly, the big data framework leverages advances in storage technology in order to actually store the data and process them. However, as the subsequent discussion will show, even if the data can be explicitly stored, it is often not easy to process and extract insights from them. This is because an increasing size of the data implies that a distributed file system must be used in order to store the information, and distributed processing techniques are required in order to ensure sufficient scalability. The challenge here is that if large segments of the data are available on different machines, it is often too expensive to shuffle the data across different machines in order to extract integrated insights from them. Thus, as in all distributed infrastructures, it is desirable to exchange intermediate insights, so as to minimize communication costs. For an application programmer, this can sometimes create challenges in terms of keeping track of where different parts of the data are stored, and the precise ordering of communications in order to minimize the costs.

In this context, Google's *MapReduce* framework [28] provides an effective method for analysis of large amounts of data, especially when the nature of the computations involves linearly computable statistical functions over the elements of the data streams. One desirable aspect of this framework is that it abstracts out the precise details of where different parts of the data are stored to the application programmer. As stated in [28]: "*The run-time system takes care of the details of partitioning the input data, scheduling the program's execution across a set of machines, handling machine failures, and managing the required inter-machine communication. This allows programmers without any experience with parallel and distributed systems to easily utilize the resources of a large distributed system*" p. 107. Many clustering algorithms such as k-means are naturally linear in terms of their scalability with the size of the data. A primer on the *MapReduce* framework implementation on *Apache Hadoop* may be found in [84]. The key idea here is to use a *Map* function to distribute the work across the different machines, and then provide an automated way to shuffle out much smaller data in (key,value) pairs containing intermediate results. The *Reduce* function is then applied to the aggregated results from the *Map* step to obtain the final results.

Google's original *MapReduce* framework was designed for analyzing large amounts of web logs and more specifically deriving linearly computable statistics from the logs. While the clustering process is not as simple as linearly computable statistics, it has nevertheless been shown [26] that many existing clustering algorithms can be generalized to the *MapReduce* framework. A proper choice of the algorithm to adapt to the *MapReduce* framework is crucial, since the framework is particularly effective for linear computations. It should be pointed out that the major attraction of the *MapReduce* framework is its ability to provide application programmers with a cleaner abstraction, which is independent of very specific run-time details of the distributed system. It should not, however, be assumed that such a system is somehow inherently superior to existing methods for distributed parallelization from an *effectiveness* or *flexibility* perspective, especially if an application programmer is willing to design such details from scratch. A detailed discussion of clustering algorithms for big data is provided in Chapter 11.

1.3 Data Types Studied in Cluster Analysis

The specific data type has a tremendous impact on the particular choice of the clustering algorithm. Most of the earliest clustering algorithms were designed under the implicit assumption that the data attributes were numerical. However, this is not true in most real scenarios, where the data could be drawn from any number of possible types such as discrete (categorical), temporal, or structural. This section discusses the impact of data types on the clustering process. A brief overview of the different data types is provided in this section.

1.3.1 Clustering Categorical Data

Categorical data is fairly common in real data sets. This is because many attributes in real data such as sex, race, or zip code are inherently discrete and do not take on a natural ordering. In many cases, the data sets may be *mixed*, in which some attributes such as salary are numerical, whereas other attributes such as sex or zip code are categorical. A special form of categorical data is market basket data, in which all attributes are binary.

Categorical data sets lead to numerous challenges for clustering algorithms:

- When the algorithms depends upon the use of a similarity or distance function, the standard L_k metrics can no longer be used. New similarity measures need to be defined for categorical data. A discussion of similarity measures for categorical data is provided in [32].
- Many clustering algorithms such as the k-means or k-medians methods construct clustering representatives as the means or medians of the data points in the clusters. In many cases, statistics such as the mean or median are naturally defined for numerical data but need to be appropriately modified for discrete data.

When the data is mixed, then the problem because more difficult because the different attributes now need to be treated in a heterogeneous way, and the similarity functions need to explicitly account for the underlying heterogeneity.

It should be noted that some models of clustering are more amenable to different data types than others. For example, some models depend only on the distance (or similarity) functions between records. Therefore, as long as an appropriate similarity function can be defined between records, cluster analysis methods can be used effectively. Spectral clustering is one class of methods which can be used with virtually any data type, as long as appropriate similarity functions are defined. The downside is that the the methods scale with the square of the similarity matrix size. Generative models can also be generalized easily to different data types, as long as an appropriate generative model can be defined for each component of the mixture. Some common algorithms for categorical data clustering include *CACTUS* [38], *ROCK* [40], *STIRR* [39], and *LIMBO* [15]. A discussion of categorical data clustering algorithms is provided in Chapter 12.

1.3.2 Clustering Text Data

Text data is a particularly common form of data with the increasing popularity of the web and social networks. Text data is typically represented in vector space format, in which the specific ordering is abstracted out, and the data is therefore treated as a bag-of-words. It should be noted that the methods for clustering text data can also be used for clustering set-based attributes. Text data has a number of properties which should be taken into consideration:

- The data is extremely high-dimensional and sparse. This corresponds to the fact that the text lexicon is rather large, but each document contains only a small number of words. Thus, most attributes take on zero values.
- The attribute values correspond to word frequencies and are, therefore, typically nonnegative. This is important from the perspective of many co-clustering and matrix factorization methods, which leverage this nonnegativity.

The earliest methods for text clustering such as the scatter–gather method [25, 75] use distance-based methods. Specifically, a combination of k-means and agglomerative methods is used for the clustering process. Subsequently, the problem of text clustering has often been explored in the context of topic modeling, where a soft membership matrix of words and documents to clusters is created. The EM-framework is used in conjunction with these methods. Two common methods used for generative topic modeling are *PLSI* and *LDA* [22, 45]. These methods can be considered soft versions of methods such as co-clustering [29, 30, 71] and matrix factorization [55], which cluster the rows and columns together at the same time. Spectral methods [73] are often used to perform this partitioning by creating a bipartite similarity graph, which represents the membership relations of the rows (documents) and columns (words). This is not particularly surprising since matrix factorization methods and spectral clustering are known to be closely related [57], as discussed in Chapter 8 of this book. Surveys on text clustering may be found in [12, 88]. In recent years, the popularity of social media has also lead to an increasing importance of *short* text documents. For example, the posts and tweets in social media web sites are constrained in length, both by the platform and by user preferences. Therefore, specialized methods have also been proposed recently for performing the clustering in cases where the documents are relatively short. Methods for clustering different kinds of text data are discussed in Chapter 13.

1.3.3 Clustering Multimedia Data

With the increasing popularity of social media, many forms of multimedia data may be used in conjunction with clustering methods. These include image data, audio and video. Examples of social media sites which contain large amounts of such data are *Flickr* and *Youtube*. Even the web and conventional social networks typically contain a significant amount of multimedia data of different types. In many cases, such data may occur in conjunction with other more conventional forms of text data.

The clustering of multimedia data is often a challenging task, because of the varying and heterogeneous nature of the underlying content. In many cases, the data may be multimodal, or it may be contextual, containing both behavioral and contextual attributes. For example, image data are typically contextual, in which the position of a pixel represents its context, and the value on the pixel represents its behavior. Video and music data are also contextual, because the temporal ordering of the data records provides the necessary information for understanding. The heterogeneity and contextual nature of the data can only be addressed with proper data representations and analysis. In fact, data representation seems to be a key issue in all forms of multimedia analysis, which significantly impacts the final quality of results. A discussion of methods for clustering multimedia data is provided in Chapter 14.

1.3.4 Clustering Time-Series Data

Time-series data is quite common in all forms of sensor data, stock markets, or any other kinds of temporal tracking or forecasting applications. The major aspect of time series is that the data values are not independent of one another, but they are temporally dependent on one another. Specifically, the data contain a contextual attribute (time) and a behavioral attribute (data value). There is a significant diversity in problem definitions in the time-series scenario. The time-series data can be clustered in a variety of different ways, depending upon whether correlation-based online analysis is required or shape-based offline analysis is required.

- In correlation-based online analysis, the correlations among the different time-series data streams are tracked over time in order to create online clusters. Such methods are often useful for sensor selection and forecasting, especially when streams exhibit lag correlations within the clustered patterns. These methods are often used in stock market applications, where it is desirable to maintain groups of clustered stock tickers, in an online manner, based on their correlation patterns. Thus, the distance functions between different series need to be computed continuously, based on their mutual regression coefficients. Some examples of these methods include the MUSCLES approach [86] and a large scale time-series correlation monitoring method [90].

- In shape-based offline analysis, the time-series objects are analyzed in offline manner, and the specific details about when a particular time series was created is not important. For example, for a set of ECG time series collected from patients, the precise time of when a series was collected is not important, but the overall shape of the series is important for the purposes of clustering. In such cases, the distance function between two time series is important. This is important, because the different time series may not be drawn on the same range of data values and may also show time-warping effects, in which the shapes can be matched only by elongating or shrinking portions of the time-series in the temporal direction. As in the previous case, the design of the distance function [42] holds the key to the effective use of the approach.

A particular interesting case is that of multivariate time series, in which many series are simultaneously produced over time. A classical example of this is trajectory data, in which the different coordinate directions form the different components of the multivariate series. Therefore, trajectory analysis can be viewed as a special kind of time-series clustering. As in the case of univariate time series, it is possible to perform these steps using either online analysis (trajectories moving together in real time) or offline analysis (similar shape). An example of the former is discussed in [59], whereas an example of the latter is discussed in [68]. A survey on time-series data is found in [60], though this survey is largely focussed on the case of offline analysis. Chapter 15 discusses both the online and offline aspects of time series clustering.

1.3.5 Clustering Discrete Sequences

Many forms of data create discrete sequences instead of categorical ones. For example, web logs, the command sequences in computer systems, and biological data are all discrete sequences. The contextual attribute in this case often corresponds to placement (e.g., biological data), rather than time. Biological data is also one of the most common applications of sequence clustering.

As with the case of continuous sequences, a key challenge is the creation of similarity functions between different data objects. Numerous similarity functions such as the hamming distance, edit distance, and longest common subsequence are commonly used in this context. A discussion of the similarity functions which are commonly used for discrete sequences may be found in [58]. Another key problem which arises in the context of clustering discrete sequences is that the intermediate and summary representation of a set of sequences can be computationally intensive. Unlike numerical data, where averaging methods can be used, it is much more difficult to find such representations for discrete sequences. A common representation, which provides a relatively limited level of summarization is the suffix tree. Methods for using suffix trees for sequence clustering methods have been proposed in CLUSEQ [85].

Generative models can be utilized, both to model the distances between sequences and to create probabilistic models of cluster generation [76]. A particularly common approach is the use of mixtures of HMMs. A primer on Hidden Markov Models may be found in [70]. Hidden Markov Models can be considered a special kind of mixture model in which the different components of the mixture are dependent on one another. A second level of mixture modeling can be used in order to create clusters from these different HMMs. Much of the work on sequence clustering is performed in the context of biological data. Detailed surveys on the subject may be found in [33, 49, 64]. A discussion of sequence clustering algorithms is provided in Chapter 16, though this chapter also provides a survey of clustering algorithms for other kinds of biological data.

1.3.6 Clustering Network Data

Graphs and networks are among the most fundamental (and general) of all data representations. This is because virtually every data type can be represented as a similarity graph, with similarity values on the edges. In fact, the method of spectral clustering can be viewed as the most general connection between all other types of clustering and graph clustering. Thus, as long as a similarity function can be defined between arbitrary data objects, spectral clustering can be used in order to perform the analysis. Graph clustering has been studied extensively in the classical literature, especially in the context of the problem of 2-way and multi-way graph partitioning. The most classical of all multi-way partitioning algorithms is the Kernighan-Lin method [56]. Such methods can be used in conjunction with graph coarsening methods in order to provide efficient solutions. These methods are known as multilevel graph partitioning techniques. A particularly popular algorithm in this category is METIS [53].

A number of methods are commonly used in the literature in order to create partitions from graphs:

- *Generative Models:* As discussed earlier in this chapter, it is possible to define a generative model for practically any clustering problem, as long as an appropriate method exists for defining each component of the mixture as a probability distribution. An example of a generative model for network clustering is found in [77].

- *Classical Combinatorial Algorithms:* These methods use network flow [13] or other iterative combinatorial techniques in order to create partitions from the underlying graph. It should be pointed out that even edge sampling is often known to create good quality partitions, when it is repeated many times [51]. It is often desirable to determine cuts which are well balanced across different partitions, because the cut with the smallest absolute value often contains the large majority of the nodes in a single partition and a very small number of nodes in the remaining partitions. Different kinds of objective functions in terms of creating the cut, such as the unnormalized cut, normalized cut, and ratio cut, provide different tradeoffs between cluster balance and solution quality [73]. It should be pointed out that since graph cuts are a combinatorial optimization problem, they can be formulated as integer programs.

- *Spectral Methods:* Spectral methods can be viewed as *linear programming relaxations* to the integer programs representing the optimization of graph cuts. Different objective functions can be constructed for different kinds of cuts, such as the unnormalized cut, ratio cut, and normalized cut. Thus, the continuous solutions to these linear programs can be used to create a multidimensional embedding for the nodes, on which conventional k-means algorithms can be applied. These linear programs can be shown to take on a specially convenient form, in which the generalized eigenvectors of the graph Laplacian correspond to solutions of the optimization problem.

- *Nonnegative Matrix Factorization:* Since a graph can be represented as an adjacency matrix, nonnegative matrix factorization can be used in order to decompose it into two low rank matrices. It is possible to apply the matrix factorization methods either to the node–node adjacency matrix or the node–edge adjacency matrix to obtain different kinds of insights. It is also relatively easy to augment the matrix with content to create analytical methods, which can cluster with a combination of content and structure [69].

While the aforementioned methods represent a sampling of the important graph clustering methods, numerous other objective functions are possible for the construction of graph cuts such as the use of modularity-based objective functions [24]. Furthermore, the problem becomes even more challenging in the context of social networks, where content may be available at either the nodes [89] or edges [69]. Surveys on network clustering may be found in [36, 72]. Algorithms for network clustering are discussed in detail in Chapter 17.

1.3.7 Clustering Uncertain Data

Many forms of data either are of low fidelity or the quality of the data has been intentionally degraded in order to design different kinds of network mining algorithms. This has lead to the field of probabilistic databases. Probabilistic data can be represented either in the form of attribute-wise uncertainty or in the form of a *possible worlds* model, in which only particular subsets of attributes can be present in the data at a given time [3]. The key idea here is that the incorporation of probabilistic information can improve the quality of data mining algorithms. For example, if two attributes are equally desirable to use in an algorithm in the deterministic scenario, but one of them has greater uncertainty than the other, then the attribute with less uncertainty is clearly more desirable for use. Uncertain clustering algorithms have also been extended recently to the domain of streams and graphs. In the context of graphs, it is often desirable to determine the most reliable subgraphs in the underlying network. These are the subgraphs which are the most difficult to disconnect under edge uncertainty. A discussion of algorithms for reliable graph clustering may be found in [62]. Uncertain data clustering algorithms are discussed in Chapter 18.

1.4 Insights Gained from Different Variations of Cluster Analysis

While the aforementioned methods discuss the basics of the different clustering algorithms, it is often possible to obtain enhanced insights either by using more rigorous analysis or by incorporating additional techniques or data inputs. These methods are particularly important because of the subjectivity of the clustering process, and the many different ways in which the same data set can be clustered. How do we know that a particular clustering is good or that it solves the needs of the application? There are numerous ways in which these issues can be explored. The exploration could be through interactive visualization and human interaction, external knowledge-based supervision, explicitly exploring the multiple solutions to evaluate different possibilities, combining the multiple solutions to create more robust ensembles, or trying to judge the quality of different solutions with the use of different validation criteria. For example, the presence of labels adds domain knowledge to the clustering process, which can be used in order to improve insights about the quality of the clustering. This approach is referred to as semisupervision. A different way of incorporating domain knowledge (and indirect supervision) would be for the human to explore the clusters interactively with the use of visual methods and interact with the clustering software in order to create more meaningful clusters. The following subsections will discuss these alternatives in detail.

1.4.1 Visual Insights

Visual analysis of multidimensional data is a very intuitive way to explore the structure of the underlying data, possibly incorporating human feedback into the process. Thus, this approach could be considered an informal type of supervision, when human feedback is incorporated into cluster generation. The major advantage of incorporating human interaction is that a human can often provide intuitive insights, which are not possible from an automated computer program of significant complexity. On the other hand, a computer is much faster at the detailed calculations required both for clustering and for "guessing" the most appropriate feature-specific views of high dimensional data. Thus, a combination of a human and a computer often provides clusters which are superior to those created by either.

One of the most well-known systems for visualization of high-dimensional clusters is the HD-Eye method [44], which explores different subspaces of the data in order to determine clusters in different feature-specific views of the data. Another well-known technique is the *IPCLUS* method [2]. The latter method generates feature-specific views in which the data is *well polarized.* A well-polarized view refers to a 2-dimensional subset of features in which the data clearly separates out into clusters. A kernel-density estimation method is used to determine the views in which the data is well polarized. The final clustering is determined by exploring different views of the data, and counting how the data separates out into clusters in these different views. This process is essentially an ensemble-based method, an approach which is used popularly in the clustering literature, and will be discussed in a later part of this section. Methods for both incorporating and extracting visual insights from the clustering process are discussed in Chapter 19.

1.4.2 Supervised Insights

The same data set may often be clustered in multiple ways, especially when the dimensionality of the data set is high and subspace methods are used. Different features may be more relevant to different kinds of applications and insights. Since clustering is often used as an intermediate step in many data mining algorithms, it then becomes difficult to choose a particular kind of clustering that may suit that application. The subjectiveness of clustering is highly recognized, and small changes in the underlying algorithm or data set may lead to significant changes in the underlying clusters. In many cases, this subjectiveness also implies that it is difficult to refer to one particular clustering as significantly better than another. It is here that supervision can often play an effective role, because it takes the specific goal of the analyst into consideration.

Consider a document clustering application, in which a web portal creator (analyst) wishes to segment the documents into a number of categories. In such cases, the analyst may already have an approximate idea of the categories in which he is interested, but he may not have *fully* settled on a particular set of categories. This is because the data may also contain as yet unknown categories in which the analyst is interested. In such cases, semisupervision is an appropriate way to approach the problem. A number of labeled examples are provided, which approximately represent the categories in which the analyst is interested. This is used as domain knowledge or prior knowledge, which is used in order to supervise the clustering process. For example, a very simple form of supervision would be to use seeding, in which the documents of the appropriate categories are provided as (some of the) seeds to a representative clustering algorithm such as k-means. In recent years, spectral methods have also been heavily adapted for the problem of semisupervised clustering. In these methods, Laplacian smoothing is used on the labels to generate the clusters. This allows the learning of the lower dimensional data surface in a semisupervised way, as it relates to the underlying clusters. The area of semisupervised clustering is also sometimes referred to as constrained clustering. An excellent discussion on constrained clustering algorithms may be found in [18]. A number of interesting methods for semisupervised clustering are discussed in Chapter 20, with a special focus on the graph-based algorithms.

1.4.3 Multiview and Ensemble-Based Insights

As discussed above, one of the major issues in the clustering process is that different kinds of clusters are possible. When no supervision is available, the bewildering number of possibilities in the clustering process can sometimes be problematic for the analyst, especially from the perspective of interpretability. These are referred to as alternative clusters, and technically represent the behavior from different perspectives. In many cases, the ability to provide different clustering solutions that are significantly different provides insights to the analyst about the key clustering properties of the underlying data. This broader area is also referred to as multiview clustering.

The most naive method for multiview clustering is to simply run the clustering algorithm multiple times, and then examine the different clustering solutions to determine those which are different. A somewhat different approach is to use spectral methods in order to create approximately orthogonal clusters. Recall that the eigenvectors of the Laplacian matrix represent alternative cuts in the graph and that the small eigenvectors represent the best cuts. Thus, by applying a 2-means algorithm to the embedding on each eigenvector, it is possible to create a clustering which is very different from the clustering created by other (orthogonal) eigenvectors. The orthogonality of the eigenvectors is important, because it implies that the embedded representations are very different. Furthermore, the smallest eigenvectors represent the best clusterings, whereas clusterings derived from successively larger eigenvectors represent successively suboptimal solutions. Thus, this approach not only provides alternative clusterings which are quite different from one another, but also provides a ranking of the quality of the different solutions. A discussion of alternative clustering methods is provided in Chapter 21.

In many cases, the alternative clustering methods can be combined to create more robust solutions with the use of ensemble-based techniques. The idea here is that a combination of the output of the different clusterings provides a more robust picture of how the points are related to one another. Therefore, the outputs of the different alternatives can be used as input to a meta-algorithm which combines the results from the different algorithms. Such an approach provides a more robust clustering solution. A discussion of ensemble-based methods for clustering is provided in Chapter 22.

1.4.4 Validation-Based Insights

Given a particular clustering, how do we know what the quality of the clustering really is? While one possible approach is to use synthetic data to determine the matching between the input and output clusters, it is not fully satisfying to rely on only synthetic data. This is because the results on synthetic data may often be specific to a particular algorithm and may not be easily generalizable to arbitrary data sets.

Therefore, it is desirable to use validation criteria on the basis of real data sets. The problem in the context of the clustering problem is that the criteria for quality is not quite as crisp as many other data mining problems such as classification, where external validation criteria are available in the form of labels. Therefore, the use of one or more criteria may inadvertently favor different algorithms. As the following discussion suggests, clustering is a problem in which precise quantification is often not possible because of its unsupervised nature. Nevertheless, many techniques provide a partial understanding of the underlying clusters. Some common techniques for cluster validation are as follows:

- A common method in the literature is to use case studies to illustrate the subjective quality of the clusters. While case studies provide good intuitive insights, they are not particularly effective for providing a more rigorous quantification of the quality. It is often difficult to compare two clustering methods from a quantitative perspective with the use of such an approach.
- Specific measurements of the clusters such as the cluster radius or density may be used in order to provide a measure of quality. The problem here is that these measures may favor different algorithms in a different way. For example, a k-means approach will typically be superior to a density-based clustering method in terms of average cluster radius, but a density-based method may be superior to a k-means algorithms in terms of the estimated density of the clusters. This is because there is a circularity in using a particular criterion to evaluate the algorithm, when the same criterion is used for clustering purposes. This results in a bias during the evaluation. On the other hand, it may sometimes be possible to reasonably compare two different algorithms of a very similar type (e.g., two variations of k-means) on the basis of a particular criterion.
- In many data sets, labels may be associated with the data points. In such cases, cluster quality can be measured in terms of the correlations of the clusters with the data labels. This provides an external validation criterion, if the labels have not been used in the clustering process. However, such an approach is not perfect, because the class labels may not always align with the natural clusters in the data. Nevertheless, the approach is still considered more "impartial" than the other two methods discussed above and is commonly used for cluster evaluation.

A detailed discussion of the validation methods commonly used in clustering algorithms is provided in Chapter 23.

1.5 Discussion and Conclusions

Clustering is one of the most fundamental data mining problems because of its numerous applications to customer segmentation, target marketing, and data summarization. Numerous classes of methods have been proposed in the literature, such as probabilistic methods, distance-based methods, density-based methods, grid-based methods, factorization techniques, and spectral methods. The problem of clustering has close relationships to the problems of dimensionality reduction, especially through projected clustering formulations. High-dimensional data is often challenging for analysis, because of the increasing sparsity of the data. Clustering methods can be viewed as an integration of feature selection/dimensionality reduction methods with clustering.

The increasing advances in hardware technology allow the collection of large amounts of data through in an ever-increasing number of ways. This requires dedicated methods for streaming and distributed processing. Streaming methods typically work with only one pass over the data, and explicitly account for temporal locality in the clustering methods. Big data methods develop distributed techniques for clustering, especially through the *MapReduce* framework. The diversity of different data types significantly adds to the richness of the clustering problems. Many variations and enhancements of clustering such as visual methods, ensemble methods, multiview methods, or supervised methods can be used to improve the quality of the insights obtained from the clustering process.

Bibliography

[1] C. Aggarwal. *Data Streams: Models and Algorithms*, Springer, 2007.

[2] C. C. Aggarwal. A human-computer interactive system for projected clustering, *ACM KDD Conference*, 2001.

[3] C. Aggarwal. *Managing and Mining Uncertain Data*, Springer, 2009.

[4] C. Aggarwal. Re-designing distance functions and distance-based applications for high dimensional data, *ACM SIGMOD Record*, March 2001.

[5] C. Aggarwal. Towards systematic design of distance functions for data mining applications, *KDD Conference*, 2003.

[6] R. Agrawal, J. Gehrke, D. Gunopulos, and P. Raghavan. Automatic subspace clustering of high dimensional data for data mining applications. *ACM SIGMOD Conference*, 1998.

[7] C. Aggarwal, J. Han, J. Wang, and P. Yu. A framework for clustering evolving data streams. In *VLDB Conference*, 2003.

[8] C. Aggarwal, A. Hinneburg, and D. Keim. On the surprising behavior of distance metrics in high dimensional space, *ICDT Conference*, 2001.

[9] C. Aggarwal, C. Procopiuc, J. Wolf, P. Yu, and J.-S. Park. Fast algorithms for projected clustering. In *ACM SIGMOD Conference*, 1999.

[10] C. Aggarwal and P. Yu. Finding generalized projected clusters in high dimensional spaces, *ACM SIGMOD Conference*, 2000.

[11] C. Aggarwal and P. Yu. The Igrid index: Reversing the dimensionality curse for similarity indexing in high dimensional space, *KDD Conference*, 2000.

[12] C. Aggarwal and C. Zhai. A survey of text clustering algorithms, *Mining Text Data*, Springer, 2012.

[13] R Ahuja, T. Magnanti, and J. Orlin. *Network Flows: Theory, Algorithms and Applications*, Prentice Hall, 1993.

[14] M. Anderberg. *Cluster Analysis for Applications.* Academic Press, 1973.

[15] P. Andritsos et al. LIMBO: Scalable clustering of categorical data. *EDBT Conference*, 2004.

[16] A. Baraldi and P. Blonda. A survey of fuzzy clustering algorithms for pattern recognition. I. *IEEE Transactions on Systems, Man, and Cybernetics, Part B: Cybernetics*, 29(6):778–785, 1999.

[17] A. Baraldi and P. Blonda. A survey of fuzzy clustering algorithms for pattern recognition. II. *IEEE Transactions on Systems, Man, and Cybernetics, Part B: Cybernetics*, 29(6):786–801, 1999.

[18] S. Basu, I. Davidson, and K. Wagstaff. *Constrained clustering: Advances in theory, Algorithms and applications,* Chapman and Hall, 2008.

[19] M. Belkin and P. Niyogi. Laplacian eigenmaps for dimensionality reduction and data representation, *Neural Computation*, 15(6), 2003.

[20] P. Berkhin. A survey of clustering data mining techniques. *Grouping Multidimensional Data*, pages 25–71, 2006.

[21] K. Beyer, J. Goldstein, U. Shaft, and R. Ramakrishnan. When is nearest neighbor meaningful? *ICDT Conference*, 2001.

[22] D. Blei, A. Ng, and M. Jordan. Latent Dirichlet allocation, *Journal of Machine Learning Research*, 3:993–1022, 2003.

[23] K. Chakrabarti and S. Mehrotra. Local dimensionality reduction: A new approach to indexing high dimensional spaces. *VLDB Conference Proceedings*, pages 89–100, 2000.

[24] A. Clauset, M. E. J. Newman, and C. Moore, Finding community structure in very large networks, *Physical Review E* 70:066111, 2004.

[25] D. Cutting, D. Karger, J. Pedersen, and J. Tukey. Scatter/gather: A cluster-based approach to browsing large document collections. *ACM SIGIR Conference*, pages 318–329,1992.

[26] R. Cordeiro, C. Traina Jr., A. Traina, J. López, U. Kang, and C. Faloutsos. Clustering very large multi-dimensional datasets with mapreduce. In *KDD*, pages 690–698, 2011.

[27] A. P. Dempster, N. M. Laird, and D. B. Rubin. Maximum likelihood from incomplete data via the EM algorithm, *Journal of the Royal Statistical Society*, B, 39(1):1–38, 1977.

[28] J. Dean and S. Ghemawat. MapReduce: A flexible data processing took, *Communication of the ACM*, 53: 72–77, 2010.

[29] I. Dhillon. Co-clustering documents and words using bipartite spectral graph partitioning, *ACM KDD Conference*, 2001.

[30] I. Dhillon, S. Mallela, and D. Modha. Information-theoretic co-clustering, *ACM KDD Conference*, 2003.

[31] B. Duran. *Cluster Analysis: A Survey*. Springer-Verlag, 1974.

[32] G. Das and H. Mannila. Context-based similarity measures for categorical databases. *PKDD Conference*, pages 201–210, 2000.

[33] M. B. Eisen, P. T. Spellman, P. O. Brown, and D. Botstein. Cluster analysis and display of genome-wide expression patterns. *Proceedings of the National Academy of Sciences*, 95(25):14863–14868, 1998. `http://rana.lbl.gov/EisenSoftware.htm`

[34] M. Ester, H.-P. Kriegel, J. Sander, and X. Xu. A density-based algorithm for discovering clusters in large spatial databases with noise. *ACM KDD Conference*, pages 226–231, 1996.

[35] M. Filippone, F. Camastra, F. Masulli, and S. Rovetta. A survey of kernel and spectral methods for clustering. *Pattern Recognition*, 41(1):176–190, 2008.

[36] S. Fortunato. Community detection in graphs, *Physics Reports*, 486(3–5):75–174, February 2010.

[37] G. Gan, C. Ma, and J. Wu. *Data Clustering: Theory, Algorithms, and Applications.* SIAM, Society for Industrial and Applied Mathematics, 2007.

[38] V. Ganti, J. Gehrke, and R. Ramakrishnan. CACTUS-clustering categorical data using summaries. *ACM KDD Conference*, 1999.

[39] D. Gibson, J. Kleiberg, and P. Raghavan. Clustering categorical data: An approach based on Dynamical Systems. *VLDB Conference*, 1998.

[40] S. Guha, R. Rastogi, and K. Shim. ROCK: A robust clustering algorithm for categorical attributes. *Information Systems*, 25(5):345–366, 2000.

[41] S. Guha, R. Rastogi, and K. Shim. CURE: An efficient clustering algorithm for large databases. In *ACM SIGMOD Conference*, 1998.

[42] D. Gunopulos and G. Das. Time-series similarity measures, and time series indexing, *ACM SIGMOD Conference*, 2001.

[43] A. Hinneburg, C. Aggarwal, and D. Keim. What is the nearest neighbor in high dimensional spaces, *VLDB Conference*, 2000.

[44] A. Hinneburg, D. Keim, and M. Wawryniuk. Hd-eye: Visual mining of high-dimensional data. *IEEE Computer Graphics and Applications*, 19:22–31, 1999.

[45] T. Hofmann. Probabilistic latent semantic indexing. *ACM SIGIR Conference*, 1999.

[46] A. Jain. Data clustering: 50 years beyond k-means. *Pattern Recognition Letters*, 31(8):651–666, 2010.

[47] A. Jain and R. Dubes. *Algorithms for Clustering Data.* Prentice Hall, 1988.

[48] A. Jain, M. Murty, and P. Flynn. Data clustering: A review. *ACM Computing Surveys (CSUR)*, 31(3):264–323, 1999.

[49] D. Jiang, C. Tang, and A. Zhang. Cluster analysis for gene expression data: A survey. *IEEE Transactions on Knowledge and Data Engineering*, 16(11):1370–1386, 2004.

[50] I. Jolliffe. *Principal Component Analysis*, Springer, 2002.

[51] D. R. Karger. Random sampling in cut, flow, and network design problems, *STOC*, pp. 648–657, 1994.

[52] L. Kaufman and P. Rousseeuw. *Finding Groups in Data: An Introduction to Cluster Analysis.* Wiley-Interscience, 2005.

[53] G. Karypis and V. Kumar. Multilevel k-way partitioning scheme for irregular graphs. *Journal of Parallel and Distributed Computing*, 48(1):96–129, 1998.

[54] H.-P. Kriegel, P. Kröger, and A. Zimek. Clustering high-dimensional data: A survey on subspace clustering, pattern-based clustering, and correlation clustering. *ACM Transactions on Knowledge Discovery from Data (TKDD)*, 3(1):1–58, 2009.

[55] D. D. Lee and H. S. Seung. Learning the parts of objects by non-negative matrix factorization, *Nature*, 401:788–791, 1999.

[56] B.W. Kernighan and S. Lin. An efficient heuristic procedure for partitioning graphs, *Bell System Tech. Journal*, 49:291–307, Feb. 1970.

[57] C. Ding, X. He, and H. Simon. On the equivalence of nonnegative matrix factorization and spectral clustering. *SDM Conference*, 2005.

[58] D. Gusfield. *Algorithms for Strings, Trees and Sequences*, Cambridge University Press, 1997.

[59] J.-G. Lee, J. Han, and K.-Y. Whang. Trajectory clustering: A partition-and-group framework. *SIGMOD Conference*, 593–604, 2007.

[60] T. Liao. Clustering of time series data—A survey. *Pattern Recognition*, 38(11):1857–1874, 2005.

[61] H. Liu and H. Motoda. *Computational Methods of Feature Selection*, Chapman and Hall (CRC Press), 2007.

[62] L. Liu, R. Jin, C. Aggarwal, and Y. Shen. Reliable clustering on uncertain graphs, *ICDM Conference*, 2012.

[63] U. von Luxberg. A tutorial on spectral clustering, *Statistics and Computing*, Springer, 2007.

[64] S. Madeira and A. Oliveira. Biclustering algorithms for biological data analysis: A survey. *IEEE/ACM Transactions on Computational Biology and Bioinformatics*, 1(1):24–45, 2004.

[65] F. Murtagh. A survey of recent advances in hierarchical clustering algorithms. *The Computer Journal*, 26(4):354–359, 1983.

[66] R. T. Ng and J. Han. CLARANS: A method for clustering objects for spatial data mining. *IEEE Trans. Knowl. Data Eng.*, 14(5):1003–1016, 2002

[67] L. Parsons, E. Haque, and H. Liu. Subspace clustering for high dimensional data: A review. *ACM SIGKDD Explorations*, 6(1):90–105, 2004.

[68] G. Qi, C. Aggarwal, and T. Huang. Online community detection in social sensing. *WSDM Conference*, 2013.

[69] G. Qi, C. Aggarwal, and T. Huang. Community detection with edge content in social media networks, *ICDE Coference*, 2013.

[70] L. R. Rabiner. A tutorial on hidden Markov models and selected applications in speech recognition, *Proceedings of the IEEE*, 77(2):257–285, Feb. 1989.

[71] M. Rege, M. Dong, and F. Fotouhi. Co-clustering documents and words using bipartite isoperimetric graph partitioning. *ICDM Conference*, pp. 532–541, 2006.

[72] S. Schaeffer. Graph clustering. *Computer Science Review*, 1(1):27–64, 2007.

[73] J. Shi and J. Malik. Normalized cuts and image segmentation. *IEEE Transaction on Pattern Analysis and Machine Intelligence*, 2000.

[74] B. W. Silverman. *Density Estimation for Statistics and Data Analysis*. Chapman and Hall, 1986.

[75] H. Schutze and C. Silverstein. Projections for efficient document clustering, *ACM SIGIR Conference*, 1997.

[76] P. Smyth. Clustering sequences with hidden Markov models, *Neural Information Processing*, 1997.

[77] Y. Sun, C. Aggarwal, and J. Han. Relation-strength aware clustering of heterogeneous information networks with incomplete attributes, *Journal of Proceedings of the VLDB Endowment*, 5(5):394–405, 2012.

[78] J. Tenenbaum, V. de Silva, and J. Langford. A global geometric framework for nonlinear dimensionality reduction. *Science*, 290(5500):2319–2323, 2000.

[79] P. Willett. Recent trends in hierarchical document clustering: A critical review. *Information Processing & Management*, 24(5):577–597, 1988.

[80] R. Xu and D. Wunsch. Survey of clustering algorithms. *IEEE Transactions on Neural Networks*, 16(3):645–678, 2005.

[81] R. Xu and D. Wunsch. *Clustering*. Wiley-IEEE Press, 2008.

[82] F. Wang and J. Sun. Distance metric learning in data mining, *SDM Conference (Tutorial)*, 2012.

[83] W. Wang, J. Yang, and R. Muntz. Sting: A statistical information grid approach to spatial data mining. *VLDB Conference*, 1997.

[84] T. White. Hadoop: The Definitive Guide. *Yahoo! Press*, 2011.

[85] J. Yang and W. Wang. CLUSEQ: Efficient and effective sequence clustering, *ICDE Conference*, 2003.

[86] B.-K. Yi, N. D. Sidiropoulos, T. Johnson, H. Jagadish, C. Faloutsos, and A. Biliris. Online data mining for co-evolving time sequences. *ICDE Conference*, 2000.

[87] T. Zhang, R. Ramakrishnan, and M. Livny. BIRCH: An efficient data clustering method for very large databases. In *ACM SIGMOD Conference*, 1996.

[88] Y. Zhao, G. Karypis, and U. Fayyad. Hierarchical clustering algorithms for document datasets. *Data Mining and Knowledge Discovery*, 10(2):141–168, 2005.

[89] Y. Zhou, H. Cheng, and J. X. Yu, Graph clustering based on structural/attribute similarities, *Proc. VLDB Endow.*, 2(1):718–729, 2009.

[90] Y. Zhu and D. Shasha. StatStream: Statistical monitoring of thousands of data streams in real time. *VLDB Conference*, pages 358–369, 2002.

Chapter 2

Feature Selection for Clustering: A Review

Salem Alelyani
Arizona State University
Tempe, AZ
salelyan@asu.edu

Jiliang Tang
Arizona State University
Tempe, AZ
Jiliang.Tang@asu.edu

Huan Liu
Arizona State University
Tempe, AZ
huan.liu@asu.edu

2.1 Introduction

The growth of the high-throughput technologies nowadays has led to exponential growth in the harvested data with respect to dimensionality and sample size. As a consequence, storing and processing these data becomes more challenging. Figure (2.1) shows the trend of this growth for UCI Machine Learning Repository. This augmentation made manual processing for these datasets impractical. Therefore, data mining and machine learning tools were proposed to automate pattern recognition and the knowledge discovery process. However, using data mining techniques on ore data is mostly useless due to the high level of noise associated with collected samples. Usually, data noise is either due to imperfection in the technologies that collected the data or to the nature of the source of this data. For instance, in the medical images domain, any deficiency in the imaging device will be reflected as noise in the dataset later on. This kind of noise is caused by the device itself. On the other hand, text datasets crawled from the Internet are noisy by nature because they are usually informally written and suffer from grammatical mistakes, misspelling, and improper punctuation. Undoubtedly, extracting useful knowledge from such huge and noisy datasets is a painful task.

Dimensionality reduction is one popular technique to remove noisy (i.e., irrelevant) and redundant attributes (aka features). Dimensionality reduction techniques can be categorized mainly into feature extraction and feature selection. In the feature extraction approach, features are projected into a new space with lower dimensionality. Examples of the feature extraction technique include Principle Component Analysis (PCA), Linear Discriminant Analysis (LDA), and Singular Value Decomposition (SVD), to name a few. On the other hand, the feature selection approach aims to select a small subset of features that minimize redundancy and maximize relevance to the target (i.e., class label). Popular feature selection techniques include Information Gain, Relief, Chi Squares, Fisher Score, and Lasso, to name a few.

Both dimensionality reduction approaches are capable of improving learning performance, lowering computational complexity, building better generalizable models, and decreasing required storage. However, feature selection is superior in terms of better readability and interpretability since it maintains the original feature values in the reduced space, while feature extraction transforms the data from the original space into a new space with lower dimension, which cannot be linked to the features in the original space. Therefore, further analysis of the new space is problematic since there is no physical meaning for the transformed features obtained from the feature extraction technique.

Feature selection is broadly categorized into four models: filter model, wrapper model, embedded model, and hybrid model. As mentioned above, feature selection selects subset of highly discriminant features. In other words, it selects features that are capable of discriminating samples that belong to different classes. Thus, we need to have labeled samples as training samples in order

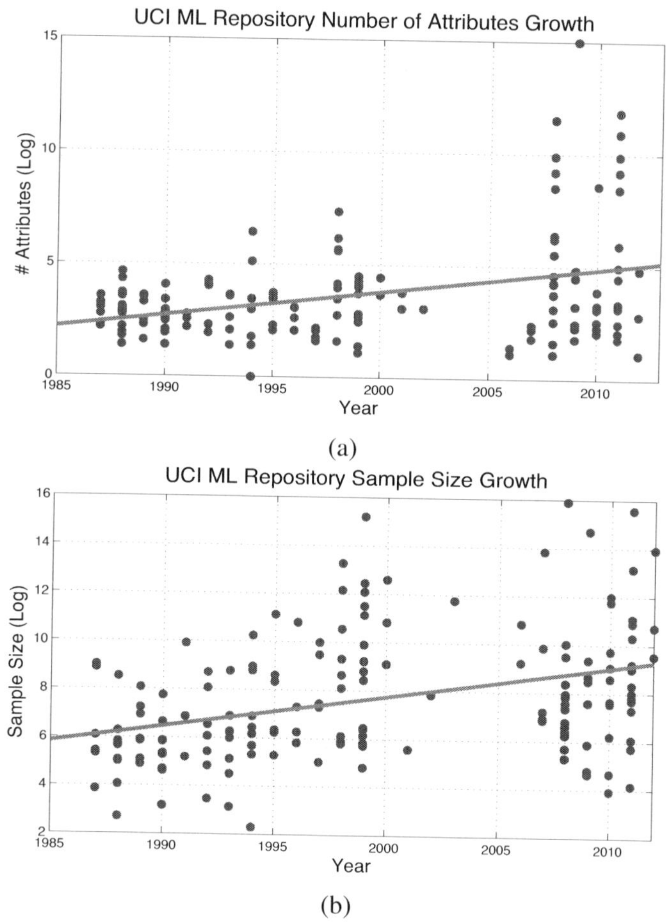

FIGURE 2.1: Plot (a) shows the dimensionality growth trend in UCI Machine Learning Repository from mid' 80s to 2012 while (b) shows the growth in the sample size for the same period.

to select these features. This kind of learning is called *supervised learning*, which means that the dataset is labeled. In supervised learning, it is easy to define what *relevant feature* means. It simply refers to the feature that is capable of distinguishing different classes. For example, a feature f_i is said to be relevant to a class c_j if f_i and c_j are highly correlated.

Unlabeled data poses yet another challenge in feature selection. In such cases, defining relevancy becomes unclear. However, we still believe that selecting subset(s) of features may help improve *unsupervised learning* in a way similar to improving the supervised learning. One of the most utilized unsupervised learning technique is *data clustering*. Data clustering is the unsupervised classification of samples into groups. In other words, it is the technique that aims to group similar samples into one group called a *cluster*. Each cluster has maximum within-cluster similarity and minimum between-cluster similarity based on certain similarity index. However, finding clusters in high-dimensional space is computationally expensive and may degrade the learning performance. Furthermore, equally good candidate of features' subsets may produce different clusters. Therefore, we demand to utilize feature selection for clustering to alleviate the effect of high-dimensionality.

We will review the literature of data clustering in Section 2.1.1 followed by general discussion about feature selection models in Section 2.1.2 and feature selection for clustering in Section 2.1.3.

2.1.1 Data Clustering

Due to the increase in data size, human manual labeling has become extremely difficult and expensive. Therefore, automatic labeling has become an indispensable step in data mining. Data clustering is one of the most popular data labeling techniques. In data clustering, we are given unlabeled data and are to put similar samples in one pile, called a cluster, and the dissimilar samples should be in different clusters. Usually, neither cluster's description nor its quantification is given in advance unless a domain knowledge exists, which poses a great challenge in data clustering.

Clustering is useful in several machine learning and data mining tasks including image segmentation, information retrieval, pattern recognition, pattern classification, network analysis, and so on. It can be seen as either an exploratory task or preprocessing step. If the goal is to explore and reveal the hidden patterns in the data, clustering becomes a stand-alone exploratory task by itself. However, if the generated clusters are going to be used to facilitate another data mining or machine learning task, clustering will be a preprocessing step.

There are many clustering methods in the literature. These methods can be categorized broadly into partitioning methods, hierarchical methods, and density-based methods. The partitioning methods use a distance-based metric to cluster the points based on their similarity. Algorithms belonging to this type produce one level partitioning and nonoverlapping spherical shaped clusters. K-means and k-medoids are popular partitioning algorithms. The hierarchical method, on the other hand, partitions the data into different levels that look like a hierarchy. This kind of clustering helps in data visualization and summarization. Hierarchical clustering can be done in either bottom-up (i.e., agglomerative) fashion or top-down (i.e., divisive) fashion. Examples of this type of clustering are BIRCH, Chameleon, AGNES, and DIANA. Unlike these two clustering techniques, density-based clustering can capture arbitrarily shaped clusters such as S-shape. Data points in dense regions will form a cluster while data points from different clusters will be separated by low density regions. DBSCAN and OPTICS are popular examples of density-based clustering methods.

2.1.2 Feature Selection

In the past thirty years, the dimensionality of the data involved in machine learning and data mining tasks has increased explosively. Data with extremely high dimensionality has presented serious challenges to existing learning methods [35], known as the curse of dimensionality [23]. With the existence of a large number of features, learning models tend to overfit and their learning performance degenerates. To address the problem of the curse of dimensionality, dimensionality reduction techniques have been studied, which form an important branch in the machine learning research area. Feature selection is one of the most used techniques to reduce dimensionality among practitioners. It aims to choose a small subset of the relevant features from the original ones according to certain relevance evaluation criterion [36, 21], which usually leads to better learning performance, e.g., higher learning accuracy, lower computational cost, and better model interpretability. Feature selection has been successfully applied in many real applications, such as pattern recognition [26, 58, 44], text categorization [74, 29, 51], image processing [26, 55], and bioinformatics [45, 56], and so forth.

According to whether the label information is utilized, different feature selection algorithms can be categorized into supervised [69, 60], unsupervised [16, 44], or semisupervised algorithms [78, 73]. With respect to different selection strategies, feature selection algorithms can also be categorized as being of either the filter [37, 13], wrapper [31], hybrid, or embedded models [12, 50]. Feature selection algorithms of the filter model are independent of any classifier. They evaluate the relevance of a feature by studying its characteristics using certain statistical criteria. Relief [59], Fisher score [15], CFS [22], and FCBF [75] are among the most representative algorithms of the filter model. On the other hand, algorithms belonging to the wrapper model utilize a classifier as a selection criteria. In other words, they select a set of features that has the most discriminative power using a given classifier, such as SVM and KNN. Some examples of the wrapper model are the FSSEM [17] and ℓ_1SVM [8]. Other examples of the wrapper model could be any combination of a preferred search strategy and given classifier. Since the wrapper model depends on a given classifier, cross-validation is usually required in the evaluation process. It is in general more computationally expensive and biased to the chosen classifier. Therefore, in real applications, the filter model is more popular, especially for problems with large datasets. However, the wrapper model has been empirically proven to be superior, in terms of classification accuracy, to those of a filter model. Due to these shortcomings in each model, the hybrid model [12, 38], was proposed to bridge the gap between the filter and wrapper models. First, it incorporates the statistical criteria, as the filter model does, to select several candidate features subsets with a given cardinality. Second, it chooses the subset with the highest classification accuracy [38]. Thus, the hybrid model usually achieves both accuracy comparable to the wrapper and efficiency comparable to the filter model. Representative feature selection algorithms of the hybrid model include BBHFS [12], and HGA [52]. Finally, the embedded model performs feature selection in the learning time. In other words, it achieves model fitting and feature selection simultaneously. Examples of embedded model include C4.5 [53], BlogReg [11], and SBMLR [11]. Based on different types of outputs, most feature selection algorithms fall into one of the three categories: subset selection [76], which returns a subset of selected features identified by the index of the feature; feature weighting [59], which returns weight corresponding to each feature; and the hybrid of subset selection and feature weighting, which returns a ranked subset of features.

Feature weighting, on the other hand, is thought of as a generalization of feature selection [70]. In feature selection, a feature is assigned a binary weight, where 1 means the feature is selected and 0 otherwise. However, feature weighting assigns a value, usually in the interval [0,1] or [−1,1], to each feature. The greater this value is, the more salient the feature will be. Feature weighting was found to outperform a feature selection in tasks where features vary in their relevance score [70], which is true in most real-world problems. Feature weighting could be, also, reduced to feature selection if a threshold is set to select features based on their weights. Therefore, most of feature selection algorithms mentioned in this chapter can be considered as using a feature weighting scheme.

Typically, a feature selection method consists of four basic steps [38], namely, subset generation, subset evaluation, stopping criterion, and result validation. In the first step, a candidate feature subset will be chosen based on a given search strategy, which is sent, in the second step, to be evaluated according to certain criterion. The subset that best fits the evaluation criterion will be chosen from all the candidates that have been evaluated after the stopping criterion are met. In the final step, the chosen subset will be validated using domain knowledge or validation set.

2.1.3 Feature Selection for Clustering

The existence of irrelevant features in the data set may degrade learning quality and consume more memory and computational time that could be saved if these features were removed. From the clustering point of view, removing irrelevant features will not negatively affect clustering accuracy while reducing required storage and computational time. Figure 2.2 illustrates this notion where (a) shows the relevant feature f_1 which can discriminate clusters. Figures 2.2(b) and (c) show that f_2 and f_3 cannot distinguish the clusters; hence, they will not add any significant information to the clustering.

In addition, different relevant features may produce different clustering. Figure 2.3(a) shows four clusters by utilizing knowledge from f_1 and f_2, while Figure 2.3(b) shows two clusters if we use f_1 only. Similarly, Figure 2.3(c) shows two different clusters if we use f_2 only.Therefore, different subsets of relevant features may result in different clustering, which greatly helps discovering different hidden patterns in the data.

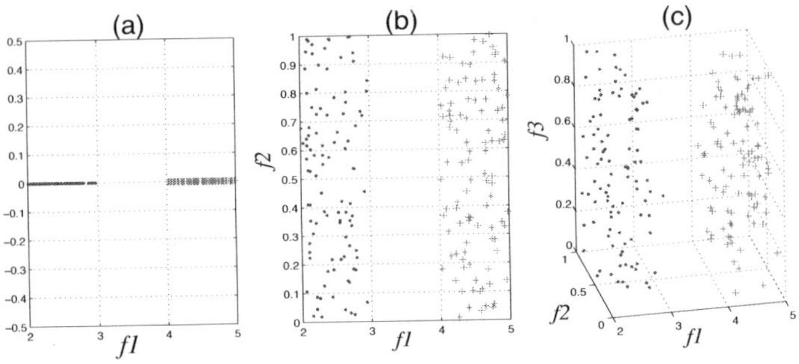

FIGURE 2.2: Feature f_1 is relevant while f_2 and f_3 are irrelevant. We are able to distinguish the two clusters from f_1 only. Thus, removing f_2 and f_3 will not affect the accuracy of clustering.

Motivated by these facts, different clustering techniques were proposed to utilize feature selection methods that eliminate irrelevant and redundant features while keeping relevant features to improve clustering efficiency and quality. For simplicity and better organization, we are going to describe different feature selection for clustering (FSC) methods based on the domain. The following sections will be organized as follows: conventional FSC, FSC in text data, FSC in streaming data, and FSC link data.

Similar to feature selection for supervised learning, methods of feature selection for clustering are categorized into filter [13] wrapper [54], and hybrid models [18]. A wrapper model evaluates the candidate feature subsets by the quality of clustering while a filter model is independent of clustering algorithm. Thus, the filter model is still preferable in terms of computational time and as unbiased toward any clustering method, while the wrapper model produces better clustering if we know the clustering method in advance. To alleviate the computational cost in the wrapper model, filtering criteria are utilized to select the candidate feature subsets in the hybrid model.

In the following subsections, we will briefly discuss feature selection for clustering methods that falls in the filter, wrapper, and hybrid models. For more about conventional methods, we refer the reader to [18].

2.1.3.1 Filter Model

Filter model methods do not utilize any clustering algorithm to test the quality of the features [18]. They evaluate the score of each feature according to certain criteria. Then, they select the features with the highest scores. It is called the filter since it filters out the irrelevant features using given criteria. Furthermore, feature evaluation could be either *univariate* or *multivariate*. Univariate means each feature is evaluated independently of the feature space. This approach is much faster

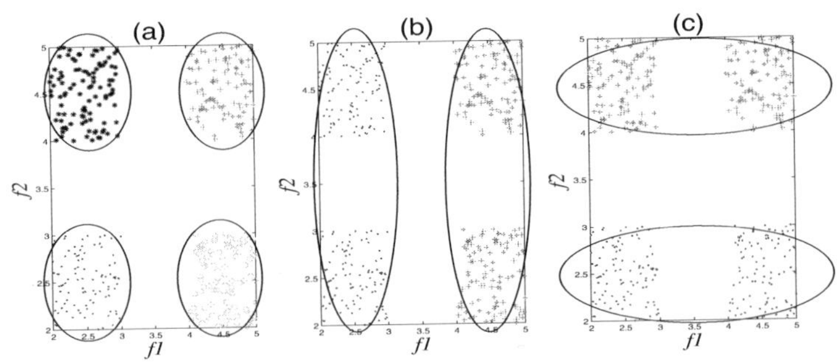

FIGURE 2.3: Different sets of features may produce different clustering.

TABLE 2.1: Nomenclature

D	Dataset
n	Sample size
m	Number of features
x_j	j^{th} sample
f_i	i^{th} feature
F	Selected feature set
l	Number of selected features
K	Number of clusters
C_k	k^{th} cluster

and more efficient than the multivariate, which evaluates features with respect to the other features. Therefore, the multivariate, unlike the univariate approach, is capable of handling redundant features. SPEC, see Section (2.2.1.1), is an example of the univariate filter model, although it was extended to the multivariate approach in [79]. Other examples of filter model criteria used in feature selection for clustering include feature dependency [62], entropy-based distance [13], and Laplacian score [24, 80].

2.1.3.2 Wrapper Model

The wrapper model utilizes a clustering algorithm to evaluate the quality of selected features. It starts by (1) finding a subset of features. Then, (2) it evaluates the clustering quality using the selected subset. Finally, it repeats (1) and (2) until the desired quality is found. Evaluating all possible subsets of features is impossible in high-dimensional datasets. Therefore, heuristic search strategy is adopted to reduce the search space. The wrapper model is very computationally expensive compared to filter model. Yet, it produces better clustering since we aim to select features that maximize the quality. It is still biased toward the clustering method used. Different wrapper feature selection methods for clustering were proposed by changing the combination of search strategy and the utilized clustering algorithm. The method proposed in [16] is an example of a wrapper that involves maximum likelihood criteria, feature selection, and a mixture of Gaussians as clustering method. Others use conventional clustering methods such as k-means and any search strategy as feature selector [30].

2.1.3.3 Hybrid Model

To overcome the drawbacks of filter and wrapper models a hybrid model is used to benefit from the efficient filtering criteria and better clustering quality from the wrapper model. A typical hybrid process goes through the following steps: (1) it utilizes filtering criteria to select different candidate subsets. Then, (2) it evaluates the quality of clustering of each candidate subsets. Finally, (3) the subset with the highest clustering quality is selected. Algorithms belonging to the hybrid model usually produce better clustering quality than those of the filter model, yet, they are less efficient. Compared to the wrapper model, the hybrid model is much more efficient.

2.2 Feature Selection for Clustering

Several feature selection for clustering methods have been proposed in the literature. Some algorithms handle text data, while others handle streaming data. Still others are capable of handling different kinds of data. In this section, we will discuss different methods with respect to data types they can handle. We will review algorithms for text, streaming, and linked data, as well as algorithms that are able to handle generic data.

2.2.1 Algorithms for Generic Data

In this section, we will discuss feature selection for clustering methods that are able to handle generic datasets. In other words, it is not necessary to be designed to handle only text, data stream, or linked data.

2.2.1.1 Spectral Feature Selection (SPEC)

Although Algorithm 1, the Spectral Feature Selection (SPEC) algorithm, is a unified framework that enables the joint study of supervised and unsupervised learning, we will use SPEC in this work as an example of filter-based unsupervised feature selection methods. SPEC [80] estimates the feature relevance by estimating feature consistency with the spectrum of a matrix derived from a similarity matrix S. SPEC uses the Radial-Basis Function (RBF) as a similarity function between two samples x_i and x_j:

$$S_{ij} = e^{-\frac{||x_i - x_j||^2}{2\sigma^2}} \tag{2.1}$$

Graph G will be constructed from S and adjacency matrix W will be constructed from G. Then, degree matrix $\bar{D}$ will be computed from W. $\bar{D}$ is the diagonal matrix where $\bar{D}_{ii} = \sum_{j=1}^{n} W_{ij}$. Given $\bar{D}$ and W, the Laplacian matrix L and the normalized Laplacian matrix $\mathcal{L}$ are computed as follows:

$$L = \bar{D} - W; \quad \mathcal{L} = \bar{D}^{-\frac{1}{2}} L \bar{D}^{-\frac{1}{2}} \tag{2.2}$$

The main idea behind SPEC is that the features consistent with the graph structure are assigned similar values to instances that are near to each other in the graph. Therefore, these features should be relevant since they behave similarly in each similar group of samples (i.e., clusters). Motivated by graph theory that states that graph structure information can be captured from its spectrum, SPEC studies how to select features according to the structure of the graph G induced from the samples similarity matrix S.

The weight of each feature f_i in SPEC is evaluated using three functions: ψ_1, ψ_2, and ψ_3. These functions were derived from the normalized cut function with the spectrum of the graph and extended to their more general forms. In this chapter, we will not explain these functions in detail, therefore, we refer the reader to [80] for more details. We assume here that each function ψ takes feature vector f_i and returns the weight based on the normalized Laplacian $\mathcal{L}$.

2.2.1.2 Laplacian Score (LS)

Laplacian Score (LS)[24] is a special case of SPEC if the ranking function used is

$$F_i \leftarrow \frac{\hat{f}_i^T L \hat{f}_i}{\hat{f}_i^T \bar{D} \hat{f}_i} \quad where \quad \hat{f}_i = f_i - \frac{f_i^T \bar{D}\mathbf{1}}{\mathbf{1}^T \bar{D}\mathbf{1}}\mathbf{1} \tag{2.3}$$

Where $\mathbf{1}$ is one vector. LS is very effective and efficient with respect to the data size. Similar to SPEC, the most time consuming step in LS is constructing the similarity matrix S. The beauty of this algorithm is that it can handle both labeled and unlabeled data.

Algorithm 1 Spectral Feature Selection (SPEC)

Input:

D: dataset

$\psi \in \{\psi_1, \psi_2, \psi_3\}$: feature weighting functions

n: number of samples

Output:

F: the ranked feature list

Construct similarity matrix S from D

Construct Graph G from S

Construct W from S

Construct $\bar{D}$ from W

Define L and $\mathcal{L}$ according to Equation (2.2)

for each feature vector f_i **do**

  $\hat{f}_i \leftarrow \frac{\bar{D}^{-\frac{1}{2}} f_i}{||\bar{D}^{-\frac{1}{2}} f_i||}$

  $F_i \leftarrow \psi(\hat{f}_i)$

end for

Rank F based on ψ

2.2.1.3 Feature Selection for Sparse Clustering

Witten and Tibshirani in [72] propose a framework for feature selection in sparse clustering. They apply Lasso-type, $\ell_1 - norm$, as a feature selection method embedded in the clustering process. This framework can be applied to any similarity-based clustering technique, yet, they used k-means clustering in [72]. The number of selected features l is chosen using gap statistics in a similar fashion to choosing the number of clusters in [67]. The proposed method attempts to minimize the following objective function with respect to the clusters $\{C_1, \ldots, C_K\}$ and the feature weight vector w :

$$\begin{aligned} \min \quad & \textstyle\sum_{j=1}^{m} w_j \Psi_j \\ \textit{subject to} \quad & ||w||^2 \leq 1, \\ & ||w||_1 \leq l, \\ & w_j \geq 0 \ \ \forall j \end{aligned}$$

Where Ψ_j is given by the following equation for k-means over j^{th} feature:

$$\Psi_j = \sum_{c=1}^{K} \frac{1}{n_c} \sum_{i,i' \in C_K} Sim(i,i',j) - \frac{1}{n} \sum_{i=1}^{n} \sum_{i'=1}^{n} Sim(i,i',j). \tag{2.4}$$

K is the number of clusters, n_c is the number of samples in cluster c, and $Sim(i,i',j)$ is the similarity index of sample i and i' using only the selected feature j. Optimizing Equation (2.4) is done using iterative algorithm by holding w fixed and optimizing Equation (2.4) with respect to the clusters $\{C_1, \ldots, C_K\}$. In this step, we apply standard k-means clustering on *n-by-n* similarity matrix using the jth feature. Then, we hold the clusters fixed and optimize Equation (2.4) with respect to w. w is set in this step to be

$$w = \frac{S(\Psi_+, \Delta)}{S(||\Psi_+, \Delta)||_2} \tag{2.5}$$

where Ψ_+ is the positive part of Ψ and $\Delta = 0$, when $||w||_1 \leq l$ and $\Delta > 0$ otherwise, so, $||w||_1 = l$. Algorithm 2 illustrates these steps of optimizing Equation (2.4).

Algorithm 2 Feature Selection for Sparse Clustering

Input:
D: dataset
l: number of selected features obtained from gab statistics-like approach
n: number of samples
Output:
the clusters and w
Initialize:
$w_1 = w_2, \ldots, w_m = \frac{1}{\sqrt{m}}$

```
while not converge do
    Hold w fixed
    Optimize Equation (2.4) with respect to C1,...,C_K
    Holding C_1,...,C_K fixed
    Optimize Equation (2.4) with respect to w by applying Equation (2.5)
end while
```

Witten and Tibshirani in [72], also, propose sparse hierarchical clustering based on lasso penalty similar to Algorithm 2. The hierarchical clustering involves *n-by-n* similarity matrix, which is optimized iteratively with w. Then, we perform hierarchical clustering on the constructed similarity matrix. For more about this algorithm we refer the reader to [72].

2.2.1.4 Localized Feature Selection Based on Scatter Separability (LFSBSS)

Li et al in [33] proposed a localized feature selection based on scatter separability (LFSBSS). This is motivated by the fact that the set of features that are relevant to one clustering result are not necessary the same set that is relevant to another clustering result. In other words, clustering dataset D using a set of feature F_1 may produce clusters $\{C_1, C_2, C_3\}$ while clustering using another set of features F_2 may lead to clusters $\{C_4, C_5\}$, where $F_1 \neq F_2$. This notion is also illustrated in Figure (2.3). Furthermore, each cluster in a clustering result may be associated to different set of relevant features. In document clustering, for instance, documents that belong to sport news are more likely to have different set of relevant terms such as: FIFA, Ball, and so on. While the set of documents that belong to technology news contains relevant terms such as: Apple, IBM, and so on. In this section we are will use the cluster set $C = \{(C_1, F_1), \ldots, (C_j, F_j), \ldots, (C_K, F_K)\}$ to refer to the clustering result where C_1 and F_1 are the first cluster and the set of selected features corresponds to the first cluster, respectively.

Li et al in [33] borrowed the notion of scatter separability from Dy and Brodley [16] and adopted it as a localized feature selection. They defined the scatter separability as

$$\Omega = tr(S_w^{-1} S_b)$$

where S_w^{-1} is the inverse of within-cluster separability and S_b is the between-cluster separability. If we need to evaluate Ω_i for cluster i, we should use the within that cluster separability $S_w^{(i)-1}$, instead. It was proven in [33] that Ω_i is monotonically increasing with dimensions as long as the clustering assignments remain the same. To mitigate this problem, separability criteria must be normalized with respect to the dimensionality for feature selection. Moreover, since localized feature selection attempts to select different sets of relevant features for each cluster, the between-cluster separability needs to be appropriately normalized as well. This is performed using cross-projecting over individual clusters. We assume that the projected cluster is $\hat{C} = \{(\hat{C}_1, \hat{F}_1), \ldots, (\hat{C}_K, \hat{F}_K)\}$. At each step of the projection, we replace the projected cluster $\hat{C}_i$ with the largest overlap and the original cluster C_j

to generate the new clustering $C^* = \{(C_1, F_1), \ldots, (\hat{C}_i, \hat{F}_i), \ldots, (C_K, F_K)\}$. Finally, we cross-project them into each other, which generates the normalized value, v, that allows us to compare two different clusters with different subspaces. Larger v indicates greater separability between clusters. More details about the projection may be found in [33].

LFSBSS reduces the impact of overlapping and unassigned data by penalizing using what they call adjusted normalized value a. This value penalizes the cross-projection if the amount of unsigned or overlap has increased in the projected cluster compared to the original clusters.

LFSBSS adopts the sequential backward feature selection. This means that the clusters are generated first using the whole feature space, then, iteratively removing irrelevant or noisy feature based on a from each cluster individually. Algorithm 3 illustrates the steps of LFSBSS.

Algorithm 3 Localized Feature Selection Based on Scatter Separability (LFSBSS)

Input:
D: dataset
l: number of selected features
n: number of samples
K: number of clusters
Output:
the clusters and their corresponding features sets
Initialize:
initialize C using all feature space F

$F'_1 = F'_2 = \cdots = F'_K = F$
while not converged **do**
  for $c = 1$ to K **do**
    Evaluate a for C_c
    Choose feature f_i to be removed based on a
    $F_c = F_c - f_i$
    Generate a new cluster set C' based on F_c
    Compare cluster in C' with clusters in C
    if *BetterClusterFound* **then**
      Replace the corresponding cluster in C
    end if
  end for
  if *Desired* **then**
    Process unassigned samples
  end if
end while

2.2.1.5 Multicluster Feature Selection (MCFS)

Similar to the motivation illustrated in Figure (2.3), Cai et al in [9] propose a multicluster feature selection (MCFS) method that is able to select the set of features that can cover all the possible clustering in the data. In MCFS, spectral analysis is used to measure the correlation between different features without needing label information. Using the top eigenvectors of graph Laplacian, spectral clustering can cluster data samples without utilizing label information. Thus, MCFS applied k-Nearest-Neighbors approach to construct the graph of the data samples, where k is a predetermined parameter. Next, the heat kernel weighting matrix W is computed as follows:

$$W_{ij} = e^{-\frac{\|x_i - x_j\|^2}{\sigma}}$$

where x_i and x_j are connected samples in the k-Nearest-Neighbors graph and σ is used as a predefined parameter. From W a degree matrix is computed as explained earlier in this chapter. Then, a graph Lapalcian matrix $L = \bar{D} - W$ is constructed. After that, MCFS solves the following eigen-problem:

$$Ly = \lambda \bar{D} y \tag{2.6}$$

Given $Y = [y_1, \ldots, y_K]$, the eigenvectors of Equation (2.6), we can find a relevant subset of features by minimizing the following objective function:

$$\begin{array}{ll} \min_{a_k} & ||y_k - X^T a_k||^2 \\ s.t. & ||a_k||_0 = l \end{array}$$

Where a_k is an $m-$dimensional vector and $||a_k||_0$ is the number of nonzero elements in a_k. Then, K sparse coefficient vectors will be chosen to correspond to each cluster. For each feature f_j, the maximum value of a_k that correspond to f_j will be chosen. Finally, MCFS will choose the top l features. MCFS shows improvement over other methods such as LS according to [9].

2.2.1.6 Feature Weighting *k*-Means

k-means clustering is one of the most popular clustering techniques. It has been extensively used in data mining and machine learning problems. A large number of k-means variations have been proposed to handle feature selection [7, 68, 28, 25, 46]. Most of these variations start by clustering the data into k clusters. Then, weight is assigned to each feature. The feature that minimizes within-cluster distance and maximizes between-cluster distance is preferred and, hence, gets higher weight. In [28], for example, an entropy weighting k-means (EWKM) was proposed for subspace clustering. It simultaneously minimizes the within-cluster dispersion and maximizes the negative weight entropy in the clustering process. EWKM calculates the weight of each feature in each cluster by including the weight entropy in the objective function of k-means. The subset of features corresponding to each cluster are, then, selected based on that weight. Thus, EWKM allows subspace clustering where the set of selected features may differ from one cluster to another.

In addition, [46] proposes feature weighting k-means clustering using generalized Fisher ratio that minimizes the ratio of the average of within-cluster distortion over the average between-cluster distortion. In this algorithm, several candidate clusterings are generated and the one with the minimal Fisher ratio is determined to be the final cluster.

Similarly, [25] proposes another variation of feature weighting k-means (W-k-means) that measures the weight of each feature based on its variance of the within-cluster distance. Algorithm 4 illustrates the process of W-k-means. It, iteratively, minimizes Equation (2.7) by fixing two parameters at each step and solving Ψ with respect to the third one. If there is no change in Ψ after the minimization, the algorithm is said to be converged.

$$\Psi(C, Z, \mathbf{w}) = \sum_{l=1}^{k} \sum_{i=1}^{n} \sum_{j=1}^{m} c_{il} w_j^{\beta} d(x_{ij}, z_{lj}) \tag{2.7}$$

Where C is a *n-by-k* partition matrix that contains binary values, $c_{il} = 1$ indicates that x_i belongs to cluster l. Z is the centroids and $d(\cdot,\cdot)$ is the distance matrix. $\mathbf{w}$ is the weight vector and β is the parameter of the attribute weight. Minimizing Ψ with respect to C and Z is straightforward. However, minimizing Ψ with respect to $\mathbf{w}$ depends of the value of β. We refer the reader to [25] for more about minimizing with respect to $\mathbf{w}$.

Algorithm 4 Feature Weighting k-means (W-k-means)

Input:
D: dataset
n: number of samples
Ψ: the objective function Equation (2.7)
Initialize:
C: apply *k-means* on D to obtain initial clusters
Z: randomly choose k centroids
$\mathbf{w}$: randomly initialize the weight of each feature so that $\sum_{i=1}^{m} w_i = 1$
t: 0
Output:
C: the clustering
Z: the centroids
$\mathbf{w}$: the features weights

```
while not stop do
    Fix Z and w and solve Ψ with respect to C.
    Stop when no changes occur on C.
    Fix C and w and solve Ψ with respect to Z.
    Stop when no changes occur on Z.
    Fix C and Z and solve Ψ with respect to w.
    Stop when no changes occur on w.
    t=t+1
end while
```

2.2.2 Algorithms for Text Data

Document clustering aims to segregate documents into meaningful clusters that reflect the content of each document. For example, in the news wire, manually assigning one or more categories for each document requires exhaustive human labor, especially with the huge amount of text uploaded online daily. Thus, efficient clustering is essential. Another problem associated with document clustering is the huge number of terms. In a matrix representation, each term will be a feature and each document is an instance. In typical cases, the number of features will be close to the number of words in the dictionary. This imposes a great challenge for clustering methods where the efficiency will be greatly degraded. However, a huge number of these words are either stop words, irrelevant to the topic, or redundant. Thus, removing these unnecessary words may help significantly reduce dimensionality.

Feature selection not only reduces computational time but also improves clustering results and provides better data interpretability [48]. In document clustering, the set of selected words that are related to a particular cluster will be more informative than the whole set of words in the documents with respect to that cluster. Different feature selection methods have been used in document clustering recently, for example, term frequency, pruning infrequent terms, pruning highly frequent terms, and entropy-based weighting. Some of these methods and others will be explained in the following subsections.

2.2.2.1 Term Frequency (TF)

Term Frequency is one of the earliest and most simple yet effective term methods. It is dated back to 1957 in [41]. Thus, it is, indeed, a conventional term selection method. In a text corpus, the documents that belong to the same topic more likely will use similar words. Therefore, these frequent terms will be a good indicator for a certain topic. We can say that a very frequent term that is normally distributed across different topics is not informative; hence, such term would be unselected. We call this technique *pruning highly frequent terms*. Similarly, very rare terms should be pruned as well and that is called *pruning infrequent terms*. Stop words most likely will be pruned due to their high frequency. Furthermore, words such as *abecedarian* will be ignored since they will not be very frequent.

TF for term f_i with respect to the whole corpus is given by

$$TF(f_i) = \sum_{j \in D_{f_i}} tf_{ij} \tag{2.8}$$

where D_{f_i} is the documents that contain the term f_i, and tf_{ij} is the frequency of f_i in document j.

2.2.2.2 Inverse Document Frequency (IDF)

TF is an effective term selection method. However, it is not effective in terms of term weighting, where all selected terms will be assigned the same weight. Also, we cannot link TF value to any document. In other words, we cannot distinguish between frequent words that appear in a small set of documents, which could have discriminative power for this set of documents, and frequent words that appear in all or most of the documents in the corpus. In order to scale the term's weight, we use, instead, the inverse document frequency (IDF). IDF measures whether the term is frequent or rare across all documents:

$$idf(f_i) = \log \frac{|D|}{|D_{f_i}|} \tag{2.9}$$

where $|D|$ is the total number of documents (i.e., sample size) and $|D_{f_i}|$ is the number of documents that contain the term f_i. The value of IDF will be high for rare terms and low for highly frequent ones.

2.2.2.3 Term Frequency-Inverse Document Frequency (TF-IDF)

We can now combine the above mentioned measures (i.e., TF and IDF) to produce weight for each term f_i in each document d_j. This measure is called TF-IDF. It is given by

$$tf\text{-}idf(f_i, d_j) = tf_{ij} * idf_(f_i) \tag{2.10}$$

tf-idf assigns greater values to terms that occur frequently in a small set of documents, thus having more discriminative power. This value gets lower when the term occurs in more documents, while the lowest value is given to terms that occur in all documents. In document clustering, terms that have higher *tf-idf* have a higher ability for better clustering.

2.2.2.4 Chi Square Statistic

Chi square (χ^2) statistic has been widely used in supervised feature selection [71]. It measures the statistical dependency between the feature and the class. χ^2 with r different values and C classes is defined as

$$\chi^2 = \sum_{i=1}^{r} \sum_{j=1}^{C} \frac{(n_{ij} - \mu_{ij})^2}{\mu_{ij}},$$

where n_{ij} is the number of samples (i.e., documents) with i^{th} feature value in the jth class and $\mu_{ij} = \frac{n_{i*} n_{*j}}{n}$ and n is the total number of documents. This equation can be interpreted using the probability as

$$\chi^2(f,c) = \frac{n(p(f,c)p(\neg f,\neg c) - p(f,\neg c)p(\neg f,c))^2}{p(f)p(\neg f)p(\neg c)p(c)} \tag{2.11}$$

where $p(f,c)$ is the probability of class c that contains the term f, and $p(\neg f, \neg c)$ is the probability of not being in class c and not containing term f and so on. Thus, χ^2 cannot be directly applied in an unsupervised learning such as clustering due to the absence of class label. Y. Li et al in [34] propose a variation of χ^2 called $r\chi^2$ that overcomes some drawbacks of the original χ^2 and is embedded in an Expectation-Maximization (EM) algorithm to be used for text clustering problems. [34] found out that χ^2 cannot determine whether the dependency between the feature and the class is negative or positive, which leads to ignoring relevant features and selecting irrelevant features sometimes. Therefore, they proposed a relevance measure (R) that can be used in the original χ^2 to overcome this limitation. This new measure R follows.

$$R(f,c) = \frac{p(f,c)p(\neg f,\neg c) - p(f,\neg c)p(\neg f,c)}{p(f)p(c)} \tag{2.12}$$

R in Equation (2.12) will be equal to 1 if there is no such dependency between the class and the feature, greater than 2 if there is a positive dependency, and less than 1 if the dependency is negative.

From Equations (2.11) and (2.12), Hoffman et al. [34] proposed a new variation of χ^2 that is able to distinguish positive and negative relevance:

$$r\chi^2(f) = \sum_{j=1}^{C} p(R(f,c_j))\chi^2(f,c_j) \tag{2.13}$$

where $p(R(f,c_j))$ is given by $p(R(f,c_j)) = \frac{R(f,c_j)}{\sum_{j=1}^{C} R(f,c_j)}$. The larger the value of $r\chi^2$ is, the more relevant the feature f will be.

As we mentioned earlier, we cannot apply a supervised feature selection in an unsupervised learning directly. Therefore, [34] embedded their proposed method given in Equation (2.13) in a clustering algorithm using an EM approach. They used k-means as the clustering algorithm and $r\chi^2$ as the feature selection method as shown in Algorithm 5.

Algorithm 5 $r\chi^2$ as a feature selector for clustering

Input:

D: dataset

k: number of clusters

α: predetermine parameter in the range of $[0,1)$

m: number of selected features

Output:

C the clusters

Initialize:

$C \leftarrow$Apply k-means on D to obtain initial clusters

1. **E-step**:
    - Apply $r\chi^2$ from Equation (2.13) using clusters C obtained from step (1) as class labels.
    - Make the weight of the top m relevant features 1 and α for the rest of the features.
    - Calculate the new k-centroid for the new space.
2. **M-step**: Compute the new clusters using k-means.
3. Repeat E-step and M-step until convergence.

One advantage of this framework is that it does not simply remove the unselected features, instead, it keeps them while reducing their weight to α so they can be reselected in coming iterations. Also, this approach outperforms other clustering techniques even with the existence of feature selection methods using *F-measure* and the *purity*. However, [34] did not investigate the convergence of Algorithm 5 which is a big concern for such an algorithm especially when we know that the selected features may change dramatically from one iteration to another. In addition, the complexity of this algorithm is not discussed. In fact, the feature selection step in Algorithm 5 is not a feature selection; instead, it is a feature weighting. In other words, the number of features in each iteration remains the same. Thus, the complexity of *k*-means will not be reduced, which is against the goals of involving feature selection in clustering.

A similar approach is proposed for the Relief algorithm by Dash and Ong in [14]. They called their method Relief-C. It is observed that if clustering is done using the whole feature space, Relief will fail miserably in the presence of a large number of irrelevant features. Thus, Relief-C starts by clustering using exactly 2 randomly selected features and using clusters as a class label passed to Relief. After that, the feature weight will be updated. These two steps are repeated until the given criteria are met. We believe Relief-C may not perform well with huge dimensionality, say, more than 1000 features, since the chance of finding the real clusters from two randomly chosen features is very slim especially if we know that the percentage of relevant features is very small. In addition, both Relief-C and $r\chi^2$ are capable of handling generic data. We include $r\chi^2$ in the text data section since it was originally applied on text domain and requires a dataset to contain discrete values.

2.2.2.5 Frequent Term-Based Text Clustering

Frequent Term-Based Text Clustering (FTC), proposed in [6], provides a natural way to reduce dimensionality in text clustering. It follows the notion of a frequent item set that forms the basis of association rule mining. In FTC, the set of documents that contains the same frequent term set will be a candidate cluster. Therefore, clusters may overlap since the document may contain different item sets. This kind of clustering can be either flat (FTC) or hierarchical (HFTC) clustering since we will have different cardinalities of item sets.

Algorithm 6 explains the FTC algorithm. First, a dataset D, predetermined minimum support min_{sup} value, and an algorithm that finds frequent item set should be available. The algorithm starts by finding the frequent item set with minimum support min_{sup}. Then, it runs until the number of documents contributing in the selected term set $|cov(STS)|$ is equivalent to the number of documents in D. In each iteration, the algorithm calculates the entropy overlap EO for each set in the remaining term set RTS, where EO is given by

$$EO_i = \sum_{D_j \in C_i} -\frac{1}{F_j} \cdot \ln(\frac{1}{F_j})$$

where D_j is the jth document, C_i is the ith cluster and F_j is the number of all frequent term sets supported by document J, with the less overlap assumed to be the better. EO equals 0 if all the documents in C_i support only on frequent item set (i.e., $F_j = 1$). This value increases with the increase of F_j. This method of overlap evaluation was found to produce better clustering quality than the standard one [6]. The best candidate set *BestSet* will be the set with a minimum amount of overlap. *BestSet* will be selected and added to the STS and excluded from RTS. In addition, the set of documents that supports the *BestSet* is removed from the dataset since they have been already clustered, which leads to dramatically reducing the number of documents. They are also removed from the documents' list of RTS which leads to reducing the number of remaining term set. This greedy approach gives the computational advantage for this algorithm.

Due to the monotonicity property of a frequent item set, which means all (k–1)-items that are subset of frequent (k)-items are also frequent, we can perform an HFTC. HFTC is based on Algorithm 6. Instead of performing FTC on the whole frequent term sets, it is performed on a single level

Algorithm 6 Frequent Term-Based Clustering (FTC)

Input:
D: dataset
min_{sup}: minimum support
$\Psi(\cdot,\cdot)$: frequent item set finder
n: number of documents
Output:
STS and $cov(STS)$
Initialize:
Selected Term Set $STS = \{\}$

```
Remaining Term Set RTS = Ψ(D, min_sup)
while |cov(STS)| ≠ n do
    for each set in RTS do
        EO_i = Calculate overlap for the set
    end for
    BestSet = RTS_i which is the set with min(EO)
    STS = STS ∪ BestSet
    RTS = RTS − BestSet
    D = D − cov(BestSet)
    cov(RTS) = cov(RTS) − cov(BestSet)
end while
```

of frequent term set, say, k-term sets. Then, the obtained clusters are further partitioned using the next level of term set, (k+1)-term sets.

Both FTC and HFTC were imperially proven to be superior to other well-known clustering methods such as bisecting k-means and 9-secting k-means in efficiency and clustering quality. They also were able to provide better description and interpretation of the generated clusters by the selected term set.

2.2.2.6 Frequent Term Sequence

Similar to FTC, a clustering based on frequent term sequence (FTS) was proposed in [32]. Unlike FTC, the sequence of the terms matters in FTS. This means that the order of terms in the document is important. The frequent terms sequence, denoted as **f**, is a set that contains the frequent terms $< f_1, f_2, \ldots, f_k >$. The sequence here means that f_2 must be after f_1, but it is not necessary to be immediately after it. There could be other nonfrequent terms between them. This is true for any f_k and f_{k-1} terms. This definition of frequent terms sequence is more adaptable to the variation of human languages [32].

Similar to FTC, FTS starts by finding frequent term sets using an association rule mining algorithm. This frequent term set guarantees to contain the frequent term sequence but not vice versa. Hence, we do not need to search the whole term space for the frequent term sequence. We can search in only the frequent term set space, which is a dramatic dimension reduction. After that, FTS builds a generalized suffix tree (GST), which is a well-known data structure for sequence pattern matching, using the documents after removing the nonfrequent terms. From the suffix nodes in GST, we obtain the cluster candidates. These cluster candidates may contain subtopics that may be eligible to be merged together to create more general topics. Therefore, a merging step takes place.

The authors of [32] chose to merge cluster candidates into more general topic clusters using *k–mismatch* instead of the similarity. An example of using the *k–mismatch* concept is when we have $FS_i = \{feature, selection, clustering\}$ and $FS_j = \{feature, selection, \textit{classification}\}$, where they have one mismatch. Therefore, we can merge these two clusters if the tolerance parameter $k \geq 1$.

In [32], FTS adopted Landau–Vishkin (LV) to test three types of mismatches: insertion, deletion, substitution. Insertion means that all we need to insert is k, or fewer, terms into the FS_j in order to match FS_i. Deletion, in contrast, means we need to delete. While substitution means we need to substitute terms from FS_j with terms from FS_i.

These merged clusters are prone to overlap. Accordingly, more merging will be performed after measuring the amount of overlap using the *Jaccard Index*:

$$J(C_i,C_j) = \frac{|C_i \cup C_j|}{|C_i \cap C_j|}$$

The larger $J(C_i,C_j)$ is, the more the overlap would be. The merge takes place here when the overlap exceeds a user defined threshold, δ. However, the final number of clusters could be predefined too. In this case, this merging step would be repeated until the number of clusters meet the user's demand.

Algorithm 7 Frequent Term Sequence (FTS)

Input:
D: dataset
min_{sup}: minimum support
$\Psi(\cdot,\cdot)$: frequent item set finder
n : number of documents
K: number of clusters
δ: overlap threshold
Output:
$Clusters$
Initialize:
$Clusters = \{\}$
Frequent 2-Term Set $FS = \Psi(D, min_{sup})$
$\hat{D}$ = Reduce D by removing every term $f \notin FS$.
$G = GST(\hat{D})$
for each node j in G **do**
  if node j has Frequent term set FS_j with documents id ID_j **then**
    $C_j = \{FS_j, ID_j\}$
    $Clusters = Clusters \cup C_j$
  end if
end for
Use Landau−Vishkin to find k−mismatch
$Clusters \leftarrow$Merge $Clusters$ according to Landau−Vishkin results
while number of $Clusters > K$ **do**
  $Clusters \leftarrow$ Combine overlapped $Clusters$ based on *Jaccard Index* and δ
end while

In Algorithm 7, the most time consuming element is constructing GST, yet, it is still linear element with the number of terms. In [32], the terms reduction after finding the frequent term sets is huge; it exceeds 90% in all cases.

Similar to FTS, [32] proposed another frequent term set algorithm (FTMS) that is based on the synonymous set (synsets) of terms instead of the term itself. This is more adaptable with human language where we may use different terms to refer to the same meaning. This proposed algorithm used a WordNet dictionary to retrieve the synsets of each word and replace each with the original term. Each two synsets that intersect in at least one term will be merged into one synset. Thus, the number of synsets will be further reduced. Finally, FTS will be applied to these documents that contain the synsets.

These two proposed algorithms (i.e., FTS and FTMS) were empirically proved to be superior to other algorithms, such as bisect *k*-means and hierarchical frequent item-based clustering. Yet, FTMS is more capable of capturing more precise clustering topics than FTS. This is due to using the terms synsets instead of just the word itself.

There are several clustering techniques based on frequent item sets besides the ones mentioned in this chapter [19, 61, 77, 4, 47]. However, they all follow the same notion of reducing dimensionality by finding the frequent term sets. They differ in the parameters or the underlying utilized methods. For example, the number of terms in the frequent set may be set to a specific number or undefined. Hence, the maximum number will be found. They also differ in the way the final cluster is smoothed or merged.

2.2.3 Algorithms for Streaming Data

Streaming Data are continuous and rapid data records. Formally defined as a set of multidimensional records $X_1, \ldots, X_n, \ldots$ that come in time stamps $T_1, \ldots, T_n, \ldots$. Each record X_i is m-dimensional. Usually these samples have a huge dimensionality and arrive very fast. Therefore, they require scalable and efficient algorithms. Also, the underlying cluster structure is changing, so we need to capture this change and keep selected feature sets up to date. In this section, we introduce what we believe to be required characteristics of a good algorithm that handles streaming data:

- **Adaptivity:** The algorithm should be able to adjust features' weights or even reselect the set of features, so it is able to handle the data drift, aka dataset shift.
- **Single scan:** The algorithm should be able to cluster the incoming stream in one scan, since another scan is usually impossible or at least costly.

The following algorithms represent the literature of feature selection for data stream clustering. In fact, this area still needs great attention from the researchers due to the lack of work in this area.

2.2.3.1 Text Stream Clustering Based on Adaptive Feature Selection (TSC-AFS)

It is natural for clusters and data categories to evolve overtime. For example, the breaking news that dominates the worlds' media today may change dramatically tomorrow. Therefore, using the same set of features to process and cluster data stream may lead to unsatisfactory learning results over time. Gong et al in [20] propose using the cluster quality threshold γ to test the quality of newly arrived sample's clustering with respect to old clusters. TSC-AFS starts by applying feature selection on training data D. Then, it clusters the samples of D with respect to the selected features only.

After that, it starts to receive data streams and assign them to the closest cluster. TSC-AFS evaluates a validity index ρ for each cluster using the Davies–Bouldin D–B index. If γ is less than ρ, then TSC-AFS should reselect the set of features. Otherwise, the algorithm keeps the current feature set and clusters based on them and accepts new streams to come. Otherwise, TSC-AFS will reevaluate ρ while considering the new data stream being added the closest cluster. If the validity index cannot satisfy the threshold requirements, a tolerance parameter κ is set to allow this sample to be clustered and initialize the algorithm again. These parameters, κ and γ, are simply determined by the cross-validation approach.

This algorithm utilizes a word variance-based selection method as feature selection since the domain in [20] is text data stream. However, any other appropriate algorithm may be used. In addition, *k*-means was used as a clustering method. Algorithm 8 shows the pseudocode for TSC-AFS.

TSC-AFS, shown in Algorithm 8, is arguable in terms of performance and applicability in the

Algorithm 8 Text Stream Clustering Based on Adaptive Feature Selection (TSC-AFS)

Input:
γ: cluster quality threshold
κ: tolerance parameter
t : time stamp
K: number of clusters
$\Xi(\cdot)$: feature selection method
Output: $Clusters$
Initialize: $t = \rho = 0$

while We have more stream **do**
    $D \leftarrow$ Load new training set
    Select features $F_t = \Xi(D)$
    $\hat{D} \leftarrow$ remove unselected features from D
    Apply k-means clustering on $\hat{D}$
    $\mu \leftarrow$ the centroid of each cluster
    while $\rho < \kappa$ **do**
        $t = t + 1$
        $S_t \leftarrow$ new stream
        Evaluate the validity index ρ
        $Flag = true$
        while $Flage$ **do**
            if $\rho > \gamma$ **then**
                Assign S_t to the nearest μ
                $\rho \leftarrow$ Reevaluate the validity index
            else
                Assign S_t to the nearest μ
            end if
        end while
    end while
    $Clusters \leftarrow \{\mu, F\}$
end while

current form. However, it has a nice underlying property that is the adaptivity with the drifting features. We believe that this approach needs more attention to improve the adaptivity step.

2.2.3.2 High-Dimensional Projected Stream Clustering (HPStream)

Aggarwal et al in [2] propose a data streams clustering technique that involves a feature selection step. This method is called HPStream. Projected clustering [3] is a subset of data points P with a subset of dimensions F such that the points in P are closely clustered with respect to F.

Since data quality in data streams may decay, the stream clustering method should assign a greater level of importance to recent data points. Motivated from this belief, fading data structure was proposed in [2] to keep the clustering contemporary. Assuming, $X_1, \ldots, X_i, \ldots$ are multi-dimensional data samples arriving in time stamps $T_1, \ldots, T_i, \ldots$, each data sample X_i contains m dimensions $(x_i^1, \ldots, x_i^m)$. The fading $f(t)$ function for each data point will be according to the arrival time t. The fading function is assumed to be monotonically exponentially decreasing with respect to the time t [2].

Fading function $f(t) = 2^{-\lambda t}$, where $\lambda = 0.5$ is the decay rate. After defining the fading function, [2] defines the fading cluster structure at time t as $\Omega(P,t) = (\Psi^2(P,t), \Psi(P,t), w(t))$, where $\Psi^2(P,t)$

and $\Psi(P,t)$ for each jth dimension are given by $\sum_{i=1}^{n} f(t-T_i)\cdot(x_i^j)^2$ and $\sum_{i=1}^{n} f(t-T_i)\cdot(x_i^j)$, respectively, and $w(t)$ is the sum of all the weights of the data points at time t, $\sum_{i=1}^{n} f(t-T_i)$.

Algorithm 9 systematically describes the HPStream algorithm. HPStream is initialized offline using k-means to cluster a portion of the data D_{tr} using full dimension. Then, the least spread dimensions within each cluster will be chosen to form the initial fading cluster structure Ω. These two steps will be repeated, using the selected features until convergence. After initialization, incoming data stream X is temporally added to each cluster with the corresponding selected features of that cluster for determination of a new set of selected features. Final assignment of X will be to the closest cluster. After that, the limiting radius of each cluster will be calculated. Any data point that lies outside the limiting radii will form a cluster by itself. If the number of generated clusters exceeds the predetermined number of clusters K, the oldest cluster(s) are removed, so that the number of clusters equals K.

Algorithm 9 High-Dimensional Projected Stream Clustering (HPStream)

Input:
D_{tr}: training data set
K: number of clusters
l: number of selected features
Output: Ω
Initialize:
Perform clustering on D_{tr}
$F = \{F_1, \ldots, F_K\}$: sets of bit vectors
Ω : fading cluster structure

while $X \leftarrow$We have more incoming stream **do**
  for each cluster C_i **do**
    Add the new stream point X into C_i
    Update $\Omega(C_i,t)$
    for each dimension d **do**
      Compute the radii of $\Omega(C_i,t)$
    end for
    Pick the dimensions in $\Omega(C_i,t)$ with minimum radii
    Create selected features set F_i for C_i
  end for
  for each cluster C_i **do**
    for each selected feature F_i **do**
      $dis_i \leftarrow$ Average distance between X and the centroid of C_i
    end for
    $index = \text{argmin}_i\{\text{dis}_i\}$
    $s \leftarrow$ The radius of C_{index}
    if $dis_{index} > s$ **then**
      Add new cluster C_{K+1}
    else
      Add X to C_{index}
    end if
    Remove clusters with zero dimensions from Ω
    if Number of current clusters $> K$ **then**
      Delete the least recently added cluster from Ω
    end if
  end for
end while

There are a few more things regarding HPStream worth mentioning. First, the data set should be normalized for meaningful comparison between different dimensions. Also, Manhattan Segmental Distance is used to find the distance along the projected dimensions. This is a normalized version of Manhattan Distance that can compute the distance of different dimensionality. In terms of the empirical evaluation of HPStream, [2] conducted several experiments on real-world and synthetic datasets. HPStream was able to out perform CluStream [1], as a clustering method for evolving data streams. Also, it was shown to be very stable in terms of processing speed and scalable in terms of dimensionality.

2.2.4 Algorithms for Linked Data

Linked data has become ubiquitous in real-world applications such as tweets in Twitter [1] (tweets linked through hyperlinks), social networks in Facebook[2] (people connected by friendships), and biological networks (protein interaction networks). Typically linked data have the following three characteristics: (1) high-dimensional such as tens of thousands of terms for tweets; (2) linked, providing an important source of information beyond attributes, i.e., link information; and (3) unlabeled due to the large-scale size and the expensive cost for labeling. Such properties pose challenges to the clustering task, and feature selection is an effective way to prepare high-dimensional data for effective and efficient clustering [35]. In this subsection, we first introduce the challenges and opportunities of linked data for traditional feature selection, then present an embedded unsupervised feature selection framework for linked data, and finally follow with a discussion about future work of feature selection for linked data.

2.2.4.1 Challenges and Opportunities

Linked instances are related to each other via different types of links (e.g., hyperlinks, friendships, and interactions). Thus linked data is distinct from traditional attribute value data (or "flat" data). Figure 2.4 illustrates a typical example of linked data and its two representations. Figure 2.4(a) shows 8 linked instances (u_1 to u_8), while Figure 2.4(b) is a conventional representation of attribute-value data: rows are instances and columns are features. As mentioned above, besides attributes, linked data provide an extra source in the form of links, represented as in Figure 2.4(c). These differences present both challenges and opportunities for traditional feature selection and machine learning [43, 57].

Linked data is patently not independent and identically distributed (i.i.d.), which is among the most enduring and deeply buried assumptions of traditional machine learning methods [27, 66]. Due to the absence of class labels that guide the search for relevant information, unsupervised feature selection is particularly difficult [9]. The linked property further exacerbates the difficulty of unsupervised feature selection for linked data. On the other hand, linked data provides more information than attribute value data and the availability of link information presents unprecedented opportunities to advance research. For example, linked data allows collective inference, inferring various interrelated values simultaneously [43], and enables relational clustering, finding a more accurate pattern [39].

Many linked data-related learning tasks are proposed such as collective classification [43, 57] and relational clustering [40, 39], but the task of feature selection for linked data is rarely touched due to its unique challenges for feature selection: (1) how to exploit relations among data instances and (2) how to take advantage of these relations for feature selection. Until recently, feature selection for linked data has attracted attention. In [63], a supervised feature selection algorithm, LinkedFS, is proposed for linked data in social media. Various relations, (coPost, coFollowing, coFollowed, and Following) are extracted following social correlation theories. LinkedFS significantly improves

[1] http://www.twitter.com/

[2] https://www.facebook.com/

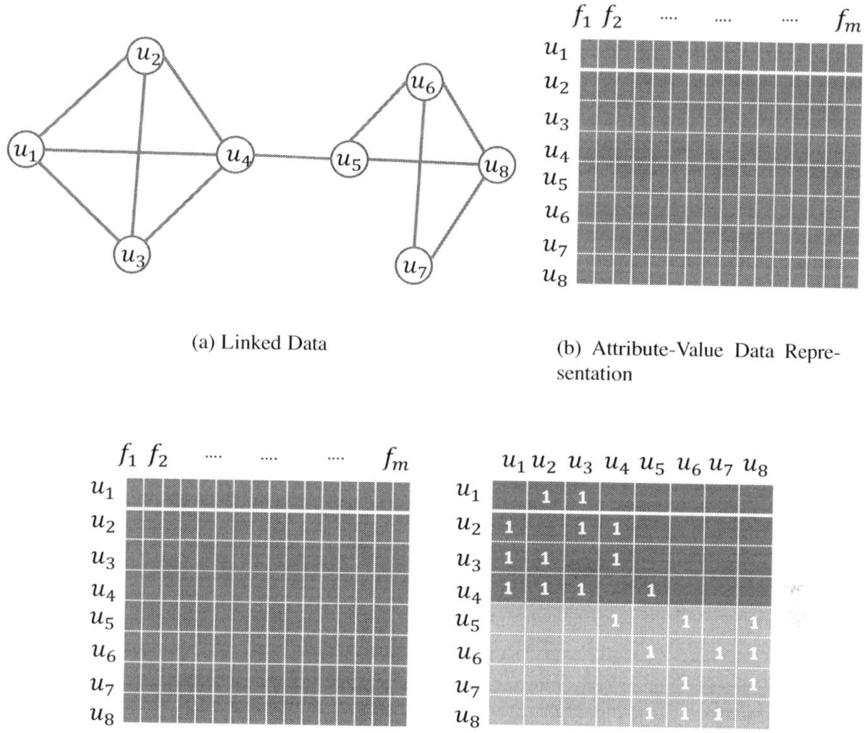

(a) Linked Data

(b) Attribute-Value Data Representation

(c) Linked Data Representation

FIGURE 2.4: A simple example of linked data.

the performance of feature selection by incorporating these relations into feature selection. In [64], an unsupervised feature selection framework, LUFS, is developed for linked data [65]. More details about LUFS will he presented in the following subsection.

2.2.4.2 LUFS: An Unsupervised Feature Selection Framework for Linked Data

LUFS is an unsupervised feature selection framework for linked data and in essence, LUFS investigates how to exploit and take advantage of link information of linked data for unsupervised feature selection. In general, the goal of feature selection is to select a subset of features to be consistent with some constraints [80]. For supervised learning, label information plays the role of constraint. Without label information, LUFS introduces the concept of pseudo-class label to guide unsupervised learning. Particularly, LUFS assumes that there is a mapping matrix $\mathbf{W} \in \mathbb{R}^{m \times c}$, assigning each data point with a pseudo-class label where m is the number of original features and c is the number of pseudo-class labels. The pseudo-class label indicator matrix is $\mathbf{Y} = \mathbf{W}^\top \mathbf{X} \in \mathbb{R}^{c \times n}$ where n is the number of data points. Each column of $\mathbf{Y}$ has only one nonzero entity, i.e., $\|\mathbf{Y}(:,i)\|_0 = 1$ where $\|\cdot\|_0$ is the vector zero norm, counting the number of nonzero elements in the vector. Then LUFS seeks pseudo-class label information by extracting constraints from both linked and attribute-value data. The constraints from link information and attribute-value parts are obtained through social dimension regularization and spectral analysis, respectively.

Social Dimension Regularization: In [64], social dimension is introduced to improve the performance of relational learning. LUFS employs social dimensions to exploit the interdependency among linked data. It first adopts Modularity Maximization [49] to extract social dimension indicator matrix $\mathbf{H}$. Since social dimensions can be considered as a type of affiliations, according to Linear Discriminant Analysis, three matrices, i.e., within, between, and total social dimension scatter matrixes $\mathbf{S}_w$, $\mathbf{S}_b$, and $\mathbf{S}_t$, are defined as follows:

$$\begin{aligned} \mathbf{S}_w &= \mathbf{Y}\mathbf{Y}^\top - \mathbf{Y}\mathbf{F}\mathbf{F}^\top\mathbf{Y}^\top, \\ \mathbf{S}_b &= \mathbf{Y}\mathbf{F}\mathbf{F}^\top\mathbf{Y}^\top, \\ \mathbf{S}_t &= \mathbf{Y}\mathbf{Y}^\top, \end{aligned} \tag{2.14}$$

where $\mathbf{F} = \mathbf{H}(\mathbf{H}^\top\mathbf{H})^{-\frac{1}{2}}$ is the weighted social dimension indicator matrix.

Instances from different social dimensions are dissimilar while instances in the same social dimension are similar. Finally the constraint, social dimension regularization, from link information can be obtained via the following maximization problem:

$$\max_{\mathbf{W}} Tr\big((\mathbf{S}_t)^{-1}\mathbf{S}_b\big). \tag{2.15}$$

Spectral Analysis: To take advantage of information from the attribute-value part, LUFS obtains the constraint from the attribute-value part through spectral analysis [42] as

$$\min \; Tr(\mathbf{Y}\mathbf{L}\mathbf{Y}^\top) \tag{2.16}$$

where $\mathbf{L}$ is a Laplacian matrix.

Considering constraints from both link information and the attribute-value part, LUFS, is equivalent to solving the following optimization problem:

$$\begin{aligned} \min_{\mathbf{W},\mathbf{s}} \; & Tr(\mathbf{Y}\mathbf{L}\mathbf{Y}^\top) - \alpha Tr\big((\mathbf{S}_t)^{-1}\mathbf{S}_b\big) \\ s.t. \; & \|\mathbf{Y}(:,i)\|_0 = 1, \;\; 1 \le i \le n \end{aligned} \tag{2.17}$$

With the spectral relaxation for label indicator matrix [42], LUFS is eventually used to solve the following optimization problem:

$$\begin{aligned} \min_{\mathbf{W}} \; & f(\mathbf{W}) = Tr(\mathbf{W}^\top\mathbf{A}\mathbf{W}) + \beta\|\mathbf{W}\|_{2,1} \\ s.t. \; & \mathbf{W}^\top\mathbf{B}\mathbf{W} = \mathbf{I}_c \end{aligned} \tag{2.18}$$

where $\mathbf{A} = \mathbf{X}\mathbf{L}\mathbf{X}^\top + \alpha\mathbf{X}(\mathbf{I}_n - \mathbf{F}\mathbf{F}^\top)\mathbf{X}^\top$ is a symmetric and positive semidefinite matrix and $\mathbf{B} = \mathbf{X}\mathbf{X}^\top + \lambda\mathbf{I}$ is a symmetric and positive matrix. Since the problem in Equation (2.18) is convex, an optimal solution $\mathbf{W}$ can be guaranteed for LUFS [65]. The detailed algorithm for LUFS is shown in Algorithm 10.

In Algorithm 10, social dimension extraction and weighted social dimension indicator construction are from line 1 to line 2. The iterative algorithm to optimize Equation (2.18) is presented from line 8 to line 13.

2.2.4.3 Conclusion and Future Work for Linked Data

In this subsection, we first analyzed the differences between linked data and attribute-view data, then presented the challenges and opportunities posed by linked data for feature selection, and finally introduced a recent proposed unsupervised feature selection framework, LUFS, for linked data in detail. To the best of our knowledge, LUFS is the first to study feature selection for linked

Algorithm 10 LUFS

Input: $\{\mathbf{X}, \mathbf{R}, \alpha, \beta, \lambda, c, K, k\}$

Output: k most relevant features

Obtain the social dimension indicator matrix $\mathbf{H}$
Set $\mathbf{F} = \mathbf{H}(\mathbf{H}^\top \mathbf{H})^{-\frac{1}{2}}$
Construct $\mathbf{S}$ through RBF kernel
Set $\mathbf{L} = \mathbf{D} - \mathbf{S}$
Set $\mathbf{A} = \mathbf{X}\mathbf{L}\mathbf{X}^\top + \alpha\mathbf{X}(\mathbf{I}_n - \mathbf{F}\mathbf{F}^\top)\mathbf{X}^\top$
Set $\mathbf{B} = \mathbf{X}\mathbf{X}^\top + \lambda\mathbf{I}$
Set $t = 0$ and initialize $\mathbf{D}_0$ as an identity matrix
while not convergent **do**
    Set $\mathbf{C}_t = \mathbf{B}^{-1}(\mathbf{A} + \beta\mathbf{D}_t)$
    Set $\mathbf{W}_t = [q_1, \ldots, q_c]$ where $q_1, \ldots, q_c$ are the eigenvectors of $\mathbf{C}_t$ corresponding to the first c smallest eigenvalues
    Update the diagonal matrix $\mathbf{D}_{t+1}$, where the i-th diagonal element is $\frac{1}{2\|\mathbf{W}_t(i,:)\|_2}$;
    Set $t = t + 1$
end while
Sort each feature according to $\|\mathbf{W}(i,:)\|_2$ in **descending** order and select the top-k ranked ones;

data in an unsupervised scenario, and many works are needed to further exploit linked data for feature selection.

Since LUFS can be categorized as an embedded method, analogous to conventional data, how to develop filter, wrapper, and hybrid models for linked data is an interesting problem for further research. Furthermore, LUFS employs social dimension to capture link information, a further investigation into sophisticated methods of exploiting link will lead to novel approaches and help us develop a deeper understanding of how to improve the performance of feature selection. Finally, the availability of various link formation can lead to links with heterogeneous strengths. Therefore how to incorporate automatic link selection into feature selection is another promising direction for linked data.

2.3 Discussions and Challenges

Here are several challenges and concerns that we need to mention and discuss briefly in this chapter about feature selection for clustering.

2.3.1 The Chicken or the Egg Dilemma

In feature selection for clustering, there is a dilemma in choosing which one to start with or which one serves the demands of the other. Do we utilize feature selection to improve clustering quality? Or, do we use clustering as an indicator of relevant features? This, in fact, leads us to the chicken or the egg dilemma. Which one comes first, clustering or feature selection? If the answer to the latter is yes, that means feature selection is a goal in itself. However, this is not the case in feature selection literature. We usually use feature selection, as mentioned earlier, to improve learning quality, reduce computational time, and reduce required storage. Thus, the answer to the first question is, indeed, yes. We use feature selection to improve clustering quality. Accordingly, we should select the features that preserve cluster structure. However, some current methods initialize clusters using all features. Then, they apply the supervised feature selection method using the initial clusters as class labels. We believe that such techniques will not achieve the goal of feature selection since the initial clusters may not be the real ones or even close. If these clusters are claimed to be real ones, why bother doing feature selection at all? Some other methods apply the same process but in an iterative manner using an EM-like approach. Although this might lead to better results, we still believe that there are several drawbacks associated with this approach. For example, if the dimensionality in the original space is huge, such an approach may take a very long time to converge. Therefore, we prefer utilizing feature selection that does not depend on any clustering input to define the relevancy of the the feature. In other words, we may utilize the clustering method to guide the searching process but we do not use the clustering as a class label. This means we apply the feature selection method on the whole dataset and then use the selected features to construct the clusters. If we are not satisfied with the clustering quality, we may use this as an indicator that this set of features is not satisfactory. This is exactly the *wrapper approach*.

2.3.2 Model Selection: K and l

Selecting the number of clusters K or the number of selected features l is an open problem. In real-world problems, we have limited knowledge about each domain. Consequently, determining optimal K or l is almost impossible. Although this problem seems not to be a big issue with some approaches (such as frequent term set-based feature selection methods), we still need to determine other sensitive parameters such as the minimum support, which leads us to another problem. Choosing different l for the same problem usually results in different equally good subsets of features and hence, different clustering results. Similarly, choosing different K leads to merging totally different clusters into one or splitting one cluster into smaller ones. In other cases, this may lead to changing the set of selected features which results in losing potential clusters and/or constructing less important ones. Therefore, picking nearly optimal parameters is desirable for better clustering quality. Some work has been done to quantify such parameters. For instance, [10] uses the notion of false nearest neighbor to find the number of selected features for clustering. On the other hand, [67] uses gap statistics to estimate the number of clusters in a dataset. However, the effort in finding better parameters for clustering is still limited. We believe it is necessary to pay more attention to these two parameters for better clustering results.

2.3.3 Scalability

With the tremendous growth of dataset sizes, the scalability of current algorithms may be in jeopardy, especially with those domains that require online clustering. For example, streaming data or data that cannot be loaded into the memory require a single data scan where the second pass is either unavailable or very expensive. Using feature selection methods for clustering may reduce the issue of scalability for clustering. However, some of the current methods that involve feature selection in the clustering process require keeping full dimensionality in the memory to observe any evolving in the cluster structure. If a change in data structure is observed, the clustering process should be restarted. Furthermore, other methods require an iterative process where each sample is visited more than once until convergence.

On the other hand, the scalability of feature selection algorithms is a big problem. Usually, they require a sufficient number of samples to obtain, statically, good enough results. It is very hard to observe a feature relevance score without considering the density around each sample. Some methods try to overcome this issue by memorizing only samples that are important or a summary, say, the mean of each cluster. In conclusion, we believe that the scalability of clustering and feature selection methods should be given more attention to keep pace with the growth and fast streaming of the data.

2.3.4 Stability

In supervised learning such as classification, stability of feature selection algorithms has gained increasing attention the last few years. Stability of feature selection algorithms or selection stability is the sensitivity of the selection toward data perturbation [5]. An example of data perturbation would be new data sample(s). Therefore, with a small amount of perturbation introduced to the data, we do not expect dramatic change in the selected features. Otherwise, the algorithm is considered to be unstable. To the best to our knowledge, selection stability in unsupervised learning has not yet been studied.

Studying stability in supervised learning is much easier than in the unsupervised. Due to the absence of class label in the latter, we cannot maintain enough knowledge about the underlying clusters within the data. For this reason, we cannot be confident enough whether the new sample(s) belong to any existing clusters or they form one or more new clusters. However, in supervised learning, we have a limited number of classes. Thus, a sample that does not belong to any existing class would be considered an outlier and we do not need to modify our selected set to obey outliers. So, to address the issue of feature selection stability in clustering, we discuss it with respect to a different situation.

Predefined Clusters:

In case we have enough domain knowledge to predefine the possible clusters, the stability issue will be identical to the stability in supervised learning. Therefore, we expect the feature selection algorithm to be stable with the existence of a small amount of perturbation. However, if the topics do not change while the data change, we might consider unstable selection.

Undefined Clusters:

In contrast to the first case, if we do not have enough knowledge about the clusters of the data, which is true in real-world situations, we generally can say that this case depends on the domain and the practitioner's desire. In streaming data, for example, we usually expect some kind of dataset shift or drift where the distribution and the clustering topics of the data may evolve over time. Therefore, we should have an adaptive feature selection process that adjusts the selected features in a way that can capture the evolving structure of the data. However, we might desire a stable selection for a period of time and an unstable one for another period. For example, in a news data clustering, we wish to have stable selection while we do not have breaking news that takes the distribution to different topic.

On the other hand, subspace and projected clustering may not promote stable selection since they search for all possible sets of features that define clusters. In some cases, we might want to maintain stable selection on some clusters but unstable on others. For instance, if we are performing projected clustering on news data, we may want to have stable selection on clusters that do not evolve rapidly while having unstable selection on rapidly evolving clusters.

Bibliography

[1] C.C. Aggarwal, J. Han, J. Wang, and P.S. Yu. A framework for clustering evolving data streams. In *Proceedings of the 29th International Conference on Very Large Data Bases—Volume 29*, pages 81–92. VLDB Endowment, 2003.

[2] C.C. Aggarwal, J. Han, J. Wang, and P.S. Yu. A framework for projected clustering of high dimensional data streams. In *Proceedings of the Thirtieth International Conference on Very Large Data Bases-Volume 30*, pages 852–863. VLDB Endowment, 2004.

[3] C.C. Aggarwal, J.L. Wolf, P.S. Yu, C. Procopiuc, and J.S. Park. Fast algorithms for projected clustering. *ACM SIGMOD Record*, 28(2):61–72, 1999.

[4] T.M. Akhriza, Y. Ma, and J. Li. Text clustering using frequent contextual termset. In *2011 International Conference on Information Management, Innovation Management and Industrial Engineering (ICIII)*, Volume 1, pages 339–342. IEEE, 2011.

[5] S. Alelyani, L. Wang, and H. Liu. The effect of the characteristics of the dataset on the selection stability. In *Proceedings of the 23rd IEEE International Conference on Tools with Artificial Intelligence*, 2011.

[6] F. Beil, M. Ester, and X. Xu. Frequent term-based text clustering. In *Proceedings of the Eighth ACM SIGKDD International Conference on Knowledge Discovery and Data Mining*, pages 436–442. ACM, 2002.

[7] C. Boutsidis, M.W. Mahoney, and P. Drineas. Unsupervised feature selection for the *k*-means clustering problem. *Advances in Neural Information Processing Systems*, 22:153–161, 2009.

[8] P.S. Bradley and O. L. Mangasarian. Feature selection via concave minimization and support vector machines. pages 82–90. Morgan Kaufmann, 1998.

[9] D. Cai, C. Zhang, and X. He. Unsupervised feature selection for multi-cluster data. In *Proceedings of the 16th ACM SIGKDD International Conference on Knowledge Discovery and Data Mining*, pages 333–342. ACM, 2010.

[10] A.C. Carvalho, R.F. Mello, S. Alelyani, H. Liu, et al. Quantifying features using false nearest neighbors: An unsupervised approach. In *Proceedings of the 23rd IEEE International Conference on Tools with Artificial Intelligence (ICTAI)*, pages 994–997. IEEE, 2011.

[11] G. C. Cawley, N. L. C. Talbot, and M. Girolami. Sparse multinomial logistic regression via Bayesian L1 regularisation. In *NIPS*, 2006.

[12] S. Das. Filters, wrappers and a boosting-based hybrid for feature selection. In *ICML '01: Proceedings of the Eighteenth International Conference on Machine Learning*, pages 74–81, San Francisco, CA, USA, 2001. Morgan Kaufmann Publishers Inc.

[13] M. Dash, K. Choi, P. Scheuermann, and H. Liu. Feature selection for clustering—A filter solution. In *Proceedings of the Second International Conference on Data Mining*, pages 115–122, 2002.

[14] M. Dash and Y.S. Ong. Relief-c: Efficient feature selection for clustering over noisy data. In *23rd IEEE International Conference on Tools with Artificial Intelligence (ICTAI)*, pages 869–872. IEEE, 2011.

[15] R.O. Duda, P.E. Hart, and D.G. Stork. *Pattern Classification*, 2nd edition. John Wiley & Sons, New York, 2001.

[16] J. G. Dy and C. E. Brodley. Feature selection for unsupervised learning. *Journal of Machine Learning Research*, 5:845–889, 2004.

[17] J. G. Dy and C. E. Brodley. Feature subset selection and order identification for unsupervised learning. In *Proceedings of the 17th International Conference on Machine Learning*, pages 247–254. Morgan Kaufmann, 2000.

[18] J.G. Dy. Unsupervised feature selection. *Computational Methods of Feature Selection*, pages 19–39, 2008.

[19] B.C.M. Fung, K. Wang, and M. Ester. Hierarchical document clustering using frequent itemsets. In *Proceedings of the SIAM International Conference on Data Mining*, Volume 30, pages 59–70, 2003.

[20] L. Gong, J. Zeng, and S. Zhang. Text stream clustering algorithm based on adaptive feature selection. *Expert Systems with Applications*, 38(3):1393–1399, 2011.

[21] I. Guyon and A. Elisseeff. An introduction to variable and feature selection. *Journal of Machine Learning Research*, 3:1157–1182, 2003.

[22] M. A. Hall. Correlation-based feature selection for machine learning. Technical report, 1999.

[23] T. Hastie, R. Tibshirani, and J. Friedman. *The Elements of Statistical Learning*. Springer, 2001.

[24] X. He, D. Cai, and P. Niyogi. Laplacian score for feature selection. *Advances in Neural Information Processing Systems*, 18:507, 2006.

[25] J.Z. Huang, M.K. Ng, H. Rong, and Z. Li. Automated variable weighting in *k*-means type clustering. *IEEE Transactions on Pattern Analysis and Machine Intelligence*, 27(5):657–668, 2005.

[26] A. Jain and D. Zongker. Feature selection: Evaluation, application, and small sample performance. *IEEE Transactions on Pattern Analysis and Machine Intelligence*, 19:153–158, 1997.

[27] D. Jensen and J. Neville. Linkage and autocorrelation cause feature selection bias in relational learning. In *ICML*, pages 259–266, 2002.

[28] L. Jing, M.K. Ng, and J.Z. Huang. An entropy weighting k-means algorithm for subspace clustering of high-dimensional sparse data. *IEEE Transactions on Knowledge and Data Engineering*, 19(8):1026–1041, 2007.

[29] T. Joachims. Text categorization with support vector machines: Learning with many relevant features, 1997.

[30] Y.S. Kim, W.N. Street, and F. Menczer. Evolutionary model selection in unsupervised learning. *Intelligent Data Analysis*, 6(6):531–556, 2002.

[31] R. Kohavi and G. H. John. Wrappers for feature subset selection, *Artificial Intelligence* 97(1997):273–324.

[32] Y. Li, S.M. Chung, and J.D. Holt. Text document clustering based on frequent word meaning sequences. *Data & Knowledge Engineering*, 64(1):381–404, 2008.

[33] Y. Li, M. Dong, and J. Hua. Localized feature selection for clustering. *Pattern Recognition Letters*, 29(1):10–18, 2008.

[34] Y. Li, C. Luo, and S.M. Chung. Text clustering with feature selection by using statistical data. *IEEE Transactions on Knowledge and Data Engineering*, 20(5):641–652, 2008.

[35] H. Liu and H. Motoda, editors. *Computational Methods of Feature Selection*. Chapman and Hall/CRC Press, 2007.

[36] H. Liu and H. Motoda. *Feature Selection for Knowledge Discovery and Data Mining*. Kluwer Academic Publishers, Boston,1998.

[37] H. Liu and R. Setiono. A probabilistic approach to feature selection—A filter solution. In *Proceedings of the 13th International Conference on Machine Learning* (ICML '96) July 1996, pages 319–327. Bari, Italy.

[38] H. Liu and L. Yu. Toward integrating feature selection algorithms for classification and clustering. *Knowledge and Data Engineering, IEEE Transactions on*, 17(4):49–502, April 2005.

[39] B. Long, Z.M. Zhang, and P.S. Yu. A probabilistic framework for relational clustering. In *Proceedings of the 13th ACM SIGKDD International Conference on Knowledge Discovery and Data Mining*, pages 470–479. ACM, 2007.

[40] B. Long, Z.M. Zhang, X. Wu, and P.S. Yu. Spectral clustering for multi-type relational data. In *Proceedings of the 23rd international Conference on Machine Learning*, pages 585–592. ACM, 2006.

[41] H.P. Luhn. A statistical approach to mechanized encoding and searching of literary information. *IBM Journal of Research and Development*, 1(4):309–317, 1957.

[42] U. Luxburg. A tutorial on spectral clustering. *Statistics and Computing*, 17(4):395–416, 2007.

[43] S.A. Macskassy and F. Provost. Classification in networked data: A toolkit and a univariate case study. *Journal of Machine Learning Research*, 8:935–983, 2007.

[44] P. Mitra, C. A. Murthy, and S. K. Pal. Unsupervised feature selection using feature similarity. *IEEE Transactions on Pattern Analysis and Machine Intelligence*, 24:301–312, 2002.

[45] F. Model, P, Adorjan, A. Olek, and C. Pierpenbrock. Feature selection for DNA methylation based cancer classification. *Bioinformatics*, 17 Suppl. 1:S157–S164, 2001.

[46] D.S. Modha and W.S. Spangler. Feature weighting in *k*-means clustering. *Machine Learning*, 52(3):217–237, 2003.

[47] K. Morik, A. Kaspari, M. Wurst, and M. Skirzynski. Multi-objective frequent termset clustering. *Knowledge and Information Systems*, 30(3)175–738, 2012.

[48] M. Mugunthadevi, M. Punitha, and M. Punithavalli. Survey on feature selection in document clustering. *International Journal*, 3, 2011.

[49] M. E. J. Newman and M. Girvan. Finding and evaluating community structure in networks. *Physical Review E*, 69(2):26113, 2004.

[50] A. Y. Ng. On feature selection: Learning with exponentially many irrelevant features as training examples. In *Proceedings of the Fifteenth International Conference on Machine Learning*, pages 404–412. Morgan Kaufmann, 1998.

[51] K. Nigam, A. Mccallum, S. Thrun, and T. Mitchell. Text classification from labeled and unlabeled documents using Em. In *J. Machine Learning*, 39(2–3):103–134, 1999.

[52] I.S. Oh, J.S. Lee, and B.R. Moon. Hybrid genetic algorithms for feature selection. *IEEE Transactions on Pattern Analysis and Machine Intelligence*, 26(11):1424–1437, 2004.

[53] J. R. Quinlan. *C4.5: Programs for Machine Learning*. Morgan Kaufmann, 1993.

[54] V. Roth and T. Lange. Feature selection in clustering problems. *Advances in Neural Information Processing Systems*, 16, 2003.

[55] Y. Rui and T. S. Huang. Image retrieval: Current techniques, promising directions and open issues. *Journal of Visual Communication and Image Representation*, 10:39–62, 1999.

[56] Y. Saeys, I. Inza, and P. Larranaga. A review of feature selection techniques in bioinformatics. *Bioinformatics*, 23(19):2507–2517, Oct. 2007.

[57] P. Sen, G. Namata, M. Bilgic, L. Getoor, B. Galligher, and T. Eliassi-Rad. Collective classification in network data. *AI Magazine*, 29(3):93, 2008.

[58] W. Siedlecki and J. Sklansky. On automatic feature selection. In *Handbook of Pattern Recognition & Computer Vision*, 63–87, 1993.

[59] M. R. Sikonja and I. Kononenko. Theoretical and empirical analysis of Relief and ReliefF. *Machine Learning*, 53:23–69, 2003.

[60] L. Song, A. Smola, A. Gretton, K. Borgwardt, and J. Bedo. Supervised feature selection via dependence estimation. In *International Conference on Machine Learning*, 2007.

[61] C. Su, Q. Chen, X. Wang, and X. Meng. Text clustering approach based on maximal frequent term sets. In *2009. SMC 2009. IEEE International Conference on Systems, Man and Cybernetics*, pages 1551–1556. IEEE, 2009.

[62] L. Talavera. Feature selection as a preprocessing step for hierarchical clustering. In *Machine Learning—International Workshop then Conference*, pages 389–397, 1999.

[63] J. Tang and H. Liu. Feature selection with linked data in social media. In *SDM*, 2012.

[64] L. Tang and H. Liu. Relational learning via latent social dimensions. In *Proceedings of the 15th ACM SIGKDD International Conference on Knowledge Discovery and Data Mining*, pages 817–826. ACM, 2009.

[65] J. Tang and H. Liu. Unsupervised feature selection for linked social media data. In *KDD*, 2012.

[66] B. Taskar, P. Abbeel, M.F. Wong, and D. Koller. Label and link prediction in relational data. In *Proceedings of the IJCAI Workshop on Learning Statistical Models from Relational Data*. Citeseer, 2003.

[67] R. Tibshirani, G. Walther, and T. Hastie. Estimating the number of clusters in a data set via the gap statistic. *Journal of the Royal Statistical Society: Series B (Statistical Methodology)*, 63(2):411–423, 2001.

[68] C.Y. Tsai and C.C. Chiu. Developing a feature weight self-adjustment mechanism for a *k*-means clustering algorithm. *Computational Statistics & Data analysis*, 52(10):4658–4672, 2008.

[69] J. Weston, A. Elisseff, B. Schoelkopf, and M. Tipping. Use of the zero norm with linear models and kernel methods. *Journal of Machine Learning Research*, 3:1439–1461, 2003.

[70] D. Wettschereck, D. W. Aha, and T. Mohri. A review and empirical evaluation of feature weighting methods for a class of lazy learning algorithms. *Artificial Intelligence Review*, 11:273–314, 1997.

[71] I.H. Witten and E. Frank. *Data Mining: Practical machine learning tools and techniques*. Morgan Kaufmann, 2005.

[72] D.M. Witten and R. Tibshirani. A framework for feature selection in clustering. *Journal of the American Statistical Association*, 105(490):713–726, 2010.

[73] Z. Xu, R, Jin, J. Ye, M. R. Lyu, and I. King. Discriminative semi-supervised feature selection via manifold regularization. In *IJCAI' 09: Proceedings of the 21th International Joint Conference on Artificial Intelligence*, 2009.

[74] Y. Yang and J. O. Pedersen. A comparative study on feature selection in text categorization. pages 412–420. Morgan Kaufmann, 1997.

[75] L. Yu and H. Liu. Efficient feature selection via analysis of relevance and redundancy. *Journal of Machine Learning Research (JMLR)*, 5(Oct):1205–1224, 2004.

[76] L. Yu and H. Liu. Feature selection for high-dimensional data: A fast correlation-based filter solution. In T. Fawcett and N. Mishra, editors, *Proceedings of the 20th International Conference on Machine Learning (ICML-03)*, pages 856–863, Washington, D.C., August 21-24, 2003. Morgan Kaufmann.

[77] W. Zhang, T. Yoshida, X. Tang, and Q. Wang. Text clustering using frequent itemsets. *Knowledge-Based Systems*, 23(5):379–388, 2010.

[78] Z. Zhao and H. Liu. Semi-supervised feature selection via spectral analysis. In *Proceedings of SIAM International Conference on Data Mining (SDM)*, 2007.

[79] Z. Zhao and H. Liu. *Spectral Feature Selection for Data Mining*. Chapman & Hall/CRC Data Mining and Knowledge Discovery. Taylor & Francis, Boca Raton, FL, 2011.

[80] Z. Zhao and H. Liu. Spectral feature selection for supervised and unsupervised learning. In *ICML '07: Proceedings of the 24th International Conference on Machine Learning*, pages 1151–1157, New York, 2007. ACM.

Chapter 3

Probabilistic Models for Clustering

Hongbo Deng
University of Illinois at Urbana-Champaign
Urbana, IL
hbdeng@illinois.edu

Jiawei Han
University of Illinois at Urbana-Champaign
Urbana, IL
hanj@illinois.edu

3.1 Introduction

Probabilistic model-based clustering techniques have been widely used and have shown promising results in many applications, ranging from image segmentation [71, 15], handwriting recognition [60], document clustering [36, 81], topic modeling [35, 14] to information retrieval [43]. Model-based clustering approaches attempt to optimize the fit between the observed data and some mathematical model using a probabilistic approach. Such methods are often based on the assumption that the data are generated by a mixture of underlying probability distributions. In practice, each cluster can be represented mathematically by a parametric probability distribution, such as a Gaussian or a Poisson distribution. Thus, the clustering problem is transformed into a parameter estimation problem since the entire data can be modeled by a mixture of K component distributions.

Data points (or objects) that belong most likely to the same distribution can then easily be defined as clusters.

In this chapter, we introduce several fundamental models and algorithms for probabilistic clustering, including mixture models [45, 48, 25], EM algorithm [23, 46, 53, 16], and probabilistic topic models [35, 14]. For each probabilistic model, we will introduce its general framework of modeling, the probabilistic explanation, the standard algorithms to learn the model, and its applications. Mixture models are probabilistic models which are increasingly used to find the clusters for univariate and multivariate data. We therefore begin our discussion of mixture models in Section 3.2, in which the values of the discrete latent variables can be interpreted as the assignments of data points to specific components (i.e., clusters) of the mixture. To find maximum likelihood estimations in mixture models, a general and elegant technique, the Expectation-Maximization (EM) algorithm, is introduced in Section 3.3. We first use the Gaussian mixture model to motivate the EM algorithm in an informal way, and then give a more general view of the EM algorithm, which is a standard learning algorithm for many probabilistic models. In Section 3.4, we present two popular probabilistic topic models, i.e., probabilistic latent semantic analysis (PLSA) [35] and latent Dirichlet allocation (LDA) [14], for document clustering and analysis. Note that some of the methods (e.g., EM/mixture models) may be more appropriate for quantitative data, whereas others such as topic models, PLSI, and LDA, are used more commonly for text data. Finally, we give the conclusions and summary in Section 3.5.

3.2 Mixture Models

Mixture models for cluster analysis [75, 45, 48, 25, 47] have been addressed in a number of ways. The underlying assumption is that the observations to be clustered are drawn from one of several components, and the problem is to estimate the parameters of each component so as to best fit the data. Inferring the parameters of these components and identifying which component produced each observation leads to a clustering of the set of observations. In this section, we first give an overview of mixture modeling, then introduce two most common mixtures, Gaussian mixture model and Bernoulli mixture model, and finally discuss several issues about model selection.

3.2.1 Overview

Suppose we have a set of data points $\mathbf{X} = \{\mathbf{x}_1, \ldots, \mathbf{x}_N\}$ consisting of N observations of a D-dimensional random variable $\mathbf{x}$. The random variable $\mathbf{x}_n$ is assumed to be distributed according to a mixture of K components. Each component (i.e., cluster) is mathematically represented by a parametric distribution. An individual distribution used to model a specific cluster is often referred to as a component distribution. The entire data set is therefore modeled by a mixture of these distributions. Formally, the mixture distribution, or probability density function, of $\mathbf{x}_n$ can be written as

$$p(\mathbf{x}_n) = \sum_{k=1}^{K} \pi_k p(\mathbf{x}_n|\theta_k) \tag{3.1}$$

where $\pi_1, \ldots, \pi_K$ are the *mixing probabilities* (i.e., mixing coefficients or weights), each θ_k is the set of parameters specifying the kth component, and $p(\mathbf{x}_n|\theta_k)$ is the component distribution. In order to be valid probabilities, the mixing probabilities $\{\pi_k\}$ must satisfy

$$0 \leq \pi_k \leq 1 \ (k = 1, \ldots, K), \ \mathit{and} \ \sum_{k=1}^{K} \pi_k = 1.$$

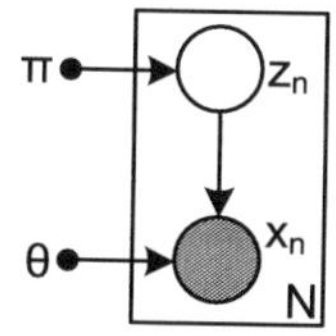

FIGURE 3.1: Graphical representation of a mixture model. Circles indicate random variables, and shaded and unshaded shapes indicate observed and latent (i.e., unobserved) variables.

An obvious way of generating a random sample $\mathbf{x}_n$ with the mixture model, given by (3.1), is as follows. Let z_n be a categorical random variable taking on the values $1,\ldots,K$ with probabilities $p(z_n = k) = \pi_k$ (also denoted as $p(z_{nk} = 1) = \pi_k$). Suppose that the conditional distribution of $\mathbf{x}_n$ given $z_n = k$ is $p(\mathbf{x}_n|\theta_k)$. Then the marginal distribution of $\mathbf{x}_n$ is obtained by summing the joint distribution over all possible values of z_n to given $p(\mathbf{x}_n)$ as (3.1). In this context, the variable z_n can be thought of as the component (or cluster) label of the random sample $\mathbf{x}_n$. Instead of using a single categorical variable z_n, we introduce a K-dimensional binary random vector $\mathbf{z}_n$ to denote the component label for $\mathbf{x}_n$. The K-dimensional random variable $\mathbf{z}_n$ has a 1-of-K representation, in which one of the element $z_{nk} = (\mathbf{z}_n)_k$ equals to 1, and all other elements equal to 0, denoting the component of origin of $\mathbf{x}_n$ is equal to k. For example, if we have a variable with $K = 5$ clusters and a particular observation $\mathbf{x}_n$ of the variable happens to correspond to the cluster where $z_{n4} = 1$, then $\mathbf{z}_n$ will be represented by $\mathbf{z}_n = (0,0,0,1,0)^T$. Note that the values of z_{nk} satisfy $z_{nk} \in \{0,1\}$ and $\sum_{k=1}^{K} z_{nk} = 1$. Because $\mathbf{z}_n$ uses a 1-of-K representation, the marginal distribution over $\mathbf{z}_n$ is specified in terms of *mixing probabilities* π_k, such that

$$p(\mathbf{z}_n) = \pi_1^{z_{n1}} \pi_2^{z_{n2}} \ldots \pi_K^{z_{nK}} = \prod_{k=1}^{K} \pi_k^{z_{nk}}. \tag{3.2}$$

Similarly, the conditional distribution of $\mathbf{x}_n$ given $\mathbf{z}_n$ can be written in the form

$$p(\mathbf{x}_n|\mathbf{z}_n) = \prod_{k=1}^{K} p(\mathbf{x}_n|\theta_k)^{z_{nk}}. \tag{3.3}$$

The joint distribution is given by $p(\mathbf{z}_n)p(\mathbf{x}_n|\mathbf{z}_n)$, and the marginal distribution of $\mathbf{x}_n$ is obtained as

$$p(\mathbf{x}_n) = \sum_{\mathbf{z}_n} p(\mathbf{z}_n)p(\mathbf{x}_n|\mathbf{z}_n) = \sum_{k=1}^{K} \pi_k p(\mathbf{x}_n|\theta_k). \tag{3.4}$$

Thus, the above marginal distribution of $\mathbf{x}$ is an equivalent formulation of the mixture model involving an explicit latent variable. The graphical representation of the mixture model is shown in Figure 3.1. From the generative process point of view, a given set of data points could have been generated by repeating the following procedure N times, once for each data point $\mathbf{x}_n$:

(a) Choose a hidden component (i.e., cluster) label $\mathbf{z}_n \sim Mult_K(1,\pi)$. This selects the kth component from which to draw point $\mathbf{x}_n$.

(b) Sample a data point $\mathbf{x}_n$ from the kth component according to the conditional distribution $p(\mathbf{x}_n|\theta_k)$.

Because we have represented the marginal distribution in the form $p(\mathbf{x}_n) = \sum_{\mathbf{z}_n} p(\mathbf{x}_n, \mathbf{z}_n)$, it follows that for every observed data point $\mathbf{x}_n$ there is a corresponding latent variable $\mathbf{z}_n$. Using Bayes' theorem, we can obtain the conditional probability of $z_{nk} = 1$ given $\mathbf{x}_n$ as

$$p(z_{nk} = 1|\mathbf{x}_n) = \frac{p(z_{nk} = 1)p(\mathbf{x}_n|z_{nk} = 1)}{\sum_{j=1}^{K} p(z_{nj} = 1)p(\mathbf{x}_n|z_{nj} = 1)} = \frac{\pi_k p(\mathbf{x}_n|\theta_k)}{\sum_{j=1}^{K} \pi_j p(\mathbf{x}_n|\theta_j)}, \tag{3.5}$$

where π_k (i.e., $p(z_{nk} = 1)$) is the *prior probability* that data point $\mathbf{x}_n$ was generated from component k, and $p(z_{nk} = 1|\mathbf{x}_n)$ is the *posterior probability* that the observed data point $\mathbf{x}_n$ came from component k. In the following, we shall use $\gamma(z_{nk})$ to denote $p(z_{nk} = 1|\mathbf{x}_n)$, which can also be viewed as the *responsibility* that component k takes for explaining the observation $\mathbf{x}_n$.

In this formulation of the mixture model, we need to infer a set of parameters from the observation, including the *mixing probabilities* $\{\pi_k\}$ and the parameters for the *component distributions* $\{\theta_k\}$. The number of components K is considered fixed, but of course in many applications, the value of K is unknown and has to be inferred from the available data [25]. Thus, the overall parameter of the mixture model is $\Theta = \{\pi_1, \ldots, \pi_K, \theta_1, \ldots, \theta_K\}$. If we assume that the data points are drawn independently from the distribution, then we can write the probability of generating all the data points in the form

$$p(\mathbf{X}|\Theta) = \prod_{n=1}^{N} \sum_{k=1}^{K} \pi_k p(\mathbf{x}_n|\theta_k), \tag{3.6}$$

or in a logarithm form

$$\log p(\mathbf{X}|\Theta) = \sum_{n=1}^{N} \log \sum_{k=1}^{K} \pi_k p(\mathbf{x}_n|\theta_k). \tag{3.7}$$

In statistics, maximum likelihood estimation (MLE) [23, 8, 41] is an important statistical approach for parameter estimation,

$$\Theta_{ML} = \arg\max_{\Theta} \{\log p(\mathbf{X}|\Theta)\},$$

which considers the best estimate as the one that maximizes the probability of generating all the observations. Sometimes we have a priori information $p(\Theta)$ about the parameters, and it can be incorporated into the mixture models. Thus, the maximum a posteriori (MAP) estimation [70, 27] is used instead,

$$\Theta_{MAP} = \arg\max_{\Theta} \{\log p(\mathbf{X}|\Theta) + \log p(\Theta)\},$$

which considers the best estimate as the one that maximizes the posterior probability of Θ given the observed data. MLE and MAP give a unified approach to parameter estimation, which is well-defined in the case of the normal distribution and many other problems.

As mentioned previously, each cluster is mathematically represented by a parametric distribution. In principle, the mixtures can be constructed with any types of components, and we could still have a perfectly good mixture model. In practice, a lot of effort is given to parametric mixture models, where all components are from the same parametric family of distributions, but with different parameters. For example, they might all be Gaussians with different means and variances, or all Poisson distributions with different means, or all power laws with different exponents. In the following section, we will introduce the two most common mixtures, mixture of Gaussian (continuous) and mixture of Bernoulli (discrete) distributions.

3.2.2 Gaussian Mixture Model

The most well-known mixture model is the Gaussian mixture model (GMM), where each component is a Gaussian distribution. Recently, GMM has been widely used for clustering in many applications, such as speaker identification and verification [62, 61], image segmentation [15, 56], and object tracking [71].

The Gaussian, also known as the normal distribution, is a widely used model for the distribution of continuous variables. In the case of a single variable x, the Gaussian distribution can be written in the form

$$\mathcal{N}(x|\mu, \sigma^2) = \frac{1}{(2\pi\sigma^2)^{1/2}} \exp\{-\frac{1}{2\sigma^2}(x-\mu)^2\}, \tag{3.8}$$

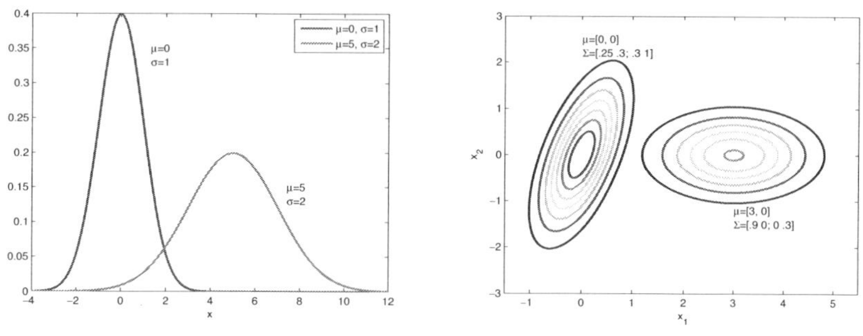

(a) (b)

FIGURE 3.2 (See color insert): (a) Plots of the univariate Gaussian distribution given by (3.8) for various parameters of μ and σ, and (b) contours of the multivariate (2-D) Gaussian distribution given by (3.9) for various parameters of μ and Σ.

where μ is the mean and σ^2 is the variance. Figure 3.2(a) shows plots of the Gaussian distribution for various values of the parameters. For a D-dimensional vector $\mathbf{x}$, the multivariate Gaussian distribution takes the form

$$\mathcal{N}(\mathbf{x}|\mu,\Sigma) = \frac{1}{(2\pi)^{D/2}}\frac{1}{|\Sigma|^{1/2}}\exp\{-\frac{1}{2}(\mathbf{x}-\mu)^T\Sigma^{-1}(\mathbf{x}-\mu)\}, \tag{3.9}$$

where μ is a D-dimensional mean vector, Σ is a $D \times D$ covariance matrix, and $|\Sigma|$ denotes the determinant of Σ. Figure 3.2(b) shows contours of the Gaussian distribution for various values of the parameters.

In the Gaussian mixture model, each component is represented by the parameters of a multivariate Gaussian distribution $p(\mathbf{x}_k|\theta_k) = \mathcal{N}(\mathbf{x}_n|\mu_k,\Sigma_k)$. Based on (3.1), the Gaussian mixture distribution can be written as a linear superposition of Gaussians in the form

$$p(\mathbf{x}_n|\Theta) = p(\mathbf{x}_n|\pi,\mu,\Sigma) = \sum_{k=1}^{K}\pi_k\mathcal{N}(\mathbf{x}_n|\mu_k,\Sigma_k). \tag{3.10}$$

For a given set of observations $\mathbf{X}$, the log-likelihood function is given by

$$l(\Theta) = \log p(\mathbf{X}|\Theta) = \sum_{n=1}^{N}\log p(\mathbf{x}_n|\Theta) = \sum_{n=1}^{N}\log\sum_{k=1}^{K}\pi_k\mathcal{N}(\mathbf{x}_n|\mu_k,\Sigma_k). \tag{3.11}$$

To find maximum likelihood solutions that are valid at local maxima, we compute the derivatives of $\log p(\mathbf{X}|\pi,\mu,\Sigma)$ with respect to π_k, μ_k, and Σ_k, respectively. The derivative with respect to the mean μ_k is given by

$$\frac{\partial l}{\partial \mu_k} = \sum_{n=1}^{N}\frac{\pi_k\mathcal{N}(\mathbf{x}_n|\mu_k,\Sigma_k)}{\sum_{j=1}^{K}\pi_j\mathcal{N}(\mathbf{x}_n|\mu_j,\Sigma_j)}\Sigma_k^{-1}(\mathbf{x}_n-\mu_k) = \sum_{n=1}^{N}\gamma(z_{nk})\Sigma_k^{-1}(\mathbf{x}_n-\mu_k),$$

where we have used (3.9) and (3.5) for the Gaussian distribution and the responsibilities (i.e., posterior probabilities), respectively. Setting this derivative to zero and multiplying by Σ_k, we obtain

$$\mu_k = \frac{\sum_{n=1}^{N}\gamma(z_{nk})\mathbf{x}_n}{\sum_{n=1}^{N}\gamma(z_{nk})} \tag{3.12}$$

and

$$\gamma(z_{nk}) = \frac{\pi_k \mathcal{N}(\mathbf{x}_n|\mu_k, \Sigma_k)}{\sum_{j=1}^{K} \pi_j \mathcal{N}(\mathbf{x}_n|\mu_j, \Sigma_j)}. \tag{3.13}$$

We can interpret that the mean μ_k for the kth Gaussian component is obtained by taking a weighted mean of all the points in the data set, in which the weighting factor corresponds to the posterior probability $\gamma(z_{nk})$ that component k was responsible for generating $\mathbf{x}_n$.

Similarly, we set the derivative of $\log p(\mathbf{X}|\pi, \mu, \Sigma)$ with respect to Σ_k to zero, and we obtain

$$\Sigma_k = \frac{\sum_{n=1}^{N} \gamma(z_{nk})(\mathbf{x}_n - \mu_k)(\mathbf{x}_n - \mu_k)^T}{\sum_{n=1}^{N} \gamma(z_{nk})}. \tag{3.14}$$

Similar to (3.12), each data point is weighted by the conditional probability generated by the corresponding component and with the denominator given by the effective number of points associated with the corresponding component.

The derivative of $\log p(\mathbf{X}|\pi, \mu, \Sigma)$ with respect to the mixing probabilities π_k requires a little more work, because the values of π_k are constrained to be positive and sum to one. This constraint can be handled using a Lagrange multiplier and maximizing the following quantity

$$\log p(\mathbf{X}|\pi, \mu, \Sigma) + \lambda(\sum_{k=1}^{K} \pi_k - 1).$$

After simplifying and rearranging we obtain

$$\pi_k = \frac{\sum_{n=1}^{N} \gamma(z_{nk})}{N} \tag{3.15}$$

so that the mixing probabilities for the kth component are given by the average responsibility that the component takes for explaining the data points.

It is worth noting that the Equations (3.12), (3.14), and (3.15) are not the closed-form solution for the parameters of the mixture model. The reason is that these equations are intimately coupled with Equation (3.13). More specifically, the responsibilities $\gamma(z_{nk})$ given by Equation (3.13) depend on all the parameters of the mixture model, while all the results of (3.12), (3.14), and (3.15) rely on $\gamma(z_{nk})$. Therefore, maximizing the log likelihood function for a Gaussian mixture model turns out to be a very complex problem. An elegant and powerful method for finding maximum likelihood solutions for models with latent variables is called the *Expectation-Maximization* algorithm or EM algorithm [23].

However, the above equations do suggest an iterative solution for the Gaussian mixture model, which turns out to be an instance of the EM algorithm for the particular case of the Gaussian mixture model. Here we shall give such an iterative algorithm in the context of Gaussian mixture model, and later we shall give a general framework of EM algorithm in Section 3.3. This algorithm is started by initializing with guesses about the parameters, including the means, covariances, and mixing probabilities. Then we alternative between two updating steps, the *expectation* step and *maximization* step. In the *expectation* step, or E-step, we use the current parameters to calculate the responsibilities (i.e., posterior probabilities) according to (3.13). In the *maximization* step, or M-step, we maximize the log-likelihood with the updated responsibilities, and reestimate the means, covariances, and mixing coefficients using (3.12), (3.14), and (3.15). In Section 3.3, we shall show that each iteration of the EM algorithm is guaranteed to increase the log-likelihood. In practice, the EM algorithm is converged when the change in the log-likelihood or the parameter values fall below some threshold. We summarize the EM algorithm for Gaussian mixtures in Algorithm 11.

As illustrated in Figure 3.3, the EM algorithm for a mixture of two Gaussian components is applied to a random generated data set. Plot (a) shows the data points together with the random initialization of the mixture model in which the two Gaussian components are shown. Plot (b) shows

Algorithm 11 EM for Gaussian Mixtures

Given a set of data points and a Gaussian mixture model, the goal is to maximize the log-likelihood with respect to the parameters.

1: Initialize the means μ_k^0, covariances Σ_k^0, and mixing probabilities π_k^0.
2: **E-step**: Calculate the responsibilities $\gamma(z_{nk})$ using the current parameters based on Equation (3.13).
3: **M-step**: Update the parameters using the current responsibilities. Note that we first update the new means using (3.12), then use these new values to calculate the covariances using (3.14), and finally reestimate the mixing probabilities using (3.15).
4: Compute the log-likelihood using (3.10) and check for convergence of the algorithm. If the convergence criterion is not satisfied, then repeat steps 2–4; otherwise, return the final parameters.

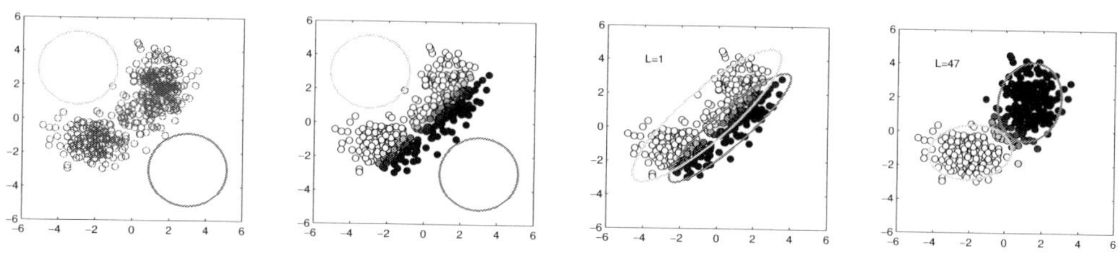

(a) (b) (c) (d)

FIGURE 3.3: Illustration of the EM algorithm for two Gaussian components.

the result of the initial E-step, in which each data point is depicted using a proportion of white ink and black ink according to the posterior probability of having been generated by the corresponding component. Plot (c) shows the situation after the first M-step, in which the means and covariances of both components have changed. Plot (d) shows the results after 47 cycles of EM, which is close to convergence.

Compared with K-means algorithm, the EM algorithm for GMM takes many more iterations to reach convergence [8, 55]. There will be multiple local maxima of the log-likelihood which depends on different initialization, and EM is not guaranteed to find the largest of these maxima. In order to find a suitable initialization and speed up the convergence for a Gaussian mixture model, it is common to run the K-means algorithm [8] and choose the means and covariances of the clusters, and the fractions of data points assigned to the respective clusters, for initializing μ_k^0, Σ_k^0 and π_k^0, respectively. Another problem of the EM algorithm for GMM is the singularity of the likelihood function in which a Gaussian component collapses onto a particular data point. It is reasonable to avoid the singularities by using some suitable heuristics [8, 55], for instance, by detecting when a Gaussian component is collapsing and resetting its mean and covariance, and then continuing with the optimization.

3.2.3 Bernoulli Mixture Model

Although most research in mixture models has focused on distributions over continuous variables described by mixtures of Gaussians, there are many clustering tasks for which binary or discrete mixture models are better fitted. We now discuss mixtures of discrete binary variables described by Bernoulli distributions, which is also known as *latent class analysis* [42, 48].

A Bernoulli mixture model [18] is a special case of mixture models, where each component k has a D-dimensional Bernoulli distribution with parameters $\mu_k = \{\mu_{k1}, \ldots, \mu_{kD}\}^T$,

$$p(\mathbf{x}|\mu_k) = \prod_{i=1}^{D} \mu_{ki}^{x_i}(1-\mu_{ki})^{1-x_i}. \tag{3.16}$$

Let K denote the number of components, a mixture model with K Bernoulli components can be written in the form

$$p(\mathbf{x}|\mu,\pi) = \sum_{k=1}^{K} \pi_k p(\mathbf{x}|\mu_k), \tag{3.17}$$

where $\mu = \{\mu_1, \ldots, \mu_K\}$ and $\pi = \{\pi_1, \ldots, \pi_K\}$.

Given a data set $\mathbf{X} = \{\mathbf{x}_1, \ldots, \mathbf{x}_N\}$, the log-likelihood for the model is given by

$$\log p(\mathbf{X}|\mu,\pi) = \sum_{n=1}^{N} \log\left\{\sum_{k=1}^{K} \pi_k p(\mathbf{x}_n|\mu_k)\right\} \tag{3.18}$$

$$= \sum_{n=1}^{N} \log\left\{\sum_{k=1}^{K} \pi_k \prod_{i=1}^{D} \mu_{ki}^{x_{ni}}(1-\mu_{ki})^{1-x_{ni}}\right\}. \tag{3.19}$$

Note that the summation appears inside the logarithm, so that there is no close-form result for maximum likelihood solution. EM algorithm is a standard learning algorithm for estimating the parameters of this model. In Section 3.3 we will show the general EM algorithm, and the solution for Bernoulli mixture model will be shown and revisited in Section 3.3.2.

Here let us discuss a real application of the Bernoulli mixture model by using it to model handwritten digits. Imagine that we are given an $M \times M$ binary (black-and-white) image that is known to be a scan of a handwritten digit between 0 and 9, but we do not know which digit is written. We can create a mixture model with $K = 10$ different components, where each component is a vector of size M^2 of Bernoulli distributions (one per pixel). Such a model can be trained with the EM algorithm on an unlabeled set of handwritten digits and will effectively cluster the images according to the digit being written.

3.2.4 Model Selection Criteria

With the model-based approach to clustering, issues such as the selection of the number of clusters or the assessment of the validity of a given model can be addressed in a principled and formal way. In mixture modeling, an important issue is the selection of the number of components, which is assumed to be fixed in previous subsections. However, like the usual trade-off in model selection problems, the mixture with too many components may overfit the data, while a mixture with too few components may not be flexible enough to approximate the true underlying model. Thus, we need to adopt some statistical procedures besides the parameter estimation to infer the number of components.

To illustrate the problem of model selection, let us denote M_k as the class of all possible k-component mixtures built from a certain type of distribution. The MLE criterion cannot be directly used to estimate the number of mixture components k because these classes are nested, i.e., $M_k \in M_{k+1}$, and the maximized likelihood function $\log p(\mathbf{X}|\Theta_{ML})$ is a nondecreasing function of k. Therefore, MLE criterion is useless to estimate the number of components, and some advanced model selection methods have been proposed.

From a computational point of view, there are two main types of model selection methods: deterministic and stochastic methods. Deterministic methods usually start by obtaining a set of candidate models for a range of values of k (from k_{min} to k_{max}) which is assumed to contain the optimal k. The number of components is then selected according to

$$k^* = \arg\min_k \{C(\Theta_k, k), k = k_{min}, \ldots, k_{max}\}, \tag{3.20}$$

where Θ_k is an estimate of the mixture parameters with k components, and $C(\Theta_k, k)$ is some model selection criteria. A very common criterion can be expressed in the form

$$C(\Theta_k, k) = -\log p(\mathbf{X}|\Theta_k) + \mathcal{P}(k),$$

where $\mathcal{P}(k)$ is an increasing function penalizing higher values of k. There are many examples of such criteria that have been used for mixture models, including Bayesian approximation criteria, such as Laplace-empirical criterion (LEC) [65, 48] and Schwarz's Bayesian inference criterion (BIC) [67, 26], and information-theoretic approaches [17], such as Rissanen's minimum description length (MDL) [64, 57], the minimum message length (MML) criterion [54, 79, 25], and Akaike's information criterion (AIC) [17]. In addition, several model section criteria have been proposed based on the classification likelihood, such as the normalized entropy criterion (NEC) [21, 4] and the integrated classification likelihood (ICL) criterion [5]. For finite mixture models, a comprehensive review of model selection approaches is given in [48]. Besides deterministic methods, stochastic methods based on Markov chain Monte Carlo (MCMC) can be used for model inference, by either implementing model selection criteria [3, 66], or modeling in fully Bayesian way and sampling from the full a posteriori distribution with k considered unknown [63, 59]. Moreover, cross-validation approaches [69] have also been used to estimate the number of mixture components.

3.3 EM Algorithm and Its Variations

In most applications, parameters of a mixture model are estimated by maximizing the likelihood and the standard approach to MLE for mixture models and other probabilistic models is the EM algorithm [23]. In this section, we first present a general view of the EM algorithm, then revisit the EM algorithm for mixture models, and finally discuss the variations and applications of EM algorithm. We will also illustrate the EM algorithm for topic models in Section 3.4.

3.3.1 The General EM Algorithm

The EM (Expectation-Maximization) algorithm is a popular iterative method for finding MLE or MAP estimations for probabilistic models with latent variables [23].

Suppose we have a set of observed data points (or objects) $\mathbf{X} = \{\mathbf{x}_1, \ldots, \mathbf{x}_N\}$ and a set of latent variables $\mathbf{Z} = \{\mathbf{z}_1, \ldots, \mathbf{z}_N\}$, in which each data point $\mathbf{x}_n$ is associated with a latent variable $\mathbf{z}_n$, and we use Θ to denote the set of all model parameters. To use the method of maximum likelihood, the likelihood (or likelihood function) of a set of parameters given the observed data is defined as the probability of all the observations given those parameters values $p(\mathbf{X}|\Theta)$. In practice, it is often more convenient to optimize the log-likelihood function for the EM algorithm, which is given by

$$l(\Theta) = \log p(\mathbf{X}|\Theta) = \sum_{n=1}^{N} \log p(\mathbf{x}_n|\Theta) = \sum_{n=1}^{N} \log \sum_{\mathbf{z}_n} p(\mathbf{x}_n, \mathbf{z}_n|\Theta).$$

Here we are assuming $\mathbf{z}_n$ is discrete, although the discussion is identical if $\mathbf{z}_n$ comprises continuous

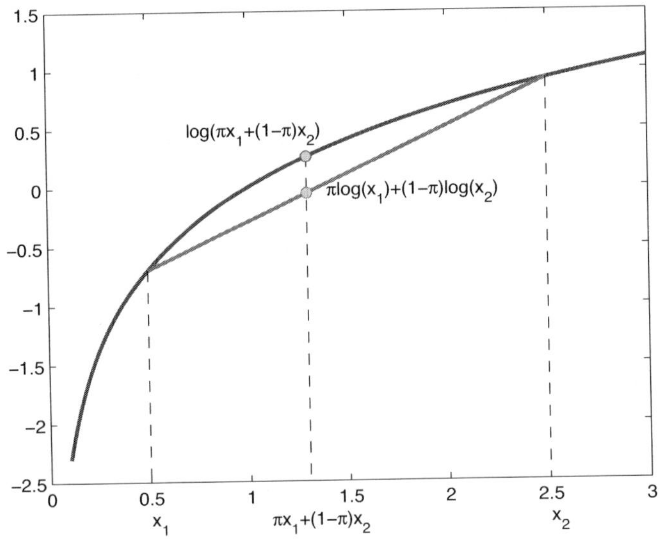

FIGURE 3.4: For a concave function $\log(x)$, the secant line consists of weighted means of the concave function that lies on or below the function.

variables with summation replaced by integration as appropriate. Note that the summation over the latent variables appears inside the logarithm, which prevents the logarithm from acting directly on the joint distribution, resulting in complicated expressions for the maximum likelihood solution. The essential trick of the EM algorithm is to maximize a *lower bound* on the log-likelihood instead of the log-likelihood.

Jensen's inequality is often used to bound the logarithm of a summation of terms. Given K non-negative numbers $\pi_1, \ldots, \pi_K$ that add up to one (i.e., mixing probabilities), and K arbitrary numbers $x_1, \ldots, x_K$, for a convex function f, Jensen's inequality can be stated as

$$f(\sum_{k=1}^{K} \pi_k x_k) \leq \sum_{k=1}^{K} \pi_k f(x_k),$$

and the inequality is reversed if f is a concave function, which is

$$f(\sum_{k=1}^{K} \pi_k x_k) \geq \sum_{k=1}^{K} \pi_k f(x_k).$$

Let us take the logarithm function, a concave function, as an example. As shown in Figure 3.4, for two points x_1 and x_2, the concave function of the weighted means $\log(\pi x_1 + (1-\pi)x_2)$ is larger than the weighted means of the concave function $\pi \log(x_1) + (1-\pi)\log(x_2)$.

We can introduce an arbitrary distribution $q(\mathbf{z}_n)$ defined over the latent variables ($\sum_{\mathbf{z}_n} q(\mathbf{z}_n) = 1$, $q(\mathbf{z}_n) \geq 0$), and we observe that, for any choice of $q(\mathbf{z}_n)$, the following decomposition holds based on the *Jensen's inequality*:

$$\begin{aligned} l(\Theta) = \log p(\mathbf{X}|\Theta) &= \sum_{n=1}^{N} \log \sum_{\mathbf{z}_n} p(\mathbf{x}_n, \mathbf{z}_n|\Theta) \\ &= \sum_{n=1}^{N} \log \sum_{\mathbf{z}_n} q(\mathbf{z}_n) \frac{p(\mathbf{x}_n, \mathbf{z}_n|\Theta)}{q(\mathbf{z}_n)} \\ &\geq \sum_{n=1}^{N} \sum_{\mathbf{z}_n} q(\mathbf{z}_n) \log \frac{p(\mathbf{x}_n, \mathbf{z}_n|\Theta)}{q(\mathbf{z}_n)} \\ &\equiv \mathcal{L}(q, \Theta). \end{aligned} \tag{3.21}$$

It is obvious that $\mathcal{L}(q,\Theta)$ is a lower bound on $\log p(\mathbf{X}|\Theta)$. Note that $\mathcal{L}(q,\Theta)$ is a function of the distribution $q(\mathbf{z})$ and a function of the parameters Θ. It is worth studying carefully that it will be easier to maximize this lower bound on the log-likelihood than the actual log-likelihood, and the lower bound is reasonably tight. In terms of tightness, suppose $q(\mathbf{z}_n) = p(\mathbf{z}_n|\mathbf{x}_n,\Theta)$; we can simplify $\mathcal{L}(q,\Theta)$ to be

$$\begin{aligned} \mathcal{L}(q,\Theta)|_{q(\mathbf{z}_n)=p(\mathbf{z}_n|\mathbf{x}_n,\Theta)} &= \sum_{n=1}^{N} \sum_{\mathbf{z}_n} p(\mathbf{z}_n|\mathbf{x}_n,\Theta) \log \frac{p(\mathbf{x}_n,\mathbf{z}_n|\Theta)}{p(\mathbf{z}_n|\mathbf{x}_n,\Theta)} \\ &= \sum_{n=1}^{N} \sum_{\mathbf{z}_n} p(\mathbf{z}_n|\mathbf{X},\Theta) \log p(\mathbf{x}_n|\Theta) \\ &= \sum_{n=1}^{N} \log p(\mathbf{x}_n|\Theta) = l(\Theta). \end{aligned} \tag{3.22}$$

Therefore, with that choice of q, the lower bound $\mathcal{L}(q,\Theta)$ will equal the log-likelihood, which indicates that the lower bound is tight. Moreover, since $\mathcal{L}(q,\Theta) \leq l(\Theta)$, this choice of q maximizes $\mathcal{L}(q,\Theta)$ for fixed Θ.

The EM algorithm is a two-stage iterative optimization framework for finding maximum likelihood solutions. Now we shall demonstrate how the EM algorithm does indeed maximize the log-likelihood. Suppose that the current estimates of the parameters are $\Theta^{(t)}$. In the E-step, the lower bound $\mathcal{L}(q,\Theta^{(t)})$ is maximized with respect to $q(\mathbf{z}_n)$ while holding $\Theta^{(t)}$ fixed, which can be stated in the form

$$q^{(t)}(\mathbf{z}_n) = \arg\max_q \mathcal{L}(q,\Theta^{(t)}) = p(\mathbf{z}_n|\mathbf{x}_n,\Theta^{(t)}).$$

The solution to this maximization problem is easily seen by noting that the value of $\log p(\mathbf{X}|\Theta^{(t)})$ does not depend on $q(\mathbf{z}_n)$ when $q(\mathbf{z}_n)$ is equal to the posterior distribution $p(\mathbf{z}_n|\mathbf{x}_n,\Theta^{(t)})$. In this case, the lower bound will equal the log-likelihood given by Equation (3.22). In other words, the maximization in the E-step is just computing the posterior probabilities $p(\mathbf{z}_n|\mathbf{x}_n,\Theta^{(t)})$.

In the following M-step, the distribution $q(\mathbf{z}_n)$ is held fixed and the lower bound $\mathcal{L}(q^{(t)},\Theta)$ is maximized with respect to Θ to obtain a new estimate $\Theta^{(t+1)}$, which can be stated in the form

$$\Theta^{(t+1)} = \arg\max_\Theta \mathcal{L}(q^{(t)},\Theta). \tag{3.23}$$

If we substitute $q^{(t)}(\mathbf{z}_n) = p(\mathbf{z}_n|\mathbf{x}_n,\Theta^{(t)})$ into $\mathcal{L}(q^{(t)},\Theta)$, the lower bound can be expressed as

$$\mathcal{L}(q^{(t)},\Theta) = Q(\Theta,\Theta^{(t)}) - \sum_{n=1}^{N} \sum_{\mathbf{z}_n} p(\mathbf{z}_n|\mathbf{x}_n,\Theta^{(t)}) \log p(\mathbf{z}_n|\mathbf{x}_n,\Theta^{(t)}) \tag{3.24}$$

where

$$Q(\Theta,\Theta^{(t)}) = \sum_{n=1}^{N} \sum_{\mathbf{z}_n} p(\mathbf{z}_n|\mathbf{x}_n,\Theta^{(t)}) \log p(\mathbf{x}_n,\mathbf{z}_n|\Theta), \tag{3.25}$$

Algorithm 12 The General EM Algorithm

Given a statistical model consisting of a set of observed variables $\mathbf{X}$, a set of unobserved latent variables $\mathbf{Z}$, and a vector of unknown parameters Θ, the goal is to maximize the log-likelihood with respect to the parameters Θ.

1: Start with an initial guess for the parameters $\Theta^{(0)}$ and compute the initial log-likelihood $\log p(\mathbf{X}|\Theta^{(0)})$.
2: **E-step**: Evaluate $q^{(t)} = \arg\max_q \mathcal{L}(q, \Theta^{(t)}) = p(\mathbf{z}_n|\mathbf{x}_n, \Theta^{(t)})$.
3: **M-step**: Update the parameters $\Theta^{(t+1)} = \arg\max_\Theta Q(\Theta, \Theta^{(t)})$.
4: Compute the log-likelihood $\log p(\mathbf{X}|\Theta^{(t+1)})$ and check for convergence of the algorithm. If the convergence criterion is not satisfied, then repeat steps 2-4, otherwise, return the final parameters.

and the second sum is simply the negative entropy (constant) of the q distribution, and is therefore independent of Θ. Thus, in the M-step, maximizing $\mathcal{L}(q^{(t)}, \Theta)$ is equivalent to maximizing $Q(\Theta, \Theta^{(t)})$,

$$\Theta^{(t+1)} = \arg\max_\Theta Q(\Theta, \Theta^{(t)}). \tag{3.26}$$

The above quantity that is being maximized is the expectation of the complete data log-likelihood.

The convergence properties of the EM algorithm are discussed in detail in [23] and [46]. Here we discuss the general convergence of the algorithm. We have seen that both the E and the M steps of the EM algorithm are about maximizing $\mathcal{L}(q, \Theta)$. It is easy to see that, after E-step,

$$l(\Theta^{(t)}) = \mathcal{L}(q^{(t)}, \Theta^{(t)}) \geq \mathcal{L}(q^{(t-1)}, \Theta^{(t)})$$

and that, after the M-step,

$$\mathcal{L}(q^{(t)}, \Theta^{(t+1)}) \geq \mathcal{L}(q^{(t)}, \Theta^{(t)}).$$

Putting these two inequalities together, we obtain $\mathcal{L}(q^{(t+1)}, \Theta^{(t+1)}) \geq \mathcal{L}(q^{(t)}, \Theta^{(t)})$ and $\log p(\mathbf{X}|\Theta^{(t+1)}) \geq \log p(\mathbf{X}|\Theta^{(t)})$. It is clear that each EM cycle can increase the lower bound on the log-likelihood function and will change the model parameters in such a way as to increase the actual log-likelihood, which is guaranteed to reach a local maximum of the log-likelihood function. The general EM algorithm is summarized in Algorithm 12.

Maximum A Posterior Estimation: The EM algorithm can also be used to find MAP solutions for models in which a prior $p(\Theta)$ over the parameters is introduced. We note that as a function of Θ, $p(\Theta|\mathbf{X}) = \frac{p(\mathbf{X}|\Theta)p(\Theta)}{p(\mathbf{X})}$, based on Bayes' theorem. By making use of the decomposition (3.21), we have

$$\begin{aligned} \log p(\Theta|\mathbf{X}) &= \log p(\mathbf{X}|\Theta) + \log p(\Theta) - \log p(\mathbf{X}) \\ &\geq \mathcal{L}(q, \Theta) + \log p(\Theta) - \log p(\mathbf{X}), \end{aligned} \tag{3.27}$$

where $\log p(\mathbf{X})$ is a constant. Again we can maximize the right-hand side alternatively with respect to q and Θ. In this case the E-step remains the same as in the maximum likelihood case as q only appears in $\mathcal{L}(q, \Theta)$, whereas in the M-step the quantity to be maximized is given by $\mathcal{L}(q, \Theta) + \log p(\Theta)$, which typically requires only a small modification to the standard maximum likelihood M-step.

Furthermore, a number of variants of the EM algorithm for clustering have been studied for when the original EM algorithm is difficult to compute in either the E-step or the M-step. Here we list some of them.

Generalized EM: Often in practice, the solution to the M-step exists in closed form, but there are many cases for which it is intractable to find the value of Θ that maximizes the function

$Q(\Theta, \Theta^{(t)})$. In such situations, a generalized EM algorithm (GEM) [23, 53, 16] is defined for which the M-step requires $\Theta^{(t+1)}$ to be chosen such that

$$Q(\Theta^{(t+1)}, \Theta^{(t)}) \geq Q(\Theta^{(t)}, \Theta^{(t)})$$

holds. That is, one chooses a better value $\Theta^{(t+1)}$ to increase the Q-function, $Q(\Theta, \Theta^{(t)})$, rather than to find the local maximum of it. The convergence can still be guaranteed using GEM, and it has been widely used for the estimation of topic modeling with graph-based regularization, such as NetPLSA [50] and TMBP [24].

Variational EM: The variational EM algorithm addresses the problem of an intractable E-step, which is one of the approximate algorithms used in LDA [14]. The idea is to find a set of variational parameters with respect to hidden variables that attempts to obtain the tightest possible lower bound in E-step and to maximize the lower bound in M-step. The variational parameters are chosen in a way that simplifies the original probabilistic model and are thus easier to calculate. There is no guarantee that variational EM will find a local optimum.

Stochastic EM: In some situations, the EM algorithm involves computing an integral that may not be tractable, such as LDA [14]. The idea of stochastic EM (SEM) [20, 46] is then to replace this tedious computation with a stochastic simulation. The E-step is computed with Monte Carlo sampling. This introduces randomness into the optimization, but asymptotically it will converge to a local optimum.

3.3.2 Mixture Models Revisited

With our general definition of the EM algorithm, let us revisit the case of estimating the parameters for Gaussian mixture model and Bernoulli mixture model, respectively. The EM algorithm is an iterative algorithm that has two main steps. In the E-step, it tries to guess the values of the hidden variable z_{nk}, while in the M-step, it updates the parameters of the model based on the guesses of the E-step.

Gaussian Mixture Model: The parameters of a Gaussian mixture model include the mean vectors μ, covariance matrices Σ, and mixing weights π. In the E-step, we calculate the posterior probability, or responsibility, of the hidden variable z_{nk}, given the current setting of parameters. Using Bayes' theorem, we obtain

$$q(z_{nk} = 1) = p(z_{nk} = 1 | \mathbf{x}_n, \mu_k, \Sigma_k) = \frac{\pi_k \mathcal{N}(\mathbf{x}_n | \mu_k, \Sigma_k)}{\sum_{j=1}^{K} \pi_j \mathcal{N}(\mathbf{x}_n | \mu_j, \Sigma_j)} = \gamma(z_{nk}), \tag{3.28}$$

which is just the responsibility of the component k for data point $\mathbf{x}_n$. In the M-step, we need to maximize the quantity $Q(\Theta, \Theta^{(t)})$. From the independence assumption, the expectation of the complete data log-likelihood (3.25) can be rewritten as

$$Q(\Theta, \Theta^{(t)}) = \sum_{n=1}^{N} \sum_{k=1}^{K} \gamma(z_{nk})^{(t)} \log \pi_k \mathcal{N}(\mathbf{x}_n | \mu_k, \Sigma_k). \tag{3.29}$$

Compared with the log-likelihood function (3.11) for the incomplete data, the logarithm now acts directly on the Gaussian distributions. By keeping the responsibilities fixed, we can maximize $Q(\Theta, \Theta^{(t)})$ with respect to μ_k, Σ_k, and π_k. This leads to closed-form solutions for updating μ_k, Σ_k, and π_k, given by (3.12), (3.14), and (3.15). This is precisely the EM algorithm for Gaussian mixtures as derived earlier.

Relationship of EM with the K-means Algorithm: Given a set of data, the goal of K-means is to partition the data set into some number K of clusters, such that the sum of the square distances of each data points to the center of its cluster is a minimum. We introduce a binary responsibility variable, r_{nk}, to describe the ownership of a data point that belongs to cluster k if $r_{nk} = 1$, where

$\sum_{k=1}^{K} = 1$, and a set of vectors μ_k to denote the mean vector associated with the kth cluster. The cost function of K-means algorithm can be defined as

$$J = \sum_{n=1}^{N} \sum_{k=1}^{K} r_{nk} \|x_n - \mu_k\|^2, \tag{3.30}$$

which represents the sum of the square distances of each data point to its associated vector μ_k. Our goal is to find values for the $\{r_{nk}\}$ and the $\{\mu_k\}$ so as to minimize J. We can solve K-means through an iterative procedure in which each iteration involves two successive steps corresponding to EM algorithm as follows,

> Choose some initial values for mean vectors $\{\mu_k\}$, then iterate between the following two phases until convergence:
>
> (E-step): Assign each data point to the nearest cluster by minimizing J with respect to the r_{nk} and keeping the μ_k fixed,
>
> (M-step): Update mean vectors for clusters by minimizing J with respect to the μ_k and keeping r_{nk} fixed.

More formally, we calculate r_{nk} in the E-step,

$$r_{nk} = \begin{cases} 1 & if\ k = \arg\min_j \|x_n - \mu_j\|^2 \\ 0 & otherwise, \end{cases}$$

and update μ_k in the M-step,

$$\mu_k = \frac{\sum_n r_{nk} x_n}{\sum_n r_{nk}}.$$

The two phases of reassigning data points to clusters and recomputing the cluster means are repeated until there is no further change in the assignments or some other convergence criteria are satisfied.

As described above, the K-means algorithm can be solved through iterative EM algorithm, which shows that there is a close similarity between the K-means algorithm and the EM algorithm. Actually, we can derive the K-means algorithm as a particular case of EM for Gaussian mixtures. The major difference is that the K-means algorithm performs a hard assignment of data points to clusters, while the EM algorithm makes a soft assignment based on the posterior probabilities. Consider a Gaussian mixture model with common covariance matrices given by εI, where ε is a variance parameter that is shared by all of the components, and I is the identity matrix. The Gaussian distribution can be written as

$$p(x|\mu_k, \Sigma_k) = \frac{1}{(2\pi\varepsilon)^{1/2}} \exp\left\{-\frac{1}{2\varepsilon}\|x - \mu_k\|^2\right\}. \tag{3.31}$$

Based on Equation (3.28), the responsibility $\gamma(z_{nk})$ is given by

$$\gamma(z_{nk}) = \frac{\pi_k \exp\{-\|x_n - \mu_k\|^2/2\varepsilon\}}{\sum_j \pi_j \exp\{-\|x_n - \mu_j\|^2/2\varepsilon\}}.$$

If we consider the limit $\varepsilon \to 0$, the responsibilities $\gamma(z_{nk})$ for the data point x_n all go to zero except for the cluster j with the smallest $\|x_n - \mu_j\|^2$, such that the responsibility becomes binary. Thus, in this limit, we obtain a hard assignment of data points to cluster, just as in the K-means algorithm, so that $\gamma(z_{nk}) \to r_{nk} \in \{0, 1\}$. In this case, the EM re-estimation equation for the μ_k given by Equation (3.12) boils down to the K-means result. Therefore, as we can see in this limit, the EM algorithm for Gaussian mixtures is precisely equivalent to the K-means algorithm.

Bernoulli Mixture Model: We now derive the EM algorithm for maximizing the likelihood function for estimating the parameters of the Bernoulli mixture model $\Theta = \{\mu, \pi\}$. As in the case of the Gaussian mixture, let us introduce a latent variable $\mathbf{z}$ associated with each instance of $\mathbf{x}$. The expectation of the complete data log-likelihood with respect to the posterior distribution of the latent variable can be given by

$$Q(\Theta, \Theta^{(t)}) = \sum_{n=1}^{N}\sum_{k=1}^{K} \gamma(z_{nk}) \log \pi_k p(\mathbf{x}_n|\mu_k) \tag{3.32}$$

$$= \sum_{n=1}^{N}\sum_{k=1}^{K} \gamma(z_{nk}) \left\{ \log \pi_k + \sum_{i=1}^{D} [x_{ni} \log \mu_{ki} + (1 - x_{ni}) \log(1 - \mu_{ki})] \right\}$$

where $\gamma(z_{nk})$ is the posterior probability, or responsibility, of component k, given $\mathbf{x}_n$ and the current setting of parameters.

The EM algorithm proceeds iteratively in two steps. In the E-step, the posterior probability, or responsibility, of component k, given $\mathbf{x}_n$ and the current setting of parameters, is estimated using Bayes' theorem, which can be written in the form

$$\gamma(z_{nk}) = \frac{\pi_k p(\mathbf{x}_n|\mu_k)}{\sum_{j=1}^{K} \pi_j p(\mathbf{x}_n|\mu_j)}.$$

In the M-step, we maximize the above log-likelihood function with respect to the parameters μ_{ki} and π_k. By setting the derivative of (3.32) with respect to μ_{ki} to zero and rearrange the terms, we cab get

$$\mu_{ki} = \frac{\sum_{n=1}^{N} \gamma(z_{nk}) x_{ni}}{\sum_{n=1}^{N} \gamma(z_{nk})}.$$

For the maximization with respect to π_k, we introduce a Lagrange multiplier to enforce the constraint $\sum_k \pi_k = 1$, and then we obtain

$$\pi_k = \frac{\sum_{n=1}^{N} \gamma z_{nk}}{N}$$

To start the EM algorithm, initial values for the parameters are required. Provided that a nonpathological starting point is used, the EM algorithm always increases the value of the log-likelihood, and the algorithm is guaranteed to converge to a local maximum. Generally, the mixing probabilities are initialized to $\pi_k = 1/K$, and the parameters μ_{ki} are set to random values chosen uniformly in the range $(0.25, 0.75)$ and then normalized to satisfy the constraint that $\sum_i \mu_{ki} = 1$.

3.3.3 Limitations of the EM Algorithm

The popularity of the EM algorithm is due to its simple implementation together with the guaranteed monotone increase of the likelihood of the training set during optimization. However, there are several well-known limitations associated with the EM algorithm.

The EM algorithm converges to a local maximum and its solution is highly dependent on initialization, which consequently produces suboptimal maximum likelihood estimates. One simple strategy is to use multiple random starts and choose the final estimate with the highest likelihood [33, 46, 48, 65], and the other one is to initialize by clustering algorithms [33, 46, 48, 8]. In [6], several simple methods to choose sensible starting values for the EM algorithm to get maximum likelihood parameter estimation in mixture models are compared. In addition, many adaptations and extensions of the EM algorithm have been proposed in order to address the problems of convergence to a local optimum and the initialization issue, such as deterministic annealing (DA) versions of EM algorithm [76], split-and-merge EM algorithm [77], component-wise EM algorithm [19, 25], and genetic-based EM algorithms [44, 57].

Moreover, EM is known to converge slowly in some situations, which has received much attention recently. Many algorithms aiming to speed up the convergence of EM while preserving its simplicity have been proposed [48, 19]. Additionally, the EM algorithm assumes that the number of components for modeling the distributions is known. In many practical applications, the optimal number of components is unknown and has to be determined. In such cases, a set of candidate models is established by applying the algorithm for a different number of components, and the best model is selected according to a model selection criterion as discussed in Section 3.2.4.

3.3.4 Applications of the EM Algorithm

The EM algorithm turns out to be a general way of maximizing the likelihood when some variables are unobserved and, hence, is useful for other applications besides mixture models. There are two main applications of the EM algorithm. The first occurs when the data indeed has missing values, due to problems with or limitations of the observation process. The second occurs when optimizing the likelihood function is analytically intractable but when the likelihood function can be simplified by assuming the existence of and values for additional but missing (or hidden) parameters. The latter application is more common in the computational pattern recognition community. But the range of potentially useful applications is much broader than presented here, including, for instance, specialized variance components models, problems of missing values in general parametric models, and models with discrete or continuous latent variables. A complementary overview of the applications is studied in [23, 7].

Here we selectively review some applications of the EM algorithm, including parameter estimation for Hidden Markov Models and the problem of missing data. The Hidden Markov Model (HMM) is a probabilistic model of the joint probability of a collection of random variables with both observations and hidden states. Generally, an HMM can be considered as a generalization of a mixture model where the hidden variables are related through a Markov process rather than independent of each other. Given a set of observed feature vectors or sequences, the parameter learning task in HMMs is to find the best set of state transition and output probabilities, which is usually to derive the maximum likelihood estimate of the unknown parameters of an HMM. A local maximum likelihood can be derived efficiently using the Baum-Welch algorithm [2, 7], and it is a particular case of a generalized EM algorithm. The general EM algorithm involves incomplete data and, therefore, includes the problem of accidental or unintended missing data. Three situations of the problem of missing data have been studied in [23], namely, the multinomial model, the normal linear, model and the multivariate normal model. In addition, the EM algorithm is frequently used for data clustering in machine learning. With the ability to deal with missing data and observe unidentified variables, EM is becoming a useful tool to price and manage risk of a portfolio [49]. The EM algorithm is also widely used in medical image reconstruction [38], especially in positron emission tomography [40, 30] and single photon emission computed tomography [68].

3.4 Probabilistic Topic Models

Probabilistic topic models [35, 14, 72, 9, 10] are a type of statistical model for discovering the abstract topics that occur in a collection (or corpus) of documents, where each document may be viewed as a mixture of various topics. In the view of probabilistic clustering, each topic can be interpreted as a cluster, such that each document usually belongs to multiple clusters. In this section, we introduce two key topic models, PLSA (also called PLSI) [35] and LDA [14], for document analysis.

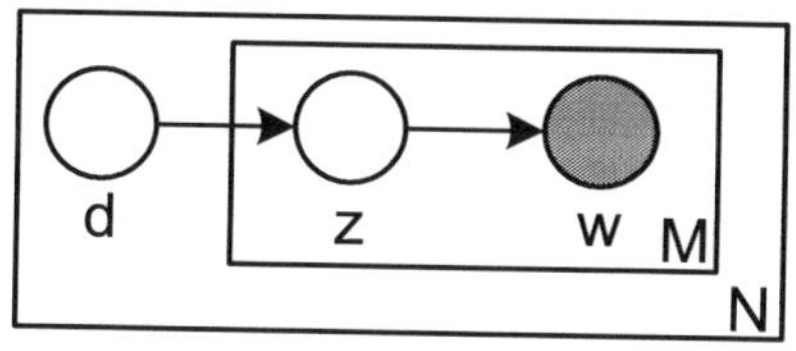

FIGURE 3.5: Graphical representation of PLSA. Circles indicate random variables, and shaded and unshaded shapes indicate observed and latent (i.e., unobserved) variables, respectively. The boxes are plates representing replicates. The outer plate represents documents, while the inner plate represents the repeated choice of topics and words within a document.

3.4.1 Probabilistic Latent Semantic Analysis

Probabilistic Latent Semantic Analysis (PLSA) [35, 36] is a statistical model for the analysis of co-occurrence data, which has been widely used in many applications, such as information retrieval, natural language processing, and machine learning from text.

The basic idea of PLSA is to treat the co-occurrences of words and documents as observations from a mixture model where each component (i.e., aspect or topic) is a distribution over words, and the proportions of different components correspond to the mixing probabilities. In PLSA, an unobserved latent variable $z_k \in \{z_1, \ldots, z_K\}$ is associated with the occurrence of a word $w_j \in \{w_1, \ldots, w_M\}$ in a particular document $d_i \in \{d_1, \ldots, d_N\}$. Let us introduce the following probabilities: $p(d_i)$ denotes the probability that a word occurrence will be observed in a specific document d_i, $p(w_j|z_k)$ is used to denote the conditional probability of a particular word w_j conditioned on the unobserved aspect variable z_k, and finally $p(z_k|d_i)$ denotes the mixing probability for document d_i to choose the kth component z_k such that $\sum_{k=1}^{K} p(z_k|d_i) = 1$. Formally, the joint probability of generating an observation pair (d_i, w_j) can be expressed in the form

$$p(d_i, w_j) = p(d_i)p(w_j|d_i) = p(d_i)\sum_{k=1}^{K} p(w_j|z_k)p(z_k|d_i). \tag{3.33}$$

Note that there is a conditional independence assumption, i.e., d_i and w_j are independent conditioned on the latent variable. The graphical representation of the PLSA model is shown in Figure 3.5. From the generative process point of view, an observation pair $p(d_i, w_j)$ could be generated by the following procedure:

(a) select a document d_i with probability $p(d_i)$,

(b) pick a latent component (i.e., aspect) z_k with probability $p(z_k|d_i)$,

(c) generate a word w_j with probability $p(w_j|z_k)$.

For a collection of documents C, we make the assumption that all co-occurrences of d_i and w_j in the collection are independent and identically distributed. Following the likelihood principle, these parameters can be determined by maximizing the log-likelihood of a collection as follows:

$$\begin{aligned} l(C) &= \sum_{i=1}^{N}\sum_{j=1}^{M} n(d_i, w_j)\log p(d_i, w_j) \\ &= \sum_{i=1}^{N} n(d_i)\log p(d_i) + \sum_{i=1}^{N}\sum_{j=1}^{M} n(d_i, w_j)\log \sum_{k=1}^{K} p(w_j|z_k)p(z_k|d_i), \end{aligned} \tag{3.34}$$

where $n(d_i) = \sum_{j=1}^{M} n(d_i, w_j)$ refers to the document length. The parameters $\{p(d_i)\}$ can be estimated trivially. By taking the derivative with respect to $p(d_i)$ and using a Lagrange multiplier, we obtain

$$p(d_i) = \frac{n(d_i)}{\sum_{i=1}^{N} n(d_i)}.$$

The other parameters Θ include $\{p(w_j|z_k)\}$ and $\{p(z_k|d_i)\}$, which can be estimated by using the standard EM algorithm [23, 36]. The total number of parameters is equal to $N + KN + MK$. Since the trivial estimate $p(d_i)$ can be carried out independently, we will only consider the second term in the following estimation.

From (3.21), the lower bound $\mathcal{L}(q,\Theta)$ on the log-likelihood (the second term of (3.34)) can be written in the form

$$\mathcal{L}(q,\Theta) = \sum_{i=1}^{N}\sum_{j=1}^{M} n(d_i,w_j) \sum_{k=1}^{K} q(z_k) \log \frac{p(w_j|z_k)p(z_k|d_i)}{q(z_k)},$$

where $q(z_k)$ is an arbitrary distribution defined over the latent variables. In the E-step, as described in Section 3.3.1, the lower bound is maximized with respect to $q(z_k)$ while holding Θ fixed, and the solution is to choose the posterior probabilities $p(z_k|d_i,w_j)$ as $q(z_k)$. Therefore, based on the current estimates of the parameters, we can simply apply Bayes' formula to compute the posterior probabilities

$$p(z_k|d_i,w_j) = \frac{p(w_j|z_k)p(z_k|d_i)}{\sum_{l=1}^{K} p(w_j|z_l)p(z_l|d_i)}. \tag{3.35}$$

In the M-step, parameters are updated by maximizing the expectation of the complete data log-likelihood $Q(\Theta,\Theta^{(t)})$, which depends on the posterior probabilities computed in the E-step, where

$$Q(\Theta,\Theta^{(t)}) = \sum_{i=1}^{N}\sum_{j=1}^{M} n(d_i,w_j) \sum_{k=1}^{K} p(z_k|d_i,w_j) \log\left\{p(w_j|z_k)p(z_k|d_i)\right\}. \tag{3.36}$$

In order to take account of the probability constraints, (3.36) has to be augmented by appropriate Language multipliers λ_k and τ_i,

$$\hat{Q} = Q(\Theta,\Theta^{(t)}) + \sum_{k=1}^{K} \lambda_k \left(1 - \sum_{j=1}^{M} p(w_j|z_k)\right) + \sum_{i=1}^{N} \tau_i \left(1 - \sum_{k=1}^{K} p(z_k|d_i)\right).$$

We maximize $\hat{Q}$ with respect to the probability functions $p(w_j|z_k)$ and $p(z_k|d_i)$, leading to the following equations, respectively.

$$\sum_{i=1}^{N} n(d_i,w_j)p(z_k|d_i,w_j) - \lambda_k p(w_j|z_k) = 0,$$

$$\sum_{j=1}^{M} n(d_i,w_j)p(z_k|d_i,w_j) - \tau_i p(z_k|d_i) = 0.$$

After eliminating λ_k and τ_i, we obtain the M-step reestimation equations

$$P(w_j|z_k) = \frac{\sum_{i=1}^{N} n(d_i,w_j)p(z_k|d_i,w_j)}{\sum_{j'=1}^{M}\sum_{i=1}^{N} n(d_i,w_{j'})p(z_k|d_i,w_{j'})}, \tag{3.37}$$

$$P(z_k|d_i) = \frac{\sum_{j=1}^{M} n(d_i,w_j)p(z_k|d_i,w_j)}{n(d_i)}. \tag{3.38}$$

The EM algorithm alternates between the E-step and the M-step until a local maximum of the log-likelihood is achieved or the convergence condition is satisfied.

Topic 1 (DB)	Topic 2 (DM)	Topic 3 (IR)	Topic 4 (AI)
data	data	information	problem
database	mining	retrieval	algorithm
systems	learning	web	paper
query	based	based	reasoning
system	clustering	learning	logic
databases	classification	knowledge	based
management	algorithm	text	time
distributed	image	search	algorithms
queries	analysis	system	search
relational	detection	language	show

FIGURE 3.6: The representative terms generated by PLSA. The terms are selected according to the probability $P(w|z)$.

vspace*-12pt

Figure 3.6 illustrates example inference with PLSA using a subset of the DBLP records [24]. The dataset includes four areas: database, data mining, information retrieval, and artificial intelligence, and contains 28,569 documents. The most representative terms generated by PLSA are shown in Figure 3.6. We can observe that the extracted topics could automatically reveal the DBLP subset is a mixture of four areas.

In clustering models for documents [37], one typically associates a latent class variable with each document in the collection. It can be shown that the hidden components z_k (i.e., aspects or topics) extracted by the topic modeling approaches can be regarded as clusters. The posterior probability $P(z_k|d_i)$ is used to infer the cluster label for each document, indicating the posterior probability of document d_i belonging to cluster z_k. In principle, this model is a soft clustering algorithm. In order to obtain a hard clustering, documents are often then assigned to the cluster z_k to which they most likely belong.

Although PLSA provides a good probabilistic model for document clustering, it has two problems. First, the number of parameters grows linearly with the number of documents, so that it tends to overfit the training data. Second, although PLSA is a generative model of the training documents on which it is estimated, it is not a generative model of new documents, and there is no natural way to estimate the probability of a previously unseen document.

3.4.2 Latent Dirichlet Allocation

Latent Dirichlet allocation (LDA) [14, 10] extends PLSA and overcomes both problems of PLSA by further considering priors on the parameters. As we will see in the following, LDA is a well-defined generative model and generalizes easily to new documents. Furthermore, the $K+KM$ parameters in a K-topic LDA model do not grow with the size of the training documents.

Similar to PLSA, LDA is a generative probabilistic model of a collection of documents. The basic idea is that documents are represented as mixtures of latent topics, where each topic is characterized by a distribution over words. Technically, the model assumes that the topics are specified before any data has been generated. From the generative process point of view, each document d_n in a collection C could be generated by the following two-stage process.

1. Randomly choose a distribution over topics $\theta_n \sim Dir(\alpha)$.
2. For each word w_{nj} in the document d_n
   (a) Randomly choose a topic $z_{nj} \sim Multinomial(\theta_n)$,
   (b) Randomly choose a word w_{nj} with probability $p(w_{nj}|z_{nj},\beta)$.

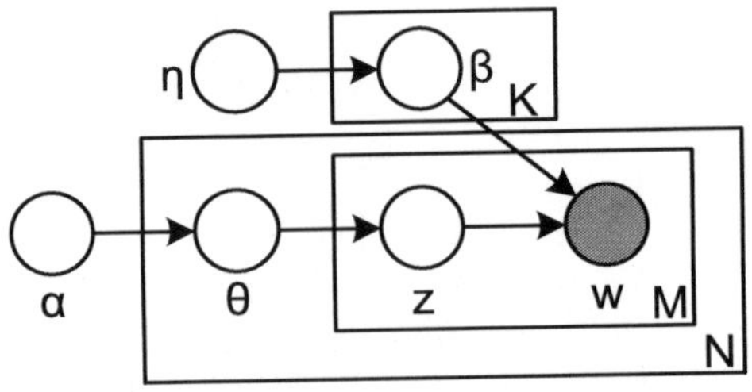

FIGURE 3.7: Graphical representation of LDA. Circles denote random variables, and edges denote dependence between random variables. Shaded circles, the words of the documents, are observed variables, while unshaded circles, like the topic proportions, assignments, and topics, are latent (i.e., unobserved) variables. The rectangular boxes are plates representing replication. The M plate represents documents, while the N plate represents the repeated choice of topics and words within a document.

The generative process of LDA can also be described in a graphical model representation as shown in Figure 3.7. The observed variables are the words of the documents, and the hidden variables are the topic proportions of documents, topic assignments of words within documents, and topics. The above generative process defines a joint probabilistic distribution over both the observed and hidden random variables. With the joint distribution, we can compute the posterior distribution of the hidden variables given the observed variables.

Now let us formally define some notations. The topic proportions for the nth document are θ_n, and the topics are β. Note that the word probabilities of topics are parameterized by a $K \times M$ matrix β where $\beta_{kj} = p(w_j = 1|z_j = k)$, which are treated as fixed quantity to be estimated. In an LDA model, a conjugate prior α is added to the multinomial distribution θ over topics, where θ follows a Dirichlet distribution $\theta \sim Dir(\alpha)$ and α is a K-dimensional parameter vector. In addition, another Dirichlet prior $\eta \sim Dir(\beta)$ [14, 31] can be further added to the multinomial distribution β over words, serving as a smoothing functionality over words, where η is an M-dimensional parameter vector and M is the size of the vocabulary. The topic assignments for the nth document are z_n, where z_{nj} is the topic assignment for the jth word in document d_n. Finally, the observed words for document d_n are $\mathbf{w}_n$, where w_{nj} is the jth word in document d_n. With this notation, the probability of observing all the words in a single document d_n can be expressed as

$$p(\mathbf{w}_n|\alpha,\beta) = \int p(\theta_n|\alpha)\left(\prod_{j=1}^{M}\sum_{z_{nj}} p(z_{nj}|\theta_n)p(w_{nj}|z_{nj},\beta)\right)d\theta_n.$$

For a collection of documents C, the joint probability of observing all the documents is given by

$$p(C|\alpha,\eta) = \prod_{n=1}^{N}\int p(\mathbf{w}_n|\alpha,\beta)p(\beta|\eta)d\beta. \tag{3.39}$$

As illustrated in Figure 3.7, there are three levels to the LDA representation. The parameters α, η, and β are collection-level parameters, assumed to be sampled once in the process of generating a collection. The variables θ_n are document-level variables, sampled once per document. Finally, the variables z_{nj} and w_{nj} are word-level variables and are sampled once for each word in each document.

To estimate the parameters of LDA, there are two types of approximate algorithms, variational algorithms and sampling-based algorithms. Variational methods are used as approximation methods in a wide variety of settings [39, 78]. The application of variational methods converts the original complex problem into a simplified problem in which inference is efficient. The simplified model is generally characterized by introducing additional parameters and ignoring some troublesome dependencies. A good descriptions of variational inference algorithm for LDA is presented in [14].

Based on that, Expectation–Propagation [52] is proposed that leads to higher accuracy at comparable cost, and a much faster online algorithm [34] is presented that easily handles millions of documents. Sampling-based methods [28, 31, 8] are a stochastic alternative which attempt to collect samples from the posterior to approximate it with an empirical distribution. Gibbs sampling [31], a form of Markov chain Monte Carlo [29, 1], is the most commonly used sampling algorithm for topic modeling. Gibbs sampling generates samples until the sampled values approximate the target distribution, and then estimates the parameters using the statistics of the distribution according to the samples. It was shown in [31] that Gibbs sampling is more efficient than other LDA learning algorithms including variational EM [14] and Expectation-Propagation [52]. Please refer to [31, 72] for a detailed description of Gibbs sampling for LDA, and a fast collapsed Gibbs sampling for LDA is proposed in [58].

3.4.3 Variations and Extensions

The simple PLSA and LDA models provide powerful tools for discovering and exploiting the hidden thematic structure in large collections of documents, which have been extended and adapted in many ways.

One active area of topic modeling research is to relax and extend the statistical assumptions, so as to uncover more sophisticated structure in the texts. For example, [80] developed a topic model to relax the bag-of-words assumption by assuming that the topics generate words conditional on the previous words, and [32] developed a topic model that switches between LDA and a standard HMM. These models expand the parameter space significantly but show improved language modeling performance. Another assumption is that the order of documents does not matter, but long-running collections (e.g., *The New York Times* collections and *Science* magazine) span years and centuries, and the topics may change over time. The dynamic topic model [12] is proposed that respects the ordering of the documents and gives a richer posterior topical structure. A third assumption about topic modeling is that the number of topics is assumed known and fixed. The Bayesian nonparametric topic model [74, 11] provides an elegant solution to determine the number of topics by the collection during posterior inference.

In traditional topic models, documents are considered independent of each other. However, the documents could be correlated with each other based on additional information, such as author and links, which should be taken into account when fitting a topic model. The author-topic model [73] is proposed to consider the relations between authors and documents: Papers with multiple authors are assumed to attach each word to an author, drawn from a topic from his topic proportions. In such way, the model allows for inferences about authors as well as documents. In addition, many documents are linked; for example, scientific papers are linked by citation or web pages are linked by hyperlink. We can assume linked webpages tend to be similar with each other, a paper cites another paper indicates the two papers are somehow similar. NetPLSA [50] is proposed to incorporate topic modeling with homogeneous networks, and TMBP [24] is another extended algorithm that could handle heterogeneous networks. Other models that incorporate meta-data into topic models includes relational topic model [22], supervised topic model [13], and Dirichlet-multinomial regression models [51]. A comprehensive overview of current research and future directions in topic modeling is given in [10].

3.5 Conclusions and Summary

This chapter has introduced the most frequently used probabilistic models for clustering, which include mixture models, such as Gaussian mixture model and Bernoulli mixture model, and probabilistic topic models, such as PLSA and LDA. Some of the methods, like mixture models, may be more appropriate for quantitative data, whereas others, such as topic models, PLSI, and LDA, are used more commonly for text data. The goals of learning algorithms for these probabilistic models are to find MLE or MAP estimations for parameters in these models. There are no closed form solutions for most of the models. The EM algorithm provides a general and elegant framework to learn mixture models and topic models. For more complicated cases, some variations of the EM algorithm, such as generalized EM and variational EM, can be employed to solve these problems.

Bibliography

[1] C. Andrieu, N. de Freitas, A. Doucet, and M. I. Jordan. An introduction to MCMC for machine learning. *Machine Learning*, 50(1–2):5–43, 2003.

[2] L. Baum, T. Petrie, G. Soules, and N. Weiss. A maximization technique occurring in the statistical analysis of probabilistic functions of Markov chains. *The Annals of Mathematical Statistics*, 41(1):164–171, 1970.

[3] H. Bensmail, G. Celeux, A. Raftery, and C. Robert. Inference in model-based cluster analysis. *Statistics and Computing*, 7(1):1–10, 1997.

[4] C. Biernacki, G. Celeux, and G. Govaert. An improvement of the nec criterion for assessing the number of clusters in a mixture model. *Pattern Recognition Letters*, 20(3):267–272, 1999.

[5] C. Biernacki, G. Celeux, and G. Govaert. Assessing a mixture model for clustering with the integrated completed likelihood. *IEEE Transactions on Pattern Analysis and Machine Intelligence*, 22(7):719–725, 2000.

[6] C. Biernacki, G. Celeux, and G. Govaert. Choosing starting values for the EM algorithm for getting the highest likelihood in multivariate Gaussian mixture models. *Computational Statistics & Data Analysis*, 41(3):561–575, 2003.

[7] J. Bilmes. A gentle tutorial of the EM algorithm and its application to parameter estimation for Gaussian mixture and hidden Markov models. *International Computer Science Institute*, 4:126, 1998.

[8] C. M. Bishop. *Pattern Recognition and Machine Learning*. Springer, 2006.

[9] D. Blei and J. Lafferty. Topic models. In A. Srivastava and M. Sahami, editors, *Text Mining: Classification, Clustering, and Applications*. Chapman & Hall/CRC Data Mining and Knowledge Discovery Series, Boca Raton, FL, 2009.

[10] D. M. Blei. Probabilistic topic models. *Communications of the ACM*, 55(4):77–84, 2012.

[11] D. M. Blei, T. L. Griffiths, and M. I. Jordan. The nested chinese restaurant process and Bayesian nonparametric inference of topic hierarchies. *Journal of the ACM*, 57(2), 2010.

[12] D. M. Blei and J. D. Lafferty. Dynamic topic models. In *ICML*, pages 113–120, 2006.

[13] D. M. Blei and J. D. McAuliffe. Supervised topic models. In *NIPS*, 2007.

[14] D. M. Blei, A. Y. Ng, and M. I. Jordan. Latent Dirichlet allocation. *Journal of Machine Learning Research*, 3:993–1022, 2003.

[15] K. Blekas, A. Likas, N. Galatsanos, and I. Lagaris. A spatially constrained mixture model for image segmentation. *IEEE Transactions on Neural Networks*, 16(2):494–498, 2005.

[16] S. Borman. The expectation maximization algorithm—a short tutorial 2004. Unpublished paper available at http://www. seanborman. com/publications

[17] K. Burnham and D. Anderson. *Model Selection and Multimodel Inference: A Practical Information-Theoretic Approach*. Springer Verlag, 2002.

[18] M. Carreira-Perpiñán and S. Renals. Practical identifiability of finite mixtures of multivariate Bernoulli distributions. *Neural Computation*, 12(1):141–152, 2000.

[19] G. Celeux, S. Chrétien, F. Forbes, and A. Mkhadri. A component-wise EM algorithm for mixtures. *Journal of Computational and Graphical Statistics*, 10(4):697–712, 2001.

[20] G. Celeux and G. Govaert. A classification EM algorithm for clustering and two stochastic versions. *Computational Statistics & Data Analysis*, 14(3):315–332, 1992.

[21] G. Celeux and G. Soromenho. An entropy criterion for assessing the number of clusters in a mixture model. *Journal of Classification*, 13(2):195–212, 1996.

[22] J. Chang and D. M. Blei. Relational topic models for document networks. *Journal of Machine Learning Research—Proceedings Track*, 5:81–88, 2009.

[23] A. Dempster, N. Laird, and D. Rubin. Maximum likelihood from incomplete data via the em algorithm. *Journal of the Royal Statistical Society. Series B (Methodological)*, 39(1):1–38, 1977.

[24] H. Deng, J. Han, B. Zhao, Y. Yu, and C. X. Lin. Probabilistic topic models with biased propagation on heterogeneous information networks. In *KDD*, pages 1271–1279, 2011.

[25] M. A. T. Figueiredo and A. K. Jain. Unsupervised learning of finite mixture models. *IEEE Transactions on Pattern Analysis and Machine Intelligence*, 24(3):381–396, 2002.

[26] C. Fraley and A. Raftery. How many clusters? Which clustering method? Answers via model-based cluster analysis. *The Computer Journal*, 41(8):578–588, 1998.

[27] J. Gauvain and C. Lee. Maximum a posteriori estimation for multivariate Gaussian mixture observations of Markov chains. *IEEE Transactions on Speech and Audio Processing*, 2(2):291–298, 1994.

[28] A. Gelfand and A. Smith. Sampling-based approaches to calculating marginal densities. *Journal of the American Statistical Association*, 35(410):398–409, 1990.

[29] W. Gilks, S. Richardson, and D. Spiegelhalter. *Markov Chain Monte Carlo in Practice*. Chapman & Hall/CRC, Boca Raton, FL, 1996.

[30] P. Green. Bayesian reconstructions from emission tomography data using a modified EM algorithm. *IEEE Transactions on Medical Imaging*, 9(1):84–93, 1990.

[31] T. Griffiths and M. Steyvers. Finding scientific topics. *Proceedings of the National Academy of Sciences*, 101(Suppl 1):5228, 2004.

[32] T. Griffiths, M. Steyvers, D. Blei, and J. Tenenbaum. Integrating topics and syntax. *Advances in Neural Information Processing Systems*, 17:537–544, 2005.

[33] T. Hastie and R. Tibshirani. Discriminant analysis by Gaussian mixtures. *Journal of the Royal Statistical Society. Series B (Methodological)*, 58(1):155–176, 1996.

[34] M. D. Hoffman, D. M. Blei, and F. R. Bach. Online learning for Latent Dirichlet allocation. In *NIPS*, pages 856–864, 2010.

[35] T. Hofmann. Probabilistic latent semantic analysis. In *UAI*, pages 289–296, 1999.

[36] T. Hofmann. Unsupervised learning by probabilistic latent semantic analysis. *Machine Learning*, 42(1/2):177–196, 2001.

[37] T. Hofmann, J. Puzicha, and M. I. Jordan. Learning from dyadic data. In *NIPS*, pages 466–472, 1998.

[38] H. Hudson and R. Larkin. Accelerated image reconstruction using ordered subsets of projection data. *IEEE Transactions on Medical Imaging*, 13(4):601–609, 1994.

[39] M. I. Jordan, Z. Ghahramani, T. Jaakkola, and L. K. Saul. An introduction to variational methods for graphical models. *Machine Learning*, 37(2):183–233, 1999.

[40] L. Kaufman. Implementing and accelerating the EM algorithm for positron emission tomography. *IEEE Transactions on Medical Imaging*, 6(1):37–51, 1987.

[41] D. Kleinbaum and M. Klein. Maximum likelihood techniques: An overview. In *Logistic Regression*, pages 103–127, Springer Science + Business Media, 2010.

[42] P. Lazarsfeld and N. Henry. *Latent Structure Analysis*. Houghton, Mifflin, 1968.

[43] C. D. Manning, P. Raghavan, and H. Schütze. *Introduction to Information Retrieval*. Cambridge University Press, 2008.

[44] A. Martınez and J. Vitria. Learning mixture models using a genetic version of the EM algorithm. *Pattern Recognition Letters*, 21(8):759–769, 2000.

[45] G. McLachlan and K. Basford. Mixture models: Inference and applications to clustering. *Statistics: Textbooks and Monographs*, Marcel Dekker, New York, 1988.

[46] G. McLachlan and T. Krishnan. *The EM Algorithm and Extensions*, volume 274. Wiley, New York, 1997.

[47] G. McLachlan, S. Ng, and R. Bean. Robust cluster analysis via mixture models. *Austrian Journal of Statistics*, 35(2):157–174, 2006.

[48] G. McLachlan and D. Peel. *Finite Mixture Models*. Wiley, New York, 2000.

[49] A. McNeil, R. Frey, and P. Embrechts. *Quantitative Risk Management: Concepts, Techniques and Tools*. Princeton University Press, 2005.

[50] Q. Mei, D. Cai, D. Zhang, and C. Zhai. Topic modeling with network regularization. In *Proceedings of the 17th International Conference on World Wide Web*, pages 101–110. ACM, 2008.

[51] D. M. Mimno and A. McCallum. Topic models conditioned on arbitrary features with Dirichlet-multinomial regression. In *UAI*, pages 411–418, 2008.

[52] T. P. Minka and J. D. Lafferty. Expectation-propogation for the generative aspect model. In *UAI*, pages 352–359, 2002.

[53] R. Neal and G. Hinton. A view of the EM algorithm that justifies incremental, sparse, and other variants. In M. I. Jordan, editor, *Learning in Graphical Models*, pages 355–368. Kluwer Academic Publishers, 1998.

[54] J. J. Oliver, R. A. Baxter, and C. S. Wallace. Unsupervised learning using MML. In *ICML*, pages 364–372, 1996.

[55] H. Park and T. Ozeki. Singularity and slow convergence of the EM algorithm for Gaussian mixtures. *Neural Processing Letters*, 29(1):45–59, 2009.

[56] H. Permuter, J. Francos, and I. Jermyn. A study of Gaussian mixture models of color and texture features for image classification and segmentation. *Pattern Recognition*, 39(4):695–706, 2006.

[57] F. Pernkopf and D. Bouchaffra. Genetic-based EM algorithm for learning Gaussian mixture models. *IEEE Transactions on Pattern Analysis and Machine Intelligence*, 27(8):1344–1348, 2005.

[58] I. Porteous, D. Newman, A. T. Ihler, A. U. Asuncion, P. Smyth, and M. Welling. Fast collapsed Gibbs sampling for latent Dirichlet allocation. In *KDD*, pages 569–577, 2008.

[59] C. Rasmussen. The infinite gaussian mixture model. *Advances in Neural Information Processing Systems*, 12(5.2):2, 2000.

[60] M. Revow, C. Williams, and G. Hinton. Using generative models for handwritten digit recognition. *IEEE Transactions on Pattern Analysis and Machine Intelligence*, 18(6):592–606, 1996.

[61] D. Reynolds, T. Quatieri, and R. Dunn. Speaker verification using adapted Gaussian mixture models. *Digital signal processing*, 10(1–3):19–41, 2000.

[62] D. Reynolds and R. Rose. Robust text-independent speaker identification using Gaussian mixture speaker models. *IEEE Transactions on Speech and Audio Processing*, 3(1):72–83, 1995.

[63] S. Richardson and P. Green. On Bayesian analysis of mixtures with an unknown number of components. *Journal of the Royal Statistical Society: Series B*, 59(4):731–792, 1997.

[64] J. Rissanen. A universal prior for integers and estimation by minimum description length. *The Annals of Statistics*, 11(2):416–431, 1983.

[65] S. Roberts, D. Husmeier, I. Rezek, and W. Penny. Bayesian approaches to gaussian mixture modeling. *IEEE Transactions on Pattern Analysis and Machine Intelligence*, 20(11):1133–1142, 1998.

[66] K. Roeder and L. Wasserman. Practical Bayesian density estimation using mixtures of normals. *Journal of the American Statistical Association*, 92(439):894–902, 1997.

[67] G. Schwarz. Estimating the dimension of a model. *The Annals of Statistics*, 6(2):461–464, 1978.

[68] L. Shepp and Y. Vardi. Maximum likelihood reconstruction for emission tomography. *IEEE Transactions on Medical Imaging*, 1(2):113–122, 1982.

[69] P. Smyth. Model selection for probabilistic clustering using cross-validated likelihood. *Statistics and Computing*, 10(1):63–72, 2000.

[70] H. Sorenson. *Parameter Estimation: Principles and Problems*. Marcel Dekker, New York, 1980.

[71] C. Stauffer and W. Grimson. Adaptive background mixture models for real-time tracking. In *CVPR*, volume 2. IEEE, 1999.

[72] M. Steyvers and T. Griffiths. Probabilistic topic models. In S. D. T. Landauer, D McNamara, and W. Kintsch, editors, *Latent Semantic Analysis: A Road to Meaning*. Laurence Erlbau, 2006.

[73] M. Steyvers, P. Smyth, M. Rosen-Zvi, and T. L. Griffiths. Probabilistic author-topic models for information discovery. In *KDD*, pages 306–315, 2004.

[74] Y. W. Teh, M. I. Jordan, M. J. Beal, and D. M. Blei. Sharing clusters among related groups: Hierarchical dirichlet processes. In *NIPS*, 2004.

[75] D. Titterington, A. Smith, and U. Makov. *Statistical Analysis of Finite Mixture Distributions*, volume 38. John Wiley & Sons, 1985.

[76] N. Ueda and R. Nakano. Deterministic annealing EM algorithm. *Neural Networks*, 11(2):271–282, 1998.

[77] N. Ueda, R. Nakano, Z. Ghahramani, and G. Hinton. SMEM algorithm for mixture models. *Neural Computation*, 12(9):2109–2128, 2000.

[78] M. J. Wainwright and M. I. Jordan. Graphical models, exponential families, and variational inference. *Foundations and Trends in Machine Learning*, 1(1–2):1–305, 2008.

[79] C. Wallace and D. Dowe. Minimum message length and Kolmogorov complexity. *The Computer Journal*, 42(4):270–283, 1999.

[80] H. Wallach. Topic modeling: Beyond bag-of-words. In *Proceedings of the 23rd international Conference on Machine Learning*, pages 977–984. ACM, 2006.

[81] S. Zhong and J. Ghosh. Generative model-based document clustering: A comparative study. *Knowledge and Information Systems*, 8(3):374–384, 2005.

Chapter 4

A Survey of Partitional and Hierarchical Clustering Algorithms

Chandan K. Reddy
Wayne State University
Detroit, MI
reddy@cs.wayne.edu

Bhanukiran Vinzamuri
Wayne State University
Detroit, MI
bhanukiranv@wayne.edu

4.1 Introduction

The two most widely studied clustering algorithms are partitional and hierarchical clustering. These algorithms have been heavily used in a wide range of applications primarily due to their simplicity and ease of implementation relative to other clustering algorithms. Partitional clustering algorithms aim to discover the groupings present in the data by optimizing a specific objective function and iteratively improving the quality of the partitions. These algorithms generally require certain user parameters to choose the prototype points that represent each cluster. For this reason they are also called *prototype-based clustering* algorithms.

Hierarchical clustering algorithms, on the other hand, approach the problem of clustering by developing a binary tree-based data structure called the dendrogram. Once the dendrogram is constructed, one can automatically choose the right number of clusters by splitting the tree at different levels to obtain different clustering solutions for the same dataset without rerunning the clustering algorithm again. Hierarchical clustering can be achieved in two different ways, namely, bottom-up and top-down clustering. Though both of these approaches utilize the concept of dendrogram while clustering the data, they might yield entirely different sets of results depending on the criterion used during the clustering process.

Partitional methods need to be provided with a set of initial seeds (or clusters) which are then improved iteratively. Hierarchical methods, on the other hand, can start off with the individual data points in single clusters and build the clustering. The role of the distance metric is also different in both of these algorithms. In hierarchical clustering, the distance metric is initially applied on the data points at the base level and then progressively applied on subclusters by choosing *absolute* representative points for the subclusters. However, in the case of partitional methods, in general, the representative points chosen at different iterations can be *virtual* points such as the centroid of the cluster (which is nonexistent in the data).

This chapter is organized as follows. In Section 4.2, the basic concepts of partitional clustering are introduced and the related algorithms in this field are discussed. More specifically, Subsections 4.2.1-4.2.3 will discuss the widely studied K-Means clustering algorithm and highlight the major factors involved in these partitional algorithms such as initialization methods and estimating the number of clusters K. Subsection 4.2.4 will highlight several variations of the K-Means clustering. The distinctive features of each of these algorithms and their advantages are also featured. In Section 4.3 the fundamentals of hierarchical clustering are explained. Subsections 4.3.1 and 4.3.2 will discuss the agglomerative and divisive hierarchical clustering algorithms, respectively. We will also highlight the differences between the algorithms in both of these categories. Subsection 4.3.3 will briefly discuss the other prominent hierarchical clustering algorithms. Finally, in Section 4.4, we will conclude our discussion highlighting the merits and drawbacks of both families of clustering algorithms.

4.2 Partitional Clustering Algorithms

The first partitional clustering algorithm that will be discussed in this section is the K-Means clustering algorithm. It is one of the simplest and most efficient clustering algorithms proposed in the literature of data clustering. After the algorithm is described in detail, some of the major factors that influence the final clustering solution will be highlighted. Finally, some of the widely used variations of K-Means will also be discussed in this section.

4.2.1 K-Means Clustering

K-means clustering [33, 32] is the most widely used partitional clustering algorithm. It starts by choosing K representative points as the initial centroids. Each point is then assigned to the closest centroid based on a particular proximity measure chosen. Once the clusters are formed, the centroids for each cluster are updated. The algorithm then iteratively repeats these two steps until the centroids do not change or any other alternative relaxed convergence criterion is met. K-means clustering is a *greedy* algorithm which is guaranteed to converge to a local minimum but the minimization of its score function is known to be NP-Hard [35]. Typically, the convergence condition is relaxed and a weaker condition may be used. In practice, it follows the rule that the iterative procedure must be continued until 1% of the points change their cluster memberships. A detailed proof of the mathematical convergence of K-means can be found in [45].

Algorithm 13 K-Means Clustering

```
1: Select K points as initial centroids.
2: repeat
3:     Form K clusters by assigning each point to its closest centroid.
4:     Recompute the centroid of each cluster.
5: until convergence criterion is met.
```

Algorithm 13 provides an outline of the basic K-Means algorithm. Figure 4.1 provides an illustration of the different stages of the running of **3**-means algorithm on the Fisher Iris dataset. The first iteration initializes three random points as centroids. In subsequent iterations the centroids change positions until convergence. A wide range of proximity measures can be used within the K-means algorithm while computing the closest centroid. The choice can significantly affect the centroid assignment and the quality of the final solution. The different kinds of measures which can be used here are Manhattan distance (L_1 norm), Euclidean distance (L_2 norm), and Cosine similarity. In general, for the K-means clustering, Euclidean distance metric is the most popular choice. As mentioned above, we can obtain different clusterings for different values of K and proximity measures. The objective function which is employed by K-means is called the Sum of Squared Errors (SSE) or Residual Sum of Squares (RSS). The mathematical formulation for SSE/RSS is provided below.

Given a dataset $D=\{x_1,x_2,\ldots,x_N\}$ consists of N points, let us denote the clustering obtained after applying K-means clustering by $C=\{C_1,C_2,\ldots,C_k\ldots,C_K\}$. The SSE for this clustering is defined in the Equation (4.1) where c_k is the centroid of cluster C_k. The objective is to find a clustering that minimizes the SSE score. The iterative assignment and update steps of the K-means algorithm aim to minimize the SSE score for the given set of centroids.

$$SSE(C) = \sum_{k=1}^{K} \sum_{x_i \in C_k} \| x_i - c_k \|^2 \tag{4.1}$$

$$c_k = \frac{\sum_{x_i \in C_k} x_i}{|C_k|} \tag{4.2}$$

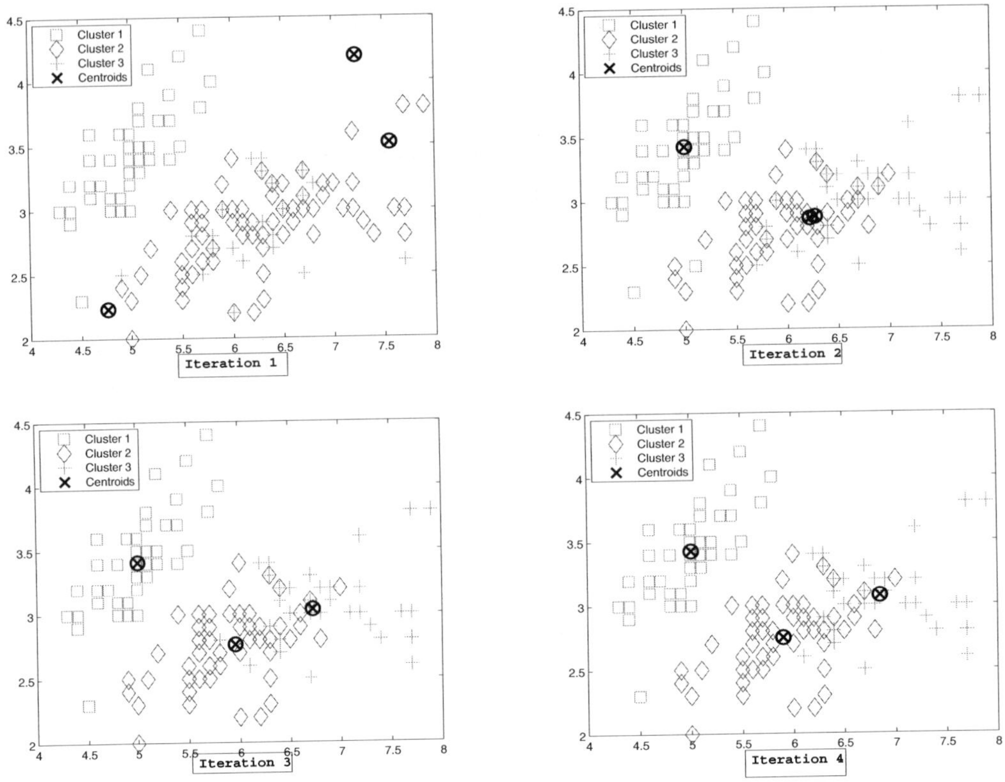

FIGURE 4.1: An illustration of 4 iterations of K-means over the Fisher Iris dataset.

4.2.2 Minimization of Sum of Squared Errors

K-means clustering is essentially an optimization problem with the goal of minimizing the SSE objective function. We will mathematically prove the reason behind choosing the mean of the data points in a cluster as the prototype representative for a cluster in the K-means algorithm. Let us denote C_k as the kth cluster, x_i is a point in C_k, and c_k is the mean of the kth cluster. We can solve for the representative of C_j which minimizes the SSE by differentiating the SSE with respect to c_j and setting it equal to zero.

$$SSE(C) = \sum_{k=1}^{K} \sum_{x_i \in C_k} (c_k - x_i)^2 \tag{4.3}$$

$$\begin{aligned}
\frac{\partial}{\partial c_j} SSE &= \frac{\partial}{\partial c_j} \sum_{k=1}^{K} \sum_{x_i \in C_k} (c_k - x_i)^2 \\
&= \sum_{k=1}^{K} \sum_{x_i \in C_j} \frac{\partial}{\partial c_j} (c_j - x_i)^2 \\
&= \sum_{x_i \in C_j} 2 * (c_j - x_i) = 0
\end{aligned}$$

$$\sum_{x_i \in C_j} 2 * (c_j - x_i) = 0 \Rightarrow |C_j| \cdot c_j = \sum_{x_i \in C_j} x_i \Rightarrow c_j = \frac{\sum_{x_i \in C_j} x_i}{|C_j|}$$

Hence, the best representative for minimizing the SSE of a cluster is the mean of the points in the cluster. In K-means, the SSE monotonically decreases with each iteration. This monotonically decreasing behaviour will eventually converge to a local minimum.

4.2.3 Factors Affecting K-Means

The major factors that can impact the performance of the K-means algorithm are the following:

1. Choosing the initial centroids.
2. Estimating the number of clusters K.

We will now discuss several methods proposed in the literature to tackle each of these factors.

4.2.3.1 Popular Initialization Methods

In his classical paper [33], MacQueen proposed a simple initialization method which chooses K seeds at random. This is the simplest method and has been widely used in the literature. The other popular K-means initialization methods which have been successfully used to improve the clustering performance are given below.

1. *Hartigan and Wong* [19]: Using the concept of nearest neighbor density, this method suggests that the points which are well separated and have a large number of points within their surrounding multi-dimensional sphere can be good candidates for initial points. The average pair-wise Euclidean distance between points is calculated using Equation (4.4). Subsequent points are chosen in the order of their decreasing density and simultaneously maintaining the separation of d_1 from all previous seeds. Note that for the formulae provided below we continue using the same notation as introduced earlier.

$$d_1 = \frac{1}{N(N-1)} \sum_{i=1}^{N-1} \sum_{j=i+1}^{N} \| x_i - x_j \| \tag{4.4}$$

2. *Milligan* [37]: Using the results of agglomerative hierarchical clustering (with the help of the dendrogram), this method uses the results obtained from the Ward's method. Ward's method chooses the initial centroids by using the sum of squared errors to evaluate the distance between two clusters. Ward's method is a greedy approach and keeps the agglomerative growth as small as possible.

3. *Bradley and Fayyad* [5]: Choose random subsamples from the data and apply K-means clustering to all these subsamples using random seeds. The centroids from each of these subsamples are then collected, and a new dataset consisting of only these centroids is created. This new dataset is clustered using these centroids as the initial seeds. The minimum SSE obtained guarantees the best seed set chosen.

4. *K-Means++* [1]: The K-means++ algorithm carefully selects the initial centroids for K-means clustering. The algorithm follows a simple probability-based approach where initially the first centroid is selected at random. The next centroid selected is the one which is farthest from the currently selected centroid. This selection is decided based on a weighted probability score. The selection is continued until we have K centroids and then K-means clustering is done using these centroids.

4.2.3.2 Estimating the Number of Clusters

The problem of estimating the correct number of clusters (K) is one of the major challenges for the K-means clustering. Several researchers have proposed new methods for addressing this challenge in the literature. We will briefly describe some of the most prominent methods.

1. *Calinski–Harabasz Index* [6]: The *Calinski–Harabasz* index is defined by Equation (4.5):

$$CH(K) = \frac{\frac{B(K)}{(K-1)}}{\frac{W(K)}{N-K}} \tag{4.5}$$

   where N represents the number of data points. The number of clusters is chosen by maximizing the function given in Equation (4.5). Here $B(K)$ and $W(K)$ are the between and within cluster sum of squares, respectively (with K clusters).

2. *Gap Statistic* [48]: In this method, B different datasets each with the same range values as the original data are produced. The within cluster sum of squares is calculated for each of them with a different number of clusters. $W_b^*(K)$ is the within cluster sum of squares for the bth uniform dataset.

$$Gap(K) = \frac{1}{B} \times \sum_b log(W_b^*(K)) - log(W(K)) \tag{4.6}$$

   The number of clusters chosen is the smallest value of K which satisfies Equation (4.7):

$$Gap(K) \geq Gap(K+1) - s_{k+1} \tag{4.7}$$

   where S_{k+1} represents the estimate of standard deviation of $log(W_b^*(K+1))$.

3. *Akaike Information Criterion (AIC)* [52]: AIC has been developed by considering the log-likelihood and adding additional constraints of Minimum Description Length (MDL) to estimate the value of K. M represents the dimensionality of the dataset. SSE (Equation 4.1) is the sum of squared errors for the clustering obtained using K. K-means uses a modified AIC as given below.

$$KMeans_{AIC} : K = argmin_K[SSE(K) + 2MK] \tag{4.8}$$

4. *Bayesian Information Criterion (BIC)* [39]: BIC serves as an asymptotic approximation to a transformation of the Bayesian posterior probability of a candidate model. Similar to AIC, the computation is also based on considering the logarithm of likelihood (L). N represents the number of points. The value of K that minimizes the BIC function given below will be used as the initial parameter for running the K-means clustering.

$$BIC = \frac{-2 * ln(L)}{N} + \frac{K * ln(N)}{N} = \frac{1}{N} \times ln(\frac{N^K}{L^2}) \tag{4.9}$$

5. *Duda and Hart* [11]: This is a method for estimation that involves stopping the hierarchical clustering process by choosing the correct cut-off level in the dendrogram. The methods which are typically used to cut a dendrogram are the following: (i) Cutting the dendrogram at a pre-specified level of similarity where the threshold has been specified by the user. (ii) Cutting the dendrogram where the gap between two successive merges is the largest. (iii) Stopping the process when the density of the cluster that results from the merging is below a certain threshold.

6. *Silhouette Coefficient* [26]: This is formulated by considering both the intra- and inter-cluster distances. For a given point x_i, first the average of the distances to all points in the same cluster is calculated. This value is set to a_i. Then for each cluster that does not contain x_i, the average distance of x_i to all the data points in each cluster is computed. This value is set to b_i. Using these two values the silhouette coefficient of a point is estimated. The average of all the silhouettes in the dataset is called the average silhouettes width for all the points in the dataset. To evaluate the quality of a clustering one can compute the average silhouette coefficient of all points.

$$S = \frac{\sum_{i=1}^{N} \frac{b_i - a_i}{max(a_i, b_i)}}{N} \tag{4.10}$$

7. *Newman and Girvan* [40]: In this method, the dendrogram is viewed as a graph, and a betweenness score (which will be used as a dissimilarity measure between the edges) is proposed. The procedure starts by calculating the betweenness score of all the edges in the graph. Then the edge with the highest betweenness score is removed. This is followed by recomputing the betweenness scores of the remaining edges until the final set of connected components is obtained. The cardinality of the set derived through this process serves as a good estimate for K.

8. *ISODATA* [2]: ISODATA was proposed for clustering the data based on the nearest centroid method. In this method, first K-means is run on the dataset to obtain the clusters. Clusters are then merged if their distance is less than a threshold ϕ or if they have fewer than a certain number of points. Similarly, a cluster is split if the within cluster standard deviation exceeds that of a user defined threshold.

4.2.4 Variations of K-Means

The simple framework of the K-means algorithm makes it very flexible to modify and build more efficient algorithms on top of it. Some of the variations proposed to the K-means algorithm are based on (i) Choosing different representative prototypes for the clusters (K-medoids, K-medians, K-modes), (ii) choosing better initial centroid estimates (Intelligent K-means, Genetic K-means), and (iii) applying some kind of feature transformation technique (Weighted K-means, Kernel K-means). In this section, we will discuss the most prominent variants of K-means clustering that have been proposed in the literature of partitional clustering.

4.2.4.1 K-Medoids Clustering

K-medoids is a clustering algorithm which is more resilient to outliers compared to K-means [38]. Similar to K-means, the goal of K-medoids is to find a clustering solution that minimizes a predefined objective function. The K-medoids algorithm chooses the actual data points as the prototypes and is more robust to noise and outliers in the data. The K-medoids algorithm aims to minimize the absolute error criterion rather than the SSE. Similar to the K-means clustering algorithm, the K-medoids algorithm proceeds iteratively until each representative object is actually the medoid of the cluster. The basic K-medoids clustering algorithm is given in Algorithm 14.

In the K-medoids clustering algorithm, specific cases are considered where an arbitrary random point x_i is used to replace a representative point m. Following this step the change in the membership of the points that originally belonged to m is checked. The change in membership of these points can occur in one of the two ways. These points can now be closer to x_i (new representative point) or to any of the other set of representative points. The cost of swapping is calculated as the absolute

Algorithm 14 K-Medoids Clustering

```
Select K points as the initial representative objects.
repeat
    Assign each point to the cluster with the nearest representative object.
    Randomly select a nonrepresentative object x_i.
    Compute the total cost S of swapping the representative object m with x_i.
    If S < 0, then swap m with x_i to form the new set of K representative objects.
until Convergence criterion is met.
```

error criterion for K-medoids. For each reassignment operation this cost of swapping is calculated and this contributes to the overall cost function.

To deal with the problem of executing multiple swap operations while obtaining the final representative points for each cluster, a modification of the K-medoids clustering called Partitioning Around Medoids (PAM) algorithm is proposed [26]. This algorithm operates on the dissimilarity matrix of a given dataset. PAM minimizes the objective function by swapping all the nonmedoid points and medoids iteratively until convergence. K-medoids is more robust compared to K-means but the computational complexity of K-medoids is higher and hence is not suitable for large datasets. PAM was also combined with a sampling method to propose the Clustering LARge Application (CLARA) algorithm. CLARA considers many samples and applies PAM on each one of them to finally return the set of optimal medoids.

4.2.4.2 K-Medians Clustering

The K-medians clustering calculates the median for each cluster as opposed to calculating the mean of the cluster (as done in K-means). K-medians clustering algorithm chooses K cluster centers that aim to minimize the sum of a distance measure between each point and the closest cluster center. The distance measure used in the K-medians algorithm is the L_1-norm as opposed to the square of the L_2-norm used in the K-means algorithm. The criterion function for the K-medians algorithm is defined as follows:

$$S = \sum_{k=1}^{K} \sum_{x_i \in C_k} |x_{ij} - med_{kj}| \tag{4.11}$$

where x_{ij} represents the jth attribute of the instance x_i and med_{kj} represents the median for the jth attribute in the kth cluster C_k. K-medians is more robust to outliers compared to K-means. The goal of the K-Medians clustering is to determine those subsets of median points which minimize the cost of assignment of the data points to the nearest medians. The overall outline of the algorithm is similar to that of K-means. The two steps that are iterated until convergence are (i) All the data points are assigned to their nearest median and (ii) the medians are recomputed using the median of the each individual feature.

4.2.4.3 K-Modes Clustering

One of the major disadvantages of K-means is its inability to deal with nonnumerical attributes [51, 3]. Using certain data transformation methods, categorical data can be transformed into new feature spaces, and then the K-means algorithm can be applied to this newly transformed space to obtain the final clusters. However, this method has proven to be very ineffective and does not produce good clusters. It is observed that the SSE function and the usage of the mean are not appropriate when dealing with categorical data. Hence, the K-modes clustering algorithm [21] has been proposed to tackle this challenge.

K-modes is a nonparametric clustering algorithm suitable for handling categorical data and optimizes a matching metric (L_0 loss function) without using any explicit distance metric. The loss function here is a special case of the standard L_p norm where p tends to zero. As opposed to the L_p norm which calculates the distance between the data point and centroid vectors, the loss function in K-modes clustering works as a *metric* and uses the number of mismatches to estimate the similarity between the data points. The K-modes algorithm is described in detail in Algorithm 15. As with K-means, this is also an optimization problem and this method cannot guarantee a global optimal solution.

Algorithm 15 K-Modes Clustering

1: Select K initial modes.
2: **repeat**
3: Form K clusters by assigning all the data points to the cluster with the nearest mode using the matching metric.
4: Recompute the modes of the clusters.
5: **until** Convergence criterion is met.

4.2.4.4 Fuzzy *K*-Means Clustering

This is also popularly known as Fuzzy C-Means clustering. Performing hard assignments of points to clusters is not feasible in complex datasets where there are overlapping clusters. To extract such overlapping structures, a fuzzy clustering algorithm can be used. In fuzzy C-means clustering algorithm (FCM) [12, 4], the membership of points to different clusters can vary from 0 to 1. The SSE function for FCM is provided in Equation (4.12):

$$SSE(C) = \sum_{k=1}^{K} \sum_{x_i \in C_k} w_{xik}^{\beta} \parallel x_i - c_k \parallel^2 \tag{4.12}$$

$$w_{xik} = \frac{1}{\sum_{j=1}^{K} \left(\frac{x_i - c_k}{x_i - c_j}\right)^{\frac{2}{\beta-1}}} \tag{4.13}$$

$$c_k = \frac{\sum_{x_i \in C_k} w_{xik}^{\beta} x_i}{\sum_{x_i \in C_k} w_{xik}} \tag{4.14}$$

Here w_{xik} is the membership weight of point x_i belonging to C_k. This weight is used during the update step of fuzzy C-means. The weighted centroid according to the fuzzy weights for C_k is calculated (represented by c_k). The basic algorithm works similarly to K-means where the algorithm minimizes the SSE iteratively followed by updating w_{xik} and c_k. This process is continued until the convergence of centroids. As in K-means, the FCM algorithm is sensitive to outliers and the final solutions obtained will correspond to the local minimum of the objective function. There are further extensions of this algorithm in the literature such as *Rough C-means* [34] and *Possibilistic C-means* [30].

4.2.4.5 *X*-Means Clustering

X-means [42] is a clustering method that can be used to efficiently estimate the value of K. It uses a method called *blacklisting* to identify those sets of centroids among the current existing ones that can be split in order to fit the data better. The decision making here is done using the Akaike or Bayesian Information Criterion. In this algorithm, the centroids are chosen by initially reducing the search space using a heuristic. K values for experimentation are chosen between a selected lower and upper bound value. This is followed by assessing the goodness of the model for different K in the bounded space using a specific model selection criterion. This model selection criterion is developed using the Gaussian probabilistic model and the maximum likelihood estimates. The best K value corresponds to the model that scores the highest on the model score. The primary goal of this algorithm is to estimate K efficiently and provide a scalable K-means clustering algorithm when the number of data points becomes large.

4.2.4.6 Intelligent K-Means Clustering

Intelligent K-means (IK-means) clustering [38] is a method which is based on the following principle: *the farther a point is from the centroid, the more interesting it becomes.* IK-means uses the basic ideas of principal component analysis (PCA) and selects those points farthest from the centroid that correspond to the maximum data scatter. The clusters derived from such points are called as *anomalous pattern* clusters. The IK-means clustering algorithm is given in Algorithm 16.

In Algorithm 16, line 1 initializes the centroid for the dataset as c_g. In line 3, a new centroid is created which is farthest from the centroid of the entire data. In lines 4–5, a version of 2-means clustering assignment is made. This assignment uses the center of gravity of the original dataset cluster c_g and that of the new anomalous pattern cluster s_g as the initial centroids. In line 6, the centroid of the dataset is updated with the centroid of the anomalous cluster. In line 7, a threshold condition is applied to discard small clusters being created because of outlier points. Lines 3–7 are run until one of the stopping criteria is met: (i) Centroids converge or (ii) prespecified K number of clusters has been obtained or (iii) the entire data have been clustered.

Algorithm 16 IK-Means Clustering

1: Calculate the center of gravity for the given set of data points c_g.
2: **repeat**
3: Create a centroid c farthest from c_g.
4: Create a cluster S_{iter} of data points that is closer to c compared to c_g by assigning all the remaining data points x_i to S_{iter} if $d(x_i, c) < d(x_i, c_g)$.
5: Update the centroid of S_{iter} as s_g.
6: Set $c_g = s_g$.
7: Discard small clusters (if any) using a prespecified threshold.
8: **until** Stopping criterion is met.

There are different ways by which we can select the K for IK-means, some of which are similar to choosing K in K-means described earlier. A structural based approach which compares the internal cluster cohesion with between-cluster separation can be applied. Standard hierarchical clustering methods which construct a dendrogram can also be used to determine K. K-means is considered to be a *nondeterministic* algorithm whereas IK-means can be considered a *deterministic* algorithm.

IK-means can be very effective in extracting clusters when they are spread across the dataset rather than being compactly structured in a single region. IK-means clustering can also be used for initial centroid seed selection before applying K-means. At the end of the IK-means we will be left with only the good centroids for further selection. Small anomalous pattern clusters will not contribute any candidate centroids as they have already been pruned.

4.2.4.7 Bisecting K-Means Clustering

Bisecting K-means clustering [47] is a divisive hierarchical clustering method which uses K-means repeatedly on the parent cluster C to determine the best possible split to obtain two child clusters C_1 and C_2. In the process of determining the best split, bisecting K-means obtains uniform-sized clusters. The algorithm for bisecting K-means clustering is given in Algorithm 17.

Algorithm 17 Bisecting K-Means Clustering

```
1: repeat
2:    Choose the parent cluster to be split C.
3:    repeat
4:       Select two centroids at random from C.
5:       Assign the remaining points to the nearest subcluster using a prespecified distance mea-
         sure.
6:       Recompute centroids and continue cluster assignment until convergence.
7:       Calculate inter-cluster dissimilarity for the 2 subclusters using the centroids.
8:    until I iterations are completed.
9:    Choose those centroids of the subclusters with maximum inter-cluster dissimilarity.
10:   Split C as C1 and C2 for these centroids.
11:   Choose the larger cluster among C1 and C2 and set it as the parent cluster.
12: until K clusters have been obtained.
```

In line 2, the parent cluster to be split is initialized. In lines 4–7, a 2-means clustering algorithm is run I times to determine the best split which maximizes the Ward's distance between C_1 and C_2. In lines 9–10, the best split obtained will be used to divide the parent cluster. In line 11, the larger of the split clusters is made the new parent for further splitting. The computational complexity of the bisecting K-means is much higher compared to the standard K-means.

4.2.4.8 Kernel K-Means Clustering

In Kernel K-means clustering [44], the final clusters are obtained after projecting the data onto the high-dimensional kernel space. The algorithm works by initially mapping the data points in the input space onto a high-dimensional feature space using the kernel function. Some important kernel functions are polynomial kernel, Gaussian kernel, and sigmoid kernel. The formula for the SSE criterion of kernel K-means along with that of the cluster centroid is given in Equation (4.15). The formula for the kernel matrix K for any two points $x_i, x_j \in C_k$ is also given below.

$$SSE(C) = \sum_{k=1}^{K} \sum_{x_i \in C_k} ||\phi(x_i) - c_k||^2 \tag{4.15}$$

$$c_k = \frac{\sum_{x_i \in C_k} \phi(x_i)}{|C_k|} \tag{4.16}$$

$$K_{x_i x_j} = \phi(x_i) \cdot \phi(x_j) \tag{4.17}$$

The difference between the standard K-means criteria and this new kernel K-means criteria is only in the usage of projection function ϕ. The Euclidean distance calculation between a point and the centroid of the cluster in the high-dimensional feature space in kernel K-means will require the knowledge of only the kernel matrix K. Hence, the clustering can be performed without the actual individual projections $\phi(x_i)$ and $\phi(x_j)$ for the data points x_i, $x_j \in C_k$. It can be observed that the computational complexity is much higher than K-means since the kernel matrix has to be generated from the kernel function for the given data. A weighted version of the same algorithm called *Weighted Kernel K-means* has also been developed [10]. The widely studied spectral clustering can be considered as a variant of kernel K-means clustering.

4.2.4.9 Mean Shift Clustering

Mean shift clustering [7] is a popular nonparametric clustering technique which has been used in many areas of pattern recognition and computer vision. It aims to discover the modes present in the data through a convergence routine. The primary goal of the mean shift procedure is to determine the local maxima or modes present in the data distribution. The Parzen window kernel density estimation method forms the basis for the mean shift clustering algorithm. It starts with each point and then performs a gradient ascent procedure until convergence. As the mean shift vector always points toward the direction of maximum increase in the density, it can define a path leading to a stationary point of the estimated density. The local maxima (or modes) of the density are such stationary points. This mean shift algorithm is one of the widely used clustering methods that fall into the category of mode-finding procedures.

We provide some of the basic mathematical formulation involved in the mean shift clustering Algorithm 18. Given N data points x_i, where $i = 1,\ldots,N$ on a d-dimensional space R^d, the multivariate Parzen window kernel density estimate $f(x)$ is obtained with kernel $K(x)$ and window radius h. It is given by

$$f(x) = \frac{1}{Nh^d}\sum_{i=1}^{N} K\left(\frac{x-x_i}{h}\right) \tag{4.18}$$

$$m_h(x) = \frac{\sum_{i=1}^{N} x_i \cdot g(\| \frac{x-x_i}{h} \|^2)}{\sum_{i=1}^{N} g(\| \frac{x-x_i}{h} \|^2)} \tag{4.19}$$

Algorithm 18 Mean Shift Clustering

1: Select K random points as the modes of the distribution.
2: **repeat**
3:     For each given mode x calculate the mean shift vector $m_h(x)$.
4:     Update the point $x = m_h(x)$.
5: **until** Modes become stationary and converge.

More detailed information about obtaining the gradient from the kernel function and the exact kernel functions being used can be obtained from [7]. A proof of convergence of the modes is also provided in [8].

4.2.4.10 Weighted K-Means Clustering

Weighted K-means (WK-means) Algorithm 19 [20] introduces a feature weighting mechanism into the standard K-means. It is an iterative optimization algorithm in which the weights for different features are automatically learned. Standard K-means ignores the importance of a particular feature and considers all of the features to be equally important. The modified SSE function optimized by the WK-means clustering algorithm is given in Equation (4.20). Here the features are numbered from $v = 1,\ldots,M$ and the clusters are numbered from $k = 1,\ldots,K$. A user-defined parameter β which employs the impact of the feature weights on the clustering is also

Algorithm 19 Weighted K-Means Clustering

1: Choose K random centroids and set up M feature weights such that they sum to 1.
2: **repeat**
3: Assign all data points x_i to the closest centroid by calculating $d(x_i, c_k)$.
4: Recompute centroids of the clusters after completing assignment.
5: Update weights using w_v.
6: **until** Convergence criterion has been met.

used. The clusters are numbered $C = C_1, \ldots, C_k, \ldots, C_K$, c_k is the M-dimensional centroid for cluster C_k, and c_{kv} represents the vth feature value of the centroid. Feature weights are updated in WK-means according to w_v. D_v is the sum of within cluster variances of feature v weighted by cluster cardinalities.

$$SSE(C,w) = \sum_{k=1}^{K} \sum_{x_i \in C_k} \sum_{v=1}^{M} s_{xik} w_v^{\beta} (x_{iv} - c_{kv})^2 \tag{4.20}$$

$$w_v = \frac{1}{\sum_{u \in V} \left[\frac{D_v}{D_u}\right]^{\frac{1}{\beta - 1}}} \tag{4.21}$$

$$d(x_i, c_k) = \sum_{v=1}^{M} w_v^{\beta} (x_{iv} - c_{kv})^2 \tag{4.22}$$

$$\begin{cases} s_{xik} \in (0,1) \\ \sum_{k=1}^{K} s_{xik} = 1 \\ \sum_{v=1}^{M} w_v = 1 \end{cases} \tag{4.23}$$

The WK-means clustering algorithm runs similar to K-means clustering but the distance measure is also weighted by the feature weights. In line 1, the centroids and weights for M features are initialized. In lines 3–4, points are assigned to their closest centroids and the weighted centroid is calculated. This is followed by a weight update step such that the sum of weights is constrained as shown in Equation (4.23). These steps are continued until the centroids converge. This algorithm is computationally more expensive compared to K-means. Similar to K-means, this algorithm also suffers from convergence issues. Intelligent K-means (IK-means) [38] can also be integrated with WK-means to yield the Intelligent Weighted K-means algorithm.

4.2.4.11 Genetic K-Means Clustering

K-means suffers from the problem of converging to a local minimum. To tackle this problem, stochastic optimization procedures which are good at avoiding the convergence to a local optimal solution can be applied. Genetic algorithms (GA) are proven to converge to a global optimum. These algorithms evolve over *generations*, where during each generation they produce a new population from the current one by applying a set of genetic operators such as natural selection, crossover, and mutation. They develop a fitness function and pick up the fittest individual based on the probability score from each generation and use them to generate the next population using the mutation operator. The problem of local optima can be effectively solved by using GA and this gives rise to the Genetic K-means algorithm (GKA) [29]. The data is initially converted using a string of group numbers coding scheme and a population of such strings forms the initial dataset for GKA. The following are the major steps involved in the GKA algorithm:

1. **Initialization**: Select a random population initially to begin the algorithm. This is analogous to the random centroid initialization step in K-means.

2. **Selection**: Using the probability computation given in Equation (4.24), identify the fittest individuals in the given population.

$$P(s_i) = \frac{F(s_i)}{\sum_{j=1}^{N} F(s_j)} \tag{4.24}$$

where $F(s_i)$ represents the fitness value of a string s_i in the population. A fitness function is further developed to assess the goodness of the solution. This fitness function is analogous to the SSE of K-means.

3. **Mutation**: This is analogous to the K-means assignment step where points are assigned to their closest centroids followed by updating the centroids at the end of iteration. The selection and mutation steps are applied iteratively until convergence is obtained.

The pseudocode of the exact GKA algorithm is discussed in detail in [29] and a proof of convergence of the GA is given [43].

4.2.5 Making K-Means Faster

It is believed that the K-means clustering algorithm consumes a lot of time in its later stages when the centroids are close to their final locations but the algorithm is yet to converge. An improvement to the original Lloyd's K-means clustering using a kd-tree data structure to store the data points was proposed in [24]. This algorithm is called the *filtering algorithm* where for each node a set of candidate centroids is maintained similar to a normal kd-tree. These candidate set centroids are pruned based on a distance comparison which measures the proximity to the midpoint of the cell. This filtering algorithm runs faster when the separation between the clusters increases. In the K-means clustering algorithm, usually there are several redundant calculations that are performed. For example, when a point is very far from a particular centroid, calculating its distance to that centroid may not be necessary. The same applies for a point which is very close to the centroid as it can be directly assigned to the centroid without computing its exact distance. An optimized K-means clustering method which uses the triangle inequality metric is also proposed to reduce the number of distance metric calculations [13]. The mathematical formulation for the lemma used by this algorithm is as follows. Let x be a data point and let b and c be the centroids.

$$d(b,c) \geq 2d(x,b) \rightarrow d(x,c) \geq d(x,b) \tag{4.25}$$

$$d(x,c) \geq max\{0, d(x,b) - d(b,c)\} \tag{4.26}$$

This algorithm runs faster than the standard K-means clustering algorithm as it avoids both kinds of computations mentioned above by using the lower and upper bounds on distances without affecting the final clustering result.

4.3 Hierarchical Clustering Algorithms

Hierarchical clustering algorithms [23] were developed to overcome some of the disadvantages associated with flat or partitional-based clustering methods. Partitional methods generally require a user predefined parameter K to obtain a clustering solution and they are often *nondeterministic* in nature. Hierarchical algorithms were developed to build a more deterministic and flexible mechanism for clustering the data objects. Hierarchical methods can be categorized into *agglomerative* and *divisive* clustering methods. Agglomerative methods start by taking singleton clusters (that contain only one data object per cluster) at the bottom level and continue merging two clusters at a time to build a *bottom-up* hierarchy of the clusters. Divisive methods, on the other hand, start with all the data objects in a huge macro-cluster and split it continuously into two groups generating a *top-down* hierarchy of clusters.

A cluster hierarchy here can be interpreted using the standard binary tree terminology as follows. The root represents all the sets of data objects to be clustered and this forms the apex of the hierarchy (level 0). At each level, the child entries (or nodes) which are subsets of the entire dataset correspond to the clusters. The entries in each of these clusters can be determined by traversing the tree from the current cluster node to the base singleton data points. Every level in the hierarchy corresponds to some set of clusters. The base of the hierarchy consists of all the singleton points which are the leaves of the tree. This cluster hierarchy is also called a *dendrogram*. The basic advantage of having a hierarchical clustering method is that it allows for cutting the hierarchy at any given level and obtaining the clusters correspondingly. This feature makes it significantly different from partitional clustering methods in that it does not require a predefined user specified parameter k (number of clusters). We will discuss more details of how the dendrogram is cut later in this chapter.

In this section, we will first discuss different kinds of agglomerative clustering methods which primarily differ from each other in the similarity measures that they employ. The widely studied algorithms in this category are the following: *single link* (nearest neighbour), *complete link* (diameter), *group average* (average link), *centroid similarity*, and *Ward's criterion* (minimum variance). Subsequently, we will also discuss some of the popular divisive clustering methods.

4.3.1 Agglomerative Clustering

The basic steps involved in an agglomerative hierarchical clustering algorithm are the following. First, using a particular proximity measure a dissimilarity matrix is constructed and all the data points are visually represented at the bottom of the dendrogram. The closest sets of clusters are merged at each level and then the dissimilarity matrix is updated correspondingly. This process of agglomerative merging is carried on until the final maximal cluster (that contains all the data objects in a single cluster) is obtained. This would represent the apex of our dendrogram and mark the completion of the merging process. We will now discuss the different kinds of proximity measures which can be used in agglomerative hierarchical clustering. Subsequently, we will also provide a complete version of the agglomerative hierarchical clustering algorithm in Algorithm 20.

4.3.1.1 Single and Complete Link

The most popular agglomerative clustering methods are single link and complete link clusterings. In *single link clustering* [36, 46], the similarity of two clusters is the similarity between their most similar (nearest neighbor) members. This method intuitively gives more importance to the regions where clusters are closest, neglecting the overall structure of the cluster. Hence, this method falls under the category of a *local* similarity-based clustering method. Because of its local behavior, single linkage is capable of effectively clustering nonelliptical, elongated shaped groups of data objects. However, one of the main drawbacks of this method is its sensitivity to noise and outliers in the data.

Complete link clustering [27] measures the similarity of two clusters as the similarity of their most dissimilar members. This is equivalent to choosing the cluster pair whose merge has the smallest diameter. As this method takes the cluster structure into consideration it is nonlocal in behavior and generally obtains compact shaped clusters. However, similar to single link clustering, this

	1	2	3	4
1	0.0	0.20	0.15	0.30
2	0.20	0.0	0.40	0.50
3	0.15	0.40	0.0	0.10
4	0.30	0.50	0.10	0.0

(a) Dissimilarity Matrix

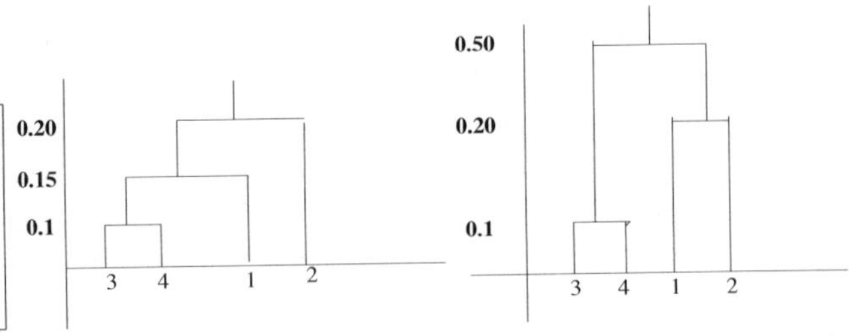

(b) Single Link (c) Complete Link

FIGURE 4.2: An illustration of agglomerative clustering. (a) A dissimilarity matrix computed for four arbitrary data points. The corresponding dendrograms obtained using (b) single link and (c) complete link hierarchical clustering methods.

method is also sensitive to outliers. Both single link and complete link clustering have their graph-theoretic interpretations [16], where the clusters obtained after single link clustering would correspond to the connected components of a graph and those obtained through complete link would correspond to the maximal cliques of the graph.

Figure 4.2 shows the dissimilarity matrix and the corresponding two dendrograms obtained using single link and complete link algorithms on a toy dataset. In the dendrograms, the X-axis indicates the data objects and the Y-axis indicates the dissimilarity (distance) at which the points were merged. The difference in merges between both the dendrograms occurs due to the different criteria used by single and complete link algorithms. In single link, first data points 3 and 4 are merged at 0.1 as shown in (b). Then, based on the computations shown in Equation (4.27), cluster (3,4) is merged with data point 1 at the next level; at the final level cluster (3,4,1) is merged with 2. In complete link, merges for cluster (3,4) are checked with points 1 and 2 and as d(1,2) = 0.20, points 1 and 2 are merged at the next level. Finally, clusters (3,4) and (1,2) are merged at the final level. This explains the difference in the clustering in both the cases.

$$\begin{aligned}
d_{min}((3,4),1) &= min(d(3,1),d(4,1)) = 0.15 \\
d_{min}((3,4,1),2) &= min(d(3,2),d(4,2),d(1,2)) = 0.20 \\
d_{max}((3,4),1) &= max(d(3,1),d(4,1)) = 0.30 \\
d_{max}((3,4),2) &= max(d(3,2),d(4,2)) = 0.50 \\
d_{max}((3,4),(1,2)) &= max(d(3,1),d(3,2),d(4,1),d(4,2)) = 0.50
\end{aligned} \quad (4.27)$$

4.3.1.2 Group Averaged and Centroid Agglomerative Clustering

Group Averaged Agglomerative Clustering (GAAC) considers the similarity between all pairs of points present in both the clusters and diminishes the drawbacks associated with single and complete link methods. Before we look at the formula let us introduce some terminology. Let two clusters C_a and C_b be merged so that the resulting cluster is $C_{a \cup b} = C_a \cup C_b$. The new centroid for this cluster is $c_{a \cup b} = \frac{N_a c_a + N_b c_b}{N_a + N_b}$, where N_a and N_b are the cardinalities of the clusters C_a and C_b, respectively. The similarity measure for GAAC is calculated as follows:

$$S_{GAAC}(C_a, C_b) = \frac{1}{(N_a + N_b)(N_a + N_b - 1)} \sum_{i \in C_a \cup C_b} \sum_{j \in C_a \cup C_b, i \neq j} d(i,j) \quad (4.28)$$

We can see that the distance between two clusters is the average of all the pair-wise distances between the data points in these two clusters. Hence, this measure is expensive to compute especially when the number of data objects becomes large. *Centroid-based agglomerative clustering*, on the other hand, calculates the similarity between two clusters by measuring the similarity between their centroids. The primary difference between GAAC and Centroid agglomerative clustering is that, GAAC considers all pairs of data objects for computing the average pair-wise similarity, whereas centroid-based agglomerative clustering uses only the centroid of the cluster to compute the similarity between two different clusters.

4.3.1.3 Ward's Criterion

Ward's criterion [49, 50] was proposed to compute the distance between two clusters during agglomerative clustering. This process of using Ward's criterion for cluster merging in agglomerative clustering is also called as *Ward's agglomeration*. It uses the K-means squared error criterion to determine the distance. For any two clusters, C_a and C_b, the Ward's criterion is calculated by measuring the increase in the value of the SSE criterion for the clustering obtained by merging them into $C_a \cup C_b$. The Ward's criterion is defined as follows:

$$\begin{aligned} W(C_{a\cup b}, c_{a\cup b}) - W(C,c) &= \frac{N_a N_b}{N_a + N_b}\sum_{v=1}^{M}(c_{av} - c_{bv})^2 \\ &= \frac{N_a N_b}{N_a + N_b} d(c_a, c_b) \end{aligned} \tag{4.29}$$

So the Ward's criterion can be interpreted as the squared Euclidean distance between the centroids of the merged clusters C_a and C_b weighted by a factor that is proportional to the product of cardinalities of the merged clusters.

4.3.1.4 Agglomerative Hierarchical Clustering Algorithm

In Algorithm 20, we provide a basic outline of an agglomerative hierarchical clustering algorithm. In line 1, the dissimilarity matrix is computed for all the points in the dataset. In lines 3–4, the closest pairs of clusters are repeatedly merged in a bottom-up fashion and the dissimilarity matrix is updated. The rows and columns pertaining to the older clusters are removed from the dissimilarity matrix and are added for the new cluster. Subsequently, merging operations are carried out with this updated dissimilarity matrix. Line 5 indicates the termination condition for the algorithm.

Algorithm 20 Agglomerative Hierarchical Clustering

1: Compute the dissimilarity matrix between all the data points.
2: **repeat**
3:     Merge clusters as $C_{a\cup b} = C_a \cup C_b$. Set new cluster's cardinality as $N_{a\cup b} = N_a + N_b$.
4:     Insert a new row and column containing the distances between the new cluster $C_{a\cup b}$ and the remaining clusters.
5: **until** Only one maximal cluster remains.

4.3.1.5 Lance–Williams Dissimilarity Update Formula

We have discussed many different proximity measures that are used in agglomerative hierarchical clustering. A convenient formulation in terms of dissimilarity which embraces all the hierarchical methods mentioned so far is the Lance–Williams dissimilarity update formula [31]. If points i and j are agglomerated into cluster $i \cup j$, then we will have to specify just the new dissimilarity

TABLE 4.1: Values of the Coefficients for the Lance–Williams Dissimilarity Update Formula for Different Hierarchical Clustering Algorithms.

Name of the Method	Lance–Williams Dissimilarity Update Formula
Single Link	$\alpha_i = 0.5$; $\beta = 0$; and $\gamma = -0.5$
Complete Link	$\alpha_i = 0.5$; $\beta = 0$; and $\gamma = 0.5$
GAAC	$\alpha_i = \frac{\lvert i \rvert}{\lvert i \rvert + \lvert j \rvert}$; $\beta = 0$; and $\gamma = 0$
Centroid	$\alpha_i = \frac{\lvert i \rvert}{\lvert i \rvert + \lvert j \rvert}$; $\beta = -\frac{\lvert i \rvert \lvert j \rvert}{(\lvert i \rvert + \lvert j \rvert)^2}$; and $\gamma = 0$
Ward's	$\alpha_i = \frac{\lvert i \rvert + \lvert k \rvert}{\lvert i \rvert + \lvert j \rvert + \lvert k \rvert}$; $\beta = -\frac{\lvert k \rvert}{\lvert i \rvert + \lvert j \rvert + \lvert k \rvert}$; and $\gamma = 0$

between the cluster and all other points. The formula is given as follows:

$$d(i \cup j, k) = \alpha_i d(i,k) + \alpha_j d(j,k) + \beta d(i,j) + \gamma |d(i,k) - d(j,k)| \tag{4.30}$$

Here, α_i, α_j, β, and γ define the agglomerative criterion. The coefficient values for the different kinds of methods we have studied so far are provided in Table 4.1.

4.3.2 Divisive Clustering

Divisive hierarchical clustering is a *top-down* approach where the procedure starts at the root with all the data points and recursively splits it to build the dendrogram. This method has the advantage of being more efficient compared to agglomerative clustering especially when there is no need to generate a complete hierarchy all the way down to the individual leaves. It can be considered as a *global approach* since it contains the complete information before splitting the data.

4.3.2.1 Issues in Divisive Clustering

We will now discuss the factors that affect the performance of divisive hierarchical clustering.

1. *Splitting criterion*: The Ward's K-means square error criterion is used here. The greater reduction obtained in the difference in the SSE criterion should reflect the goodness of the split. Since the SSE criterion can be applied to numerical data only, Gini index (which is widely used in decision tree construction in classification) can be used for handling the nominal data.

2. *Splitting method*: The splitting method used to obtain the binary split of the parent node is also critical since it can reduce the time taken for evaluating the Ward's criterion. The Bisecting K-means approach can be used here (with $K = 2$) to obtain good splits since it is based on the same criterion of maximizing the Ward's distance between the splits.

3. *Choosing the cluster to split*: The choice of cluster chosen to split may not be as important as the first two factors, but it can still be useful to choose the most appropriate cluster to further split when the goal is to build a compact dendrogram. A simple method of choosing the cluster to be split further could be done by merely checking the square errors of the clusters and splitting the one with the largest value.

4. *Handling noise*: Since the noise points present in the dataset might result in aberrant clusters, a threshold can be used to determine the termination criteria rather splitting the clusters further.

4.3.2.2 Divisive Hierarchical Clustering Algorithm

In Algorithm 21, we provide the steps involved in divisive hierarchical clustering. In line 1, we start with all the points contained in the maximal cluster. In line 3, the Bisecting K-means approach is used to determine the uniform splitting mechanism to obtain C_1 and C_2. In line 4, we use the heuristic mentioned above and choose the cluster with higher squared error for splitting as the next parent. These steps (lines 3–4) are run repeatedly until the complete dendrogram (up to the individual) has been constructed. As mentioned above, we can use the threshold to handle noise during the construction of the dendrogram.

Algorithm 21 Basic Divisive Hierarchical Clustering

1: Start with the root node consisting all the data points
2: **repeat**
3: Split parent node into two parts C_1 and C_2 using Bisecting K-means to maximize Ward's distance $W(C_1,C_2)$.
4: Construct the dendrogram. Among the current, choose the cluster with the highest squared error.
5: **until** Singleton leaves are obtained.

4.3.2.3 Minimum Spanning Tree-Based Clustering

In a weighted graph, a minimum spanning tree is an acyclic subgraph that covers all the vertices with the minimum edge weights. Prim's and Kruskal's algorithms [9] are used for finding the minimum spanning tree (MST) in a weighted graph. In a Euclidean minimum spanning tree (EMST), the data points represent the vertices and the edge weights are computed using the Euclidean distance between two data points. Each edge in an EMST represents the shortest distance between those two points. Using this EMST a divisive clustering method can be developed which removes the largest weighted edge to get two clusterings and subsequently removes the next largest edge to get three clusterings and so on. This process of removing edges from an EMST gives rise to an effective divisive clustering method. The major advantage of this method is that it is able to detect clusters with nonspherical shapes effectively.

A basic EMST clustering algorithm proceeds by taking a user supplied parameter K where the edges present in the graph are sorted in descending order. This is followed by removing the edges with the top $(K-1)$ weights one by one to get the K connected components. This is similar to the process of divisive clustering where finer clustering partitions are obtained after each split. Subsequently, we can also use the EMST to build a clustering algorithm which continuously prunes the inconsistent edges present in the graph. An inconsistent edge is the one whose edge weight is much higher than the average weight of the edges in the neighbourhood of that edge. Algorithm 22 describes the minimum spanning tree-based clustering algorithm that was originally proposed by Zahn [53].

Algorithm 22 Zahn Minimum Spanning Tree-Based Divisive Clustering

Create the EMST using Prim's/Kruskal's algorithm on all the data points.
repeat
  Remove edge with highest inconsistency measure.
until No more inconsistent edges can be removed.

4.3.3 Other Hierarchical Clustering Algorithms

The agglomerative and divisive hierarchical clustering methods are successful in capturing convex shaped clusters effectively. As mentioned above, agglomerative methods, especially single link and complete link, suffer from the "chaining problem" and are ineffective at capturing arbitrarily shaped clusters. Hence, to capture arbitrarily shaped clusters, algorithms such as CURE [17] and CHAMELEON [25] have been proposed in the literature. Some of the popular extensions of hierarchical algorithms are discussed below.

1. **CURE** (Clustering Using REpresentatives) [17] is an algorithm which incorporates a novel feature of representing a cluster using a set of well-scattered representative points. The distance between two clusters is calculated by looking at the minimum distance between the representative points chosen. In this manner, CURE incorporates features of both the Single link and GAAC hierarchical clustering methods. Choosing scattered points helps CURE capture clusters of arbitrary shapes also. In addition, CURE employs a shrinking factor α in the algorithm, where the points are shrunk toward the centroid by a factor α. α shrinking has a greater effect in the case of outliers compared to normal points. This makes CURE more robust to outliers. Similar to this approach, an algorithm called ROCK [18] was also proposed to handle categorical data. This algorithm uses the concept of common links and determines the Jaccard coefficient between candidate clusters for hierarchical clustering.

2. **CHAMELEON** [25] is a clustering algorithm which uses graph partitioning methods on the K-nearest neighbor graph of the data. These initial partitions are then used as the seed clusters for the agglomerative hierarchical clustering process. The algorithm uses two metrics based on the relative inter-connectivity and relative closeness of clusters to merge the clusters. These metrics capture the local information of the clusters during the clustering process thus enabling this algorithm to behave like a *dynamic* framework. CHAMELEON is one of the best hierarchical clustering algorithms and is extremely effective in capturing arbitrarily shaped clusters which is primarily due to the dynamic behavior. A detailed comparison between the clustering results of CURE and CHAMELEON for synthetic datasets with clusters of varying shapes can also be found in [25].

3. **COBWEB** [15]: This is a conceptual clustering algorithm that works incrementally by updating the clusters object by object. Probabilistically described clusters are arranged as a tree to form a hierarchical clustering known as probabilistic categorization tree. It handles uncertainty associated with categorical attributes in clustering through a probabilistic framework that is similar to Naive Bayes. The dendrogram in this algorithm is also called a classification tree and the nodes are referred to as concepts.

4. **Self-Organizing Maps (SOM)** [28] were developed on the same lines of Artificial Neural Networks and are useful for hierarchical representation. It is an efficient data visualization technique. Similar to K-means, data points are assigned to their closest centroids. The difference arises in the centroid update step where, when a centroid is updated, those in its neighborhood which are close to this centroid are also updated. The final output is an SOM neural network which can be explored to understand the relationships between different objects involved in the clustering.

4.4 Discussion and Summary

A major advantage of partitional clustering algorithms is that they can gradually improve the clustering quality through an iterative optimization process [3]. This cannot be done in standard hierarchical clustering since the dendrogram cannot revisit the merges (or splits) that were already completed. Partitional algorithms are also effective in detecting compact spherical-shaped clusters and are easy to implement and use in practice [22]. K-means is also a computationally efficient algorithm compared to hierarchical clustering. Although there is no consensus, it is believed that K-means is better than hierarchical clustering algorithms [35] in terms of the quality of the final clustering solution.

Hierarchical clustering methods can potentially overcome some of the critical problems associated with flat (partitional) clustering methods. One of the major advantages of hierarchical algorithms is the generation of the visual dendrograms which can assist the end-users during the clustering process. In such applications, generally a user will label the clusters to understand more about them. This is also called cluster labeling. Hierarchical methods are also *deterministic* compared to the nondeterministic behavior experienced with the basic K-means algorithm.

Despite these advantages, it is observed that in hierarchical clustering methods the merge or split decisions once made at any given level in the hierarchy cannot be undone [3]. This is considered to be a weakness for such hierarchical algorithms since it reduces the flexibility. To overcome this problem, [14] proposes an iterative optimization strategy that keeps modifying the created dendrogram until the optimal solution is obtained. The run-time complexity of these hierarchical algorithms is quadratic which is not desirable especially for large-scale problems. Parallel hierarchical clustering methods [41] have also been proposed to reduce the complexity to linear time.

In spite of the numerous advances made in the field of data clustering in the past two decades, both partitional and hierarchical clustering algorithms form a solid foundation for data clustering. Many of the newly proposed data clustering algorithms (to be discussed in the next few chapters) typically compare their performance to these fundamental clustering algorithms. In addition, due to their simplicity and ease of usage, these algorithms are heavily used in several other application domains such as bioinformatics, information retrieval, text mining, imaging, climate science, and astronomy. The development of new variants of both partitional and hierarchical clustering algorithms is still an active area of research.

Bibliography

[1] D. Arthur and S. Vassilvitskii. K-means++: The advantages of careful seeding. In *Proceedings of the Eighteenth Annual ACM-SIAM Symposium on Discrete Algorithms*, pages 1027–1035. Society for Industrial and Applied Mathematics, 2007.

[2] G. H. Ball and D. J. Hall. ISODATA, a novel method of data analysis and pattern classification. Technical report, DTIC Document, 1965.

[3] P. Berkhin. A survey of clustering data mining techniques. In *Grouping Multidimensional Data*, J. Kogan, C. Nicholas, and M. Teoulle, Eds., Springer, Berlin Heidelberg, pages 25–71, 2006.

[4] J. C. Bezdek. *Pattern recognition with fuzzy objective function algorithms*. Kluwer Academic Publishers, 1981.

[5] P. S. Bradley and U. M. Fayyad. Refining initial points for k-means clustering. In *Proceedings of the Fifteenth International Conference on Machine Learning*, volume 66. San Francisco, CA, USA, 1998.

[6] T. Caliński and J. Harabasz. A dendrite method for cluster analysis. *Communications in Statistics—Theory and Methods*, 3(1):1–27, 1974.

[7] Y. Cheng. Mean shift, mode seeking, and clustering. *IEEE Transactions on Pattern Analysis and Machine Intelligence*, 17(8):790–799, 1995.

[8] D. Comaniciu and P. Meer. Mean shift: A robust approach toward feature space analysis. *IEEE Transactions on Pattern Analysis and Machine Intelligence*, 24(5):603–619, 2002.

[9] T. H. Cormen. *Introduction to Algorithms*. MIT Press, 2001.

[10] I. S. Dhillon, Y. Guan, and B. Kulis. Kernel k-means: Spectral clustering and normalized cuts. In *Proceedings of the Tenth ACM SIGKDD International Conference on Knowledge Discovery and Data Mining (KDD)*, pages 551–556. ACM, 2004.

[11] R. O. Duda and P. E. Hart. *Pattern Classification and Scene Analysis*, Vol. 3, Wiley, New York, 1973.

[12] J. C. Dunn. A fuzzy relative of the ISODATA process and its use in detecting compact well-separated clusters. *Journal of Cybernetics*, 3(3):32–57, 1973.

[13] C. Elkan. Using the triangle inequality to accelerate k-means. In *Proceedings of International Conference on Machine Learning (ICML)*, pages 147–153, 2003.

[14] D. Fisher. Optimization and simplification of hierarchical clusterings. In *Proceedings of the 1st International Conference on Knowledge Discovery and Data Mining (KDD)*, pages 118–123, 1995.

[15] D. H. Fisher. Knowledge acquisition via incremental conceptual clustering. *Machine Learning*, 2(2):139–172, 1987.

[16] J. C. Gower and G. J. S. Ross. Minimum spanning trees and single linkage cluster analysis. *Journal of the Royal Statistical Society. Series C (Applied Statistics)*, 18(1):54–64, 1969.

[17] S. Guha, R. Rastogi, and K. Shim. Cure: An efficient clustering algorithm for large databases. In *ACM SIGMOD Record*, volume 27, pages 73–84. ACM, 1998.

[18] S. Guha, R. Rastogi, and K. Shim. ROCK: A robust clustering algorithm for categorical attributes. In *Proceedings of the 15th International Conference on Data Engineering*, pages 512–521. IEEE, 1999.

[19] J. A. Hartigan and M. A. Wong. Algorithm as 136: A k-means clustering algorithm. *Journal of the Royal Statistical Society. Series C (Applied Statistics)*, 28(1):100–108, 1979.

[20] J. Z. Huang, M. K. Ng, H. Rong, and Z. Li. Automated variable weighting in k-means type clustering. *IEEE Transactions on Pattern Analysis and Machine Intelligence*, 27(5):657–668, 2005.

[21] Z. Huang. Extensions to the k-means algorithm for clustering large data sets with categorical values. *Data Mining and Knowledge Discovery*, 2(3):283–304, 1998.

[22] A. K. Jain. Data clustering: 50 years beyond k-means. *Pattern Recognition Letters*, 31(8):651–666, 2010.

[23] A. K. Jain, M. N. Murty, and P. J. Flynn. Data clustering: A review. *ACM Computing Surveys (CSUR)*, 31(3):264–323, 1999.

[24] T. Kanungo, D. M. Mount, N. S. Netanyahu, C. D. Piatko, R. Silverman, and A. Y. Wu. An efficient k-means clustering algorithm: Analysis and implementation. *IEEE Transactions on Pattern Analysis and Machine Intelligence*, 24(7):881–892, 2002.

[25] G. Karypis, E. H. Han, and V. Kumar. CHAMELEON: Hierarchical clustering using dynamic modeling. *Computer*, 32(8):68–75, 1999.

[26] L. Kaufman, P.J. Rousseeuw, et al. *Finding Groups in Data: An Introduction to Cluster Analysis*, volume 39. Wiley Online Library, 1990.

[27] B. King. Step-wise clustering procedures. *Journal of the American Statistical Association*, 62(317):86–101, 1967.

[28] T. Kohonen. The self-organizing map. *Proceedings of the IEEE*, 78(9):1464–1480, 1990.

[29] K. Krishna and M. N. Murty. Genetic k-means algorithm. *IEEE Transactions on Systems, Man, and Cybernetics, Part B: Cybernetics*, 29(3):433–439, 1999.

[30] R. Krishnapuram and J. M. Keller. The possibilistic C-means algorithm: Insights and recommendations. *IEEE Transactions on Fuzzy Systems*, 4(3):385–393, 1996.

[31] G. N. Lance and W. T. Williams. A general theory of classificatory sorting strategies II. Clustering systems. *The Computer Journal*, 10(3):271–277, 1967.

[32] S. Lloyd. Least squares quantization in PCM. *IEEE Transactions on Information Theory*, 28(2):129–137, 1982.

[33] J. MacQueen. Some methods for classification and analysis of multivariate observations. In *Proceedings of the Fifth Berkeley Symposium on Mathematical Statistics and Probability*, volume 1, pages 281–297, Berkeley, CA, USA, 1967.

[34] P. Maji and S. K. Pal. Rough set based generalized fuzzy c-means algorithm and quantitative indices. *IEEE Transactions on Systems, Man, and Cybernetics, Part B*, 37(6):1529–1540, 2007.

[35] C. D. Manning, P. Raghavan, and H. Schutze. *Introduction to Information Retrieval*, volume 1. Cambridge University Press, Cambridge, 2008.

[36] L. L. McQuitty. Elementary linkage analysis for isolating orthogonal and oblique types and typal relevancies. *Educational and Psychological Measurement* 17(2):207–229, 1957.

[37] G. W. Milligan. A Monte Carlo study of thirty internal criterion measures for cluster analysis. *Psychometrika*, 46(2):187–199, 1981.

[38] B. G. Mirkin. *Clustering for Data Mining: A Data Recovery Approach*, volume 3. CRC Press, Boca Raton, FL, 2005.

[39] R. Mojena. Hierarchical grouping methods and stopping rules: An evaluation. *The Computer Journal*, 20(4):359–363, 1977.

[40] M. E. J. Newman and M. Girvan. Finding and evaluating community structure in networks. *Physical Review E*, 69(2):026113+, 2003.

[41] C. F. Olson. Parallel algorithms for hierarchical clustering. *Parallel Computing*, 21(8):1313–1325, 1995.

[42] D. Pelleg and A. Moore. X-means: Extending k-means with efficient estimation of the number of clusters. In *Proceedings of the Seventeenth International Conference on Machine Learning*, pages 727–734, San Francisco, CA, USA, 2000.

[43] G. Rudolph. Convergence analysis of canonical genetic algorithms. *IEEE Transactions on Neural Networks*, 5(1):96–101, 1994.

[44] B. Schölkopf, A. Smola, and K. R. Müller. Nonlinear component analysis as a kernel eigenvalue problem. *Neural Computation*, 10(5):1299–1319, 1998.

[45] S. Z. Selim and M. A. Ismail. *K*-means-type algorithms: A generalized convergence theorem and characterization of local optimality. *IEEE Transactions on Pattern Analysis and Machine Intelligence*, 6(1):81–87, 1984.

[46] P. H. A. Sneath and R. R. Sokal. Numerical taxonomy. *Nature*, 193:855–860, 1962.

[47] M. Steinbach, G. Karypis, and V. Kumar. A comparison of document clustering techniques. In *KDD Workshop on Text Mining*, volume 400, pages 525–526. Boston, MA, USA, 2000.

[48] R. Tibshirani, G. Walther, and T. Hastie. Estimating the number of clusters in a data set via the gap statistic. *Journal of the Royal Statistical Society: Series B (Statistical Methodology)*, 63(2):411–423, 2001.

[49] J. Ward. Hierarchical grouping to optimize an objective function. *Journal of the American Statistical Association*, 58(301):236–244, 1963.

[50] D. Wishart. An algorithm for hierarchical classifications. *Biometrics*, 25(1):165–170, 1969.

[51] R. Xu and D. Wunsch. Survey of clustering algorithms. *IEEE Transactions on Neural Networks*, 16(3):645–678, 2005.

[52] K.Y. Yeung, C. Fraley, A. Murua, A.E. Raftery, and W.L. Ruzzo. Model-based clustering and data transformations for gene expression data. *Bioinformatics*, 17(10):977–987, 2001.

[53] C.T. Zahn. Graph-theoretical methods for detecting and describing gestalt clusters. *IEEE Transactions on Computers*, 20(1):68–86, 1971.

Chapter 5

Density-Based Clustering

Martin Ester
Simon Fraser University
British Columbia, Canada
ester@cs.sfu.ca

5.1 Introduction

Many of the well-known clustering algorithms make, implicitly or explicitly, the assumption that data are generated from a probability distribution of a given type, e.g., from a mixture of k Gaussian distributions. This is the case in particular for EM (Expectation Maximization) clustering and for k-means. Due to this assumption, these algorithms produce spherical clusters and cannot deal well with datasets in which the actual clusters have nonspherical shapes. Nonspherical clusters occur naturally in spatial data, i.e., data with a reference to some two- or three-dimensional concrete space corresponding to our real world. Spatial data include points, lines, and polygons and support a broad range of applications. Clusters in spatial data may have arbitrary shape, i.e., they are often drawn-out, linear, elongated etc., because of the constraints imposed by geographic entities such as mountains and rivers. In geo-marketing, one may want to find clusters of homes with a given characteristic, e.g., high-income homes, while in crime analysis one of the goals is to detect crime hot-spots, i.e., clusters of certain types of crimes. Even in higher-dimensional data the assumption of a certain number of clusters of a given shape is very strong and may often be violated. In this case, algorithms such as k-means will break up or merge the actual clusters, leading to inaccurate results. The objective to minimize the average squared distances of points from their corresponding cluster center leads to a partitioning of the dataset that is equivalent to the Voronoi diagram of the cluster centers, irrespective of the shape of the actual clusters. Figure 5.1 illustrates this weakness on a small 2-dimensional dataset, showing that k-means with $k = 3$ breaks up and merges the three horizontal clusters.

This observation motivates the requirement to discover clusters of arbitrary shape. The increasingly large sizes of real-life databases require scalability to large databases, i.e., efficiency on databases of up to millions of points or more. Finally, the clustering of large databases requires the

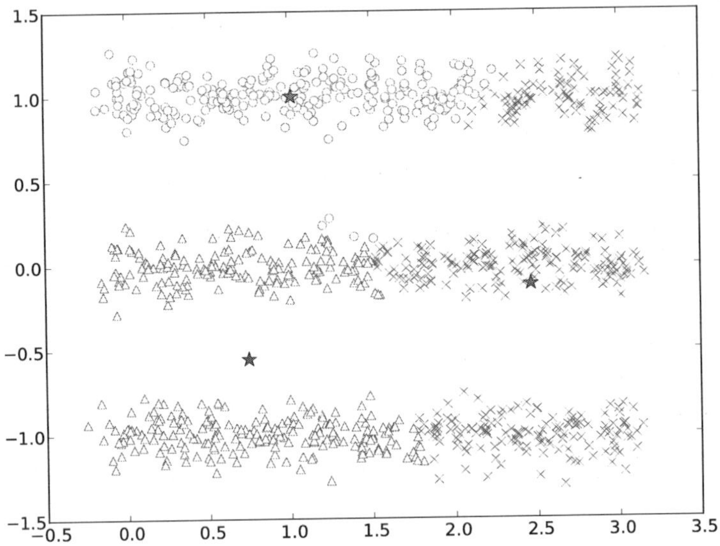

FIGURE 5.1: k-means with $k = 3$ on a sample 2-dimensional dataset.

ability to detect and remove noise and outliers. The paradigm of density-based clustering has been proposed to address all of these requirements. Density-based clustering can be considered as a non-parametric method, as it makes no assumptions about the number of clusters or their distribution.

Density-based clusters are connected, dense areas in the data space separated from each other by sparser areas. Furthermore, the density within the areas of noise is assumed to be lower than the density in any of the clusters. Due to their local nature, dense connected areas in the data space can have arbitrary shape. Given an index structure that supports region queries, density-based clusters can be efficiently computed by performing at most one region query per database object. Sparse areas in the data space are treated as noise and are not assigned to any cluster.

It is worth noting that the algorithms that are being referred to in the literature as density-based clustering have various predecessors that have already explored some of the ideas. In particular, Wishart [33] explored ways to avoid the so-called chaining effect in single-link clustering, which is caused by a small number of noisy data points that connect sets of points that should in principle form separate clusters. First, nondense points, that have fewer than k neighbors within a distance of r, are removed. Second, single-link is employed to cluster the remaining points. Finally, nondense points may be allocated to one of the clusters according to some criterion. We would also like to point out the relationship of the paradigms of density-based clustering and mean-shift clustering [8]. The mean-shift procedure is an iterative procedure that replaces each point by the weighted mean of its neighboring points, where the neighborhood and weights are determined by the chosen kernel function, and it converges to the nearest stationary point of the underlying density function. Mean-shift clustering employs the mean-shift procedure as density-estimator. In mean-shift clustering, very narrow kernels create singleton clusters, very wide kernels create one cluster, and intermediate kernels create a natural number of clusters. As opposed to density-based clustering, in mean-shift clustering, the neighborhood membership is weighted (instead of Boolean), the minimum number of points does not need to be specified, and there is no guarantee that clusters are connected.

A density-based clustering algorithm needs to answer several key design questions:

- How is the density estimated?

- How is connectivity defined?
- Which data structures support the efficient implementation of the algorithm?

In the following, we will present the main density-based clustering algorithms and discuss the ways in which they answer these questions.

5.2 DBSCAN

DBSCAN [11] estimates the density by counting the number of points in a fixed-radius neighborhood and considers two points as connected if they lie within each other's neighborhood. A point is called *core point* if the neighborhood of radius *Eps* contains at least *MinPts* points, i.e., the density in the neighborhood has to exceed some threshold. A point q is directly density-reachable from a core point p if q is within the *Eps*-neighborhood of p, and density-reachability is given by the transitive closure of direct density-reachability. Two points p and q are called density-connected if there is a third point o from which both p and q are density-reachable. A cluster is then a set of density-connected points which is maximal with respect to density-reachability. Noise is defined as the set of points in the database not belonging to any of its clusters. The task of density-based clustering is to find all clusters with respect to parameters *Eps* and *MinPts* in a given database.

In the following we provide more formal definitions. Let D be a set (database) of data points. The definition of density-based clusters assumes a distance function $dist(p,q)$ for pairs of points. The Eps-neighborhood of a point p, denoted by $NEps(p)$, is defined by $NEps(p) = \{q \in D | dist(p,q) \leq Eps\}$. A point p is *directly density-reachable* from a point q with respect to *Eps*, *MinPts* if (1) $p \in NEps(q)$ and (2) $|NEps(q)| \geq MinPts$. A point p is *density-reachable* from a point q with respect to *Eps* and *MinPts* if there is a chain of points $p_1, ..., p_n, p_1 = q, p_n = p$ such that p_{i+1} is directly density-reachable from p_i. Density-reachability is a canonical extension of direct density-reachability. Since this relation is not transitive, another relation is introduced. A point p is *density-connected* to a point q with respect to *Eps* and *MinPts* if there is a point o such that both p and q are density-reachable from o with respect to *Eps* and *MinPts*. Figure 5.2 illustrates these concepts. While p is density-reachable from q, q is not density-reachable from p. a and c are density-connected via b.

Intuitively, a density-based cluster is a maximal set of density-connected points. Formally, a cluster C with respect to *Eps* and *MinPts* is a nonempty subset of D satisfying the following two conditions:

1. $\forall p,q$ if $p \in C$ and q is density-reachable from p with respect to Eps and $MinPts$, then $q \in C$. (maximality)

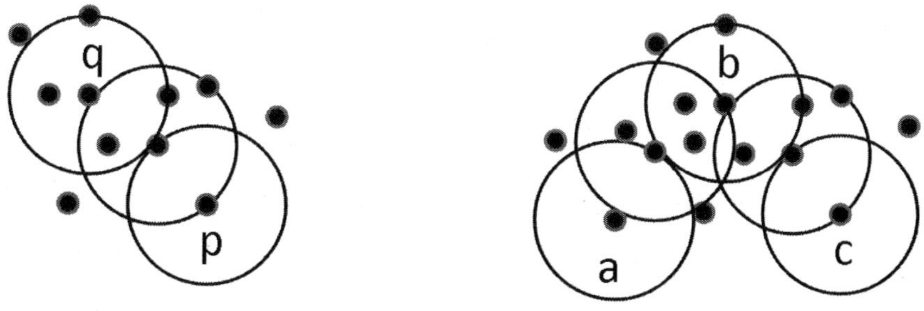

FIGURE 5.2: Density-reachability and connectivity.

2. $\forall p,q \in C$: p is density-connected to q with respect to Eps and $MinPts$. (connectivity)

Let $C_1,\ldots,C_k$ be the clusters of the database D with respect to *Eps* and *MinPts*. Then the *noise* is defined as the set of points in D not belonging to any cluster C_i, i.e., $noise = \{p \in D | p \notin C_i \forall i\}$. Density-based clustering distinguishes three different types of points (see Figure 5.2):

- core points, i.e., points with a dense neighborhood ($|NEps(p)| \geq MinPts$),
- border points, i.e., points that belong to a cluster, but whose neighborhood is not dense, and
- noise points, i.e., points which do not belong to any cluster.

In Figure 5.2, e.g., q and b are core points, and p, a and c are border points.

Density-based clusters have two important properties that allow their efficient computation. Let p be a core point in D. Consider the set O of all points drawn from D, which are density-reachable from p with respect to Eps and $MinPts$. This set O is a cluster with respect to *Eps* and *MinPts*. Let C be a cluster in D. Each point in C is density-reachable from any of the core points of C and, therefore, a cluster C contains exactly the points which are density-reachable from an arbitrary core point of C. Thus, a cluster C with respect to *Eps* and *MinPts* is uniquely determined by any of its core points. This is the foundation of the DBSCAN algorithm.

To find a cluster, DBSCAN starts with an arbitrary database point p and retrieves all points density-reachable from p with respect to *Eps* and *MinPts*, performing region queries first for p and if necessary for p's direct and indirect neighbors. If p is a core point, this procedure yields a cluster with respect to *Eps* and *MinPts*. If p is not a core point, no points are density-reachable from p and DBSCAN assigns p to the noise and applies the same procedure to the next database point. If p is actually a border point of some cluster C, it will later be reached when collecting all the points density-reachable from some core point of C and will then be (re-)assigned to C. The algorithm terminates when all points have been assigned to a cluster or to the noise.

Standard DBSCAN implementations are based on a spatial index such as an R-tree [14] or X-tree [4], which provides efficient support of region queries that retrieve the *Eps*-neighborhood of a given point. In the worst case, DBSCAN performs one region query per database point. This leads to a runtime complexity of $O(nlogn)$ for DBSCAN, where n denotes the number of database points. Unfortunately, spatial indexes degenerate for high-dimensional data, i.e., the performance of region queries degenerates from $O(logn)$ to $O(n)$, and the runtime complexity of DBSCAN becomes $O(n^2)$ for such data. On the other hand, if a grid-based data structure is available that supports $O(1)$ region queries, the runtime complexity of DBSCAN decreases to $O(n)$. Note that a runtime complexity of $O(nlogn)$ is considered to be scalable to large datasets.

An incremental version of DBSCAN can further improve its efficiency in dynamic databases with insertions and deletions. [10] shows that a density-based clustering can be updated incrementally without having to rerun the DBSCAN algorithm on the updated database. It examines which part of an existing clustering is affected by an update of the database and presents algorithms for incremental updates of a clustering after insertions and deletions. Due to the local nature of density-based clusters, the portion of affected database objects tends to be small which makes the incremental algorithm very efficient.

The basic idea of density-based clusters can be generalized in several ways [30]. First, any notion of a neighborhood can be employed instead of a distance-based *Eps*-neighborhood as long as the definition of the neighborhood is based on a predicate $NPred(p,q)$ which is symmetric and reflexive. The neighborhood N of p is then defined as the set of all points q satisfying $NPred(p,q)$. Second, instead of simply counting the elements in a neighborhood we can as well use a more general predicate $MinWeight(N)$ to determine whether the neighborhood N is dense, if *MinWeight* is monotone in N, i.e., if *MinWeight* is satisfied for all supersets of sets that satisfy N. Finally, not only point-like objects but also spatially extended objects such as polygons can be clustered. When clustering polygons, for example, the following predicates are more natural than the *Eps*-neighborhood and the *MinPts* cardinality constraint: $NPred(X,Y)$ iff $intersect(X,Y)$ and $MinWeight(N)$ iff $\sum_{p \in N} population(p) \geq MinPop$. The GDBSCAN algorithm [30] for finding generalized density-based clusters is a straightforward extension of the DBSCAN algorithm.

5.3 DENCLUE

DENCLUE [16] takes another approach to generalize the notion of density-based clusters, based on the concept of *influence functions* that mathematically model the influence of a data point in its neighborhood. The density at some point is estimated by the sum of the influences of all data points. A point is said to be *density-attracted* to a so-called density-attractor, if they are connected through a path of high-density points. An influence function should be symmetric, continuous, and differentiable, and typical examples of influence functions are square wave functions or Gaussian functions. The *density function* at a point x is computed as the sum of the influence functions of all data points at point x. *Density-attractors* are points that correspond to local maxima of the density function. A point p is density-attracted to a density-attractor q if q can be reached from p through a path of points that lie within a distance of *Eps* from each other in the direction of the gradient. An *arbitrary-shape cluster* for a set of density-attractors X is then defined as the set of all points that are density-attracted to one of the density-attractors x from X where the density function at x exceeds a threshold ξ. In addition, all pairs of density-attractors need to be connected to each other via paths of points whose density meets the same threshold.

More precisely, the *influence function* of a data point $y \in F^d$ is a function $f_B^y(x)$ defined in terms of a basic influence function f_B, i.e., $f_B^y(x) = f_B(x,y)$. A simple example of an influence function is the Square Wave Influence Function:

$$f_B(x,y) = \begin{array}{l} 1 \ \text{if } d(x,y) \leq \sigma, \\ 0 \ \text{otherwise.} \end{array}$$

Given a database of points $D = \{x_1, \ldots, x_N\} \subset F^d$, the *density function* is defined as $f_B^D(x) = \sum_{i=1}^N f_B^{x_i}(x)$. The gradient of the density function is defined as $\Delta f_B^D(x) = \sum_{i=1}^N (x_i - x) f_B^{x_i}(x)$. A point x^* is called a *density-attractor*, if x^* is a local maximum of the density function $f_B^D(x)$. A point x is *density-attracted* to density-attractor x^*, if there is a sequence of points $x^k, d(x^k, x^*) \leq \varepsilon$ for some distance function d, with $x^i = x^{i-1} + \delta \frac{\Delta f_B^D(x^{i-1})}{\|\Delta f_B^D(x^{i-1})\|}$.

Given a set of density-attractors X, an *arbitrary-shape cluster* with respect to σ and ξ is a subset $C \subseteq D$, where

1. $\forall x \in C \; \exists x^* \in X : f_B^D(x^*) \geq \xi$ and x is density-attracted to x^*, and

2. $\forall x_1^*, x_2^* \in X \; \exists$ path $P \subset F^d$ from x_1^* to x_2^* with $\forall p \in P : f_B^D(p) \geq \xi$.

The parameter σ, employed by the basic influence function, determines the reach of the influence of a point, while parameter ξ specifies when a density-attractor is significant. Note that the DENCLUE clusters become identical to the DBSCAN clusters when choosing the Square Wave Influence Function with $\sigma = Eps$ and $\xi = MinPts$.

The efficient implementation of the DENCLUE algorithm is based on the observation that most data points do not contribute to the density function at any given point of the data space. This can be exploited by computing only a local density function, while guaranteeing tight error bounds. To efficiently access neighboring points, a so-called *map* data structure is created, a d-dimensional grid structure of grid length 2σ. Only grid cells that actually contain points are determined and are mapped to one-dimensional keys, which are indexed, e.g., in a B^+-tree. To determine the density-attractors for each point in a grid cell, a hill-climbing procedure based on the local density function and its gradient is used. After determining the density-attractor x^* for a point x, the point x is assigned to the cluster including x^*. Some heuristics allows the algorithm to assign nearby points without having to apply the hill-climbing procedure.

5.4 OPTICS

In many real-life databases the intrinsic cluster structure cannot be characterized by global density parameters, and very different local densities may be needed to reveal clusters in different regions of the data space. In principle, one could apply a density-based clustering algorithm with different parameter settings, but there are an infinite number of possible parameter values. The basic idea of the OPTICS algorithm [3] to address this challenge is to create not one explicit clustering, but to produce a novel cluster-ordering of the database points with respect to its density-based clustering structure containing the information about every clustering level of the data set up to a generating distance *Eps*. This ordering is visualized graphically to support interactive analysis of the cluster structure.

For a constant *MinPts*-value, density-based clusters with respect to a higher density (i.e., a lower value for *Eps*) are completely contained in clusters with respect to a lower density (i.e., a higher value for *Eps*). Consequently, the DBSCAN algorithm could be extended to simultaneously cluster a database for several *Eps* values. However, points which are density-reachable with respect to the lowest *Eps* value would always have to be processed first to guarantee that clusters with respect to higher density are finished first. OPTICS works in principle like an extended DBSCAN algorithm for an infinite number of distance parameters Eps_i which are smaller than a generating distance *Eps*. The only difference is that it does not assign cluster memberships, but stores the order in which the points are processed (the clustering order) and the following two pieces of information which would be used by an extended DBSCAN algorithm to assign cluster memberships. The *core-distance* of a point p is the smallest distance Eps' between p and a point in its *Eps*-neighborhood such that p would be a core point with respect to Eps' if this neighbor is contained in $NEps(p)$. The *reachability-distance* of a point p with respect to another point o is the smallest distance such that p is directly density-reachable from o if o is a core point. The clustering structure of a data set can be visualized by a *reachability plot* (see Figure 5.3) that shows the reachability-distance values r for all points sorted according to the clustering order. Valleys in the reachability plot correspond to clusters, which can be hierarchically nested.

5.5 Other Algorithms

The work in [31] introduces a statistical, grid-based index structure called STING (Statistical INformation Grid) to efficiently process region queries on databases of points. STING divides the data space into rectangular cells at different levels of resolution, and these cells form a tree structure. A cell at a high level contains a number of cells of the next lower level. The following statistical information is calculated and stored for each cell: (1) the number of points in the cell and (2) for each attribute, the mean, standard deviation, minimum value, and maximum value of the attribute for the points in this cell, as well as the type of distribution of these attribute values. To process a region

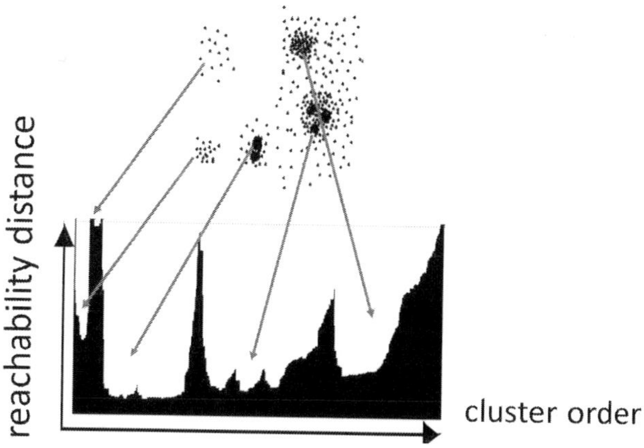

FIGURE 5.3: Reachability plot for sample dataset.

query, the STING index is explored, starting with the root. At the current cell, the likelihood that this cell is relevant to the query at some confidence level is calculated using the statistical information of this cell. Only children of likely relevant cells are recursively explored. The algorithm terminates when the lowest level of the index structure has been searched. The runtime complexity of a region query is $O(K)$, where K is the number of grid cells at the lowest level. Usually, $K << N$, where N is the number of data points. While STING is more efficient than deterministic index structures discussed above, its probabilistic nature implies a loss of accuracy in query processing. STING can provide efficient support for density-based clustering algorithms which employ region queries as basic operations.

The restriction of k-means-like algorithms to find spherical clusters can be traced back to the representation of a cluster by only one point. Density-based clustering algorithms, on the other hand, represent a cluster by the set of all of its points, which enables them to discover arbitrary shape clusters. CURE (Clustering Using REreprosentatives) [12] adopts a compromise, representing each cluster by a certain fixed number of points that are generated by selecting well- scattered points from the cluster and shrinking them toward the center of the cluster by a specified fraction. Having more than one representative point per cluster allows CURE to adjust well to the geometry of nonspherical shapes, and the shrinking helps to dampen the effects of outliers. In order to scale to large databases, CURE employs a combination of random sampling and partitioning. First, a random sample drawn from the data set is partitioned, and each partition is partially clustered. Then the partial clusters are clustered to obtain the final clusters.

The clustering algorithm CHAMELEON [21] also aims to discover arbitrary shape clusters. As in CURE, the approach is not density-based, but is distance-based. CHAMELEON is a hierarchical algorithm. The key idea is to consider both the inter-connectivity and closeness of clusters when identifying the most similar pair of clusters. These two clusters are merged only if the inter-connectivity and closeness between the two clusters are comparable to the internal inter-connectivity of the clusters and closeness of points within the clusters. This is different from the CURE algorithm which ignores the information about the aggregate inter-connectivity of points in two clusters. CHAMELEON finds the clusters in two phases. During the first phase, CHAMELEON uses a graph partitioning algorithm to cluster the data items into several relatively small subclusters. In the second phase, it creates clusters by applying hierarchical clustering to the subclusters from the first phase.

While DBSCAN can find clusters of arbitrary shapes, it cannot handle data with clusters of different densities, due to a single density threshold. OPTICS (see Section 5.4) provides an approach to address this challenge, but it only visualizes the cluster structure without actually determining clusters. [9] presents an approach based on shared nearest neighbors (SNN). The SNN similarity of two points is defined as the cardinality of the intersection of the sets of the k nearest neighbors of the two points. The SNN graph is created by linking a pair of points if and only if they have each other in their k-nearest neighbor lists. The shared nearest neighbor graph keeps the links in regions of uniform density and breaks links in transition regions, i.e., it keeps links in a region of any density, high or low, as long as the region has relatively uniform density. This property supports the detection of clusters of different densities. A point is called core point if the number of points that have an SNN similarity of *Eps* or greater is at least *MinPts*. With these definitions, the DBSCAN algorithm (see Section 5.2) can be applied.

5.6 Subspace Clustering

In high-dimensional datasets, clusters tend to reside in lower-dimensional subspaces rather than in the full-dimensional space, which has motivated the task of subspace clustering, as discussed in Chapter 9. The CLIQUE algorithm [2], a prominent subspace clustering algorithm, is density-based. It discretizes the data space through a grid and estimates the density by counting the number of points in a grid cell. Two grid cells are considered to be connected if they share some edge or face. A subspace cluster is then a set of neighboring dense cells in an arbitrary subspace. Note that CLIQUE discovers not only the subspace clusters, but also some minimal descriptions of the clusters. Figure 5.4 illustrates a two-dimensional dataset without full-space clusters, that contains a one-dimensional subspace cluster in the "age" dimension ($18 \leq age \leq 24$).

More formally, let $S = A_1 \times A_2 \times \cdots \times A_d$ be a d-dimensional numerical space, and $D = v_1, v_2, \ldots, v_m$ a set of d-dimensional points. The data space S is partitioned into units by partitioning every dimension into ξ intervals of equal length, and each unit u is the intersection of one interval from each dimension. The *selectivity* of a unit is defined to be the fraction of total data points contained in the unit. We call a unit u *dense* if *selectivity*(u) is greater than density threshold τ. Similarly, units are defined in all subspaces of the original d-dimensional space. Two k-dimensional

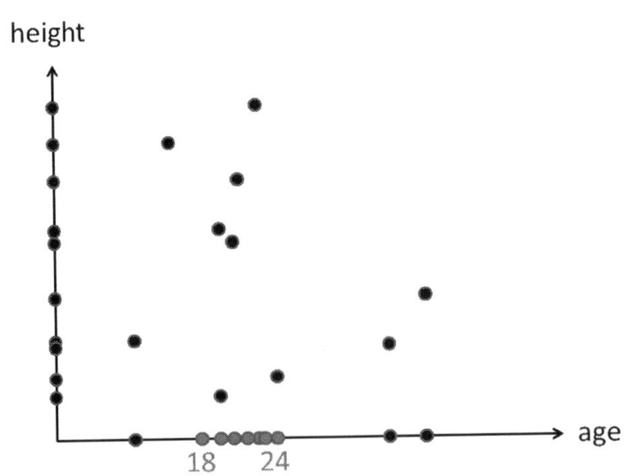

FIGURE 5.4: Sample dataset with subspace clusters.

units u_1, u_2 are connected if they have a common face or if there exists another k-dimensional unit u_3 such that u_1 is connected to u_3 and u_2 is connected to u_3. A *cluster* is then defined as a maximal set of connected dense units in k dimensions. A *region* in k dimensions is an axis-parallel rectangular k-dimensional set, expressed as a DNF expression on intervals of the domains A_i. A region R contained in a cluster C is said to be *maximal* if no proper superset of R is contained in C. A *minimal description* of a cluster is a nonredundant covering of the cluster with maximal regions. Given a set of data points and the input parameters, ξ and τ, the task of subspace clustering is to find all clusters in all subspaces of the original data space together with a minimal description of each cluster.

The CLIQUE algorithm decomposes the task of subspace clustering into three subtasks:

1. identification of subspaces that contain clusters,

2. identification of clusters, and

3. generation of minimal descriptions.

We discuss only the first subtask, which is the most challenging. Its algorithmic solution is based on the following monotonicity property: If a set of points S is a cluster in a k-dimensional space, then S is also part of a cluster in any $(k-1)$-dimensional projections of this space. Due to this property, dense units (and therefore subspaces with clusters) can be efficiently discovered in a level-wise manner, starting with 1-dimensional units, and extending dense units by one dimension at every level. The determination of the density of all candidate units at a given level requires one database scan. The algorithm terminates when no more candidates are generated at the current level.

If a dense unit exists in k dimensions, then all of its projections in any of the $O(2k)$ subsets of the k dimensions are also dense. The running time of the CLIQUE algorithm is therefore exponential in the highest dimensionality of any dense unit. While it can be shown that the candidate generation procedure produces the minimal number of candidates that can guarantee that all dense units will be found, a heuristic inspired by the Minimal Description Length principle was proposed to prune all subspaces and their dense units that are not "interesting." To measure the interestingness of a subspace, its coverage is computed, i.e., the fraction of the database that is covered by the dense units in that subspace. Only subspaces with large coverage are selected, and the remaining ones are pruned.

In CLIQUE, as in all grid-based approaches, the quality of the results crucially depends on the appropriate choice of the number and width of the partitions and grid cells. To address this problem, [17] suggests the OptiGrid method. Good cutting planes should partition the dataset in areas of low density and should discriminate clusters as much as possible. Guided by these requirements, the OptiGrid algorithm works as follows. In each step, it partitions the dataset into multiple subsets. A cutting plane has to be orthogonal to at least one projection. The density at a cutting plane is bound by the density of the orthogonal projection of the cutting plane in the projected space. q cutting planes with minimum density are chosen. Partitions containing at least one cluster are processed recursively. The recursion stops if no good cutting plane can be found for a given partition. It can be shown analytically that the OptiGrid algorithm finds all center-defined clusters, which roughly correspond to clusters generated by a Gaussian distribution.

To avoid the drawbacks of grid-based approaches, researchers have investigated approaches similar to DBSCAN that define the neighborhood of a point through region queries. In principle, one could apply the DBSCAN algorithm as is to all subspaces to detect all density-based subspace clusters, but this approach is infeasible due to the exponential number of subspaces. SUBCLU [26] discovers the same clusters as the naive application of DBSCAN to all subspaces, but achieves an efficient solution by employing the following monotonicity property: while density-based clusters are not anti-monotone, density-connected sets are, i.e., they are density-connected in any subset of the set of their dimensions.

More precisely, let A denote the set of all attributes or dimensions of dataset D. Different from

DBSCAN, the *Eps*-neighborhood of a point is defined relative to a subspace $S \subseteq A$ based on the distance of two points being projected to the dimensions of S.

$$NEps(p) = \{q \in D | dist(\pi_s(p), \pi_s(q)) \leq Eps\}$$

All other DBSCAN notation can be transferred to subspaces in a straightforward way. A density-connected set in subspace $S \subseteq A$ is also a density-connected set in any subspace $T \subseteq S$. The SUBCLU algorithm starts by applying DBSCAN to each 1-dimensional subspace. Let S_k denote the set of all k-dimensional subspaces containing at least one cluster. For each detected cluster with k (initially, $k = 1$) dimensions, it is checked whether (part of) this cluster still exists in higher-dimensional subspaces. Therefore, candidate $k+1$ dimensional subspaces are generated by joining two k-dimensional subspaces that have $k-1$ attributes in common. Candidate subspaces having at least one k-dimensional subspace not included in S_k are pruned. The algorithm terminates when no more clusters are detected at the current dimensionality.

As DBSCAN is applied to different subspaces, an index structure for the full-dimensional data space A is not applicable. Instead, inverted lists are employed to support the region queries that retrieve the *Eps*-neighborhood of a point. Inverted lists provide $O(logn)$ region queries for individual attributes, where n denotes the number of data points. For region queries on multiple attributes, region queries are performed separately on each of the attributes, and these intermediate results are intersected to obtain the final result.

Many algorithms for subspace clustering or projected clustering of high-dimensional data have been proposed. However, the relevant densities in different subspaces may vary greatly. In particular, the threshold density that can distinguish between different clusters in one subspace may be equal to the density of the noise in another subspace. To address this issue, visualization-based, interactive methods have been presented as a way to involve the user with his cognitive abilities and domain knowledge. Hinneburg et al. [18] develop the visual tool HD-Eye that enables the user to interact with clusters in lower-dimensional subspaces. Predefined projections (subspaces) are visualized and combined iteratively by the user. Aggarwal [1] presents the IPCLUS system providing more active guidance to the user. In each iteration, the system proposes a well-polarized projection to the user, i.e., a projection in which several clusters can be clearly distinguished. A visual profile of the data density in that projection is presented to the user, who then manually separates the clusters in the subspace. The system terminates as soon as most points have been assigned to at least one cluster in one of the subspaces considered.

5.7 Clustering Networks

Most density-based clustering algorithms, as well as other clustering algorithms, work on data points or records with a fixed number of attributes having one value each. However, network data is becoming increasingly common, e.g., social networks, the web, or biological interaction networks. Networks can naturally be modeled as graphs. Let $G = (V, E)$ be a graph with a set of vertices V and a set of edges E. Vertices represent objects, and edges represent relationships between pairs of objects. In this section, we discuss density-based clustering methods for networks and refer to the Chapter 17 on Network Clustering for other methods.

The SCAN algorithm [34] transfers the concepts of DBSCAN from point data to network data, assuming that network clusters are dense graph components. SCAN discovers not only clusters, but also hubs connecting several clusters and outliers not belonging to any cluster. Figure 5.5 shows a sample network dataset with two clusters (vertices 1 to 6, and vertices 7 to 10), one hub (vertex 11) and one outlier (vertex 12).

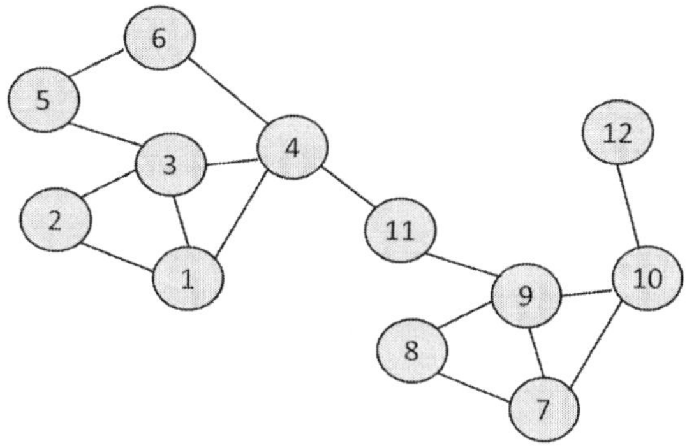

FIGURE 5.5: Sample network dataset.

Intuitively, vertices sharing a lot of neighbors should belong to the same cluster. To formalize this intuition, a similarity function for pairs of vertices v and w, denoted by $sim(v,w)$, is defined as follows based on the intersection of their sets of neighbors:

$$sim(v,w) = \frac{|\Gamma(v) \cap \Gamma(w)|}{\sqrt{|\Gamma(v)| \cdot |\Gamma(w)|}},$$

where $\Gamma(v)$ denotes the set of all (direct) neighbors of vertex v, i.e., $\Gamma(v) = \{w|(v,w) \in E\} \cup \{v\}$. The *ε-neighborhood* of a vertex v is given by the set of all neighbors whose similarity exceeds the threshold of ε, i.e.,

$$N_{\varepsilon}(v) = \{w | w \varepsilon \Gamma(v) \wedge sim(v,w) \geq \varepsilon\}.$$

A vertex v is called a *core*, if its ε-neighborhood has a cardinality of at least μ. Based on these definitions, the concepts of direct structure-reachability, structure-reachability, structure-connectivity, and (structure-connected) cluster with respect to ε and μ can be introduced in analogy to the DBSCAN concepts.

If a vertex is not a member of any cluster, it is either a hub or an outlier, depending on its neighborhood. It is a hub, if it has neighbors in at least two different clusters, and it is an outlier otherwise, i.e., if all neighbors belong to the same cluster or not to any cluster.

The SCAN algorithm is completely analogous to the DBSCAN algorithm. The crucial operation is the retrieval of the neighborhood of a given vertex, which can be efficiently supported using an *adjacency list*, a data structure where each vertex is associated with a list of directly neighboring vertices. With this data structure, the cost of a neighborhood query is proportional to the number of neighbors, that is, the degree of the query vertex. SCAN performs at most one neighborhood query per vertex. Therefore, the total running time of SCAN is $O(m)$, where m denotes the number of all edges, which scales well to large datasets.

SCAN, like most other methods of network clustering such as graph partitioning and quasi-clique finding work on graph data only. However, in many applications more informative graphs are given, where feature vectors are associated with vertices representing object properties such as demographic attributes of customers and expression data of genes. Often features and edges contain complementary information, i.e., neither can the relationships be derived from the attributes nor vice-versa. In such scenarios the simultaneous use of both data types promises more meaningful and accurate results. Figure 5.6 shows an example of a social network of computer science students

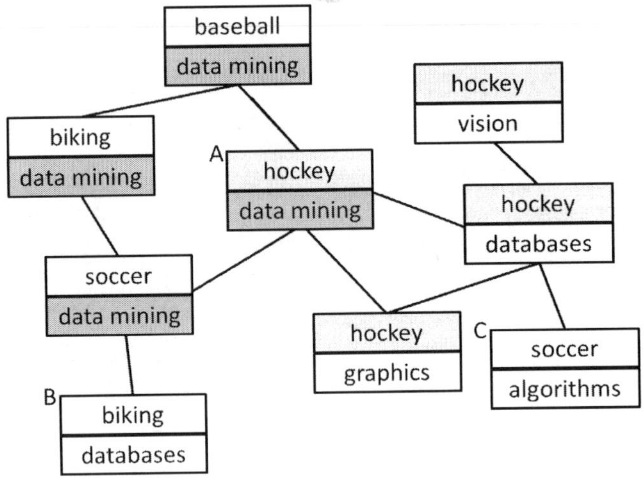

FIGURE 5.6: Sample social network dataset with feature vectors.

where students are associated with two features, their favorite sports and their research areas. The first cluster consists of hockey players, united by their favorite sports, and the second one contains students who share data mining as their research area. Note that student A belongs to both clusters, while students B and C are not part of any of the clusters, because they do not share any interest (research or sports) with their friends.

Integrating the concepts of dense subgraphs and subspace clusters, Moser et al. [28] introduced the problem of finding cohesive patterns. A *cohesive pattern* is defined as a subgraph which

1. is connected,
2. has a density exceeding the threshold of α, and
3. has homogeneous values in at least d features, for example, the values of each of these features may have at most a specified variance within the vertices of the pattern.

Different from graph partitioning methods and similar to frequent pattern mining methods, cohesive patterns can overlap and do not have to cover the entire dataset. Furthermore, the number of patterns does not need to be specified in advance. Integrating constraints on the feature vectors reduces the number of patterns substantially and adds more meaning to the identified patterns. In order to reduce the potentially large number of cohesive patterns, the search is restricted to finding only maximal cohesive patterns, i.e., those whose supergraphs do not form cohesive patterns.

The problem of discovering all maximal cohesive patterns is NP-hard, but it becomes tractable in many practical cases by exploiting the following monotonicity properties. A constraint is called *anti-monotone* if the satisfaction of the constraint by a pattern of size k implies the satisfaction of the constraint by all subpatterns of size $k-1$, and a constraint is called *loose anti-monotone* if the implication holds for at least one subpattern of size $k-1$. The homogeniety constraint is anti-monotone. The density constraint and the connectivity constraint are loose anti-monotone. Finally, the simultaneous satisfaction of all three constraints is loose anti-monotone for $\alpha \geq 0.5$. The CoPaM algorithm enumerates subgraphs in a levelwise, bottom-up fashion, starting with subgraphs of size 2. At level k, subgraphs of size k that do not satisfy all of the three constraints are pruned, while the remaining graphs are extended by one neighboring vertex to form candidates for the next level $k+1$. The algorithm terminates at a level where no cohesive patterns are discovered.

To further reduce the number of cohesive patterns, Güennemann et al. [13] introduce a notion of redundancy among such patterns, based on their density, size, and number of relevant dimensions, and present the Gamer algorithm for discovering redundancy-free sets of patterns.

While all of the above methods cluster static snapshots of a network, [22] proposes a method for clustering dynamic networks, employing the framework of temporal smoothness. This framework assumes that the structure of clusters does not significantly change in a very short time and tries to smooth clusterings over time. The framework aims for a trade-off between the snapshot quality, i.e., the strength of the cluster structure in one snapshot, and the history quality, i.e., the similarity of the clusterings of a snapshot and the previous snapshot. A density-based approach similar to that of SCAN is adopted for clustering the static snapshots, determining smoothed local clusters of high quality using a cost-embedding technique and optimal modularity.

5.8 Other Directions

A data stream consists of a sequence of data points that arrive continuously and rapidly and are too large to be stored permanently, so that an algorithm can take only one look at each data point to analyze the data stream. Stream data is becoming available in many applications, e.g., sensor data and click stream data. For density-based clustering of data streams, [6] introduces the concepts of core-micro-clusters, potential core-micro-clusters and outlier micro-clusters to maintain the relevant statistical information. A novel pruning strategy is designed based on these concepts, which allows accurate clustering in the context of limited memory. [7] presents the D-Stream algorithm, which uses an online component to map input data into a grid and an offline component to compute the grid cell density and cluster the cells based on the density. The algorithm adopts a density decaying technique to capture the dynamic changes of a data stream. Sporadic grid cells, resulting from outliers in the data stream, are detected and removed.

Recently, uncertain data has become more common in scenarios such as mobile services, where the locations of moving objects are transmitted at discrete time stamps and can only be estimated with some uncertainty between these time stamps. Essentially, attribute values are generalized to probability distributions over the attribute domain. For density-based clustering of uncertain data, the FDBSCAN algorithm [24] introduces a probabilistic concept of density-based clusters generalizing the deterministic concept of DBSCAN. In particular, the distance function is replaced by a distance density function and a distance distribution function for pairs of uncertain data points, and the Boolean core point property is replaced by a core point probability. The main challenge of the implementation is the efficient computation of the core point probability. A Monte-Carlo sampling method is developed in order to efficiently approximate this probability. A similar generalization (FOPTICS) has been proposed [25] for OPTICS, supporting the hierarchical density-based clustering of uncertain data. See Chapter 18 on clustering uncertain data for other clustering methods.

Januzaj et al. [19] investigate density-based clustering of distributed databases. The databases are clustered locally, and suitable representatives from the clusters are determined. These representatives are sent to a global server where the overall clustering is computed based on the local representatives. In [20], the same authors improve the scalability of this approach, allowing the user to specify the trade-off between clustering quality and the number of transmitted objects from the different local databases to the global server.

Various researchers have explored the use of constraints in the context of density-based clustering. Wang et al. [32] propose the spatial clustering method DBRS+ to cluster spatial data in the presence of both obstacles and facilitators. In [29], the authors enhance the density-based algorithm

DBSCAN with *Must-Link* and *Cannot-Link* constraints for pairs of points that must or cannot belong to the same cluster. Lelis and Sander [27] demonstrate how labeled objects can be employed to help the DBSCAN algorithm detecting suitable density parameters.

Outliers are observations that deviate significantly from the other observations. The task of outlier detection is quite closely related to that of clustering, since points in a cluster can be considered normal cases and points not assigned to any cluster can be considered outliers. However, clustering algorithms treat outliers only as by-products and do not focus on their discovery. Knorr and Ng [23] introduce the notion of distance-based outliers, i.e., points for which at least p percent of the database points have a distance larger than D. It is shown that this definition generalizes the standard definition of outliers from a Gaussian distribution. We would like to point out that the concept of distance-based outliers can also be understood as density-based: points are considered outliers if their neighborhood of radius D does not contain at least a fraction of $1-p$ of all database points. Distance-based outliers are global in the sense that the number of points that should be further away than D is specified as a fraction of the database cardinality. Breunig et al. [5] motivate the importance of local outliers, points that deviate strongly from their local neighborhoods, particularly with respect to the densities of the neighborhoods. Different from other methods, an outlier factor is introduced measuring the degree to which a point p is an outlier. Adopting the notions of OPTICS (see Section 5.4), the local outlier factor is defined as the average of the ratio of the local reachability-density of p and those of the *MinPts* nearest neighbors of p. Similar to DBSCAN and distance-based outlier detection, the detection of local outliers requires one region query per database point, i.e., the runtime complexity is $O(n \cdot \log(n))$ if an effective index structure is available and $O(n^2)$ otherwise.

5.9 Conclusion

The paradigm of density-based clustering has been proposed to address the requirements of (1) discovery of arbitrary shape clusters, (2) scalability to large databases, and (3) the ability to detect and remove noise and outliers. Density-based clusters are connected, dense areas in the data space separated from each other by sparser areas. Furthermore, the density within the areas of noise is assumed to be lower than the density in any of the clusters. Due to their local nature, dense connected areas in the data space can have arbitrary shape. Given an index structure that supports region queries, density-based clusters can be efficiently computed by performing at most one region query per data point. Sparse areas in the data space are treated as noise and are not assigned to any cluster. Different from clustering algorithms that optimize a certain objective function, the number of clusters does not need to be specified by the user.

In this chapter, we have reviewed the major concepts and algorithms for density-based clustering. In particular, we have discussed the key aspects of density estimation, connectivity definition, and data structures for efficient implementation. We have also covered advanced density-based approaches to subspace clustering and the clustering of network data as well as the clustering of data streams and uncertain data. Density-based clustering has been successfully applied in many practical applications. The density-based clustering algorithms DBSCAN and OPTICS have been implemented in WEKA [15], the leading public domain data mining toolkit.

Bibliography

[1] Charu C. Aggarwal. A human-computer cooperative system for effective high dimensional clustering. In *KDD*, pages 221–226, 2001.

[2] Rakesh Agrawal, Johannes Gehrke, Dimitrios Gunopulos, and Prabhakar Raghavan. Automatic subspace clustering of high dimensional data for data mining applications. In *SIGMOD Conference*, pages 94–105, 1998.

[3] Mihael Ankerst, Markus M. Breunig, Hans-Peter Kriegel, and Jörg Sander. OPTICS: Ordering points to identify the clustering structure. In *SIGMOD Conference*, pages 49–60, 1999.

[4] Stefan Berchtold, Daniel A. Keim, and Hans-Peter Kriegel. The X-tree: An index structure for high-dimensional data. In *VLDB*, pages 28–39, 1996.

[5] Markus M. Breunig, Hans-Peter Kriegel, Raymond T. Ng, and Jörg Sander. LOF: Identifying density-based local outliers. In *SIGMOD Conference*, pages 93–104, 2000.

[6] Feng Cao, Martin Ester, Weining Qian, and Aoying Zhou. Density-based clustering over an evolving data stream with noise. In *Proceedings of the SIAM Conference on Data Mining*, pages 328–339, 2006.

[7] Yixin Chen and Li Tu. Density-based clustering for real-time stream data. In *KDD*, pages 133–142, 2007.

[8] Dorin Comaniciu and Peter Meer. Mean shift: A robust approach toward feature space analysis. *IEEE Trans on Pattern Analysis and Machine Intelligence*, 24(5):603–619, 2002.

[9] Levent Ertöz, Michael Steinbach, and Vipin Kumar. Finding clusters of different sizes, shapes, and densities in noisy, high dimensional data. In *Proceedings of the Second SIAM Conference on Data Mining*, pages 47–58, 2003.

[10] Martin Ester, Hans-Peter Kriegel, Jörg Sander, Michael Wimmer, and Xiaowei Xu. Incremental clustering for mining in a data warehousing environment. In *VLDB*, pages 323–333, 1998.

[11] Martin Ester, Hans-Peter Kriegel, Jörg Sander, and Xiaowei Xu. A density-based algorithm for discovering clusters in large spatial databases with noise. In *KDD*, pages 226–231, 1996.

[12] Sudipto Guha, Rajeev Rastogi, and Kyuseok Shim. CURE: An efficient clustering algorithm for large databases. *Inf. Syst.*, 26(1):35–58, 2001.

[13] Stephan Günnemann, Ines Färber, Brigitte Boden, and Thomas Seidl. Subspace clustering meets dense subgraph mining: A synthesis of two paradigms. In *ICDM*, pages 845–850, 2010.

[14] Antonin Guttman. R-trees: A dynamic index structure for spatial searching. In *SIGMOD Conference*, pages 47–57, 1984.

[15] Mark Hall, Eibe Frank, Geoffrey Holmes, Bernhard Pfahringer, Peter Reutemann, and Ian H. Witten. The WEKA data mining software: An update. *SIGKDD Explorations*, 11(1):10–18, 2009.

[16] Alexander Hinneburg and Daniel A. Keim. An efficient approach to clustering in large multimedia databases with noise. In *KDD*, pages 58–65, 1998.

[17] Alexander Hinneburg and Daniel A. Keim. Optimal grid-clustering: Towards breaking the curse of dimensionality in high-dimensional clustering. In *VLDB*, pages 506–517, 1999.

[18] Alexander Hinneburg, Daniel A. Keim, and Markus Wawryniuk. HD-Eye: Visual mining of high-dimensional data. *IEEE Computer Graphics and Applications*, 19(5):22–31, 1999.

[19] Eshref Januzaj, Hans-Peter Kriegel, and Martin Pfeifle. DBDC: Density based distributed clustering. In *EDBT*, pages 88–105, 2004.

[20] Eshref Januzaj, Hans-Peter Kriegel, and Martin Pfeifle. Scalable density-based distributed clustering. In *PKDD*, pages 231–244, 2004.

[21] George Karypis, Eui-Hong Han, and Vipin Kumar. Chameleon: Hierarchical clustering using dynamic modeling. *IEEE Computer*, 32(8):68–75, 1999.

[22] Min-Soo Kim and Jiawei Han. A particle-and-density based evolutionary clustering method for dynamic networks. *PVLDB*, 2(1):622–633, 2009.

[23] Edwin M. Knorr and Raymond T. Ng. Algorithms for mining distance-based outliers in large datasets. In *VLDB*, pages 392–403, 1998.

[24] Hans-Peter Kriegel and Martin Pfeifle. Density-based clustering of uncertain data. In *KDD*, pages 672–677, 2005.

[25] Hans-Peter Kriegel and Martin Pfeifle. Hierarchical density-based clustering of uncertain data. In *ICDM*, pages 689–692, 2005.

[26] Peer Kröger, Hans-Peter Kriegel, and Karin Kailing. Density-connected subspace clustering for high-dimensional data. In *SDM*, 2004.

[27] Levi Lelis and Jörg Sander. Semi-supervised density-based clustering. In *ICDM*, pages 842–847, 2009.

[28] F. Moser, R. Colak, A. Rafiey, and M. Ester. Mining cohesive patterns from graphs with feature vectors. In *SIAM Data Mining Conference (SDM)*, pages 593–604, 2009.

[29] Carlos Ruiz, Myra Spiliopoulou, and Ernestina Menasalvas Ruiz. C-DBSCAN: Density-based clustering with constraints. In *RSFDGrC*, pages 216–223, 2007.

[30] Jörg Sander, Martin Ester, Hans-Peter Kriegel, and Xiaowei Xu. Density-based clustering in spatial databases: The algorithm GDBSCAN and its applications. *Data Mining and Knowledge Discovery*, 2(2):169–194, 1998.

[31] Wei Wang, Jiong Yang, and Richard R. Muntz. STING: A statistical information grid approach to spatial data mining. In *VLDB*, pages 186–195, 1997.

[32] Xin Wang, Camilo Rostoker, and Howard J. Hamilton. Density-based spatial clustering in the presence of obstacles and facilitators. In *PKDD*, pages 446–458, 2004.

[33] David Wishart. Mode analysis: A generalization of nearest neighbor which reduces chaining effects. In *Numerical Taxonomy*. Academic Press, 1969.

[34] Xiaowei Xu, Nurcan Yuruk, Zhidan Feng, and Thomas A. J. Schweiger. SCAN: A structural clustering algorithm for networks. In *KDD*, pages 824–833, 2007.

Chapter 6

Grid-Based Clustering

Wei Cheng
University of North Carolina at Chapel Hill
Chapel Hill, NC 27599
chengw02@gmail.com

Wei Wang
University of California, Los Angeles
Los Angeles, CA 90095
weiwang@cs.ucla.edu

Sandra Batista
Duke University
Durham, NC 27710
sandraleebatitsa@yahoo.com

6.1 Introduction

Grid-based clustering algorithms are efficient in mining large multidimensional data sets. These algorithms partition the data space into a finite number of cells to form a grid structure and then form clusters from the cells in the grid structure. Clusters correspond to regions that are more dense in data points than their surroundings. Grids were initially proposed by Warnekar and Krishna [30] to organize the feature space, e.g., in GRIDCLUS [25], and increased in popularity after STING [28], CLIQUE [1], and WaveCluster [27] were introduced. The great advantage of grid-based clustering is a significant reduction in time complexity, especially for very large data sets. Rather than clustering the data points directly, grid-based approaches cluster the neighborhood surrounding the data points represented by cells. In most applications since the number of cells is significantly smaller than the number of data points, the performance of grid-based approaches is significantly improved. Grid-based clustering algorithms typically involve the following five steps [9, 10]:

1. Creating the grid structure, i.e., partitioning the data space into a finite number of cells.
2. Calculating the cell density for each cell.
3. Sorting of the cells according to their densities.
4. Identifying cluster centers.
5. Traversal of neighbor cells.

Since cell density often needs to be calculated in order to sort cells and select cluster centers, most grid-based clustering algorithms may also be considered density-based. Some grid-based clustering algorithms also combine hierarchical clustering or subspace clustering in order to organize cells based on their density. Table 6.1 lists several representative grid-based algorithms which also use hierarchical clustering or subspace clustering.

Grid-based clustering is susceptible to the following data challenges:

1. *Non-Uniformity*: Using a single inflexible, uniform grid may not be sufficient to achieve desired clustering quality or efficiency for highly irregular data distributions.
2. *Locality*: If there are local variations in the shape and density of the distribution of data points, the effectiveness of grid-based clustering is limited by predefined cell sizes, cell borders, and the density threshold for significant cells.
3. *Dimensionality*: Since performance depends on the size of the grid structures and the size of grid structures may increase significantly with more dimensions, grid-based approaches may not be scalable for clustering very high-dimensional data. In addition, there are aspects of the "curse of dimensionality" including filtering noise and selecting the most relevant attributes that are increasingly difficult with more dimensions in a grid-based clustering approach.

To overcome the challenge of nonuniformity, *adaptive* grid-based clustering algorithms that divide the feature space at multiple resolutions, e.g., AMR [15] and MAFIA [21], were proposed. The varying grid sizes can cluster data well with nonuniform distributions. For example, as illustrated

TABLE 6.1: Grid-Based Algorithms That Use Hierarchical Clustering or Subspace Clustering

hierarchical clustering	GRIDCLUS, BANG-clustering, AMR, STING, STING+
subspace clustering	MAFIA, CLIQUE, ENCLUS

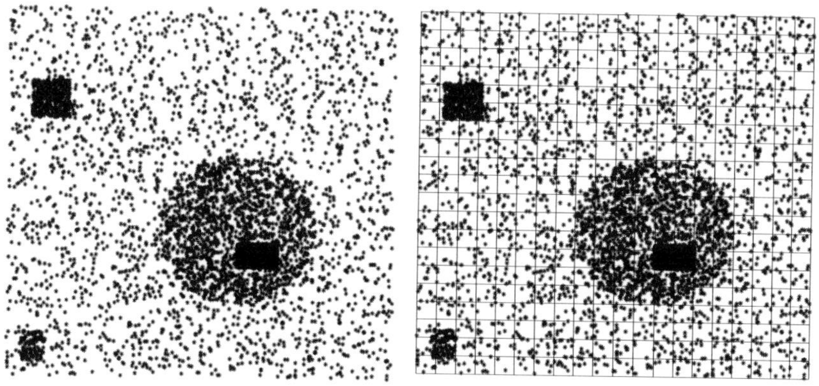

(a) Original data (b) Uniform grid

FIGURE 6.1: Nonuniformity example with nested clusters.

in Figure 6.1(a), the data is dispersed throughout the spatial domain with several more dense nested regions in the shape of a circle, square, and rectangle. A single resolution uniform grid would have difficulty identifying those more dense, nested regions as clusters as shown in Figure 6.1(b). In contrast, an adaptive algorithm, such as AMR, that permits higher resolution throughout the space can recognize those nested, more dense clusters with centers at the most clear, dense shapes. (Figure 6.1 is adapted from Figure 1 from Liao et al. [15] and is only illustrative, not based on real data.)

To address locality, *axis-shifting* algorithms were introduced. These methods adopt axis-shifted partitioning strategies to identify areas of high density in the feature space. For instance, in Figure 6.2(a), traditional grid-based algorithms will have difficulty adhering to the border and continuity of the most dense regions because of the predefined grids and the threshold of significant cells. The clustering from using a single uniform grid, shown in Figure 6.2(b), demonstrates that some clusters are divided into several smaller clusters because the continuity of the border of the dense regions is disturbed by cells with low density. To remedy this, axis-shifting algorithms, such as ASGC [17], shift the coordinate axis by half a cell width in each dimension creating a new grid structure. This shifting yields a clustering that recognizes more dense regions adjacent to lower density cells as shown in Figure 6.2(c). By combining the clustering from both axes, algorithms can recognize dense regions as clusters as shown in Figure 6.2(d). (Figures 6.2(a), 6.2(b), 6.2(c), and 6.2(d) are adapted from Figures 11, 12, 14, and 18, respectively, from Lin et al. [17] and are only illustrative, not based on real data or real clustering algorithm results.)

For handling high dimensional data, there are several grid-based approaches. For example, the CLIQUE algorithm selects appropriate subspaces rather than the whole feature space for finding the dense regions. In contrast, the OptiGrid algorithm uses density estimations. A summary of grid-based algorithms that address these three challenges is presented in Table 6.2.

TABLE 6.2: Grid-based Algorithms Addressing Nonuniformity (Adaptive), Locality (Axis-Shifting), and Dimensionality

Adaptive	MAFIA, AMR
Axis-shifting	NSGC, ADCC, ASGC, GDILC
High-dimension	CLIQUE, MAFIA, ENCLUS, OptiGrid, O-cluster, CBF

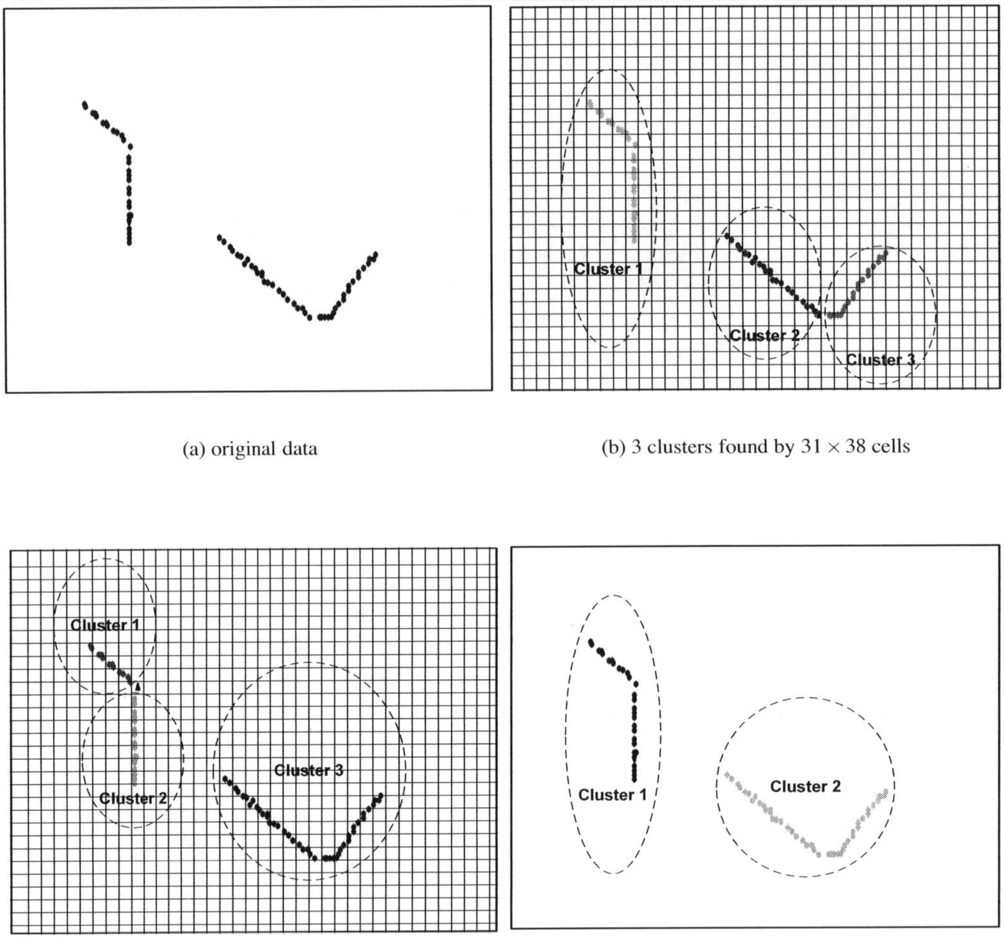

(a) original data

(b) 3 clusters found by 31 × 38 cells

(c) 3 clusters found by 31 × 38 cells after axis-shifting

(d) gr/final clusters by combining *b* and *c*

FIGURE 6.2: Locality Example: Axis-shifting grid-based clustering.

In the remainder of this chapter we survey classical grid-based clustering algorithms as well as those algorithms that directly address the challenges of nonuniformity, locality, and high-dimensionality. First, we discuss some classical grid-based clustering algorithms in Section 6.2. These classical grid-based clustering algorithms include the earliest approaches: GRIDCLUS, STING, WaveCluster, and variants of them. We present an adaptive grid-based algorithm, AMR, in Section 6.3. Several axis-shifting algorithms are evaluated in Section 6.4. In Section 6.5, we discuss high dimensional grid-based algorithms, including CLIQUE, OptiGrid, and their variants. We offer our conclusions and summary in Section 6.6.

6.2 The Classical Algorithms

In this section, we introduce three classical grid-based clustering algorithms together with their variants: GRIDCLUS, STING, and WaveCluster.

6.2.1 Earliest Approaches: GRIDCLUS and BANG

Schikuta [25] introduced the first GRID-based hierarchical CLUStering algorithm called GRIDCLUS. The algorithm partitions the data space into a grid structure composed of disjoint d-dimensional hyper rectangles or blocks. Data points are considered points in d-dimensional space and are designated to blocks in the grid structure such that their topological distributions are maintained. Once the data is assigned to blocks, clustering is done by a neighbor search algorithm. In some respects, GRIDCLUS is the canonical grid-based clustering algorithm and its basic steps coincide with those given for grid-based algorithms in Section 6.1. Namely, GRIDCLUS inserts points into blocks in its grid structure, calculates the resultant density of the blocks, sorts the blocks according to their density, recognizes the most dense blocks as cluster centers, and constructs the rest of clusters using a neighbor search on the blocks.

The grid structure has a scale for each dimension, a grid directory, and the set of data blocks. Each scale is used to partition the entire d-dimensional space and this partitioning is stored in the grid directory. The data blocks contain the data points and there is an upper bound on the number of points per block. The blocks must be nonempty, cover all the data points, and not have any data points in common. Hinrichs offers a more thorough discussion of the grid file structure used [13].

The density index of a block, B, is defined as the number of points in the block divided by the spatial volume of the block, i.e.,

$$D_B = \frac{p_B}{V_B}, \tag{6.1}$$

where p_B is the number of data points in the block B and V_B is the spatial volume of the block B, i.e.,

$$V_B = \prod_{i=1}^{d} e_{B_i}, \tag{6.2}$$

where d is the number of dimensions and e_{B_i} is the extent of the block in the i dimension.

GRIDCLUS sorts the blocks according to their density and those with the highest density are chosen as the cluster centers. The blocks are clustered in order of descending density iteratively to create a nested sequence of nonempty, disjoint clusters. Starting from cluster centers only neighboring blocks are merged into clusters. The neighbor search is done recursively starting at the cluster center, checking for adjacent blocks that should be added to the cluster, and for only those neighboring blocks added to the cluster, continuing the search. The GRIDCLUS algorithm is described in Algorithm 23 and the function *NEIGHBOR_SEARCH* is the recursive procedure described in Algorithm 24 [10, 25].

While no explicit time complexity analysis is given for GRIDCLUS in the original paper, the algorithm may not have time complexity much better than other hierarchical clustering algorithms in the worst case. The number of blocks in the worst case is $O(n)$ where n is the number of data points and sorting the blocks by density is $O(n \log n)$. However, this complexity would still be better than hierarchical clustering. The problem is that step 4 can also require $O(n)$ if all the blocks have different densities and step 7 can also require $O(n)$ if all the blocks have the same density. In addition, while the number of neighbors of any block is a function of the number of dimensions, the depth of the recursive calls to the neighbor search function can also be $O(n)$. This can occur if the blocks are adjacent in a single place analogous to a spanning tree that is a straight line. Without any

Algorithm 23 GRIDCLUS Algorithm

1: Set $u := 0$, $W[] := \{\}$,$C[] := \{\}\{\text{initialization}\}$;
2: Create the grid structure and calculate the block density indices;
3: Generate a sorted block sequence $B_{1'}, B_{2'}, ..., B_{b'}$ and mark all blocks "not active" and "not clustered";
4: **while** a "not active" block exists **do**
5: $\quad u \Leftarrow u + 1$;
6: $\quad$ mark first $B_{1'}, B_{2'}, ..., B_{j'}$ with equal density index "active";
7: $\quad$ **for** each "not clustered" block $B_{l'} := B_{1'}, B_{2'}, ..., B_{j'}$ **do**
8: $\quad\quad$ Create a new cluster set $C[u]$;
9: $\quad\quad W[u] \Leftarrow W[u] + 1, C[u, W[u]] \leftarrow B_{l'}$;
10: $\quad\quad$ Mark block $B_{l'}$ clustered;
11: $\quad\quad$ *NEIGHBOR_SEARCH*$(B_{l'}, C[u, W[u]])$;
12: $\quad$ **end for**
13: $\quad$ **for** each "not active" block B **do**
14: $\quad\quad W[u] \Leftarrow W[u] + 1, C[u, W[u]] \leftarrow B$;
15: $\quad$ **end for**
16: $\quad$ Mark all blocks "not clustered";
17: **end while**

Algorithm 24 Procedure NEIGHBOR_SEARCH(B,C)

1: **for** each "active" and "not clustered" neighbor B' of B **do**
2: $\quad C \leftarrow B'$;
3: $\quad$ Mark block B' "clustered";
4: $\quad$ *NEIGHBOR_SEARCH*(B', C);
5: **end for**

discriminatory density thresholds, the pathological case of step 7 could also apply and the complexity would be $O(n^2)$. (Granted average case complexity for several distributions may be significantly better (i.e., $O(n)$) and that may be a better analysis to consider.)

The BANG algorithm introduced by Schikuta and Erhart [26] is an extension of the GRIDCLUS algorithm. It addresses some of the inefficiencies of the GRIDCLUS algorithm in terms of grid structure size, searching for neighbors, and managing blocks by their density. BANG also places data points in blocks and uses a variant of the grid directory called a BANG structure to maintain blocks. Neighbor search and processing the blocks in decreasing order of density are also used for clustering blocks. Nearness of neighbors is determined by the maximum dimensions shared by a common face between blocks. A binary tree is used to store the grid structure, so that neighbor searching can be done more efficiently. From this tree in the grid directory and the sorted block densities, the dendrogram is calculated. Centers of clusters are still the most highly dense blocks in the clustering phase. The BANG algorithm is summarized in Algorithm 25 [10]. While both GRIDCLUS and BANG can discern nested clusters efficiently, BANG has been shown to be more efficient than GRIDCLUS on large data sets because of its significantly reduced growth of grid structure size [26].

6.2.2 STING and STING+: The Statistical Information Grid Approach

Wang et al. [28] proposed a STatistical INformation Grid-based clustering method (STING) to cluster spatial databases and to facilitate region-oriented queries. STING divides the spatial area into rectangular cells and stores the cells in a hierarchical grid structure tree. Each cell (except leaves in

Algorithm 25 BANG-Clustering Algorithm

1: Partition the feature space into rectangular blocks which contain up to a maximum of p_{max} data points.
2: Build a binary tree to maintain the populated blocks, in which the partition level corresponds to the node depth in the tree.
3: Calculate the dendrogram in which the density indices of all blocks are calculated and sorted in decreasing order.
4: Starting with the highest density index, all neighbor blocks are determined and classified in decreasing order. BANG-clustering places the found regions in the dendrogram to the right of the original blocks.
5: Repeat step 4 for the remaining blocks of the dendrogram.

the tree) is partitioned into 4 child cells at the next level with each child corresponding to a quadrant of the parent cell. A parent cell is the union of its children; the root cell at level 1 corresponds to the whole spatial area. The leaf level cells are of uniform size, determined globally from the average density of objects. For each cell, both attribute-dependent and attribute-independent parameters of the statistical information are maintained. These parameters are defined in Table 6.3.

STING maintains summary statistics for each cell in its hierarchical tree. As a result, statistical parameters of parent cells can easily be computed from the parameters of child cells. Note that the distribution types may be normal, uniform, exponential, and none. Value of *dist* may be either assigned by the user or obtained by hypothesis tests such as the χ^2 test. Even though measures of these statistical parameters are calculated in a bottom-up fashion from any leaf node, the STING algorithm adopts a top-down approach for clustering and query by starting from the root of its hierarchical grid structure tree. The algorithm is summarized in Algorithm 26 [10, 28].

The tree can be constructed in O(N) time, where N is the total number of data points. Dense cells are identified and clustered by examining the density of these cells in a similar vein to the density-based DBSCAN algorithm [7]. If the cell tree has K leaves, then the complexity of spatial querying and clustering for STING is O(K), which is $O(N)$ in the worst case since cells that would be empty never need to be materialized and stored in the tree. A common misconception is that K would be $O(2^d)$ where d is the number of dimensions and that this would be problematic in high dimensions. STING may have problems with higher dimensional data common to all grid-based algorithms (e.g., handling noise and selecting most relevant attributes) [11], but scalability of the grid structure is not one of them.

There are several advantages of STING. First, it is a query-independent approach since the statistical information exists independent of queries. The computational complexity of STING for clustering is O(K), and this is quite efficient in clustering large data sets especially when $K \ll N$. The algorithm is readily parallelizable and allows for multiple resolutions for examining the data in its hierarchical grid structure. In addition, incremental data updating is supported, so there is lower overhead for incorporating new data points.

TABLE 6.3: Statistical Information in STING

n	number of objects (points) in the cell
mean	mean of each dimension in this cell
std	standard deviation of each dimension in this cell
min	the minimum value of each dimension in this cell
max	the maximum value of each dimension in this cell
dist	the distribution of points in this cell

Algorithm 26 STING Algorithm

1: Determine a level at which to begin.
2: For each cell of this level, we calculate the confidence interval (or estimated range) of probability that this cell is relevant to the query.
3: From the interval calculated above, label the cell as *relevant* or *not relevant*.
4: If this level is the leaf level, go to Step 6; otherwise, go to Step 5.
5: Go down the hierarchy structure by one level. Go to Step 2 for those cells that form the relevant cells of the higher level.
6: If the specification of the query is met, go to Step 8; otherwise, go to Step 7.
7: Retrieve those data that fall into the *relevant* cells and do further processing. Return the result that meets that the requirement of the query. Go to Step 9.
8: Find the regions of *relevant* cells. Return those regions that meet the requirement of the query. Go to Step 9.
9: Stop.

Wang et al. extended STING to STING+ [29] so that it is able to process dynamically evolving spatial databases. In addition, STING+ enables active data mining by supporting user-defined trigger conditions.

6.2.3 WaveCluster: Wavelets in Grid-Based Clustering

Sheikholeslami et al. [27] proposed a grid-based and density-based clustering approach that uses wavelet transforms: WaveCluster. This algorithm applies wavelet transforms to the data points and then uses the transformed data to find clusters. A wavelet transform is a signal processing technique that decomposes a signal into different frequency subbands. The insight to using the wavelet transforms is that data points are considered d-dimensional signals where d is the number of dimensions. The high-frequency parts of a signal correspond to the more sparse data regions such as boundaries of the clusters whereas the low-frequency high-amplitude parts of a signal correspond to the more dense data regions such as cluster interiors [3]. By examining different frequency subbands, clustering results may be achieved at different resolutions and scales from fine to coarse. Data are transformed to preserve relative distance between objects at different levels of resolution. A hat-shaped filter is used to emphasize regions where points cluster and to suppress weaker information in their boundaries. This makes natural clusters more distinguishable and eliminates outliers simultaneously. As input parameters the algorithm requires the number of grid cells for each dimension, the wavelet, and the number of applications of the wavelet transform. This algorithm is summarized in Algorithm 27 [27].

Algorithm 27 WaveCluster Algorithm

INPUT: Multidimensional data objects' feature vectors
OUTPUT: cluster objects

1: First bin the feature space, then assign objects to the units, and compute unit summaries.
2: Apply wavelet transform on the feature space.
3: Find connected components (clusters) in the subbands of transformed feature space, at multiple levels.
4: Assign labels to the units in the connected components.
5: Make the lookup table.
6: Map the objects to the clusters.

WaveCluster offers several advantages. The time complexity is O(N) where N is the number of data points; this is very efficient for large spatial databases. The clustering results are insensitive to outliers and the data input order. The algorithm can accurately discern arbitrarily shaped clusters such as those with concavity and nesting. The wavelet transformation permits multiple levels of resolution, so that clusters may be detected more accurately. This algorithm is primarily only suited for low dimensional data. However, in the case of very high-dimensional data, PCA may be applied to the data to reduce the number of dimensions, so that $N > m^f$ where m is the number of intervals in each dimension and f is the number of dimensions selected after PCA. After this, WaveCluster may be applied to the data to cluster it and still achieve linear time efficiency [27].

6.3 Adaptive Grid-Based Algorithms

When a single inflexible, uniform grid is used, it may be difficult to achieve the desired clustering quality or efficiency for highly irregular data distributions. In such instances, adaptive algorithms that modify the uniform grid may be able to overcome this weakness in uniform grids. In this section, we introduce an adaptive grid-based clustering algorithm: AMR. Another adaptive algorithm (MAFIA) will be discussed in Section 6.5.2.2.

6.3.1 AMR: Adaptive Mesh Refinement Clustering

Liao et al. [15] proposed a grid-based clustering algorithm AMR using an Adaptive Mesh Refinement technique that applies higher resolution grids to the localized denser regions. Different from traditional grid-based clustering algorithms, such as CLIQUE and GRIDCLUS, which use a single resolution mesh grid, AMR divides the feature space at multiple resolutions. While STING also offers multiple resolutions, it does so over the entire space, not localized regions. AMR creates a hierarchical tree constructed from the grids at multiple resolutions. Using this tree, this algorithm can discern clusters, especially nested ones, that may be difficult to discover without clustering several levels of resolutions at once. AMR is very fit for data mining problems with highly irregular data distributions.

The AMR clustering algorithm mainly contains two steps summarized here from [14]:

1. Grid Construction: First grids are created at multiple resolutions based on regional density. The grid hierarchy tree contains nested grids of increasing resolution since the grid construction is done recursively. The construction of the AMR tree starts with a uniform grid covering the entire space, and for those cells that exceed a density threshold, the grid is refined into higher resolution grids at each recursive step. The new child grids created as part of the refinement step are connected in the tree to parent grid cells whose density exceeds the threshold.

2. Clustering: To create clusters, each leaf node is considered to be the center of an individual cluster. The algorithm recursively assigns objects in the parent nodes to clusters until the root node is reached. Cells are assigned to clusters based on the minimum distance to the clusters under the tree branch.

The overall complexity for constructing the AMR tree is $O(dtN\frac{1-p^h}{1-p} + (dtk+6^d)r\frac{1-q^h}{1-q})$, where N is the number of data points, d is the dimensionality, t is the number of attributes in each dimension, h is the AMR tree height, p represents the average percentage of data points to be refined at each level, r is the mesh size at the root, and q is the average ratio of mesh sizes between two grid levels [15].

As with most grid-based methods, AMR exhibits insensitivity to the order of input data. The AMR clustering algorithm may be applied to any collection of attributes with numerical values even those with very irregular or very concentrated data distributions. However, like GDILC, it cannot be scaled to high-dimensional databases because of its overall complexity.

6.4 Axis-Shifting Grid-Based Algorithms

The effectiveness of a grid-based clustering algorithm is seriously limited by the size of the predefined grids, the borders of the cells, and the density threshold of the significant cells in the face of local variations in shape and density in a data space. These challenges motivate another kind of grid-based algorithm: axis-shifting algorithms. In this section, we introduce four axis-shifting algorithms: NSGC, ADCC, ASGC, and GDILC.

6.4.1 NSGC: New Shifting Grid Clustering Algorithm

Fixed grids may suffer from the boundary effect. To alleviate this, Ma and Chow [18] proposed a New Shifting Grid Clustering algorithm (NSGC). NSGC is both density-based and grid-based. To form its grid structure, the algorithm divides each dimension of the space into an equal number of intervals. NSGC shifts the whole grid structure and uses the shifted grid along with the original grid to determine the density of cells. This reduces the influence of the size and borders of the cells. It then clusters the cells rather than the points. Specifically, NSGC consists of four main steps summarized in Algorithm 28 [18].

Algorithm 28 NSGC Algorithm

1: Cell construction: It divides each dimension of the space into $2w$ intervals, where w is the number of iterations.

2: Cell assignment: It first finds the data points belonging to a cell, then shifts by half cell-size of the corresponding dimension, and finds the data points belonging to shifted cells.

3: Cell density computation: It uses both the density of the cell itself and its nearest neighborhood to obtain a descriptive density profile.

4: Group assignment(clustering): It starts when the considered cell or one of its neighbor cells has no group assigned. Otherwise, the next cell is considered until all non-empty cells are assigned.

NSGC repeats the steps above until the result of the previous iteration and that of the current iteration are smaller than a specified accepted error threshold. The complexity of NSGC is $O((2w)^d)$, where d is the dimensionality and w is the number of iterations of the algorithm. While it is claimed that this algorithm is nonparametric, its performance is dependent upon the choice of the number of iterations, w, and the accepted error threshold. If w is set too low (or high) or the error threshold too high, then clustering results may not be accurate; there is no a priori way to know the best values of these parameters for specific data. NSGC is susceptible to errors caused by cell sizes that are too small also. As the size of cells decreases (and the number of iterations increases), the total number of cells and the number of clusters reported both increase. The reported clusters may be too small and not correspond to clusters in the original data. The strongest advantage of NSGC is that its grid shifting strategy permits it to recognize clusters of very arbitrary boundary shapes with great accuracy.

6.4.2 ADCC: Adaptable Deflect and Conquer Clustering

The clustering quality of grid-based clustering algorithms often depends on the size of the predefined grids and the density threshold. To reduce their influence, Lin et al. adopted "deflect and conquer techniques" to propose a new grid-based clustering algorithm ADCC (Adaptable Deflect and Conquer Clustering) [16]. Very similar to NSGC, the idea of ADCC is to utilize the predefined grids and predefined threshold to identify significant cells. Nearby cells that are also significant can be merged to develop a cluster. Next, the grids are deflected half a cell size in all directions and the significant cells are identified again. Finally, the newly generated significant cells and the initial set of significant cells are merged to improve the clustering of both phases. Specifically, ADCC is summarized in Algorithm 29.

Algorithm 29 ADCC Algorithm

1: Generate the first grid structure.
2: Identify the significant cells.
3: Generate the first set of clusters.
4: Transform the grid structure.
5: Generate the second set of clusters.
6: Revise the original clusters. In this case, the first and second sets of clusters are combined recursively.
7: Generate the final clustering result.

The overall complexity of ADCC is $O(m^d + N)$, where m is the number of intervals in each dimension, d is the dimensionality of data, and N is the number of data points. While ADCC is very similar to NSGC in its axis-shifting strategy, it is quite different in how it constructs clusters from the sets of grids. Rather than examining a neighborhood of the two grids at once as NSGC does, ADCC examines the two grids recursively looking for consensus in the significance of cells in both clusterings especially those that overlap a previous clustering to make a determination about the final clustering. This step can actually help to separate clusters more effectively especially if there is only a small distance with very little data between them. Both methods are suspectible to errors caused by small cell sizes, but can for the most part handle arbitrary borders and shapes in clusters very well. ADCC is not dependent on many parameters to determine its termination. It is only dependent on the choice of the number of intervals per dimension, m.

6.4.3 ASGC: Axis-Shifted Grid-Clustering

Another attempt by Lin et al. [17] to minimize the impact of the size and borders of the cells is ASGC (Axis-Shifted Grid-Clustering) (also referred to as $ACICA^+$). After creating an original grid structure and initial clustering from that grid structure, the original grid structure is shifted in each dimension and another clustering is done. The shifted grid structure can be translated an arbitrary distance to be specified. The effect of this is to implicitly change the size of the original cells. It also offers greater flexibility to adjust to boundaries of clusters in the original data and minimize the effect of the boundary cells. The clusters generated from this shifted grid structure can be used to revise the original clusters. Specifically, the ASGC algorithm involves 7 steps, and is summarized in Algorithm 30 from [14].

The complexity of ASGC is the same as that of ADCC, which is $O(m^d + N)$, where N is the number of data points, d is the dimensionality of the data, and m is the number of intervals in each dimension. The main difference between ADCC and ASGC is that the consensus method to revise clusters is bidirectional in ASGC: using the overlapping cells the clusters from the first phase can be used to modify the clusters of the second phase and vice versa. When a cluster of the first clustering overlaps a cluster of the second clustering, the combined cluster of union of

Algorithm 30 ASGC Algorithm

1: Generate the first grid structure: the entire feature space is divided into non overlapping cells thus forming the first grid structure.
2: Identify the significant cells: These are cells whose density is more than a predefined threshold.
3: Generate the first set of clusters: all neighboring significant cells are grouped together to form clusters.
4: Transform the grid structure: the original coordinate origin is shifted by distance ξ in each dimension of the feature space to obtain a new grid structure.
5: Generate the second set of clusters: new clusters are generated using steps 2 and 3.
6: Revise the original clusters: the clusters generated from the shifted grid structures can be used to revise the clusters generated from the original grid structure.
7: Generate the final clustering result.

both can then be modified in order to generate the final clustering. This permits great flexibility in handling arbitrary shapes of clusters in the original data and minimizes the extent to which either grid structure will separate clusters. By essentially translating the original grid structure an arbitrary distance to create the second grid and overlapping it with the original grid structure, a different resolution (and implicitly different cell size) is also achieved by this translation. While this method is less susceptible to the effects of cell sizes and cell-density thresholds than other axis-shifting grid-clustering methods, it still requires careful initial choice of cell size and cell-density threshold.

6.4.4 GDILC: Grid-Based Density-IsoLine Clustering Algorithm

Zhao and Song [32] proposed a Grid-based Density-IsoLine Clustering algorithm (GDILC) to perform clustering by making use of the density-isoline figure. It assumes that all data samples have been normalized. All attributes are numerals and are in the range of $[0,1]$. This is for the convenience of distance and density calculation. GDILC first implicitly calculates a density-isoline figure, the contour figure of the density of data points. Then clusters are discovered from the density-isoline figure.

GDILC computes the density of a data point by counting the number of points in its neighbor region. Specifically, the density of a data point x is defined as follows:

$$Density(x) = |\{y : Dist(x,y) \leq T\}|, \tag{6.3}$$

where T is a given distance threshold and $Dist(x,y)$ is a distance function (e.g., Euclidean distance) used to measure the dissimilarity between data points x and y. The density-isoline figure is never drawn, but is obtained from the density vectors. The density vectors are computed by counting the elements of each row of the distance matrix that are less than the radius of the neighbor region, T. To avoid enumerating all data points for calculating the density vector, GDILC employs a grid-based method. The grid-based method first partitions each dimension into several intervals creating hyper-rectangular cells. Then, to calculate the density of data point x, GDILC considers only data points in the same cell with x and those data points in its neighbor cells; this is identical to axis shifting. The GDILC algorithm is shown in Algorithm 31 [10].

For many data sets, this grid-based method significantly reduces the search space of calculating the point pair distances; the complexity may appear nearly linear. In the worst case the time complexity of GDILC remains $O(N^2)$ (i.e., for the pathological case when all the points cluster in the neighborhood of a constant number of cells). However, it cannot be scaled to high-dimensional data because the space is divided into m^d cells, where m is the number of intervals in each dimension, and d is the dimensionality. When the dimensionality d is very large, m^d is significantly large, and the data points in each cell are very sparse, the GDILC algorithm will no longer work. (There will be difficulty computing any distances or thresholds.)

Algorithm 31 GDILC Algorithm

1: Cells are initialized by dividing each dimension into m intervals.
2: The distances between sample points and those in neighboring cells are calculated. The distance threshold T is computed.
3: The density vector and density threshold τ are computed.
4: At first, GDILC takes each data point whose density is more than the density threshold τ as a cluster. Then, for each data point x, check, for every data point whose density is more than the density threshold τ in the neighbor cells of x, whether its distance to x is less than the distance threshold T. If so, GDILC then combines the two clusters containing those two data points. The algorithm continues until all point pairs have been checked.
5: Outliers are removed.

There are two significant advantages to this algorithm. First, it can handle outliers explicitly and this can be refined as desired. Second, it computes necessary thresholds such as those for density and distance directly from the data. These can be fine-tuned as needed (i.e., they don't need to be guessed at any point). In essence this algorithm dynamically learns the data distribution of samples while learning the parameters for thresholds in addition to discerning the clustering in the data.

6.5 High-Dimensional Algorithms

The scalability of grid-based approaches is a significant problem in higher dimensional data because of the increase in the size of the grid structure and the resultant time complexity increase. Moreover, inherent issues in clustering high dimensional data such as filtering noise and identifying the most relevant attributes or dimensions that represent the most dense regions must be addressed inherently in the grid structure creation as well as the actual clustering algorithm. In this section, we examine carefully a subspace clustering approach presented in CLIQUE and a density estimation approach presented by OptiGrid. This section is greatly influenced by the survey of Berkhin [3] with additional insights on the complexity, strengths, and weaknesses of each algorithm presented.

6.5.1 CLIQUE: The Classical High-Dimensional Algorithm

Agrawal et al. [1] proposed a hybrid density-based, grid-based clustering algorithm, CLIQUE (CLustering In QUEst), to find automatically subspace clustering of high-dimensional numerical data. It locates clusters embedded in subspaces of high-dimensional data without much user intervention to discern significant subclusters. In order to present the clustering results in an easily interpretable format, each cluster is given a minimal description as a disjunctive normal form (DNF) expression.

CLIQUE first partitions its numerical space into units for its grid structure. More specifically, let $\mathcal{A} = \{A_1, A_2, ..., A_d\}$ be a set of bounded, totally ordered domains (attributes) and $\mathcal{S} = A_1 \times A_2 \times ..., \times A_d$ be a d-dimensional numerical space. By partitioning every dimension A_i $(1 \leq i \leq d)$ into m intervals of equal length, CLIQUE divides the d-dimensional data space into m^d nonoverlapping rectangular units. A d-dimensional data point, v, is considered in a unit, u, if the value of v in each attribute, is greater than or equal to the left boundary of that attribute in u and less than the right boundary of that attribute in u. The selectivity of a unit is defined to be the fraction of total data points in the unit. Only units whose selectivity is greater than a parameter τ are viewed as dense and retained. The definition of dense units applies to all subspaces of the original d-dimensional space.

To identify dense units to retain and subspaces that contain clusters, CLIQUE considers projections of the subspaces from the bottom up (i.e., the least dimensional subspaces to those of increasing dimension). Given a projection subspace, $A_{t_1} \times A_{t_2} \times ... \times A_{t_p}$, where $p < d$ and $t_i < t_j$ if $i < j$, a unit is the intersection of an interval in each dimension. By leveraging the *Apriori* algorithm, CLIQUE employs a bottom-up scheme because monotonicity holds: if a collection of points is a cluster in a p-dimensional space, then this collection of points is also part of a cluster in any $(p-1)$-dimensional projections of this space. In CLIQUE, the recursive step from $(p-1)$-dimensional units to p-dimensional units involves a self-join of the $p-1$ units sharing *first* common $(p-2)$-dimensions [3]. To reduce the time complexity of the *Apriori* process, CLIQUE prunes the pool of candidates, only keeping the set of dense units to form the candidate units in the next level of the dense unit generation algorithm. To prune the candidates, all the subspaces are sorted by their coverage, i.e., the fraction of the database that is covered by the dense units in it. The less covered subspaces are pruned. The cut point between retained and pruned subspaces is selected based on the MDL [24] principle in information theory.

CLIQUE then forms clusters from the remaining candidate units. Two p-dimensional units u_1, u_2 are connected if they have a common face or if there exists another p-dimensional unit u_s such that u_1 is connected to u_s and u_2 is connected to u_s. A cluster is a maximal set of connected dense units in p-dimensions. Finding clusters is equivalent to finding connected components in the graph defined to represent the dense units as the vertices and edges between vertices existing if and only if the units share a common face. In the worst case, this can be done in quadratic time in the number of dense units. After finding all the clusters, CLIQUE uses a DNF expression to specify a finite set of maximal segments (*regions*) whose union is the cluster. Finding the minimal descriptions for the clusters is equivalent to finding an optimal cover of the clusters; this is NP-hard. In light of this, instead, CLIQUE adopts a greedy approach to cover the cluster in regions and then discards redundant regions.

By integrating density-based, grid-based, and subspace clustering, CLIQUE discovers clusters embedded in subspaces of high-dimensional data without requiring users to select subspaces of interest. The DNF expressions for the clusters give a clear representation of clustering results. The time complexity of CLIQUE is $O(c^p + pN)$, where p is the highest subspace dimension selected, N is the number of input points, and c is a constant; this grows exponentially with respect to p. The algorithm offers an effective, efficient method of pruning the space of dense units in order to counter the inherent exponential nature of the problem. However, there is a trade-off for the pruning of dense units in the subspaces with low coverage. While the algorithm is faster, there is an increased likelihood of missing clusters. In addition, while CLIQUE does not require users to select subspaces of interest, its susceptibility to noise and ability to identify relevant attributes is highly dependent on the user's choice of unit intervals, m, and sensitivity threshold, τ.

6.5.2 Variants of CLIQUE

There are two aspects of the CLIQUE algorithm that can be improved. The first one is the criterion for the subspace selection. The second is the size and resolution of the grid structure. The former is addressed by the ENCLUS algorithm by using entropy as subspace selection criterion. The latter is addressed by the MAFIA algorithm by using adaptive grids for fast subspace clustering.

6.5.2.1 ENCLUS: Entropy-Based Approach

The algorithm ENCLUS (ENtropy-based CLUStering) [6] is an adaptation of CLIQUE that uses a different, entropy-based criterion for subspace selection. Rather than using the fraction of total points in a subspace as a criterion to select subspaces, ENCLUS uses an entropy criteria and only those subspaces spanned by attributes $A_1,...,A_p$ with entropy $H(A_1,...,A_p) < \varpi$(a threshold) are selected for clustering. A low-entropy subspace corresponds to a more dense region of units.

An analogous monotonicity condition or *Apriori* property also exists in terms of entropy. If a p-dimensional subspace has low entropy, then so does any $(p-1)$-dimensional projections of this subspace:

$$H(A_1,...,A_{p-1}) = H(A_1,...,A_p) - H(A_p|A_1,...,A_{p-1}) < \varpi. \tag{6.4}$$

A significant limitation of ENCLUS is its extremely high computational cost, especially in terms of computing the entropy of subspaces. However, this cost also yields the benefit that this approach has increased sensitivity to detect clusters especially extremely dense small ones.

6.5.2.2 MAFIA: Adaptive Grids in High Dimensions

MAFIA (Merging of Adaptive Finite IntervAls) proposed by Nagesh et al. [21] is a descendant of CLIQUE. Instead of using a fixed size cell grid structure with an equal number of bins in each dimension, MAFIA constructs adaptive grids to improve subspace clustering and also uses parallelism on a shared-nothing architecture to handle massive data sets. MAFIA proposes an adaptive grid of bins in each dimension. Then using an *Apriori* algorithm, dense intervals are merged to create clusters in the higher dimensional space. The adaptive grid is created by partitioning each dimension independently based on the distribution (i.e., the histogram) observed in that dimension, merging intervals that have the same observed distribution, and pruning those intervals with low density. This pruning during the construction of the adaptive grid reduces the overall computation of the clustering step. The steps of MAFIA are summarized from [3] in Algorithm 32.

Algorithm 32 MAFIA Algorithm

1: Do one scan of the data to construct *adaptive grids* in each dimension.
2: Compute the histograms by reading blocks of data into memory using bins.
3: Using the histograms to merge bins into a smaller number of adaptive variable-size bins, combine adjacent bins with similar histogram values to form larger bins. The bins that have low density of data are pruned.
4: Select bins that are α-times (α is a parameter called the cluster dominance factor) more densely populated than average as p ($p = 1$ now) CDUs.
5: Iteratively scan data for higher dimensions, construct new p-CDU from two $(p-1)$-CDUs if they share any $(p-2)$-face, and merge adjacent CDUs into clusters.
6: Generate minimal DNF expressions for each cluster.

If p is the highest dimensionality of a candidate dense unit (CDU), N is the number of data points, and m is a constant, the algorithm's complexity is O $(m^p + pN)$, still exponential in the dimension as CLIQUE also is. However, the performance results on real data sets show that MAFIA is 40 to 50 times faster than CLIQUE because of the use of adaptive grids and their ability to select a smaller set of interesting CDUs [6]. Parallel MAFIA further offers the ability to obtain a highly scalable clustering for large data sets. Since the adaptive grid permits not only variable resolution because of the variable bin size, but also variable, adaptive grid boundaries, MAFIA yields with greater accuracy cluster boundaries that are very close to grid boundaries and are readily expressed as minimal DNF expressions.

6.5.3 OptiGrid: Density-Based Optimal Grid Partitioning

Hinneburg and Keim proposed OptiGrid (OPTimal GRID-Clustering) [12] to address several aspects of the "curse of dimensionality": noise, scalability of the grid construction, and selecting relevant attributes by optimizing the density function over the data space. OptiGrid uses density estimations to determine the centers of clusters as the clustering was done for the DENCLUE algorithm [11]. A cluster is a region of concentrated density centered around a strong density attractor or local maximum of the density function with density above the noise threshold. Clusters may also have multiple centers if the centers are strong density attractors and there exists a path between them above the noise threshold. By recursively partitioning the feature space into multidimensional grids, OptiGrid creates an optimal grid-partition by constructing the best cutting hyperplanes of the space. These cutting planes cut the space in areas of low density (i.e., local minima of the density function) and preserve areas of high density or clusters, specifically the cluster centers (i.e., local maxima of the density function). The cutting hyperplanes are found using a set of *contracting* linear projections of the feature space. The contracting projections create upper bounds for the density of the planes orthogonal to them. Namely, for any point, x, in a contracting projection, P, then for any point y such that $P(y) = x$, the density of y is at most the density of x.

To define the grid more precisely, we present the definitions offered in [12] as summarized in [10]. A cutting plane is a $(d-1)$-dimensional hyperplane consisting of all points y that satisfy the equation $\sum_{i=1}^{d} w_i y_j = 1$. The cutting plane partitions $\mathbb{R}^d$ into two half spaces. The decision function $H(x)$ determines the half space, where a point $x \in \mathbb{R}$ is located:

$$H(x) = \begin{cases} 1 & if \quad \sum_{i=1}^{d} w_i x_j \geq 1, \\ 0 & otherwise. \end{cases} \tag{6.5}$$

Then, a multi-dimensional grid G for the feature space S is defined to be a set $H = \{H_1, H_2, ..., H_k\}$ of $(d-1)$-dimensional cutting planes. The coding function $c^G : S \to \mathbb{N}$ is defined as

$$x \in S, c(x) = \sum_{i=1}^{k} 2^i \cdot H_i(x). \tag{6.6}$$

OptiGrid uses a density function to determine the best cutting places and to locate clusters. The density function $\hat{f}^D$ is defined as

$$\hat{f}^D = \frac{1}{nh} \sum_{i=1}^{n} KD(\frac{x - x_i}{h}), \tag{6.7}$$

where D is a set of N d-dimensional points, h is the smoothness level, and KD is the kernel density estimator. Clusters are defined as the maxima of the density function, which are above a certain noise level ξ.

A center-based cluster for a maximum x^* of the density function $\hat{f}^D$ is the subset $C \subseteq D$, with $x \in C$ being density-attracted by x^* and $\hat{f}^D(x^*) \geq \xi$. Points $x \in D$ are called outliers if they are density-attracted by a local maximum x_0^* with $\hat{f}^D(x_0^*) < \xi$.

OptiGrid selects a set of contracting projections. These projections are then used to find the optimal cutting planes. The projections are useful because they concentrate the density of points, and the cutting planes, in contrast, will have low density. Each cutting plane is selected to have minimal point density and to separate two dense half spaces. After each step of constructing a multi-dimensional grid defined by the best cutting planes, OptiGrid finds the clusters using the density function. The algorithm is then applied recursively to the clusters. In each round of recursion, OptiGrid maintains only data objects in the dense grids from the previous round of recursion. The algorithm OptiGrid is described in Algorithm 33 [10, 12].

The time complexity of OptiGrid is $O(d \cdot N \cdot \log N)$, where N is the number of data points and d is the dimensionality of the data. This is very efficient for clustering large high-dimensional databases. However, it may perform poorly in locating clusters embedded in a low-dimensional subspace of a very high-dimensional database, because its recursive method only reduces the dimensions by one at every step [10]. In addition, it suffers sensitivity to parameter choice and does not efficiently handle grid sizes that exceed available memory [20]. Moreover, OptiGrid requires very careful selection of the projections, density estimate, and determination of what constitutes a best or optimal cutting

Algorithm 33 OptiGrid Algorithm

INPUT: data set D, q, min_cut_score

1: Determine a set of contracting projections $P = \{P_0, P_1, ..., P_k\}$ and calculate all the projections of the data set $D : P_i(D), i = 1, 2, ..., k$;
2: Initialize a list of cutting planes $BEST\_CUT \Leftarrow \Phi$, $CUT \Leftarrow \Phi$;
3: **for** $i = 0$ to k **do**
4: $CUT \Leftarrow$best local cuts $P_i(D)$;
5: $CUT_SCORE \Leftarrow$Score best local cuts $P_i(D)$;
6: Insert all the cutting planes with a score $\geq min\_cut\_score$ into $BEST\_CUT$;
7: **if** $BEST\_CUT = \Phi$ **then**
8: **return** D as a cluster;
9: **else**
10: Select the q cutting planes of the highest score from $BEST\_CUT$ and construct a multidimensional grid G using the q cutting planes;
11: Insert all data points in D into G and determine the highly populated grid cells in G; add these cells to the set of clusters C;
12: Refine C;
13: **for all** clusters C_i in C **do**
14: Do the same process with data set C_i;
15: **end for**
16: **end if**
17: **end for**

plane from users. The difficulty of this is only determined on a case by case basis on the data being studied. However, a special case of applying this algorithm can be considered a more efficient variant of CLIQUE and MAFIA. Namely, if the projections used are the projection maps in each dimension, the density estimate is uniform, and there are sufficient cutting planes to separate each density attractor on each dimension, then a more efficient and accurate clustering can be achieved that circumvents the difficulties of CLIQUE, i.e, the time complexity is no longer exponential in the dimensions and clusters are not missed.

6.5.4 Variants of the OptiGrid Approach

In this section, we consider variants of OptiGrid that were introduced to address the issues of the scalabity of the grid structure, especially with respect to available memory, and a clear criterion for the selection of cutting planes.

6.5.4.1 O-Cluster: A Scalable Approach

Milenova and Campos proposed an O-Cluster (Orthogonal partitioning CLUSTERing) [20] to address three limitations of OptiGrid: scalability in terms of data relative to memory size, lack of clear criterion to determine if a cutting plane is optimal or not, and sensitivity to threshold parameters for noise and cut plane density. O-Clusters addresses the first limitation by using a random sampling technique on the original data and a small buffer size. Only partitions that are not resolved (i.e., ambiguous) have data points maintained in the buffer. As a variant of OptiGrid, O-Cluster uses an axis-parallel partitioning strategy to locate high density areas in the data. To do so, O-Cluster uses contracting projections, but also proposes the use of a statistical test to validate the quality of a cutting plane. The statistical test checks for statistical significance between the difference in the density of the peaks and a valley when the valley separates the two peaks using a standard χ^2 test. If statistical significance is found, the cutting plane would then be through such a valley.

O-Cluster is also a recursive method. After testing the splitting points for all projections in a partition, the optimal one is chosen to partition the data. The algorithm then searches for cutting planes in the new partitions. A hierarchical tree structure is used to divide the data into rectangular regions. The main steps are summarized in Algorithm 34.

Algorithm 34 O-Cluster Algorithm

1: Load data buffer.
2: Compute histograms for active partitions.
3: Find "best" splits for active partitions.
4: Flag ambiguous and "frozen" partitions.
5: Split active partitions.
6: Reload buffer.

The time complexity of O-Cluster is approximated to be $O(Nd)$, where N is the number of data points and d is the number of dimensions. However, Hinneburg and Keim claim a superlinear lower bound for the time complexity of clustering high dimensional data with noise [12]. Their proof sketch addresses the issue that given an $O(N)$ amount of noise in data that has been read and must be searched, there is not a constant time way to do that even for random noise. Moreover, the time complexity of the OptiGrid algorithm is dominated by the insertion time into the grid structure. They assert that the insertion time for axis parallel planes is $O(NqI)$ where q is the number of planes and I is the insertion time and that the insertion time is the minimum of q and $\log N$. Since the O-Cluster algorithm uses a binary clustering tree to load the buffer, it is possible that its running time is dominated by this and is $O(Ns)$ where s is the depth of the binary clustering tree, but clear worst case analysis of that depth was not offered. However, empirically O-Cluster has shown good scalability results even using a small buffer size.

While O-Cluster handles uniform noise well, its performance degrades with increasing noise. O-Cluster, unlike OptiGrid, also has a tendency to split clusters because it tends to oversplit the data space and uses histograms that do not sufficiently smooth the distribution density. However, like OptiGrid, it may have difficulty with clusters embedded in a low-dimensional subspace of a very high-dimensional feature space because of its recursion step [10].

6.5.4.2 CBF: Cell-Based Filtering

CBF (Cell-Based Filtering) [4] proposed by Chang and Jin focuses on the scalability of the grid structure, handling large data sets in memory, and the efficiency of insertion and retrieval of clusters from the grid structure. It also offers a clear criteria for a cutting plane.

CBF creates its grid structure by splitting each dimension into a set of partitions using a split index. Once a dimension is split into classes, the split index is equal to the sum of the relative density squared of each class. CBF finds the optimal split section in each dimension by repeatedly examining the the split value index for partitions along the potential split points in the dimension until the maximum value of a split index for a partitioning is found. Cells are created from the overlapping regions of the partitions in each dimension. This cell creation algorithm results in many fewer cells in the grid structure than other high dimensional grid-based algorithms we have discussed.

Cells with higher density than a given threshold are inserted into the clusters in the index structure. There is a second layer of the index structure built upon the cluster information file that provides approximation information of the clusters by grouping them into sections. For sections of clusters, if the density of the section is greater than a threshold, then this secondary index is set to true for that section and false otherwise. By using this approximation technique, the index structure is able to filter queries by first examining the section information and then accessing the cell and cluster information as necessary. This filtering-based index structure achieves good retrieval performance on clusters by minimizing the I/O access to a cluster information file. While CBF is very efficient in its cell grid creation and query retrieval time, the trade-off for such efficiency is lower precision for clustering.

6.6 Conclusions and Summary

The efficiency of grid-based clustering algorithms comes from how data points are grouped into cells and clustered collectively rather than individually. This results in drastic time complexity improvements because often data is grouped into far few cells than there are data points. The general approach of these algorithms is to divide the data space into grid data structures, summarize each cell in the grids by a statistic such as density, and then cluster the grid cells. The greatest challenge for using these algorithms is then determining the best strategy of constructing the grid structure. The size of the grid structure and the time for its construction largely determine the time complexity of grid-based algorithms. The size of the grid structure is directly influenced by the size of the cell, and the size of the cell determines the resolution at which data may be clustered. The resolution of the grid determines the clustering accuracy and the ability to recognize diverse data cluster boundaries.

We have presented classical grid-based algorithms that use uniform grids and several variant classes that deal with the specific challenges of using uniform grid structures. Adaptive algorithms permit grid structures that offer finer resolutions over some regions of the data space in order to deal with data that has highly irregular or concentrated data distributions. Axis-shifting algorithms deal with local variations in data shape clusters and density of the data by translating the original grid structure across the data space. Grid-based algorithms for high-dimensional data deal with a problem that is inherently exponential in nature; constructing a grid structure that partitions the data in each dimension will create one that is exponential in the number of dimensions. These algorithms choose ways to filter the grid structure to investigate only relevant subspaces or to filter the original subspace in order to select only the most relevant attributes and filter noise.

The time complexities of several grid-based algorithms discussed in this chapter are summarized in Table 6.4. Figure 6.3 summarizes the grid-based algorithms we have discussed in this chapter and their relationships. GCA is a low dimensional grid-based clustering algorithm that can be applied to high-dimensional data after PCA is applied to the data [33]. GCHL is a variant of GRIDCLUS that is suitable for high-dimensional data [23]. Some current research on grid-based clustering focuses on parallel implementation of grid-based clustering such as PGMCLU [31] and GridDBSCAN [19], grid-based clustering of fuzzy query results [2], and domain-specific applications such as grid-based clustering on data streams such as SGCS, D-Stream , and DUCstream [22, 5, 8].

TABLE 6.4: Time Complexity of Grid-Based Algorithms

Algorithm Name	Time Complexity
STING	$O(K)$, K is the number of cells at bottom layer
WaveCluster	$O(N)$, N is the number of data objects
NSGC	$O((2w)^d)$, d is the dimensionality, w is number of iterations
ADCC,ASGC	$O(m^d)+O(N)$, m is the number of intervals in each dimension
GDILC	$O(N)$
CLIQUE,MAFIA	$O(c^p+pN)$, p is the highest subspace dimension selected, c is a constant
OptiGrid	$O(d \cdot N \cdot \log N)$
O-Cluster	$\Omega(N \cdot d)$

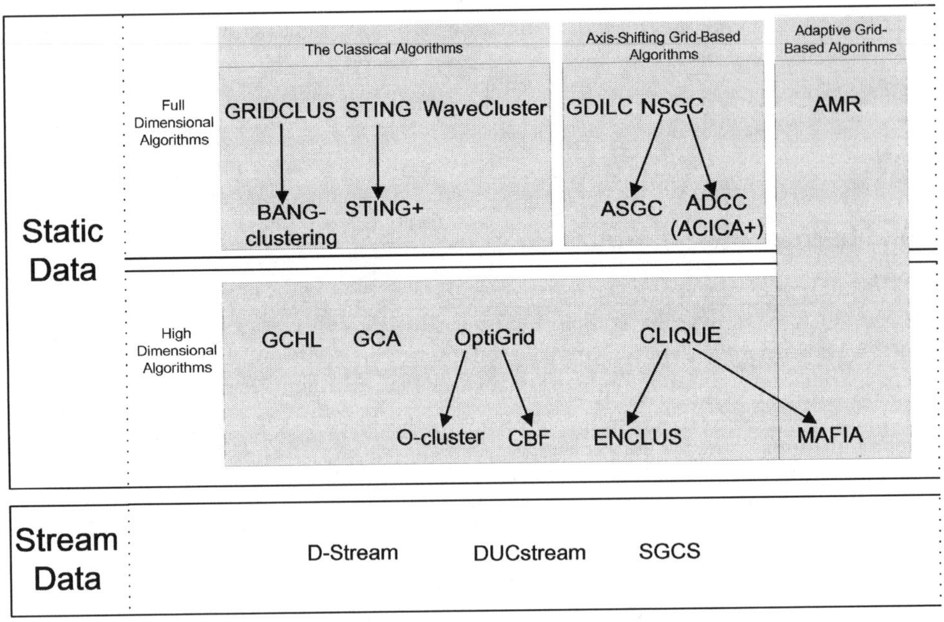

FIGURE 6.3: Relationships among grid-based algorithms.

Bibliography

[1] Rakesh Agrawal, Johannes Gehrke, Dimitrios Gunopulos, and Prabhakar Raghavan. Automatic subspace clustering of high dimensional data for data mining applications. In *Proceedings of the 1998 ACM SIGMOD International Conference on Management of Data*, SIGMOD '98, pages 94–105, New York, USA, 1998. ACM.

[2] Mounir Bechchi, Amenel Voglozin, Guillaume Raschia, and Noureddine Mouaddib. Multidimensional grid-based clustering of fuzzy query results. Rapports de recherche, INRIA, 2008. doi: HAL: http://hal.archives-ouvertes.fr/inria-00346540/en/

[3] Pavel Berkhin. Survey of clustering data mining techniques. Technical report, Accrue Software, Inc., San Jose, CA, 2002.

[4] Jae-Woo Chang and Du-Seok Jin. A new cell-based clustering method for large, high-dimensional data in data mining applications. In *Proceedings of the 2002 ACM Symposium on Applied Computing*, SAC '02, pages 503–507. ACM, 2002.

[5] Yixin Chen and Li Tu. Density-based clustering for real-time stream data. In *Proceedings of the 13th ACM SIGKDD International Conference on Knowledge Discovery and Data Mining*, KDD '07, pages 133–142, 2007.

[6] Chun-Hung Cheng, Ada Wai-chee, and Fu Yi Zhang. ENCLUS: Entropy-based subspace clustering for mining numerical data. In *Proceedings of ACM SIGKDD International Conference on Knowledge Discovery and Data Mining*, (KDD-99), 1999.

[7] Martin Ester, Hans-Peter Kriegel, Jorg Sander, and Xiaowei Xu. A density-based algorithm for discovering clusters in large spatial databases with noise. In *KDD'96*, pages 226–231, 1996.

[8] Jing Gao, Jianzhong Li, Zhaogong Zhang, and Pang-Ning Tan. An incremental data stream clustering algorithm based on dense units detection. In *Proceedings of the 9th Pacific-Asia Conference on Advances in Knowledge Discovery and Data Mining*, PAKDD'05, pages 420–425, Berlin, Heidelberg, 2005. Springer-Verlag.

[9] Peter Grabusts and Arkady Borisov. Using grid-clustering methods in data classification, In *Proceedings of the International Conference on Parallel Computing in Electrical Engineering*, pages 425–426, 2002.

[10] Chaoqun Ma Guojun Gan and Jianhong Wu. *Data Clustering: Theory, Algorithms, and Applications*. SIAM, 2007.

[11] Alexander Hinneburg, and Daniel A. Keim. An efficient approach to clustering in large multimedia databases with noise. In *Proceedings of the 4th International Conference on Knowledge Discovery and Data Mining*, pages 58–65. AAAI Press, 1998.

[12] Alexander Hinneburg and Daniel A. Keim. Optimal grid-clustering: Towards breaking the curse of dimensionality in high-dimensional clustering. In *VLDB'99*, pages 506–517. Morgan Kaufmann, 1999.

[13] Klaus Helmer Hinrichs. The grid file system: Implementation and case studies of applications. Technical report, ETH Zurich, 1985.

[14] M.R. Ilango and Dr V. Mohan. A survey of grid based clustering algorithms. *International Journal of Engineering Science and Technology*, 2(8):3441–3446, 2010.

[15] Wei-keng Liao, Ying Liu, and Alok Choudhary. A grid-based clustering algorithm using adaptive mesh refinement, 2004.

[16] Nancy P. Lin, Chung-I Chang, and Chao-Lung Pan. An adaptable deflect and conquer clustering algorithm. In *Proceedings of the 6th International Conference on Applied Computer Science—Volume 6*, ACOS'07, pages 155–159, 2007.

[17] Nancy P. Lin, Chung-I Chang, Nien-yi Jan, Hao-En Chueh, Hung-Jen Chen, and Wei-Hua Hao. An axis-shifted crossover-imaged clustering algorithm. *WTOS*, 7(3):175–184, March 2008.

[18] Eden W.M. Ma and Tommy W.S. Chow. A new shifting grid clustering algorithm. *Pattern Recognition*, 37(3):503–514, 2004.

[19] S. Mahran and K. Mahar. Using grid for accelerating density-based clustering. In *8th IEEE International Conference on Computer and Information Technology*, pages 35–40, 2008.

[20] Boriana L. Milenova and Marcos M. Campos. O-cluster: Scalable clustering of large high dimensional data sets. In *Proceedings from the IEEE International Conference on Data Mining*, pages 290–297, 2002.

[21] H. Nagesh, S. Goil, and A. Choudhary. MAFIA: Efficient and scalable subspace clustering for very large data sets, Technical Report No. CPDC-TR-9906-010 ©1999 Center for Parallel and Distributed Computing.

[22] Nam Hun Park and Won Suk Lee. Statistical grid-based clustering over data streams. *SIGMOD Record*, 33(1):32–37, March 2004.

[23] Abdol Hamid Pilevar and M. Sukumar. GCHL: A grid-clustering algorithm for high-dimensional very large spatial data bases. *Pattern Recognition Letters*, 26(7): 999–1010, 2005.

[24] Jorma Rissanen. *Stochastic Complexity in Statistical Inquiry Theory.* World Scientific Publishing Co., Inc., River Edge, NJ, USA, 1989.

[25] E. Schikuta. Grid-clustering: an efficient hierarchical clustering method for very large data sets. *Proceedings of the 13th International Conference on Pattern Recognition*, 2:101–105, 1996.

[26] Erich Schikuta and Martin Erhart. The BANG-clustering system: Grid-based data analysis. In *Proceedings of the 2nd International Symposium on Advances in Intelligent Data Analysis, Reasoning about Data. IDA-97*, pages 513–524. Springer-Verlag, 1997.

[27] Gholamhosein Sheikholeslami, Surojit Chatterjee, and Aidong Zhang. WaveCluster: A multi-resolution clustering approach for very large spatial databases. In *VLDB'98*, pages 428–439, 1998.

[28] Wei Wang, Jiong Yang, and Richard R. Muntz. STING: A statistical information grid approach to spatial data mining. In *VLDB'97, Proceedings of 23rd International Conference on Very Large Data Bases*, Athens, Greece, pages 186–195. Morgan Kaufmann, 1997.

[29] Wei Wang, Jiong Yang, and Richard R. Muntz. STING+: An approach to active spatial data mining. In *ICDE'99, Proceedings of the 15th International Conference on Data Engineering*, Sydney, Austrialia, pages 116-125, 1999.

[30] C. S. Warnekar and G. Krishna. A heuristic clustering algorithm using union of overlapping pattern-cells. *Pattern Recognition*, 11(2):85–93, 1979.

[31] Chen Xiaoyun, Chen Yi, Qi Xiaoli, Yue Min, and He Yanshan. PGMCLU: A novel parallel grid-based clustering algorithm for multi-density datasets. In *SWS '09: 1st IEEE Symposium on Web Society*, pages 166–171, 2009.

[32] Yanchang Zhao and Junde Song. GDILC: A grid-based density-isoline clustering algorithm. In *International Conferences on Info-tech and Info-net*, 3: 140 –145, 2001.

[33] Zhiwen Yu and Hau-San Wong. GCA: A real-time grid-based clustering algorithm for large data set. In *ICPR* (2): 740–743. IEEE Computer Society, 2006.

Chapter 7

Nonnegative Matrix Factorizations for Clustering: A Survey

Tao Li
Florida International University
Miami, FL
taoli@cs.fiu.edu

Chris Ding
University of Texas at Arlington
Arlington, TX
chqding@uta.edu

7.1 Introduction

Recently there has been significant development in the use of nonnegative matrix factorization (NMF) methods for various clustering tasks. NMF factorizes an input nonnegative matrix into two nonnegative matrices of lower rank. Although NMF can be used for conventional data analysis, the recent overwhelming interest in NMF is due to the newly discovered ability of NMF to solve challenging data mining and machine learning problems. In particular, NMF with the sum of squared error cost function is equivalent to a relaxed K-means clustering, the most widely used unsupervised learning algorithm. In addition, NMF with the I-divergence cost function is equivalent to probabilistic latent semantic indexing, another unsupervised learning method popularly used in text analysis. Many other data mining and machine learning problems can be reformulated as an NMF problem. This chapter aims to provide a comprehensive review of nonnegative matrix factorization methods for clustering. In particular, we outline the theoretical foundations on NMF for clustering, provide an overview of different variants on NMF formulations, and examine several practical issues in NMF algorithms. We also summarize recent advances on using NMF-based methods for solving many other clustering problems including co-clustering, semisupervised clustering, and consensus clustering and discuss some future research directions.

7.1.1 Background

Originally proposed for parts-of-whole interpretation of matrix factors, NMF has attracted a lot of attention recently and has been shown to be useful in a variety of applied settings, including environmetrics [88], chemometrics [128], pattern recognition [70], multimedia data analysis [26], text mining [130, 91], document summarization [89, 118], DNA gene expression analysis [11], financial data analysis [43], and social network analysis [138, 123, 19]. NMF can be traced back to the 1970s (Notes from G. Golub) and is studied extensively by Paatero and Tapper [88]. The work of Lee and Seung [68, 67] brought much attention to NMF in machine learning and data mining fields. Algorithmic extensions of NMF have been developed to accommodate a variety of objective functions [29] and a variety of data analysis problems, including classification [97] and collaborative filtering [106]. A number of studies have focused on further developing computational methodologies for NMF. In addition, various extensions and variations of NMF and the complexity proof that NMF is NP-hard have been proposed recently [56, 7, 37, 39, 111, 97, 135, 113, 38].

The true power of NMF, however, is NMF's ability to solve challenging data mining and pattern recognition problems. In fact, it has been shown [35, 37] that NMF with the sum of squared error cost function is equivalent to a relaxed K-means clustering, the most widely unsupervised learning algorithm. In addition, NMF with the I-divergence cost function is equivalent to probabilistic latent semantic indexing [48, 39, 40], another unsupervised learning method popularly used in text analysis. Thanks to the clustering capability, NMF has attracted a lot of recent attentions in data mining.

In this chapter, we provide a comprehensive review of nonnegative matrix factorization methods for clustering. To appeal to a broader audience in the data mining community, our review focuses more on conceptual formulation and interpretation rather than detailed mathematical derivations. The reader should be cautioned, however, that NMF is such a large research area that truly comprehensive surveys are almost impossible and, thus, that our overview may be a little eclectic. An interested reader is encouraged to consult with other papers for further reading. The reader should be cautioned also that in our presentation many mathematical descriptions are modified so that they adapt to data mining problems.

The rest of the chapter is organized as follows: the remainder of Section 7.1 introduces the basic formulations of NMF; Section 7.2 outlines the theoretical foundations on NMF for clustering and presents the equivalence results between NMF and various clustering methodologies; Section 7.3 demonstrates the NMF clustering capabilities and analyzes the advantages of NMF in clustering analysis; Section 7.4 provides an outline on the NMF algorithm development and discusses several practical issues in NMF algorithms; Section 7.5 provides an overview of many different NMF variants; and Section 7.6 summarizes recent advances on using NMF-based methods for solving many other clustering problems including co-clustering, semisupervised clustering, and consensus clustering. Finally, Section 7.7 concludes the chapter and discusses future research directions.

7.1.2 NMF Formulations

Let the input data matrix $X = (\mathbf{x}_1, \cdots, \mathbf{x}_n)$ contain the collection of n data column vectors. Generally, we factorize X into two matrices,

$$X \approx FG^T, \tag{7.1}$$

where $X \in \mathbb{R}^{p\times n}$, $F \in \mathbb{R}^{p\times k}$, and $G \in \mathbb{R}^{n\times k}$. Generally, $p < n$ and the rank of matrices F, G is much lower than the rank of X, i.e., $k \ll \min(p,n)$. F, G are obtained by minimizing a cost function. The most common cost function is the **sum of squared errors**,

$$\min_{F,G\geq 0} J_{\text{sse}} = \|X - FG^T\|^2. \tag{7.2}$$

In this chapter, the matrix norm is implicitly assumed to be the Frobenius norm. A rank nondeficiency condition is assumed for F, G.

Another cost function is the so-called **I-divergence**:

$$\min_{F,G\geq 0} J_{\text{ID}} = \sum_{i=1}^{m}\sum_{j=1}^{n}\left[X_{ij}\log\frac{X_{ij}}{(FG^T)_{ij}} - X_{ij} + (FG^T)_{ij}\right]. \tag{7.3}$$

It is easy to show that the inequality $I(x) = x\log x - x + 1 \geq 0$ holds when $x \geq 0$; the equality holds when $x = 1$. The quantity $I(u,v) = (u/v)\log(u/v) - u/v + 1$ is called I-divergence,

7.2 NMF for Clustering: Theoretical Foundations

Although NMF can be used for conventional data analysis [88], the overwhelming interest in NMF is the newly discovered ability of NMF to solve challenging clustering problems [35, 37]. Here we outline the recent results on (1) the relationship between NMF with least square objective and K-means clustering; (2) the relationship between NMF using I-divergence and PLSI; and (3) the relationship between NMF and spectral clustering. These results establish the theoretical foundations for NMF to solve unsupervised learning problems. We also present the boundedness theorem which offers the theoretical foundation for the normalization of factor matrices.

7.2.1 NMF and K-Means Clustering

The K-means clustering algorithm is one of the most popularly used data clustering methods. Let $X = (\mathbf{x}_1, \cdots, \mathbf{x}_n)$ be n data points. We partition them into K mutually disjoint clusters. The K-means clustering objective can be written as

$$J_{\text{Kmeans}} = \sum_{i=1}^{n}\min_{1\leq k\leq K}\|\mathbf{x}_i - \mathbf{f}_k\|^2 = \sum_{k=1}^{K}\sum_{i\in C_k}\|\mathbf{x}_i - \mathbf{f}_k\|^2.$$

The following theorem shows that NMF is inherently related to the K-means clustering algorithm [35].

Theorem 1 *G-orthogonal NMF,*

$$\min_{F\geq 0, G\geq 0} \|X - FG^T\|^2, \ \ s.t. \ G^T G = I. \tag{7.4}$$

is equivalent to K-means clustering. This holds even if X and F have mixed-sign entries.

We can understand this relationship in this way [35]. Let $C = (\mathbf{c}_1, \cdots, \mathbf{c}_k)$ be the cluster centroids obtained via K-means clustering. Let H be the cluster indicators; i.e., $h_{ki} = 1$ if $\mathbf{x}_i$ belongs to cluster c_k; $h_{ki} = 0$ otherwise. We can write the K-means cluster objective as $J = \sum_{i=1}^{n} \sum_{k=1}^{K} h_{ik} \|\mathbf{x}_i - \mathbf{c}_k\|^2 = \|X - CH^T\|^2$. From this analogy, in NMF F has the meaning of cluster centroids and G is the cluster indicator [35]. Thus K-means and NMF have the same objective function but with different constraints.

Indeed, the K-means objective function can be expressed in such a way: if we ignore the nonnegativity constraint while keeping the orthogonality constraint, the principal component is the solution [34, 137]. On the other hand, if we ignore the orthogonality while keeping the nonnegativity, NMF is the solution.

7.2.2 NMF and Probabilistic Latent Semantic Indexing

Probabilistic Latent Semantic Indexing (PLSI) is a unsupervised learning method based on statistical latent class models and has been successfully applied to document clustering [55]. (PLSI is further developed into a more comprehensive Latent Dirichlet Allocation model [9].) PLSI maximizes the likelihood

$$J_{\text{PLSI}} = \sum_{i=1}^{m} \sum_{j=1}^{n} X(w_i, d_j) \log P(w_i, d_j), \tag{7.5}$$

where the joint occurrence probability is factorized (i.e., parameterized or approximated) as

$$P(w_i, d_j) = \sum_k P(w_i, d_j|z_k)P(z_k) = \sum_k P(w_i|z_k)P(d_j|z_k)P(z_k), \tag{7.6}$$

assuming that w_i and d_j are conditionally independent given z_k. The following theorem shows the equivalence between NMF and PLSI [39, 40]:

Theorem 2 *PLSI is equivalent to NMF with I-divergence objective. (A) The objective function of PLSI is identical to the objective function of NMF using the I-divergence. (B) We can express* $FG^T = \bar{F}D\bar{G}^T$ *where* $\bar{F}, D, \bar{G}^T$ *satisfy the probability normalization:* $\sum_{i=1}^{m} \bar{F}_{ik} = 1, \sum_{j=1}^{n} \bar{G}_{jk} = 1, \sum_{k=1}^{K} \bar{D}_{kk} = 1$.

Therefore, the NMF update algorithm and the EM algorithm in training PLSI are alternative methods to optimize the same objective function [39]. The relationships between NMF and PLSI have also been studied in [48].

7.2.3 NMF and Kernel K-Means and Spectral Clustering

For a square symmetric matrix W, we would expect a $W \simeq HH^T$ type decomposition. The following theorem points out its usefulness for data clustering [35].

Theorem 3 *Orthogonal symmetric NMF*

$$\min_{H\geq 0} \|W - HH^T\|^2, \ \ s.t. \ H^T H = I \tag{7.7}$$

is equivalent to kernel K-means clustering.

It has also been shown that symmetric matrix factorization, a special case of nonnegative matrix factorization, is equivalent to sophisticated normalized cut spectral clustering [100]. Given the adjacent matrix W of a graph, it can be easily seen that the following matrix factorization

$$\min_{H^T H = I, H \geq 0} ||\widetilde{W} - HH^T||^2, \tag{7.8}$$

where

$$\widetilde{W} = D^{-1/2} W D^{-1/2}, \; D = \mathrm{diag}(d_1, \cdots, d_n), \; d_i = \sum_j w_{ij},$$

is equivalent to Normalized Cut spectral clustering [35, 118].

7.2.4 NMF Boundedness Theorem

NMF differs from SVD (Singular Value Decomposition) due to the absence of cancellation of plus and minus signs. But what is the fundamental significance of this absence of cancelation? It is the Boundedness Property [140, 141]. The boundedness theorem offers the theoretical foundation for the normalization of F and G in $X = FG^T$. A matrix A is bounded, if $0 \leq A_{ij} \leq 1$. Note that for any nonnegative input matrix, we can rescale it into the bounded form.

The boundedness property of NMF states: *if X is bounded, then the factor matrices F, G must also be bounded.* More precisely:

Theorem 4 (Boundedness Theorem) *Let $0 \leq X \leq 1$ be the input data matrix. F, G are the nonnegative matrices satisfying $X = FG^T$. There exists a diagonal matrix D such that when we rescale $X = FG^T = (FD)(GD^{-1})^T = F^*(G^*)^T$, the rescaled matrices satisfy $0 \leq F^*_{ij}, G^*_{ij} \leq 1$. D is constructed this way: $D = D_F^{1/2} D_G^{-1/2}$, $D_F = \mathrm{diag}(f_1, \cdots, f_k), f_k = max_p F_{kp}$ and $D_G = \mathrm{diag}(g_1, \cdots, g_k), g_k = max_p G_{kp}$. This holds when X is symmetric, i.e., $X = HH^T$, $0 \leq H_{ij} \leq 1$.*

This theorem assures the existence of an appropriate scale such that both W and H are bounded, i.e., their elements cannot exceed the magnitude of the input data matrix. We note that SVD decomposition does not have the boundedness property. In this case, even if the input data are in the range of $0 \leq X_{ij} \leq 1$, we cannot guarantee that for all i, j, $|F^*_{ij}| \leq 1$ and $|V^*_{ij}| \leq 1$.

7.3 NMF Clustering Capabilities

7.3.1 Examples

We demonstrate the clustering capabilities of NMF using several examples. Figures 7.1 and 7.2 present the computed F factors (images) on two image datasets: ORL face image dataset and Digits image dataset. Here each image is represented as a vector of pixel gray values. Note that by Theorem 1, these F factors are cluster *centroids* (representatives of clusters). Intuitively, these images are *representatives* of clusters of the original images. The examples show that NMF provides a holistic view of the datasets.

7.3.2 Analysis

In general, the advantages of NMF over the existing unsupervised learning methods can be summarized as follows. NMF can model widely varying data distributions due to the flexibility of matrix factorization as compared to the rigid spherical clusters that the K-means clustering objective

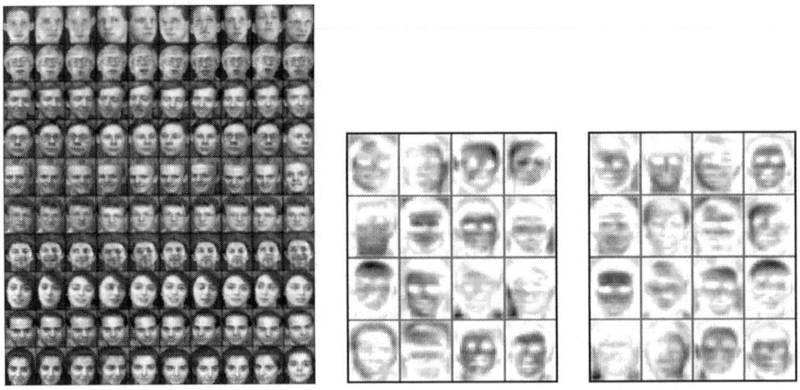

FIGURE 7.1: Left: ORL face image dataset, containing 100 images of 10 persons. Middle/Right: Computed NMF factors $F = (\mathbf{f}_1, \cdots, \mathbf{f}_{16})$ for 2 runs with random initialization.

function attempts to capture. When the data distribution is far from a spherical clustering, NMF may have advantages. Another advantage of NMF is that NMF can do both hard and soft clustering simultaneously.

Figure 7.3 gives an example. The 2-D dataset in the left figure consists of 38 data points and the G values are shown in the right figure. As seen in the figure, G values for points in regions $\{C, D, E\}$ indicate they are fractionally assigned to different clusters. As a result, NMF is able to perform soft clustering.

The third advantage is that NMF is able to perform simultaneously clustering of the rows (features) and the columns (data points) of an input data matrix. Consider the following NMF objective with orthonormal constraints on the factor matrices:

$$J_{orth} = \|X - FG^T\|^2, \ \ s.t. \ \ F \geq 0, G \geq 0, F^T F = I, G^T G = I. \tag{7.9}$$

The orthonormal constraints and the nonnegative constraints in Equation (7.9) make the resulted F and G approximate the K-means clustering results on both features and data points [35]. The fourth advantage is that many other data mining and machine learning problems can be reformulated as NMF problems.

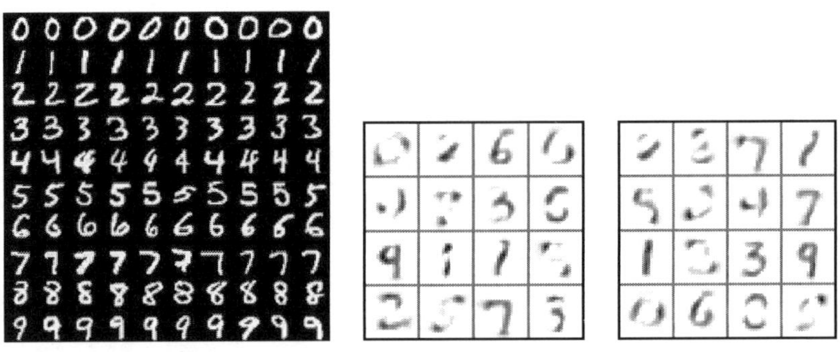

FIGURE 7.2: Left: Digits image dataset. Middle/Right: Computed NMF factors $F = (\mathbf{f}_1, \cdots, \mathbf{f}_{16})$ for 2 runs with random initialization.

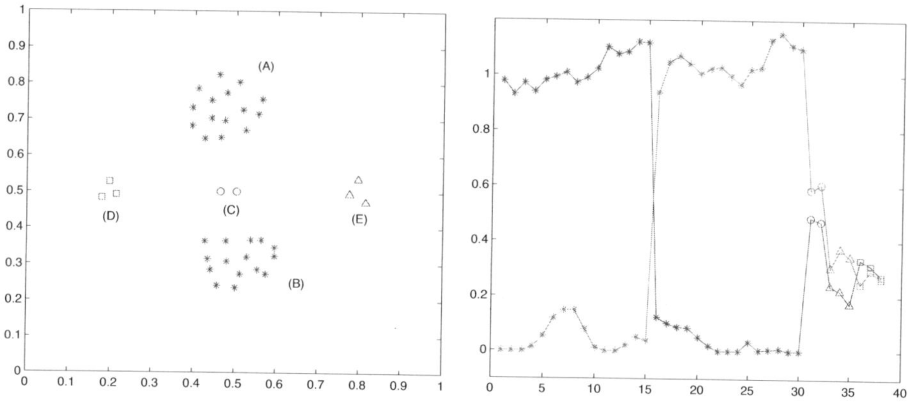

FIGURE 7.3: Left: A 2D dataset of 38 data points. Right: Their $G = (\mathbf{g}_1, \mathbf{g}_2)$ values are shown as blue and red curves. Points in regions $\{A, B\}$ are numbered 1–30, where $\mathbf{g}_1$ and $\mathbf{g}_2$ values differ significantly, indicating that these points are uniquely clustered into either cluster A or cluster B (hard clustering). Points in regions $\{C, D, E\}$ are numbered 31–38, where $\mathbf{g}_1$ and $\mathbf{g}_2$ values are close, indicating that these points are fractionally clustered into clusters A and B (soft clustering).

7.4 NMF Algorithms

7.4.1 Introduction

The algorithms for matrix factorizations are generally iterative updating procedures: updating one factor while fixing the other factors. Existing algorithms for NMF include multiplicative updates of Lee and Seung [68, 67], alternating least square (ALS) algorithms [88, 7], alternating nonnegative least squares (ANLS) [59, 60], gradient descent methods [56, 98], projected gradient methods [79], and Newton-type algorithms [134, 58]. More details on NMF algorithms can be found in [105, 7, 24]. In this section, we provide an outline on the algorithm development for orthogonal NMF and also discuss several practical issues when applying NMF for clustering.

7.4.2 Algorithm Development

Here we provide an outline on the algorithm development for orthogonal NMF. We wish to solve the optimization problem

$$\min_{F \geq 0} J(F) = ||X - FG^T||^2, \quad s.t. \; F^T F = I, \tag{7.10}$$

where $X, G \geq 0$ are fixed. We show that given an initial F, the updating algorithm

$$F_{ik} \leftarrow F_{ik} \frac{(XG)_{ik}}{(FG^T X^T F)_{ik}} \tag{7.11}$$

correctly finds a local minimum of this problem.

We can use the gradient-descent method, the current standard approach, to update F as $F \leftarrow F - \delta \tilde{\nabla}_G J$, where $\tilde{\nabla}_F J = \nabla_F J - F(\nabla_F J)^T F$ is the *natural gradient* [46] to enforce the orthogonality $F^T F = I$ (F on the Stiefel manifold). We obtain the updating rule

$$F_{ik} \leftarrow F_{ik} - \delta(-XG + FG^T X^T F)_{ik}. \tag{7.12}$$

Note that in Equation (7.12), setting the stepsize $\delta = G_{ik}/(FG^T X^T F)_{ik}$ we recover the updating rule Equation (7.11).

We can also show that the fixed point of the iteration in Equation (7.11) is in fact the KKT (Karush–Kuhn–Tucker) condition in the theory of *constrained optimization*. We introduce the Lagrangian multipliers $\Lambda \in \Re^{K\times K}$ and minimize the Lagrangian function $L(F) = ||X - FG^T||^2 + \mathrm{Tr}[\Lambda(F^T F - I)]$. The KKT complementarity slackness condition for the nonnegativity of F_{ik} gives

$$(-2XG + 2FG^T G + 2F\Lambda)_{ik} F_{ik} = 0. \tag{7.13}$$

This is the fixed point condition that any local minimum F^* must satisfy. From Equation (7.13) we obtain the Lagrangian multiplier $\Lambda = G^T X^T F^* - G^T G$. Now, we can easily check that the converged solution of the update rule Equation (7.11) also satisfies the fixed point relation Equation (7.13) with Λ substituted in. This shows that the converged solution of the update rule Equation (7.11) is a local minimum of the optimization problem.

The convergence of Equation (7.11) is guaranteed by the monotonicity theorem which can be proved using the auxiliary function approach [67, 41].

Theorem 5 *The Lagrangian function L is nonincreasing (monotonically decreasing) under the update rule Equation (7.11).*

7.4.3 Practical Issues in NMF Algorithms

In the following we discuss several practical issues when applying NMF for clustering.

7.4.3.1 Initialization

Similar to K-means clustering, initialization also plays an important role when applying NMF for clustering since the objective function may have many local minima [109, 53]. The intrinsic alternating minimization in NMF algorithms is nonconvex, even though the objective function is convex with respect to one set of variables.

A simple random initialization, where the factor matrices are initialized as random matrices, is generally not effective as it often leads to slow convergence to a local minimum. Many different techniques have been developed for improving the random initialization:

- Multiple initializations: The core idea is to perform multiple runs using different random initializations and select the best estimates from multiple runs. This method needs to perform NMF multiple times and is computationally expensive.
- Factorization-based initialization: Note that NMF is a constrained low rank matrix factorization; hence, we can use the results from alternative low rank factorization methods as the initialization [10]. Typical examples include SVD-based initialization [10] and CUR decomposition-based initialization [66].
- Clustering-based initialization: If we think of NMF as a clustering process, we can seek the initialization strategy based on the results obtained from other clustering algorithms (e.g., spherical k-means [127, 41], divergence-based K-means [131], and fuzzy clustering [142]).

A comparison of different initialization methods can be found in [66]. Note that both factorization-based initialization and clustering-based initialization methods are able to lead to rapid error reduction and faster convergence.

7.4.3.2 Stopping Criteria

Beyond a predefined number of iterations or a fixed running time, there are simple heuristics that can be used as the stopping criteria for the iterative algorithms of NMF: (1) the objective function is reduced to below a given threshold, and (2) there is little change on the factor matrices or the objective function between successive iterations. Recently, Lin employed stopping conditions from bound-constrained optimization for the iterative procedures of NMF [79] and Kim and Park tested the combined convergence criterion using the KKT optimality conditions and the convergence of positions of the largest elements in the factor matrices [59].

Based on Theorem 1, upon the completion of NMF algorithms, G corresponds to the cluster indicator. Thus, the clustering assignment can be determined by the largest element of each row in G. Clustering assignments can also be determined by finding a discrete clustering membership function that is close to the resulted G, using the strategy similar to [133].

7.4.3.3 Objective Function vs. Clustering Performance

In clustering problems, when we have the external structure information (i.e., the derived class labels of the data points), various performance measures such as purity, normalized mutual information (NMI), and accuracy [109, 53, 39] have been used as the performance measure. In practice, the clusters of a given dataset could have many different shapes and sizes. In addition, these clusters can overlap with each other and they cannot be easily identified and separated. As a result, it is difficult to effectively capture the cluster structures using a single (even the "best" if exists) clustering objective function.

Hence there is generally a gap between the objective function and the clustering performance measure [33]. In many real applications, when the clustering objective function is optimized, the quality of clustering in terms of accuracy or NMI is usually improved. But the improvement works only up to a certain point, and beyond that, further optimization of the objective function will not improve the clustering performance as the objective function does not necessarily capture the cluster structure. It is also possible that as the objective function is optimized, the clustering performance (e.g., accuracy or NMI) of clustering can be degraded.

7.4.3.4 Scalability

Many prior efforts have been reported on scaling up matrix factorization methods through delicate algorithm designs. While efficient computation algorithms for matrix factorization have been developed, they have primarily been in the context of scientific/engineering applications where sizes of matrices (tens of thousands by tens of thousands with millions of nonzero values) are typically much smaller than those for Web analysis (millions-by-millions matrices with billions of observations). Here we present several practical mechanisms for dealing with large-scale datasets for NMF:

- Shrinking scheme: The scheme first "coarsens" the original large scale problem into a small one (level by level), then solves the coarsened optimization problem, and finally refines the solution to the coarsened problem to approximate the solution to the original problem. A typical example of the shrinking scheme is the multi-level graph partitioning [31, 54, 57].

- Partitioning scheme: The scheme first divides the original problem into a series of small scale problems, and then solves those small problems, and finally combines their solutions to approximate the solution to the original problem. Typical examples of this scheme include blockwise principal component analysis [87], canopy-based clustering [85], and K-means preclustering [121].

- Online/Incremental scheme: The scheme does not require the input dataset to reside in the memory and performs factorization in an incremental fashion (i.e., one sample or a chunk of samples per step). Examples of the scheme include incremental NMF [16], online NMF by dynamic updating of the factor matrices [13], online NMF using stochastic approximation [122, 52, 120], and evolutionary NMF [125, 94].

- Parallel and distributed implementation: An orthogonal direction for improving the scalability of matrix factorization methods is to use parallel and distributed computing platforms [42, 80, 6]. MapReduce is a programming model proposed for processing and generating large data sets [26]. Although the initial purpose of MapReduce is to perform large-scale data processing, it turns out that this model is much more expressive and has been widely used in many data mining and machine learning tasks [112, 21, 15]. Liu et al. [80] successfully scaled up the basic NMF for web-scale dyadic data analysis using MapReduce, and Sun et al. [108] presented different matrix multiplication implementations and scaled up three types of nonnegative matrix factorizations on MapReduce. Recently, Graphics Processing Unit (GPU) implementation of the NMF algorithms has also been reported [65, 84, 93].

7.5 NMF Related Factorizations

Here we provide an overview on related matrix factorization methods.

1. **SVD:** The classic matrix factorization is Principal Component Analysis (PCA) which uses the singular value decomposition [45, 50], $X \approx U\Sigma V^T$, where we allow U, V to have mixed signs; the input data could have mixed signs. Absorbing Σ into U, we can write

$$\text{SVD:} \quad X_{\pm} \approx U_{\pm} V_{\pm}. \tag{7.14}$$

2. **NMF:** When the input data is nonnegative, we restrict F and G to be nonnegative. The standard NMF can be written as

$$\text{NMF:} \quad X_{+} \approx F_{+} G_{+}, \tag{7.15}$$

   using an intuitive notation for $X, F, G \geq 0$.

3. **Regularized NMF:** Additional constraints can be added to the standard NMF formulation. In general, regularized NMF can be written as the following optimization problem:

$$\min_{F,G \geq 0} \{\|X - FG^T\|^2 + \alpha J_1(F) + \beta J_2(G)\}, \tag{7.16}$$

   where the functions $J_1(F)$ and $J_2(G)$ are penalty terms to enforce certain constraints and α and β are the corresponding regularization parameters [90]. Many different penalty terms have been used in the literature to have different effects on the computed solutions such as the smoothness constraint (e.g., the matrix norm) [90], the sparsity constraints (e.g., L_1-norm regularization) [59, 56], the geometric constraints (e.g., manifold structures) [12], and the local learning regularizations based on neighborhoods [51].

4. **Projective NMF:** A projective NMF model aims to find an approximate projection of the input matrix [132, 24]. It can be formulated as follows:

$$X \approx FF^T X. \tag{7.17}$$

   The idea of the projective NMF is similar to subspace clustering.

5. **Semi-NMF:** When the input data has mixed signs, we can restrict G to be nonnegative while placing no restriction on the signs of F. This is called semi-NMF [37, 36]:

$$\text{semi-NMF:} \quad X_{\pm} \approx F_{\pm} G_{+}. \tag{7.18}$$

6. **Convex-NMF:** In general, the basis vectors $F = (\mathbf{f}_1, \cdots, \mathbf{f}_k) \in \Re_+^{n\times k}$ are a much larger space than the input space spanned by the columns of $X = (\mathbf{x}_1, \cdots, \mathbf{x}_n)$, while according to Theorem 1, F has the meaning of cluster centroids. To enforce this geometry meaning, we restrict F to be a convex combination of the input data points, i.e., F lies in the input space, $\mathbf{f}_l = w_{1l}\mathbf{x}_1 + \cdots + w_{nl}\mathbf{x}_n = X\mathbf{w}_l$, or $F = XW, w_{il} \geq 0$. We call this restricted form of factorization Convex-NMF [37, 36]. Convex-NMF applies to both nonnegative and mixed-sign input data:
$$X_\pm \approx X_\pm W_+ G_+^T. \tag{7.19}$$
Recently, ***Convex-Hull NMF*** has been proposed to extend Convex-NMF by restricting the convexity on the columns of both F and G, thus leading to the factorization where each input data point is expressed as a convex combination of convex hull data points [110]. Convex-Hull NMF can be expressed as
$$X \approx CG^T, \tag{7.20}$$
where C consists of a set of appropriate points $c_i \in conv(X)$, and $conv(X)$ is the convex hull of X.

7. **Cluster NMF:** In Convex-NMF, we require the columns of F to be convex combinations of input data. Suppose now that we interpret the entries of G as posterior cluster probabilities. In this case the cluster centroids can be computed as $\mathbf{f}_k = X\mathbf{g}_k/n_k$, or $F = XGD_n^{-1}$, where $D_n = \text{diag}(n_1, \cdots, n_k)$. The extra degree of freedom for F is not necessary. Therefore, the pair of desiderata—(1) F encodes centroids and (2) G encodes posterior probabilities—motivates a factorization $X \approx XGD_n^{-1}G^T$. We can absorb D_n^{-1} into G and solve for
$$\text{Cluster-NMF}: \; X \approx XG_+G_+^T. \tag{7.21}$$
We call this factorization *Cluster-NMF* because the degree of freedom in this factorization is the cluster indicator G, as in a standard clustering problem [37]. The objective function is $J = \|X - XGG^T\|^2$.

8. **Tri-Factorization:** To simultaneously cluster the rows and the columns of the input data matrix X, we consider the following nonnegative 3-factor decomposition [41]:
$$X \approx FSG^T. \tag{7.22}$$
Note that S provides additional degrees of freedom such that the low-rank matrix representation remains accurate while F gives row clusters and G gives column clusters. More precisely, we solve
$$\min_{F\geq 0, G\geq 0, S\geq 0} \|X - FSG^T\|^2, \; s.t. \; F^TF = I, \; G^TG = I. \tag{7.23}$$
This form gives a good framework for simultaneously clustering the rows and columns of X [30, 136].

An important special case is that the input X contains a matrix of pairwise similarities: $X = X^T = W$. In this case, $F = G = H$. We call it symmetric NMF, which optimizes
$$\min_{W\geq 0, S\geq 0} \|X - HSH^T\|^2, \quad \min_{W\geq 0, S\geq 0} \|X - HSH^T\|^2, \; s.t. \; H^TH = I.$$

9. **Kernel NMF:** Consider a mapping such as those used in support vector machines:
$$\mathbf{x}_i \to \phi(\mathbf{x}_i), \text{ or }, \; X \to \phi(X) = (\phi(\mathbf{x}_1), \cdots, \phi(\mathbf{x}_n)).$$
Similar to the concept proposed for convex-NMF, we restrict F to be a convex combination of transformed input data points:
$$\phi(X) \simeq \phi(X)WG^T, \tag{7.24}$$

rather than a standard NMF like $\phi(X) \approx FG^T$, which would be difficult since F, G will depend explicitly on the mapping function $\phi(\cdot)$. It is easy to see that the minimization objective:

$$||\phi(X) - \phi(X)WG^T||^2 = \mathrm{Tr}[\phi(X)^T\phi(X) - 2G^T\phi^T(X)\phi(X)W + W^T\phi^T(X)\phi(X)WG^TG], \quad (7.25)$$

depends only on the kernel $K = \phi^T(X)\phi(X)$. This kernel extension of NMF is similar to kernel-PCA and kernel$K$-means .

10. **Multi-layer NMF:** In multi-layer NMF, the basic factor matrix is replaced by a set of cascaded matrices using a sequential decomposition [22, 32, 23]. Given an input matrix X, it is first decomposed as $F_{(1)}G^T_{(1)}$. Then in the second step, $G^T_{(1)}$ is further decomposed to $F_{(2)}G^T_{(2)}$ and the process is repeated for a number of times. The multi-layer NMF model can be written as follows:

$$X \approx F_{(1)}F_{(2)}\cdots F_{(t)}G^T_{(t)} = FG^T. \quad (7.26)$$

It has been shown that the multi-layer NMF can improve the performance of most NMF algorithms and address the problem of local minima due to the distributed and multi-stage nature and the sequential decomposition with different initial conditions [24].

11. **Binary NMF:** When the input data X is binary, binary NMF factorizes X into two binary matrices thus conserving the most important integer property of X [141, 140, 71]. The binary NMF model can be written as

$$\text{Binary NMF:} \quad X_{0-1} \approx F_{0-1}G^T_{0-1}. \quad (7.27)$$

A special case of binary NMF is the boolean factorization [86]

$$X_{0-1} \approx W_{0-1} \oplus H_{0-1}.$$

12. **Weighted Feature Subset NMF:** In weighed NMF, weights are incorporated to indicate the importance of the corresponding rows and columns. If we consider only the feature importance, this leads to the feature subset NMF (FS-NMF):

$$\min_{W \geq 0, F \geq 0, G \geq 0} ||X - FG^T||^2_W, s.t. \ \sum_j W^\alpha_j = 1, \quad (7.28)$$

where $W \in \mathbb{R}^{m\times m}_+$ is a diagonal matrix indicating the weights of the rows (keywords or features) in X, and α is a parameter [116, 117]. In general, we can also assign different weights to different samples. This leads to the weighted FS-NMF:

$$\min_{W \geq 0, F \geq 0, G \geq 0} ||X - FG^T||^2_W,$$

where we set $W_{ij} = a_i b_j$. This becomes

$$\min_{W \geq 0, F \geq 0, G \geq 0} (X - FG^T)^2_{ij} a_i b_j,$$
$$s.t. \ \sum_i a^\alpha_i = 1, \sum_j b^\beta_j = 1, \quad (7.29)$$

where α, β are two parameters with $0 < \alpha < 1, 0 < \beta < 1$.

13. **Robust NMF:** Recently a robust NMF by using $L_{2,1}$-norm loss function has been proposed in [63]. The error function is

$$||X - FG^T||_{2,1} = \sum_i \sqrt{\sum_j (X - FG^T)^2_{ij}}. \quad (7.30)$$

The proposed robust NMF formulation can handle outliers and noises in a better way than standard NMF.

In summary, various newly proposed NMF formulations are collectively presented as follows:

SVD:	$X_\pm \approx U_\pm V_\pm$
NMF:	$X_+ \approx F_+ G_+^T$
Regularized NMF:	$\min_{F,G\geq 0} \{\|X - FG^T\|^2 + \alpha J_1(F) + \beta J_2(G)\}$
Projective NMF:	$X \approx FF^T X$
Semi-NMF:	$X_\pm \approx F_\pm G_+^T$
Convex-NMF:	$X_\pm \approx X_\pm W_+ G_+^T$
Convex-hull NMF:	$X \approx CG^T, \; c_i \in conv(X)$
Cluster-NMF:	$X \approx XG_+ G_+^T$
Kernel-NMF:	$\phi(X_\pm) \approx \phi(X_\pm) W_+ G_+^T$
Tri-Factorization:	$X_+ \approx F_+ S_+ G_+^T$
Symmetric-NMF:	$W_+ \approx H_+ S_+ H_+^T$
Multi-layer NMF:	$X \approx F_{(1)} F_{(2)} \cdots F_{(t)} G_{(t)}^T$
Binary NMF:	$X_{0-1} \approx F_{0-1} G_{0-1}^T$
Weighted Feature Subset NMF:	$\min \|\|X - FG^T\|\|_W^2$
Robust NMF:	$\min \|\|X - FG^T\|\|_{2,1}^2$

Note that there are other NMF formulations that are not included in the above discussion, such as convolutive NMF for a set of nonnegative matrices [24, 103] and Bayesian NMF with the incorporation of Bayesian techniques to NMF [14, 95, 96].

7.6 NMF for Clustering: Extensions

We have shown that NMF provides a general framework for unsupervised learning. NMF can model widely varying data distributions and can do both hard and soft clustering simultaneously. In fact, many other clustering problems such as co-clustering, consensus clustering, semisupervised clustering, and graph clustering can be reformulated as an NMF problems.

7.6.1 Co-Clustering

In many real world applications, a typical task often involves more than one type of data points and the input data are association data relating different types of data points. For example, in document analysis, there are *terms* and *documents*. In DNA micro-array data, rows represent *genes* and columns represent *samples*. Co-clustering algorithms aim at clustering different types of data simultaneously by making use of the dual relationship information such as the term-document matrix and the gene-sample matrix [30, 136, 20, 82, 83].

To simultaneously cluster the rows and the columns of the input data matrix X, tri-factorization has been proposed for 3-factor nonnegative matrix decomposition [41], which aims to solve

$$\min_{F\geq 0, G\geq 0, S\geq 0} \|X - FSG^T\|^2, \; s.t. \; F^T F = I, \; G^T G = I. \tag{7.31}$$

Note that S provides additional degrees of freedom such that the low-rank matrix representation remains accurate while F gives row clusters and G gives column clusters. Tri-factorization provides a nice co-clustering framework. Recently, a fast tri-factorization extension has been proposed in [126] for large-scale data co-clustering by restricting the factor matrices to be cluster indicator matrices (a special type of nonnegative matric).

7.6.2 Semisupervised Clustering

In many situations when we discover new patterns using clustering, there exists some prior, partial, incomplete knowledge about the problem. We wish to incorporate that knowledge into the clustering algorithm. Semisupervised clustering refers to the situation where the clustering is done with many prespecified must-link constraints (two data points must be clustered into the same cluster) and cannot-link constraints (two data points cannot be clustered into the same cluster) [5, 8, 4, 115, 129, 61].

Specifically the above constraints are formulated as follows [115]: (1) Must-link constraints. $A = \{(i_1, j_1), \cdots, (i_a, j_a)\}, a = |A|$, contains pairs of data points, where $\mathbf{x}_{i_1}, \mathbf{x}_{j_1}$ are considered similar and must be clustered into the same cluster. (2) Cannot-link constraints. $B = \{(i_1, j_1), \cdots, (i_b, j_b)\}, b = |B|$, where each pair of points are considered dissimilar and cannot be clustered into the same clusters. A, B can also be viewed as symmetric matrices containing $\{0, 1\}$. Using cluster indicator H, the must-link of (i_1, j_1) implies that $\mathbf{x}_{i_1}, \mathbf{x}_{j_1}$ should have significant nonzero posterior probability at the same cluster k, i.e., the overlap $\sum_{k=1}^{K} h_{i_1k}h_{j_1k} = (HH^T)_{i_1j_1}$ should be maximized. Thus, the must-link condition is $\max_H \sum_{(ij)\in A}(HH^T)_{ij} = \sum_{ij} A_{ij}(HH^T)_{ij} = \mathrm{Tr}H^TAH$. Similarly, the cannot-link constraints can be formulated as $\min_H \mathrm{Tr}H^TBH$. Putting these constraint conditions together, the semisupervised clustering problem can be cast as the following optimization problem:

$$\max_{H^TH=I, H\geq 0} \mathrm{Tr}[H^TWH + \alpha H^TAH - \beta H^TBH], \tag{7.32}$$

where parameter α controls the weight for must-link constraints in A, and β controls the weight of cannot-link constraints in B. Weights α, β allow certain levels of uncertainties so that must-link and cannot-link constraints are not necessarily always vigorously enforced.

Let $W^+ = W + \alpha A \geq 0, \quad W^- = \beta B \geq 0$. The semisupervised clustering problem can be reformulated as an NMF problem [73]

$$\min_{H^TH=I, H\geq 0} ||(W^+ - W^-) - HH^T||^2. \tag{7.33}$$

Thus, the semisupervised clustering problem is equivalent to a semi-NMF problem [37, 72].

Chen et al. formulated semisupervised clustering with the instance-level constraints using symmetric nonnegative tri-factorization [17]. Zhu et al. [143] noted that must-link and cannot-link constraints play different roles in clustering and proposed a constrained NMF method where must-link constraints are used to control the distance of the data in the compressed form, and cannot-link constraints are used to control the encoding factor.

There are also some other research efforts on incorporating the (partial) class label information into the matrix factorization framework. For example, Lee et al. [69] presented semisupervised NMF (SSNMF) which jointly incorporates the data matrix and the (partial) class label matrix into NMF. Liu and Wu [81] proposed a form of constrained NMF by requiring that the data points sharing the same label have the same coordinate in the new representation space.

7.6.3 Semisupervised Co-Clustering

In co-clustering two types of objects, sometimes we have partial knowledge on x-type objects and also partial knowledge on y-type objects. Semisupervised co-clustering aims to incorporate the knowledge in co-clustering. As in Section 7.6.2, we can formulate the partial knowledge as must-link and cannot-link constraints on both x-type and y-type objects. Let A^x contain the must-link pairs for x-type objects (A^y for y-type objects), and B^x contain the cannot-link pairs for x-type objects (B^y for y-type objects). Then, the semisupervised co-clustering problem can be formulated as

$$\min_{F\geq 0, G\geq 0} J = \|X - FSG^T\|^2 + \mathrm{Tr}[aF^T(A^x - B^x)F + bG^T(A^y - B^y)G], \tag{7.34}$$

where a, b are parameters to control the effects of different types of constraints [124]. Another semisupervised co-clustering method has been proposed in [18] using symmetric nonnegative tri-factorization.

Recently, Li et al. [74, 78] proposed several constrained nonnegative tri-factorization knowledge transformation methods to use the partial knowledge (such as instance-level constraints and partial class label information) from one type of objects (e.g., words) to improve the clustering of another type of objects (e.g., documents). Their models bring together semisupervised clustering/co-clustering and learning from labeled features [44, 101, 102].

7.6.4 Consensus Clustering

Consensus clustering, also called aggregation of clustering, refers to the situation where a number of clustering results on the same dataset are already obtained and the task is to find a clustering which is closest to those already obtained clusterings [49, 107, 47, 75].

Formally let $X = \{x_1, x_2, \cdots, x_n\}$ be a set of n data points. Suppose we are given a set of T clusterings (or partitionings) $\mathcal{P} = \{P^1, P^2, \cdots, P^T\}$ of X. Note that the number of clusters could be different for different clusterings. Let us define the *connectivity* matrix $M_{ij}(P^t)$ for a single partition P^t as

$$M_{ij}(P^t) = \begin{cases} 1 & (i,j) \text{ belong to the same cluster} \\ 0 & \text{otherwise} \end{cases} \tag{7.35}$$

Consensus clustering looks for a consensus partition (consensus clustering) P^* which is the closest to all the given partitions:

$$\min_{P^*} J = \frac{1}{T}\sum_{t=1}^{T}\sum_{i,j=1}^{n}[M_{ij}(P^t) - M_{ij}(P^*)]^2 = \frac{1}{T}\sum_{t=1}^{T}\|M(P^t) - M(P^*)\|_F^2.$$

Let the average association between i and j be $\widetilde{M}_{ij} = \frac{1}{T}\sum_{t=1}^{T} M_{ij}(P^t)$. We have

$$J = \frac{1}{T}\sum_{t=1}^{T}\|M(P^t) - \widetilde{M}\|_F^2 + \|\widetilde{M} - M(P^*)\|_F^2.$$

The first term is a constant which measures the average difference from the consensus association $\widetilde{M}$. The smaller this term is, the closer to each other the partitions are.

We therefore minimize the second term. The optimal clustering solution P^* can be specified by clustering indicators $H = \{0,1\}^{n\times k}$, with the constraint that in each row of H there can be only one "1" and the rest must be zeros. The key connection here is that the connectivity matrix $M(P^*) = HH^T$. With this, the consensus clustering problem becomes

$$\min_{H} \|\widetilde{M} - HH^T\|^2 \;\; s.t. \;\; H \text{ is a cluster indicator.} \tag{7.36}$$

We can relax the constraint. Clearly $(H^TH) = D = \mathrm{diag}(n_1, \cdots, n_k)$ where $n_k = |C_k|$. However, before we solve the problem, we have no way to know D and thus no way to impose the constraints. A slight reformulation resolves the problem. We define $\widetilde{H} = H(H^TH)^{-1/2}$. Thus,

$HH^T = \widetilde{H}D\widetilde{H}^T$, $\widetilde{H}^T\widetilde{H} = H(H^TH)^{-1}H = I$. Therefore, the consensus clustering becomes the following optimization problem:

$$\min_{\widetilde{H}^T\widetilde{H}=I,\ \widetilde{H},D\geq 0} \quad ||\widetilde{M} - \widetilde{H}D\widetilde{H}^T||^2, \ s.t.\ D \text{ is diagonal.} \tag{7.37}$$

Now, $\widetilde{H}$ and D are new variables. We do not need to prespecify the cluster sizes. Thus, the consensus clustering problem is equivalent to a symmetric NMF problem [73].

7.6.5 Graph Clustering

Three NMF models (Symmetric NMF, Asymmetric NMF, and Joint NMF) have been proposed in [123] to identify communities in three different types of networks (undirected, directed, and compound). Among their proposed models, symmetric NMF and asymmetric NMF are special cases of tri-factorization. Joint NMF involves multiple and possibly heterogeneous networks. For example, in music recommendation, we are given three networks: (1) the user network U showing the relationships among users, (2) the music network D showing the relationship among music songs, and (3) the user–music network M showing user preferences. Joint NMF is the problem of finding a latent matrix G, which reflects some "intrinsic" characteristics of the user–music network M, such that the following three objectives are minimized simultaneously: $||M-G||$, $||U-GG^T||$, $||D-G^TG||$. Formally, joint NMF aims to solve

$$\min_G ||M-G||^2 + \alpha||U-GG^T||^2 + \beta||D-G^TG||^2 \ s.t.\ \mathbf{G} \in \mathbb{R}_+^{n\times m}, \tag{7.38}$$

where $\alpha > 0$ and $\beta > 0$ are constants to trade-off the importance between different terms. Recently, an efficient Newton-like algorithm has been proposed for graph clustering using symmetric NMF [64].

7.6.6 Other Clustering Extensions

NMF methods have also been developed for many other clustering extensions. Badea has proposed an NMF model which simultaneously factorizes two linked nonnegative matrices with a shared factor matrix, aiming to uncover the common characteristics between the two input datasets [2]. Saha and Sindhwani have proposed a framework for online topic detection to handle streaming nonnegative data matrices with possibly growing number of components [94]. Li et al. [77, 76] have proposed constrained nonnegative matrix tri-factorizations for cross-domain transfer clustering with the input matrices in both the source and target domains. The proposed constrained matrix factorization framework naturally incorporates document labels via a least squares penalty incurred by a certain linear model and enables direct and explicit knowledge transfer across different domains. Wang et al. [119] have proposed a new NMF-based language model to simultaneously cluster and summarize documents by making use of both the document-term and sentence-term matrices. The proposed framework leads to a better document clustering method with more meaningful interpretation using representative sentences. Wang et al. [116, 117] have proposed weighted NMF-based approaches which combine keyword selection and document clustering (topic discovery) together by incorporating weights describing the importance of the keywords.

NMF has also been extended for analyzing multi-way data (or multi-way tensor). Multi-way data are generalizations of matrices and they appear in many applications [114, 3, 104, 62, 92, 1]. One typical type of three-way data is multiple two-way data/matrices with different time periods. For example, a series of 2-D images, 2-D text data (documents vs. terms) or 2-D microarray data (genes vs. conditions) are naturally represented as three-way data. Nonnegative Tensor Factorization (NTF) [99] is an extension of Nonnegative Matrix Factorization. The input data is a nonnegative tensor. For a three-way tensor, the standard NTF can be thought as HOSVD (high-order SVD) [27,

28] with nonnegativity constraints. Tri-factorization has also been extended as Tri-NTF (Tri-factor Nonnegative Tensor Factorization) to analyze three-way tensors [139].

7.7 Conclusions

Matrix-based methodologies are rapidly becoming a significant part of data mining as they are amenable to vigorous analysis and can benefit from the well-established knowledge in linear algebra accumulated through centuries. In particular, NMF factorizes an input nonnegative matrix into two nonnegative matrices of lower rank and has the capability to solve challenging data mining problems. Thanks to the data mining capabilities, NMF has attracted a lot of recent attention and has been used in a variety of fields. This chapter provides a comprehensive review of nonnegative matrix factorization methods for clustering by outlining the theoretical foundations on NMF for clustering and providing an overview of different variants on NMF formulations. We also examine the practical issues in NMF algorithms and summarize recent advances on using NMF-based methods for solving many other clustering problems.

There are many future research directions on NMF for clustering including the following:

1. extending NMF for better cluster representation and for dealing with more challenging clustering problems;
2. providing deeper understanding of NMF's clustering capability besides the established theoretical results;
3. developing novel and rigorous proof strategies to prove the correctness and convergence properties of the numerical algorithms;
4. studying NMF with other distance measures (such as other matrix norms and Bregman divergences);
5. improving the scalability of NMF algorithms for large-scale datasets; and
6. applying NMF to many different real-world applications and solving real problems.

Acknowledgment

The work of T. Li is supported by the National Science Foundation under grants DMS-0915110 and CCF-0830659, and by the Army Research Office under grants W911NF-10-1-0366 and W911NF-12-1-0431. The work of C. Ding is supported by the National Science Foundation under grants DMS-0915228 and CCF-0830780.

Bibliography

[1] E. Acar and B. Yener. Unsupervised multiway data analysis: A literature survey. *IEEE Transactions on Knowledge and Data Engineering*, 21(1):6–20, January 2009.

[2] L. Badea. Extracting gene expression profiles common to colon and pancreatic adenocarcinoma using simultaneous nonnegative matrix factorization. In *Pacific Symposium on Biocomputing'08*, pages 267–278, 2008.

[3] B. W. Bader, R. A. Harshman, and T. G. Kolda. Temporal analysis of semantic graphs using ASALSAN. In *Proceedings of the ICDM07*, pages 33–42, October 2007.

[4] S. Basu, A. Banerjee, and R. J. Mooney. Semi-supervised clustering by seeding. In *Proceedings of the Nineteenth International Conference on Machine Learning*, pages 27–34, 2002.

[5] S. Basu, M. Bilenko, and R. J. Mooney. A probabilistic framework for semi-supervised clustering. In *KDD '04: Proceedings of the 2004 ACM SIGKDD International Conference on Knowledge Discovery and Data Mining*, pages 59–68, New York, USA, 2004. ACM Press.

[6] E. Battenberg and D. Wessel. Accelerating non-negative matrix factorization for audio source separation on multi-core and many-core architectures. In *ISMIR'09*, pages 501–506, 2009.

[7] M. W. Berry, M. Browne, A. N. Langville, V. P. Pauca, and R. J. Plemmons. Algorithms and applications for approximate nonnegative matrix factorization. *Computational Statistics and Data Analysis*, 52(1):155–173, 2006.

[8] M. Bilenko, S. Basu, and R. J. Mooney. Integrating constraints and metric learning in semi-supervised clustering. In *Proceedings of International Conference on Machine Learning*, 2004.

[9] D. Blei, A. Ng, and M. Jordan. Latent Dirichlet allocation. *Journal of Machine Learning Research*, 3:993–1022, 2003.

[10] C. Boutsidis and E. Gallopoulos. SVD based initialization: A head start for nonnegative matrix factorization. *Pattern Recognition*, 41(4):1350–1362, April 2008.

[11] J.-P. Brunet, P. Tamayo, T.R. Golub, and J.P. Mesirov. Metagenes and molecular pattern discovery using matrix factorization. *Proceedings of National Academy of Sciences USA*, 102(12):4164–4169, 2004.

[12] D. Cai, X. He, J. Han, and T. S. Huang. Graph regularized nonnegative matrix factorization for data representation. *IEEE Transactions on Pattern and Analysis Machine Intelligence*, 33(8):1548–1560, August 2011.

[13] B. Cao, D. Shen, J. Sun, X. Wang, Q. Yang, and Z. Chen. Detect and track latent factors with online nonnegative matrix factorization. In *Proceedings of the 20th International Joint Conference on Artifical Intelligence*, IJCAI'07, pages 2689–2694, San Francisco, CA, USA, 2007. Morgan Kaufmann Publishers Inc.

[14] A. T. Cemgil. Bayesian inference for nonnegative matrix factorisation models. *Intelligent Neuroscience*, 2009:4:1–4:17, January 2009.

[15] E. Y. Chang, K. Zhu, and H. Bai. Parallel algorithms for mining large-scale datasets. CIKM tutorial, 2009.

[16] W. Chen, B. Pan, B. Fang, M. Li, and J. Tang. Incremental nonnegative matrix factorization for face recognition. *Mathematical Problems in Engineering*, 2008.

[17] Y. Chen, M. Rege, M. Dong, and J. Hua. Non-negative matrix factorization for semi-supervised data clustering. *Knowledge and Information Systems*, 17(3):355–379, November 2008.

[18] Y. Chen, L. Wang, and M. Dong. Non-negative matrix factorization for semisupervised heterogeneous data coclustering. *IEEE Transactions on Knowledge and Data Engineering*, 22(10):1459–1474, October 2010.

[19] Y. Chi, S. Zhu, Y. Gong, and Y. Zhang. Probabilistic polyadic factorization and its application to personalized recommendation. In *CIKM '08: Proceeding of the 17th ACM Conference on Information and Knowledge Management*, pages 941–950. ACM, 2008.

[20] H. Cho, I. Dhillon, Y. Guan, and S. Sra. Minimum sum squared residue co-clustering of gene expression data. In *Proceedings of The 4th SIAM Data Mining Conference*, pages 22–24, April 2004.

[21] C.-T. Chu et al. Map-Reduce for machine learning on multicore. In *NIPS*, 2006.

[22] A. Cichocki and R. Zdunek. Multilayer nonnegative matrix factorization. *Electronics Letters*, 42(16):947–948, 2006.

[23] A. Cichocki, R. Zdunek, and S. Amari. Hierarchical ALS algorithms for nonnegative matrix and 3D tensor factorization. In *In Lecture Notes on Computer Science, LNCS-4666, Proceedings of Independent Component Analysis (ICA07)*, pages 169–176. Springer, 2007.

[24] A. Cichocki, R. Zdunek, A. H. Phan, and S. Amari. *Nonnegative Matrix and Tensor Factorizations: Applications to Exploratory Multi-way Data Analysis and Blind Source Separation*. Wiley, 2009.

[25] M. Cooper and J. Foote. Summarizing video using non-negative similarity matrix factorization. In *Proceedings of the IEEE Workshop on Multimedia Signal Processing*, pages 25–28, 2002.

[26] J. Dean and S. Ghemawat. MapReduce: Simplified data processing on large clusters. In *OSDI*, 2004.

[27] L. De Lathauwer, B. De Moor and J. Vandewalle. On the best rank-1 and rank-$(R_1, R_2, \ldots, R_n)$ approximation of higher-order tensors. *SIAM Journal on Matrix Analysis and Applications*, 21(4):1324-1342, 2000

[28] L. De Lathauwer, B. De Moor and J. Vandewalle. A multilinear singular value decomposition. *SIAM Journal on Matrix Analysis and Applications*, 21(4):1253-1278, 2000

[29] I. Dhillon and S. Sra. Generalized nonnegative matrix approximations with Bregman divergences. In *Advances in Neural Information Processing Systems 17*, Cambridge, MA, MIT Press, 2005.

[30] I. S. Dhillon. Co-clustering documents and words using bipartite spectral graph partitioning. *Proceedings of the ACM International Conference on Knowledge Discovery and Data Minings on Pattern Analysis and Machine Intelligence (KDD 2001)*, 2001.

[31] I. S. Dhillon, Y. Guan, and B. Kulis. Weighted graph cuts without eigenvectors: A multilevel approach. *Proceedings of the ACM International Conference on Knowledge Discovery and Data Minings on Pattern Analysis and Machine Intelligence*, 29:2007, 2007.

[32] I. S. Dhillon and S. Sra. Generalized nonnegative matrix approximations with Bregman divergences. In *Proceedings of the Neural Information Processing Systems (NIPS) Conference*, pages 283–290, 2005.

[33] C. Ding and X. He. Cluster merging and splitting in hierarchical clustering algorithms. In *Proceedings of the 2002 IEEE International Conference on Data Mining, ICDM '02*, pages 139–146, Washington, DC, USA, 2002. IEEE Computer Society.

[34] C. Ding and X. He. K-means clustering via principal component analysis. In *Proceedings of the Twenty-First International Conference on Machine Learning, ICML '04*, pages 225–232, 2004.

[35] C. Ding, X. He, and H.D. Simon. On the equivalence of nonnegative matrix factorization and spectral clustering. *Proc. SIAM Data Mining Conf*, 2005.

[36] C. Ding, T. Li, and M. I. Jordan. Convex and semi-nonnegative matrix factorizations. *IEEE Transactions on Pattern Analysis and Machine Intelligence*, 32(1):45–55, January 2010.

[37] C. Ding, T. Li, and M. Jordan. Convex and semi-nonnegative matrix factorizations for clustering and low-dimension representation. Technical Report LBNL-60428, Lawrence Berkeley National Laboratory, University of California, Berkeley, 2006.

[38] C. Ding, T. Li, and M. I. Jordan. Nonnegative matrix factorization for combinatorial optimization: Spectral clustering, graph matching, and clique finding. In *Proceedings of the 2008 Eighth IEEE International Conference on Data Mining, ICDM '08*, pages 183–192, Washington, DC, USA, 2008. IEEE Computer Society.

[39] C. Ding, T. Li, and W. Peng. Nonnegative matrix factorization and probabilistic latent semantic indexing: Equivalence, chi-square statistic, and a hybrid method. In *Proceedings of the National Conference on Artificial Intelligence (AAAI-06)*, 2006.

[40] C. Ding, T. Li, and W. Peng. On the equivalence between non-negative matrix factorization and probabilistic latent semantic indexing. *Computational Statistics & Data Analysis*, 52(8):3913–3927, April 2008.

[41] C. Ding, T. Li, W. Peng, and H. Park. Orthogonal nonnegative matrix tri-factorizations for clustering. In *Proceedings of the Twelfth ACM SIGKDD International Conference on Knowledge Discovery and Data Mining*, 2006.

[42] C. Dong, H. Zhao, and W. Wang. Parallel nonnegative matrix factorization algorithm on the distributed memory platform. *International Journal of Parallel Programming*, 38(2):117–137, 2010.

[43] K. Drakakis, S. Rickard, R. D. Frein, and A. Cichocki. Analysis of financial data using non-negative matrix factorization. *International Mathematical Forum*, 3(38):1853–1870, 2008.

[44] G. Druck, G. Mann, and A. McCallum. Learning from labeled features using generalized expectation criteria. In *Proceedings of the 31st Annual International ACM SIGIR Conference on Research and Development in Information Retrieval, SIGIR '08*, pages 595–602, New York, USA, 2008. ACM.

[45] C. Eckart and G. Young. The approximation of one matrix by another of lower rank. *Psychometrika*, 1:183–187, 1936.

[46] A. Edelman, T. Arias, and S.T. Smith. The geometry of algorithms with orthogonality constraints. *SIAM Journal on Matrix Analysis Applications*, 20(2):303–353, 1999.

[47] X. Z. Fern and C. E. Brodley. Solving cluster ensemble problems by bipartite graph partitioning. In *ICML '04: Proceedings of the Twenty-First International Conference on Machine Learning*, page 36, 2004.

[48] E. Gaussier and C. Goutte. Relation between PLSA and NMF and implications. In *SIGIR '05: Proceedings of the 28th Annual International ACM SIGIR Conference on Research and Development in Information Retrieval*, pages 601–602, New York, USA, 2005. ACM Press.

[49] A. Gionis, H. Mannila, and P. Tsaparas. Clustering aggregation. In *ICDE*, pages 341–352, 2005.

[50] G. Golub and C. Van Loan. *Matrix Computations,* 3rd edition. Johns Hopkins, Baltimore, 1996.

[51] Q. Gu and J. Zhou. Local learning regularized nonnegative matrix factorization. In *Proceedings of the 21st International Joint Conference on Artifical Intelligence, IJCAI'09*, pages 1046–1051, San Francisco, CA, USA, 2009. Morgan Kaufmann Publishers Inc.

[52] N. Guan, D. Tao, Z. Luo, and B. Yuan. Online nonnegative matrix factorization with robust stochastic approximation. *IEEE Transactions on Neural Network Learning Systems*, 23:1087–1099, 2012.

[53] J. Han, M. Kamber, and J. Pei. *Data Mining: Concepts and Techniques.* Morgan Kaufmann Publishers, 2011.

[54] B. Hendrickson and R. Leland. A multilevel algorithm for partitioning graphs. In *Proceedings of the 1995 ACM/IEEE Conference on Supercomputing, Supercomputing '95*, New York, USA, 1995. ACM.

[55] T. Hofmann. Probabilistic latent semantic analysis. In *Proceedings of the 15th Annual Conference on Uncertainty in Artificial Intelligence (UAI-99)*, pages 289–296, 1999.

[56] P. O. Hoyer. Non-negative matrix factorization with sparseness constraints. *Journal of Machine Learning Research*, 5:1457–1469, December 2004.

[57] G. Karypis and V. Kumar. A fast and high quality multilevel scheme for partitioning irregular graphs. *SIAM Journal of Scientific Computing*, 20(1):359–392, December 1998.

[58] D. Kim, S. Sra, and I. S. Dhillon. Fast projection-based methods for the least squares nonnegative matrix approximation problem. *Statistical Analysis and Data Mining*, 1(1):38–51, February 2008.

[59] H. Kim and H. Park. Sparse non-negative matrix factorizations via alternating non-negativity-constrained least squares for microarray data analysis. *Bioinformatics*, 23(12):1495–1502, 2007.

[60] J. Kim and H. Park. Fast nonnegative matrix factorization: An active-set-like method and comparisons. *SIAM Journal on Scientific Computing*, 33(6):3261–3281, 2012.

[61] D. Klein, S. D. Kamvar, and C. D. Manning. From instance-level constraints to space-level constraints: Making the most of prior knowledge in data clustering. In *Proceedings of the Nineteenth International Conference on Machine Learning*, pages 307–314, 2002.

[62] T. G. Kolda and B. W. Bader. The TOPHITS model for higher-order web link analysis. In *Workshop on Link Analysis, Counterterrorism and Security*, 2006.

[63] D. Kong, C. Ding, and H. Huang. Robust nonnegative matrix factorization using L21-norm. In *Proceedings of the 20th ACM International Conference on Information and Knowledge Management (CIKM'11)*, pages 673–682, 2011.

[64] D. Kuang, C. Ding, and H. Park. Symmetric nonnegative matrix factorization for graph clustering. In *Proceedings of the 2012 SIAM International Conference on Data Mining (SDM 2012)*, 2012.

[65] V. Kysenko, K. Rupp, O. Marchenko, S. Selberherr, and A. Anisimov. GPU-accelerated non-negative matrix factorization for text mining. *Natural Language Processing and Information Systems*, page 158–163, 2012.

[66] A. N. Langville, C. D. Meyer, and R. Albright. Initalizations for the nonnegative matrix factorization. In *Proccedings of the Twelfth ACM SIGKDD International Conference on Knowledge Discovery and Data Mining (KDD 2006)*, 2006.

[67] D.D. Lee and H. S. Seung. Algorithms for non-negatvie matrix factorization. In T. G. Dietterich and V. Tresp, editors, *Advances in Neural Information Processing Systems*, volume 13. The MIT Press, 2001.

[68] D.D. Lee and H. S. Seung. Learning the parts of objects by non-negative matrix factorization. *Nature*, 401:788–791, 1999.

[69] H. Lee, J. Yoo, and S. Choi. Semi-supervised nonnegative matrix factorization. *IEEE Signal Processing Letters*, 17(1), 2010.

[70] S.Z. Li, X. Hou, H. Zhang, and Q. Cheng. Learning spatially localized, parts-based representation. In *Proceedings of the IEEE Conference on Computer Vision and Pattern Recognition*, pages 207–212, 2001.

[71] T. Li. A general model for clustering binary data. In *Proceedings of the Eleventh ACM SIGKDD International Conference on Knowledge Discovery in Data Mining, KDD '05*, pages 188–197, New York, USA, 2005. ACM.

[72] T. Li and C. Ding. The relationships among various nonnegative matrix factorization methods for clustering. In *In Proceedings of the 2006 IEEE International Conference on Data Mining (ICDM 2006)*, pages 362–371, 2006.

[73] T. Li, C. Ding, and M. I. Jordan. Solving consensus and semi-supervised clustering problems using nonnegative matrix factorization. In *Proceedings of the 2007 Seventh IEEE International Conference on Data Mining, ICDM '07*, pages 577–582, Washington, DC, USA, 2007. IEEE Computer Society.

[74] T. Li, C. Ding, Y. Zhang, and B. Shao. Knowledge transformation from word space to document space. In *Proceedings of the 31st Annual International ACM SIGIR Conference on Research and Development in Information Retrieval*, SIGIR '08, pages 187–194, New York, USA, 2008. ACM.

[75] T. Li, M. Ogihara, and S. Ma. On combining multiple clusterings. In *CIKM '04: Proceedings of the Thirteenth ACM International Conference on Information and Knowledge Management*, pages 294–303, 2004.

[76] T. Li, V. Sindhwani, C. Ding, and Y. Zhang. Bridging domains with words: Opinion analysis with matrix tri-factorizations. In *Proceedings of the Tenth SIAM Conference on Data Mining (SDM 2010)*, pages 293–302, 2010.

[77] T. Li, V. Sindhwani, C. Ding, and Y. Zhang. Knowledge transformation for cross-domain sentiment classification. In *Proceedings of the 32nd Annual International ACM SIGIR Conference on Research and Development in Information Retrieval (SIGIR 2009)*, pages 716–717, 2009.

[78] T. Li, Y. Zhang, and V. Sindhwani. A non-negative matrix tri-factorization approach to sentiment classification with lexical prior knowledge. In *Proceedings of the Joint Conference of the 47th Annual Meeting of the ACL and the 4th International Joint Conference on Natural Language Processing of the AFNLP: Volume 1*, ACL '09, pages 244–252, Stroudsburg, PA, USA, 2009. Association for Computational Linguistics.

[79] C. Lin. Projected gradient methods for nonnegative matrix factorization. *Neural Computation*, 19(10):2756–2779, October 2007.

[80] C. Liu, H. Yang, J. Fan, L. He, and Y. Wang. Distributed nonnegative matrix factorization for web-scale dyadic data analysis on mapreduce. In *WWW '10: Proceedings of the 19th International Conference on World Wide Web*, pages 681–690, 2010.

[81] H. Liu and Z. Wu. Non-negative matrix factorization with constraints. In *Proceedings of the Twenty-Fourth AAAI Conference on Artificial Intelligence (AAAI-10)*, 2010.

[82] B. Long, X. Wu, Z. Zhang, and P. S. Yu. Unsupervised learning on k-partite graphs. In *Proceedings of ACM SIGKDD*, pages 317–326, 2006.

[83] B. Long, Z. Zhang, and P.S. Yu. Co-clustering by block value decomposition. In *KDD '05: Proceedings of the Eleventh ACM SIGKDD International Conference on Knowledge Discovery and Data Mining*, pages 635–640, New York, USA, 2005. ACM Press.

[84] N. Lopes and B. Ribeiro. Non-negative matrix factorization implementation using graphic processing units. In *Proceedings of the 11th International Conference on Intelligent Data Engineering and Automated Learning, IDEAL'10*, pages 275–283, Berlin, Heidelberg, 2010. Springer-Verlag.

[85] A. McCallum, K. Nigam, and L. H. Ungar. Efficient clustering of high-dimensional data sets with application to reference matching. In *Proceedings of the sixth ACM SIGKDD International Conference on Knowledge Discovery and Data Mining*, KDD '00, pages 169–178, New York, USA, 2000. ACM.

[86] P. Miettinen, T. Mielikainen, A. Gionis, G. Das, and H. Mannila. The discrete basis problem. In *Proceedings of the 10th European Conference on Principle and Practice of Knowledge Discovery in Databases, PKDD'06*, pages 335–346, Berlin, Heidelberg, Springer-Verlag. 2006.

[87] K. Nishino, S. K. Nayar, and T. Jebara. Clustered blockwise pca for representing visual data. *IEEE Transactions on Pattern Analysis and Machine Intelligence*, 27(10):1675–1679, October 2005.

[88] P. Paatero and U. Tapper. Positive matrix factorization: A non-negative factor model with optimal utilization of error estimates of data values. *Environmetrics*, 5:111–126, 1994.

[89] S. Park, J. Lee, D. Kim, and C. Ahn. Multi-document summarization based on cluster using non-negative matrix factorization. In *Proceedings of the 33rd Conference on Current Trends in Theory and Practice of Computer Science, SOFSEM '07*, pages 761–770, Berlin, Heidelberg, 2007. Springer-Verlag.

[90] V. P. Pauca, J. Piper, and Robert J. Plemmons. Nonnegative matrix factorization for spectral data analysis. *Linear Algebra and Its Applications*, 416(1):29–47, 2006.

[91] V. P. Pauca, F. Shahnaz, M.W. Berry, and R.J. Plemmons. Text mining using non-negative matrix factorization. In *Proceedings of the SIAM International Conference on Data Mining*, pages 452–456, 2004.

[92] W. Peng and T. Li. Temporal relation co-clustering on directional social network and author-topic evolution. *Knowledge and Information Systems*, 26(3):467–486, March 2011.

[93] J. Platos, P. Gajdos, P. Kromer, and V. Snasel. Non-negative matrix factorization on gpu. In Filip Zavoral, Jakub Yaghob, Pit Pichappan, and Eyas El-Qawasmeh, editors, *Networked Digital Technologies*, volume 87 of *Communications in Computer and Information Science*, pages 21–30. Berlin Heidelberg, Springer, 2010.

[94] A. Saha and V. Sindhwani. Learning evolving and emerging topics in social media: A dynamic NMF approach with temporal regularization. In *Proceedings of the Fifth ACM International Conference on Web Search and Data Mining, WSDM '12*, pages 693–702, New York, USA, 2012. ACM.

[95] M. N. Schmidt and H. Laurberg. Nonnegative matrix factorization with Gaussian process priors. *Intelligent Neuroscience*, 2008:3:1–3:10, January 2008.

[96] M. N. Schmidt, O. Winther, and L. Hansen. Bayesian non-negative matrix factorization. In *International Conference on Independent Component Analysis and Signal Separation*, volume 5441 of *Lecture Notes in Computer Science (LNCS)*, pages 540–547. Springer, 2009.

[97] F. Sha, L.K. Saul, and D.D. Lee. Multiplicative updates for nonnegative quadratic programming in support vector machines. *Advances in Neural Information Processing Systems 15*: 1041–1048. 2003.

[98] F. Shahnaz, M. W. Berry, V. P. Pauca, and R. J. Plemmons. Document clustering using nonnegative matrix factorization. *Information Processing and Management*, 42(2):373–386, March 2006.

[99] A. Shashua and T. Hazan. Non-negative tensor factorization with applications to statistics and computer vision. In *ICML*, 2005.

[100] J. Shi and J. Malik. Normalized cuts and image segmentation. *IEEE Transactions on Pattern Analysis and Machine Intelligence*, 22:888–905, 2000.

[101] V. Sindhwani, J. Hu, and A. Mojsilovic. Regularized co-clustering with dual supervision. In *Proceedings of NIPS'08*, pages 1505–1512, 2008.

[102] V. Sindhwani and P. Melville. Document-word co-regularization for semi-supervised sentiment analysis. In *Proceedings of the 2008 Eighth IEEE International Conference on Data Mining, ICDM '08*, pages 1025–1030, Washington, DC, USA, 2008. IEEE Computer Society.

[103] P. Smaragdis. Non-negative matrix factor deconvolution; extraction of multiple sound sources from monophonic inputs. In *Proceedings of ICA'04*, pages 494–499, 2004.

[104] A. Smilde, R. Bro, and P. Geladi. *Multi-Way Analysis: Applications in the Chemical Sciences.* West Sussex, UK, Wiley, 2004.

[105] S. Sra and I. S. Dhillon. Nonnegative matrix approximation: algorithms and applications. Technical Report TR-06-27, UTCS, 2006.

[106] N. Srebro, J. Rennie, and T. Jaakkola. Maximum margin matrix factorization. In *Advances in Neural Information Processing Systems*, Cambridge, MA, MIT Press. 2005.

[107] A. Strehl and J. Ghosh. Cluster ensembles—A Knowledge reuse framework for combining multiple partitions. *The Journal of Machine Learning Research*, 3:583–617, March 2003.

[108] Z. Sun, T. Li, and N. Rishe. Large-scale matrix factorization using MapReduce. In *Proceedings of the 2010 IEEE International Conference on Data Mining Workshops*, ICDMW '10, pages 1242–1248, Washington, DC, USA, 2010. IEEE Computer Society.

[109] P. Tan, M. Steinbach, and V. Kumar. *Introduction to Data Mining*. Addison Wesley, 2006.

[110] C. Thurau, K. Kersting, M. Wahabzada, and C. Bauckhage. Convex non-negative matrix factorization for massive datasets. *Knowledge and Information Systems*, 29(2):457–478, November 2011.

[111] F. D. Torre and T. Kanade. Discriminative cluster analysis. In *Proceedings of the 23rd International Conference on Machine Learning (ICML 2006)*, 2006.

[112] A. Tveit. Mapreduce and hadoop algorithms in academic papers. Blog, Internet.

[113] S. A. Vavasis. On the complexity of nonnegative matrix factorization. http://arxiv.org/abs/0708.4149, 2007.

[114] M. Vichi, R. Rocci, and H. A. L. Kiers. Simultaneous component and clustering models for three-way data: Within and between approaches. *Journal of Classification*, 24(1):71–98, 2007.

[115] K. Wagsta, C. Cardie, S. Rogers, and S. Schroedl. Constrained k-means clustering with background knowledge. In *Proceedings of International Conference on Machine Learning*, pages 577–584, 2001.

[116] D. Wang, C. Ding, and T. Li. Feature subset non-negative matrix factorization and its applications to document understanding. In *Proceedings of the 33rd International ACM SIGIR Conference on Research and Development in Information Retrieval, SIGIR '10*, pages 805–806, New York, USA, 2010. ACM.

[117] D. Wang, T. Li, and C. Ding. Weighted feature subset non-negative matrix factorization and its applications to document understanding. In *Proceedings of the 2010 IEEE International Conference on Data Mining*, ICDM '10, pages 541–550, Washington, DC, USA, 2010. IEEE Computer Society.

[118] D. Wang, T. Li, S. Zhu, and C. Ding. Multi-document summarization via sentence-level semantic analysis and symmetric matrix factorization. In *Proceedings of the 31st Annual International ACM SIGIR Conference on Research and Development in Information Retrieval, SIGIR '08*, pages 307–314, New York, USA, 2008. ACM.

[119] D. Wang, S. Zhu, T. Li, Y. Chi, and Y. Gong. Integrating document clustering and multi-document summarization. *ACM Transactions on Knowledge Discovery and Data*, 5(3):14:1–14:26, August 2011.

[120] F. Wang and P. Li. Efficient nonnegative matrix factorization with random projections. In *In Proceedings of 2010 SIAM Data Mining Conference (SDM'10)*, pages 281–292, 2010.

[121] F. Wang and T. Li. Gene selection via matrix factorization. In *Proceedings of 7th IEEE Conference on Bioinformatics and Bioengineering*, pages 1046–1050.

[122] F. Wang, P. Li, and A. C. Konig. Efficient document clustering via online nonnegative matrix factorizations. In *Proceedings of the 2011 SIAM International Conference on Data Mining (SDM'11)*, pages 908–919, 2011.

[123] F. Wang, T. Li, X. Wang, S. Zhu, and C. Ding. Community discovery using nonnegative matrix factorization. *Data Mining and Knowledge Discovery*, 22(3):493–521, May 2011.

[124] F. Wang, T. Li, and C. Zhang. Semi-supervised clustering via matrix factorization. In *Proceedings of 2008 SIAM International Conference on Data Mining (SDM 2008)*, pages 1–12, 2008.

[125] F. Wang, H. Tong, and C. Lin. Towards evolutionary nonnegative matrix factorization. In *Proceedings of the Twenty-Fifth AAAI Conference on Artificial Intelligence (AAAI 2011)*, 2011.

[126] H. Wang, F. Nie, H. Huang, and F. Makedon. Fast nonnegative matrix tri-factorization for large-scale data co-clustering. In *Proceedings of the Twenty-Second International Joint Conference on Artificial Intelligence—Volume Two, IJCAI'11*, pages 1553–1558. AAAI Press, 2011.

[127] S. Wild, J. Curry, and A. Dougherty. Improving non-negative matrix factorizations through structured initialization. *Pattern Recognition*, 37(11):2217–2232, 2004.

[128] Y.-L. Xie, P.K. Hopke, and P. Paatero. Positive matrix factorization applied to a curve resolution problem. *Journal of Chemometrics*, 12(6):357–364, 1999.

[129] E. P. Xing, A. Y. Ng, M. I. Jordan, and S. Russell. Distance metric learning, with application to clustering with side-information. *Advances in Neural Information Processing Systems* 15: pages 505–512, 2003.

[130] W. Xu, X. Liu, and Y. Gong. Document clustering based on non-negative matrix factorization. In *Proceedings of ACM Conference on Research and Development in IR(SIGIR)*, pages 267–273, Toronto, Canada, 2003.

[131] Y. Xue, C. S. Tong, Y. Chen, and W. Chen. Clustering-based initialization for non-negative matrix factorization. *Applied Mathematics and Computation*, 205(2):525–536, 2008.

[132] Z. Yang and E. Oja. Linear and nonlinear projective nonnegative matrix factorization. *IEEE Transactions on Neural Networks*, 21(5):734 –749, 2010.

[133] S. X. Yu and J. Shi. Multiclass spectral clustering. In *Proceedings of the Ninth IEEE International Conference on Computer Vision—Volume 2, ICCV '03*, pages 313–319, Washington, DC, USA, 2003. IEEE Computer Society.

[134] R. Zdunek and A. Cichocki. Nonnegative matrix factorization with quadratic programming. *Neurocomputing*, 71(10–12):2309–2320, June 2008.

[135] D. Zeimpekis and E. Gallopoulos. CLSI: A flexible approximation scheme from clustered term-document matrices. In *Proceedings of SIAM Data Mining Conference*, pages 631–635, 2005.

[136] H. Zha, X. He, C. Ding, M. Gu, and H.D. Simon. Bipartite graph partitioning and data clustering. In *Proceedings of the International Conference on Information and Knowledge Management (CIKM 2001)*, 2001.

[137] H. Zha, X. He, C. Ding, M. Gu, H.D. Simon, and M. Gu. Spectral Relaxation for K-means Clustering. In *Advances in Neural Information Processing Systems 14 (NIPS'01)*, pages 1057–1064, 2001, MIT Press.

[138] S. Zhang, W. Wang, J. Ford, and F. Makedon. Learning from incomplete ratings using non-negative matrix factorization. In *Proceedings of the Sixth SIAM Conference on Data Mining(SDM)*, pages 549–553, 2006.

[139] Z. Zhang, T. Li, and C. Ding. Non-negative tri-factor tensor decomposition with applications. *Knowledge and Information Systems*, 34(2):243–265, 2012.

[140] Z. Zhang, T. Li, C. Ding, X. Ren, and X. Zhang. Binary matrix factorization for analyzing gene expression data. *Data Mininig and Knowledge Discovery*, 20(1):28–52, January 2010.

[141] Z. Zhang, T. Li, C. Ding, and X. Zhang. Binary matrix factorization with applications. In *Proceedings of the 2007 Seventh IEEE International Conference on Data Mining*, ICDM '07, pages 391–400, Washington, DC, USA, 2007. IEEE Computer Society.

[142] Z. Zheng, J. Yang, and Y. Zhu. Initialization enhancer for non-negative matrix factorization. *Engineering Applications of Artificial Intelligence*, 20(1):101–110, 2007.

[143] Y. Zhu, L. Jing, and J. Yu. Text clustering via constrained nonnegative matrix factorization. In *Proceedings of the 2011 IEEE 11th International Conference on Data Mining, ICDM '11*, pages 1278–1283, Washington, DC, USA, 2011. IEEE Computer Society.

Chapter 8

Spectral Clustering

Jialu Liu
University of Illinois at Urbana-Champaign
Urbana, IL
jliu64@illinois.edu

Jiawei Han
University of Illinois at Urbana-Champaign
Urbana, IL
hanj@illinois.edu

8.1 Introduction

In this chapter, we introduce the family of spectral clustering algorithms which have seen increasing popularity over the past few years. Starting with the seminal works in [37] and [43], a large number of papers has been published along this line of work. As opposed to "traditional clustering algorithms" such as k-means and generative mixture models which always result in clusters with convex geometric shape, spectral clustering can solve problems in much more complex scenarios, such as intertwined spirals, or other arbitrary nonlinear shapes, because it does not make assumptions on the shapes of clusters. Another disadvantage of the previous algorithms is related to the inherent challenges in the Expectation Maximization (EM) framework, which is often used to learn

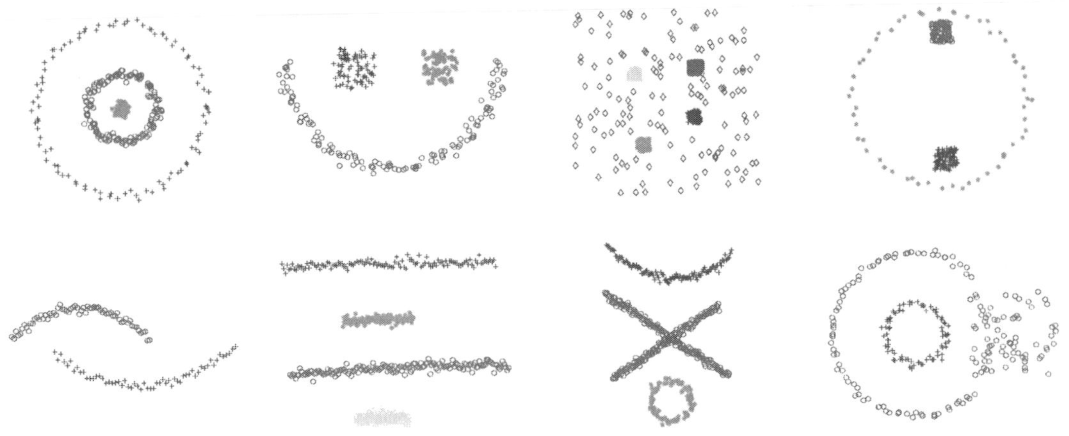

FIGURE 8.1: Self-tuning spectral clustering on toy datasets. (From Zelnik-Manor and Perona, *Advances in Neural Information Processing Systems*, 17:1601–1608, 2004.)

a mixture model for clustering. This framework is essentially an iterative process of finding local minima, and therefore multiple restarts are required to find a good solution.

Many data sets can be notoriously difficult to cluster with traditional methods. Figure 8.1 demonstrates a number of toy data sets [60], which are difficult for traditional clustering algorithms. These data sets have been constructed in order to generate clusters of different shapes. On such datasets, algorithms which implicitly assume specific shapes of clusters cannot achieve good results. For example, the Euclidian distance metrics assume a convex shape to the underlying clusters. Obviously such assumptions can impact the quality of the clustering in arbitrary data sets. As will be evident, the spectral clustering method is able to handle such data sets effectively.

The history of spectral clustering can be traced back to [14, 15], in which it was suggested that the eigenvectors of the adjacency matrix could be used in order to determine the underlying partitions. The main difference between spectral clustering algorithms is whether they use normalized or unnormalized "graph Laplacian" [52] which will be introduced in the later sections. Different versions of spectral clustering have been successfully applied to image segmentation [43], text mining [11], speech processing [1], and general purpose methods for data analysis and clustering [37, 60, 13, 12]. An excellent review on the history of spectral clustering can be found in [46].

The spectral clustering family can be viewed as a three-step algorithm:

- The first step is to construct a similarity graph for all the data points.
- The data points are embedded in a space, in which the clusters are more "obvious," with the use of the eigenvectors of the graph Laplacian.
- Finally, a classical clustering algorithm such as k-means is applied to partition the embedding.

The low-dimensional representation obtained in the second step is also referred to as "spectral embedding" and has applications beyond the clustering context, such as dimensionality reduction [2]. The word *spectral* is used to denote the fact that the clustering results are obtained by analyzing the spectrum of the graph Laplacian.

Figure 8.2 shows[1] the application of the spectral clustering on an example data set. The figure in the middle is a K-nearest neighbor (KNN) graph built in term of the Euclidean distance. The clustering result is illustrated in the adjacent figure on the right.

[1]The figure is generated by GraphDemo: http://www.ml.uni-saarland.de/GraphDemo/GraphDemo.html.

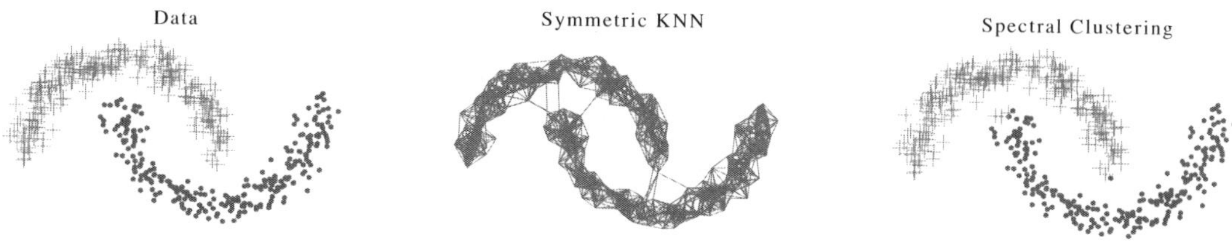

FIGURE 8.2: Example illustrating three steps of spectral clustering. (From Tenenbaum, de Silva, and Langford, *Science*, 290(5550):2319–2323, 2000.)

8.2 Similarity Graph

Let the set of data points, which we wish to partition into k subsets, be denoted by $X = \{x_1, \ldots x_n\}$ in $\mathbb{R}^m$. In order to perform spectral clustering, we first need to represent this data in the form of an undirected "similarity graph" $G = (V, E)$. Here each data point x_i is represented by a vertex v_i, and E refers to the edges between vertices. Note that x_i is a vector which denotes the data point while v_i is a vertex without any attributes. Then we can use a nonnegative weighted n by n *adjacency matrix* (or *affinity matrix*) W to describe G, where $W = \{W_{ij}\}_{i,j=1,\ldots,n}$. Note that W_{ij} equals 0 when vertices v_i and v_j are not connected. Since spectral clustering algorithms aim at partitioning the vertices to let those in the same cluster have high similarity and those in different clusters have low similarity, it becomes critical to choose an effective method to construct such an adjacency matrix.

Recent studies in spectral graph theory [8] and manifold learning theory [50, 2] have demonstrated that the adjacency matrix should model the local geometric structure of the data points. This issue will be discussed in some more detail in Section 8.7. Based on this rule, we introduce three ways to construct W [2, 31]:

1. ***K*-nearest neighbor graphs:** The idea is that v_i is connected with v_j when v_j is among the K-nearest neighbors of v_i, or v_i is among the K-nearest neighbors of v_j. The distance is computed based on the original representation of the data points x_i and x_j. Some examples include ℓ_1-norm, ℓ_2-norm, and the cosine distance. The resulting graph is usually called the *k-nearest neighbor graph*. The alternative is to connect v_i and v_j when they are mutually in the neighborhood of each other. This graph is referred to as the *mutual K-nearest neighbor graph* or *symmetric K-nearest neighbor graph*. In both cases, after adding the edges according to the neighborhood of each vertex, we can assign weights to the edges by the similarity of their endpoints or simply adopt a 0–1 weight.

2. **ε-neighborhood graph:** In this kind of graph, vertices are connected only when the pairwise distance $||x_i - x_j||^2$ is smaller than ε. However, this method often leads to graphs with disconnected components if ε is not carefully chosen.

3. **Fully connected graph:** In this case, we connect all vertices with positive similarity. Since the adjacency matrix should model the local neighborhood, the selection of the similarity function itself becomes tricky. An example is the heat kernel function:

$$W_{ij} = e^{-\frac{||x_i - x_j||^2}{\sigma^2}}$$

where σ controls the width of the neighborhoods. It is suggested in [60] that σ could be tuned locally with respect to the pair of vertices. Besides the problem of choosing an appropriate value of σ, the construction of this kind of graph also suffers from the efficiency problem. It is reported in [5] that the most commonly used approach for addressing the computational and memory challenge is to zero out some elements in the similarity matrix, or to sparsify the matrix. From this point of view, once the graph is fully connected, it will increase both the time and space complexity. In Section 8.9, we will carefully discuss this issue.

The selection of the specific method for constructing the similarity graph can sometimes be a complex and confusing problem. To the best of our knowledge, there are no theoretical studies regarding the best choice under different circumstances. Besides these three methods, some recent papers have been published to propose new graph construction techniques which will be covered in Section 8.10.

8.3 Unnormalized Spectral Clustering

Given the similarity graph G, the main step for spectral clustering is then to compute *graph Laplacian* matrices [8]. The actual construction of this very important intermediate representation is often not defined uniquely in the literature. Every author just calls "his" matrix the graph Laplacian [31]. To distinguish them, we will carefully study and discuss their properties in this and the next section. In this section, we first introduce the *unnormalized graph Laplacian*.

8.3.1 Notation

In the previous section, we defined the nonnegative weight W_{ij} for each pair of vertices v_i and v_j in the undirected similarity graph G. Since G is undirected, we have $W_{ij} = W_{ji}$.

For each vertex, its degree is then computed as the sum of the weights incident on it:

$$d_i = \sum_{j=1}^{n} W_{ij}$$

Correspondingly, the *degree matrix* D is defined as the diagonal matrix satisfying $D_{ii} = d_i$.

For a subset A of vertices V, its indicator vector is denoted by $\mathbb{1}_A = (f_1, \ldots, f_n)^T$, where $f_i = 1$ if vertex v_i belongs to A and $f_i = 0$ otherwise.

8.3.2 Unnormalized Graph Laplacian

The unnormalized graph Laplacian L is defined as follows:

$$L = D - W$$

We list several important properties of L [33, 34, 31]:

Proposition 8.3.1 *The unnormalized graph Laplacian* L *satisfies the following four properties:*

1. *For an arbitrary vector* $f \in \mathbb{R}^n$, *we have*

$$f^T L f = \frac{1}{2} \sum_{i,j=1}^{n} W_{ij}(f_i - f_j)^2 \tag{8.1}$$

2. L *is symmetric and positive semidefinite.*

3. *The smallest eigenvalue of* L *is* 0*, with eigenvector* $\mathbb{1}$.

4. L *has* n *nonnegative, real-valued eigenvalues* $0 = \lambda_1 \leq \lambda_2 \leq \ldots \leq \lambda_n$.

Proof 1 *We can verify the first property from the following equation:*

$$\begin{aligned} f^T L f &= f^T(D-W)f = \sum_{i=1}^{n} f_i^2 d_i - \sum_{i,j=1}^{n} f_i f_j W_{ij} \\ &= \frac{1}{2}\left(\sum_{i=1}^{n} f_i^2 d_i + \sum_{j=1}^{n} f_j^2 d_j - 2\sum_{i,j=1}^{n} f_i f_j W_{ij}\right) \\ &= \frac{1}{2}\sum_{i,j=1}^{n} W_{ij}(f_i - f_j)^2 \end{aligned}$$

Since $f^T L f \geq 0$ *and both* D *and* W *are symmetric,* L *is symmetric and positive semidefinite. Property 3 can be proven directly by plugging the unit vector* $\mathbb{1}$ *into Equation (8.1). Due to the fact that* L *is a real symmetric matrix, its eigenvalues should be also real numbers. Then property 4 is proved based on property 3.* ■

8.3.3 Spectrum Analysis

We first analyze the spectrum of L in the context of the unnormalized algorithm. Assume that G is an undirected similarity graph with k connected components $A_1, A_2, \ldots, A_k$. The corresponding graph Laplacians for these subsets are represented as $L_1, L_2, \ldots, L_k$. Without loss of generality, L can be represented as a block-diagonal structure:

$$L = \begin{bmatrix} L_1 & & & \\ & L_2 & & \\ & & \ddots & \\ & & & L_k \end{bmatrix}$$

Since L is a block diagonal matrix, its eigenvalues and eigenvectors are the union of eigenvalues and eigenvectors of its blocks (i.e., $L_1, L_2, \ldots, L_k$). This implies that the multiplicity of eigenvalue 0 of L should be at least k. The corresponding eigenvectors can be represented as indicator vectors $\mathbb{1}_{A_1}, \mathbb{1}_{A_2}, \ldots, \mathbb{1}_{A_k}$. As the weights W_{ij} are nonnegative, it can also be seen from Equation 8.1 that the next eigenvalue is strictly bigger than 0 because this sum can vanish if expressions $W_{ij}(f_i - f_j)^2$ for any pairs of vertices v_i and v_j vanish. So we can claim that the multiplicity k of the eigenvalue 0 of L equals the number of connected components $A_1, A_2, \ldots, A_k$.

How about the eigenvectors? Since 0 is a repeated eigenvalue in L, any other k orthogonal vectors spanning the same subspace as the eigenspace of the eigenvectors obtained above can also be the eigenvectors. Therefore, we have the following proposition [31].

Proposition 8.3.2 *The multiplicity k of eigenvalue 0 of L equals the number of connected components* $A_1, A_2, \ldots, A_k$ *in the graph. And the eigenspace of eigenvalue 0 is spanned by the indicator vectors* $\mathbb{1}_{A_1}, \mathbb{1}_{A_2}, \ldots, \mathbb{1}_{A_k}$ *of those components.*

This proposition provides theoretical foundation for the spectral clustering algorithm proposed for partitioning vertices into different subsets in the "ideal" case.

8.3.4 Unnormalized Spectral Clustering Algorithm

We will introduce the unnormalized spectral clustering algorithm based on unnormalized graph Laplacian L. Assume that $F \in \mathbb{R}^{n \times k}$ is a matrix containing the k relevant orthonormal vectors $f_1, f_2, \ldots, f_k$. It is desired to determine these orthonormal vectors $f_1, f_2, \ldots, f_k$ with the following objective function:

$$\min_F Tr(F^T L F) \quad \text{s.t. } F^T F = I \tag{8.2}$$

Here, $Tr(\cdot)$ denotes the trace of the matrix.

As we have discussed in the last subsection, these vectors are the top[2] eigenvectors of L. As discussed earlier, these eigenvectors are useful for clustering the vertices. However, unlike non-negative matrix factorization methods, these eigenvectors cannot be *directly* used for inferring the cluster labels because of the following two reasons. First, only in the ideal case where no edges exist between different connected components can the eigenspace be spanned by the indicator vectors $\mathbb{1}_{A_1}, \mathbb{1}_{A_2}, \ldots, \mathbb{1}_{A_k}$. Second, the eigenvectors could be any orthogonal transformation of the indicator vectors, which are not necessarily indicative of the cluster labels. Fortunately, if we view these eigenvectors as the low-dimensional representations of the original data points X, then any off-the-shelf clustering algorithms such as k-means and mixture models can be used to partition the embedding.

The overall algorithm for spectral clustering is illustrated below.

Unnormalized Spectral Clustering

1. Construct the similarity graph with one of the methods described in Section 8.2. Let W be the adjacency matrix and D be the degree matrix.
2. Compute the unnormalized graph Laplacian L where $L = D - W$.
3. Determine $f_1, f_2, \ldots, f_k$, the top k eigenvectors of L.
4. Construct the matrix $F \in \mathbb{R}^{n \times k}$ from $f_1, f_2, \ldots, f_k$.
5. Treat each row of F as a vertex in $\mathbb{R}^k$, partition these vertices into k clusters via any off-the-shelf method such as the k-means algorithm.

8.4 Normalized Spectral Clustering

In the literature, there are two versions of *normalized graph Laplacian*: L_{sym} and L_{rw}. The former refers to a symmetric matrix, and the latter can be explained from a random walk perspective. Close relationships exist between these two alternatives of normalized spectral clustering. These relationships will be discussed carefully in this section.

[2]Here, by *top*, we refer to the eigenvectors with the smallest eigenvalues.

8.4.1 Normalized Graph Laplacian

The two variations of the normalized graph Laplacian are defined as follows:

$$L_{sym} = D^{-\frac{1}{2}}LD^{-\frac{1}{2}} = I - D^{-\frac{1}{2}}WD^{-\frac{1}{2}}$$
$$L_{rw} = D^{-1}L = I - D^{-1}W$$

We list some properties of these two variations L_{sym} and L_{rw} below.

Proposition 8.4.1 *The normalized graph Laplacians satisfy the following six properties [31]:*

1. *For arbitrary vector* $f \in \mathbb{R}^n$, *we have*[3]

$$f^T L_{sym} f = \frac{1}{2}\sum_{i,j=1}^{n} W_{ij}\left(\frac{f_i}{\sqrt{d_i}} - \frac{f_j}{\sqrt{d_j}}\right)^2 \tag{8.3}$$

2. *λ is an eigenvalue of L_{sym} with eigenvector u if and only if λ is also an eigenvalue of L_{rm} with eigenvector w, satisfying that $u = D^{\frac{1}{2}}w$.*

3. *λ is an eigenvalue of L_{rw} with eigenvector w if and only if λ and w together solve the generalized eigen-problem $Lw = \lambda Dw$.*

4. *Both L_{sym} and L_{rw} are positive semidefinite.*

5. *The smallest eigenvalue of both L_{sym} and L_{rw} is 0, with eigenvector $D^{\frac{1}{2}}\mathbb{1}$ for L_{sym} and $\mathbb{1}$ for L_{rw}.*

6. *Both L_{sym} and L_{rw} have n nonnegative, real-valued eigenvalues $0 = \lambda_1 \leq \lambda_2 \leq \ldots \leq \lambda_n$.*

Proof 2 *Similar to the proof for Proposition 8.3.1, we can verify the first property from the following equation:*

$$\begin{aligned}
f^T L_{sym} f &= f^T(I - D^{-\frac{1}{2}}WD^{-\frac{1}{2}})f = \sum_{i=1}^{n} f_i^2 - \sum_{i,j=1}^{n} f_i f_j \frac{W_{ij}}{\sqrt{d_i d_j}} \\
&= \frac{1}{2}\left(\sum_{i=1}^{n} f_i^2 + \sum_{j=1}^{n} f_j^2 - 2\sum_{i,j=1}^{n} f_i f_j \frac{W_{ij}}{\sqrt{d_i d_j}}\right) \\
&= \frac{1}{2}\sum_{i,j=1}^{n} W_{ij}\left(\frac{f_i}{\sqrt{d_i}} - \frac{f_j}{\sqrt{d_j}}\right)^2
\end{aligned}$$

To prove property 2, we replace u with $D^{\frac{1}{2}}w$ in the eigenvalue equation of L_{sym}:

$$L_{sym}u = L_{sym}D^{\frac{1}{2}}w = D^{-\frac{1}{2}}LD^{-\frac{1}{2}}D^{\frac{1}{2}}w = D^{-\frac{1}{2}}Lw = D^{\frac{1}{2}}\lambda w = \lambda u$$

Property 3 follows directly by multiplying the eigenvalue equation of L_{rm} with D from the left:

$$\begin{aligned}
DL_{rm}w &= D\lambda w \\
Lw &= \lambda Dw
\end{aligned}$$

Since $f^T L_{sym} f \geq 0$ and both D and W are symmetric, L_{sym} is then symmetric and positive semidefinite. This property also fits L_{rm} due to property 2. Property 5 can be proved directly as $L_{rw}\mathbb{1} = 0$, and the same statement for L_{sym} follows from property 2. Due to the fact that L_{sym} and L_{rm} are both real symmetric matrices, their eigenvalues should be also real numbers. Then property 6 is proved based on property 5. ■

[3]The same case does not hold for L_{rw}.

8.4.2 Spectrum Analysis

As in the case of the spectrum analysis for unnormalized spectral clustering, we have a similar proposition [31] on the multiplicity of the 0 eigenvalues. The following proposition relies on Proposition 8.4.1:

Proposition 8.4.2 *The multiplicity k of eigenvalue 0 of both L_{sym} and L_{rw} equals the number of connected components $A_1, A_2, \dots, A_k$ in the graph. For L_{rw}, the eigenspace of eigenvalue 0 is spanned by the indicator vectors $\mathbb{1}_{A_1}, \mathbb{1}_{A_2}, \dots, \mathbb{1}_{A_k}$ of those components. For L_{sym}, the eigenspace of eigenvalue 0 is spanned by $D^{\frac{1}{2}}\mathbb{1}_{A_i}$.*

8.4.3 Normalized Spectral Clustering Algorithm

We will introduce two normalized spectral clustering algorithms based on the two normalized graph Laplacian variants. These are aimed at finding orthonormal vectors $f_1, f_2, \dots, f_k$ with the following objective functions (assume $F \in \mathbb{R}^{n\times k}$ is a matrix consisting of orthogonal vectors), respectively:

$$\min_F Tr(F^T L_{sym} F) \quad \text{s.t. } F^T F = I \tag{8.4}$$

$$\min_F Tr(F^T LF) \quad \text{s.t. } F^T DF = I \tag{8.5}$$

We first introduce the normalized spectral clustering algorithm based on L_{sym}. In this case, the eigenspace of eigenvalue 0 is spanned by $D^{\frac{1}{2}}\mathbb{1}_{A_i}$ instead of $\mathbb{1}_{A_i}$. Therefore, before applying k-means, one row normalization step is usually adopted for L_{sym} in order to make each row comparable in the Euclidean space. This is an additional step, which needs to be applied, beyond what is discussed for the case of unnormalized spectral clustering.

The algorithm is shown below.

Normalized Spectral Clustering (symmetric version)

1. Construct similarity graph by one of the methods described in Section 8.2. Let W be the adjacency matrix and D be the degree matrix.
2. Compute symmetric normalized graph Laplacian L_{sym} where $L_{sym} = D^{-\frac{1}{2}}LD^{-\frac{1}{2}}$.
3. Determine $f_1, f_2, \dots, f_k$, the top k eigenvectors of L_{sym}.
4. Construct the matrix $F \in \mathbb{R}^{n\times k}$ from $f_1, f_2, \dots, f_k$.
5. Normalize the rows of F to 1 such that $\forall i \le n, \sum_j F_{ij}^2 = 1$.
6. Treat each row of F as a vertex in $\mathbb{R}^k$, partition these vertices into k clusters via k-means algorithm.

The other version of normalized spectral clustering algorithms is based on L_{rm}. This version is more similar to the unnormalized case, because their eigenspaces are both spanned by $\mathbb{1}_{A_i}$ in the ideal case where there are k connected components. Therefore, their procedures are almost the same except for the computation step for the graph Laplacian.

Normalized Spectral Clustering (random walk version)

1. Construct similarity graph by one of the methods described in Section 8.2. Let W be the adjacency matrix and D be the degree matrix.
2. Compute unnormalized graph Laplacian L where $L = D - W$.
3. Find $f_1, f_2, \ldots, f_k$, the top k eigenvectors of the generalized eigenproblem $Lf = \lambda Df$.
4. Form the matrix $F \in \mathbb{R}^{n \times k}$ from $f_1, f_2, \ldots, f_k$.
5. Treat each row of F as a vertex in $\mathbb{R}^k$, partition these vertices into k clusters via k-means algorithm.

8.5 Graph Cut View

In the last two sections, we introduced the definitions for different formulations of graph Laplacians and specific procedures for all three kinds of spectral clustering algorithms. Although we have analyzed the spectrum of both unnormalized and normalized Laplacian matrices, the intuition behind the graph Laplacians is not completely clear. Therefore, in this section, the spectral clustering method will be presented from the perspective of the graph cut.

Given a graph $G = (V, E)$, suppose we have two subsets of vertices A_1, A_2, satisfying $A_1 \bigcap A_2 = \emptyset$ and $A_1 \bigcup A_2 \subseteq V$. We then introduce the definition of the *cut* as the sum of weights of the edges across the two subsets:

$$cut(A_1, A_2) = \sum_{v_i \in A_1, v_j \in A_2} W_{ij}$$

Here, W_{ij} is the weight of the edge between vertices i and j. It is evident that the cut describes the closeness between these two subsets in the graph. In other words, a smaller value of the cut indicates greater separability of the two subsets.

In graph theory, a *minimum cut* (MinCut) of a graph is a cut with the smallest possible value under the condition $A_1 \bigcup A_2 = V$. This is also the most direct way to construct the partition. MinCut is a classical problem in the literature and relatively easy to solve. More details of MinCut may be found in [49].

By extending the cut definition from two sets to the multi-set situation, we can reformulate the cut definition as follows:

$$cut(A_1, A_2, \ldots, A_k) = \sum_{i=1}^{k} cut(A_i, \overline{A_i})$$

where $\overline{A_i}$ stands for the set of vertices $\{v_j | v_j \notin A_i\}$.

Although MinCut is an intuitive way to partition the graph, it suffers from a critical problem, which is quite common in graph partitioning. In many cases, MinCut separates an individual vertex or a small set of vertices from the remaining graph, overlooking the balance between the sizes of different partitions. Some normalization strategies can be incorporated into the objective function of MinCut to circumvent this problem. Two examples of such normalized objective functions are the *Ratio Cut* (RatioCut) [19] and *Normalized Cut* (NCut) [43].

The RatioCut is defined as follows:

$$RatioCut(A_1,\ldots,A_k) = \sum_{i=1}^{k} \frac{cut(A_i,\overline{A_i})}{|A_i|}$$

Here, $|A_i|$ denotes the number of vertices belonging to A_i. By minimizing this measure, the partitions with more balanced cluster sizes are preferred.

The objective function for NCut is defined as follows:

$$NCut(A_1,\ldots,A_k) = \sum_{i=1}^{k} \frac{cut(A_i,\overline{A_i})}{assoc(A_i,V)}$$

where $assoc(A_i,V) = \sum_{v_j \in A_i}(d_j)$. Different from RatioCut which measures the balance by the cluster size, NCut focuses on the total degree within different clusters.

The incorporation of normalization into the objective function makes it NP-hard [53]. However, by relaxing some constraints, spectral clustering is a way to give the approximate solution. Specifically, unnormalized spectral clustering can be used to solve the relaxed RatioCut problem, and normalized spectral clusterings are designed to solve the relaxed NCut problem.

8.5.1 Ratio Cut Relaxation

Suppose we already have k vectors $\{f_i\}_{i=1}^{k}$ indicating the cluster labels for the vertices in the graph where $f_i = (f_{i1},\ldots,f_{in})^T$. Define that:

$$f_{ij} = \begin{cases} \frac{1}{\sqrt{|A_i|}} & \text{if } v_j \in A_i \\ 0 & \text{otherwise.} \end{cases} \tag{8.6}$$

Then, under this condition, for each cluster i among all k clusters, we can rewrite Equation 8.1 of the unnormalized graph Laplacian as follows:

$$\begin{aligned} f_i^T L f_i &= \frac{1}{2}\sum_{j,l=1}^{n} W_{jl}(f_{ij}-f_{il})^2 \\ &= \frac{1}{2}\Big(\sum_{v_j\in A_i, v_l\notin A_i} W_{jl}(\frac{1}{\sqrt{|A_i|}}-0)^2 + \sum_{v_j\notin A_i, v_l\in A_i} W_{jl}(0-\frac{1}{\sqrt{|A_i|}})^2\Big) \\ &= \frac{1}{2}(\frac{cut(A_i,\overline{A_i})}{|A_i|}) + \frac{1}{2}(\frac{cut(\overline{A_i},A_i)}{|A_i|}) \\ &= \frac{cut(A_i,\overline{A_i})}{|A_i|} \end{aligned}$$

The RatioCut measure is linked to the unnormalized graph Laplacian because of the following calculation:

$$\begin{aligned} \sum_{i=1}^{k} f_i^T L f_i &= Tr(F^T L F) \\ &= \sum_{i=1}^{k} \frac{cut(A_i,\overline{A_i})}{|A_i|} \\ &= RatioCut(A_1,\ldots,A_k) \end{aligned}$$

where $F \in \mathbb{R}^{n\times k}$ is formed from $f_1, f_2, \ldots, f_k$.

Moreover, according to Equation 8.6, f_i satisfies the following:

$$f_i^T f_i = \sum_{j=1}^{n} f_{ij}^2 = |A_i|\frac{1}{|A_i|} = 1$$

It is also easy to check that any two different column vectors of F are orthogonal.

Now the RatioCut problem is equivalent to the following:

$$\min_{A_1,\dots,A_k} Tr(F^T LF) \quad \text{s.t. } F^T F = I \tag{8.7}$$

Note that f_{ij} can only be 0 or $1/\sqrt{|A_i|}$. The objective function is therefore discrete, which is known to be NP-hard in the general case. A relaxation of this problem allows f_{ij} to take on arbitrary real values. This implies the following:

$$\min_{F\in\mathbb{R}^{n\times k}} Tr(F^T LF) \quad \text{s.t. } F^T F = I$$

The aforementioned objective function is exactly the same as the objective function in Equation (8.2) for unnormalized spectral clustering.

According to Rayleigh-Ritz theorem [30], to minimize the above reflexed RatioCut objective function is equivalent to minimizing Rayleigh quotient of the unnormalized spectral clustering, where the Rayleigh quotient $R(L,f)$ is defined as follows:

$$R(L,f) = \frac{f^T Lf}{f^T f}$$

The Rayleigh quotient (objective function value) reaches its minimum value (the smallest eigenvalue of L) when f is the corresponding eigenvector[4]. Therefore, the solution of Equation (8.2) is obtained by choosing the top k eigenvectors of L. Later, we need to convert the real valued matrix F back to discrete indicators. As mentioned in Section 8.3.4, the k-means approach can be the candidate algorithm to be adopted on the rows of F.

8.5.2 Normalized Cut Relaxation

Similar to RatioCut, we first define the indicator vectors $f_i = (f_{i1},\dots,f_{in})^T$ where $1 \le i \le k$ such that:

$$f_{ij} = \begin{cases} \dfrac{1}{\sqrt{assoc(A_i,V)}} & \text{if } v_j \in A_i \\ 0 & \text{otherwise.} \end{cases} \tag{8.8}$$

Under this condition, we can rewrite Equation 8.1 of the unnormalized graph Laplacian as follows:

$$\begin{aligned} f_i^T Lf_i &= \frac{1}{2}\sum_{j,l=1}^{n} W_{jl}(f_{ij}-f_{il})^2 \\ &= \frac{1}{2}\left(\sum_{v_j\in A_i, v_l\notin A_i} W_{jl}\frac{1}{assoc(A_i,V)} + \sum_{v_j\notin A_i, v_l\in A_i} W_{jl}\frac{1}{assoc(A_i,V)}\right) \\ &= \frac{1}{2}\left(\frac{cut(A_i,\overline{A_i})}{assoc(A_i,V)}\right) + \frac{1}{2}\left(\frac{cut(\overline{A_i},A_i)}{assoc(A_i,V)}\right) \\ &= \frac{cut(A_i,\overline{A_i})}{assoc(A_i,V)} \end{aligned}$$

[4]The other eigenvalues and corresponding eigenvectors of L follow the same way.

The NCut measure can be linked with the graph Laplacian, based on the following calculation:

$$\begin{aligned}\sum_{i=1}^{k} f_i^T L f_i &= Tr(F^T L F) \\ &= \sum_{i=1}^{k} \frac{cut(A_i, \overline{A_i})}{assoc(A_i, V)} \\ &= NCut(A_1, \ldots, A_k)\end{aligned}$$

The matrix $F \in \mathbb{R}^{n \times k}$ can be constructed with the k column vectors $f_1, f_2, \ldots, f_k$.

According to Equation 8.8, f_i satisfies the following:

$$f_i^T D f_i = \sum_{j=1}^{n} d_j f_{ij}^2 = \frac{\Sigma_{v_j \in A_i} d_j}{assoc(A_i, V)} = \frac{assoc(A_i, V)}{assoc(A_i, V)} = 1$$

For any pair of columns f_i, f_j from F, it is easy to show that they are orthogonal.

The NCut objective function is equivalent to the following:

$$\min_{A_1, \ldots, A_k} Tr(F^T L F) \quad \text{s.t. } F^T D F = I$$

Similar to the RatioCut problem, the value of f_{ij} is discrete, which is NP-hard in the general case. By relaxing f_{ij} to take on real values, the problem is reduced to the following:

$$\min_{F \in \mathbb{R}^{n \times k}} Tr(F^T L F) \quad \text{s.t. } F^T D F = I$$

This is exactly the same as the objective function in Equation (8.5) for normalized spectral clustering.

If we define $P = D^{\frac{1}{2}} F$ according to Proposition 8.4.1, and substitute F by P in the above relaxed objective function for NCut, we have

$$\min_{P \in \mathbb{R}^{n \times k}} Tr(P^T D^{-\frac{1}{2}} L D^{-\frac{1}{2}} P) \quad \text{s.t. } P^T P = I$$

This is exactly the same as the symmetric version of objective function in Equation (8.4) for normalized spectral clustering. The relaxed NCut problem can be solved by simply computing the first k eigenvectors of L_{rw}. The first k eigenvectors of the generalized eigenproblem $Lf = \lambda D f$, which is the method adopted in [43]. To convert the real value matrix F (the first k eigenvectors of L_{rw}) back to discrete indicators, k-means could be adopted.

We can also compute the first k eigenvectors of L_{sym} instead of L_{rw} according to [37]. For P (the first k eigenvectors of L_{sym}), we need row normalization before discretization because $P = D^{\frac{1}{2}} F$.

8.6 Random Walks View

A random walk on a graph is a *Markov chain* which can be described by an $n \times n$ square matrix M, where n is the number of vertices in the graph. The matrix M denotes the transition probabilities. Therefore, the conditional probability of the next state being vertex v_j, given the current state v_i, is given by $0 \leq P(v_i|v_j) = M_{ij} \leq 1$. If there is no edge from vertex v_i to vertex v_j, then $M_{ij} = 0$. The matrix M is a stochastic matrix. Therefore, all its entries are nonnegative and every row adds up to one.

A formal relationship analysis between NCut and random walks has been provided in [32]. Specifically, the transition probability M_{ij} is proportional to the edge weight W_{ij} and is computed by the following:

$$M_{ij} = \frac{W_{ij}}{d_i}$$

If we use matrix operations to represent this computation, we have:

$$M = D^{-1}W$$

In the following, we will use the transition matrix M to achieve a better understanding of the normalized spectral clustering algorithm. First, define $\pi = (\pi_1, \ldots, \pi_n)^T$ to be a vector with length n as follows:

$$\pi_i = \frac{d_i}{assoc(V,V)}$$

Here, $assoc(V,V)$ represents the total degrees in the graph. It is easy to verify that $M^T\pi = \pi$ and, thus, that π is a *stationary distribution* of the Markov chain.

Assume in the graph $G = (V,E)$, we have two disjoint subsets A and $\overline{A}$. Define $M_{A\overline{A}} = P(A \to \overline{A}|A)$ as the probability of the random walk transiting from set A to set $\overline{A}$ in one step if the current state is in A and the random walk is started from its stationary distribution.

$$\begin{aligned}
M_{A\overline{A}} &= \frac{\sum_{v_i \in A, v_j \in \overline{A}} \pi_i M_{ij}}{\sum_{v_i \in A} \pi_i} \\
&= \frac{\sum_{v_i \in A, v_j \in \overline{A}} \frac{W_{ij}}{assoc(V,V)}}{\sum_{v_i \in A} \frac{d_i}{accoc(V,V)}} \\
&= \frac{\sum_{v_i \in A, v_j \in \overline{A}} W_{ij}}{assoc(A,V)} \\
&= \frac{cut(A,\overline{A})}{assoc(A,V)}
\end{aligned}$$

From this, we now have:

$$NCut(A,\overline{A}) = M_{A\overline{A}} + M_{\overline{A}A}$$

If the NCut measure is small for a certain partition A and $\overline{A}$, then the probability of the walk moving from one partition to the other is small. In other words, when minimizing NCut, we are trying to partition the set V into different groups such that the random walk, once in one of the parts, tends to remain in it [32].

This random walk view gives a new and intuitive characterization of the NCut algorithm. The NCut algorithm is just another view of spectral clustering. Moreover, as the random walk is essentially a Markov chain on the graph, it provides spectral clustering algorithm with a probabilistic foundation.

8.7 Connection to Laplacian Eigenmap

The *Laplacian eigenmap* algorithm [2] is a very successful method for dimensionality reduction and is closely related to spectral clustering. Dimensionality reduction is also called subspace learning or feature extraction, aiming at extracting low-dimensional structure dimensionality from high-dimensional data. One of the most popular dimensionality reduction algorithms is Principal Component Analysis (PCA) which gives the solution by projecting the original high-dimensional data onto the low-dimensional linear subspace spanned by the leading eigenvectors of the data's covariance matrix.

In many real-world problems, the data is not always sampled from a linear subspace. For example, it is always believed that the face images are sampled from a nonlinear low-dimensional manifold which is embedded in the high-dimensional ambient space. This has motivated researchers to consider manifold-based techniques for dimensionality reduction and Laplacian eigenmap is just one of them. The basic idea of Laplacian eigenmap is also based on spectral graph theory [8]. Its objective function is similar to that in normalized spectral clustering:

$$\text{argmin}_F \, Tr(F^T LF) \quad \text{s.t. } F^T DF = I$$

where F is considered to be the new low-dimensional representation.

We can reconsider the following equation from the dimensionality reduction point of view:

$$f^T Lf = \frac{1}{2} \sum_{i,j=1}^{n} W_{ij}(f_i - f_j)^2$$

The intuition is that for any pair of data points, if they are near to each other on the high-dimensional space (a space in which a low-dimensional manifold is embedded), they should still be close in the mapped low-dimensional space. In real life, the underlying manifold is often unknown. Similar to [50], the geodesic distance between all pairs of points on the manifold can be reflected by their shortest path distance on the adjacency graph. Therefore, the minimization of the aforementioned cost function based on the graph preserves the local distance between every pair of data points. The additional constraint $F^T DF = I$ fixes the arbitrary scaling factor in the embedding.

Figure 8.3 is used in [50] to illustrate how geodesic paths work for nonlinear dimensionality reduction. In the left figure, for two arbitrary points (circled) on a nonlinear manifold, their Euclidean distance in the high-dimensional input space (length of dashed line) may not accurately reflect their intrinsic similarity, as measured by geodesic distance along the low-dimensional manifold (length of solid curve). The figure on the right shows an approximation to the true geodesic path, as the shortest path in adjacency matrix W.

We can see that the only difference between these two algorithms is that spectral clustering makes one step further to cluster the information in the new dimension. It is discussed in [3] that for the local approaches to do dimensionality reduction, although they attempt only to approximate or preserve neighborhood information, they may also be interpreted as imposing a soft clustering of the data (which may be converted to a hard clustering by a variety of heuristic techniques such as k-means). In this sense, the local approaches to dimensionality reduction impose a natural clustering on the data.

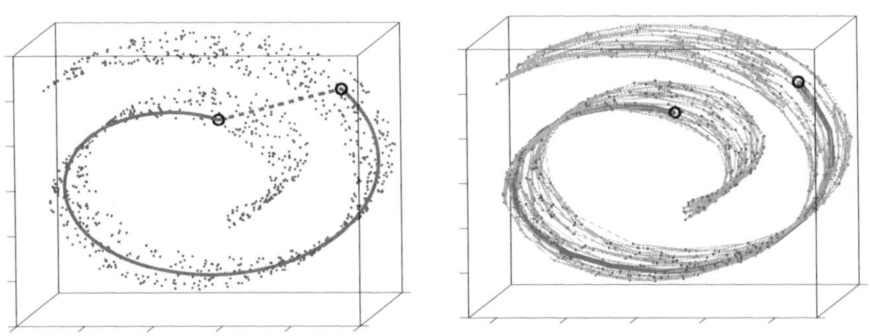

FIGURE 8.3: The "swiss roll" data set, illustrating how geodesic paths work for nonlinear dimensionality reduction.

8.8 Connection to Kernel k-Means and Nonnegative Matrix Factorization

Recent studies [12, 13] show that spectral clustering is theoretically closely related to two other popular clustering algorithms: *kernel k-means* and *nonnegative matrix factorization* [25]. It is proved that these three are unified in a simple way: they are different prescriptions of the same problem with slightly different constraints and optimization solutions.

It is known that k-means uses k centroids to characterize the data. The objective function is to minimize the sum of squared errors [13]:

$$J = \sum_{i=1}^{k} \sum_{j \in A_i} ||x_j - m_i||^2 = c - \sum_{i=1}^{k} \frac{1}{|A_i|} \sum_{j,l \in A_i} x_j^T x_l \tag{8.9}$$

where $m_i = \sum_{j \in A_i} x_j / |A_i|$ is the centroid of cluster A_i of $|A_i|$ points, and $c = \sum_j ||x_j||^2$.

We can represent the solution by k nonnegative indicator vectors:

$$F = (f_1, \cdots, f_k), \quad f_i^T f_j = \delta_{ij}$$

where

$$f_{ij} = \begin{cases} \frac{1}{\sqrt{|A_i|}} & \text{if } x_j \in A_i \\ 0 & \text{otherwise.} \end{cases}$$

We can rewrite Equation 8.9 as $J = Tr(X^T X) - Tr(F^T X^T X F)$. As the first term is a constant, minimizing J is equal to the following:

$$\max Tr(F^T G F)$$

Here, $G = X^T X$ is a pairwise similarity matrix called inner-product linear kernel matrix which can be extended to any other kernels. It is clear that if we set the kernel matrix G to be the negative of unnormalized graph Laplacian[5] $-L = W - D$, the objective function of kernel k-means clustering is equivalent to that of RatioCut in Equation 8.7. Therefore, for the same objective function, spectral clustering first relaxes F to be real value and computes it through solving an eigen-decomposition problem. The clustering labels are finally obtained by running k-means on F. Different from this "global" minimum after relaxation of F, kernel k-means uses an iterative algorithm similar to k-means to directly partition all data points, which is essentially a local search. It is worth noting that in [12], a weighted kernel k-means is proposed and the authors also prove it equivalent to spectral clustering where kernel matrix G is set to $D^{-1}WD^{-1}$ and the weight matrix for data points equals to D.

Nonnegative matrix factorization (NMF) aims at factorizing X into two nonnegative matrices,

$$X \approx UV^T$$

where $U \in \mathbb{R}_+^{n \times k}$ and $V \in \mathbb{R}_+^{m \times k}$. As $k \ll \min(m,n)$, UV^T can be viewed as a low-rank approximation of X. The clustering aspect of NMF is studied in [57, 18].

Now we consider the symmetric NMF of

$$G \approx FF^T$$

[5] If G is set to be $I - L_{sym}$, then the objective function is equivalent to NCut, which is proved in [13].

The factorizations can be obtained by the least squares minimization:

$$\begin{aligned} F &= \text{argmin}_{F\geq 0} ||G - FF^T||^2 \\ &= \text{argmin}_{F\geq 0} ||G||^2 - 2Tr(F^T GF) + ||F^T F||^2 \\ &= \text{argmax}_{F\geq 0} Tr(F^T GF) \end{aligned}$$

Again if we set G to be $-L = W - D$, the objective function of NMF is very similar to unnormalized spectral clustering except that for NMF we require F to be nonnegative while for spectral clustering F is expected to be orthonormal.

8.9 Large Scale Spectral Clustering

The general spectral clustering method needs to construct an adjacency matrix and calculate the eigen-decomposition of the corresponding Laplacian matrix. Both of these two steps are computational expensive. For a data set consisting of n data points with m dimensions, the above two steps will have time complexities of $O(n^2m)$ and $O(n^3)$, respectively. This is an unreasonable burden for large-scale applications.

Fortunately, if we only want to construct an *approximate* adjacency matrix with a few number of nearest neighbors, there is a shortcut for reducing the complexity. This is called approximate nearest neighbor search. We first introduce randomized kd-trees [45, 36]. A kd-tree partitions the space by hyperplanes that are perpendicular to the coordinate axes. At each node of the tree, the corresponding data points are partitioned into two halves with a hyperplane through a specified dimension which exhibits the greatest variance. As kd-tree's performance rapidly degrades with the number of dimensions increasing, Silpa-Anan and Hartley [45] recently have proposed an improved version of the kd-tree algorithm in which multiple randomized kd-trees are created. By comparison, the randomized trees are built by choosing the split dimension randomly from the first few dimensions on which data has the greatest variance. The other available technique is called locality-sensitive hashing (LSH) [10] which offers probabilistic guarantees of retrieving data points similar to the query. The basic idea is to compute a set of functions that reduce the dimensionality of the data points through certain mathematical operations. By concatenating several such functions to form a hash function, data points mapped in the same bin as the query are returned as candidates of nearest neighbors with probabilistic guarantees. Moreover, several hash tables can be built in the same way to further enhance the performance. A few libraries for approximate nearest neighbor search are available, including FLANN [36] and ANN [35].

For the problem of eigen-decomposition, state-of-the-art eigensolvers can be categorized into three classes: local, global, and extreme solvers [27]. The power method is the simplest local solver, which only computes the maximum eigenvalue and its associated eigenvector. QR-decomposition method is a typical global solver, which computes all the eigenvalues and eigenvectors. Extreme eigensolvers compute several extreme (smallest or largest) eigenvalues and the corresponding eigenvectors, which is obviously the most suitable for spectral clustering since the number of clusters k usually is not too large. Typical examples of extreme eigensolvers include Lanczos, preconditioned conjugate gradient (PCG), and Jacobi-Davidson (JD) methods. Interested readers can refer to [41] for details. Several software implementations of the Lanczos method are publicly available and are well maintained, e.g., ARPACK [26]. PCG and JD methods also have some implementations available [22, 48]. If the Laplacian matrix is sparse, i.e., the similarity graph is not fully connected, these state-of-art eigensolvers can benefit from this property because most elements are zero and could be removed from the computation.

Besides this, many works have been proposed to accelerate spectral clustering itself instead of focusing on the general eigensolvers. [16] adopts the classical Nyström method which was originally proposed for finding numerical approximations to eigenfunction problems. It first chooses samples randomly among the data points to obtain the small-size eigenvectors and then extrapolates the solution for the whole dataset with weights. The work in [44] reduces the original data set to a relatively small number of data points in a biased way. The first step is to apply an efficient clustering algorithm such as k-means to get the initial partitions. Then data points near the same centroid are merged and viewed as one. By some additional similarity computations for such new representations, the algorithm can operate directly on this smaller graph. Similar to this idea, in [58], all data points are collapsed into centroids through k-means or random projection trees so that the later eigen-decomposition needs to be applied only on the centroids. [42] uses random projection for another purpose which is to reduce the dimensionality of X (i.e., m mentioned at the beginning) and therefore to help save time in constructing the adjacency matrix. Additionally, random sampling is later applied to drastically reduce the size of data points within the eigen-decomposition step. [4, 27] introduce early stop strategies to speed up eigen-decomposition. It is based on observation that well-separated data points will converge to the final embedding more quickly, and hence, we can sequentially depress the scale of the Laplacian matrix in the iteration steps of conjugate-based eigensolvers without inferencing the final clustering results by much. In [6], landmark points are first selected randomly or through k-means among all the data points to serve as codebook. Later each data point can be encoded in the form of linear combinations of these landmarks. After constructing the affinity matrix by inner product using such landmark representations, acceleration can be achieved since the complexity now is related to the number of landmark points which is far less than n. Different from the above methods, authors in [21] work on resistance distance embedding which has an idea similar to spectral clustering and also has quite similar clustering capability. The proposed method does not require using any sampling technique or computing any eigenvector. Instead it uses random projection and a linear time solver of Spielman and Teng [47] to find the approximate and accurate embedding.

To address resource bottlenecks of both memory use and computation time, another promising direction is to make use of distributed system. In [5], a parallel framework is proposed which first distributes n data instances onto p distributed machine nodes. On each node, similarities between local data and the whole set are computed in a way that uses minimal disk I/O. The eigenvector matrices are stored on distributed nodes to reduce per-node memory use. Finally, together with parallel eigensolver and k-means, the proposed parallel spectral clustering achieves good speedup with large data sets. Additionally, for constructing the sparse similarity matrix, the authors choose MapReduce because this step may be the most time consuming and having a fault tolerant mechanism is essential. For eigen-decomposition and the k-means step, MPI-based open source softwares are adopted. Gao et al. [17] use a different approach to first map data points into buckets and then implement spectral clustering for each bucket simultaneously. The main idea is within the preprocess step, which is to adopt locality sensitive hashing (LSH) [10] to create signatures (i.e., M-bit binary numbers) for data points and later to project near-duplicate signatures[6] into one bucket. The hash function they use to generate the signatures falls into the family of random projection. Besides this, they propose the MapReduce algorithm for LSH and call the spectral clustering module in Apache Mahout [40] which is also implemented on top of Apache Hadoop using the MapReduce paradigm.

---

[6]Two M-bit signatures are considered to be near-duplicate if they have at least $M - \varepsilon$ bits in common.

8.10 Further Reading

The family of spectral clustering has become popular especially since the works in [37] and [43]. Since then, several works have been proposed to slightly change the original framework and achieve better clustering results. In [59], the final k-means step on the eigenvectors shared by the family of spectral clustering is explained not to be optimal. The authors clarify the role of eigenvectors obtained from graph Laplacian as a generator of all optimal solutions through orthonormal transforms. A better solution is then proposed by rotating normalized eigenvectors to obtain an optimal segmentation. Their method iterates between nonmaximum suppression and using SVD to recover the rotation. Later in [60], it is pointed out that this iterative method can easily get stuck in local minima and thus does not reliably find the optimal alignment. To find the best alignment for a set of eigenvectors, the authors define a new cost function which measures the alignment quality of a set of vectors and adopts a gradient descent scheme which provably converges. However, this paper also mentions that such alignments might only be effective when the data is not highly noisy. Another insight of this paper concerns the construction of adjacency matrix W. It suggests that in the fully connected graph, any two vertices are connected and the value of W_{ij} is defined to be $e^{-\frac{||x_i-x_j||^2}{\sigma_i\sigma_j}}$. Here σ_i is called a local scale and is set to be the distance between vertex i and its certain neighbor (7th nearest neighbor in the paper). The intuition behind this method is that in reality data resides in multiple scales (one cluster is tight and the other is sparse) and local scaling automatically finds the two scales and results in high affinities within clusters and low affinities across clusters. Later a more natural parameterization of the neighborhood density based on the empty region graphs is proposed in [9] which is claimed to be not sensitive to parameters and data perturbation.

Many papers have also been published to discover the theoretical relations of spectral clustering to other algorithms. As discussed in Section 8.8, there exist close links among spectral clustering, kernel k-means, and nonnegative matrix factorization. Besides this, relations between spectral clustering and kernel principal component analysis (KPCA) are built on the fact that the Laplacian eigenmap can be interpreted as KPCA with kernel matrix set to be the Moore-Penrose inverse of L [20]. From Section 8.7 we know that Laplacian eigenmap can be essentially viewed as a nonlinear dimensionality reduction method which is almost the same as spectral clustering except for the final k-means step.

Spectral clustering algorithms are extended to many nonstandard data types compared to those introduced at the beginning of this chapter. In [11], a co-clustering algorithm is proposed to the bipartite graph which is a kind of graph that can be divided into two sets such that no two data points in the same set are connected, e.g., document–word graph. [28] was later published to address the clustering problem through collective factorization for heterogeneous data types which is more complicated and general than the bipartite graph case. For network data where only nodes and links are observed, a spectral clustering-like algorithm [56] is designed for optimizing "modularity" measure instead of "cut" measure because the former is also popular in the community detection. Another interesting extension to incorporate is temporal information. Chi et al. [7] present two frameworks in evolutionary spectral clustering through regularizing one additional cost on temporal smoothness. If the data points are observed in a stream-like way, [51] has been proposed to update the clustering directly without evaluating the entire affinity matrix. In [38], a more general algorithm allowing similarity changes between existing data points has been developed based on incrementally updating the eigenvectors.

Besides all these, in the machine learning community, spectral clustering techniques usually appear with other learning tasks. For example, in the semisupervised scenario, pairwise constraints, i.e., must-link and cannot-link, are incorporated into spectral clustering as the user guidance for the partition [29]. Different from this, in [61], a few class labels are directly provided and a semisupervised classification function similar to spectral clustering is proposed requiring the smoothness of inference between known labeled and unlabeled data. Multi-view learning is another hot topic where multiple complementary representations of the data are presented. In [62], the spectral clustering approach is generalized to multiple views via a random walk formulation. The work in [23] approaches the problem by adopting a co-training framework such that the similarity matrix in one view is affected by the similarity estimated based on the eigenvectors of graph Laplacian in the other view. In [24], a co-regularization framework is proposed to enforce the pairwise similarities computed from the eigenvectors learned from different views to be close with each other. Opposite to multi-view algorithms which learn the latent consensus from multiple representations, the work in [39] tries to find multiple nonredundant clustering results from a single data view through penalizing for redundancy between different results. There are also extensions of spectral clustering algorithms in the field of active learning. The work in [54] formulates this problem as incrementally querying the oracle about pairwise relations mentioned above between chosen data points. In [55], the "active" part lies in querying the pairwise similarity under the assumption that the adjacency matrix W is expensive to obtain or compute.

In the last few years, many papers have been published concerning combinations of spectral clustering with different data types, learning tasks, and applications. Interested readers are encouraged to further explore the literature in these directions.

Bibliography

[1] Francis R. Bach and Michael I. Jordan. Learning spectral clustering, with application to speech separation. *Journal of Machine Learning Research*, 7:1963–2001, 2006.

[2] Mikhail Belkin and Partha Niyogi. Laplacian eigenmaps and spectral techniques for embedding and clustering. In *Advances in Neural Information Processing Systems 14*, pages 585–591, 2001.

[3] Mikhail Belkin and Partha Niyogi. Laplacian eigenmaps for dimensionality reduction and data representation. *Neural Computation*, 15(6):1373–1396, 2003.

[4] Bo Chen, Bin Gao, Tie-Yan Liu, Yu-Fu Chen, and Wei-Ying Ma. Fast spectral clustering of data using sequential matrix compression. In *Proceedings of the 17th European Conference on Machine Learning*, pages 590–597, 2006.

[5] Wen-Yen Chen, Yangqiu Song, Hongjie Bai, Chih-Jen Lin, and E. Y. Chang. Parallel spectral clustering in distributed systems. *IEEE Transactions on Pattern Analysis and Machine Intelligence*, 33(3):568–586, 2011.

[6] Xinlei Chen and Deng Cai. Large scale spectral clustering with landmark-based representation. In *Proceedings of the Twenty-Fifth AAAI Conference on Artificial Intelligence*, pages 313–318, 2011.

[7] Yun Chi, Xiaodan Song, Dengyong Zhou, Koji Hino, and Belle L. Tseng. Evolutionary spectral clustering by incorporating temporal smoothness. In *Proceedings of the 13th ACM SIGKDD International Conference on Knowledge Discovery and Data Mining*, pages 153–162, 2007.

[8] Fan R. K. Chung. *Spectral Graph Theory*. American Mathematical Society, 1994.

[9] Carlos D. Correa and Peter Lindstrom. Locally-scaled spectral clustering using empty region graphs. In *Proceedings of the 18th ACM SIGKDD International Conference on Knowledge Discovery and Data Mining*, pages 1330–1338, 2012.

[10] Mayur Datar, Nicole Immorlica, Piotr Indyk, and Vahab S. Mirrokni. Locality-sensitive hashing scheme based on p-stable distributions. In *Proceedings of the Twentieth Annual Symposium on Computational Geometry*, pages 253–262, 2004.

[11] Inderjit S. Dhillon. Co-clustering documents and words using bipartite spectral graph partitioning. In *Proceedings of the Seventh ACM SIGKDD International Conference on Knowledge Discovery and Data Mining*, pages 269–274, 2001.

[12] Inderjit S. Dhillon, Yuqiang Guan, and Brian Kulis. Kernel k-means: Spectral clustering and normalized cuts. In *Proceedings of the Tenth ACM SIGKDD International Conference on Knowledge Discovery and Data Mining*, pages 551–556, 2004.

[13] Chris Ding, Xiaofeng He, and Horst D. Simon. On the equivalence of nonnegative matrix factorization and spectral clustering. In *Proceedings of the SIAM Conference on Data Mining*, pages 606–610, 2005.

[14] Wilm E. Donath and Alan J. Hoffman. Lower bounds for the partitioning of graphs. *IBM Journal of Research and Development*, 17(5):420–425, 1973.

[15] Miroslav Fiedler. Algebraic connectivity of graphs. *Czechoslovak Mathematical Journal*, 23(98):298–305, 1973.

[16] Charless Fowlkes, Serge Belongie, Fan Chung, and Jitendra Malik. Spectral grouping using the Nyström method. *IEEE Transactions on Pattern Analysis and Machine Intelligence*, 26(2):214–225, 2004.

[17] Fei Gao, Wael Abd-Almageed, and Mohamed Hefeeda. Distributed approximate spectral clustering for large-scale datasets. In *Proceedings of the 21st International Symposium on High-Performance Parallel and Distributed Computing*, pages 223–234, 2012.

[18] Eric Gaussier and Cyril Goutte. Relation between PLSA and NMF and implications. In *Proceedings of the 28th Annual International ACM SIGIR Conference on Research and Development in Information Retrieval*, pages 601–602, 2005.

[19] Lars W. Hagen and Andrew B. Kahng. New spectral methods for ratio cut partitioning and clustering. *IEEE Transactions on Computer-Aided Design of Integrated Circuits and Systems*, 11(9):1074–1085, 1992.

[20] Jihun Ham, Daniel D. Lee, Sebastian Mika, and Bernhard Schölkopf. A kernel view of the dimensionality reduction of manifolds. In *Proceedings of the Twenty-First International Conference on Machine Learning*, pages 47–54, 2004.

[21] Nguyen Lu Dang Khoa and Sanjay Chawla. Large scale spectral clustering using resistance distance and Spielman-Teng solvers. In *Discovery Science*, 7568:7–21, 2012.

[22] Andrew V. Knyazev, Merico E. Argentati, Ilya Lashuk, and E. E. Ovtchinnikov. Block locally optimal preconditioned eigenvalue xolvers (BLOPEX) in *hypre* and PETSC. *SIAM Journal on Scientific Computing*, 29(5):2224–2239, 2007.

[23] Abhishek Kumar and Hal Daumé III. A co-training approach for multi-view spectral clustering. In *Proceedings of the 28th International Conference on Machine Learning*, pages 393–400, 2011.

[24] Abhishek Kumar, Piyush Rai, and Hal Daumé III. Co-regularized multi-view spectral clustering. *Advances in Neural Information Processing Systems*, 24:1413–1421, 2011.

[25] Daniel D. Lee and H. Sebastian Seung. Learning the parts of objects by non-negative matrix factorization. *Nature*, 401(6755):788–791, 1999.

[26] Richard B. Lehoucq, Danny C. Sorensen, and Chao Yang. *ARPACK Users Guide—Solution of Large-Scale Eigenvalue Problems with Implicitly Restarted Arnoldi Methods*. SIAM, Philadelphia, PA, 1998.

[27] Tie-Yan Liu, Huai-Yuan Yang, Xin Zheng, Tao Qin, and Wei-Ying Ma. Fast large-scale spectral clustering by sequential shrinkage optimization. In *Proceedings of the 29th European Conference on IR Research*, pages 319–330, 2007.

[28] Bo Long, Zhongfei (Mark) Zhang, Xiaoyun Wu, and Philip S. Yu. Spectral clustering for multi-type relational data. In *Proceedings of the 23rd International Conference on Machine Learning*, pages 585–592, 2006.

[29] Zhengdong Lu and Miguel A. Carreira-Perpinan. Constrained spectral clustering through affinity propagation. In *2008 IEEE Conference on Computer Vision and Pattern Recognition*, pages 1–8, 2008.

[30] Helmut Lütkepohl. *Handbook of Matrices*. Wiley, 1997.

[31] Ulrike Luxburg. A tutorial on spectral clustering. *Statistics and Computing*, 17(4):395–416, 2007.

[32] Marina Meilă and Jianbo Shi. A random walks view of spectral segmentation. In *IEEE International Conference on Artificial Intelligence and Statistics*, 2001.

[33] Bojan Mohar. The Laplacian spectrum of graphs. *Graph Theory, Combinatorics, and Applications*, 2:871–898, 1991.

[34] Bojan Mohar. Some applications of Laplace eigenvalues of graphs. *Graph Symmetry: Algebraic Methods and Applications*, 497:227–275, 1997.

[35] David M. Mount and Sunil Arya. ANN: A library for approximate nearest neighbor searching. *CGC 2nd Annual Fall Workshop on Computational Geometry*, 1997.

[36] Marius Muja and David G. Lowe. Fast approximate nearest neighbors with automatic algorithm configuration. In *International Conference on Computer Vision Theory and Application*, pages 331–340, 2009.

[37] Andrew Y. Ng, Michael Jordan, and Yair Weiss. On spectral clustering: Analysis and an algorithm. *Advances in Neural Information Processing Systems*, 14:849–856. 2001.

[38] Huazhong Ning, Wei Xu, Yun Chi, Yihong Gong, and Thomas S. Huang. Incremental spectral clustering by efficiently updating the eigen-system. *Pattern Recognition*, 43:113–127, 2010.

[39] Donglin Niu, Jennifer G. Dy, and Michael I. Jordan. Multiple non-redundant spectral clustering views. In *Proceedings of the 27th International Conference on Machine Learning*, pages 831–838, 2010.

[40] Sean Owen, Robin Anil, Ted Dunning, and Ellen Friedman. *Mahout in Action*. Manning Publications, Shelter Island, NY, 2011.

[41] Yousef Saad. *Numerical Methods for Large Eigenvalue Problems* (Second Edition). SIAM, 2011.

[42] Tomoya Sakai and Atsushi Imiya. Fast spectral clustering with random projection and sampling. In *Proceedings of the 6th International Conference on Machine Learning and Data Mining in Pattern Recognition*, pages 372–384, 2009.

[43] Jianbo Shi and Jitendra Malik. Normalized cuts and image segmentation. *IEEE Transactions on Pattern Analysis and Machine Intelligence*, 22(8):888–905, 1997.

[44] Hiroyuki Shinnou and Minoru Sasaki. Spectral clustering for a large data set by reducing the similarity matrix size. In *Proceedings of the Sixth International Conference on Language Resources and Evaluation*, pages 201–204, 2008.

[45] Chanop Silpa-Anan and Richard Hartley. Optimised kd-trees for fast image descriptor matching. In *IEEE Conference on Computer Vision and Pattern Recognition*, pages 1–8, 2008.

[46] Daniel A. Spielman and Shang-Hua Teng. Spectral partitioning works: Planar graphs and finite element meshes. In *IEEE Symposium on Foundations of Computer Science*, pages 96–105, 1996.

[47] Daniel A. Spielman and Shang-Hua Teng. Nearly-linear time algorithms for preconditioning and solving symmetric, diagonally dominant linear systems. *CoRR*, abs/cs/0607.105, 2006.

[48] Andreas Stathopoulos and James R. McCombs. PRIMME: Preconditioned iterative multimethod eigensolver—Methods and software description. *ACM Transactions on Mathematical Software*, 37(2), 2010.

[49] Mechthild Stoer and Frank Wagner. A simple min-cut algorithm. *Journal of the ACM*, 44(4): 585–591, 1997.

[50] Joshua B. Tenenbaum, Vin de Silva, and John C. Langford. A global geometric framework for nonlinear dimensionality reduction. *Science*, 290(5500):2319–2323, 2000.

[51] Christoffer Valgren and Achim J. Lilienthal. Incremental spectral clustering and seasons: Appearance-based localization in outdoor environments. In *IEEE International Conference on Robotics and Automation*, pages 1856–1861, 2008.

[52] Ulrike von Luxburg, Mikhail Belkin, and Olivier Bousquet. Consistency of spectral clustering. *Annals of Statistics*, 36(2):555–586, 2008.

[53] Dorothea Wagner and Frank Wagner. Between min cut and graph bisection. In *Proceedings of the 18th International Symposium on Mathematical Foundations of Computer Science*, pages 744–750, 1993.

[54] Xiang Wang and Ian Davidson. Active spectral clustering. In *Proceedings of the 10th IEEE International Conference on Data Mining*, pages 561–568, 2010.

[55] Fabian L. Wauthier, Nebojsa Jojic, and Michael I. Jordan. Active spectral clustering via iterative uncertainty reduction. In *Proceedings of the 18th ACM SIGKDD International Conference on Knowledge Discovery and Data Mining*, pages 1339–1347, 2012.

[56] Scott White and Padhraic Smyth. A spectral clustering approach to finding communities in graphs. In *Proceedings of SIAM International Conference on Data Mining*, 2005.

[57] Wei Xu, Xin Liu, and Yihong Gong. Document clustering based on non-negative matrix factorization. *Proceedings of the 26th Annual International ACM SIGIR Conference on Research and Development in Informaion Retrieval*, pages 267–273, 2003.

[58] Donghui Yan, Ling Huang, and Michael I. Jordan. Fast approximate spectral clustering. In *Proceedings of the 15th ACM SIGKDD International Conference on Knowledge Discovery and Data Mining*, pages 907–916, 2009.

[59] Stella X. Yu and Jianbo Shi. Multiclass spectral clustering. In *Proceedings of the Ninth IEEE International Conference on Computer Vision—Volume 2*, pages 313–319, 2003.

[60] Lihi Zelnik-Manor and Pietro Perona. Self-tuning spectral clustering. *Advances in Neural Information Processing Systems*, 17:1601–1608, 2004.

[61] Dengyong Zhou, Olivier Bousquet, Thomas Navin Lal, Jason Weston, and Bernhard Schölkopf. Learning with local and global consistency. *Advances in Neural Information Processing Systems*, 16:321–328, 2004.

[62] Dengyong Zhou and Christopher J.C. Burges. Spectral clustering and transductive learning with multiple views. *Proceedings of the 24th International Conference on Machine Learning*, pages 1159–1166, 2007.

Chapter 9

Clustering High-Dimensional Data

Arthur Zimek
University of Alberta
Edmonton, Canada
zimek@ualberta.ca

9.1 Introduction

The general definition of the task of clustering as *to find a set of groups of similar objects within a data set while keeping dissimilar objects separated in different groups or the group of noise* is very common. Although Estivill-Castro criticizes this definition for including a grouping criterion [47], this criterion (*similarity*) is exactly what is in question among many different approaches. Especially in high-dimensional data, the meaning and definition of similarity is right at the heart of the problem. In many cases, the similarity of objects is assessed within subspaces, e.g., using a subset of the dimensions only, or a combination of (a subset of) the dimensions. These are the so-called *subspace clustering* algorithms. Note that for different clusters within one and the same clustering solution usually different subspaces are relevant. Therefore subspace clustering algorithms cannot be thought of as variations of usual clustering algorithms using just a different definition of similarity. Instead, the similarity measure and the clustering solution are usually derived simultaneously and depend on each other. In this overview, we focus on these methods. The emerging field of subspace clustering is still raising a lot of open questions. Many methods have been proposed, though, putting the emphasis on different aspects of the problem.

While we give a concise overview on the problems,[1] the problem settings, and types of solutions, other overviews on the topic of subspace clustering can be found in the literature. The first survey to discuss the young field was presented by Parsons et al. [114]. This survey earned the merits of putting the community's attention to the problem and sketching a couple of early algorithms. Later on, however, the problem was studied in much more detail and categories of similar approaches have been defined [90]. A short discussion of the fundamental problems and strategies has been provided by Kröger and Zimek [94]. Vidal focused on a specific subtype of the problem from the point of view of machine learning and computer vision [128]. Assent gives an overview in the context of high-dimensional data of different provenance, including time series and text documents [17]. Sim et al. [122] discuss "enhanced" subspace clustering; i.e., they point out specific open problems in the field and discuss methods specifically addressing those problems. Kriegel et al. [91] give a concise overview and point to open questions as well. Some recent textbooks [73, 57] sketch some example algorithms and cursorily touch on some problems. Recent experimental evaluation studies have covered some selections of a specific subtype of subspace clustering, algorithms for clustering in axis-parallel subspaces [104, 108].

As a book chapter, this survey is intended to give an overview on the fundamental problems that clustering is confronted with in high-dimensional data (associated with the infamous *curse of dimensionality*—Section 9.2). Based on that, we sketch (Section 9.3) the basic problem setting for specialized algorithms (known as *subspace clustering*, *projected clustering*, and the like) as well as some fundamental algorithms (Section 9.4). Finally, we point to open questions and issues of ongoing research in this area (Section 9.5), and we summarize and conclude the discussion in Section 9.6.

9.2 The *"Curse of Dimensionality"*

9.2.1 Different Aspects of the *"Curse"*

The motivation of specialized solutions for analyzing high-dimensional data has often been given with a general reference to the so-called *curse of dimensionality*. This general term relates to a bundle of rather different problems. Some of these are elaborated in more detail in the literature concerned with index structures [30]. Here, we highlight aspects of the problem that are often seen in relation to subspace clustering although they are of different nature and importance [90, 91].

Aspect 1: Optimization Problem

Historically, the first aspect has been described by Bellman [23], who coined the term "curse of dimensionality" in the context of optimization problems. Clearly the difficulty of any global optimization approach increases exponentially with an increasing number of variables (dimensions).

[1] This chapter is based on previously published surveys [90, 91] but it provides some updates as well as occasionally more detailed explanations, and it rounds things off in a partly different way.

This problem is particularly well known in the area of pattern recognition. At a general level, this problem relates to the clustering problem. Seeking a clustering of a data set is assuming that the data are generated by several functions (statistical processes). Hence, ideally, a clustering model would enable the user to identify the functional dependencies resulting in the data set at hand and, thus, to eventually find new and interesting insights in the domain of the data set (i.e., models of the statistical processes). Those functions are more complex when several attributes contribute to the actual relationships.

Aspect 2: Concentration Effect of L_p-Norms

The second aspect of the curse of dimensionality is the *deterioration of expressiveness* of the most commonly used similarity measures, the L_p-norms, with increasing dimensionality. In their general form, for $x, y \in \mathbb{R}^d$, L_p-norms are given as

$$\|x - y\|_p = \sqrt[p]{\sum_{i=1}^{d} |x_i - y_i|^p}. \tag{9.1}$$

The choice of p is crucial in high-dimensional data according to several studies [27, 76, 11]. The key result of Beyer et al. [27] states the following: if the ratio of the variance of the length of any point vector $x \in \mathbb{R}^d$ (denoted by $\|x\|$) to the length of the mean point vector (denoted by $E[\|x\|]$) converges to zero with increasing data dimensionality, then the proportional difference between the farthest-point distance D_{max} and the closest-point distance D_{min} (the *relative contrast*) vanishes. Formally:

$$\lim_{d \to \infty} \text{var}\left(\frac{\|x\|}{E[\|x\|]}\right) = 0 \implies \frac{D_{max} - D_{min}}{D_{min}} \to 0. \tag{9.2}$$

The precondition has been described as being valid for a broad range of data distributions. Intuitively, the relative contrast of (L_p) distances does diminish as the dimensionality increases. This *concentration effect* of the distance measure reduces the utility of the measure for discrimination.

Distance Concentration and Clustering

The distance-concentration effect means that far and close neighbors have similar distances, if they belong to the same distribution (and some mild conditions apply on the nature of the distribution). As clustering is concerned with data from different distributions, the crucial question is whether these distributions are sufficiently separated. In that case, distance concentration is not an issue for the separation of different clusters.

The behavior of integer L_p-norms (i.e., $p \in \mathbb{N}$) in high-dimensional spaces has been studied by Hinneburg et al. [76]. The authors showed by means of an analytic argument that L_1 and L_2 are the only integer norms useful for higher-dimensional spaces. In addition, they studied the use of projections for discrimination, the effectiveness of which depends on localized dissimilarity measures that did not satisfy the symmetry and triangle inequality conditions of distance metrics. Fractional L_p distance measures (with $0 < p < 1$) have been studied in a similar context by Aggarwal et al. [11]. The authors observe that smaller values of p offer better results in higher-dimensional settings.[2]

These studies are often quoted as a motivation of specialized approaches in high-dimensional data. It should be noted, though, that all these studies generally assume that the complete data set follow a single data distribution, subject to certain restrictions. When the data, instead, is actually

[2]The concentration of the cosine similarity has been studied by Radovanović et al. [119].

generated by a mixture of distributions, the concentration effect is not always observed. To notice this is important here, as clustering is usually concerned with data generated by different distributions, ideally with one of the resulting clusters capturing one of the distributions.

Distances between members of different distributions may not necessarily tend to the global mean as the dimensionality increases. The fundamental differences between data following a single distribution and data following several distributions are discussed in detail by Bennett et al. [25]. The authors demonstrate that nearest-neighbor queries are both theoretically and practically meaningful when the search is limited to objects from the same cluster (distribution) as the query point, and other clusters (distributions) are well separated from the cluster in question. The key concept is that of *pairwise stability* of clusters, which is said to hold whenever the mean distance between points of different clusters dominates the mean distance between points belonging to the same cluster. When the clusters are pairwise stable, for any point belonging to a given cluster, its nearest neighbors tend to belong to the same cluster. Here, a nearest-neighbor query whose size is of the order of the cluster size *can* be considered meaningful, whereas differentiation between nearest and farthest neighbors within the same cluster may still be meaningless. Overall, this aspect of the curse of dimensionality is anything but well studied, as noted recently [54]. Nevertheless, this aspect should be considered as a serious impediment for any data mining, indexing, or similarity search application in high-dimensional data. Recent studies discuss in practical settings this aspect of the curse of dimensionality and highlight that it does not apply when the data are following different, well-separated distributions [78, 26], since in that case the precondition does not hold. Hence, regarding the separation of clusters, the following aspects are far more important in the scenario of subspace clustering.

Aspect 3: Irrelevant Attributes

A third aspect is often confused with the previous aspect but it is actually independent. In order to find rules describing some occurring phenomena, in many domains a glut of data is collected where single entities are described with many possibly related attributes. Among the features of a high-dimensional data set, for any given query object, many attributes can be expected to be irrelevant to describing that object. Irrelevant attributes can interfere with the performance of similarity queries for that object. The relevance of certain attributes may differ for different groups of objects within the same data set. Thus, since groups of data are defined by some of the available attributes only, many irrelevant attributes may interfere with the efforts to find these groups. Irrelevant attributes are often also referred to as "noise."

Distance Concentration, Clustering, and Relevance of Attributes

There is an important interplay between Aspect 3 and Aspect 2, as many additional irrelevant attributes (noisy features) can obfuscate the separation of different distributions in the high-dimensional space. This has been studied in detail by Houle et al. [78]. Additional *relevant* attributes, i.e., attributes that carry information regarding the separation of different distributions are actually helpful rather than proving to be obstacles. In that case, more dimensions will actually make the clustering easier instead of more difficult!

Aspect 4: Correlated Attributes

In a data set containing many attributes, there may be some correlations among subsets of attributes. From the point of view of feature reduction, all but one of these attributes may be redundant. However, from the point of view of a domain scientist who collected these attributes in the first place, it may be an interesting new insight that there are so far unknown connections between features [5]. As for the previous problem, this phenomenon may be differently relevant for different subgroups of a data set as correlations among attributes that are characteristic for a cluster will be different for other clusters of the data set.

With respect to spatial queries, the observation that the intrinsic dimensionality of a data set usually is lower than the embedding dimensionality (based on interdependencies among attributes) has often been attributed to overcoming the curse of dimensionality [48, 24, 113, 85]. Durrant and Kabán [45] show that the correlation between attributes is actually an important effect for avoiding the concentration of distances. Correlated attributes will probably also result in an intrinsic dimensionality that is considerably lower than the representational (embedding) dimensionality. This effect led to confronting the curse of dimensionality with the *self-similarity blessing* [85]. As opposed to the conjecture of Beyer et al. [27] and François et al. [54], however, Durrant and Kabán [45] also show that a low intrinsic dimensionality alone does not suffice to avoid the concentration, but essentially the correlation between attributes needs to be strong enough. Let us note that, actually, different aspects such as a strong cluster-structure of data [25] and low intrinsic dimensionality [85] may be merely symptoms for a strong, though possibly latent, correlation between attributes. Durrant and Kabán also identify irrelevant dimensions as the core of the problem of the curse. In essence, the ratio between noise (e.g., irrelevant attributes or additive noise masking information in relevant attributes) and (latent) correlation in the attributes of a dataset will determine whether asking for the "nearest neighbor" is actually meaningful or not.

Intrinsic Dimensionality vs. Embedding Dimensionality

An important implication of a strong correlation between several attributes is that the so-called *intrinsic dimensionality* of a data set can be considerably lower than the so-called *embedding dimensionality*, i.e., the number of features of the data set. In short, the intrinsic dimensionality (or *expansion dimension*), as opposed to the embedding dimensionality, describes how much space is actually used by the data. These concepts have been discussed, e.g., by Gupta et al. [65].

However, Aspects 1, 2, and 3 could still apply, depending on the number and nature of the remaining uncorrelated attributes.

Aspect 5: Varying Relative Volume of an ε-Hypersphere

The relative volume of an ε-hypersphere (i.e., using the L_2-norm as query distance) as compared to the volume of the data space varies extremely with the change of dimensionality. The data space is usually seen as a hypercube. Hence, the possibly maximally occurring distance (along the diagonal) is growing with the square root of the dimensionality. If we choose ε big enough to cover the complete data space in one or two dimensions, slowly but steadily the corners of the data space will grow beyond the range of the hypersphere. Figuratively speaking, we could imagine the high-dimensional data space almost consisting of only corners.

As a consequence, what may seem a reasonable value of ε for a 3-dimensional query space may in a 20-dimensional space be extremely unrealistic (i.e., the corresponding query is a priori unlikely to contain any point at all) or, vice versa, choosing ε big enough to cover some reasonable volume in a 20-dimensional space would engulf some 3-dimensional subspace completely. This is an effect to keep in mind when designing subspace clustering algorithms. This effect (among others) has been discussed in more detail, although with respect to outlier detection in high-dimensional data, by Zimek et al. [142].

9.2.2 Consequences

The third and fourth of these aspects of the curse of dimensionality have been actually the main motivation for developing specialized methods for clustering in *subspaces* of potentially high-dimensional data even though this was not always clearly recognized. The superficial reference to the second aspect that can often be found in the literature misses the point. But even so the proposed approaches, perhaps partly unintentionally, actually tackled mainly Aspect 3 and Aspect 4.

For decades, research in statistics was concerned with the effect of irrelevant (*noise*) attributes (sometimes called *masking variables* [53]) on the performance of clustering algorithms [100, 52, 59, 127]. They discuss how to identify such noise attributes that do not contribute to the cluster structure at all. The clustering task should then concentrate on the remaining attributes only. In some cases, the contribution of attributes may be weighted instead of a discrete decision (contributing or masking). This view of the problem is obviously closely related to dimensionality reduction [66, 132].

The field of subspace clustering as we sketch it here assumes a more complex view. The reasoning is that dimensionality reduction as a first step prior to clustering is not always resolving these issues since the correlation of attributes or the relevance vs. irrelevance of attributes is usually characteristic for some clusters but not for complete data sets. In other words, masking variables may be masking certain clusters but may help to discover others.

The phenomenon that different features or a different correlation of features may be relevant for different clusters within a single clustering solution has been called *local feature relevance* or *local feature correlation* [90]. Feature selection or dimensionality reduction techniques are *global* in the following sense: they generally compute only one subspace of the original data space in which the clustering can then be performed. In contrast, the problem of *local feature relevance* or *local feature correlation* describes the observation that multiple subspaces are needed because each cluster may exist in a different subspace. The critical requirement for the design of subspace clustering algorithms is hence the integration of some heuristic for deriving the characteristic subspace for a certain subgroup of a data set with some heuristic for finding a certain subgroup of a data set that exhibits a sufficient level of similarity in a certain subspace of the data space — obviously a circular dependency of two subtasks (subspace determination vs. cluster assignment) already in the most basic problem description.

Aspect 5 is a problem whenever ε-range queries of different dimensionality are compared, which is done in many of the available algorithms.

9.3 Clustering Tasks in Subspaces of High-Dimensional Data

For the design of specific clustering algorithms suitable for high-dimensional data all the aspects of the curse of dimensionality, as sketched above, are relevant. However, motivating specialized approaches to clustering in subspaces are mainly the phenomena of irrelevant attributes (Aspect 3) and correlated attributes (Aspect 4). These two phenomena result in different types of subspaces which pose different challenges to concrete algorithms.

9.3.1 Categories of Subspaces

9.3.1.1 Axis-Parallel Subspaces

The distinction between relevant and irrelevant attributes generally assumes that the variance of the occurring values for a relevant attribute over all points of the corresponding cluster is rather small compared to the overall range of attribute values whereas the variance for irrelevant attributes within a given cluster is high (or indistinguishable from the values of the same attribute for other clusters or for background noise). For example, one could assume a relevant attribute for a given cluster being normally distributed with a small standard deviation whereas irrelevant attribute values are uniformly distributed over the complete data space. The geometrical intuition of these assumptions relates to the points of a cluster being widely scattered in the direction of irrelevant axes while being densely clustered along relevant attributes. When selecting the relevant attributes only (by projection onto the corresponding subspace, i.e., the subspace spanned by these attributes), the cluster would appear as a full-dimensional cluster in this subspace. In the full-dimensional space (including also the irrelevant attributes) the cluster points form a hyperplane parallel to the irrelevant axes. Due to this geometrical appearance, this type of cluster is addressed as an *axis-parallel subspace cluster*.

9.3.1.2 Arbitrarily Oriented Subspaces

If two attributes a_i and a_j are linearly correlated for a set of points, the points will be scattered along a hyperplane defined by some linear dependency between both attributes that corresponds to the correlation. The subspace orthogonal to this hyperplane is then a subspace where the points cluster densely irrespective of the variance of combined values of a_i and a_j. This subspace is arbitrarily oriented and hence the more general case compared to axis-parallel subspaces.

9.3.1.3 Special Cases

While the aforementioned types of subspaces have a direct relationship to aspects of the curse of dimensionality as discussed above, it should be noted that from the point of view of existing clustering algorithms there are special types of subspaces. In fact, there is a large family of algorithms, addressed as "bi-clustering," "co-clustering," "two-mode clustering," or "pattern-based clustering" [90, 99].

Considering the spatial intuition of subspaces for subspace clustering, these algorithms address different kinds of subspaces such as very specific rotations only (e.g., points that form a cluster are scattered along the bisecting line) or specific half-spaces (e.g., in some subspace, only points located on one side of the bisecting line can form a cluster). Thus, the types of the subspaces considered by these methods do not directly relate to the phenomena of the curse of dimensionality discussed here but rather are the products of specific cluster models.

As these clustering methods are predominantly used in biological data analysis, refer to Chapter 16 for an overview on this family of methods. The spatial intuition that comes with these cluster models has been elaborated in detail by Kriegel et al. [90].

9.3.2 Search Spaces for the Clustering Problem

Recognizing the different categories of subspaces can guide us to recognize different categories of subspace clustering algorithms. If we describe the problem in general as *find clusters, where each cluster can reside in a different subspace*, it is obvious that the problem of finding the subspaces containing clusters is in general intractable. The number of arbitrarily oriented subspaces is infinite. Even the number of axis-parallel subspaces is exponential in the number of dimensions (we can see this as a result of the first aspect of the curse of dimensionality). Also, since different subspaces can be relevant for different clusters, global feature reduction is not a solution. Aside from the search space for interesting subspaces, the search space for clustering solutions is also challenging, as clustering in general is an NP-complete problem [125]. Even worse here: the search for clusters and the search for subspaces are depending on each other. Neither is it a good solution to, first, find the subspaces and then find the clusters in the subspaces, as this results in a quite redundant set of clusters (we will discuss this problem of redundancy in more detail later). Nor is it a viable approach to, first, find the clusters and then define the subspace for each cluster. Hence the definition of a cluster, that comes along with an algorithmic heuristic for clustering, and the heuristic to derive a subspace for each cluster are closely related for each approach. Yet how this relationship is defined in particular will be differing in the different approaches.

Many heuristics and assumptions for efficiently solving the clustering problem have been discussed in the past decades. We will focus on the subspace search problem in this chapter. However, for the sake of completeness, we should keep in mind the fact that the solutions discussed here also rely on assumptions and heuristics that consider the clustering problem (i.e., whether, for example, a partitioning or a density-based clustering paradigm is used or adapted). These assumptions and heuristics are reflected by the underlying clustering model that specifies the meaning of clusters and the algorithms to compute such clusters. One important aspect to describe the meaning of clusters that has an impact on the subspace search problem is the choice of a certain distance or similarity measure. Which data points are considered more similar than others and, thus, cluster together, crucially depends on this choice. Some approaches are based on neighborhood in Euclidean space, typically using an L_p-norm as distance measure. Others take into account a similar "behavior" of attribute values over a certain range of attributes (often called "patterns"). But different approaches also make use of the definition of similarity in different ways, based on different intuitions of the meaning of a cluster and resulting in different cluster models. If we stay with the notion of Euclidean distances (or, more general, L_p-norms), the selection of a subspace for a cluster will mean that similarity of points is defined in different ways for different clusters. In the case of an axis-parallel subspace, only a subset of the attributes is considered to compute the L_p-distance. For arbitrarily oriented subspaces, correlations of attributes are considered and combinations thereof are computed to define a rotated subspace distance (such as some variant of a Mahalanobis distance).

9.4 Fundamental Algorithmic Ideas

Different paradigms of clustering heuristics are discussed in other chapters. Here, we survey some fundamental algorithmic ideas and example approaches according to the different heuristic restrictions of the search space for subspaces.

9.4.1 Clustering in Axis-Parallel Subspaces

9.4.1.1 Cluster Model

So far, most research in this field mainly transferred the full-dimensional cluster models of different clustering techniques into some (or many)—more or less—interesting subspaces (i.e., subsets of attributes) of a data space. Hence, there is no established "subspace cluster model" describing axis-parallel subspace clusters concisely. Instead, there are only cluster models describing clusters in a subspace as if the corresponding subspace were the full-dimensional data space. The first statistically sound model for axis-parallel subspace clusters was proposed by Moise and Sander [101]. It is based on the assumption (statistically speaking: hypothesis) that values of a relevant attribute for a subspace cluster are *not* uniformly distributed over the complete attribute domain (i.e., the null hypothesis is a uniform distribution of irrelevant attributes over the complete domain) and that this hypothesis satisfies a statistical test.

9.4.1.2 Basic Techniques

The number of possible axis-parallel subspaces where clusters could reside is exponential in the dimensionality of the data space. Hence, the main task of research in the field was the development of appropriate subspace search heuristics. Starting from the pioneering approaches to axis-parallel subspace clustering, two opposite basic techniques for searching subspaces were pursued, namely, top-down search [12] and bottom-up search [14].

Top-Down Subspace Search. The rational of top-down approaches is to determine the subspace of a cluster starting from the full-dimensional space. This is usually done by determining a subset of attributes for a given set of points—potential cluster members—such that the points meet the given cluster criterion when projected onto this corresponding subspace. Obviously, the dilemma is that for the determination of the subspace of a cluster, at least some cluster members must be identified. On the other hand, in order to determine cluster memberships, the subspace of each cluster must be known. To escape this circular dependency, most of the top-down approaches rely on a rather strict assumption, which has been called the *locality assumption* [90]. It is assumed that the subspace of a cluster can be derived from the local neighborhood (in the full-dimensional data space) of the (provisional) cluster center or the (provisional) cluster members. In other words, it is assumed that even in the full-dimensional space, the subspace of each cluster can be learned from the local neighborhood of cluster representatives or cluster members—an assumption that may make sense if the dimensionality of the sought subspaces is not much smaller than the dimensionality of the complete data space.

Other top-down approaches that do not rely on the locality assumption use random sampling as a heuristic in order to generate a set of potential cluster members.

Bottom-Up Subspace Search. The exponential search space of all possible subspaces of a data space that needs to be traversed is equivalent to the search space of the frequent item-set problem in market basket analysis in the area of transaction databases. For example, an item-set may contain items A, B, C, etc. The key idea of the APRIORI algorithm [15] is to start with item-sets (called "transactions") of size 1 (containing a certain item, irrespective of other items possibly also contained in the transaction) and exclude those larger item-sets from the search that cannot be frequent anymore, given the knowledge of which smaller item-sets are frequent. For example, if a 1-item-set containing A is not frequent (i.e., we count such an item-set less than a given minimum support threshold), all 2-item-sets, 3-item-sets, etc., containing A (e.g., $\{A,B\},\{A,C\},\{A,B,C\}$) cannot be frequent either (otherwise item-sets containing A would have been frequent as well) and need not be tested for exceeding the minimum support threshold. Theoretically, the search space remains exponential, yet practically the search is usually substantially accelerated.

Transferring the item-set problem to subspace clustering, each attribute represents an item and each subspace cluster is a transaction of the items representing the attributes that span the corresponding subspace. Finding item-sets with frequency 1 then relates to finding all combinations of attributes that constitute a subspace containing at least one cluster. This observation is the rationale of many of the early bottom-up subspace clustering approaches. The subspaces that contain clusters are determined starting from all 1-dimensional subspaces that accommodate at least one cluster by employing a search strategy similar to frequent item-set mining algorithms. To apply any efficient frequent item-set mining algorithm, the cluster criterion must implement a downward closure property (also called monotonicity property):

> If subspace S contains a cluster, then any subspace $T \subseteq S$ must also contain a cluster.

For pruning (i.e., excluding specific subspaces from consideration), the antimonotonic reverse implication can be used.

> If a subspace T does not contain a cluster, then any superspace $S \supseteq T$ also cannot contain a cluster.

Let us note that there are bottom-up algorithms that do not use an APRIORI-like subspace search but instead apply other search heuristics.

Applying an efficient subspace search strategy addresses Aspect 1 of the curse of dimensionality. Selecting a subset of attributes as relevant corresponds to Aspect 3 of the curse of dimensionality. Aspect 2 needs to be considered when adapting similarity measures to local neighborhoods and is differently important in the different approaches. Aspect 5 is often the main problem for this kind of approach since the monotonicity usually requires the same distance thresholds in subspaces of different dimensionalities.

9.4.1.3 Clustering Algorithms

The two basic techniques resulted initially in two "subspace" clustering paradigms that have been named by the pioneers in this field "subspace clustering" [14] and "projected clustering" [12]. For the latter, a variant has emerged sometimes called "soft projected clustering" [82]. Many algorithms, however, do not clearly fit in one of these categories. We therefore list some examples as "hybrid approaches."

Projected Clustering. The objective of projected clustering is to find a set of tuples (O_i, D_i), where O_i is the set of objects belonging to cluster i and D_i is the subset of attributes where the points O_i are "similar" according to a given cluster criterion.

The approach introducing the task of "projected clustering" is PROCLUS [12], a k-medoid-like clustering algorithm. PROCLUS randomly determines a set of potential cluster centers M on a sample of points first. In the iterative cluster refinement phase, for each of the k current medoids the subspace is determined by minimizing the standard deviation of the distances of the points in the neighborhood of the medoids to the corresponding medoid along each dimension. Points are then assigned to the closest medoid considering the relevant subspace of each medoid. The clusters are refined by replacing bad medoids with new medoids from M as long as the clustering quality increases. A postprocessing step identifies noise points that are too far away from their closest medoids. The algorithm always outputs a partition of the data points into k clusters (each represented by its medoid) with corresponding subspaces and a (potentially empty) set of noise points.

The k-medoid-style cluster model of PROCLUS tends to produce equally sized clusters that have spherical shape in their corresponding subspaces. In addition, since the set M of possible medoids is determined in a randomized procedure, different runs of PROCLUS with the same parametrization usually result in different clusterings.

Variations of PROCLUS are, for example, FINDIT [130] and SSPC [134]. FINDIT employs additional heuristics to enhance efficiency and clustering accuracy. SSPC offers the capability of further enhancing accuracy by using domain knowledge in the form of labeled objects and/or labeled attributes.

PreDeCon [31] applies the density-based traditional (full-dimensional) clustering algorithm DBSCAN [46] using a specialized distance measure that captures the subspace of each cluster. The definition of this specialized subspace distance is based on the so-called subspace preference that is assigned to each point p, representing the maximal-dimensional subspace in which p clusters best. A dimension is considered to be relevant for the subspace preference of a point p if the variance of points in the Euclidean ε-neighborhood of p is smaller than a user-defined threshold δ. The specialized subspace distance between points is a weighted Euclidean distance where the dimensions relevant for the subspace preference of a point are weighted by a constant $\kappa \gg 1$ while the remaining dimensions are weighted by 1. PreDeCon determines the number of clusters automatically, and handles noise implicitly. In addition, its results are determinate and the clusters may exhibit any shape and size in the corresponding subspace. However, PreDeCon requires the user to specify a number of input parameters that are usually hard to guess. Especially ε, the neighborhood radius, has rather different meaning in a different dimensionality (cf. Aspect 5 of the curse of dimensionality) but is used both for the first guess of neighborhood (locality assumption) in the full-dimensional space as well as for the adapted neighborhood in some lower-dimensional space.

CLTree [96] is a method that presents an interesting variation of the theme. The basic idea is to assign a common class label to all existing points and to add additionally points uniformly distributed over the data space and labeled as a different class. Then a decision-tree is trained to separate the two classes. As a consequence, the attributes are split independently, adaptively, and in a flexible order of the attributes. However, selecting a split is based on the evaluation of information gain which is rather costly. Furthermore, the density of the superimposed artificial data can be expected to heavily influence the quality of the results. Since the distribution parameters of existing clusters are unknown beforehand, finding a suitable parametrization seems rather hard. Another problem is the merging of adjacent regions. A cluster can easily become separated if the corresponding bins do not "touch" each other.

Soft Projected Clustering. There is a rich literature concerned with so-called *soft* projected clustering [43, 55, 79, 82, 29, 42, 98]. These are usually optimization approaches derived from k-means type clustering, learning some weight vector for the different weighting of attributes. Therefore, weights w_i for attributes $i = 1, \ldots, d$ are introduced into the distance measure:

$$\|(x-y)\|_p^w = \sqrt[p]{\sum_{i=1}^{d} w_i \cdot |x_i - y_i|^p}. \tag{9.3}$$

Usually, in these approaches, all weights are restricted at least to the condition

$$0 < w_i \leq 1, \tag{9.4}$$

i.e., to ensure the applicability of optimization techniques, no weight can become 0. In terms of subspace clustering, this means that no attribute is truly omitted and, hence, the resulting clustering resides in the full-dimensional, though skewed data space. The fact that the clustering is not truly searched for in a projected space (i.e., no *hard* subspace is assigned to a specific cluster) is indicated by the term *soft*. However, these approaches are often generally named "subspace clustering," not reflecting the differences discussed here. Essentially, all these approaches can be seen as variants of EM-clustering [40].

Geometrically, the effect of these approaches is just to allow the shape of clusters to become axis-parallel ellipsoids instead of spheres. This does not necessarily account for relevance or irrelevance of different attributes. It seems more related to some sort of normalization of attributes per cluster.

An example for these methods is LAC (Locally Adaptive Clustering) [43], which starts with k centroids and k sets of d weights (for d attributes). The algorithm proceeds to approximate a set of k Gaussians by adapting the weights.

COSA [55], another prominent example, does not derive a clustering but merely a similarity matrix that can be used by an arbitrary clustering algorithm afterwards. The matrix contains weights for each point specifying a subspace preference of the points similar to PreDeCon. The weights for a point p are determined by starting with the Euclidean k-nearest neighbors of p and by computing the average distance distribution of the k-nearest neighbors along each dimension. Roughly speaking, the weight for a given attribute is computed as the ratio between the distances of the point p to all k-nearest neighbors of p in that attribute and the average distances of p to all k-nearest neighbors of p in all attributes. Thus, the higher the variance of the k-nearest neighbors along an attribute is, the higher is the weight that is assigned to that attribute. The weight, however, cannot become 0. As long as the weight vectors still change, the k-nearest neighbors are again determined using the current weights and the weights are recomputed. The number of neighbors k is an input parameter. Very different from PreDeCon, the weights can have arbitrary values rather than only two fixed values. In addition, the weighting matrix is tested using several full-dimensional clustering algorithms rather than integrating it into only one specific algorithm. Note that the effect of weight vectors is very different in density-based approaches such as PreDeCon where the shape of clusters is not defined beforehand.

As we have seen, these approaches miss the complex interrelation of different problems in high-dimensional data. For further references, the reader may therefore refer to some other examples of recent work in this direction. Although the connections and differences to the algorithms as we outline here are not made clear there, detailed sections on related work are suitable as introduction to the history of "soft" projected clustering approaches provided in some papers specialized on this direction [79, 29, 82].

Subspace Clustering. Subspace clustering in a narrower sense pursues the goal of finding all clusters in all subspaces of the entire feature space. This goal obviously is defined to correspond to the bottom-up technique used by these approaches, based on some antimonotonic property of clusters allowing the application of the APRIORI search heuristic. However, the relationship of the definition of *subspace clustering* (*find all clusters in all subspaces*) and the subspace search strategy (bottom-up, applying an APRIORI-like heuristic) is historical and hence contingent rather than essential. We consider the problem definition as guiding category, regardless of the algorithmic approach. Nevertheless, the majority of approaches actually consist of some adaptation of the APRIORI search heuristic.

The pioneer approach for finding all clusters in all subspaces coining the term "subspace clustering" for this specific task has been CLIQUE [14], using a grid-based clustering notion. The data space is partitioned by an axis-parallel grid into equi-sized units of width ξ. Only units which contain at least τ points are considered as dense. A cluster is defined as a maximal set of adjacent dense units. Since dense units satisfy the downward closure property, subspace clusters can be explored rather efficiently in a bottom-up way. Starting with all 1-dimensional dense units, $(k+1)$-dimensional dense units are computed from the set of k-dimensional dense units in an APRIORI-like procedure. If a $(k+1)$-dimensional unit contains a projection onto a k-dimensional unit that is not dense, then the $(k+1)$-dimensional unit cannot be dense either. Furthermore, a heuristic that is based on the minimum description length principle is introduced to discard candidate units within less interesting subspaces (i.e., subspaces that contain only a very small number of dense units). This way, the efficiency of the algorithm is enhanced but at the cost of incomplete results (i.e., possibly some true clusters are lost).

There are some variants of CLIQUE; let us consider three well-known examples. The method ENCLUS [34] also relies on a fixed grid but searches for subspaces that potentially contain one or more clusters rather than for dense units. Three quality criteria for subspaces are introduced, one of them implements the downward closure property. The method MAFIA [111] uses an adaptive grid. The generation of subspace clusters is similar to CLIQUE. Another variant of CLIQUE called nCluster [97] allows overlapping windows of length δ as 1-dimensional units of the grid.

In summary, all grid-based methods use a simple but rather efficient cluster model. The shape of each resulting cluster corresponds to a polygon with axis-parallel lines in the corresponding subspace. Obviously, the accuracy and the efficiency of CLIQUE and its variants primarily depend on the granularity and the positioning of the grid. A higher grid granularity results in higher runtime-requirements but will most likely produce more accurate results.

Projected Clustering vs. Subspace Clustering

The distinction between the terms "projected clustering" and "subspace clustering" is not uniform in the literature but traditionally, dating back to the pioneering approaches CLIQUE and PROCLUS, the tasks are discerned as follows:

Projected clustering aims at a partitioning of the data set where each data point belongs to exactly one cluster, and each cluster is assigned a specific subset of attributes. Projected on these attributes, the cluster adheres to the clustering criterion as defined by the applied clustering algorithm.

Subspace clustering aims at finding *all* clusters in *all* subspaces. As a consequence, each data point can belong to many clusters simultaneously (i.e., the clusters can substantially overlap in different subspaces). This approach usually results in very redundant clusters in the retrieved clustering.

Many approaches proposed in the literature do not clearly fit in one of these traditional categories and the tendency of opinion in research is that more suitable problem descriptions should lie somewhere in between these archetypical categories, i.e., allowing for overlap of clusters, but restricting the redundancy of results. Eventually, what is a suitable clustering should not be defined by the algorithm but by the domain-specific requirements.

Note that these categories are not (or only coincidentally in the early approaches) related to the categories of subspace search heuristics (top-down vs. bottom-up).

SUBCLU [84] uses the DBSCAN cluster model of density-connected sets [46]. It is shown that density-connected sets satisfy the downward closure property. This enables SUBCLU to search for density-based clusters in subspaces in an APRIORI-like style. The resulting clusters may exhibit an arbitrary shape and size in the corresponding subspaces. In fact, for each subspace SUBCLU computes all clusters that would have been found by DBSCAN applied to that subspace only. Compared to the grid-based approaches, SUBCLU achieves a better clustering quality but requires a higher runtime.

It has been observed that a global density threshold, as used by SUBCLU and the grid-based approaches, leads to a bias toward a certain dimensionality (recall Aspect 5 of the curse of dimensionality). On the one hand, a tighter threshold which is able to separate clusters from noise well in low-dimensional subspaces tends to lose clusters in higher-dimensional subspaces. On the other hand, a more relaxed threshold which is able to detect high-dimensional clusters will produce an excessive amount of low-dimensional clusters. Motivated by this problem, the dimensionality unbiased cluster model DUSC [18] (and some variants [19, 70, 105, 106]) has been proposed. DUSC is based on a density measure that is adaptive to the dimensionality. As a major drawback, this approach is lacking of antimonotonic properties and, thus, pruning the search space is not possible. A "weak density" is thus defined as a remedy, providing antimonotonic properties. This remedy, however, introduces a global density threshold again, which renders the intended solution of the motivating problem ineffective. As this example shows, the curse of dimensionality should always be expected to keep haunting the researcher.

The initial problem formulation for projected clustering, *finding all clusters in all subspaces* [14], that was adopted in many papers afterwards, is rather questionable since the information gained by retrieving such a huge set of clusters with high redundancy is not very useful. This description of the objective of subspace clustering is based on a rather naïve use of the concept of frequent item-sets in subspace clustering [129]. What constitutes a good subspace clustering result is defined here apparently in close relationship to the design of the algorithm, i.e., the desired result appears to be defined according to the expected result (as opposed to in accordance to what makes sense). The tool "frequent item-set mining" was first, and the problem of "finding *all* clusters in *all* subspaces" has apparently been defined in order to have some new problem where the tool can be applied straightforwardly. The resulting clusters are usually highly redundant and, hence, to a major extend useless.

Therefore, subsequent methods often concentrated on possibilities of concisely restricting the resulting set of clusters by somehow assessing and reducing the redundancy of clusters, for example, to keep only clusters of highest dimensionality. Related approaches aim at reporting only the most representatives of a couple of redundant subspace clusters [19, 70, 105]. A broader overview on these issues in subspace clustering has been recently presented by Sim et al. [122].

Let us note that the statistical significance of subspace clusters (as defined by Moise and Sander [101]), is not an antimonotonic property and hence does in general not allow for APRIORI-like bottom-up approaches finding only *meaningful* (i.e., significant) clusters.

Hybrid Approaches. It is our impression that recently the majority of approaches do not stick to the initial concepts any more, but pursue some hybrid approach [117, 133, 136, 87, 3, 103, 72]. In general, the result is neither a clear partitioning without overlap nor an enumeration of all clusters in all subspaces. Still the problem definition may remain unclear and each approach defines its own goal. A fair comparison of the true merits of different algorithms thus remains difficult.

DOC [117] uses a global density threshold to define a subspace cluster by means of hypercubes of fixed side-length w containing at least α points. A random search algorithm is proposed to compute such subspace clusters from a starting seed of sampled points. A third parameter β specifies the balance between the number of points and the dimensionality of a cluster. This parameter affects the dimensionality of the resulting clusters and, thus, DOC usually also has problems with subspace clusters of significantly different dimensionality. Due to the very simple clustering model, the clusters may contain additional noise points (if w is too large) or not all points that naturally belong to the cluster (if w is too small). One run of DOC may (with a certain probability) find one subspace cluster. If k clusters need to be identified, DOC has to be applied at least k times. If the points assigned to the clusters found so far are excluded from subsequent runs, DOC can be considered as a pure projected clustering algorithm because each point is uniquely assigned to one cluster or to noise (if not assigned to a cluster). On the other hand, if the cluster points are not excluded from subsequent runs, the resulting clusters of multiple runs may overlap. Usually, DOC cannot produce all clusters in all subspaces.

MINECLUS [135, 136] is based on an idea similar to DOC, but proposes a deterministic method to find an optimal projected cluster given a sample seed point. The authors transform the problem into a frequent item-set mining problem and employ a modified frequent pattern tree growth method. Further heuristics are introduced to enhance efficiency and accuracy.

HiSC [4] and the more advanced extension DiSH [3] follow a similar idea as PreDeCon but use a hierarchical clustering model. This way, hierarchies of subspace clusters can be discovered (i.e., the information that the subspace of a lower-dimensional cluster is composed of a subset of the attributes of the subspace of a higher-dimensional cluster). The distance between points and clusters reflects the dimensionality of the subspace that is spanned by combining the corresponding subspace of each cluster. As in COSA, the weighting of attributes is learned for each object, not for entire clusters. The learning of weights, however, is based on single attributes, not on the entire feature space. DiSH uses an algorithm that is inspired by the density-based hierarchical clustering algorithm OPTICS [16]. However, DiSH extends the cluster ordering computed by OPTICS in order to find hierarchies of subspace clusters with multiple inclusions (a lower-dimensional subspace cluster may be embedded in multiple higher-dimensional subspace clusters).

HARP [133] is a hierarchical clustering algorithm similar to single-link clustering [126, 121] but uses a specialized distance function between points and clusters or between clusters and clusters and does not produce a hierarchy of subspace clusters. Starting with singleton clusters, HARP iteratively merges clusters as long as the resulting cluster has a minimum number of relevant attributes. A relevance score is introduced for attributes based on a threshold that starts at some harsh value and is progressively decreased while clusters increase in size. By design, HARP has problems finding low-dimensional clusters. The resulting dendrogram can be cut at any level in order to produce a unique assignment of points to clusters.

SCHISM [120] mines interesting subspaces rather than subspace clusters; thus, it is not exactly a subspace clustering algorithm but solves a related problem: find subspaces to look for clusters. It employs a grid-like discretization of the database and applies a depth-first search with backtracking to find maximally interesting subspaces.

FIRES [87] computes 1-dimensional clusters using any clustering technique the user is most accustomed to in a first step. These 1-dimensional clusters are then merged by applying a sort of clustering of clusters. The similarity of clusters is defined by the number of intersecting points. The resulting clusters represent hyper-rectangular approximations of the true subspace clusters. In an optional postprocessing step, these approximations can be refined by applying any clustering algorithm to the points included in the approximation projected onto the corresponding subspace. Though using a bottom-up search strategy, FIRES is rather efficient because it does not employ a worst-case exhaustive search procedure but a heuristic that is scaling polynomially in the dimensionality of the data space. However, the user can expect to pay for this performance boost by a loss of clustering accuracy. It cannot be specified whether the subspace clusters produced by FIRES may overlap or not, this partly depends on the clustering algorithms used in the intermediate steps of the framework. In general, the clusters may overlap but usually FIRES will not produce all clusters in all subspaces.

P3C [102, 103] starts with 1-dimensional intervals that are likely to approximate higher-dimensional subspace clusters. These intervals are merged using an APRIORI-like bottom-up search strategy. The maximal-dimensional subspace cluster approximations resulting from this merging procedure are reported as so-called cluster cores. In a refinement step, the cluster cores are refined by using an EM-like clustering procedure. Each cluster core is taken as one initial cluster for the EM algorithm. Points are assigned to the closest cluster core using the Mahalanobis distance. The final output of P3C is a matrix that records for each data point its probability of belonging to each projected cluster. From this matrix, a disjoint partitioning of the data points into clusters can be obtained by assigning each point to the cluster with the highest probability. If overlapping clusters are allowed, each point can be assigned to all clusters with a probability larger than $1/k$. P3C does not produce all clusters in all subspaces as any APRIORI-style algorithm does but reports only the results of the final cluster computation step using EM.

Moise and Sander [101] provided a first attempt to formulate the search for statistically significant subspace clusters as an optimization problem. In addition, the authors proposed an iterative algorithm called STATPC to search locally optimized solutions for this optimization problem.

Interesting recent developments combine techniques from ensemble clustering [58] with ideas from subspace clustering, resulting in "subspace clustering ensembles" [41] or "projective clustering ensembles" [64, 63, 62].

9.4.2 Clustering in Arbitrarily Oriented Subspaces

9.4.2.1 Cluster Model

A model for correlation clusters, i.e., clusters residing in arbitrarily-oriented subspaces, can be based on a linear equation system describing the λ-dimensional hyperplane accommodating the points of a correlation cluster $C \subset \mathbb{R}^d$. This equation system will consist of $d - \lambda$ equations for d variables, and the affinity, e.g. given by the mean point $x_C = (\bar{x}_1 \cdots \bar{x}_d)^{\mathsf{T}}$ of all cluster members:

$$\begin{array}{ccccccc}
v_{1(\lambda+1)} \cdot (x_1 - \bar{x}_1) & + & v_{2(\lambda+1)} \cdot (x_2 - \bar{x}_2) & + \cdots + & v_{d(\lambda+1)} \cdot (x_d - \bar{x}_d) & = & 0 \\
v_{1(\lambda+2)} \cdot (x_1 - \bar{x}_1) & + & v_{2(\lambda+2)} \cdot (x_2 - \bar{x}_2) & + \cdots + & v_{d(\lambda+2)} \cdot (x_d - \bar{x}_d) & = & 0 \\
 & & & \vdots & & & \\
v_{1d} \cdot (x_1 - \bar{x}_1) & + & v_{2d} \cdot (x_2 - \bar{x}_2) & + \cdots + & v_{dd} \cdot (x_d - \bar{x}_d) & = & 0
\end{array} \tag{9.5}$$

where v_{ij} is the value at row i, column j in the eigenvector matrix V_C derived (e.g., by principle component analysis (PCA) [83]) from the covariance matrix of C. The first λ eigenvectors (also called *strong* eigenvectors) give the directions of high variance and span the hyperplane accommodating C. The remaining $d - \lambda$ *weak* eigenvectors span the perpendicular subspace. The corresponding linear equation system can therefore also be given by

$$\hat{V}_C^{\mathsf{T}} \cdot x = \hat{V}_C^{\mathsf{T}} \cdot x_C \tag{9.6}$$

The defect of $\hat{V}_C^{\mathsf{T}}$ gives the number of free attributes, the remaining attributes may actually be involved in linear dependencies. The equation system is by construction at least approximately fulfilled for all points $x \in C$ and hence provides an approximate quantitative model for the cluster [5]. The degree of allowed deviation of cluster members from the hyperplane and the method of assessment differs from approach to approach. Hence, also in the area of arbitrarily-oriented clustering, future research should consider refined and more concise models.

9.4.2.2 Basic Techniques and Example Algorithms

Basic techniques to find arbitrarily oriented subspaces accommodating clusters are principle component analysis (PCA) and the Hough-transform. The first approach to this *generalized projected clustering* was the algorithm ORCLUS [13], using ideas similar to the axis-parallel approach PROCLUS [12]. ORCLUS is a k-means–like approach, picking $k_c > k$ seeds at first, assigning the data objects to these seeds according to a distance function that is based on an eigensystem of the corresponding cluster assessing the distance along the *weak* eigenvectors only (i.e., the distance in the projected subspace where the cluster objects exhibit high density). The eigensystem is iteratively adapted to the current state of the updated cluster. The number k_c of clusters is reduced iteratively by merging closest pairs of clusters until the user-specified number k is reached. The closest pair of clusters is the pair with the least average distance in the projected space (spanned by the weak eigenvectors) of the eigensystem of the merged clusters. Starting with a larger value for the parameter k_c increases the effectiveness, but also the runtime.

In contrast to ORCLUS, the algorithm [32] is based on a density-based clustering paradigm [88]. Thus, the number of clusters is not decided beforehand but clusters grow from a seed as long as a density criterion is fulfilled. Otherwise, another seed is picked to start a new cluster. The density criterion is a required minimal number of points within the neighborhood of a point, where the neighborhood is ascertained based on distance matrices computed from the eigensystems of two points. The eigensystem of a point p is based on the covariance matrix of the ε-neighborhood of p in Euclidean space. A parameter δ discerns large from small eigenvalues in the eigenvalue matrix E_p; then large eigenvalues are replaced by 1 and small eigenvalues by a value $\kappa \gg 1$. Using the adapted eigenvalue matrix E'_p, a correlation similarity matrix for p is obtained by $V_p \cdot E'_p \cdot V_p^{\mathsf{T}}$. This matrix is then used to derive the distance of two points, q and p, w.r.t. p, as the general quadratic form distance:

$$\sqrt{(p-q)^{\mathsf{T}} \cdot V_p \cdot E'_p \cdot V_p^{\mathsf{T}} \cdot (p-q)}. \tag{9.7}$$

Applying this measure symmetrically to q and choosing the maximum of both distances helps to decide whether both points are connected by a similar correlation of attributes and, thus, are similar and belong to each other's correlation neighborhood. Regarding the choice of parameters, 4C suffers from similar problems as PreDeCon, especially from the hard to guess neighborhood size ε that is used in both the Euclidean full-dimensional neighborhood and the adapted subspace neighborhood (recall Aspect 5 of the curse of dimensionality).

As a hierarchical approach, HiCO [8] defines the distance between points according to their local correlation dimensionality and subspace orientation and uses hierarchical density-based clustering [16] to derive a hierarchy of correlation clusters.

COPAC [7] is based on ideas similar to 4C but disposes of some problems such as meaningless similarity matrices due to sparse ε-neighborhoods. Instead of the problematic ε-range queries, COPAC is using a fixed number k of neighbors for the computation of local data characteristics—which raises the question of how to choose a good value for k. However, at least choosing $k > \lambda$ ensures a meaningful definition of a λ-dimensional hyperplane. Essentially, choosing k large enough but not larger than the expected minimum cluster size is a reasonable guideline. Thus, the point taken in COPAC that choosing the neighborhood in terms of the number of points rather than their distance, is worth considering. The main point in COPAC, however, is a considerable speed-up by partitioning the data set based on the observation that a correlation cluster should consist of points exhibiting the same local correlation dimensionality (i.e., the same number of strong eigenvectors in the covariance matrix of the k nearest neighbors). Thus, the search for clusters involves only the points with equal *local correlation dimensionality*. By creating one partition for each occurring correlation dimensionality, the time complexity rapidly decreases on average by getting rid of a squared factor d^2 in a d-dimensional data set.

Another related algorithm is ERiC [6], also deriving a local eigensystem for a point based on the k nearest neighbors in Euclidean space. Here, the neighborhood criterion for two points in a DBSCAN-like procedure is an approximate linear dependency and the affine distance of the correlation hyperplanes as defined by the strong eigenvectors of each point. As in COPAC, the property of clusters to consist of points exhibiting an equal local correlation dimensionality is exploited for the sake of efficiency. Furthermore, the resulting set of clusters is also ordered hierarchically to provide the user with a hierarchy of subspace clusters. In finding and correctly assigning complex patterns of intersecting clusters, COPAC and ERiC improve considerably over ORCLUS and 4C.

There are further examples of algorithms based on PCA [33, 95, 139, 20]. Some similar methods can be found in the survey of Vidal [128]. Many of these methods based on PCA use the eigensystem to adapt similarity measures in a soft way and, hence, can also be seen as variants of EM-clustering [40] (as far as they employ the partitioning clustering paradigm). A general framework has been proposed to increase the robustness of PCA-based correlation clustering algorithms [89].

The algorithm CASH [2, 1] is finding arbitrarily oriented subspace clusters by means of the Hough-transform [77, 44]. The Hough-transform was originally designed to map the points from a 2-dimensional data space (also called picture space) of Euclidean coordinates (e.g., pixels of an image) into a parameter space. The parameter space represents all possible 1D lines in the original 2D data space. In principle, each point of the data space is mapped into an infinite number of points in the parameter space which is not materialized as an infinite set but instead as a trigonometric function in the parameter space. Each function in the parameter space represents all lines in the picture space crossing the corresponding point in the data space. An intersection of two curves in the parameter space indicates a line through the two corresponding points in the picture space.

The objective of a clustering algorithm is to find intersections of many curves in the parameter space representing lines through many database objects. The key feature of the Hough-transform is that the distance of the points in the original data space is not considered any more. Objects can be identified as associated to a common line even if they are far apart in the original feature space. As a consequence, the Hough-transform is a promising candidate for developing a principle for subspace analysis that does not require the locality assumption and, thus, enables a global subspace clustering approach. CASH [2, 1] follows a grid-based approach to identify dense regions in the parameter space, successively attribute-wise dividing the space and counting the functions intersecting each of the resulting hyperboxes. In a depth-first search, most promising paths in the search tree are searched first. A hyperbox is divided along one axis if it contains enough functions to allow for dense child boxes in turn. If a dense subspace is found, the algorithm is applied on the data set accounted for by the corresponding hyperbox projected on the corresponding subspace. This recursive descent allows for finding lower-dimensional subspace clusters and implicitly yields a hierarchy of arbitrarily oriented subspaces and their accommodated clusters. However, if there are no correlation clusters in the original data space and, hence, no dense regions in the parameter space (but still, the hyperboxes remain dense enough to qualify as promising candidates), the complete search space is enumerated resulting in a worst case time complexity exponential in d. Probably, some more sophisticated heuristic may make this promising idea more practical for really high-dimensional data.

The Hough-transform is an example of using a technique in analysis of high-dimensional data that was originally developed in the context of image analysis (hence for 2-dimensional data). Other examples are the usage of image-filter techniques [35]. Occasionally, RANSAC [51] or other random sampling techniques have been used [74, 75].

As mentioned, approaches to clustering in arbitrarily oriented subspaces are also known as "correlation clustering" [32, 141]. It should be noted, though, that this term is ambiguously used also for a completely different problem in machine learning where a partitioning of the data correlate as much as possible with a pairwise similarity function f learned from past data [22].

9.5 Open Questions and Current Research Directions

Some interesting current research directions have found attention in the workshop series on "Discovering, Summarizing and Using Multiple Clusterings" (MultiClust) at KDD 2010 [50], at ECML/PKDD 2011 [109], and at SDM 2012 [110]. Intriguingly, problems known in subspace clustering meet similar problems in other areas such as ensemble clustering, alternative clustering, or multiview clustering [93]. Such problems common to all of these areas are (i) how to treat diversity of clustering solutions (are diverse clustering solutions to be unified or to be presented individually?), (ii) how to effectively summarize and present diversity to a user of clustering algorithms, (iii) how to treat redundancy of clusters, and (iv) how to assess similarity between multiple clustering solutions. Also, relationships to frequent pattern mining, which was godfather at the origin of subspace clustering, have been discussed from a more recent point of view on research in both fields [129].

As we have pointed out, redundancy of subspace cluster results is a problem inherited from the bottom-up strategy of the very first approaches, borrowed from frequent pattern mining. For current research on subspace clustering, getting rid of too much redundancy is a major topic [19, 70, 105]. Research on multiview clustering [28, 36, 80, 72] seems to approach the problem from the opposite direction but eventually is aiming at the same issue, seeking to allow some redundancy at least in order to not exclude possibly interesting concepts although they might partially have a certain overlap with other concepts. A related though distinct way of tackling the problem of redundancy and distinctiveness of different clusters is to seek diverse clusterings by directly assessing a certain notion of distance between different partitions (so-called alternative clustering approaches [60, 61, 21, 38, 118, 39, 37, 116]).

A question related to the redundancy issue concerns the appropriate density level. This is a general problem also in density-based clustering [88], but for clustering in subspaces, the problem is aggravated. Setting a fixed density threshold for an APRIORI-style subspace search is not appropriate for all possible subspaces. Consider for example some CLIQUE-style grid approach: the volume of a hypercube increases exponentially with the dimensionality; hence, the density decreases rapidly. As a consequence, any chosen threshold introduces a bias to identify clusters of (up to) a certain dimensionality. This observation motivates research on adaptive density thresholds [18, 106]. When using Euclidean distance (L_2), the appropriate choice of an ε-range becomes extremely challenging as well, not only for subspace, but also for projected clustering or arbitrarily oriented clustering, due to the rather counter-intuitive behavior of the volume of the hypersphere with increasing dimensions (which we discussed as Aspect 5 of the curse of dimensionality). Choosing the size of the neighborhood in terms of objects rather than in terms of a radius (i.e., using k nearest neighbors instead of an ε-range query) has been advocated as a workaround for this problem [7], to solve at least certain aspects such as having a well-defined (nonempty) set of objects for the density estimation or for the estimation of spatial characteristics of the neighborhood.

A rather unclear problem in arbitrarily oriented clustering is the significance of arbitrarily oriented clusters. (As discussed above, for axis-parallel subspace clustering at least some first steps have been taken [101].) Also for overlapping clusters, a common result in subspace clustering but also in other domains, a proper evaluation procedure is not known. Different approaches to evaluation have different weaknesses and all neglect certain requirements. This problem is sketched and discussed in more detail by Färber et al. [49]. Only some preliminary first steps toward improved evaluation methodology can be found in the literature [115, 92, 69, 9].

As a consequence, a comprehensive study experimentally comparing the merits of a broad selection of the different algorithms is still missing in the literature. There are only two preliminary attempts to compare a certain selection of subspace clustering algorithms [104, 108], where the applied evaluation procedures are not entirely convincing. Some of the algorithms discussed in this chapter are available in a unified implementation in the ELKI[3] framework [9].

Despite the unresolved issues in the basic techniques for real-valued feature vectors, there are already attempts to generalize subspace clustering for applications to more complex data. So-called 3D data add a third component (e.g., time) to objects and attributes. This means, instead of a data matrix, a data tensor of 3rd order is mined [140, 123, 81, 124]. While 3D data applications do have complete access to the third component (such as different points in time), applications to dynamic or stream data usually do not. These require a clustering algorithm to grasp the (subspace-) clustering structure with a single scan of the stream (and possibly some postprocessing on a summary structure). There are also first attempts to address subspace clustering in dynamic or stream data [10, 138, 86, 112]. Other challenges are given when the data are not of (purely) numeric nature but are, completely or partially, categorical [137, 107], or when the data are uncertain [71]. When some objects are labeled, this information can be used for semisupervised subspace clustering [134, 131, 56]. Lately, some subspace clustering approaches can use additionally available graph information simultaneously with the clustering process [68, 67]. A recent survey discusses such problems and approaches as "enhanced subspace clustering" [122].

9.6 Conclusion

Research in data mining and related disciplines such as statistics, pattern recognition, machine learning, and also applied sciences (e.g., bioinformatics) has led to a large variety of clustering techniques and also addressed the specialized problem of clustering high-dimensional data. New approaches to that problem are proposed in numerous conferences and journals every year. However, many researchers agree that there is no such thing as a general clustering technique suitable to all problems and universally applicable to arbitrary data sets. The aim of the concrete task of data analysis influences the choice of the clustering algorithm and obviously also the interpretation of the results of the clustering process. This is true for clustering in general, but even more so when facing tasks involving high-dimensional data.

In this chapter, in order to guide the readers to the literature most appropriate to the task they are facing, we distinguished different problem settings, namely, axis-parallel subspace clustering with the families of "subspace clustering" (in a narrower sense), "(soft) projected clustering," and an increasing number of "hybrid" approaches; as well as clustering in arbitrarily oriented subspaces (also called "correlation clustering"). Among these fields, one can find different problem settings addressed by "bi-clustering" approaches. All these different families address different aspects of the so-called "curse of dimensionality." The curse of dimensionality, however, is nothing we could resolve. There are different aspects of this "curse" and, usually, if one of these aspects is taken into consideration by some approach, other aspects will keep haunting the researcher.

Different approaches come with a different bias. The domain expert should decide which bias—if any—is most meaningful in a given application. For example, it could be meaningful in a given application to specifically search for axis-parallel subspace clusters. This means, in turn, the use of

[3] `http://elki.dbs.ifi.lmu.de/`

algorithms (e.g., P3C or of "soft" projected clustering algorithms) is discouraged in such a scenario or otherwise the results are to be closely inspected, since some approaches can possibly result in arbitrarily oriented clusters although the algorithms are designed to search for axis-parallel ones. Hence, the choice of a clustering approach as adequate to the problem at hand should be based on knowledge of the basic principles and heuristic restrictions upon which the particular clustering algorithm is based. Similarly, the interpretation of clustering results should be guided by the knowledge of the kinds of patterns a particular algorithm can or cannot find.

The family of axis-parallel subspace and projected clustering algorithms assumes that data objects belonging to the same cluster are located near each other in Euclidean space but allows assess to the corresponding distance of objects w.r.t. subsets of the attributes due to the problem of irrelevant attributes. Pattern-based approaches often disregard the assumption that a cluster consists of objects that are near each other in the Euclidean space or some Euclidean subspace and, instead, aim at collecting objects following a similar behavioral pattern over a subset of attributes. These patterns relate to simple positive correlations among the considered attributes. Correlation clustering algorithms generalize this approach to arbitrary complex positive or negative correlations but often assume, again, a certain density of the points in Euclidean space, too. Finally, let us note that the different notions of similarity employed by the different classes of algorithms usually cannot be used interchangeably. Rather the algorithms of each class are more or less tailored to the class-specific notions of similarity. The user should not ask: *"Which is the best clustering algorithm ever?"* but rather: *"Which clustering algorithm (with its assumptions and restrictions) is best suited for my problem?"* With this chapter it was intended to give some useful guidelines to this end.

Bibliography

[1] E. Achtert, C. Böhm, J. David, P. Kröger, and A. Zimek. Global correlation clustering based on the Hough transform. *Statistical Analysis and Data Mining*, 1(3):111–127, 2008.

[2] E. Achtert, C. Böhm, J. David, P. Kröger, and A. Zimek. Robust clustering in arbitrarily oriented subspaces. In *Proceedings of the 8th SIAM International Conference on Data Mining (SDM)*, Atlanta, GA, pages 763–774, 2008.

[3] E. Achtert, C. Böhm, H.-P. Kriegel, P. Kröger, I. Müller-Gorman, and A. Zimek. Detection and visualization of subspace cluster hierarchies. In *Proceedings of the 12th International Conference on Database Systems for Advanced Applications (DASFAA)*, Bangkok, Thailand, pages 152–163, 2007.

[4] E. Achtert, C. Böhm, H.-P. Kriegel, P. Kröger, I. Müller-Gorman, and A. Zimek. Finding hierarchies of subspace clusters. In *Proceedings of the 10th European Conference on Principles and Practice of Knowledge Discovery in Databases (PKDD)*, Berlin, Germany, pages 446–453, 2006.

[5] E. Achtert, C. Böhm, H.-P. Kriegel, P. Kröger, and A. Zimek. Deriving quantitative models for correlation clusters. In *Proceedings of the 12th ACM International Conference on Knowledge Discovery and Data Mining (SIGKDD)*, Philadelphia, PA, pages 4–13, 2006.

[6] E. Achtert, C. Böhm, H.-P. Kriegel, P. Kröger, and A. Zimek. On exploring complex relationships of correlation clusters. In *Proceedings of the 19th International Conference on Scientific and Statistical Database Management (SSDBM)*, Banff, Canada, page 7, 2007.

[7] E. Achtert, C. Böhm, H.-P. Kriegel, P. Kröger, and A. Zimek. Robust, complete, and efficient correlation clustering. In *Proceedings of the 7th SIAM International Conference on Data Mining (SDM)*, Minneapolis, MN, 2007.

[8] E. Achtert, C. Böhm, P. Kröger, and A. Zimek. Mining hierarchies of correlation clusters. In *Proceedings of the 18th International Conference on Scientific and Statistical Database Management (SSDBM)*, Vienna, Austria, pages 119–128, 2006.

[9] E. Achtert, S. Goldhofer, H.-P. Kriegel, E. Schubert, and A. Zimek. Evaluation of clusterings—Metrics and visual support. In *Proceedings of the 28th International Conference on Data Engineering (ICDE)*, Washington, DC, pages 1285–1288, 2012.

[10] C. C. Aggarwal, J. Han, J. Wang, and P. S. Yu. A framework for projected clustering of high dimensional data streams. In *Proceedings of the 30th International Conference on Very Large Data Bases (VLDB)*, Toronto, Canada, pages 852–863, 2004.

[11] C. C. Aggarwal, A. Hinneburg, and D. Keim. On the surprising behavior of distance metrics in high dimensional space. In *Proceedings of the 8th International Conference on Database Theory (ICDT)*, London, UK, pages 420–434, 2001.

[12] C. C. Aggarwal, C. M. Procopiuc, J. L. Wolf, P. S. Yu, and J. S. Park. Fast algorithms for projected clustering. In *Proceedings of the ACM International Conference on Management of Data (SIGMOD)*, Philadelphia, PA, pages 61–72, 1999.

[13] C. C. Aggarwal and P. S. Yu. Finding generalized projected clusters in high dimensional space. In *Proceedings of the ACM International Conference on Management of Data (SIGMOD)*, Dallas, TX, pages 70–81, 2000.

[14] R. Agrawal, J. Gehrke, D. Gunopulos, and P. Raghavan. Automatic subspace clustering of high dimensional data for data mining applications. In *Proceedings of the ACM International Conference on Management of Data (SIGMOD)*, Seattle, WA, pages 94–105, 1998.

[15] R. Agrawal and R. Srikant. Fast algorithms for mining association rules. In *Proceedings of the 20th International Conference on Very Large Data Bases (VLDB)*, Santiago de Chile, Chile, pages 487–499, 1994.

[16] M. Ankerst, M. M. Breunig, H.-P. Kriegel, and J. Sander. OPTICS: Ordering points to identify the clustering structure. In *Proceedings of the ACM International Conference on Management of Data (SIGMOD)*, Philadelphia, PA, pages 49–60, 1999.

[17] I. Assent. Clustering high dimensional data. *Wiley Interdisciplinary Reviews: Data Mining and Knowledge Discovery*, 2(4):340–350, 2012.

[18] I. Assent, R. Krieger, E. Müller, and T. Seidl. DUSC: Dimensionality unbiased subspace clustering. In *Proceedings of the 7th IEEE International Conference on Data Mining (ICDM)*, Omaha, NE, pages 409–414, 2007.

[19] I. Assent, E. Müller, S. Günnemann, R. Krieger, and T. Seidl. Less is more: Non-redundant subspace clustering. In *MultiClust: 1st International Workshop on Discovering, Summarizing and Using Multiple Clusterings Held in Conjunction with KDD 2010*, Washington, DC, 2010.

[20] M. S. Aziz and C. K. Reddy. A robust seedless algorithm for correlation clustering. In *Proceedings of the 14th Pacific-Asia Conference on Knowledge Discovery and Data Mining (PAKDD)*, Hyderabad, India, pages 28–37, 2010.

[21] E. Bae and J. Bailey. COALA: A novel approach for the extraction of an alternate clustering of high quality and high dissimilarity. In *Proceedings of the 6th IEEE International Conference on Data Mining (ICDM)*, Hong Kong, China, pages 53–62, 2006.

[22] N. Bansal, A. Blum, and S. Chawla. Correlation clustering. *Machine Learning*, 56:89–113, 2004.

[23] R. Bellman. *Adaptive Control Processes. A Guided Tour*. Princeton University Press, 1961.

[24] A. Belussi and C. Faloutsos. Estimating the selectivity of spatial queries using the "correlation" fractal dimension. In *Proceedings of the 21st International Conference on Very Large Data Bases (VLDB)*, Zurich, Switzerland, pages 299–310, 1995.

[25] K. P. Bennett, U. Fayyad, and D. Geiger. Density-based indexing for approximate nearest-neighbor queries. In *Proceedings of the 5th ACM International Conference on Knowledge Discovery and Data Mining (SIGKDD)*, San Diego, CA, pages 233–243, 1999.

[26] T. Bernecker, M. E. Houle, H.-P. Kriegel, P. Kröger, M. Renz, E. Schubert, and A. Zimek. Quality of similarity rankings in time series. In *Proceedings of the 12th International Symposium on Spatial and Temporal Databases (SSTD)*, Minneapolis, MN, pages 422–440, 2011.

[27] K. Beyer, J. Goldstein, R. Ramakrishnan, and U. Shaft. When is "nearest neighbor" meaningful? In *Proceedings of the 7th International Conference on Database Theory (ICDT)*, Jerusalem, Israel, pages 217–235, 1999.

[28] S. Bickel and T. Scheffer. Multi-view clustering. In *Proceedings of the 4th IEEE International Conference on Data Mining (ICDM)*, Brighton, UK, pages 19–26, 2004.

[29] C. Bouveyron, S. Girard, and C. Schmid. High-dimensional data clustering. *Computational Statistics and Data Analysis*, 52:502–519, 2007.

[30] C. Böhm, S. Berchtold, and D. A. Keim. Searching in high-dimensional spaces: Index structures for improving the performance of multimedia databases. *ACM Computing Surveys*, 33(3):322–373, 2001.

[31] C. Böhm, K. Kailing, H.-P. Kriegel, and P. Kröger. Density connected clustering with local subspace preferences. In *Proceedings of the 4th IEEE International Conference on Data Mining (ICDM)*, Brighton, UK, pages 27–34, 2004.

[32] C. Böhm, K. Kailing, P. Kröger, and A. Zimek. Computing clusters of correlation connected objects. In *Proceedings of the ACM International Conference on Management of Data (SIGMOD)*, Paris, France, pages 455–466, 2004.

[33] K. Chakrabarti and S. Mehrotra. Local dimensionality reduction: A new approach to indexing high dimensional spaces. In *Proceedings of the 26th International Conference on Very Large Data Bases (VLDB)*, Cairo, Egypt, pages 89–100, 2000.

[34] C. H. Cheng, A. W.-C. Fu, and Y. Zhang. Entropy-based subspace clustering for mining numerical data. In *Proceedings of the 5th ACM International Conference on Knowledge Discovery and Data Mining (SIGKDD)*, San Diego, CA, pages 84–93, 1999.

[35] R. L. F. Cordeiro, A. J. M. Traina, C. Faloutsos, and C. Traina Jr. Finding clusters in subspaces of very large, multi-dimensional datasets. In *Proceedings of the 26th International Conference on Data Engineering (ICDE)*, Long Beach, CA, pages 625–636, 2010.

[36] Y. Cui, X. Z. Fern, and J. G. Dy. Non-redundant multi-view clustering via orthogonalization. In *Proceedings of the 7th IEEE International Conference on Data Mining (ICDM)*, Omaha, NE, pages 133–142, 2007.

[37] X. H. Dang and J. Bailey. Generation of alternative clusterings using the CAMI approach. In *Proceedings of the 10th SIAM International Conference on Data Mining (SDM)*, Columbus, OH, pages 118–129, 2010.

[38] I. Davidson and Z. Qi. Finding alternative clusterings using constraints. In *Proceedings of the 8th IEEE International Conference on Data Mining (ICDM)*, Pisa, Italy, pages 773–778, 2008.

[39] I. Davidson, S. S. Ravi, and L. Shamis. A SAT-based framework for efficient constrained clustering. In *Proceedings of the 10th SIAM International Conference on Data Mining (SDM)*, Columbus, OH, pages 94–105, 2010.

[40] A. P. Dempster, N. M. Laird, and D. B. Rubin. Maximum likelihood from incomplete data via the EM algorithm. *Journal of the Royal Statistical Society: Series B (Statistical Methodology)*, 39(1):1–31, 1977.

[41] C. Domeniconi. Subspace clustering ensembles (invited talk). In *3rd MultiClust Workshop: Discovering, Summarizing and Using Multiple Clusterings Held in Conjunction with SIAM Data Mining 2012*, Anaheim, CA, 2012.

[42] C. Domeniconi, D. Gunopulos, S. Ma, B. Yan, M. Al-Razgan, and D. Papadopoulos. Locally adaptive metrics for clustering high dimensional data. *Data Mining and Knowledge Discovery*, 14(1):63–97, 2007.

[43] C. Domeniconi, D. Papadopoulos, D. Gunopulos, and S. Ma. Subspace clustering of high dimensional data. In *Proceedings of the 4th SIAM International Conference on Data Mining (SDM)*, Lake Buena Vista, FL, pages 517–521, 2004.

[44] R. O. Duda and P. E. Hart. Use of the Hough transformation to detect lines and curves in pictures. *Communications of the ACM*, 15(1):11–15, 1972.

[45] R. J. Durrant and A. Kabán. When is "nearest neighbour" meaningful: A converse theorem and implications. *Journal of Complexity*, 25(4):385–397, 2009.

[46] M. Ester, H.-P. Kriegel, J. Sander, and X. Xu. A density-based algorithm for discovering clusters in large spatial databases with noise. In *Proceedings of the 2nd ACM International Conference on Knowledge Discovery and Data Mining (KDD)*, Portland, OR, pages 226–231, 1996.

[47] V. Estivill-Castro. Why so many clustering algorithms—A position paper. *ACM SIGKDD Explorations*, 4(1):65–75, 2002.

[48] C. Faloutsos and I. Kamel. Beyond uniformity and independence: Analysis of R-trees using the concept of fractal dimension. In *Proceedings of the ACM International Conference on Management of Data (SIGMOD)*, Minneapolis, MN, pages 4–13, 1994.

[49] I. Färber, S. Günnemann, H.-P. Kriegel, P. Kröger, E. Müller, E. Schubert, T. Seidl, and A. Zimek. On using class-labels in evaluation of clusterings. In *MultiClust: 1st International Workshop on Discovering, Summarizing and Using Multiple Clusterings Held in Conjunction with KDD 2010*, Washington, DC, 2010.

[50] X. Z. Fern, I. Davidson, and J. G. Dy. MultiClust 2010: Discovering, summarizing and using multiple clusterings. *ACM SIGKDD Explorations*, 12(2):47–49, 2010.

[51] M. A. Fischler and R. C. Bolles. Random sample consensus: A paradigm for model fitting with applications to image analysis and automated cartography. *Communications of the ACM*, 24(6):381–395, 1981.

[52] E. B. Fowlkes, R. Gnanadesikan, and J. R. Kettenring. Variable selection in clustering. *Journal of Classification*, 5:205–228, 1988.

[53] E. B. Fowlkes and C. L. Mallows. A method for comparing two hierarchical clusterings. *Journal of the American Statistical Association*, 78(383):553–569, 1983.

[54] D. François, V. Wertz, and M. Verleysen. The concentration of fractional distances. *IEEE Transactions on Knowledge and Data Engineering*, 19(7):873–886, 2007.

[55] J. H. Friedman and J. J. Meulman. Clustering objects on subsets of attributes. *Journal of the Royal Statistical Society: Series B (Statistical Methodology)*, 66(4):825–849, 2004.

[56] É. Fromont, A. Prado, and C. Robardet. Constraint-based subspace clustering. In *Proceedings of the 9th SIAM International Conference on Data Mining (SDM)*, Sparks, NV, pages 26–37, 2009.

[57] G. Gan, C. Ma, and J. Wu. *Data Clustering. Theory, Algorithms, and Applications*. Philadelphia, PA: Society for Industrial and Applied Mathematics (SIAM), 2007.

[58] J. Ghosh and A. Acharya. Cluster ensembles. *Wiley Interdisciplinary Reviews: Data Mining and Knowledge Discovery*, 1(4):305–315, 2011.

[59] R. Gnanadesikan, J. R. Kettenring, and S. L. Tsao. Weighting and selection of variables for cluster analysis. *Journal of Classification*, 12(1):113–136, 1995.

[60] D. Gondek and T. Hofmann. Non-redundant clustering with conditional ensembles. In *Proceedings of the 11th ACM International Conference on Knowledge Discovery and Data Mining (SIGKDD)*, Chicago, IL, pages 70–77, 2005.

[61] D. Gondek and T. Hofmann. Non-redundant data clustering. In *Proceedings of the 4th IEEE International Conference on Data Mining (ICDM)*, Brighton, UK, pages 75–82, 2004.

[62] F. Gullo, C. Domeniconi, and A. Tagarelli. Advancing data clustering via projective clustering ensembles. In *Proceedings of the 17th ACM International Conference on Knowledge Discovery and Data Mining (SIGKDD)*, San Diego, CA, pages 733–744, 2011.

[63] F. Gullo, C. Domeniconi, and A. Tagarelli. Enhancing single-objective projective clustering ensembles. In *Proceedings of the 10th IEEE International Conference on Data Mining (ICDM)*, Sydney, Australia, pages 833–838, 2010.

[64] F. Gullo, C. Domeniconi, and A. Tagarelli. Projective clustering ensembles. In *Proceedings of the 9th IEEE International Conference on Data Mining (ICDM)*, Miami, FL, pages 794–799, 2009.

[65] A. Gupta, R. Krauthgamer, and J. R. Lee. Bounded geometries, fractals, and low-distortion embeddings. In *Proceedings of the 44th Annual IEEE Symposium on Foundations of Computer Science (FOCS)*, Cambridge, MA, pages 534–543, 2003.

[66] I. Guyon and A. Elisseeff. An introduction to variable and feature selection. *Journal of Machine Learning Research*, 3:1157–1182, 2003.

[67] S. Günnemann, B. Boden, and T. Seidl. Finding density-based subspace clusters in graphs with feature vectors. *Data Mining and Knowledge Discovery*, 25(2):243–269, 2012.

[68] S. Günnemann, I. Färber, B. Boden, and T. Seidl. Subspace clustering meets dense subgraph mining: A synthesis of two paradigms. In *Proceedings of the 10th IEEE International Conference on Data Mining (ICDM)*, Sydney, Australia, pages 845–850, 2010.

[69] S. Günnemann, I. Färber, E. Müller, I. Assent, and T. Seidl. External evaluation measures for subspace clustering. In *Proceedings of the 20th ACM Conference on Information and Knowledge Management (CIKM)*, Glasgow, UK, pages 1363–1372, 2011.

[70] S. Günnemann, I. Färber, E. Müller, and T. Seidl. ASCLU: Alternative subspace clustering. In *MultiClust: 1st International Workshop on Discovering, Summarizing and Using Multiple Clusterings Held in Conjunction with KDD 2010*, Washington, DC, 2010.

[71] S. Günnemann, H. Kremer, and T. Seidl. Subspace clustering for uncertain data. In *Proceedings of the 10th SIAM International Conference on Data Mining (SDM)*, Columbus, OH, pages 385–396, 2010.

[72] S. Günnemann, E. Müller, I. Färber, and T. Seidl. Detection of orthogonal concepts in subspaces of high dimensional data. In *Proceedings of the 18th ACM Conference on Information and Knowledge Management (CIKM)*, Hong Kong, China, pages 1317–1326, 2009.

[73] J. Han, M. Kamber, and J. Pei. *Data Mining: Concepts and Techniques*, 3rd edition. Morgan Kaufmann, 2011.

[74] R. Haralick and R. Harpaz. Linear manifold clustering. In *Proceedings of the 4th International Conference on Machine Learning and Data Mining in Pattern Recognition (MLDM)*, Leipzig, Germany, pages 132–141, 2005.

[75] R. Harpaz and R. Haralick. Mining subspace correlations. In *Proceedings of the IEEE Symposium on Computational Intelligence and Data Mining (CIDM)*, Honolulu, HI, pages 335–342, 2007.

[76] A. Hinneburg, C. C. Aggarwal, and D. A. Keim. What is the nearest neighbor in high dimensional spaces? In *Proceedings of the 26th International Conference on Very Large Data Bases (VLDB)*, Cairo, Egypt, pages 506–515, 2000.

[77] P. V. C. Hough. Methods and means for recognizing complex patterns. U.S. Patent 3069654, December 18, 1962.

[78] M. E. Houle, H.-P. Kriegel, P. Kröger, E. Schubert, and A. Zimek. Can shared-neighbor distances defeat the curse of dimensionality? In *Proceedings of the 22nd International Conference on Scientific and Statistical Database Management (SSDBM)*, Heidelberg, Germany, pages 482–500, 2010.

[79] J. Z. Huang, M. K. Ng, H. Rong, and Z. Li. Automated variable weighting in k-means type clustering. *IEEE Transactions on Pattern Analysis and Machine Intelligence*, 27(5):657–668, 2005.

[80] P. Jain, R. Meka, and I. S. Dhillon. Simultaneous unsupervised learning of disparate clusterings. *Statistical Analysis and Data Mining*, 1(3):195–210, 2008.

[81] L. Ji, K.-L. Tan, and A. K. H. Tung. Mining frequent closed cubes in 3D datasets. In *Proceedings of the 32nd International Conference on Very Large Data Bases (VLDB)*, Seoul, Korea, pages 811–822, 2006.

[82] L. Jing, M. K. Ng, and J. Z. Huang. An entropy weighting k-means algorithm for subspace clustering of high-dimensional sparse data. *IEEE Transactions on Knowledge and Data Engineering*, 19(8):1026–1041, 2007.

[83] I. T. Jolliffe. *Principal Component Analysis*. 2nd edition, Springer, 2002.

[84] K. Kailing, H.-P. Kriegel, and P. Kröger. Density-connected subspace clustering for high-dimensional data. In *Proceedings of the 4th SIAM International Conference on Data Mining (SDM)*, Lake Buena Vista, FL, pages 246–257, 2004.

[85] F. Korn, B.-U. Pagel, and C. Faloutsos. On the "dimensionality curse" and the "self-similarity blessing." *IEEE Transactions on Knowledge and Data Engineering*, 13(1):96–111, 2001.

[86] H.-P. Kriegel, P. Kröger, I. Ntoutsi, and A. Zimek. Density based subspace clustering over dynamic data. In *Proceedings of the 23rd International Conference on Scientific and Statistical Database Management (SSDBM)*, Portland, OR, pages 387–404, 2011.

[87] H.-P. Kriegel, P. Kröger, M. Renz, and S. Wurst. A generic framework for efficient subspace clustering of high-dimensional data. In *Proceedings of the 5th IEEE International Conference on Data Mining (ICDM)*, Houston, TX, pages 250–257, 2005.

[88] H.-P. Kriegel, P. Kröger, J. Sander, and A. Zimek. Density-based clustering. *Wiley Interdisciplinary Reviews: Data Mining and Knowledge Discovery*, 1(3):231–240, 2011.

[89] H.-P. Kriegel, P. Kröger, E. Schubert, and A. Zimek. A general framework for increasing the robustness of PCA-based correlation clustering algorithms. In *Proceedings of the 20th International Conference on Scientific and Statistical Database Management (SSDBM)*, Hong Kong, China, pages 418–435, 2008.

[90] H.-P. Kriegel, P. Kröger, and A. Zimek. Clustering high dimensional data: A survey on subspace clustering, pattern-based clustering, and correlation clustering. *ACM Transactions on Knowledge Discovery from Data (TKDD)*, 3(1):1–58, 2009.

[91] H.-P. Kriegel, P. Kröger, and A. Zimek. Subspace clustering. *Wiley Interdisciplinary Reviews: Data Mining and Knowledge Discovery*, 2(4):351–364, 2012.

[92] H.-P. Kriegel, E. Schubert, and A. Zimek. Evaluation of multiple clustering solutions. In *2nd MultiClust Workshop: Discovering, Summarizing and Using Multiple Clusterings Held in Conjunction with ECML PKDD 2011*, Athens, Greece, pages 55–66, 2011.

[93] H.-P. Kriegel and A. Zimek. Subspace clustering, ensemble clustering, alternative clustering, multiview clustering: What can we learn from each other? In *MultiClust: 1st International Workshop on Discovering, Summarizing and Using Multiple Clusterings Held in Conjunction with KDD 2010*, Washington, DC, 2010.

[94] P. Kröger and A. Zimek. Subspace clustering techniques. In L. Liu and M. T. Özsu, editors, *Encyclopedia of Database Systems*, pages 2873–2875. Springer, 2009.

[95] J. Li, X. Huang, C. Selke, and J. Yong. A fast algorithm for finding correlation clusters in noise data. In *Proceedings of the 11th Pacific-Asia Conference on Knowledge Discovery and Data Mining (PAKDD)*, Nanjing, China, pages 639–647, 2007.

[96] B. Liu, Y. Xia, and P. S. Yu. Clustering through decision tree construction. In *Proceedings of the 9th ACM Conference on Information and Knowledge Management (CIKM)*, Washington, DC, pages 20–29, 2000.

[97] G. Liu, J. Li, K. Sim, and L. Wong. Distance based subspace clustering with flexible dimension partitioning. In *Proceedings of the 23rd International Conference on Data Engineering (ICDE)*, Istanbul, Turkey, pages 1250–1254, 2007.

[98] Y. Lu, S. Wang, S. Li, and C. Zhou. Particle swarm optimizer for variable weighting in clustering high-dimensional data. *Machine Learning*, 82(1):43–70, 2010.

[99] S. C. Madeira and A. L. Oliveira. Biclustering algorithms for biological data analysis: A survey. *IEEE/ACM Transactions on Computational Biology and Bioinformatics*, 1(1):24–45, 2004.

[100] G. W. Milligan. An examination of the effect of six types of error perturbation on fifteen clustering algorithms. *Psychometrika*, 45(3):325–342, 1980.

[101] G. Moise and J. Sander. Finding non-redundant, statistically significant regions in high dimensional data: A novel approach to projected and subspace clustering. In *Proceedings of the 14th ACM International Conference on Knowledge Discovery and Data Mining (SIGKDD)*, Las Vegas, NV, pages 533–541, 2008.

[102] G. Moise, J. Sander, and M. Ester. P3C: A robust projected clustering algorithm. In *Proceedings of the 6th IEEE International Conference on Data Mining (ICDM)*, Hong Kong, China, pages 414–425, 2006.

[103] G. Moise, J. Sander, and M. Ester. Robust projected clustering. *Knowledge and Information Systems (KAIS)*, 14(3):273–298, 2008.

[104] G. Moise, A. Zimek, P. Kröger, H.-P. Kriegel, and J. Sander. Subspace and projected clustering: Experimental evaluation and analysis. *Knowledge and Information Systems (KAIS)*, 21(3):299–326, 2009.

[105] E. Müller, I. Assent, S. Günnemann, R. Krieger, and T. Seidl. Relevant subspace clustering: Mining the most interesting non-redundant concepts in high dimensional data. In *Proceedings of the 9th IEEE International Conference on Data Mining (ICDM)*, Miami, FL, pages 377–386, 2009.

[106] E. Müller, I. Assent, R. Krieger, S. Günnemann, and T. Seidl. DensEst: Density estimation for data mining in high dimensional spaces. In *Proceedings of the 9th SIAM International Conference on Data Mining (SDM)*, Sparks, NV, pages 173–184, 2009.

[107] E. Müller, I. Assent, and T. Seidl. HSM: heterogeneous subspace mining in high dimensional data. In *Proceedings of the 21st International Conference on Scientific and Statistical Database Management (SSDBM)*, New Orleans, LA, pages 497–516, 2009.

[108] E. Müller, S. Günnemann, I. Assent, and T. Seidl. Evaluating clustering in subspace projections of high dimensional data. In *Proceedings of the 35th International Conference on Very Large Data Bases (VLDB)*, Lyon, France, pages 1270–1281, 2009.

[109] E. Müller, S. Günnemann, I. Assent, and T. Seidl, editors. *2nd MultiClustWorkshop: Discovering, Summarizing and Using Multiple Clusterings, Held in Conjunction with ECML PKDD 2011*, Athens, Greece, 2011.

[110] E. Müller, T. Seidl, S. Venkatasubramanian, and A. Zimek, editors. *3rd MultiClust Workshop: Discovering, Summarizing and Using Multiple Clusterings, Held in Conjunction with SIAM Data Mining 2012*, Anaheim, CA, 2012.

[111] H. S. Nagesh, S. Goil, and A. Choudhary. Adaptive grids for clustering massive data sets. In *Proceedings of the 1st SIAM International Conference on Data Mining (SDM)*,New York, N,Y., pages 1–17, 2001.

[112] I. Ntoutsi, A. Zimek, T. Palpanas, P. Kröger, and H.-P. Kriegel. Density-based projected clustering over high dimensional data streams. In *Proceedings of the 12th SIAM International Conference on Data Mining (SDM)*, Anaheim, CA, 2012.

[113] B.-U. Pagel, F. Korn, and C. Faloutsos. Deflating the dimensionality curse using multiple fractal dimensions. In *Proceedings of the 16th International Conference on Data Engineering (ICDE)*, San Diego, CA, pages 589–598, 2000.

[114] L. Parsons, E. Haque, and H. Liu. Subspace clustering for high dimensional data: A review. *ACM SIGKDD Explorations*, 6(1):90–105, 2004.

[115] A. Patrikainen and M. Meila. Comparing subspace clusterings. *IEEE Transactions on Knowledge and Data Engineering*, 18(7):902–916, 2006.

[116] J. M. Phillips, P. Raman, and S. Venkatasubramanian. Generating a diverse set of high-quality clusterings. In *2nd MultiClust Workshop: Discovering, Summarizing and Using Multiple Clusterings Held in Conjunction with ECML PKDD 2011*, Athens, Greece, pages 80–91, 2011.

[117] C. M. Procopiuc, M. Jones, P. K. Agarwal, and T. M. Murali. A Monte Carlo algorithm for fast projective clustering. In *Proceedings of the ACM International Conference on Management of Data (SIGMOD)*, Madison, WI, pages 418–427, 2002.

[118] Z. J. Qi and I. Davidson. A principled and flexible framework for finding alternative clusterings. In *Proceedings of the 15th ACM International Conference on Knowledge Discovery and Data Mining (SIGKDD)*, Paris, France, pages 717–726, 2009.

[119] M. Radovanović, A. Nanopoulos, and M. Ivanović. On the existence of obstinate results in vector space models. In *Proceedings of the 33rd International Conference on Research and Development in Information Retrieval (SIGIR)*, Geneva, Switzerland, pages 186–193, 2010.

[120] K. Sequeira and M. J. Zaki. SCHISM: A new approach to interesting subspace mining. *International Journal of Business Intelligence and Data Mining*, 1(2):137–160, 2005.

[121] R. Sibson. SLINK: An optimally efficient algorithm for the single-link cluster method. *The Computer Journal*, 16(1):30–34, 1973.

[122] K. Sim, V. Gopalkrishnan, A. Zimek, and G. Cong. A survey on enhanced subspace clustering. *Data Mining and Knowledge Discovery*, 26(2):332–397, 2013.

[123] K. Sim, J. Li, V. Gopalkrishnan, and G. Liu. Mining maximal quasi-bicliques to co-cluster stocks and financial ratios for value investment. In *Proceedings of the 6th IEEE International Conference on Data Mining (ICDM)*, Hong Kong, China, pages 1059–1063, 2006.

[124] K. Sim, A. K. Poernomo, and V. Gopalkrishnan. Mining actionable subspace clusters in sequential data. In *Proceedings of the 10th SIAM International Conference on Data Mining (SDM)*, Columbus, OH, pages 442–453, 2010.

[125] J. L. Slagle, C. L. Chang, and S. L. Heller. A clustering and data-reorganization algorithm. *IEEE Transactions on Systems, Man, and Cybernetics*, 5(1):121–128, 1975.

[126] P. H. A. Sneath. The application of computers to taxonomy. *Journal of General Microbiology*, 17:201–226, 1957.

[127] D. Steinley and M. J. Brusco. Selection of variables in cluster analysis: An empirical comparison of eight procedures. *Psychometrika*, 73(1):125–144, 2008.

[128] R. Vidal. Subspace clustering. *IEEE Signal Processing Magazine*, 28(2):52–68, 2011.

[129] J. Vreeken and A. Zimek. When pattern met subspace cluster—A relationship story. In *2nd MultiClust Workshop: Discovering, Summarizing and Using Multiple Clusterings Held in Conjunction with ECML PKDD 2011*, Athens, Greece, pages 7–18, 2011.

[130] K.-G. Woo, J.-H. Lee, M.-H. Kim, and Y.-J. Lee. FINDIT: A fast and intelligent subspace clustering algorithm using dimension voting. *Information and Software Technology*, 46(4):255–271, 2004.

[131] B. Yan and C. Domeniconi. Subspace metric ensembles for semi-supervised clustering of high dimensional data. In *Proceedings of the 17th European Conference on Machine Learning (ECML)*, Berlin, Germany, pages 509–520, 2006.

[132] L. Yang. Distance-preserving dimensionality reduction. *Wiley Interdisciplinary Reviews: Data Mining and Knowledge Discovery*, 1(5):369–380, 2011.

[133] K. Y. Yip, D. W. Cheung, and M. K. Ng. HARP: A practical projected clustering algorithm. *IEEE Transactions on Knowledge and Data Engineering*, 16(11):1387–1397, 2004.

[134] K. Y. Yip, D. W. Cheung, and M. K. Ng. On discovery of extremely low-dimensional clusters using semi-supervised projected clustering. In *Proceedings of the 21st International Conference on Data Engineering (ICDE)*, Tokyo, Japan, pages 329–340, 2005.

[135] M. L. Yiu and N. Mamoulis. Frequent-pattern based iterative projected clustering. In *Proceedings of the 3rd IEEE International Conference on Data Mining (ICDM)*, Melbourne, FL, pages 689–692, 2003.

[136] M. L. Yiu and N. Mamoulis. Iterative projected clustering by subspace mining. *IEEE Transactions on Knowledge and Data Engineering*, 17(2):176–189, 2005.

[137] M. J. Zaki, M. Peters, I. Assent, and T. Seidl. CLICKS: An effective algorithm for mining subspace clusters in categorical datasets. *Data & Knowledge Engineering*, 60(1):51–70, 2007.

[138] Q. Zhang, J. Liu, and W. Wang. Incremental subspace clustering over multiple data streams. In *Proceedings of the 7th IEEE International Conference on Data Mining (ICDM)*, Omaha, NE, pages 727–732, 2007.

[139] X. Zhang, F. Pan, and W. Wang. CARE: finding local linear correlations in high dimensional data. In *Proceedings of the 24th International Conference on Data Engineering (ICDE)*, Cancun, Mexico, pages 130–139, 2008.

[140] L. Zhao and M. J. Zaki. TRICLUSTER: An effective algorithm for mining coherent clusters in 3D microarray data. In *Proceedings of the ACM International Conference on Management of Data (SIGMOD)*, Baltimore, MD, pages 694–705, 2005.

[141] A. Zimek. Correlation clustering. *ACM SIGKDD Explorations*, 11(1):53–54, 2009.

[142] A. Zimek, E. Schubert, and H.-P. Kriegel. A survey on unsupervised outlier detection in high-dimensional numerical data. *Statistical Analysis and Data Mining*, 5(5):363–387, 2012.

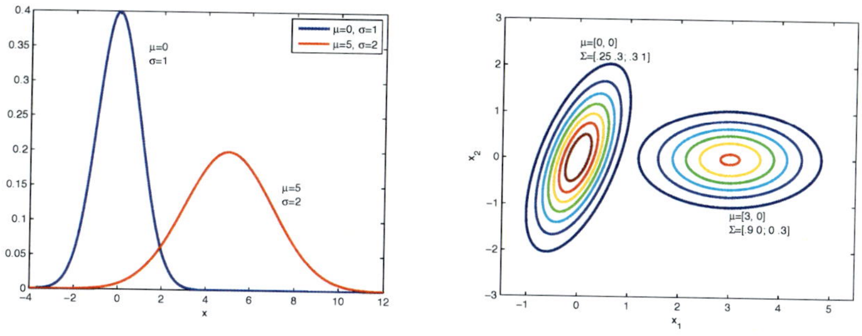

(a) (b)

FIGURE 3.2: (a) Plots of the univariate Gaussian distribution given by (3.8) for various parameters of μ and σ, and (b) contours of the multivariate (2-D) Gaussian distribution given by (3.9) for various parameters of μ and Σ.

FIGURE 12.4: Example of Hamming distances on the *zoo* categorical dataset.

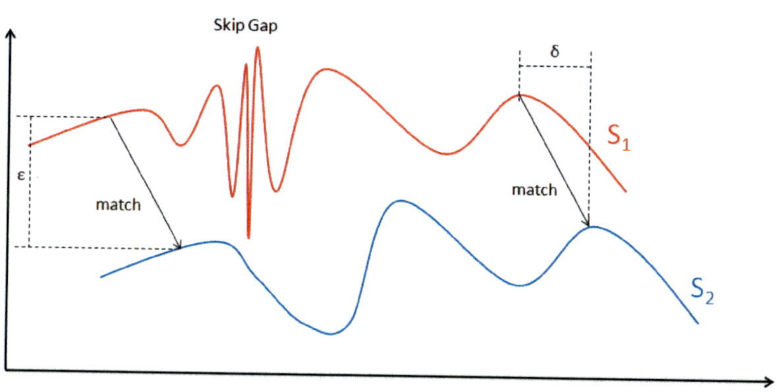

FIGURE 15.2: LCSS distance, thresholds δ and ε

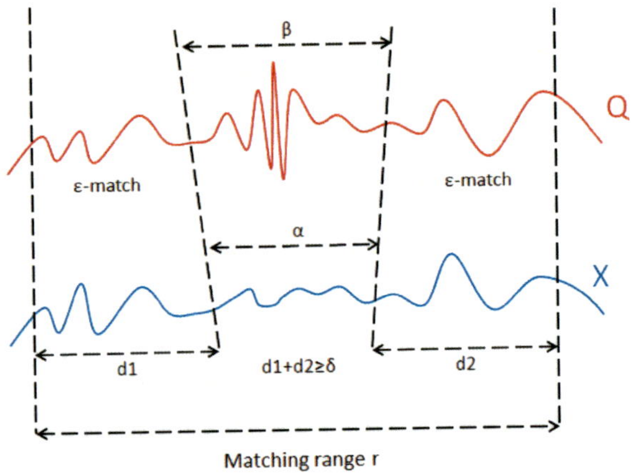

FIGURE 15.3: SMBGT distance between a query sequence Q and a database sequence X

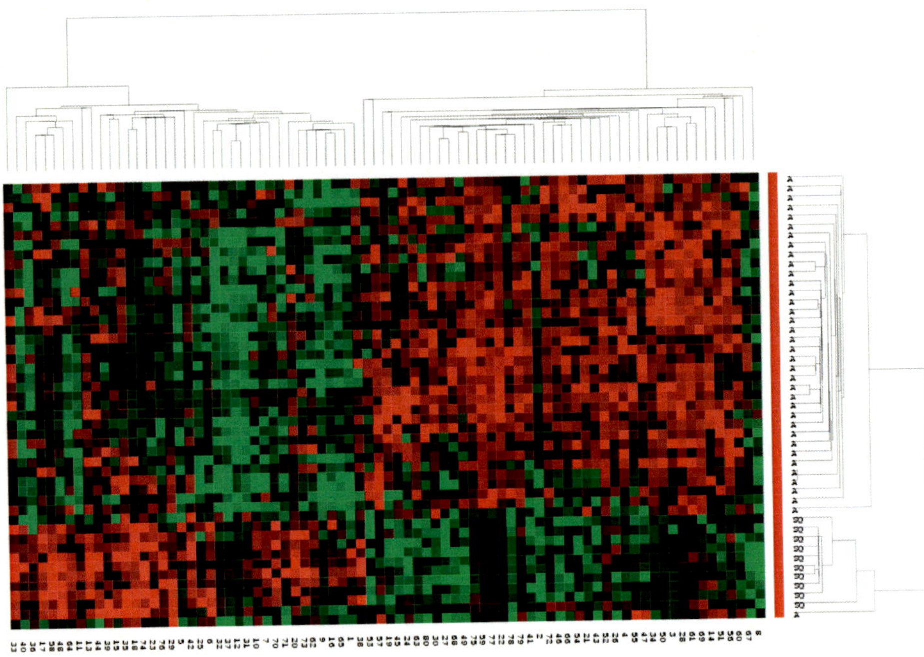

FIGURE 16.1: A simple dendrogram based on hierarchical clustering of rows and columns for gene expression data. The figure has been adapted from [39]. Here, rows correspond to the genes and columns correspond to the conditions. The color scale ranges from saturated green for log ratios -3.0 and below to saturated red for log ratios 3.0 and above.

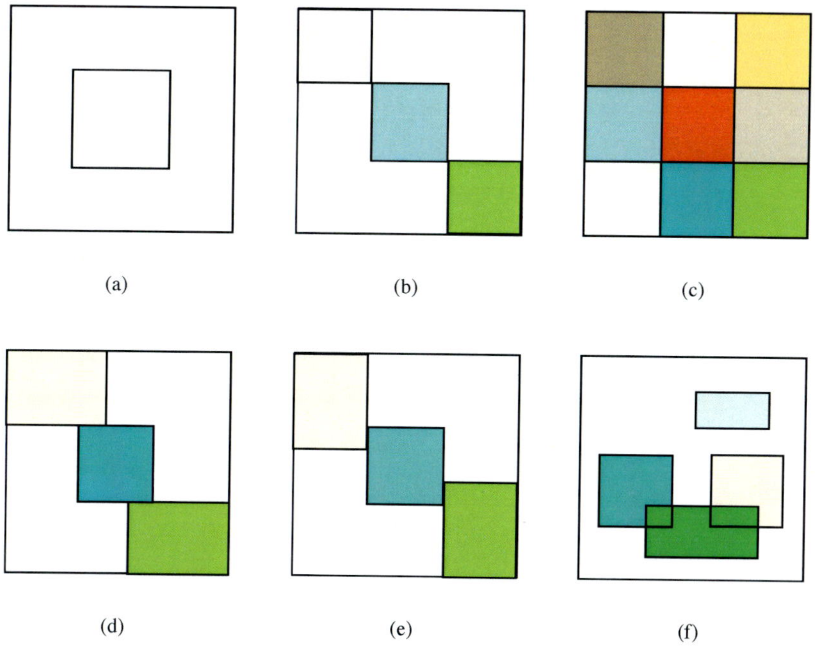

FIGURE 16.4: Examples of bicluster structures. (a) Single bicluster. (b) Exclusive row and column biclusters. (c) Checkerboard pattern biclusters. (d) Exclusive row biclusters. (e) Exclusive column biclusters. (f) Arbitrarily positioned overlapping biclusters.

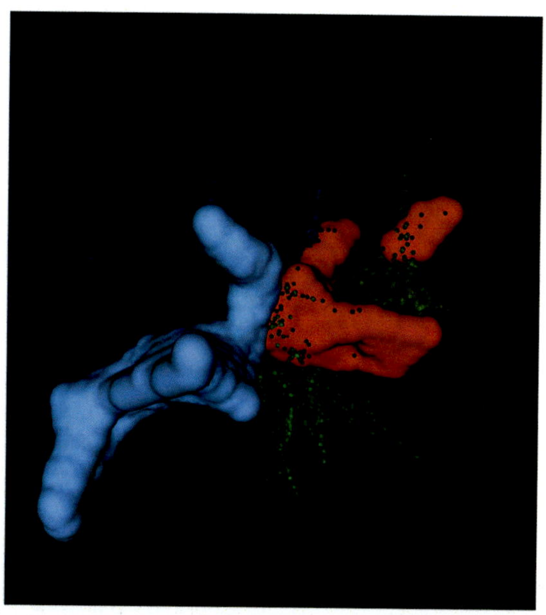

FIGURE 19.6: Two selected three-dimensional clusters shown with their hull. Note that the red cluster consists of two separate regions.

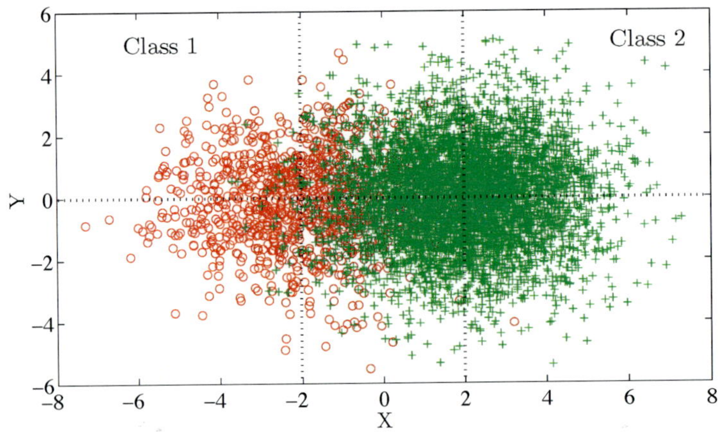

FIGURE 23.1: A Simulated Data Set ($n_1 = 1000$, $\sigma^2 = 2.5$).

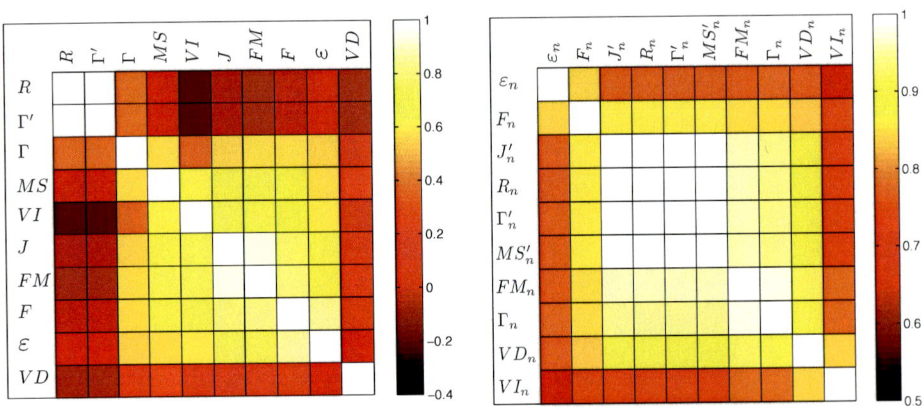

(a) Unnormalized Measures.

(b) Normalized Measures.

FIGURE 23.4: Correlations of the Measures.

Chapter 10

A Survey of Stream Clustering Algorithms

Charu C. Aggarwal
IBM T. J. Watson Research Center
Yorktown Heights, NY
charu@us.ibm.com

10.1 Introduction

In recent years, advances in hardware technology have allowed us to automatically record transactions and other pieces of information of everyday life at a rapid rate. Such processes generate huge amounts of online data which grow at an unlimited rate. These kinds of online data are referred to as *data streams*. The issues on management and analysis of data streams have been researched extensively in recent years because of their emerging, imminent, and broad applications [1].

Many important problems such as clustering and classification have been widely studied in the data mining community. The problem has been investigated classically in the context of a wide variety of methods such as the k-means, k-medians, density-based methods, and probabilistic clustering methods. A detailed discussion of different kinds of clustering methods may be found in [42, 44]. Most of the classical methods in the literature are not necessarily designed in the context of very large data sets and data streams. The stream scenario brings a unique set of challenges with it, which cannot be addressed by most of the classical methods proposed in [42, 44].

The specific challenges in the context of stream scenario are as follows:

- Streams typically have *massive volume*, and it is often not possible to store the data explicitly on disk. Therefore, the data needs to be processed in a *single pass*, in which all the summary information required for the clustering process needs to be stored and maintained. The time needed to process each record must be small and constant. Otherwise, the model construction process would never be able to catch up with the stream.

- The patterns in the data stream may continuously evolve over time [3]. From a stream mining perspective, this implies that the underlying cluster models need to be continuously updated. A usable model must be available at any time, because the end of stream computation may never be reached, and an analyst may require results at any point in time.

- Different domains of data may pose different challenges to data stream clustering. For example, in a *massive domain* of discrete attributes, it may not be possible to store summary representations of the clusters effectively without increasing the computational complexity of the problem significantly. Therefore, space-efficient methods need to be designed for massive-domain clustering of data streams.

Clearly, the issue of scalability is a primary one from the perspective of stream processing, but by no means is it the only issue. In the context of stream processing, temporal locality is also quite important, because the underlying patterns in the data may evolve, and therefore, the clusters in the past history may no longer remain relevant to the future. The issue of scalability also arises in the context of data clustering of *very large data sets* [31, 37, 51, 63]. In such cases, the major constraint is that the algorithm should require no more than one (or at least a small constant number of) pass(es) over the data. This is because I/O operations are traditionally quite expensive in databases, as compared to main memory operations. Therefore, the optimum performance is often achieved, when the underlying algorithm is designed to minimize the number of passes over the data, rather than minimize the number of CPU operations. In the context of data streams, temporal locality issues are also very important and should not be ignored in the clustering process. Therefore, a variety of stream clustering algorithms attempt to take such temporal issues into account with the use of snapshot-based methods, decay-based techniques, windowing, etc. We will make an effort to point out such techniques where they are used.

This chapter is organized as follows. In the next section, we will study clustering methods which are based on the k-means or the k-medians methodology. In Section 10.3, we will study density-based methods for stream clustering. Section 10.4 discusses probabilistic algorithms for clustering data streams. High dimensional streaming algorithms are introduced in Section 10.5. Methods for discrete and categorical stream clustering introduced in Section 10.6. Methods for clustering text streams are discussed in Section 10.7. Other scenarios for data stream clustering are discussed in Section 10.8. The conclusions and summary are presented in Section 10.9.

10.2 Methods Based on Partitioning Representatives

A number of methods for clustering are based on partitioning representatives. These include methods such as the k-means– and k-medians–based methods. In these techniques, the clusters are defined by a set of data points, which are drawn either directly from the data set or otherwise. The cluster membership of the remaining points are defined by assigning them to their closest representative. In the classical literature, such methods are iterative and require multiple passes over the data in order to estimate the representatives accurately. However, in the stream scenario, multiple passes are not required because a sufficient amount of data is available for estimating the representatives efficiently in even one pass of the data. In this section, we will first discuss the stream clustering methods, which are based on the partitioning methodology.

10.2.1 The STREAM Algorithm

The *STREAM* framework [36, 53] is based on the k-medians clustering methodology. The core idea is to break the stream into *chunks*, each of which is of manageable size and fits into main memory. Thus, for the original data stream $\mathcal{D}$, we divide it into chunks $\mathcal{D}_1 \ldots \mathcal{D}_r \ldots$, each of which contains at most m data points. The value of m is defined on the basis of a predefined memory budget.

Since each chunk fits in main memory, a variety of more complex clustering algorithms can be used for each chunk. The methods in [36, 53] use a variety of different k-medians–style algorithms for this purpose. The choice of subroutine is crucial to the quality of the underlying algorithm, and will be discussed in detail below. In k-medians algorithms, we pick a set $\mathcal{S}$ of k representatives from each chunk $\mathcal{D}_i$, so that each point in $\mathcal{D}_i$ is assigned to its closest representatives. The goal is to pick the representatives in such a way, so as to minimize the *Sum of Squared Error (SSE)* of the assigned data points from these representatives. For a set of m data points $\overline{X_1} \ldots \overline{X_m}$ in $\mathcal{S}$, and a set of k representatives $\mathcal{Y} = \overline{Y_1} \ldots \overline{Y_k}$, the objective function is defined as follows:

$$Objective(\mathcal{S}, \mathcal{Y}) = \sum_{\overline{X_i} \in \mathcal{S}, \overline{X_i} \Leftarrow \overline{Y_{j_i}}} dist(\overline{X_i}, \overline{Y_{j_i}}) \tag{10.1}$$

The assignment operator is denoted by $\Leftarrow$ in the above. The sum of squared distances between a pair of records is denoted by $dist(\cdot,\cdot)$.

After the first chunk has been processed, we now have a set of k medians, which are stored away. The number of points assigned to each representative is stored as a "weight" for that representative. Such representatives are considered *level-1* representatives. The next chunk is independently processed in order to find its k optimal median representatives. Thus, at the end of processing the second chunk, we will have $2 \cdot k$ such representatives. Thus, the memory requirement for storing the representatives also increases with time, and after processing r chunks, we will have a total of $r \cdot k$ representatives. When the number of representatives exceeds m, a second level of clustering is applied to this set of $r \cdot k$ points, except that the stored weights on the representatives are also used in the clustering process. The resulting representatives are stored as *level-2* representatives. In general, when the number of representatives of level-p reaches m, they are converted to k level-$(p+1)$ representatives. Thus, the process will result in increasing the number of representatives at all levels, though the number of representatives in higher levels will increase exponentially slower than the lower levels. At the end of processing the entire data stream (or when a specific need for the clustering result arises), all remaining representatives of different levels are clustered together in one final application of the k-medians subroutine.

Clearly, the choice of the particular algorithm which is used for the k-medians problem is critical in providing an effective solution. The other factor which impacts the quality of the final output is the effect of the problem decomposition into chunks followed by hierarchical clustering. How does such a problem decomposition affect the final quality of the output? It has been shown in [53], that the final quality of the output cannot be arbitrarily worse than the particular subroutine which is used at the intermediate stage for k-medians clustering.

Lemma 10.2.1 *Let the subroutine used for k-medians clustering in the* STREAM *algorithm have an approximation factor of c. Then, the* STREAM *algorithm will have an approximation factor of no worse than* $5 \cdot c$.

The work in [53] extends the original work in [36] by designing a more effective subroutine for the k-medians problem. This solution is based on the problem of facility location. We note that the *Lagrangian relaxation* of the clustering problem can be modeled as a facility location problem, wherein we relax the constraint on the number of representatives and incorporate it into the objective function with a Lagrangian multiplier. These representative are analogous to facilities in the context of the facility location problem, where the cost associated with a representative, can be physically visualized as the cost of "building" a facility, in order to service its assigned clients (data points). Thus, we add a cost λ (Lagrangian parameter) for each facility included. The cost of assignments is the same as previously. Thus, the new objective function in terms of the set of facilities $\mathcal{Y}$ may be expressed as follows:

$$Objective(\mathcal{S}, \mathcal{Y}) = \sum_{\overline{X_i} \in \mathcal{S}, \overline{X_i} \Leftarrow \overline{Y_{j_i}}} dist(\overline{X_i}, \overline{Y_{j_i}}) + \lambda \cdot |\mathcal{Y}| \tag{10.2}$$

Unlike the previous case, the cardinality of $\mathcal{Y}$ may not necessarily be k at the optimal value of the objective function, unless the value of the Lagrangian parameter (or facility cost) λ is picked appropriately. In order to use this approach to solve the k-medians subroutine, two steps are needed [53]:

- Given a particular value of λ, how do we determine the optimal objective function value?
- How do we determine the value of λ in such a way that the number of facilities in the optimal solution is k?

The latter step of the solution is easy. As long as the optimal number of facilities changes monotonically with λ, one can use binary search on λ in order to determine the appropriate value. It is shown in [53] that the optimal number of facilities indeed changes monotonically with k.

In order to solve the first part of the problem, a local search algorithm called *LSEARCH* is proposed. In this algorithm, we start off with a current set of open facilities O, and a particular assignment of data points to these facilities (which is not necessarily optimal). Then, we repeatedly examine a randomly selected facility x, which is currently not included in O, and compute the improvement in the objective function by adding x to O and closing other facilities. This improvement in objective function is referred to as *gain*(x). Certain rules of reassignment of data points to facilities and closing facilities are used for this purpose, as discussed below.

Specifically, any data point is allowed to be reassigned to x, when x is added to O (if there is an improvement), and it is also allowed to close a facility y and assign *all* of its points to x. The latter case would result in a gain, as long as the cost of the reassignment together to x is no greater than the savings from closing y. The cost of opening x is always subtracted from the overall gains. Thus, the value of $gain(x)$ may be positive or negative. The facility x is added to O, only if $gain(x) > 0$. This process is repeated over all facilities in random order, and the final set of open facilities and reassignments represents one possible locally optimal solution. This process would need to be repeated $\Omega(\log(m))$ times in order to provide guarantees on the underlying quality.

A major limitation of the *STREAM* algorithm is that it is not particularly sensitive to evolution in the underlying data stream. In many cases, the patterns in the underlying stream may evolve and change significantly. Therefore, it is critical for the clustering process to be able to adapt to such changes and provide insights over different time horizons. In this sense, the *CluStream* algorithm is able to provide significantly better insights at differing levels of temporal granularity.

10.2.2 CluStream: The Microclustering Framework

Since stream data naturally imposes a one-pass constraint on the design of the algorithms, it becomes more difficult to provide such a flexibility in computing clusters over different kinds of time horizons using conventional algorithms. For example, a direct extension of the stream-based k-means algorithm in [53] to such a case would require a simultaneous maintenance of the intermediate results of clustering algorithms over all possible time horizons. Such a computational burden increases with progression of the data stream and can rapidly become a bottleneck for online implementation. Furthermore, in many cases, an analyst may wish to determine the clusters at a previous moment in time, and compare them to the current clusters. This requires even greater bookkeeping and can rapidly become unwieldy for fast data streams.

Therefore, a natural design to stream clustering would separate out the process into an online microclustering component and an offline macroclustering component. The online micro-clustering component requires a very efficient process for storage of appropriate summary statistics in a fast data stream. The offline component uses these summary statistics in conjunction with other user input in order to provide the user with a quick understanding of the clusters whenever required.

It is assumed that the data stream consists of a set of multidimensional records $\overline{X}_1 \ldots \overline{X}_k \ldots$ arriving at time stamps $T_1 \ldots T_k \ldots$. Each $\overline{X_i}$ is a multidimensional record containing d dimensions which are denoted by $\overline{X_i} = (x_i^1 \ldots x_i^d)$.

The micro-clustering framework is designed to capture summary information about the data stream, in order to facilitate clustering and analysis over different time horizons. This summary information is defined by the following structures:

- **Microclusters:** We maintain statistical information about the data locality in terms of microclusters. These microclusters are defined as a temporal extension of the *cluster feature vector* [63]. The additivity property of the microclusters makes them a natural choice for the data stream problem.

- **Pyramidal Time Frame:** The microclusters are stored at snapshots in time which follow a pyramidal pattern. This pattern provides an effective trade-off between the storage requirements and the ability to recall summary statistics from different time horizons.

The summary information in the microclusters is used by an offline component which is dependent upon a wide variety of user inputs such as the time horizon or the granularity of clustering. We will first begin by defining the concept of microclusters and pyramidal time frame more precisely.

10.2.2.1 Microcluster Definition

Microclusters are defined as follows.

Definition 10.2.1 *A microcluster for a set of d-dimensional points $X_{i_1} \ldots X_{i_n}$ with time stamps $T_{i_1} \ldots T_{i_n}$ is the $(2 \cdot d + 3)$ tuple $(\overline{CF2^x}, \overline{CF1^x}, CF2^t, CF1^t, n)$, wherein $\overline{CF2^x}$ and $\overline{CF1^x}$ each correspond to a vector of d entries. The definition of each of these entries is as follows:*

- *For each dimension, the sum of the squares of the data values is maintained in $\overline{CF2^x}$. Thus, $\overline{CF2^x}$ contains d values. The p-th entry of $\overline{CF2^x}$ is equal to $\sum_{j=1}^{n} (x_{i_j}^p)^2$.*

- *For each dimension, the sum of the data values is maintained in $\overline{CF1^x}$. Thus, $\overline{CF1^x}$ contains d values. The p-th entry of $\overline{CF1^x}$ is equal to $\sum_{j=1}^{n} x_{i_j}^p$.*

- *The sum of the squares of the time stamps $T_{i_1} \dots T_{i_n}$ is maintained in $CF2^t$.*
- *The sum of the time stamps $T_{i_1} \dots T_{i_n}$ is maintained in $CF1^t$.*
- *The number of data points is maintained in n.*

The data stream clustering algorithm proposed in [5] can generate approximate clusters in any user-specified length of history from the current instant. This is achieved by storing the microclusters at particular moments in the stream which are referred to as *snapshots*. At the same time, the current snapshot of microclusters is always maintained by the algorithm. Consider for example, the case when the current clock time is t_c and the user wishes to find clusters in the stream based on a history of length h. Then, the macroclustering algorithm will use some of the additive properties of the microclusters stored at snapshots t_c and $(t_c - h)$ in order to find the higher level clusters in a history or *time horizon* of length h. Of course, since it is not possible to store the snapshots at each and every moment in time, it is important to choose particular instants of time at which it is possible to store the state of the microclusters so that clusters in any user specified time horizon $(t_c - h, t_c)$ can be approximated. This was achieved in [5] with the use of the concept of a pyramidal time frame.

10.2.2.2 Pyramidal Time Frame

In this technique, the snapshots are stored at differing levels of granularity depending upon the recency. Snapshots are classified into different *orders* which can vary from 1 to $\log(T)$, where T is the clock time elapsed since the beginning of the stream. The order of a particular class of snapshots defines the level of granularity in the time at which the snapshots are maintained. The snapshots of different order are maintained as follows:

- Snapshots of the i-th order occur at time intervals of α^i, where α is an integer and $\alpha \geq 1$. Specifically, each snapshot of the i-th order is taken at a moment in time when the clock value is exactly divisible by α^i.
- At any given moment in time, only the last $\alpha^l + 1$ snapshots of order i are stored.

The above definition allows for considerable redundancy in storage of snapshots. For example, the clock time of 8 is divisible by 2^0, 2^1, 2^2, and 2^3 (where $\alpha = 2$). Therefore, the state of the microclusters at a clock time of 8 simultaneously corresponds to order 0, order 1, order 2, and order 3 snapshots. From an implementation point of view, a snapshot needs to be maintained only once. The following observations are true:

- For a data stream, the maximum order of any snapshot stored at T time units since the beginning of the stream mining process is $\log_\alpha(T)$.
- For a data stream the maximum number of snapshots maintained at T time units since the beginning of the stream mining process is $(\alpha^l + 1) \cdot \log_\alpha(T)$.
- For any user specified time window of h, at least one stored snapshot can be found within $(1 + 1/\alpha^{l-1})$ units of the current time.

While the first two results are quite easy to see, the last one needs to be proven formally [5]. We summarize this result as follows:

Lemma 10.2.2 *Let h be a user-specified time horizon, t_c be the current time, and t_s be the time of the last stored snapshot of any order just before the time $t_c - h$. Then $t_c - t_s \leq (1 + 1/\alpha^{l-1}) \cdot h$.*

Order of Snapshots	Clock Times (Last 5 Snapshots)
0	55 54 53 52 51
1	54 52 50 48 46
2	52 48 44 40 36
3	48 40 32 24 16
4	48 32 16
5	32

TABLE 10.1: An example of snapshots stored for $\alpha = 2$ and $l = 2$

Proof: See [5].

For larger values of l, the time horizon can be approximated as closely as desired. For example, by choosing $l = 10$, it is possible to approximate any time horizon within 0.2%, while a total of only $(2^{10}+1)\cdot\log_2(100*365*24*60*60) \approx 32343$ snapshots are required for 100 years. Since historical snapshots can be stored on disk and only the current snapshot needs to be maintained in main memory, this requirement is quite feasible from a practical point of view. It is also possible to specify the pyramidal time window in accordance with user preferences corresponding to particular moments in time such as beginning of calendar years, months, and days.

In order to clarify the way in which snapshots are stored, let us consider the case when the stream has been running starting at a clock time of 1, and a use of $\alpha = 2$ and $l = 2$. Therefore $2^2 + 1 = 5$ snapshots of each order are stored. Then, at a clock time of 55, snapshots at the clock times illustrated in Table 10.1 are stored.

We note that a large number of snapshots are common among different orders. From an implementation point of view, the states of the microclusters at times of 16, 24, 32, 36, 40, 44, 46, 48, 50, 51, 52, 53, 54, and 55 are stored. It is easy to see that for more recent clock times, there is less distance between successive snapshots (better granularity). We also note that the storage requirements estimated in this section do not take this redundancy into account. Therefore, the requirements which have been presented so far are actually worst-case requirements.

10.2.2.3 Online Clustering with CluStream

The microclustering phase is the online statistical data collection portion of the algorithm. The aim is to maintain statistics at a sufficiently high level of (temporal and spatial) granularity so that it can be effectively used by the offline components such as horizon-specific macroclustering as well as evolution analysis. The algorithm works in an iterative fashion, by always maintaining a current set of microclusters. It is assumed that a total of q microclusters are stored at any moment by the algorithm. We will denote these microclusters by $\mathcal{M}_1 \ldots \mathcal{M}_q$. Associated with each microcluster i, we create a unique *id* whenever it is first created. If two microclusters are merged (as will become evident from the details of our maintenance algorithm), a *list* of *id*s is created in order to identify the constituent microclusters. The value of q is determined by the amount of main memory available in order to store the microclusters. Therefore, typical values of q are significantly larger than the natural number of clusters in the data but are also significantly smaller than the number of data points arriving in a long period of time for a massive data stream. These microclusters represent the current snapshot of clusters which change over the course of the stream as new points arrive. Their status is stored away on disk whenever the clock time is divisible by α^i for any integer i. At the same time any microclusters of order r which were stored at a time in the past more remote than α^{l+r} units are deleted by the algorithm.

Whenever a new data point $\overline{X_{i_k}}$ arrives, the microclusters are updated in order to reflect the changes. Each data point needs to be either absorbed by a microcluster or put in a cluster of its own. The first preference is to absorb the data point into a currently existing microcluster. The distance of each data point to the microcluster centroids $\mathcal{M}_1 \dots \mathcal{M}_q$ is determined. The distance value of the data point $\overline{X_{i_k}}$ to the centroid of the microcluster $\mathcal{M}_j$ is denoted by $dist(\mathcal{M}_j, \overline{X_{i_k}})$. Since the centroid of the microcluster is available in the cluster feature vector, this value can be computed relatively easily. This distance is used to compute the distance of the cluster $\mathcal{M}_p$ to the data point $\overline{X_{i_k}}$. In many cases, the point $\overline{X_{i_k}}$ does not naturally belong to the cluster $\mathcal{M}_p$. These cases are as follows:

- The data point $\overline{X_{i_k}}$ corresponds to an outlier.
- The data point $\overline{X_{i_k}}$ corresponds to the beginning of a new cluster because of evolution of the data stream.

While the two cases above cannot be distinguished until more data points arrive, the data point $\overline{X_{i_k}}$ needs to be assigned a (new) microcluster of its own with a unique *id*. In order to make this decision, the cluster feature vector of $\mathcal{M}_p$ is used to decide if this data point falls within the *maximum boundary* of the microcluster $\mathcal{M}_p$. If so, then the data point $\overline{X_{i_k}}$ is added to the microcluster $\mathcal{M}_p$ using the CF additivity property. The maximum boundary of the microcluster $\mathcal{M}_p$ is defined as a factor of t of the RMS deviation of the data points in $\mathcal{M}_p$ from the centroid. We define this as the *maximal boundary factor*. We note that the RMS deviation can be defined only for a cluster with more than one point. For a cluster with only one previous point, the maximum boundary is defined in a heuristic way. Specifically, it is chosen to be r times that of the next closest cluster.

If the data point does not lie within the maximum boundary of the nearest microcluster, then a new microcluster must be created containing the data point X_{i_k}. This newly created microcluster is assigned a new id which can identify it uniquely at any future stage of the data steam process. However, in order to create this new microcluster, the number of other clusters must be reduced by one in order to create memory space. This can be achieved by either deleting an old cluster or joining two of the old clusters. Our maintenance algorithm first determines if it is safe to delete any of the current microclusters as outliers. If not, then a merge of two microclusters is initiated.

The first step is to identify if any of the old microclusters are possibly outliers which can be safely deleted by the algorithm. While it might be tempting to simply pick the microcluster with the fewest number of points as the microcluster to be deleted, this may often lead to misleading results. In many cases, a given microcluster might correspond to a point of considerable cluster presence in the past history of the stream, but may no longer be an active cluster in the recent stream activity. Such a microcluster can be considered an outlier from the current point of view. An ideal goal would be to estimate the average timestamp of the last m arrivals in each microcluster, and delete the microcluster with the least recent timestamp. While the above estimation can be achieved by simply storing the last m points in each microcluster, this increases the memory requirements of a microcluster by a factor of m. Such a requirement reduces the number of microclusters that can be stored by the available memory and therefore reduces the effectiveness of the algorithm.

It is also necessary to approximate the average timestamp of the last m data points of the cluster $\mathcal{M}$. This is achieved by using the data about the timestamps stored in the microcluster $\mathcal{M}$. We note that the timestamp data allows the calculation of the mean and standard deviation[1] of the arrival times of points in a given microcluster $\mathcal{M}$. Let these values be denoted by $\mu\mathcal{M}$ and $\sigma\mathcal{M}$ respectively. Then, the time of arrival of the $m/(2 \cdot n)$-th percentile of the points in $\mathcal{M}$ is computed under the assumption that the timestamps are normally distributed. This timestamp is used as the approximate value of the recency. We shall call this value the *relevance stamp* of cluster $\mathcal{M}$. When the smallest such stamp of any microcluster is below a user-defined threshold δ, it can be eliminated

[1] The mean is equal to $CF1^t/n$. The standard deviation is equal to $\sqrt{CF2^t/n - (CF1^t/n)^2}$.

and a new microcluster can be created with a unique *id* corresponding to the newly arrived data point $\overline{X_{i_k}}$.

In some cases, none of the microclusters can be readily eliminated. This happens when all relevance stamps are sufficiently recent and lie above the user-defined threshold δ. In such a case, the closest microclusters are merged. The new microcluster no longer corresponds to one *id*. Instead, an *idlist* is created which is a union of the the *id*s in the individual microclusters. Thus, any microcluster which is the result of one or more merging operations can be identified in terms of the individual microclusters merged into it.

While the above process of updating is executed at the arrival of each data point, an additional process is executed at each clock time which is divisible by α^i for any integer i. At each such time, the current set of microclusters is stored on disk, together with its id list, and indexed by its time of storage. The least recent snapshot of order i is deleted, if $\alpha^l + 1$ snapshots of such order have already been stored on disk, and if the clock time for this snapshot is not divisible by α^{i+1}. In the latter case, the snapshot continues to be a viable snapshot of order $(i+1)$. These microclusters can then be used to form higher level clusters or an evolution analysis of the data stream.

It should be pointed out that the microclustering model can be used in conjunction with fast indexing structures in order to allow *anytime* stream mining. This is particularly important in the context of data streams, since the stream speed is not known on an a priori basis. A particular approach along this direction is the *ClusTree* method [46], which allows the adaptation of the granularity of the cluster model to the stream speed. The broader principle is that it is possible to follow the anytime paradigm to spend as much (or as little) time as dynamically available to digest new events.

While the use of snapshots is a natural way for examining the evolving stream at a variety of different granularities, other methods are possible for capturing the evolution of the data stream by incorporating decay into the microclusters [6], or by using sliding windows in conjunction with an exponential histogram of the temporal cluster feature vector [65]. These methods are generally preferable, if the level of evolution in the data stream is known in advance. These different methods have differing trade-offs between memory requirements and flexibility, but their goals are similar in terms of capturing the evolution of the data stream.

10.3 Density-Based Stream Clustering

Density-based methods [18, 30] construct a density profile of the data for clustering purposes. Typically, kernel density estimation methods [58] are used in order to construct a smooth density profile of the underlying data. Subsequently, the data are separated out into density-connected regions. These density connected regions may be of different shapes and sizes. One of the advantages of density-based algorithms is that an implicit shape is not assumed for the clusters. For example, when Euclidian distance functions are used, it is always assumed that the clusters have spherical shapes. Similarly, the Manhattan metric assumes that the clusters are of a diamond shape. In density-based clustering, connected regions of high density may often have arbitrary shapes. Another aspect of density-based clustering is that it does not predecide the number of clusters. Rather, a threshold on the density is used in order to determine the connected regions. Of course, this changes the nature of the parameter which needs to be presented to the algorithm (density threshold instead of number of clusters), but it does not necessarily make the approach parameter-free.

The main challenge in the stream scenario is to construct density-based algorithms which can be efficiently executed in a single pass of the data, since the process of density estimation may be computationally intensive. There are two broad classes of techniques:

- The first class of techniques extends the microclustering technique to this case, by relaxing the constraint on the number of microclusters, and imposing a constraint on the radius and "weight" of each microcluster. The dense regions are generated by connecting together the dense microclusters which satisfy a condition on connectivity similar to that in [30].

- The second class of techniques divides the data space into grids and then determines the dense grids. The dense regions in the data are reconstructed by piecing together the connected dense grids.

We will discuss both these classes of techniques in this section.

10.3.1 DenStream: Density-Based Microclustering

The *DenStream* algorithm [24] approach combines microclustering with a density-estimation process for effective clustering. The first step is to define a *core object*, which is defined as an object, in the ε-neighborhood of which the weight of the data points is at least μ. A *density area* is defined as the union of the ε neighborhoods of the core objects.

In the context of streaming data, it is difficult to determine these dense regions naturally. Therefore, dense regions of smaller granularity are defined in the form of core microclusters. A core microcluster is defined a set of data points with weight at least μ and for which the radius of the microcluster about its center is less than ε. We note that the weight of a data point is based on a decay weighted function of the time that it last arrived. Therefore, if δt is the time since a data point arrived, its weight is given by:

$$f(\delta t) = 2^{-\delta t} \tag{10.3}$$

Since the radius of the microcluster is constrained to be less than ε, it implies that the number of microclusters is much larger than the number of natural clusters in the data for small values of ε. At the same time, the number of core microclusters is much smaller than the number of points in the data stream, since each cluster contains a weight of at least μ. We note that the key difference from the standard microclustering definition is that the number of microclusters is not constrained, though the radius of each microcluster is constrained. Thus, this approach approaches a different parameter set to the underlying application. The core microcluster is also referred to as a c-microcluster.

One immediate observation is that when a microcluster is first formed by such an algorithm, it is unlikely to contain the requisite weight of data points required to be defined as a microcluster. Therefore, the concepts of *potential core microcluster* and *outlier microcluster* are defined in [24]. In the former case, the microcluster contains a weight of at least $\beta \cdot \mu$ (for some $\beta \in (0,1)$), and in the latter case, it contains a weight less than $\beta \cdot \mu$. Furthermore, since the weights of points decay over time, a cluster may also change from being a p-microcluster to an o-microcluster, if sufficient data points are not added to it in order to compensate for the decay. Thus, during its lifecycle, a microcluster may move from being an outlier microcluster to being a potential core microcluster, and finally to the stage of being a core microcluster. These two kinds of microclusters are referred to as p-microclusters and o-microclusters respectively.

When a new data point arrives, the following can be the possibilities in terms of what is done with it:

- The first step is to try to insert it into a p-microcluster, as long as it is possible to do so, without violating the radius constraint.

- If the first step is not possible, the next step is to try to insert it into an o-microcluster, as long as this can be done without violating the radius constraint.

- If the first and second steps are not possible, then a new o-microcluster is created containing this data point.

One challenge of this approach is that the number of o-microclusters will increase with time, as new o-microclusters are created, and some of the p-microclusters decay back to o-microclusters. Therefore, from time to time, we purge some of the o-microclusters, which will allow potential for becoming p-microclusters. The larger the time δt that has elapsed since the creation of an o-microcluster, the larger its weight is expected to be. Therefore, every T_p time periods, we prune all those microclusters, whose weight is less than the threshold $\psi(\delta t)$, where:

$$\psi(\delta t) = \frac{2^{-\lambda \cdot (\delta t + T_p)} - 1}{2^{-\lambda \cdot (T_p)} - 1} \tag{10.4}$$

This process continues in order to maintain the microclusters dynamically. We note that the individual microclusters can be reconstructed into density-connected regions in order to create the final set of clusters of arbitrary shape. The overall approach for creating the clusters of arbitrary shape is discussed in detail in in [24].

10.3.2 Grid-Based Streaming Algorithms

Grid-based methods are a class of density-based streaming algorithms, in which a grid structure is used in order to quantify the density at each point in the data. The core idea is that the data is discretized into ranges along each dimension, and this also results in dividing the data into cells along different dimensions. The number of points in each cell defines the density of that cell. Then, the dense cells in the data can be aggregated in order to determine the dense regions for the clustering.

10.3.2.1 D-Stream Algorithm

A method called D-Stream for real-time density based clustering of streaming data was proposed in [25]. The algorithm has many similarities with [30] in terms of trying to determine fine-grained regions of high density. The main difference at the conceptual level is that this is done with the use of grids rather than microclusters. As in the previous case, a decay function $f(\delta t(\overline{X}))$ is used to denote the weight of a point $\overline{X}$ since the time of its arrival $\delta t(\overline{X}, t_c)$ units ago from the current time t_c:

$$f(\delta t(\overline{X}, t_c)) = \mu^{-\delta t(\overline{X}, t_c)} \tag{10.5}$$

Here, we assume that $\mu > 1$, which is slightly different from the notations in [25]. We note that this decay function is identical to that proposed in [6, 30], by defining the relationship with respect to the parameter λ and assuming that $\mu = 2^{\lambda}$. Under the assumption of [25] that exactly one record arrives at each timestamp, it can be shown that the sum of the weights of all data points is no larger than $\mu/(\mu - 1)$.

The grids are defined by discretizing these ranges along each dimension. For the i-th dimension, the discretization is performed into p_i different ranges along the ith dimension. This discretization naturally defines a total of $\eta = \prod_i p_i$ d-dimensional cells. For each cell S, its weight $W(S, t_c)$ is defined as the current time t_c as follows:

$$W(S, t_c) = \sum_{\overline{X} \in S} f(\delta t(\overline{X}), t_c) \tag{10.6}$$

We note that the grid is essentially analogous to the radius constrained microcluster defined in the *DenStream* algorithm. Thus, as in the case of *DenStream*, the density of a grid is constantly changing over time. However, it is never necessary to update any decay-based statistics either in grids or microclusters in each instance [6, 25]. This is because all grids decay at the same proportional rate, and the update can be lazily performed only when the density value in the grid is updated with the addition of a new data point.

The next step is to define what it means for a grid to be dense. Since the total density over all grids is no larger than $\mu/(\mu-1)$, it follows that the *average* density of each grid is no larger than $\mu/(\eta\cdot(\mu-1))$. Therefore, a grid is defined as *dense*, when its density is a constant times larger than this factor. This is essentially analogous to the concept of a c-microcluster defined in [30]. Analogous to the concept of an o-microcluster, and a p-microcluster, the work in [25] divides the nondense grid-cells into *sparse grid cells* and *transitional grid cells*, with the use of a smaller threshold on the density. Grid cells can change between the different states of being sparse, transitional, or dense, both because of the addition of new data points and also because of decay.

One observation is that the number of grid-cells $\eta=\prod_{i=1}^{d}p_i$ is exponentially dependent upon the dimensionality d. However, in practice, most of these grid-cells are empty, and the information for empty grids need not be stored. This may not be sufficient in many real applications, where many outlier points may sporadically appear in the data. The work in [25] designs methods for identifying and removing such sporadic grids from the data. The maintained dense grids can then be consolidated into larger dense regions in the data. As in the case of [30], this is defined in the form of density-connected grids, where adjacency of two dense grids is treated as a density-connection. For the precise details of the dense region construction, we refer the reader to [25]. Thus, it is evident that grid-based and microcluster-based density clusters share a number of conceptual similarities at various stages of the algorithm.

One weakness of the approach is that a significant number of nonempty grid cells need to be discarded in order to keep the memory requirements in check. In many cases, such grid-cells occur at the borders of the clusters. The discarding of such cells may lead to degradation in cluster quality. Therefore, a method has been proposed in [43] to design a variation of the *D-Stream* method (known as *DD-Stream*), which includes the data points at the borders into adjacent denser grids, which are retained by the algorithm. It has been shown that such an approach leads to some improvements in cluster quality.

10.3.2.2 Other Grid-Based Algorithms

The method in [35] updates a full-dimensional grid of the incoming data stream. The clusters are discovered from the updated density values in this grid. At any given time, the density values in the grid can be used in order to determine the updated and most effective clusters.

An important point to be kept in mind is that grid-based algorithms require the discretization of the data along the different dimensions in order to create the grid cells. The granularity of the discretization is a critical design choice at the very beginning of the algorithm. Unfortunately, at the beginning of the stream arrival, very little may be known about how the underlying data points are distributed, and therefore, it is difficult to choose the level of discretization along each dimension in an optimal way. Furthermore, the appropriate level of discretization for each data locality may be different, and a single global level of discretization may be suboptimal over different data localities.

Therefore, the work in [56] uses a dynamic approach called *Statsgrid* for the discretization process, wherein the data cells are recursively partitioned based on their local density. The algorithm starts off with cells of equal size. As data points are added to the cells, and the number of points in a cell becomes sufficiently large, the algorithm partitions the cell into two along one of the dimensions. This process can of course be repeated recursively each time any of the children cells become dense. At some point, the maximum level of allowed granularity is reached, and such a cell is called a *unit cell*, which cannot be further divided. We note that this approach naturally leads to a hierarchical clustering of the data, which can be very useful in many applications. Nevertheless, the work does not use temporal decay and, therefore, does not adjust very well to an evolving data stream.

This method has therefore been extended to the *CellTree* method [55], which allows decay in the statistics. It explicitly uses a *CellTree* data structure in order to maintain the hierarchical relationships among the grid cells. Furthermore, when the data cells decay with time, it may be possible to merge adjacent cells. Therefore, the method in [55] provides greater flexibility, than discussed in the original algorithm.

10.4 Probabilistic Streaming Algorithms

One of the common methods for probabilistic clustering is that of *mixture modeling*, in which the data is assumed to be generated by a mixture of known distributions such as the Gaussian distribution. The parameters of this distribution are then typically learned with an EM algorithm from the actual data records [59]. The main argument in [59] is that probability density-based clustering algorithms are much more efficient than applying EM on the entire data set. On the other hand, it has been shown that it is possible to use the EM-algorithm in order to provide an efficient update process for newly arriving data. Nevertheless, since the EM algorithm requires one to learn a large number of parameters, such an approach is unlikely to be effective when the underlying data is evolving rapidly.

Another area in which probabilistic models are commonly used for clustering is for the case of text. A common technique which is used to create a soft clustering of the data for the case of text is *topic modeling* [41]. In this technique, soft clusters are associated with the data in which words and documents are probabilistically assigned to the different partitions. Since the topic modeling approach uses an EM algorithm, it can sometimes be slow in practice. Therefore, a method has been proposed in [20] for topic modeling over text streams. The work in [20] proposes online variants of three common batch algorithms for topic modeling. These correspond to the Latent Dirichlet Allocation (LDA) [22], Dirichlet Compound Multinomial (DCM) mixtures [29] and von-Mises Fisher (vMF) mixture models [21]. It is shown in [20] that the online variant of the vMF approach provides the best results. A detailed study of these topic modeling algorithms is beyond the scope of this survey. Interested readers are referred to [20].

10.5 Clustering High-Dimensional Streams

In many circumstances, data stream are very high-dimensional because a large number of features are available for the mining process. The high-dimensional case presents a special challenge to clustering algorithms even in the traditional domain of static data sets. This is due to the sparsity of the data in the high-dimensional case. In high-dimensional space, all pairs of points tend to be almost equidistant from one another. As a result, it is often unrealistic to define distance-based clusters in a meaningful way. Some recent work on high-dimensional data uses techniques for *projected clustering* which can determine clusters for a specific subset of dimensions [10, 17]. In these methods, the definitions of the clusters are such that each cluster is specific to a particular group of dimensions. This alleviates the sparsity problem in high-dimensional space to some extent. Even though a cluster may not be meaningfully defined on all the dimensions because of the sparsity of the data, some subset of the dimensions can always be found on which particular subsets of points form high quality and meaningful clusters. Of course, these subsets of dimensions may vary over the different clusters. Such clusters are referred to as *projected clusters* [10].

10.5.1 The HPSTREAM Method

The microclustering method can also be extended to the case of high-dimensional projected stream clustering. The algorithm is referred to as *HPSTREAM*. In [6, 7], methods have been proposed for high-dimensional projected clustering of data streams. The basic idea is to use an (incremental) algorithm in which we associate a set of dimensions with each cluster. This corresponds to the standard microclustering method as discussed in [6], with the main difference that the distances from data points to clusters are computed on the basis of dimension-specific clusters. Therefore, additional information needs to be associated with a cluster in order to capture the set of dimensions associated with it. The set of dimensions is represented as a d-dimensional bit vector $\mathcal{B}(C_i)$ for each cluster structure in $\mathcal{FCS}$. This bit vector contains a 1 bit for each dimension which is included in cluster C_i. In addition, the maximum number of clusters k and the average cluster dimensionality l is used as an input parameter. The average cluster dimensionality l represents the average number of dimensions used in the cluster projection. An iterative approach is used in which the dimensions are used to update the clusters and vice-versa. The structure in $\mathcal{FCS}$ uses a decay-based mechanism in order to adjust for evolution in the underlying data stream. For a data point, which arrived δt units ago, its weight is assumed to be $2^{-\lambda \cdot \delta t}$, where λ is the decay rate.

Therefore, the microclusters for the *HPSTREAM* algorithm contain decay-based statistics, wherein the microclusters are similar to the *CluStream* algorithm. Furthermore, the overall framework of the *HPSTREAM* algorithm is quite similar, because data points are assigned to microclusters on the basis of their projected distances. The main difference is that each component of the additive microcluster statistics is a decay-based weight. In addition, the bit vector corresponding to the choice of dimensions is stored with the microcluster statistics. Clearly, a number of changes need to be incorporated into the *CluStream* approach in order to account for these changes:

- We note that the decay-based statistics ensure that the weights of the microclusters change in each timestamp. However, it is not necessary to update the statistics at each timestamp, since all the microclusters decay at the same rate. Rather, we perform the update only when a new point is inserted into the data. When a new point is inserted, and δx is the time interval since the last time, then all microcluster statistics are multiplied by $2^{-\lambda \cdot \delta x}$ before adding a data point to the microcluster statistics.

- The average distance to each cluster is now computed on the basis of the projected dimensions specific to that cluster. The bit vector in the microcluster statistics is used in order to decide on the exact set of dimensions to use.

- The projected dimensions in the different clusters are updated periodically, so that the most compact dimensions are retained for each cluster. The corresponding bit vector in the microcluster statistics is updated.

It has been shown in [6], that the incorporation of projections can significantly improve the effectiveness of the approach. Details are discussed in [6].

10.5.2 Other High-Dimensional Streaming Algorithms

A variety of other high-dimensional streaming clustering algorithms have been proposed after the first *HPSTREAM* framework. In particular, a grid-based algorithm was proposed in [49] for high-dimensional projected clustering of data streams. High-dimensional projected clustering has also been applied to other domains such as uncertain data. For example, methods for high- dimensional projected clustering of uncertain data streams have been proposed in [4, 60].

Most of the methods discussed in the literature use a k-means-type approach, which fixes the number of clusters in the data. Furthermore, a k-means-type approach also makes implicit assumptions about the *shapes* of the underlying clusters. The work in [52] proposes a density-based method for high-dimensional stream clustering. Such an approach has the virtue of recognizing the fact that the number and shape of the clusters in the stream may vary over time. The work in [52] proposes HDDSTREAM, which is a method for high-dimensional stream clustering. It generalizes the density-based approach proposed in [30] in order to incorporate subspace analysis in the clustering process. Since the method in [30] was originally designed to handle variations in the number of clusters in the stream, as well as different shapes of clusters, these virtues are inherited by the HDDSTREAM method of [52] as well.

A number of methods have also been proposed for high-dimensional projected clustering of dynamic data [47, 62]. While these methods can be effectively used for dynamic data, they do require access to the raw data for clustering purposes. Therefore, these methods are not streaming techniques in the strict sense.

10.6 Clustering Discrete and Categorical Streams

Many data streams are defined on a domain of *discrete values* in which the attributes are unordered and take on one of a very large number of possible discrete values. In such cases, the cluster feature vector of the microclustering approach does not represent the values in the underlying data well. Therefore, methods are required in order to perform stream clustering algorithms in such scenarios. The simplest case of categorical stream clustering is that of *binary data* in which the data take on values from $\{0,1\}$. Binary data can be considered both quantitative and symbolic, and therefore, almost all stream algorithms can be used directly for this case.

10.6.1 Clustering Binary Data Streams with k-Means

The simplest case of categorical data is binary data, in which the discrete attributes may take on only one of two possible values. Binary data is also a special case of quantitative data, because an ordering can always be assigned to discrete values as long as there are only two of them. Therefore, virtually all of the streaming algorithms can be used for binary data. Nevertheless, it can sometimes be useful to leverage a method which is specifically designed for the case of binary data.

An algorithm for utilizing optimizations of k-means algorithms for data streams is proposed in [54]. The main observation in [54] is that the binary transactions are often *sparse*. This can be used in order to greatly speed up the distance computations. Since distance computations form the bottleneck operation for such algorithms, a speedup of the distance computations also greatly speeds up the underlying algorithm. From a practical point of view, this means that only a small fraction of the features in the transaction take on the 1-value. Therefore, the general approach used in [54] is that first the distance of the null transaction to each of the centroids is computed. Subsequently, for each position in the transaction, the effect of that position on the distances is computed. Since many distance functions are separable functions across different dimensions, this can be achieved quite effectively. It has been shown in [54] that these speedups can be implemented very effectively at no reduction in quality of the underlying results.

10.6.2 The StreamCluCD Algorithm

The *Squeezer* algorithm is a one-pass algorithm for clustering categorical data [39]. The *StreamCluCD* approach [40] is the streaming extension of this framework. The essential idea behind the algorithm is that when a data point comes in, it is placed into a cluster of its own. For subsequent incoming points, we compute their similarity to the existing clusters in the data. If the incoming points are sufficiently similar to one of the existing clusters, then they are placed in that cluster. Otherwise, the incoming data point is placed in a cluster of its own. A key operation in the *StreamCluCD* algorithm is to maintain the frequency counts of the attribute values in each cluster. The lossy counting approach introduced in [50] is used for this purpose. The motivation for this is to reduce the memory footprint for maintaining the frequency counts. The issue of memory requirements becomes especially important when the number of possible discrete values of each attribute increases. This is referred to as the *massive-domain* scenario and will be discussed in the next section.

10.6.3 Massive-Domain Clustering

Massive-domains are those data domains in which the number of possible values for one or more attributes is very large. Examples of such domains are as follows:

- In network applications, many attributes such as IP-addresses are drawn over millions of possibilities. In a multidimensional application, this problem is further magnified because of the multiplication of possibilities over different attributes.
- Typical credit-card transactions can be drawn from a universe of millions of different possibilities depending upon the nature of the transactions.
- Supermarket transactions are often drawn from a universe of millions of possibilities. In such cases, the determination of patterns which indicate different kinds of classification behavior may become infeasible from a space- and computational-efficiency perspective.

The massive domain size of the underlying data restricts the *computational approach* which may be used for discriminatory analysis. Thus, this problem is significantly more difficult than the standard clustering problem in data streams. Space-efficiency is a special concern in the case of data streams, because it is desirable to hold most of the data structures in main memory in order to maximize the processing rate of the underlying data. Smaller space requirements ensure that it may be possible to hold most of the intermediate data in fast caches, which can further improve the efficiency of the approach. Furthermore, it may often be desirable to implement stream clustering algorithms in a wide variety of space-constrained architectures such as mobile devices, sensor hardware, or cell processors. Such architectures present special challenges to the massive-domain case if the underlying algorithms are not space-efficient.

The problem of clustering can be extremely challenging from a space and time perspective in the massive-domain case. This is because one needs to retain the discriminatory characteristics of the most relevant clusters in the data. In the massive-domain case, this may entail storing the frequency statistics of a large number of possible attribute values. While this may be difficult to do explicitly, the problem is further intensified by the large volume of the data stream which prevents easy determination of the importance of different attribute-values. The work in [2] proposes a sketch-based approach in order to keep track of the intermediate statistics of the underlying clusters. These statistics are used in order to make approximate determinations of the assignment of data points to clusters. A number of probabilistic results are provided in [2], which indicate that these approximations are sufficiently accurate to provide similar results to an infinite-space clustering algorithm with high probability.

The data stream $\mathcal{D}$ contains d-dimensional records denoted by $\overline{X_1} \ldots \overline{X_N} \ldots$. The attributes of record $\overline{X_i}$ are denoted by $(x_i^1 \ldots x_i^d)$. It is assumed that the attribute value x_i^k is drawn from the unordered domain set $J_k = \{v_1^k \ldots v_{M^k}^k\}$. The value of M^k denotes the domain size for the kth attribute. The value of M^k can be very large and may range in the order of millions or billions. From the point of view of a clustering application, this creates a number of challenges, since it is no longer possible to hold the cluster statistics in a space-limited scenario.

```
Algorithm CSketch(Labeled Data Stream: 𝒟,
              NumClusters: k)
begin
  Create k sketch tables of size w·h each;
  Initialize k sketch tables to null counts;
  repeat
    Receive next data point X̄ from 𝒟;
    Compute the approximate dot product of incoming data
     point with each cluster-centroid with the use of the
     sketch tables;
    Pick the centroid with the largest approximate
      dot product to incoming point;
    Increment the sketch counts in the chosen table
     for all d dimensional value strings;
  until(all points in 𝒟 have been processed);
end
```

FIGURE 10.1: The sketch-based clustering algorithm (CSketch Algorithm).

The work in [2] uses the count-min sketch [26] for the problem of clustering massive-domain data streams. In the count-min sketch, a hashing approach is utilized in order to keep track of the attribute-value statistics in the underlying data. We use $w = \lceil \ln(1/\delta) \rceil$ pairwise independent hash functions, each of which maps onto uniformly random integers in the range $h = [0, e/\varepsilon]$, where e is the base of the natural logarithm. The data structure itself consists of a 2-dimensional array with $w \cdot h$ cells with a length of h and width of w. Each hash function corresponds to one of w 1-dimensional arrays with h cells each. In standard applications of the count-min sketch, the hash functions are used in order to update the counts of the different cells in this 2-dimensional data structure. For example, consider a 1-dimensional data stream with elements drawn from a massive set of domain values. When a new element of the data stream is received, we apply each of the w hash functions to map onto a number in $[0 \ldots h-1]$. The count of each of the set of w cells is incremented by 1. In order to *estimate* the count of an item, we determine the set of w cells to which each of the w hash-functions map and compute the minimum value among all these cells. Let c_t be the true value of the count being estimated. We note that the estimated count is at least equal to c_t, since we are dealing with nonnegative counts only, and there may be an overestimation due to collisions among hash cells. As it turns out, a probabilistic upper bound to the estimate may also be determined. It has been shown in [26], that for a data stream with T arrivals, the estimate is at most $c_t + \varepsilon \cdot T$ with probability at least $1 - \delta$.

The *CSketch* algorithm (Figure 10.1) uses number of clusters k and data stream $\mathcal{D}$ as input to the algorithm. The clustering algorithm is partition based and assigns incoming data points to the most similar cluster centroid. The frequency counts for the different attribute values in the cluster centroids are incremented with the use of the sketch table. These frequency counts can be maintained only approximately because of the massive domain size of the underlying attributes in the data stream. Similarity is measured with the computation of the dot-product function between the incoming point and the centroid of the different clusters. This computation can be performed only approximately in the massive-domain case, since the frequency counts for the values in the different dimensions cannot be maintained explicitly. For each cluster, we maintain the frequency sketches of the records which are assigned to it. Specifically, for each cluster, the algorithm maintains a separate sketch table containing the counts of the values in the incoming records. The same hash function is used for each table. The algorithm starts off by initializing the counts in each sketch table to 0.

Subsequently, for each incoming record, we will update the counts of each of the cluster-specific hash tables. In order to determine the assignments of data points to clusters, the dot products of the d-dimensional incoming records to the frequency statistics of the values in the different clusters are computed. This is sometimes not possible to do explicitly, since the precise frequency statistics are not maintained. Let $q_r^j(x_i^r)$ represents the frequency of the value x_i^r in the jth cluster. Let m_i be the number of data points assigned to the jth cluster. Then, the d-dimensional statistics of the record $(x_i^1 \ldots x_i^d)$ for the jth cluster are given by $(q_1^j(x_i^1) \ldots q_d^j(x_i^d))$. Then, the frequency-based dot product $D^j(\overline{X_i})$ of the incoming record with statistics of cluster j is given by the dot product of the fractional frequencies $(q_1^j(x_i^1)/m_j \ldots q_d^j(x_i^d)/m_j)$ of the attribute values $(x_i^1 \ldots x_i^d)$ with the frequencies of these same attribute values within record $\overline{X_i}$. We note that the frequencies of the attribute values with the record $\overline{X_i}$ are unit values corresponding to $(1, \ldots 1)$. Therefore, the corresponding dot product is the following:

$$D^j(\overline{X_i}) = \sum_{r=1}^{d} q_r^j(x_i^r)/m_j \tag{10.7}$$

The incoming record is assigned to the cluster for which the estimated dot product is the largest. We note that the value of $q_r^j(x_i^r)$ cannot be known exactly, but can only be estimated approximately due to the massive-domain constraint. There are two key steps which use the sketch table during the clustering process:

- Updating the sketch table and other required statistics for the corresponding cluster for each incoming record.
- Comparing the similarity of the incoming record to the different clusters with the use of the corresponding sketch tables.

First, we discuss the process of updating the sketch table, once a particular cluster has been identified for assignment. For each record, the sketch-table entries corresponding to the attribute values on the different dimensions are incremented. For each incoming record $\overline{X_i}$, the w hash functions are applied to the strings corresponding to the attribute values in it. Let m be the index of the cluster to which the data point is assigned. Then, exactly $d \cdot w$ entries in the sketch table for cluster m are updated by applying the w hash functions to each of the d strings which are denoted by $x_i^1 \oplus 1 \ldots x_i^d \oplus d$. The corresponding entries are incremented by one unit each.

In order to pick a cluster for assignment, the approximate dot products across different clusters need to be computed. The record is converted to its sketch representation by applying the w hash functions to each of these d different attribute values. We retrieve the corresponding d sketch-table entries for each of the w hash functions and each cluster. For each of the w hash functions for the sketch table for cluster j, the d counts are simply estimates of the value of $q_1^j(x_i^1) \ldots q_d^j(x_i^d)$. Specifically, let the count for the entry picked by the lth hash function corresponding to the rth dimension of record $\overline{X_i}$ in the sketch table for cluster j be denoted by c_{ijlr}. Then, we estimate the dot product D_{ij} between the record $\overline{X_i}$ and the frequency statistics for cluster j as follows:

$$D_{ij} = \min_l \sum_{r=1}^{d} c_{ijlr}/m_j \tag{10.8}$$

The value of D_{ij} is computed over all clusters j, and the cluster with the largest dot product to the record $\overline{X_i}$ is picked for assignment.

It has been shown in [2] that this assignment process approximates an infinite space clustering algorithm quite well. This is characterized in terms of *assignment errors* because of the approximation process. An important point to be kept in mind is that in many cases an incoming data point may match well with many of the clusters. Since similarity functions such as the dot product are heuristically defined, small errors in ordering (when b_q is small) are not very significant for the clustering process. Similarly, some of the clusters correspond to outlier points and are sparsely populated. Therefore, it is more important to ensure that assignments to "significant clusters" (for which f_p is above a given threshold) are correct. Therefore, the work in [2] defines the concept of an (f,b)-significant assignment error as follows:

Definition 10.6.1 *An assignment error which results from the estimation of dot products is said to be (f,b)-significant, if a data point is incorrectly assigned to cluster index p which contains at least a fraction f of the points, and the correct cluster for assignment has index q which satisfies the following relationship:*

$$D_{ip} \geq D_{iq} + b \tag{10.9}$$

It has been shown in [2] that it is possible to bound the probability of an (f,b)-significant error in a given assignment.

Lemma 10.6.1 *The probability of an (f,b)-significant error in an assignment is at most equal to $k \cdot (d^2/(b \cdot f \cdot h))^w$.*

This can be used to further bound the probability that there is no (f,b)-significant error over the entire course of the clustering process of N data points.

Lemma 10.6.2 *Let us assume that $N \cdot k \cdot (d^2/(b \cdot f \cdot h))^w < 1$. The probability that there is at least one (f,b)-significant error in the clustering process of N data points is given by at most $\frac{N \cdot k}{(b \cdot f \cdot h/d^2)^w}$.*

In addition, the experimental results in [2] show that the clustering process can be replicated almost exactly in practice with the use of this approximation process. Thus, the work in [2] proposes a fast and space-efficient method for clustering massive-domain data streams.

10.7 Text Stream Clustering

The problem of streaming text clustering is particularly challenging in the context of text data due to the clusters needing to be continuously maintained in real time. One of the earliest methods for streaming text clustering was proposed in [64]. This technique is referred to as the *Online Spherical k-Means (OSKM)* Algorithm, which reflects the broad approach used by the methodology. This technique divides the incoming stream into small segments, each of which can be processed effectively in main memory. A set of k-means iterations is applied to each such data segment in order to cluster them. The advantage of using a segment-wise approach for clustering is that since each segment can be held in main memory, we can process each data point multiple times as long as it is held in main memory. In addition, the centroids from the previous segment are used in the next iteration for clustering purposes. A decay factor is introduced in order to age-out the old documents, so that the new documents are considered more important from a clustering perspective. This approach has been shown in [64] to be extremely effective in clustering massive text streams.

A different method for clustering massive text and categorical data streams is discussed in [12]. This method uses an approach which examines the relationship between outliers, emerging trends, and clusters in the underlying data. Old clusters may become inactive and eventually get replaced by new clusters. Similarly, when newly arriving data points do not naturally fit in any particular cluster, these need to be initially classified as outliers. However, as time progresses, these new points may create a distinctive pattern of activity which can be recognized as a new cluster. The temporal locality of the data stream is manifested by these new clusters. For example, the first web page belonging to a particular category in a crawl may be recognized as an outlier but may later form a cluster of documents of its own. On the other hand, the new outliers may not necessarily result in the formation of new clusters. Such outliers are true short-term abnormalities in the data since they do not result in the emergence of sustainable patterns. The approach discussed in [12] recognizes new clusters by first recognizing them as outliers. This approach works with the use of a summarization methodology, in which we use the concept of *condensed droplets* [12] in order to create concise representations of the underlying clusters.

As in the case of the OSKM algorithm, we ensure that recent data points are given greater importance than older data points. This is achieved by creating a time-sensitive weight for each data point. It is assumed that each data point has a time-dependent weight defined by the function $f(t)$, which is also referred to as the *fading function*. The fading function $f(t)$ is a nonmonotonic decreasing function which decays uniformly with time t. The aim of defining a half-life is to quantify the rate of decay of the importance of each data point in the stream clustering process. The *decay rate* is defined as the inverse of the half-life of the data stream. We denote the decay rate by $\lambda = 1/t_0$. We denote the weight function of each point in the data stream by $f(t) = 2^{-\lambda \cdot t}$. From the perspective of the clustering process, the weight of each data point is $f(t)$. It is easy to see that this decay function creates a half-life of $1/\lambda$. It is also evident that by changing the value of λ, it is possible to change the rate at which the importance of the historical information in the data stream decays.

When a cluster is created during the streaming process by a newly arriving data point, it is allowed to remain as a trend-setting outlier for at least one half-life. During that period, if at least one more data point arrives, then the cluster becomes an active and mature cluster. On the other hand, if no new points arrive during a half-life, then the trend-setting outlier is recognized as a true anomaly in the data stream. At this point, this anomaly is removed from the list of current clusters. We refer to the process of removal as *cluster death*. Thus, a new cluster containing one data point dies when the (weighted) number of points in the cluster is 0.5. The same criterion is used to define the death of mature clusters. A necessary condition for this criterion to be met is that the inactivity period in the cluster has exceeded the half-life $1/\lambda$. The greater the number of points in the cluster, the greater the level by which the inactivity period would need to exceed its half-life in order to meet the criterion. This is a natural solution, since it is intuitively desirable to have stronger requirements (a longer inactivity period) for the death of a cluster containing a larger number of points.

The statistics of the data points are captured in summary statistics, which are referred to as *condensed droplets*. These represent the word distributions within a cluster and can be used to compute the similarity of an incoming data point to the cluster. The overall algorithm proceeds as follows. At the beginning of algorithmic execution, we start with an empty set of clusters. As new data points arrive, unit clusters containing individual data points are created. Once a maximum number k of such clusters have been created, we can begin the process of online cluster maintenance. Thus, we initially start off with a trivial set of k clusters. These clusters are updated over time with the arrival of new data points.

When a new data point $\overline{X}$ arrives, its similarity to each cluster droplet is computed. In the case of text data sets, the cosine similarity measure between $\overline{DF1}$ and $\overline{X}$ is used. The similarity value $S(\overline{X}, C_j)$ is computed from the incoming document $\overline{X}$ to every cluster C_j. The cluster with the maximum value of $S(\overline{X}, C_j)$ is chosen as the relevant cluster for data insertion. Let us assume that this cluster is C_{mindex}. We use a threshold denoted by *thresh* in order to determine whether the incoming data point is an outlier. If the value of $S(\overline{X}, C_{mindex})$ is larger than the threshold *thresh*, then the point $\overline{X}$ is assigned to the cluster C_{mindex}. Otherwise, we check if some inactive cluster exists in the current set of cluster droplets. If no such inactive cluster exists, then the data point $\overline{X}$ is added to C_{mindex}. On the other hand, when an inactive cluster does exist, a new cluster is created containing the solitary data point $\overline{X}$. This newly created cluster replaces the inactive cluster. We note that this new cluster is a potential true outlier or the beginning of a new trend of data points. Further understanding of this new cluster may only be obtained with the progress of the data stream.

In the event that $\overline{X}$ is inserted into the cluster C_{mindex}, we update the statistics of the cluster to reflect the insertion of the data point and temporal decay statistics. Otherwise, we replace the most inactive cluster by a new cluster containing the solitary data point $\overline{X}$. In particular, the replaced cluster is the least recently updated cluster among all inactive clusters. This process is continuously performed over the life of the data stream, as new documents arrive over time. The work in [12] also presents a variety of other applications of the stream clustering technique such as evolution and correlation analysis.

A different way of utilizing the temporal evolution of text documents in the clustering process is described in [38]. Specifically, the work in [38] uses *bursty features* as markers of new topic occurrences in the data stream. This is because the semantics of an up-and-coming topic are often reflected in the frequent presence of a few distinctive words in the text stream. At a given time, the nature of relevant topics could lead to bursts in specific features of the data stream. Clearly, such features are extremely important from a clustering perspective. Therefore, the method discussed in [38] uses a new representation, which is referred to as the *bursty feature representation* for mining text streams. In this representation, a time-varying weight is associated with the feature depending upon its burstiness. This also reflects the varying importance of the feature to the clustering process. Thus, it is important to remember that a particular document representation is dependent upon the particular instant in time at which it is constructed.

Another issue which is handled effectively in this approach is an implicit reduction in dimensionality of the underlying collection. Text is inherently a high-dimensional data domain, and the preselection of some of the features on the basis of their burstiness can be a natural way to reduce the dimensionality of document representation. This can help in both the effectiveness and efficiency of the underlying algorithm.

The first step in the process is to identify the bursty features in the data stream. In order to achieve this goal, the approach uses Kleinberg's 2-state finite automaton model [45]. Once these features have been identified, the bursty features are associated with weights which depend upon their level of burstiness. Subsequently, a bursty feature representation is defined in order to reflect the underlying weight of the feature. Both the identification and the weight of the bursty feature are dependent upon its underlying frequency. A standard k-means approach is applied to the new representation in order to construct the clustering. It is shown in [38] that the approach of using burstiness improves the cluster quality. One criticism of the work in [38] is that it is mostly focused on the issue of improving effectiveness with the use of temporal characteristics of the data stream, and does not address the issue of efficient clustering of the underlying data stream.

In general, it is evident that feature extraction is important for all clustering algorithms. While the work in [38] focuses on using temporal characteristics of the stream for feature extraction, the work in [48] focuses on using *phrase extraction* for effective feature selection. This work is also related to the concept of topic modeling, which will be discussed in detail in Chapter 13 because the different topics in a collection can be related to the clusters in a collection. The work in [48] uses topic-modeling techniques for clustering. The core idea in the work of [48] is that individual words are not very effective for a clustering algorithm because they miss the context in which the word is used. For example, the word *star* may refer either to a celestial body or to an entertainer. On the other hand, when the phrase *fixed star* is used, it becomes evident that the word *star* refers to a celestial body. The phrases which are extracted from the collection are also referred to as *topic signatures*.

The use of such phrasal clarification for improving the quality of the clustering is referred to as *semantic smoothing* because it reduces the noise which is associated with semantic ambiguity. Therefore, a key part of the approach is to extract phrases from the underlying data stream. After phrase extraction, the training process determines a translation probability of the phrase to terms in the vocabulary. For example, the word *planet* may have high probability of association with the phrase *fixed star*, because both refer to celestial bodies. Therefore, for a given document, a rational probability count may also be assigned to all terms. For each document, it is assumed that all terms in it are generated either by a topic-signature model or a background collection model.

The approach in [48] works by modeling the soft probability $p(w|C_j)$ for word w and cluster C_j. The probability $p(w|C_j)$ is modeled as a linear combination of two factors: (a) a maximum likelihood model which computes the probabilities of generating specific words for each cluster and (b) an indirect (translated) word-membership probability which first determines the maximum likelihood probability for each topic signature and then multiplies with the conditional probability of each word, given the topic signature. We note that we can use $p(w|C_j)$ in order to estimate $p(d|C_j)$ by using the product of the constituent words in the document. For this purpose, we use the frequency $f(w,d)$ of word w in document d:

$$p(d|C_j) = \prod_{w \in d} p(w|C_j)^{f(w,d)} \tag{10.10}$$

We note that in the static case, it is also possible to add a background model in order to improve the robustness of the estimation process. This is, however, not possible in a data stream because of the fact that the background collection model may require multiple passes in order to build effectively. The work in [48] maintains these probabilities in online fashion with the use of a *cluster profile*, that weights the probabilities with the use of a fading function. We note that the concept of cluster profile is analogous to the concept of condensed droplet introduced in [12]. The key algorithm (denoted by OCTS) is to maintain a dynamic set of clusters into which documents are progressively assigned with the use of similarity computations. It has been shown in [48] how the cluster profile can be used to efficiently compute $p(d|C_j)$ for each incoming document. This value is then used to determine the similarity of the documents to the different clusters and assign the documents to their closest cluster. We note that the methods in [12, 48] share a number of similarities in terms of (a) maintenance of cluster profiles, (b) use of cluster profiles (or condensed droplets) to compute similarity and assignment of documents to most similar clusters, and (c) the rules used to decide when a new singleton cluster should be created or one of the older clusters should be replaced.

The main difference between the two algorithms is the technique which is used in order to compute cluster similarity. The OCTS algorithm uses the probabilistic computation $p(d|C_j)$ to compute cluster similarity, which takes the phrasal information into account during the computation process. One observation about OCTS is that it may allow for very similar clusters to coexist in the current set. This reduces the space available for distinct cluster profiles. A second algorithm called OCTSM is also proposed in [48], which allows for merging of very similar clusters. Before each assignment, it checks whether pairs of similar clusters can be merged on the basis of similarity. If this is the case, then we allow the merging of the similar clusters and their corresponding cluster profiles. Detailed experimental results on the different clustering algorithms and their effectiveness are presented in [48]. A comprehensive review of text stream clustering algorithms may be found in the chapters on data clustering and streaming algorithms in [15].

10.8 Other Scenarios for Stream Clustering

Data stream clustering is a fundamental problem, and numerous other domains arise, in which the data occurs naturally in the streaming context. In this section, we provide a brief introduction to these different data domains, along with pointers to the relevant literature.

10.8.1 Clustering Uncertain Data Streams

Uncertain data streams arise quite often in the context of sensor data or other hardware collection technology such as RFID in which there are significant errors in the data collection process. In many of these cases, the errors in the data can be approximated either in terms of statistical parameters, such as the standard deviation, or in terms of probability density functions (pdfs). Such statistical information increases the richness of the noisy data because it provides information about the parts of data which are less reliable, and should therefore be emphasized less in the mining process.

In this context, a method called *UMicro* for clustering uncertain data streams is proposed in [13]. This method enhances the microclusters with additional information about the uncertainty of the data points in the clusters. This information is used to improve the quality of the distance functions for the assignments. This approach has been further improved for the case of projected clustering of uncertain data streams [4, 60]. We are not providing a detailed discussion of these methods, since they are discussed in detail in the chapter on uncertain data clustering.

10.8.2 Clustering Graph Streams

Graph streams are created by edge-centered activity in numerous social and information networks. Many different kinds of streaming and clustering models are possible, depending on the application scenario. These different models are as follows:

- The stream is a sequence of *objects*, each of which is a small graph containing a subset of nodes and edges. We wish to determine similar *objects* in the stream based on the similarities between the nodes and edges. For example, the DBLP bibliographic network has an incoming stream of graph objects corresponding to the coauthored papers. It may be desirable to determine clusters of objects with similar structure. An algorithm for this problem, known as *Gmicro* is proposed in [14]. The microclustering representation is extended to handle edges, and a sketch based compression is used on the edges in order to reduce the impact of the massive domain of the edges.

- The stream is a sequence of *objects*, each of which is a small graph containing a subset of nodes and edges. We wish to determine node sets which co-occur frequently in these objects, and are also densely connected together. A method is proposed in [8] with the use of min-hash-based graph stream summarization.

- The stream is a sequence of either *objects* or *edges*. It is desirable to determine dense node clusters in the graph.

The last problem is particularly general and is also related to general problem of dynamic community detection. In this context, a method was proposed for creating dynamic partitions of graphs with the use of structural reservoir sampling of edge streams [16]. While the work in [16] is targeted to outlier detection, a key intermediate step in the process is the dynamic generation of node clusters from the edge stream. The work in [28] has further refined the structural reservoir sampling techniques of [16] in order to provide more effective methods for node clustering in graph streams.

In many cases, such as social networks, content information is associated with the structure in the network. For example, the tweets in a *Twitter* stream have both structure and content. Such streams are referred to as *social streams*. The clustering of such streams requires the use of both structural information and content in the process. The work in [9] has designed methods for clustering social streams. Sketch-based methods are used in order to summarize the content in the social stream and use it for the clustering process. In addition, it has been shown in [9], how this approach may be used for event detection.

In the conventional streaming model, it is assumed that only one pass is allowed over the data, and the amount of memory available to the application is constant, irrespective of stream length. A technique for determining the densest subgraph has been proposed in [19] in a *weakly streaming model*, where a limited number of passes (more than one) are allowed over the data, and the amount of available memory is sublinear in the size of the input. This approach is able to determine the densest subgraph in passes which grow logarithmically with the graph size.

10.8.3 Distributed Clustering of Data Streams

In the context of sensor networks, the stream data is often available only in a *distributed setting*, in which large volumes of data are collected separately at the different sensors. A natural approach for clustering such data is to transmit all of the data to a centralized server. The clustering can then be performed at the centralized server to determine the final results. Unfortunately, such an approach is extremely expensive in terms of its communication costs due to the large volume of the data which must be transmitted to the centralized server. Therefore, it is important to design a method which can reduce the communication costs among the different processors. A method proposed in [27] performs local clustering at each node and merges these different clusters into a single global clustering at low communication cost. Two different methods are proposed in this work. The first method determines the cluster centers by using a *furthest point algorithm*, on the current set of data points at the local site. In the furthest point algorithm, the center of a cluster is picked as a furthest point to the current set of centers. For any incoming data point, it is assigned to its closest center, as long as the distance is within a certain factor of an optimally computed radius. Otherwise, a reclustering is forced by applying the furthest point algorithm on current set of points. After the application of the furthest point algorithm, the centers are transmitted to the central server, which then computes a global clustering from these local centers over the different nodes. These global centers can then be transmitted to the local nodes if desired. One attractive feature of the method is that an approximation bound is proposed on the quality of the clustering. A second method for distributed clustering proposed in [27] is the *parallel guessing algorithm*.

A variety of other methods have also been proposed for distributed clustering of data streams. For example, techniques for distributed data stream clustering with the use of the k-medians approach are proposed in [61]. Another method for distributed sensor stream clustering which reduces the dimensionality and communication cost by maintaining an online discretization may be found in [57]. Finally, a method for distributed clustering of data streams with the use of the EM approach is proposed in [66].

10.9 Discussion and Conclusions

The problem of clustering is fundamental to a wide variety of streaming applications, because of its natural applicability to summarizing and representing the data in a very small space. A wide variety of techniques have been proposed in the literature for stream clustering, such as partitioning methods, density-based methods, and probabilistic methods. The stream clustering method has also been extended to other data domains such as categorical, text, and uncertain data. Finally, methods have also been proposed for clustering graph streams.

Many further directions of research are likely for this problem. These are as follows:

- Heterogeneous data streams are becoming extremely common because of applications such as social networks, in which large amounts of heterogeneous content are posted over time. It is desirable to determine clusters among these groups of heterogeneous objects.

- It is desirable to use a combination of links and content in order to determine clusters from heterogeneous activity streams. This direction of research has also found increasing interest in the community. This is a further layer of complexity over the graph streaming scenario.

Furthermore, it would also be interesting to perform clustering streams from multiple sources, while simultaneously learning the nature of the dynamic relationships between the different streams.

Bibliography

[1] C. Aggarwal. *Data Streams: Models and Algorithms*, Springer, 2007.

[2] C. Aggarwal. A Framework for clustering massive-domain data streams. In *ICDE Conference*, pages 102–113, 2009.

[3] C. Aggarwal. On change diagnosis in evolving data streams. In *IEEE TKDE*, 17(5), pages 587–600, 2005.

[4] C. Aggarwal. On high-dimensional projected clustering of uncertain data streams, *IEEE ICDE Conference*, pages 1152–1154, 2009.

[5] C. Aggarwal, J. Han, J. Wang, and P. Yu. A framework for clustering evolving data streams. In *VLDB Conference*, pages 81–92, 2003.

[6] C. Aggarwal, J. Han, J. Wang, and P. Yu. A framework for projected clustering of high dimensional data streams. In *VLDB Conference*, pages 852–863, 2004.

[7] C. Aggarwal, J. Han, J. Wang, and P. Yu. On high dimensional projected clustering of data streams, *Data Mining and Knowledge Discovery Journal*, 10(3):251–273, 2005.

[8] C. Aggarwal, Y. Li, P. Yu, and R. Jin. On dense pattern mining in graph streams, *VLDB Conference*, pages 975–984, 2010.

[9] C. Aggarwal and K. Subbian. Event detection in social streams, *SIAM Conference on Data Mining*, pages 624–635, 2012.

[10] C. Aggarwal, C. Procopiuc, J. Wolf, P. Yu, and J.-S. Park. Fast algorithms for projected clustering. In *ACM SIGMOD Conference*, pages 61–72, 1999.

[11] C. Aggarwal and P. Yu. Finding generalized projected clusters in high dimensional spaces. In *ACM SIGMOD Conference*, pages 70–81, 2000.

[12] C. Aggarwal and P. Yu. A framework for clustering massive text and categorical data streams. In *SIAM Conference on Data Mining*, pages 479–483, 2006.

[13] C. Aggarwal and P. Yu. A framework for clustering uncertain data streams, *IEEE ICDE Conference*, pages 150–159, 2008.

[14] C. Aggarwal and P. Yu. On clustering graph streams, *SDM Conference*, pages 478–489, 2010.

[15] C. Aggarwal and C. X. Zhai, *Mining Text Data*, Springer, 2012.

[16] C. Aggarwal, Y. Zhao, and P. Yu. Outlier detection in graph streams, *ICDE Conference*, pages 399–409, 2011.

[17] R. Agrawal, J. Gehrke, D. Gunopulos, and P. Raghavan. Automatic subspace clustering of high dimensional data for data mining applications. In *ACM SIGMOD Conference*, pages 94–105, 1998.

[18] M. Ankerst, M. Breunig, H.-P. Kriegel, and J. Sander. OPTICS: Ordering points to identify the clustering structure. In *ACM SIGMOD Conference*, pages 49–60, 1999.

[19] B. Bahamani, R. Kumar, and V. Vassilvitskii. Densest subgraph mining in streaming and MapReduce, *VLDB Conference*, pages 454–465, 2012.

[20] A. Banerjee and S. Basu. Topic Models over text streams: A Study of Batch and Online Unsupervised Learning, *SIAM Conference*, pages 431–436, 2007.

[21] A. Banerjee, I. Dhillon, J. Ghosh, and S. Sra. Clustering on the unit hypersphere using von Mises-Fisher distributions. In *Journal of Machine Learning Research*, 6:1345–1382, 2005.

[22] D. Blei, A. Ng, and M. Jordan. Latent Dirichlet allocation, *Journal of Machine Learning Research*, 3:993–1022, 2003.

[23] P. Bradley, U. Fayyad, and C. Reina. Scaling Clustering Algorithms to Large Databases. In *ACM KDD Conference*, pages 9–15, 1998.

[24] F. Cao, M. Ester, W. Qian, A. Zhou. Density-based clustering over an evolving data stream with noise. In *SDM Conference*, pages 326–337, 2006.

[25] Y. Chen and L. Tu. Density-based clustering for real time stream data. In *ACM KDD Conference*, pages 133–142, 2007.

[26] G. Cormode and S. Muthukrishnan. An improved data-stream summary: The count-min sketch and its applications. *Journal of Algorithms*, 55(1):58–75, 2005.

[27] G. Cormode, S. Muthukrishnan, and W. Zhuang. Conquering the divide: Continuous clustering of distributed data streams. *ICDE Conference*, pages 1036–1045, 2007.

[28] A. Eldawy, R. Khandekar, and K. L. Wu. Clustering streaming graphs, *ICDCS Conference*, pages 466–475, 2012.

[29] C. Elkan. Clustering documents with an exponential-family approximation of the Dirichlet compund multinomial distribution. *ICML Conference*, pages 289–296, 2006.

[30] M. Ester, H.-P. Kriegel, J. Sander, and X. Xu. A density-based algorithm for discovering clusters in large spatial databases with noise. *ACM KDD Conference*, pages 226–231, 1996.

[31] F. Farnstrom, J. Lewis, and C. Elkan. Scalability for clustering algorithms revisited. *ACM SIGKDD Explorations*, 2(1):51–57, 2000.

[32] D. Fisher. Knowledge Acquisition via incremental conceptual clustering. *Machine Learning*, 2:139–172, 1987.

[33] M. Franz, T. Ward, J. McCarley, and W.-J. Zhu. Unsupervised and supervised clustering for topic tracking. *ACM SIGIR Conference*, pages 310–317, 2001.

[34] G. P. C. Fung, J. X. Yu, P. Yu, and H. Lu. Parameter free bursty events detection in text streams, *VLDB Conference*, pages 181–192, 2005.

[35] J. Gao, J. Li, Z. Zhang, and P.-N. Tan. An incremental data stream clustering algorithm based on dense units detection. In *PAKDD Conference*, pages 420–425, 2005.

[36] S. Guha, A. Meyerson, N. Mishra, R. Motwani, and L. O'Callaghan. Clustering data streams: Theory and practice. In *IEEE FOCS Conference*, pages 515–528, 2000.

[37] S. Guha, R. Rastogi, and K. Shim. CURE: An efficient clustering algorithm for large databases. In *ACM SIGMOD Conference*, pages 73–84, 1998.

[38] Q. He, K. Chang, E.-P. Lim, and J. Zhang. Bursty feature representation for clustering text streams. *SDM Conference*, pages 491–496, 2007.

[39] Z. He, X. Xu, and S. Deng. Squeezer: An efficient algorithm for clustering categorical data, *Journal of Computer Science and Technology (JCST)*, 17(5):611–624, 2002.

[40] Z. He, X. Xu, S. Deng, and J. Huang. Clustering categorical data streams, *Published Online on Arxiv*, December 2004.
`http://arxiv.org/ftp/cs/papers/0412/0412058.pdf`

[41] T. Hoffman. Probabilistic latent semantic indexing, *ACM SIGIR Conference*, pages 50–57, 1999.

[42] A. Jain and R. Dubes. *Algorithms for Clustering Data*, Prentice Hall, Upper Saddle River, NJ, 1998.

[43] C. Jia, C. Tan, and A. Yong. A grid and density-based clustering algorithm for processing data stream, *International Conference on Genetic and Evolutionary Computing*, pages 517–521, 2008.

[44] L. Kaufman and P. Rousseuw. *Finding Groups in Data—An Introduction to Cluster Analysis*. Wiley Series in Probability and Math Sciences, 1990.

[45] J. Kleinberg. Bursty and hierarchical structure in streams, *ACM KDD Conference*, pages 91–101, 2002.

[46] P. Kranen, I. Assent, C. Baldauf, and T. Seidl. The ClusTree: Indexing micro-clusters for anytime stream mining. *Knowledge and Information Systems*, 29(2):249–272, 2011. http://link.springer.com/article/10.1007

[47] H.-P. Kriegel, P. Kroger, I. Ntoutsi, and A. Zimek. Density based subspace clustering over dynamic data. *SSDBM Conference*, pages 387–404, 2011.

[48] Y.-B. Liu, J.-R. Cai, J. Yin, and A. W.-C. Fu. Clustering text data streams, *Journal of Computer Science and Technology*, Vol. 23(1):112–128, 2008.

[49] Y. Lu, Y. Sun, G. Xu, and G. Liu. A grid-based clustering approach for high-dimensional data streams, *Advanced Data Mining and Applications*, 384:824–831, 2005.

[50] G. Manku and R. Motwani. Approximate frequency counts over data streams, *VLDB Conference*, pages 346–359, 2002.

[51] R. Ng and J. Han. Efficient and effective clustering methods for spatial data mining. In *Very Large Data Bases Conference*, pages 144–155, 1994.

[52] I. Ntoutsi, A. Zimek, T. Palpanas, H.-P. Kriegel, and A. Zimek. Density-based projected clustering over high dimensional data streams, In *SDM Conference*, pages 987–998, 2012.

[53] L. O'Callaghan, N. Mishra, A. Meyerson, S. Guha, and R. Motwani. Streaming-data algorithms for high-quality clustering. In *ICDE Conference*, pages 685–694, 2002.

[54] C. Ordonez. Clustering binary data streams with K-means. In *Data Mining and Knowledge Discovery Workshop*, pages 12–19, 2003.

[55] N. H. Park and W. S. Lee. Cell trees: An adaptive synopsis structure for clustering multi-dimensional on-line data streams. *Data and Knowledge Engineering*, 63(2):528–549, 2007.

[56] N. H. Park and W. S. Lee. Statistical grid-based clustering over data streams, *ACN SIGMOD Record*, 33(1), March, 2004. `http://www09.sigmod.org/sigmod/record/issues/0403/A15.park-lee.pdf`

[57] P. Rodrigues, J. Gama, and L. Lopes. Clustering distributed sensor data streams, *PKDD Conference*, 5212:282–297, 2008.

[58] B. W. Silverman. *Density Estimation for Statistics and Data Analysis*. Boca Raton, FL, Chapman and Hall, 1986.

[59] M. Song and H. Wang. Highly efficient incremental estimation of Gaussian Mixture Models for online data stream clustering, *SPIE*, 5803:174–183, 2005.

[60] C. Zhang, M. Gao, and A. Zhou. Tracking high quality clusters over uncertain data streams, *ICDE Conference*, pages 1641–1648, 2009.

[61] Q. Zhang, J. Liu, and W. Wang. Approximate clustering on distributed data streams, *ICDE Conference*, pages 1131–1139, 2008.

[62] Q. Zhang, J. Liu, and W. Wang. Incremental subspace clustering over multiple data streams. *ICDM Conference*, pages 727–732, 2007.

[63] T. Zhang, R. Ramakrishnan, and M. Livny. BIRCH: An efficient data clustering method for very large databases. In *ACM SIGMOD Conference*, pages 103–114, 1996.

[64] S. Zhong. Efficient streaming text clustering. *Neural Networks*, 18(5–6):790–798, 2005.

[65] A. Zhou, F. Cao, W. Qian, and C. Jin. Tracking clusters in evolving data streams over sliding windows, *Knowledge and Information Systems*, 15(2):181–214, 2008.

[66] A. Zhou, F. Cao, Y. Yan, C. Sha, and X. He. Distributed clustering of data streams: A fast EM-based approach, *ICDE Conference*, pages 736–745, 2007.

Chapter 11

Big Data Clustering

Hanghang Tong
IBM T. J. Watson Research Center
Yorktown Heights, NY
htong@us.ibm.com

U Kang
KAIST
Seoul, Korea
ukang@cs.cmu.edu

11.1 Introduction

With the advance of Web2.0, the data size is increasing explosively. For example, Twitter data spans several terabytes; Wikipedia data (e.g., articles and authors) is of similar size; web click-through data is reported to reach petabyte scale [36]; Yahoo! web graph in 2002 has more than 1 *billion* nodes and almost *7 billion* edges [27].

On the other hand, many data clustering algorithms have a high intrinsic time complexity. For example, the classic k-means clustering is NP-hard even when $k = 2$. The normalized cut (NCut), a representative spectral clustering algorithm, is also NP-hard [43]. Therefore, a key challenge for data clustering lies in its scalability, that is, how we can speed up/scale up the clustering algorithms with the minimum sacrifice to the clustering quality.

At the high level, many data clustering algorithms have the following procedure: after some initialization (e.g., randomly choose the k cluster centers in k-means), it takes some iterative process until some convergence criteria is met. In each iteration, it adjusts/updates the cluster membership for each data point (e.g., assign, it to the closest cluster centers in k-means). Therefore, in order to speed up/scaleup such clustering algorithms, we have three basic ways: (1) by reducing the iteration number (e.g., one-pass algorithm) [47, 22], (2) by reducing the access to the data points (e.g., by randomized techniques) [25, 38], and (3) by distributing/parallelizing the computation [28, 13]. In this chapter, we will review some representative algorithms for each of these categories.

Notice that another important aspect related to the scalability is for the stream data, that is, how we can adopt the batch-mode data clustering algorithms in the case the data arrives in the stream mode, so that ideally the algorithms scale well with respect to the size of the new arrival data as opposed to that of the entire data set [16, 4, 35]. These issues will be discussed independently in the chapter on density-based clustering.

11.2 One-Pass Clustering Algorithms

In this section, we will review some earlier, classic clustering algorithms which are designed for large-scale data sets. CLARANS [39, 40] is one of the earliest algorithms to use randomized search to fight with the exponential search space in the K-medoid clustering problem. In BIRCH [47], a new data structure called clustering feature tree (CF-tree) was proposed to reduce the I/O cost when the input data set exceeds the memory size. Finally, CURE [22] uses multiple representative data points for each cluster in order to capture the irregular clustering shapes and it further adopts sampling to speed up the hierarchical clustering procedure. Thanks to these techniques, we can often largely reduce the iteration number in the clustering procedure. In some extreme cases, they often generate pretty good clustering results by one pass over the input data set. We collectively call these algorithms as one-pass clustering algorithms.

11.2.1 CLARANS: Fighting with Exponential Search Space

Let us start with one of the earliest representative clustering algorithms, i.e., Partitioning Around Medoids (PAM) [32]. The basic idea of PAM is to choose one representative point from each cluster (e.g., the most central data point in the cluster), which is referred to as a *medoid*. If we want to find k clusters for n data points, the algorithm essentially tries to find the k best medoids. Once such optimal k-medoid is found, the remaining (nonmedoid) data points are assigned to one of the selected medoids according to their distance to the k-medoid. To this end, it follows an iterative procedure where in each iteration, it tries to replace one of the current medoids by one of the nonmedoid data points. It is easy to see that the computational cost for PAM is high: first of all, in each iteration, we need to check all the $(n-k)\times k$ possible pairs; second, the overall search space is exponential, i.e., we have $\binom{n}{k}$ possible k-medoid assignment. To address this issue, Clustering Large Applications (CLARA) [32] relies on *sampling* to reduce the search space. Instead of trying to find the optimal k-medoid from the original *entire* data set as in PAM, CLARA first takes a sample of k from the original n data points and then calls PAM on these $O(k)$ sampled data points to find k-medoid. Once the optimal k-medoid is found, the remaining (nonsampled) data points are assigned to one of these k clusters based on their distance to the k-medoid.

Conceptually, both PAM and CLARA can be regarded as graph-searching problems. In this graph, each node is a possible clustering solution (i.e., a k-medoid), and the two nodes are linked to each other if they only differ in 1-out-of-k medoids. PAM starts with one of the randomly chosen

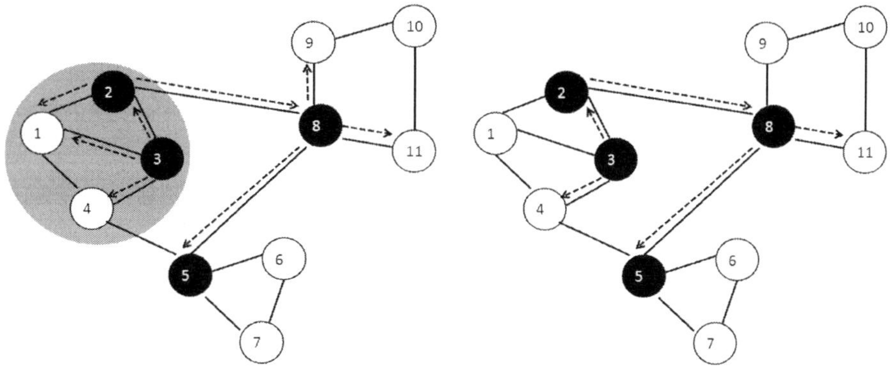

(a) PAM and CLARA (b) CLARANS

FIGURE 11.1: Each node in the graph is a possible clustering assignment (i.e., k-medoid); and each link means that two nodes differ in only one medoid. PAM (a) starts from a randomly selected node (e.g., node 3, dark), and greedily moves to the next best neighbor (e.g., node 2). In each iteration, it searches all the possible neighbors (dashed arrows). It tries to search over the entire graph. CLARA first samples a subgraph (shadowed area) and restricts the search procedure within this subgraph. CLARANS also searches the entire graph as in PAM. But in each iteration, it only searches only randomly sampled neighbors of the current node (dashed arrow).

nodes in the conceptual graph, and it greedily moves to one of its neighbors until it cannot find a better neighbor. CLARA aims to reduce the search space by only searching a subgraph that is induced by the sampled $O(k)$ data points.

Based on this observation, Clustering Large Applications based on Randomized Sampling (CLARANS) [39, 40] has been proposed to further improve the efficiency. As in PAM, CLARANS aims to find a local optimal solution by searching the entire graph. But unlike in PAM, in each iteration, it checks only a sample of the neighbors of the current node in the graph. Notice that both CLARA and CLARANS use sampling techniques to reduce the search space. But they conduct the sampling in the different ways. In CLARA, the sampling is done at the beginning stage to restrict the entire search process within a particular subgraph; whereas in CLARANS, the sampling is conducted *dynamically* at each iteration of the search process. The empirical evaluation in [39] shows that such dynamic sampling strategy in CLARANS leads to further efficiency improvement over CLARA. Figure 11.1 provides a pictorial comparison between PAM, CLARA and CLARANS.

11.2.2 BIRCH: Fighting with Limited Memory

When the data size exceeds the available memory amount, the I/O cost may dominate the in-memory computational time, where CLARANS and its earlier versions (e.g., PAM, CLARA) suffer. Balanced Iterative Reducing and Clustering using Hierarchies (BIRCH) [47] was one of the first methods to address this issue, and explicitly tries to reduce the I/O costs given the limited amount of memory.

The key of BIRCH is to introduce a new data structure called *clustering feature* (CF) as well as CF-tree. Therefore, before we present the BIRCH algorithm, let us take a short detour to introduce these concepts.

CF can be regarded as a concise summary of each cluster. This is motivated by the fact that not every data point is equally important for clustering and we cannot afford to keep every data point in the main memory given that the overall memory is limited. On the other hand, for the purpose of

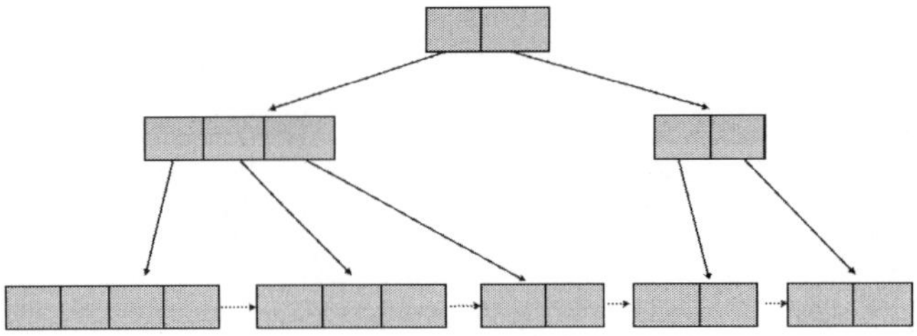

FIGURE 11.2: An illustration of the CF-tree. Each node represents a cluster which consists of up to B subclusters. Due to the additive property of CF, the CF of the parent node is simply the sum of the CFs of its children. All the leaf nodes are chained together for efficient scanning. The diameter of the subcluster in the leaf node is bounded by the threshold T. The bigger T is, the smaller the tree is.

clustering, it is often enough to keep up to the second order of data moment. In order words, CF is not only efficient, but also sufficient to cluster the entire data set.

To be specific, CF is a triple $< N, LS, SS >$ which contains the number of the data points in the cluster (i.e., the zero-order moment N), the linear sum of the data points in the cluster (i.e., the first-order moment LS), and the square sum of the data points in the cluster (i.e., the second order moment SS). It is easy to check that CF satisfies the additive property; that is, if we want to merge two existing clusters, the CF for the merged cluster is simply the sum of the CFs of the two original clusters. This feature is critical as it allows us to easily merge two existing clusters without accessing the original data set.

CF-tree is a height-balanced tree which keeps track of the hierarchical clustering structure for the entire data set. Each node in the CF-tree represents a cluster which is in turn made up of at most B subclusters, where B is the so-called balancing factor. All the leaf nodes are chained together for the purpose of efficient scanning. In addition, for each subcluster in the leaf node, its diameter is upper-bounded by a second parameter T (the threshold). Apparently, the larger T is, the smaller the CF-tree is in terms of the tree height. The insertion on CF-tree can be performed in a similar way as the insertion in the classic B-tree. Figure 11.2 presents a pictorial illustration of the CF-tree.

There are two main steps in the BIRCH algorithm. First, as it scans the input data points, it builds a CF-tree by inserting the data points with a default threshold $T = 0$. By setting the threshold $T = 0$, we treat each data point as an individual cluster at the leaf node. Thus, the size of the resulting CF-tree might be very large, which might lead to an out-of-memory issue in the middle of the scanning process. If such a situation happens, we will need to increase the threshold T and rebuild the CF-tree by reinserting all the leaf nodes from the old CF-tree. By grouping close data points together, the resulting CF-tree builds an initial summary of the input data set which reflects its rough clustering structure.

Due to the skewed input order and/or the splitting effect by the page size, the clustering structure from the initial CF-tree might not be accurate enough to reflect the real underlying clustering structure. To address this issue, the second key step ("global clustering") tries to cluster all the subclusters in the leaf nodes. This is done by converting a subcluster with n' data points n' times at the centroid and then running either an agglomerative hierarchical clustering algorithm or a modified clustering algorithm.

Between these two main steps, there is an optional step to refine the initial CF-tree by reinserting its leaf entries ("tree condensing"). Usually, this leads to a much more compact CF-tree. After the global clustering step, there is another optional step ("clustering refinement"), which reassigns all the data points based on the cluster centroid produced by the global clustering step. Figure 11.3 summarizes the overall flowchart of the BIRCH algorithm.

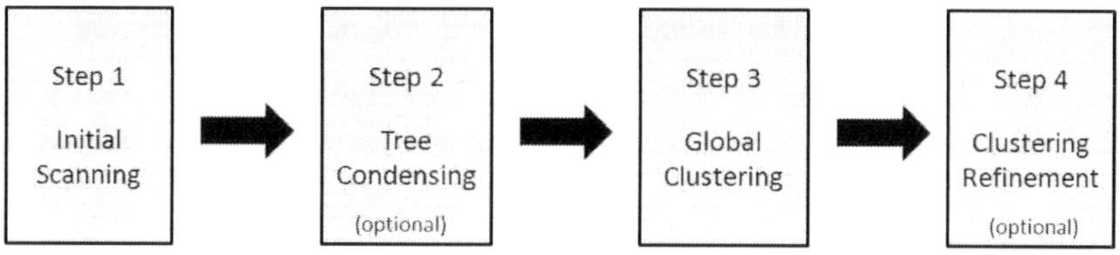

FIGURE 11.3: The flowchart of BIRCH algorithm.

The empirical evaluation in [47] indicates that BIRCH consistently outperforms the previous method CLARANS in terms of both time and space efficiency. It is also more robust to the ordering of the input data sequence. BIRCH offers some additional advantages, e.g., being robust to outliers and being more easily parallelized.

11.2.3 CURE: Fighting with the Irregular Clusters

In both CLARANS and BIRCH, we use one single data point to represent a cluster. Conceptually, we implicitly assume that each cluster has a spherical shape, which may not be the case in some real applications where the clusters can exhibit more complicated shapes. At the other extreme, we can keep all the data points within each cluster, whose computational as well as space cost might be too high for large data sets. To address this issue, clustering using representatives (CURE) [22] proposes to use a set of well-scattered data points to represent a cluster.

CURE is essentially a hierarchical clustering algorithm. It starts by treating each data point as a single cluster and then recursively merges two existing clusters into one until we have only k clusters. In order to decide which two clusters to be merged at each iteration, it computes the minimum distance between all the possible pairs of the representative points from the two clusters. CURE uses two major data structures to enable the efficient search. First, it uses a heap to track the distance of each existing cluster to its closest cluster. Additionally, it uses k-d tree to store all the representative points for each cluster.

In order to speed up the computation, CURE first draws a sample of the input data set and runs the above procedure on the sampled data. The authors further use Chernoff bound to analyze the necessary sample size. When the original data set is large, the different clusters might overlap each other, which in turn requires a large sample size. To alleviate this issue, CURE further uses partitions to speed up. To be specific, it first partitions the n' sampled data points into p partitions. Within each partition, it then runs a *partial* hierarchical clustering algorithm until either a predefined number of clusters is reached or the distance between the two clusters to be merged exceeds some threshold. After that, it runs another clustering pass on all the partial clusters from all the p partitions ("global clustering"). Finally, each nonsampled data point is assigned to a cluster based on its distance to the representative point ("labeling"). Figure 11.4 summarizes the flowchart of the CURE algorithm.

The empirical evaluation in [22] shows that CURE achieves lower execution time compared to BIRCH. In addition, as with BIRCH, CURE is also robust to outliers by shrinking the representative points to the centroid of the cluster with a constant factor.

11.3 Randomized Techniques for Clustering Algorithms

In the previous section, we have already seen how sampling can be used to speed up the clustering algorithms, e.g, to reduce the search space as in CLARA and CLARANS and to do pre-

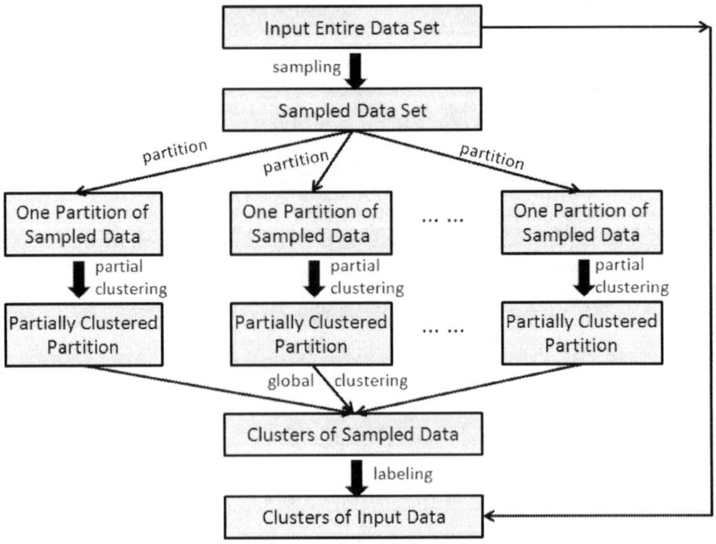

FIGURE 11.4: The flowchart of CURE algorithm.

clustering/partial clustering as in BIRCH and CURE. In this section, let us take one step further to see how randomized techniques can be used to address the scalability issue in clustering.

Suppose that we want to cluster n data points in d dimensional space into k clusters. We can represent these data points by an $n \times d$ data matrix A; and each row of this data matrix is a data point in d dimensional space. The basic idea of various randomized techniques is to reduce the scale of the input data matrix A, e.g., by transforming (embedding, projecting, etc.) it into a lower t-dimensional space ($t \ll d$) and then performing clustering on this reduced space. In this section, we will review two major types of such techniques.

11.3.1 Locality-Preserving Projection

In the locality-preserving projection (also referred to as random projection) methods, after we project the original d-dimensional data set into a lower dimensional space, we want the *pair-wise distance* to be preserved in a probabilistic sense. In many clustering algorithms, it is mainly the pair-wise distance measure that determines the clustering structure. Thus we would expect that the clustering results in the projected subspace would provide a reasonably good approximation to that in the original space.

In addition to clustering, random projection itself has a broad applicability in various data mining tasks (e.g., similarity search, near-duplicate detection). Here, we will first provide some general introduction to the typical random projection algorithms and then introduce how to apply them in the clustering setting.

A classic result for the locality-preserving projection comes from Johnson and Lindenstrauss [25], which can be formally stated as follows:

Lemma 11.3.1 *Assume* $\varepsilon > 0$, *and* n *is an integer. Let* t *be a positive integer such that* $t \geq t_0 = O(log(n)/\varepsilon^2)$. *For any set* P *with* n *data points in* R^d *space, there exist* $f : R^d \rightarrow R^t$ *such that for all* $u, v \in P$, *we have*

$$(1-\varepsilon)\|u-v\|^2 \leq \|f(u)-f(v)\|^2(1+\varepsilon)\|u-v\|^2$$

At the algorithmic level, such random projection (also referred to as "JL-embedding") can be performed by a linear transformation of the original data matrix A. Let R be a $d \times t$ rotation matrix with its elements $R(i,j)$ being independent random variables. Then $\tilde{A} = A \cdot R$ is the projected data matrix,[1] and each of its row is a vector in t dimensional space which is the projection of the corresponding data point in d dimensional space.

Different random projection algorithms differ in terms of different ways to construct such a rotation matrix R. Earlier methods suggest $R(i,j)$ as being independent Normal random variables with mean 0 and variance 1. In [1], the authors propose two database-friendly ways to construct the rotation matrix R. The first method sets $R(i,j) = \pm 1$ with equal probabilities of 0.5. Alternatively, we can also set $R(i,j) = \pm\sqrt{3}$ with equal probabilities of $1/6$ and $R(i,j) = 0$ with the probability of $2/3$.

Once we have projected the original data matrix A into a low-dimensional space, the clustering can be performed in the projected subspace $\tilde{A}$. In [18], the authors assume that the projected data points form a Gaussian-mixture model and further propose using the expectation-maximization (EM) algorithm to find the soft clusters; that is, each data point i is assigned a cluster membership probability $p(l|i,\theta)$, where l is the lth cluster and θ is a parameter for the underlying Gaussian-mixture model. The rationality of this approach can be traced back to [14], which found that by random projection, the original high-dimensional distribution tends to be more like Gaussian distribution in the projected subspace. It was also found in [10] that eccentric clusters can be reshaped to be more spherical by random projection.

Due to its random nature, the clustering result generated by the above procedure (RP+EM) might be highly unstable; that is, different runs of random projections might lead to very different clustering assignment despite the fact that each of them might *partially* reveal the true clustering structure. Based on this observation, the authors in [18] further propose running this RP+EM procedure multiple times followed up by an ensemble step. To be specific, after we run RP+EM multiple times (say T in total), we construct an $n \times n$ similarity/affinity matrix P, where each element indicates the average probability that the corresponding two data points are assigned to the same cluster, i.e., $P(i,j) = \frac{1}{T}\sum_{t=1}^{T}\sum_{l=1}^{k} p(l|i,\theta_t) \times p(l|j,\theta_t)$, where l and t are the indices for clusters and different runs, respectively. Then, an agglomerative clustering algorithm is run based on this affinity matrix P to generate the final k clusters. The empirical evaluations show that such ensemble framework leads to much more robust clustering results. Figure 11.5 summarizes the overall flowchart of this clustering ensemble framework based on RP+EM.

Given that the k-means clustering is NP-hard even if $k = 2$, many recent research works have focused on the so-called γ–approximation algorithms. Let us introduce an $n \times k$ cluster indictor matrix X, where each row of X has only one nonzero element; and $X(i,j) \neq 0$ indicates that the data point i belongs to the jth cluster.[2] Optimal k-means searches for an indicator matrix X_{opt} such that $X_{opt} = argmin_{\mathcal{X}} \|A - XX^T A\|_{fro}^2$, where $\mathcal{X}$ is the set of all valid $n \times k$ indictor matrices.

For any $\gamma > 1$ and the failure probability $0 \leq \delta_\gamma < 1$, a γ–approximation algorithm finds an indicator matrix X_γ such that with the probability at least $1 - \delta_\gamma$, we have $\|A - XX^T A\|_{fro}^2 \leq \gamma \cdot min_{\mathcal{X}} \|A - XX^T A\|_{fro}^2$. In [33], the authors propose the first *linear* time γ–approximation algorithm by sampling. More recently, the authors in [3] propose further speeding up the computation by running such a γ–approximation algorithm on the projected subspace.

---

[1] Depending on the way we construct the rotation matrix, we might need to do a constant scaling on the elements of the projected matrix $\tilde{A}$.

[2] The actual value of such non-zero elements is determined by the size of the clusters. If we have n_j data points in the jth cluster, we have $X(i,j) = 1/\sqrt{n_j}$.

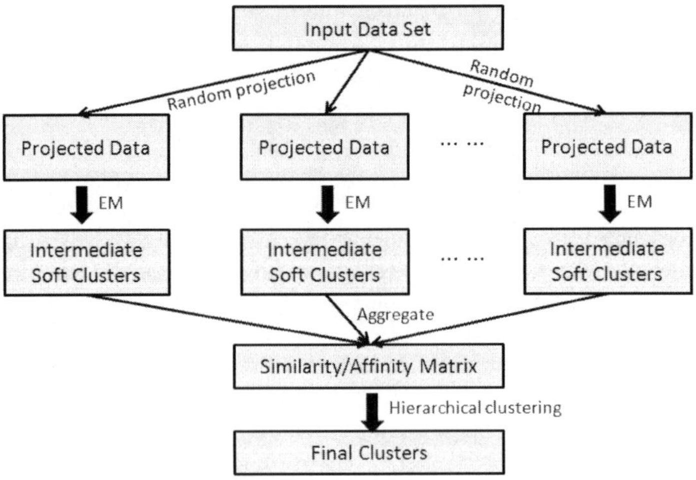

FIGURE 11.5: The flowchart of clustering ensemble based on Random Projection and EM algorithm.

11.3.2 Global Projection

In the locality-preserving projection, for any *two* data points, we want their pair-wise distance approximately unchanged before and after the projection. In global projection, for *each* data point, we want its projection to be as close as possible to the original data point. If we use the Euclidian distance, it is equivalent to saying that we want to minimize the Frobenius norm of $(\tilde{A} - A)$, where $\tilde{A}$ is the approximation of the original data matrix A.

We can represent a global projection using the notation of matrix low-rank approximation. Formally, a rank-c approximation of matrix A is a matrix $\tilde{A}$; $\tilde{A} = L \cdot M \cdot R$ where L, M, and R are of rank-c; and $\|\tilde{A} - A\|$ is small. Such a low-rank approximation often provides a powerful tool for clustering. For example, for the spatial data, the left matrix L often provides a good indicator for the cluster-membership; and the row of the right matrix R provides the description of the corresponding cluster. For the bipartite graph data, we can represent it by its adjacency matrix A; then the left and right matrices are often good indicators for row and column cluster memberships; and the middle matrix M indicates the interaction between the row and the column clusters. We refer the readers to Chapter 7 for more details on the clustering algorithms based on matrix low-rank approximation. Here, we introduce how to get such low-rank approximation efficiently so that the overall clustering procedure is efficient and scalable.

Depending on the properties of those matrices (L, M, and R), many different approximations have been proposed in the literature. For example, in singular value decomposition (SVD) [21], L and R are orthogonal matrices whose columns/rows are singular vectors and M is a diagonal matrix whose diagonal entries are singular values. Among all the possible rank-c approximations, SVD gives the best approximation in terms of squared error. However, SVD is usually dense, even if the original matrix is sparse. Furthermore, the singular vectors are abstract notions of best orthonormal basis, which is not intuitive for the interpretation.

To address the computational challenges in SVD, sparsification was proposed in [1]. The basic idea is to randomly set a significant portion of the entries in A as zeros and rescale the remaining entries; and to run SVD on the resulting sparse matrix instead. For instance, with uniform sampling, each entry $A(i,j)$ is set as zero with the probability of $1 - 1/s (s > 1)$ and is scaled as $sA(i,j)$ with the probability of $1/s$.[3] The resulting sparse matrix $\bar{A}$ can be viewed as a perturbated version of the

[3] The authors in [1] also proposed a non-uniform sampling procedure.

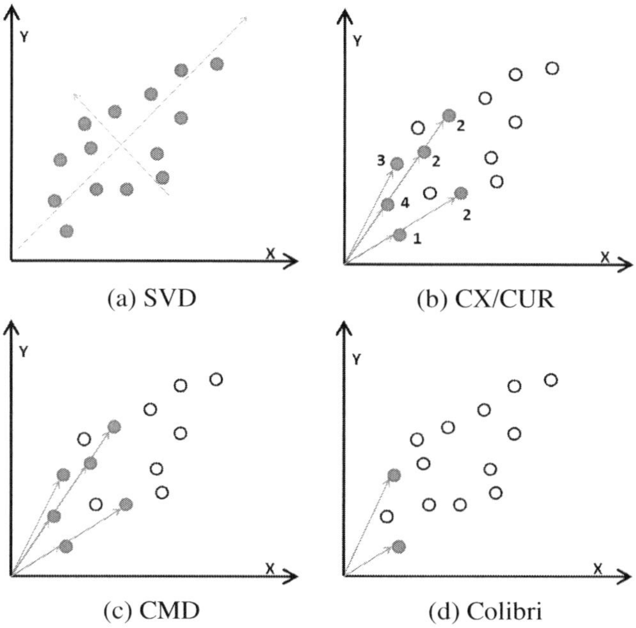

FIGURE 11.6: An illustrative comparison of SVD, CX/CUR, CMD, and Colibri. Each dot is a data point in 2-D space. SVD (a) uses all the available data points to generate optimal projection directions. CX/CUR (b) uses actual sampled data points (dark dots) as projection directions and there are a lot of duplicates. CMD (c) removes duplicate sampled data points. Colibri (d) further removes all the linearly correlated data points from sampling.

original matrix A by adding a random matrix E to it. From the random matrix theory [20], it is well known that the norm of a random matrix is well-bounded. As a result, the SVD on this sparse matrix $\tilde{A}$ provides a *near-optimal* low-rank approximation of the original data matrix A, at the benefit of significantly speeding up the computation.

Notice that the sparse SVD provides computational gain compared with the original exact SVD, but it does not address the issue of the space cost or the interpretability of the resulting low-rank approximation. For example, as shown in [44], these methods often need a huge amount of space. More recently, the so-called *example-based low-rank approximations* have started to appear, such as CX/CUR [15], CMD [44] and Colibri [45], which use the actual columns and rows of the data matrix A to form L and R. The benefit is that they provide an intuitive as well as a sparse representation, since L and R are directly sampled from the original data matrix. However, the approximation is often sub-optimal compared to SVD and the matrix M is no longer diagonal, which means a more complicated interaction.

In CX decomposition, we first randomly sample a set of columns C from the original data matrix A, and then project the A into the column space of C, i.e., $\tilde{A} = CC^{\dagger}A$, where $C^{\dagger}$ is the Moore-Penrose pseudo inverse of C. In CUR decomposition, we also do sampling on the row side to further reduce the space cost. One drawback of the original CX/CUR is the repeated sampling, i.e., some sampled columns/rows are duplicate. Based on this observation, CMD tries to remove such duplicate rows and columns before performing the projection. In [45], the authors further propose removing all the linearly correlated sampled columns. Since it does not affect the column/row space if we remove the linearly correlated columns/rows, these methods (CMD and Colibri) lead to the same approximation accuracy as CX/CUR, but require much less space and cost. Figure 11.6 provides an illustrative comparison of these different projection methods.

11.4 Parallel and Distributed Clustering Algorithms

In this section, we survey parallel and distributed clustering algorithms to handle big data. In contrast to the typical single machine clustering, parallel and distributed algorithms use multiple machines to speed up the computation and increase the scalability. The parallel and distributed algorithms are divided into two groups: traditional memory-based algorithms [13, 19, 5, 6, 41] and modern disk-based algorithms [11, 9, 29, 26].

The traditional memory-based algorithms assume that the data fit in the memory of multiple machines. The data is distributed over multiple machines, and each machine loads the data into memory.

Many of the modern disk-based algorithms use MapReduce [11], or its open-source counterpart Hadoop, to process disk resident data. MapReduce is a programming framework for processing huge amounts of data in a massively parallel way. MapReduce has two major advantages: (a) the programmer is oblivious of the details of the data distribution, replication, load balancing, etc., and (b) the programming concept is familiar, i.e., the concept of functional programming. Briefly, the programmer needs to provide only two functions, a *map* and a *reduce*. The typical framework is as follows [34]: (a) the *map* stage sequentially passes over the input file and outputs (key, value) pairs; (b) the *shuffling* stage groups all values by key, and (c) the *reduce* stage processes the values with the same key and outputs the final result. Optionally, *combiners* can be used to aggregate the outputs from local mappers. MapReduce has been used widely for large scale data mining [29, 26].

We describe both the traditional and modern parallel and distributed clustering algorithms. We first describe the general principle of parallel and distributed algorithms. Then we survey traditional algorithms including DBDC and ParMETIS . Finally we survey modern, disk-based MapReduce algorithms including PKMeans, DisCo, and BoW.

11.4.1 General Framework

Most parallel and distributed clustering algorithms follow the general framework depicted in Figure 11.7.

1. **Partition.** Data are partitioned and distributed over machines.
2. **Local Clustering.** Each machine performs local clustering on its partition of the data.
3. **Global Clustering.** The cluster information from the previous step is aggregated globally to produce global clusters.
4. **Refinement of Local Clusters.** Optionally, the global clusters are sent back to each machine to refine the local clusters.

Some algorithms (e.g. PKMeans in Section 11.4.4 and DisCo in Section 11.4.5) iteratively perform steps 3 and 4 until the quality of clusters becomes reasonably good. One of the main challenges of a parallel and distributed clustering algorithm is to minimize data traffic between the steps 2 and 3. Minimizing the traffic is important since the advantage of a parallel and distributed clustering algorithms comes from the fact that the output from the local clustering step is much smaller than the raw data. We will see how different algorithms use different strategies to minimize the data traffic.

Another issue in parallel and distributed clustering is the lower accuracy compared to the serial counterpart. There are two main reasons for the lower accuracy. First, each machine performing the local clustering might use a different clustering algorithm than the serial counterpart due to heavy communication costs. Second, even though the same algorithm is used for the local clustering step

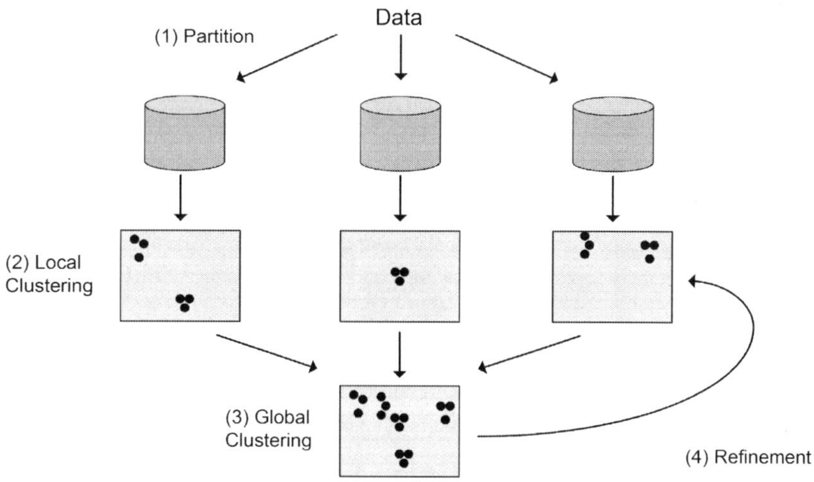

FIGURE 11.7: The general framework of most parallel and distributed clustering algorithms. (1) Data is partitioned. (2) Data is locally clustered. (3) Local clusters are merged to make global clusters. (4) Optionally, local clusters are refined using the global clusters.

as well as the serial algorithm, the divided data might change the final aggregated clusters. We will survey the clustering quality, too, in the following.

11.4.2 DBDC: Density-Based Clustering

DBDC [24] is a density-based distributed clustering algorithm. Density-based clustering aims to discover clusters of arbitrary shape. Each cluster has a density of points which is considerably higher than outside of the cluster. Also, the density within the areas of noise is lower than the density in any of the clusters.

DBDC is an exemplary algorithm that follows the general framework given in Section 11.4.1. Initially the data is partitioned over machines. At the local clustering step, each machine performs a carefully designed clustering algorithm to output a set of a small number of representatives which has an accurate description of local clusters. The representatives are merged in the global clustering step using DBSCAN [17], a single-machine density-based clustering algorithm. Then the global clusters are sent back to all clients sites which relabel all objects located on their site independently of each other.

The experimental results clearly show the advantage of the distributed clustering. The running time of DBDC is more than 30 times faster than the serial clustering counterpart. Moreover, DBDC yields almost the same clustering quality as the serial algorithm.

11.4.3 ParMETIS: Graph Partitioning

ParMETIS [31] is a parallel graph partitioning algorithm. Given a vertex and edge weighted graph, k-way graph partitioning algorithm partitions the vertices into k subsets so that the sum of the weight of the vertices in each subset is roughly the same, and the weight of the edges whose incident edges belong to different subsets is small. For example, Figure 11.8 shows an example of 3-way graph partitioning of a graph with 14 vertices. The graph partitioning is essentially a clustering problem where the goal is to find good clusters of vertices.

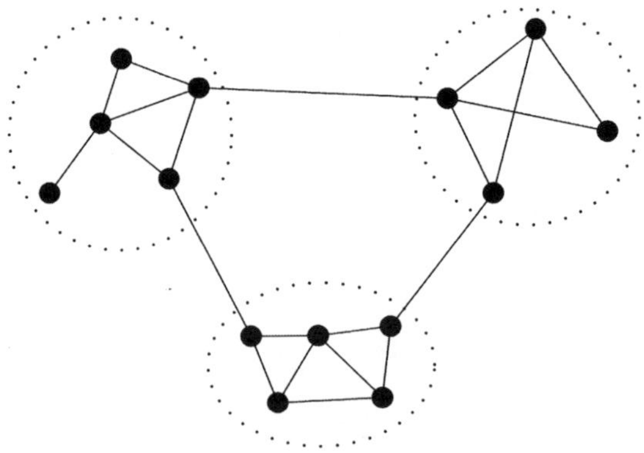

FIGURE 11.8: The k-way graph partitioning problem partitions the vertices into k subsets so that the sum of the weight of the vertices in each subset is roughly the same, and the weight of the edges whose incident edges belong to different subsets is small. The figure shows 3-way partitioning of a graph with 14 vertices. A dotted circle represents a partition.

There have been many works on the graph partitioning. One of the state-of-the-art graph partitioning algorithms is METIS [30]. ParMETIS is a parallel algorithm of METIS for distributed memory system.

METIS is a multilevel partitioning algorithm. There are three main phases in the METIS. In the first *coarsening* phase, maximal matching on the original graph is performed. Then, the matched vertices are collapsed together to create a smaller graph. This process is repeated until the number of vertices is small enough. In the second *partitioning* phase, k-way partitioning of the coarsened graph is performed by a multilevel recursive bisection algorithm. In the third *uncoarsening* phase, the partitioning from the second phase is projected back to the original graph by a greedy refinement algorithm.

ParMETIS performs the three phases of METIS using memory-based distributed systems. ParMETIS does not follow the general framework in Section 11.4.1, since the clustering is mainly based on the coarsening and the uncoarsening operations which are graph operations different from clustering operations. Initially each processor receives an equal number of vertices. At the coarsening phase, ParMETIS first computes a coloring of a graph and then computes a global graph incrementally matching only vertices of the same color one at a time. At the partitioning phase, the coarsened graph is broadcasted to all the machines. Each machine performs recursive bisection by exploring only a single path of the recursive bisection tree. At the uncoarsening phase, vertices of the same color are considered for a movement; subsets of the vertices that lead to a reduction in the edge-cut are moved. When moving vertices, the vertices themselves do not move until the final step: only the partition number associated with each vertex moves to minimize communication cost.

Experiments were performed on a 128-processor Cray T3D parallel computer with distributed memories. In terms of partition quality, the edge-cut produced by the parallel algorithm is quite close to that produced by the serial algorithm. The running time of the ParMETIS using 128 processors was from 14 to 35 times faster than the serial algorithm.

11.4.4 PKMeans: K-Means with MapReduce

MapReduce has been used for many clustering algorithms. Here we describe PKMeans [48], a k-means [37, 2, 23, 46] clustering algorithm on MapReduce. Given a set of n objects and the number of cluster k, the k-means algorithm partitions the objects so that the objects within a cluster are more similar to each other than the objects in different clusters. Initially, k objects are randomly selected as the center of clusters. Then the k-means algorithm performs the following two tasks iteratively:

- **Step 1.** Assign each object to the cluster whose center is the closest to the object.
- **Step 2.** For each cluster, update the center with the mean of the objects in the cluster.

PKMeans uses MapReduce to distribute the computation of k-means. PKMeans follows the general framework shown in Section 11.4.1. The partitioning is implicitly performed when the data is uploaded to the distributed file system (e.g., GFS or HDFS) of MapReduce. The local clustering, which corresponds to Step 1 above, is performed in the mapper. The global clustering, which corresponds to Step 2 above, is performed in the combiner and the reducer.

Specifically, the objects data $x_1 ... x_n$, for which we assume d-dimensional points, are distributed in the HDFS. The centers $y_1 ... y_k$, which are also d-dimensional points, of clusters are given to the mapper in the format of $\{i, y_i\}$ by the parameter passing mechanism of Hadoop. The details of the mapper, the combiner, and the reducer are as follows:

- Mapper: Read the object data x_i, and find the center y_j which is closest to x_i. That is, $j = \text{argmin}_l ||x_i - y_l||_2$. Emit $< j, x_i >$.
- Combiner: Take $< j, \{x_i\} >$ and emit $< j, \sum x_i, num >$ where num is the number of objects that the combiner received with the key j.
- Reducer: Take $< j, \{(\sum x_i, num)\} >$, and compute the new center. Emit the new center $< j, y_j >$ of the cluster.

The outputs of the reducer are the centers of clusters which are fed into the mapper of the next iteration. We note that the combiner greatly decreases the amount of intermediate data by emitting only the sum of the input data.

Experimental results show that PKMeans provides good *speed up*, *scale up*, and *size up*. The speed up is evaluated by the ratio of running time while keeping the dataset constant and increasing the number of machines in the system. PKMeans shows near-linear speed up. The scale up measures whether x-times larger system can perform x-times larger job in the same run-time as the original system. PKMeans shows a good scale up, better than 0.75 for 4 machines. In addition, PKMeans has a linear size up: given the fixed number of machines, the running time grows linearly with the data size. Finally, PKMeans is an exact algorithm, and thus the quality of the clustering is the same as that of the serial k-means.

11.4.5 DisCo: Co-Clustering with MapReduce

DisCo [42] is a distributed co-clustering algorithm with MapReduce. Given a data matrix, co-clustering groups the rows and columns so that the resulting permuted matrix has concentrated nonzero elements. For example, co-clustering on a documents-to-words matrix finds document groups as well as word groups. For an $m \times n$ input matrix, co-clustering outputs the row and column labeling vector $\mathbf{r} \in \{1, 2, ..., k\}^m$ and $\mathbf{c} \in \{1, 2, ..., l\}^n$, respectively, and $k \times l$ group matrix G where k and l are the number of desired row and column partitions, respectively. Searching for an optimal cluster is NP-hard [12], and thus co-clustering algorithms perform a local search. In the local search, each row is iterated to be assigned to the best group that gives the minimum cost while the column group assignments are fixed. Then in the same fashion each column is iterated while the row group assignments are fixed. This process continues until the cost function stops decreasing.

There are two important observations that affect the design of the distributed co-clustering algorithm on MapReduce:

- The numbers k and l are typically small, and thus the $k \times l$ matrix G is small. Also, the row and the column labeling vectors r and c can fit in the memory.
- For each row (or column), finding the best group assignment requires only r, c, and G. It does not require other rows.

Exploiting the above two characteristics, G, r, and c are broadcast to mappers via parameter passing mechanism of MapReduce. Then each row (or column) can be independently processed to be assigned to the best group that minimizes the cost. Each row (or column) iteration requires a map and a reduce stage, and it follows the general framework shown in Section 11.4.1. The mapper reads each row and performs local clustering. The reducer performs global clustering by gathering local cluster information and outputs the updated G matrix and the r row-label vector (c column-label vector for the column iteration). DisCo minimizes network traffic by transferring only the label vectors and the G matrix between the mapper and the reducer; the matrix data are not transferred.

Experiments were performed on a 39-node Hadoop cluster on matrix data up to 2.5 million by 1.3 million. The performance, measured by the aggregated throughput, increased linearly with the number of machines. The quality of DisCo is the same as that of the serial algorithm since DisCo is an exact algorithm.

11.4.6 BoW: Subspace Clustering with MapReduce

BoW [7] is a distributed subspace clustering algorithm on MapReduce. BoW provides two subspace clustering algorithms on MapReduce: Parallel Clustering (ParC) and Sample-and-Ignore (SnI). ParC has three steps as illustrated in Figure 11.9: (1) partition the input data using mappers so that data with the same partition are aggregated to a reducer, (2) reducers find clusters in their assigned partitions using any subspace clustering algorithm (e.g., [8]) as a plug-in), and (3) merge the clusters from the previous step to get final clusters. Steps (1) and (2) are performed in a map and a reduce stage, respectively, and step (3) is done serially in a machine.

SnI uses sampling to minimize network traffic. SnI comprises two phases as depicted in Figure 11.10. In the first phase, (1) mappers randomly sample data (e.g., 10% of the data) and send

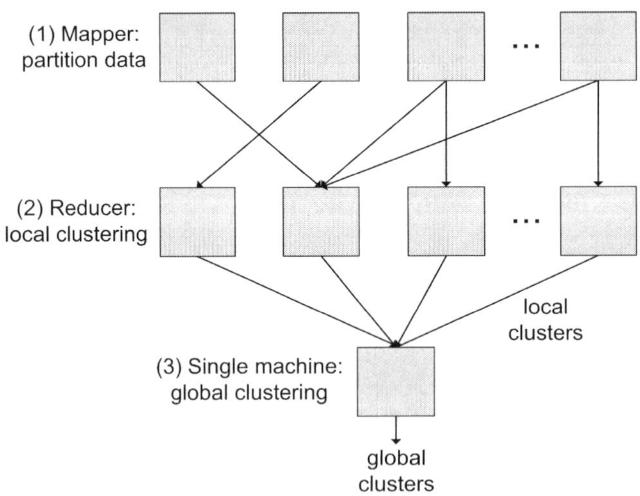

FIGURE 11.9: The ParC algorithm for subspace clustering. (1) Mappers partition the data. (2) Each reducer finds clusters. (3) A single machine collects and merges all the clusters from the output of the reducers to make final clusters.

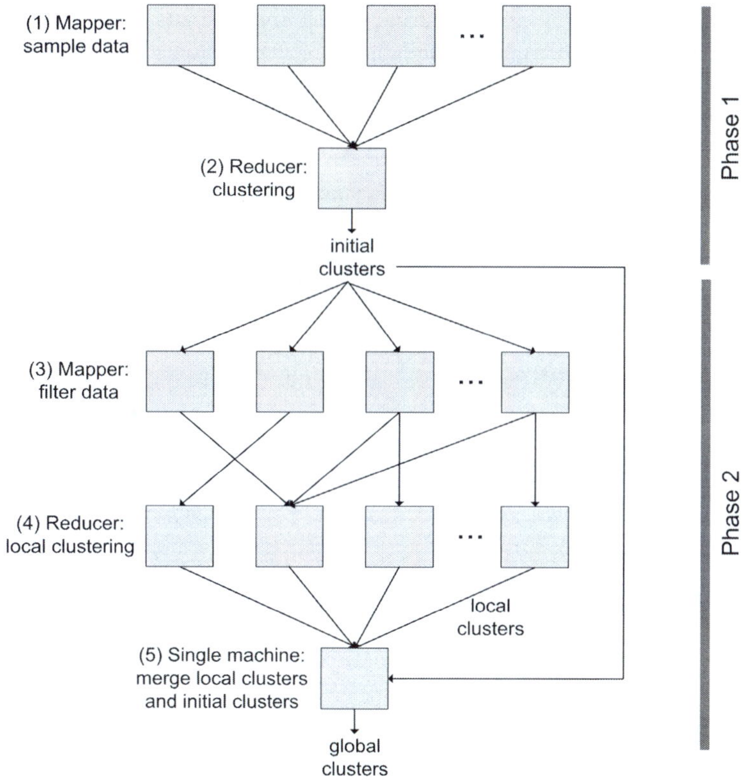

FIGURE 11.10: The SnI algorithm for subspace clustering. (1) Mappers sample the data. (2) A reducer finds initial clusters from the sampled data. (3) Mappers sends to reducers only the data which do not belong to the initial clusters from the second step. (4) Reducers find clusters. (5) A single machine combines the clusters from the second and the fourth steps to make final clusters.

results to a reducer. (2) The reducer runs a subspace clustering algorithm to find initial clusters. In the second phase, (3) mappers send to reducers only the data points which do not belong to the initial clusters found in the first phase, (4) the reducers find clusters, and (5) a single machine combines the clusters from the reducer with the initial clusters from the first phase to make final clusters.

ParC and the second phase of SnI follow the general framework described in Section 11.4.1: data is partitioned, data is locally clustered, local clusters are aggregated to make global clusters. The first phase of SnI is a preprocessing step to minimize network traffic.

ParC and SnI have their own pros and cons. In terms of the number of disk I/Os, ParC is better since it requires only one map-and-reduce step while SnI requires two map-and-reduce steps. However, in terms of the network traffic SnI is better due to the sampling in the first phase and the filtering in the second phase.

The main question is which algorithm runs faster? There are many parameters to consider to answer the question: e.g., number of reducers used, data size, disk speed, network speed, start-up cost, the plug-in algorithm cost, ratio of data transferred in the shuffling stage, and the sampling rate. BoW derives the cost (running time) as a function of the parameters to determine the algorithm that gives the minimum running time.

Experimental results show that BoW correctly chooses the best algorithm for different numbers of reducers. BoW also shows linear scalability on the data size. Finally, we note that both ParC and

SnI are not exact algorithms due to the clustering on the divided data. However, the quality of the clustering from either ParC or SnI is comparable to the serial version of the algorithm.

11.5 Conclusion

Given that (1) data size keeps growing explosively and (2) the intrinsic complexity of a clustering algorithm is often high (e.g., NP-hard), the scalability seems to be a "never-ending" challenge in clustering. In this chapter, we have briefly reviewed three basic techniques to speed up/scale up a data clustering algorithm, including (a) "one-pass" algorithms to reduce the iteration number in clustering procedure, (b) randomized techniques to reduce the complexity of the input data size, and (c) distributed and parallel algorithms to speed up/scale up the computation. A future trend is to integrate all these available techniques to achieve even better scalability.

Bibliography

[1] Dimitris Achlioptas and Frank McSherry. Fast computation of low-rank matrix approximations. *Journal of the ACM*, 54(2), 2007.

[2] Sanghamitra Bandyopadhyay, Chris Giannella, Ujjwal Maulik, Hillol Kargupta, Kun Liu, and Souptik Datta. Clustering distributed data streams in peer-to-peer environments. *Information Sciences*, 176(14):1952–1985, 2006.

[3] Christos Boutsidis, Anastasios Zouzias, and Petros Drineas. Random projections for k-means clustering. In *NIPS*, pages 298–306, 2010.

[4] Moses Charikar, Chandra Chekuri, Tomás Feder, and Rajeev Motwani. Incremental clustering and dynamic information retrieval. In *STOC*, pages 626–635, 1997.

[5] Wen-Yen Chen, Yangqiu Song, Hongjie Bai, Chih-Jen Lin, and Edward Y. Chang. Parallel spectral clustering in distributed systems. *IEEE Transactons on Pattern Analysis and Machine Intelligence*, 33(3):568–586, 2011.

[6] Cheng-Tao Chu, Sang Kyun Kim, Yi-An Lin, YuanYuan Yu, Gary R. Bradski, Andrew Y. Ng, and Kunle Olukotun. Map-Reduce for machine learning on multicore. In *NIPS*, pages 281–288, 2006.

[7] Robson Leonardo Ferreira Cordeiro, Caetano Traina Jr., Agma Juci Machado Traina, Julio López, U. Kang, and Christos Faloutsos. Clustering very large multi-dimensional datasets with MapReduce. In *KDD*, pages 690–698, 2011.

[8] Robson Leonardo Ferreira Cordeiro, Agma J. M. Traina, Christos Faloutsos, and Caetano Traina Jr. Finding clusters in subspaces of very large, multi-dimensional datasets. In *ICDE*, pages 625–636, 2010.

[9] Abhinandan Das, Mayur Datar, Ashutosh Garg, and ShyamSundar Rajaram. Google news personalization: Scalable online collaborative filtering. In *WWW*, pages 271–280, 2007.

[10] Sanjoy Dasgupta. Experiments with random projection. In *UAI*, pages 143–151, 2000.

[11] J. Dean and S. Ghemawat. MapReduce: Simplified data processing on large clusters. *OSDI*, pages 139–149, 2004.

[12] Inderjit S. Dhillon, Subramanyam Mallela, and Dharmendra S. Modha. Information-theoretic co-clustering. In *KDD*, pages 89–98, 2003.

[13] Inderjit S. Dhillon and Dharmendra S. Modha. A data-clustering algorithm on distributed memory multiprocessors. In *Large-Scale Parallel Data Mining*, pages 245–260, 1999.

[14] P. Diaconis and D. Freedman. Asymptotics of graphical projection pursuit. *Annals of Statistics*, 12(3):793–813, 1984.

[15] Petros Drineas, Ravi Kannan, and Michael W. Mahoney. Fast Monte Carlo algorithms for matrices III: Computing a compressed approximate matrix decomposition. *SIAM Journal of Computing*, 36(1):132–157, 2005.

[16] Martin Ester, Hans-Peter Kriegel, Jörg Sander, Michael Wimmer, and Xiaowei Xu. Incremental clustering for mining in a data warehousing environment. In *VLDB*, pages 323–333, 1998.

[17] Martin Ester, Hans-Peter Kriegel, Jörg Sander, and Xiaowei Xu. A density-based algorithm for discovering clusters in large spatial databases with noise. In *KDD*, pages 226–231, 1996.

[18] Xiaoli Zhang Fern and Carla E. Brodley. Random projection for high dimensional data clustering: A cluster ensemble approach. In *ICML*, pages 186–193, 2003.

[19] George Forman and Bin Zhang. Distributed data clustering can be efficient and exact. *SIGKDD Explorations*, 2(2):34–38, 2000.

[20] Z. Fredi and J. Komlos. The eigenvalues of random symmetric matrices. *Combinatorica*, 1:233–241, 1981.

[21] G. H. Golub and C. F. Van-Loan. *Matrix Computations*, 2nd edition, The Johns Hopkins University Press, Baltimore, 1989.

[22] Sudipto Guha, Rajeev Rastogi, and Kyuseok Shim. CURE: An efficient clustering algorithm for large databases. *Information Systems*, 26(1):35–58, 2001.

[23] Geetha Jagannathan and Rebecca N. Wright. Privacy-preserving distributed k-means clustering over arbitrarily partitioned data. In *KDD*, pages 593–599, 2005.

[24] Eshref Januzaj, Hans-Peter Kriegel, and Martin Pfeifle. DBDC: Density based distributed clustering. In *EDBT*, pages 88–105, 2004.

[25] W. B. Johnson and J. Lindenstrauss. Extensions of Lipschitz mappings into a Hilbert space. *Contemporary Mathematics*, 26:189–206, 1984.

[26] U. Kang, Evangelos E. Papalexakis, Abhay Harpale, and Christos Faloutsos. Gigatensor: Scaling tensor analysis up by 100 times—Algorithms and discoveries. In *KDD*, pages 316–324, 2012.

[27] U. Kang, Hanghang Tong, Jimeng Sun, Ching-Yung Lin, and Christos Faloutsos. GBASE: A scalable and general graph management system. In *KDD*, pages 1091–1099, 2011.

[28] U. Kang, Charalampos E. Tsourakakis, and Christos Faloutsos. Pegasus: A peta-scale graph mining system. In *ICDM*, pages 229–238, 2009.

[29] U. Kang, Charalampos E. Tsourakakis, and Christos Faloutsos. Pegasus: Mining peta-scale graphs. *Knowledge and Information Systems*, 27(2):303–325, 2011.

[30] George Karypis and Vipin Kumar. Multilevel k-way partitioning scheme for irregular graphs. *Journal of Parallel and Distributed Computing*, 48(1):96–129, 1998.

[31] George Karypis and Vipin Kumar. Parallel multilevel k-way partitioning for irregular graphs. *SIAM Review*, 41(2):278–300, 1999.

[32] Leonard Kaufman and Peter J. Rousseeuw. *Finding Groups in Data: An Introduction to Cluster Analysis*. John Wiley & Sons, 1990.

[33] Amit Kumar, Yogish Sabharwal, and Sandeep Sen. A simple linear time (1+)-approximation algorithm for k-means clustering in any dimensions. In *FOCS*, pages 454–462, 2004.

[34] Ralf Lämmel. Google's MapReduce programming model—Revisited. *Science of Computer Programming*, 70:1–30, 2008.

[35] Jessica Lin, Michail Vlachos, Eamonn J. Keogh, and Dimitrios Gunopulos. Iterative incremental clustering of time series. In *EDBT*, pages 106–122, 2004.

[36] Chao Liu, Fan Guo, and Christos Faloutsos. BBM: Bayesian browsing model from petabyte-scale data. In *KDD*, pages 537–546, 2009.

[37] J.B. MacQueen. Some methods for classification and analysis of multivariate observations. In *Proceedings of 5th Berkeley Symposium on Mathematical Statistics and Probability*, pages 281–297. University of California Press.

[38] Andrew McCallum, Kamal Nigam, and Lyle H. Ungar. Efficient clustering of high-dimensional data sets with application to reference matching. In *KDD*, pages 169–178, 2000.

[39] Raymond T. Ng and Jiawei Han. CLARANS: A method for clustering objects for spatial data mining. *IEEE Transactions on Knowledge and Data Engineering*, 14(5):1003–1016, 2002.

[40] Raymond T. Ng and Jiawei Han. Efficient and effective clustering methods for spatial data mining. In *VLDB*, pages 144–155, 1994.

[41] Clark F. Olson. Parallel algorithms for hierarchical clustering. *Parallel Computing*, 21(8):1313–1325, 1995.

[42] Spiros Papadimitriou and Jimeng Sun. DisCo: Distributed co-clustering with Map-Reduce: A case study towards petabyte-scale end-to-end mining. In *ICDM*, pages 512–521, 2008.

[43] Jianbo Shi and Jitendra Malik. Normalized cuts and image segmentation. *IEEE Transactions on Pattern Analysis and Machine Intelligence*, 22(8):888–905, 2000.

[44] Jimeng Sun, Yinglian Xie, Hui Zhang, and Christos Faloutsos. Less is more: Compact matrix decomposition for large sparse graphs. In *SDM*, 2007.

[45] Hanghang Tong, Spiros Papadimitriou, Jimeng Sun, Philip S. Yu, and Christos Faloutsos. Colibri: Fast mining of large static and dynamic graphs. In *KDD*, pages 686–694, 2008.

[46] Jaideep Vaidya and Chris Clifton. Privacy-preserving k-means clustering over vertically partitioned data. In *KDD*, pages 206–215, 2003.

[47] Tian Zhang, Raghu Ramakrishnan, and Miron Livny. BIRCH: An efficient data clustering method for very large databases. In *SIGMOD Conference*, pages 103–114, 1996.

[48] Weizhong Zhao, Huifang Ma, and Qing He. Parallel k-means clustering based on MapReduce. In *CloudCom*, pages 674–679, 2009.

Chapter 12

Clustering Categorical Data

Bill Andreopoulos

Lawrence Berkeley National Laboratory
Berkeley, CA
billandreo@gmail.com

12.1 Introduction

A growing number of clustering algorithms for categorical data have been proposed in recent years, along with interesting applications, such as partitioning large software systems [8, 9] and protein interaction data [13, 21, 38, 77].

A categorical dataset with m attributes is viewed as an m-dimensional "cube", offering a spatial density basis for clustering. A cell of the cube is mapped to the number of objects having values equal to its coordinates. Clusters in such a cube are regarded as *subspaces* of high object density and are separated by subspaces of low object density. Clustering the cube poses several challenges:

(i) Since there is no ordering of attribute values, the cube cells have no ordering either. The search for dense subspaces might have to consider several orderings of each dimension of the cube to identify the best clustering (unless all the attributes have binary values).

(ii) The density of a subspace is often defined relative to a user-specified value, such as a radius. However, different radii are preferable for different subspaces of the cube [17]. In dense subspaces where no information should be missed, the search is more accurately done "cell by cell" with a low radius of 1. In sparse subspaces, a higher radius may be preferable to aggregate information. The cube search could start from a low radius and gradually move to higher radii. Although the term *radius* is derived from geometrical analogies that assume circular constructs, with categorical data *radius* is not a Euclidean distance.

Figure 12.1 illustrates a 3-dimensional cube with subspaces that may be clusters. Figure 12.2 shows examples of creating and expanding clusters in a 3-D dataset. The radius is the maximum number of dimensions by which neighbors can differ.

Categorical clustering algorithms have often been motivated by density-based clustering, such as CLIQUE [4], CLICKS [75], CACTUS [32], COOLCAT [19], DBSCAN [29, 68], OPTICS [17],

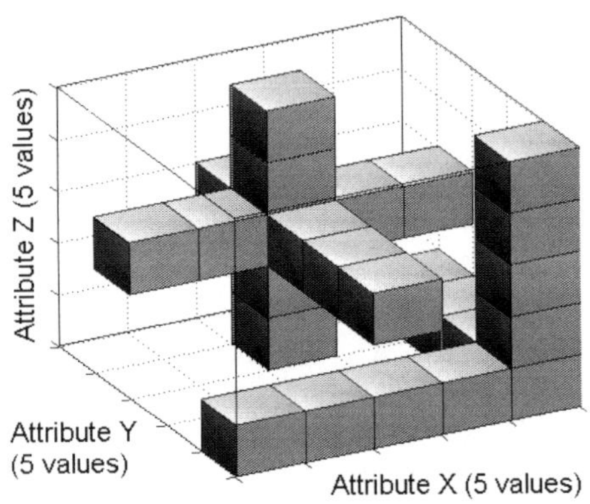

FIGURE 12.1: Two "subspaces" in a 3-D cube, for r=1.

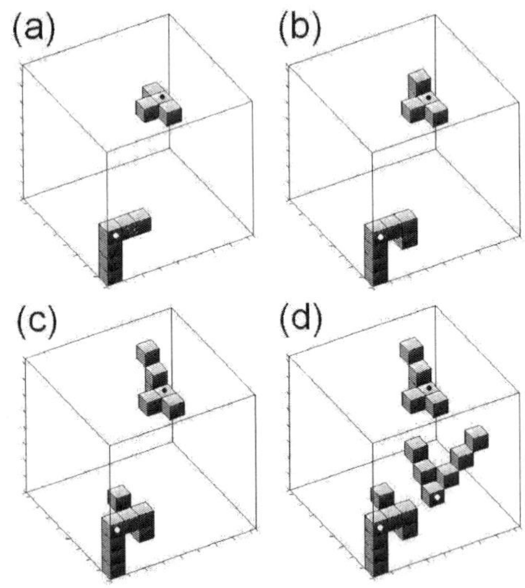

FIGURE 12.2: A cluster is a dense subspace with a "central" cell marked with a dot. (a) radius=1, two new clusters. (b) radius=1, clusters expand. (c) radius=2, clusters expand. (d) radius=2, one new cluster.

CHAMELEON [53], ROCK [39], and DENCLUE [44], and others. Although most of these approaches are efficient and relatively accurate, many of these algorithms require the user to specify input parameters (with wrong parameter values resulting in a bad clustering), may return too many clusters or too many outliers, often have difficulty finding clusters within clusters or subspace clusters, or are sensitive to the order of object input [21, 35, 38, 75].

Hierarchical algorithms for categorical clustering take the approach of building a hierarchy representing a dataset's entire underlying cluster structure. Hierarchical algorithms require few user-specified parameters and are insensitive to object ordering. Such approaches offer the cluster structure of a dataset as a hierarchy, which is usually built independent of user-specified parameters or object ordering. A user can cut its branches and study the cluster structure at different levels of granularity and detect subclusters within clusters. Though some hierarchical algorithms are slow achieving quadratic runtimes, they inspire faster simplifications that are useful for finding the rich cluster structure of a dataset.

The objectives of this chapter are to survey important clustering applications for categorical (discrete) datasets and to explain benefits and drawbacks of existing categorical clustering algorithms. This chapter is organized as follows. Section 12.2 presents the goals of categorical clustering algorithms in general. Section 12.3 gives an overview of similarity measures for categorical data. Section 12.4 describes previous categorical clustering algorithms from the literature and discusses the scalability of the algorithms. Finally we discuss open problems for future work in Section 12.5.

12.2 Goals of Categorical Clustering

Categorical clustering algorithms have various features, which make them suitable for applications with different requirements. To evaluate a categorical clustering algorithm's suitability for a problem, we use a general set of desirable features [15, 41, 42, 73]:

- Scalability: the runtime and memory requirements should not explode on large (high-dimensional) datasets [28].
- Robustness: it should have ability to detect outliers that are distant from the rest of the objects. Outliers may indicate objects that belong to a different population of the samples [39].
- Order insensitivity: a clustering algorithm should not be sensitive to the ordering of the input objects. Reordering the objects in a dataset should not result in different clusters. Order insensitivity is important for every application, as it is key to ensure reproducibility of results.
- Minimum user-specified input: parameters, such as the number of clusters, will affect the result [21, 38].
- Mixed datatypes: objects may have numerical descriptive attributes, such as a set of genes expressed at different levels over time, and discrete (categorical) descriptive attributes, such as genes with Gene Ontology annotations [1].
- Point proportion admissibility: duplicating objects and re-clustering should not change the result [37].

Evaluation of clustering quality is application-dependent, with choices for quality measures including

- Precision/recall to a gold standard [33],

- Entropy of the clusters [65],
- Reproducibility [59],
- Hubert-Arabie Indices, the number of pairs of objects that are correctly placed in the same or different clusters divided by all pairs of objects [49].

12.2.1 Clustering Road Map

Categorical clustering partitions N objects into k clusters. Object o has m attributes, $\{o_1, \cdots, o_m\}$ (usually $N >> m$). Attribute o_i, $i = 1 \cdots m$, has a domain D_i of a categorical or boolean datatype. Figure 12.3 depicts inheritance relationships between categorical clustering algorithms. Root approaches are separated into algorithms with different features: partitioning, hierarchical, density-based, model-based [41, 42, 73]. Refinement algorithms improve upon a root approach, inheriting the approach's features, while possibly introducing drawbacks. Table 12.1 compares categorical clustering algorithms' features and shows whether they are recommended for an application. The

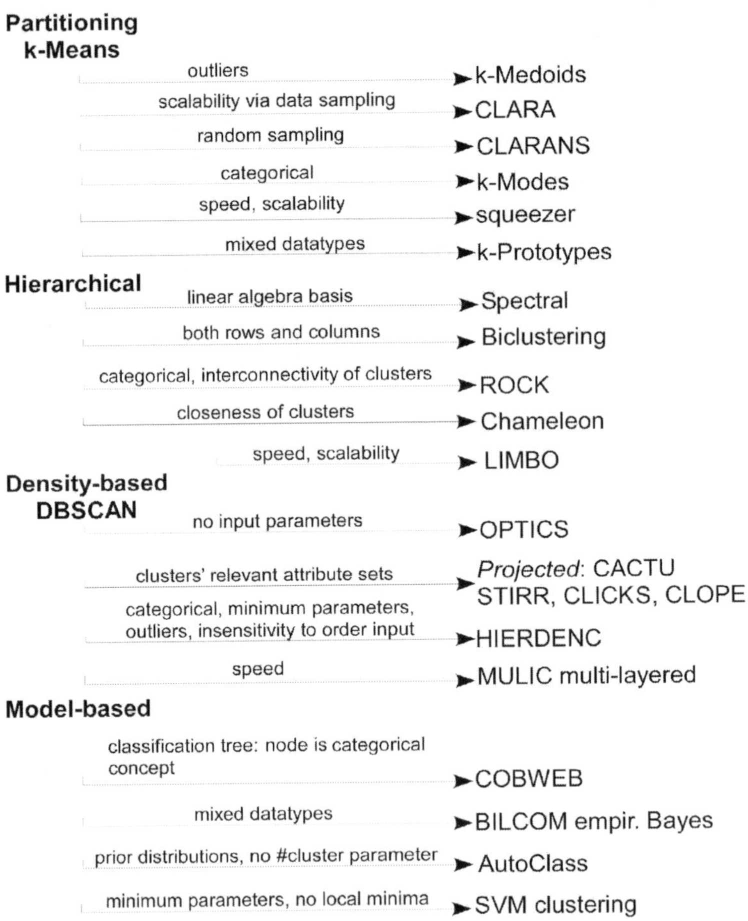

FIGURE 12.3: Classification of categorical clustering algorithms.

TABLE 12.1: Properties of Different Categorical Data Clustering Algorithms

Algorithm	Complexity	Robust to outliers	Order independence	User input	Mixed datatypes	Point prop. admiss.	Availability	Since
Partitioning (*k*-Means)								
k-Modes [47]	$O(tkN)$	no	no	1	no	no	MATLAB, Weka, R	1998
k-Prototypes [46]	$O(tkN)$	no	no	1	yes	no	MATLAB, Weka	1997
Fuzzy *k*-Modes [48]	$O(tkN)$	no	no	1	no	no	MATLAB, Weka	1999
Squeezer [43]	$O(kN)$	no	no	13	yes	no	Code	2006
COOLCAT [19]	$O(N^2)$	no	no	1	no	no	-	2002
Hierarchical								
ROCK [39]	$O(kN^2)$	no	yes	1,13	yes	no	C	2000
Chameleon [53]	$O(N^2)$	yes	yes	13	no	yes	Code, web service	1999
COBWEB [31]	$O(Nd^2)$	yes	no	-	no	no	Weka	1987
LIMBO [16]	$O(NlogN)$	yes	yes	14	no	no	C++	2004
Density-based								
HIERDENC [12, 14]	$O(N)$	yes	yes	-	yes	no	Code	2007
MULIC [13, 12, 14]	$O(N^2)$	yes	no	-	yes	no	Code, web service, Cytoscape	2006
CACTUS [32]	"scalable"	no	yes	1,4	no	no	Code	1999
CLICKS [76]	"scalable"	no	yes	-	no	no	Code	2005
STIRR [34]	"scalable"	no	no	12	no	no	-	1998
CLOPE [74]	$O(kdN)$	no	yes	-	no	no	-	2002
Model-based								
BILCOM [10]	$O(N^2)$	yes	no	5	yes	no	Code	2006
AutoClass (ExpMax) [71]	$O(kd^2Nt)$	yes	yes	-	yes	no	Weka, C++, R mclust	1995
SVM clustering [72]	$O(N^{1.8})$	no	no	-	yes	yes	MATLAB, C++, Java, SVMlight	2007

$O(.)$ notation describes how the size of the dataset affects an algorithm's runtimes; higher values are slower. In the next sections, we discuss these algorithms.

12.3 Similarity Measures for Categorical Data

The notion of similarity or distance for categorical data is not as intuitive as it is for continuous numerical data. One of the main characteristics of categorical data is that a categorical attribute takes discrete values that do not have any inherent ordering, unlike numerical attributes. Although two categorical attribute values may be identical or different, they are not directly comparable as in numerical data (by the less than $\leq$ or greater than $\geq$ relationships).

12.3.1 The Hamming Distance in Categorical and Binary Data

The simplest categorical similarity measure is the Hamming distance, which measures the overlap between two categorical data objects by counting the number of matching attributes (in which the objects have identical values). For a fixed length m, the Hamming distance (HD) is a metric on the vector space of the words of that length. Figure 12.4 shows an example of HDs in the *zoo* dataset [60]. The serpent *tuatara* is within a relatively small HD from the other serpents; the maximum distance is $HD(tuatara \leftrightarrow seasnake) = 5$. On the other hand, $HD(tuatara \leftrightarrow gorilla) = 8$, and gorilla is unlikely to belong to the class of serpents. For binary strings a and b, the HD is equivalent to the number of ones in $a\ xor\ b$. The metric space of length-m binary strings, with the HD, is known as the Hamming cube.

One obvious drawback of the Hamming distance is that all matches and mismatches are treated equally since HD does not distinguish between the different values taken by an attribute. The Hamming distance is too simplistic as it gives equal weight to all matches and mismatches. Although categorical data values do not have any inherent ordering, there is other information in categorical data that can be used to define what is more or less similar. A combination of attribute values may co-occur frequently or rarely in the dataset. This observation leads to similarity measures for categorical attributes, which take into account the frequency distribution of different attribute values in a given dataset. These measures may use probability or information theory to define similarity between categorical attribute values. Several such categorical similarity measures are described next.

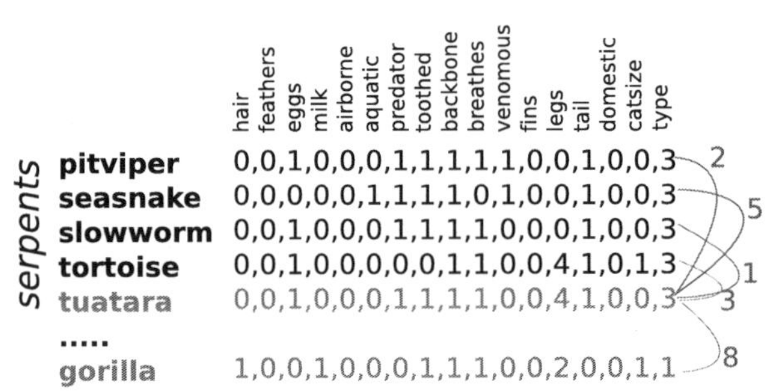

FIGURE 12.4 (See color insert): Example of Hamming distances on the *zoo* categorical dataset.

12.3.2 Probabilistic Measures

Probabilistic approaches take into account the probability of a match between attribute values taking place. The following measures are probabilistic:

Goodall: This measure assigns a higher similarity to a match if the value is infrequent than if the value is frequent. This similarity measure essentially considers the probability that the similarity value would be observed in a random sample of two points from the dataset. The maximum similarity is attained when the matching value X_k occurs twice and all other possible A_k values occur more than twice. The minimum similarity is attained when attribute A_k has only one value. The Goodall similarity formula is

$$S_k(X_k,Y_k) = 1 - \sum_{q \in Q} p_k^2(q) \ \ if \ \ X_k = Y_k, \ \ 0 \ \ otherwise.$$

The original Goodall similarity measure combines similarities in multivariate categorical data, by considering dependencies between attributes; however, this procedure is computationally expensive to compute and Boriah et al. have proposed 4 simpler variants [36, 22].

Smirnov: The Smirnov measure is a probabilistic measure that considers both matches and mismatches. This measure considers both a value's frequency as well as the distribution of the other values taken by the same attribute. The similarity is higher for a match, if the matching values are infrequent, but the other values for the attribute are frequently occurring; the maximum similarity is attained if the matching value X_k occurs only twice and there is just one other value for A_k, which occurs $N-2$ times. The minimum similarity for a matching value is attained if X_k is the only value for A_k occurring N times. For a mismatch, the maximum similarity is attained if X_k and Y_k occur once each and A_k takes just one more value occurring $N-2$ times. For a mismatch, the minimum similarity is attained when X_k, Y_k are the only two possible values for A_k [70].

Anderberg: This similarity measure also takes into account both matches and mismatches. Using this approach, matches on rare values indicate a strong similarity, while mismatches on rare values should be treated distinctly and should indicate lower similarity. It assigns higher similarity for matches on rare values and lower similarity for matches on frequent values. As for mismatches, it decreases the similarity for mismatches on rare values (decreasing it less for mismatches on frequent value) [6].

12.3.3 Information-Theoretic Measures

Information-theoretic approaches incorporate the information content of a particular attribute value with respect to the data set. Usually, these measures are inspired by information theory, where attribute values that are rarely observed are considered more informative. The following measures are information-theoretic:

Burnaby: This measure assigns higher similarity to mismatches on frequent values and lower similarity to mismatches on rare values. For mismatching values, the range of $S_k(X_k,Y_k)$ is $[\frac{Nlog(1-frac1N)}{Nlog(1-frac1N)-log(N-1)}, 1]$. For matching values, this similarity measure returns $S_k(X_k,Y_k) = 1$. This formula returns the minimum similarity when all the values for attribute A_k are infrequent (each occurring only once) and the maximum value when X_k, Y_k are frequent (each occurring $N/2$ times) [25].

Lin: This measure gives higher similarity to matches on frequent values and lower similarity to mismatches on infrequent values. The basic Lin similarity formula is [58]:

$$S_k(X_k,Y_k) = 2 \ log \ p_k(X_k) \ \ if \ \ X_k = Y_k, \quad 2log(p_k(X_k) + p_k(Y_k)) \ \ otherwise.$$

12.3.4 Context-Based Similarity Measures

Context-based measures evaluate similarity between two objects by examining the contexts in which they appear. For example, a context-based measure may evaluate the distance between two soft drink customers, considering one is a Coke customer and the other a Pepsi customer. So, the question becomes how to define the distance between two vectors of boolean or categorical attributes in m-space, which correspond to Coke and Pepsi customers.

The context-based metric proposed by Das and Mannila follows an iterative approach [26]. Given distances between all pairs of m attributes in the dataset, this approach iteratively computes distances between subrelations (e.g. Coke and Pepsi customers), and then it computes distances between all attributes again. After several iterations, a stable set of distances between all pairs of attributes is produced.

An overview of this similarity measure is given next. We are given a set R of m attributes. Let r be a boolean 0/1 relation over R. Initially, all distances between attribute pairs $A, B \in R$ are initialized to random values. Then, subrelation centers are computed as follows. For each $t \in r$, all attributes $A_1, \cdots, A_k$ to which t is similar (according to a criterion) are used to compute a vector $f(t)$ consisting of all the pairwise distances between t and A_i. Then, the subrelation center for each attribute $A \in R$ is the vector c_A on R, defined by the average of all vectors $f(t)$ from the previous step where t and A are similar. Next, for each pair of attributes $A, B \in R$, the distance is defined by computing distance between subrelation centers c_A and c_B. This iteration is repeated until the distance between $A, B \in R$ converges. Given attribute distances, it is easy to compute object distances and subrelation distances.

12.4 Descriptions of Algorithms

In this section, a description of the key classes of categorical data clustering algorithms is provided.

12.4.1 Partition-Based Clustering

In the partitioning approach, objects are partitioned and may change clusters based on dissimilarity. Partitioning methods are useful for bioinformatics applications where a fixed number of clusters is desired, such as small gene expression datasets [24]. A drawback is that the user typically specifies the number of clusters as an input parameter.

12.4.1.1 *k*-Modes

The k-Modes algorithm represents a cluster by a summary of the attribute-value frequencies of objects classified under the node. Iteratively, objects are assigned to a cluster and the summaries are reevaluated. Usually, the algorithm converges after a finite (small) number of iterations. The number of clusters is user-specified.

The k-Modes algorithm is an adaptation of k-Means to categorical datasets. K-Modes removes the numerical-only limitation of the k-Means algorithm but maintains its efficiency in clustering large categorical datasets. K-Modes makes the following modifications: 1) it uses a dissimilarity measure for categorical objects; 2) it replaces the means of clusters with the modes; and 3) it uses a frequency-based method to find the modes. It has similar runtime, benefits, and drawbacks as k-Means [47]. The Hamming Distance can be used for finding an object's nearest cluster mode.

The mode of a cluster is used to assign an unassigned object to the closest cluster. Let $\mathbf{X} = \{X_1, X_2, \cdots, X_n\}$ be a set of n categorical objects described by categorical attributes, $A_1, A_2, \cdots, A_m$. A mode of $\mathbf{X}$ is a vector $Q = [q_1, q_2, \cdots, q_m]$ that minimizes

$$D(\mathbf{X}, Q) = \sum_{i=1}^{n} d_1(X_i, Q)$$

where d_1 is a distance function, such as Hamming Distance. It was proven that the function $D(\mathbf{X}, Q)$ is minimized if and only if every position j of the mode Q contains the most frequently occurring value in attribute A_j, such that $fr(A_j = q_j|\mathbf{X}) \geq fr(A_j = c_{i,j}|\mathbf{X})$ for $q_j \neq c_{i,j}$ for all $j = 1, \cdots, m$. Note that the mode vector Q is not necessarily an element of X.

The mode of cluster c is a vector $\mu_c = \{\mu_{c1} \cdots \mu_{cm}\}$, where μ_{cj} is the most frequent value in c for the jth attribute. The mode of c is determined by setting μ_{cj} to the most frequent value for the jth category in c. To find a mode for a set, let $n_{ci,j}$ be the number of objects having the ith value $c_{i,j}$ in attribute A_j and $fr(A_j = c_{i,j}|\mathbf{X}) = n_{ci,j}/n$ be the relative frequency of value $c_{i,j}$ in $\mathbf{X}$.

The basic algorithm evaluates the total cost against the whole dataset, according to the cost function $P(\underline{X}, Q) = \sum_{l=1}^{k} \sum_{i=1}^{n} \sum_{j=1}^{m} \delta(x_{i,j}, q_{l,j})$, each time a new mode is obtained over the k clusters, n objects, and m categorical attributes. To make the computation more efficient, the following algorithm can be used instead in practice:

1. Select k initial modes, one for each cluster.
2. Assign an object to the cluster with the nearest mode, according to a distance measure. Update the mode of the cluster after each object allocation.
3. After all objects have been assigned to clusters, retest the dissimilarity of objects against the current modes. If an object is found such that its nearest mode belongs to another cluster rather than its current one, reassign the object to that cluster and update the modes of both clusters.
4. Repeat until no object changes clusters.

In the original published implementation of the k-Modes algorithm two initial mode selection methods were described. The first and simplest method selects the first k distinct records from the dataset as the initial k modes. The second method evaluates the frequencies of all values for all categorical attributes and stores them in an array in descending order of frequency; then it assigns the most frequent values to the initial k modes.

The main disadvantage of k-Modes is that (like the k-Means algorithm) it produces locally optimal solutions that are heavily dependent on the initial modes and the order of objects in the dataset.

12.4.1.2 k-Prototypes (Mixed Categorical and Numerical)

An extension of k-Modes called k-Prototypes handles mixed datatypes [46]. The k-Prototypes algorithm that is used to cluster mixed-type objects integrates the k-Means and k-Modes algorithms. The k-Prototypes algorithm is practically more useful because frequently encountered objects in real world databases are mixed-type objects. k-Prototypes uses a distance metric that weighs the contribution of the numerical versus categorical attributes. Like the k-Means and k-Modes algorithms, k-Prototypes iterates until few objects change clusters.

The dissimilarity between two mixed-type objects X and Y, which are described by p numerical and $m-p$ categorical attributes $A_1^r, \cdots, A_p^r, A_{p+1}^c \ldots A_m^c$, can be measured by

$$d_2(X, Y) = \sum_{j=1}^{p} (x_j - y_j)^2 + \gamma \sum_{j=p+1}^{m} \delta(x_j, y_j)$$

where the first term is the squared Euclidean distance measure on the numerical attributes and the second term is the simple matching dissimilarity measure on the categorical attributes. The weight γ is used to avoid favoring either type of attribute.

12.4.1.3 Fuzzy *k*-Modes

Fuzzy *k*-Modes extends fuzzy *k*-Means to categorical data [48]. Fuzzy *k*-Modes is based on the same ideas as *k*-Modes, involving iterative assignment of objects to the closest modes and re-evaluation of modes, until convergence. However, objects' assignments to clusters involve degrees of membership between 0 and 1. The degree of membership of object $1 \leq i \leq n$ in cluster $1 \leq l \leq k$ is represented by the weight $0 \leq w_{li} \leq 1$. An object's total membership across all clusters must equal 1, represented as $\sum_{l=1}^{k} w_{li} = 1$, for an object $1 \leq i \leq n$. The fuzzy *k*-Modes algorithm evaluates the total cost against the whole dataset $\mathbf{X}$, according to the cost function $F(\mathbf{W},\mathbf{X},Z) = \sum_{l=1}^{k}\sum_{i=1}^{n} w_{li}^{\alpha} d(X_i,Z_l)$, where $\alpha \geq 1$ is a weighting exponent and $Z_l = [z_{l,1},\cdots,z_{l,m}]$ is the mode of the lth cluster. The weights $0 \leq w_{li} \leq 1$ are set at each iteration, such that the ith object will tend to join the lth cluster with the least dissimilar mode Z_l. In turn, the mode Z_l of cluster $1 \leq l \leq k$, represented by $[z_{l,1},\cdots,z_{l,m}]$, has each position $z_{l,j}$ set to the value for the attribute A_j that will minimize the cost function. The cost function is the sum of the weighted distances of all objects against the cluster mode. It was proven that the fuzzy *k*-Means algorithm will converge in a finite number of iterations. Its main disadvantage is that it often terminates at a local minimum.

12.4.1.4 Squeezer

Squeezer is a one-pass algorithm that is based on summarizing clusters like *k*-Modes. However, Squeezer improves upon the iteration-bound speed of *k*-Modes [43]. Squeezer reads objects one-by-one. The first tuple forms a cluster alone. Next objects are either put into an existing cluster or rejected by all to form a new cluster.

The Squeezer algorithm accepts as input the dataset of objects and a value s for the similarity threshold. The algorithm then fetches objects in an iterative fashion. Initially, the first object is read in and a new Clustering Structure is established, which includes a summary and the cluster itself. For each subsequent object, the similarity between any existing cluster C and the object is computed. The maximal similarity value (denoted by sim-max) and the corresponding index of a cluster (denoted by index) are evaluated from the above computing results. Then, if the sim-max is larger than the input threshold s, the object is assigned to the selected cluster. Otherwise, a new Clustering Structure is constructed, consisting of the cluster and summary. Finally, outliers are handled and the clustering results are returned.

One of the disadvantages of Squeezer is that on some datasets, it may not produce accurate clusters. Squeezer is considered efficient with a complexity of $O(kN)$.

12.4.1.5 COOLCAT

COOLCAT is based on similar ideas as *k*-Modes, but instead of modes (summaries) of clusters it uses objects as cluster centers. COOLCAT attempts to deal with *k*-Modes' sensitivity to the initial cluster modes problem. Clusters are created by reducing their entropy [19]. COOLCAT finds a set of k maximally dissimilar objects to create initial clusters. All remaining objects are placed in one of the clusters, such that the increase in entropy is minimized. The name of COOLCAT comes from the notion of reducing the entropy of the clusters, thereby "cooling" them. Entropy is the measure of information and uncertainty of a random variable. If X is a random variable, $S(X)$ the set of values that X can take, and $p(x)$ the probability function of X, the entropy $E(X)$ is defined as $E(X) = -\sum_{x \in S(X)} p(x) log(p(x))$. Entropy is sometimes referred to as a measure of the amount of

TABLE 12.2: Three Different Clusterings of a Zoo Dataset

Cluster #	*Clustering 1*		*Clustering 2*		*Clustering 3*	
	members	E	members	E	members	E
Cluster 0	{ tail, 4-legs }, { tail, 2-legs }	1.0	{ tail, 4-legs }, { no-tail, 0-legs }	2.0	{ tail, 4-legs }	0
Cluster 1	{ no-tail, 0-legs }	0	{ tail, 2-legs }	0	{ tail, 2-legs }, { no-tail, 0-legs }	2.0
Exp. E		0.66		1.33		1.33

Note: Table has three objects (shown in brackets) and two attributes: tail (yes/no) and legs (zero/two/four). As shown, clustering 1 minimizes the expected entropy of the two clusters.

"disorder" in a system. The authors argue that as a measure of the similarity between two vectors, the use of entropy is equivalent to that of other widely used similarity coefficients.

COOLCAT starts with a sample of points. The initial step selects the k most dissimilar objects from the sample set, such that the pairwise entropy of the chosen objects is maximized. These serve as initial representatives of the k clusters. All remaining objects of the dataset are placed in one of the clusters such that, at each step, the increase in the entropy of the resulting clustering is minimized. After the initialization, COOLCAT proceeds to process the remaining objects of the dataset (not selected to seed the clusters initially) incrementally, finding a suitable cluster for each object. This is done by computing the expected entropy that results after placing the object in each of the clusters and selecting the cluster for which that expected entropy is the minimum. Table 12.2 compares three clusterings and selects the one that minimizes the expected entropy.

Disadvantages of COOLCAT include its sensitivity to the order of object selection, as well as its quadratic complexity of $O(N^2)$.

12.4.2 Hierarchical Clustering

Hierarchical clustering algorithms partition the objects into a tree of nodes, where each node represents a potential cluster [56, 61]. Hierarchical clustering methods applied to categorical data usually cluster the data in an agglomerative (bottom-up) fashion, where the most similar objects are gradually placed in clusters at different levels of the resulting tree. For choosing a similarity metric, there are many choices, such as Hamming distance. Alternatively, a classification tree can be created in a divisive (top-down) fashion, where every node at a level consists of a statistical summary of the resulting cluster and a new object is placed in the most suitable cluster.

Disadvantages of hierarchical methods for categorical data include their quadratic runtime and often slower speed. The resulting clustering may be sensitive to the ordering by which objects are presented. Errors in merging clusters cannot be undone and will affect the result. If large clusters are merged then interesting local cluster structure may be lost. Next, we discuss several hierarchical clustering algorithms for categorical data.

12.4.2.1 ROCK

ROCK is an agglomerative (bottom-up) hierarchical algorithm [39]. ROCK handles categorical data clustering by building a tree; at each tree level clusters are merged in such a way that the resulting intra-cluster similarity is maximized, where similarity is evaluated by the number of similar object pairs within a resulting cluster. ROCK assumes a special similarity measure between objects and defines a "link" between two objects the similarity of which exceeds a threshold.

The motivation for ROCK was to develop a global clustering approach that considers the links between objects. For this purpose, ROCK uses common neighbors to define links. If object A neighbors object C and object B neighbors object C, then the objects A and B are linked (even if they are not themselves neighbors). Two objects belonging to the same cluster should have many common neighbors, while objects belonging to different clusters should have few common neighbors. Initially, each object is assigned to a separate cluster. Then, clusters are merged repeatedly according to their "closeness," defined by the sum of the number of links between all pairs of objects between two clusters.

In order to decide which objects are "neighbors," ROCK defines a similarity function, $sim(A,C)$, which encodes the level of similarity ("closeness") between two objects. The similarity is normalized, such that $sim(A,C)$ is one when A equals C and zero when they are completely dissimilar. Objects A and C are considered to be "neighbors" if $sim(A,C) \geq \theta$, where θ is a user-provided parameter. Then $link(A,B)$ is defined to be the number of common neighbors between A and B. The similarity function can be any metric, such as Euclidean distance or the Jaccard coefficient.

Then, a hierarchical clustering algorithm that uses links is applied to the samples. It iteratively merges the clusters C_i, C_j that maximize the goodness function

$$g(C_i,C_j) = \frac{total\#crosslinks}{expected\#crosslinks} = \frac{link[C_i,C_j]}{(n_i+n_j)^{1+2f(\theta)} - n_i^{1+2f(\theta)} - n_j^{1+2f(\theta)}}.$$

It stops merging once there are no more links between clusters or the required number of clusters has been reached.

The best set of clusters is characterized through the use of a criterion function, E_l, such that the best set of clusters is the one maximizing E_l. The most common approach is to maximize the number of links between pairs of points in each cluster: $E_l = \sum_{i=1}^{k} \sum_{p_q,p_r \in C_i} link(p_q,p_r)$. This criterion keeps points that share many links in the same cluster, but it does not force points with few links to split into different clusters. This criterion approach may end up with a large cluster, or it may result in relatively weak clusters.

Disadvantages of ROCK include its cubic complexity in N, which makes it generally unsuitable for large datasets [38, 76].

12.4.2.2 COBWEB

COBWEB creates a hierarchical clustering in the form of a classification tree. COBWEB incrementally organizes objects into a classification tree. A classification tree differs from decision trees, which label branches rather than nodes and use logical rather than probabilistic descriptions. Sibling nodes at a classification tree level form a partition [31].

Each node in a classification tree represents a class (concept) and is labeled by a probabilistic concept that summarizes the attribute-value distributions of objects classified under the node. This classification tree can be used to predict the class of a new object or missing attributes. COBWEB integrates objects incrementally into an existing classification tree by classifying the object along a path of best matching nodes.

There are four operations COBWEB employs in building the classification tree. Which operation is selected for a step depends on the category utility of the classification tree achieved by applying it:

1. Insert a new node: a node is created corresponding to the object being inserted into the tree.

2. Merge two nodes: replace two nodes by a node whose children is the union of the original nodes' sets of children. The new node summarizes the attribute-value distributions of all objects classified under it.

3. Split a node: a node is split by replacing it with its children.

4. Pass an object down the hierarchy: the COBWEB algorithm is called on the object and the subtree rooted in the node.

A benefit of COBWEB is that it can adjust the number of clusters in a partition, without the user specifying this input parameter. A disadvantage of COBWEB is that it assumes categorical attributes are independent and it may assume correlated attributes are independent.

12.4.2.3 LIMBO

LIMBO handles the categorical clustering problem by building a tree, where a node contains statistics about the objects that are members of the corresponding cluster. New objects are added to the tree in a top-down fashion by finding the best matching node and possibly breaking it into a new node, thereby extending the tree. LIMBO employs Entropy and the Information Bottleneck (IB) framework for quantifying the relevant information preserved when clustering [16]. LIMBO scales to large datasets using a memory bound summary for the data, thereby improving on the scalability of other hierarchical clustering algorithms. The approach most similar to LIMBO is the COOLCAT algorithm, a nonhierarchical algorithm also based on entropy minimization, which was presented previously.

LIMBO summarizes a cluster of objects using a Distributional Cluster Feature (DCF). For a cluster c, $DCF(c) = (p(c), p(A|c))$, where $p(c)$ is the probability of cluster c, and $p(A|c)$ is the conditional probability distribution of the attribute values given the cluster c. The information in DCFs is used to compute the distance between two clusters or between a cluster and an object. The LIMBO algorithm proceeds in three phases. In the first phase, the DCF tree is constructed to summarize the data. In the second phase, the DCFs of the tree leaves are merged to produce a chosen number of clusters. In the third phase, each object is associated with the DCF to which the object is closest.

Phase 1 starts by reading and inserting objects into the DCF tree one by one. A tree is created, where a node contains one or more DCF entries. Object o is converted into DCF(o). Then, starting from the root, a path downward in the DCF tree is traced. When at a nonleaf node, the distance between DCF(o) and each DCF entry of the node is computed, finding the closest DCF entry to DCF(o). The child pointer of this entry to the next level of the tree is followed. When at a leaf node, DCF(c) denotes the DCF entry in the leaf node that is closest to DCF(o). DCF(c) is the summary of some cluster c. At this point, the decision is made whether o will be merged into the cluster c or not. If there is an empty entry in the leaf node that contains DCF(c) (according to a space bound) then DCF(o) is placed in that entry. If there is no empty leaf entry and there is sufficient free space, then the leaf node is split into two leaves. The two DCFs in the leaf node that are farthest apart are used as seeds for the new leaves. The remaining DCFs and DCF(o) are placed in the leaf that contains the closest seed DCF.

Phase 2 clusters the leafs of the DCF tree. After the construction of the DCF tree, the leaf nodes represent the DCFs of a clustering C of the objects. Each DCF(c) corresponds to a cluster $c \in C$ and contains statistics for computing probability p(c). The Agglomerative Information Bottleneck (AIB) algorithm is employed to cluster the DCFs in the leaves and produce clusterings of the DCFs. Any clustering algorithm is applicable at this phase of the algorithm.

Phase 3 associates the objects with clusters. For a chosen value of k, Phase 2 produced k DCFs that are representatives of k clusters. In the final phase 3, a scan over the dataset assigns each object to the cluster whose representative DCF is closest to the object.

12.4.3 Density-Based Clustering

Density-based approaches use a local density criterion for clustering categorical data; clusters are subspaces in which the objects are dense and are separated by subspaces of low density. Density-based methods are useful in bioinformatics for finding the densest subspaces in networks, typically involving cliques [13, 18]. Advantages of many density-based algorithms include time efficiency and the ability to find clusters of arbitrary shapes. A challenge of applying density-based clustering to categorical datasets is that the "cube" of attribute values has no ordering defined. Some density-based algorithms take user-specified input parameters, though they usually do not require the number of clusters k. Sometimes they cannot identify clusters of varying densities. With some density-based algorithms the central subspace of a cluster cannot be distinguished from the rest of the cluster based on a higher density [35, 38, 76]. Some density-based approaches are also grid-based, since a histogram is constructed by partitioning the dataset into a number of nonoverlapping regions.

12.4.3.1 Projected (Subspace) Clustering

Projected clustering is motivated by high-dimensional datasets, where clusters exist only in specific attribute subsets. Clusters are subspaces of high-dimensional datasets, determined by the subset of attributes most relevant to each cluster. The object membership in a cluster is defined by a specific range of values for the relevant attributes, while objects of other clusters are less likely to have such values. The drawback of projected (subspace) methods is that clustering depends on user parameters for determining the attributes relevant to each cluster; such parameters are the number of clusters or the average number of dimensions for each cluster. Projected clustering may distinguish the center of a cluster based on a higher frequency of values for the relevant attributes [4, 2, 3]. Next, we present several subspace clustering algorithms.

12.4.3.2 CACTUS

CACTUS is a projected clustering method, which assumes that a cluster is identified by a unique set of attribute values that seldom occur in other clusters [32]. It searches for the minimal set of relevant attribute sets that are sufficient to define a cluster.

Assume all attributes $A_1, \ldots, A_m$ are independent and all values in an attribute are equally likely. Then, the measure $\sigma(a_i, a_j)$ indicates the co-occurrence (and the similarity) of attribute values a_i and a_j. The values a_i and a_j are strongly connected if their co-occurrence $\sigma(a_i, a_j)$ is higher by a user-specified factor $\alpha > 1$ than the value expected under the attribute-independence assumption.

A set of attribute values $C = \{a_1, \ldots, a_n\}$ over the attributes $\{A_1, \ldots, A_n\}$ is a cluster if the set C is a set of strongly connected attribute values. In other words, the condition should be satisfied that all pairs of attribute values in C are strongly connected and their co-occurrence $\sigma(a_i, a_j) > \alpha$ for $i \neq j$. Cluster C is also called a subspace cluster. C is a subcluster if there is another value for one of attributes $\{A_1, \ldots, A_n\}$ that is not included in C, but is strongly connected to the other attribute values in $C = \{a_1, \ldots, a_n\}$.

The CACTUS algorithm collects inter-attribute summaries and intra-attribute summaries on categorical attributes. The inter-attribute summaries consist of all strongly connected attribute value pairs where each pair has attribute values from different attributes. The intra-attribute summaries consist of similarities between attribute values of the same attribute. Then, the CACTUS algorithm consists of three phases: summarization, clustering, and validation. The summarization phase computes the summary information from the dataset. The clustering phase uses the summary information to discover a set of candidate clusters. The validation phase determines the actual set of clusters from the set of candidate clusters.

A drawback of CACTUS is that the assumption of a cluster being identified by a unique set of attribute values that seldom occur in other clusters may be unnatural for clustering some real world datasets. CACTUS may also return too many clusters [76, 35, 38].

12.4.3.3 CLICKS

CLICKS is another subspace clustering method, standing for "Subspace Clustering of Categorical data via maximal K-partite cliques." CLICKS creates a graph representation of the dataset, where nodes are categorical attribute values and an edge is a co-occurrence of values in an object.

CLICKS models a categorical dataset D as a graph where the nodes are attribute values that form disjoint sets, one set per attribute. Edges between nodes in different partitions indicate dense relationships. A k-partite clique is a subgraph C consisting of two disjoint sets of nodes, where every pair of nodes from two sets is connected by an edge. The k-partite clique C is a maximal k-partite clique if it has no k-partite superset. C is dense if the number of objects in the dataset that have the values defined by C exceeds a user-specified threshold.

A k-subspace $C = (C_1 \times \cdots \times C_k)$ is a (subspace) cluster over attributes $A_1 \cdots A_k$ if and only if it is a maximal, dense, and a strongly connected k-partite clique in D, such that all or most pairs of nodes are connected by an edge.

CLICKS uses a three-step approach to mine all subspace clusters in D: In the preprocessing step it creates the k-partite graph from the input database D. In the clique detection step, it enumerates maximal k-partite cliques in the graph. The approach of this step is based on a backtracking search that tries to expand the current clique to ensure maximality. In the postprocessing phase it verifies the density property for the detected cliques, given a user-specified threshold. A maximal clique may fail the density test, whereas one of its subcliques may be dense.

Disadvantages include that CLICKS may return too many clusters or too many outliers [76].

12.4.3.4 STIRR

STIRR stands for "Sieving Through Iterated Relational Reinforcement." STIRR is an iterative approach for assigning and propagating weights on the categorical values; this allows a similarity measure for categorical data, which is based on the co-occurrence of values in the dataset. STIRR looks for relationships between all attribute values to detect a potential cluster and converges to clusters of highly correlated values between different categorical attribute fields [34].

Each possible value in a categorical attribute is represented by a node and the data is represented as a set D of objects. Each object $d \in D$ is represented as a set of nodes, consisting of one node from each attribute field. STIRR assigns a weight w_v to each node v. The weight configuration is referred to as w. A normalization function $N(w)$ rescales the weights of the nodes associated with an attribute such that their squares add to 1.

A function f is repeatedly applied to a set of nodes (values) until a fixed point u is reached for which $f(u) = u$. The function f maps the node weights to a new configuration. So, the purpose is to converge to a point that remains the same under repeated applications of f. The function f updates a weight w_v for a node v by applying an operator $\oplus$ to all objects that contain value v (as well as $m-1$ other values) and summing the results, as follows:

$$for\ each\ object\ o = \{v, u_1 \cdots, u_{m-1}\}\ that\ contains\ value\ v$$

$$x_r \leftarrow \oplus(u_1 \cdots, u_{m-1})$$

$$w_v \leftarrow \sum_r x_r.$$

The operator $\oplus$ may be a simple multiplication or addition operator. The function f then normalizes the set of weights using $N()$. After several iterations of yielding a new configuration $f(w)$ the system is expected to converge to a point where $f(w) = w$. Then, the nodes with large positive weights and the nodes with extreme negative weights represent dense regions in the dataset that are separated and have few interconnections, possibly defining clusters.

Disadvantages of STIRR include its sensitivity to the initial object ordering. It also lacks a definite convergence. The notion of weights is nonintuitive and several parameters are user-specified. The final detected clusters are often incomplete [76].

12.4.3.5 CLOPE

CLOPE stands for "CLustering with SLOPE." It is useful for clustering large transactional databases with high dimensions, such as market-basket data and web server logs. CLOPE is considered to be fast and scalable on transactional databases and high-dimensional datasets. CLOPE uses a heuristic of increasing the height-to-width ratio of the cluster histogram [74]. Given a cluster C, CLOPE finds all the distinct items in the cluster, with their respective occurrences, i.e., the number of transactions containing that item. D(C) is the set of distinct items, and Occ(i, C) the occurrence of item i in cluster C. CLOPE then draws the histogram of a cluster C, with items on the X-axis decreasingly ordered by their occurrences, and occurrence as the Y-axis.

Figure 12.5 shows an example of two clusterings with two clusters each. For each cluster, the occurrence of every distinct item is counted and the results are plotted as histograms. Then the height (H) and width (W) of the histogram for each cluster is obtained. Using the height-to-width ratio of the cluster histograms, CLOPE defines a global criterion function for clustering. This example shows that a larger height-to-width ratio of the histogram means better intra-cluster similarity.

The main disadvantage of CLOPE clustering is that the histogram used may produce suboptimal clusters for some datasets.

12.4.3.6 HIERDENC: Hierarchical Density-Based Clustering

The HIERDENC algorithm for "hierarchical density-based clustering of categorical data" offers a probabilistic basis for designing categorical clustering algorithms, as well as indexing methods for categorical data. Benefits of HIERDENC are its insensitivity to order of object input, no re-clustering needed when new objects are presented, no user-specified input parameters required, and the ability to find clusters within clusters and outliers [12]. HIERDENC clusters the m-dimensional *cube* representing the spatial density of a set of objects with m categorical attributes. To find its dense subspaces, HIERDENC considers an object's neighbors to be all objects that are within a *radius* of *maximum dissimilarity*. Clusters start from the densest subspaces of the cube. Clusters expand

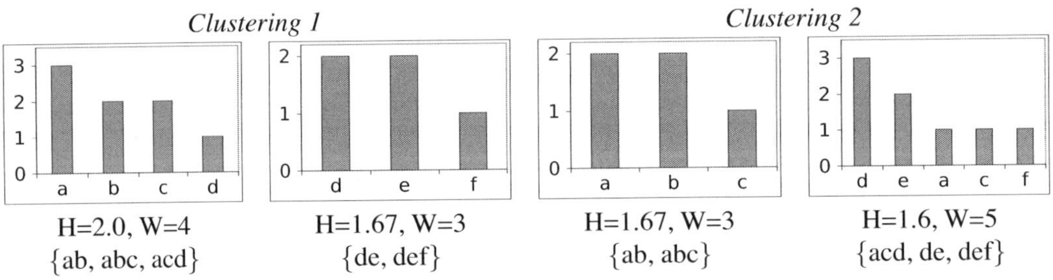

FIGURE 12.5: Histograms of two clusterings (with two clusters each) for a small dataset: (1) {{ab, abc, acd}, {de, def}} and (2) {{ab, abc}, {acd, de, def}}. The height-to-width of the clusters is used to select the best clustering. While the two histograms for clusters {de, def} and {ab, abc} are identical, the histograms for the other two clusters are of different quality. The first cluster {ab, abc, acd} has the occurrences of a:3, b:2, c:2, and d:1. The histogram for cluster {ab, abc, acd} has 4 items for 8 blocks with H=2.0 and W=4 (H/W=0.5). The histogram for the second cluster {acd, de, def} has 5 items for 8 blocks with H=1.6 and W=5 (H/W=0.32). The first clustering is preferable since a larger height-to-width ratio means more overlap among transactions in the same cluster.

outward from a dense subspace, by connecting nearby dense subspaces. Object neighborhoods are insensitive to attribute or value ordering.

We are given a dataset of objects S (which might contain duplicates) with m categorical attributes, $X_1, \cdots, X_m$. Each attribute X_i has a domain D_i with a finite number of d_i possible values. The space S^m can be viewed as an m-dimensional "cube" with $\prod_{i=1}^{m} d_i$ cells (positions). A cell $\mathbf{x} = (x_1, \cdots, x_m) \in S^m$ represents a number of objects (given by function $freq()$) with all m attribute values equal to $(x_1, \cdots, x_m)$. We define the HIERDENC *hyper-cube* $C(\mathbf{x_0}, r) \subset S^m$, centered at cell $\mathbf{x_0}$ with radius r, as follows:

$$C(\mathbf{x_0}, r) = \{\mathbf{x} : \mathbf{x} \in S^m \; and \; dist(\mathbf{x}, \mathbf{x_0}) \leq r \; and \; freq(\mathbf{x}) > 0\}.$$

The $dist(\cdot)$ is a distance function, which may be the Hamming distance.

Figure 12.1 illustrates two HIERDENC hyper-cubes in a 3-dimensional cube. Since r=1, the hyper-cubes are visualized as "crosses" in 3D and are not shown as actually having a cubic shape. A cube's cells for which $freq()$ is 0 do not represent any objects in the dataset. Normally, a hyper-cube will equal a subspace of S^m.

The *density* of a subspace $X \subset S^m$, where X could equal a hyper-cube $C(\mathbf{x_0}, r) \subset S^m$, involves the sum of $freq()$ evaluated over all cells of X:

$$density(X) = \sum_{\mathbf{c} \in X} \frac{freq(\mathbf{c})}{|S|}.$$

This density can also be viewed as the likelihood that a hyper-cube contains a random object from S, where $|S|$ is the size of S. HIERDENC seeks the densest hyper-cube $C(\mathbf{x_0}, r) \subset S^m$. This is the hyper-cube centered at $\mathbf{x_0}$ that has the maximum likelihood of containing a random object from S.

Figure 12.6 shows the HIERDENC clustering algorithm. G_k represents the kth cluster. The remainder set, $R = \{\mathbf{x} : \mathbf{x} \in S^m \; and \; \mathbf{x} \notin G_i, i = 1, \cdots, k\}$, is the collection of unclustered cells after the formation of k clusters.

Step 1 retrieves the *densest* hyper-cube $C \subset S^m$ of radius r. Step 1 checks that the densest hyper-cube represents more than one object ($density(C(\mathbf{x_0}, r)) > \frac{1}{|S|}$), since otherwise the cluster will not expand, ending up with one object. If the hyper-cube represents zero or one object, then r is incremented. *Step 2* creates a new *leaf* cluster at level $r \geq 1$. Starting from an existing leaf cluster, *step 3* tries to move to the densest hyper-cube of radius r nearby. If a dense hyper-cube is found near the cluster, then in *step 4* the cluster expands by collecting the hyper-cube's cells. This is repeated for a cluster until no such connection can be made. New objects are clustered until $r = m$, or $density(R) \leq 1\%$ and the unclustered cells are identified as outliers (*step 5*). For many datasets, most objects are likely to be clustered long before $r = m$.

Initially $r = 1$ by default, since most datasets contain subsets of similar objects. Such subsets are used to initially identify dense hyper-cubes. When r is incremented, a special process *merges* clusters that are connected relative to r. Although the initial $r = 1$ value may result in many clusters, similar clusters are merged gradually. A merge is represented as a *link* between two or more *leaf* clusters or links, created at a level $r \geq 1$. A link represents a group of merged clusters. This process gradually constructs one or more cluster tree structures, resembling hierarchical clustering [50, 61]. The user specifies a cut-off level (e.g., $r = 3$) to cut tree branches; links at the cut-off level are extracted as merged clusters. *Step 5* checks if a newly formed cluster is connected to another cluster relative to r and if so links them at level r. *Step 6* continues linking existing clusters into a tree, until $r = m$. By allowing r to reach m, an entire tree is built. At the top of the tree, there is a single cluster containing all objects of the dataset.

12.4.3.7 MULIC: Multiple Layer Incremental Clustering

MULIC stands for "MUltiple Layer Incremental Clustering" of categorical data. MULIC is a faster simplification of HIERDENC. MULIC balances clustering accuracy with time efficiency fo-

Input: space S^m.
Output: a hierarchy of clusters.
Method:

$r = 1$. //radius of hyper-cubes
$R = S^m$. //set of unclustered cells
$k = 0$. //number of leaf clusters
$k_r = 0$. //number of clusters at level r
$G_k = null$. //kth cluster
$U = null$. //set of hyper-cube centers

Step 1: Find $\mathbf{x_0} \in R$ such that $\max_{\mathbf{x_0}} density(C(\mathbf{x_0}, r))$.

If $density(C(\mathbf{x_0}, r)) \leq \frac{1}{|S|}$, then:
(1) $r = r + 1$.
(2) If $k_{r-1} > 1$, then:
(3) Merge clusters that are connected relative to r.
(4) $k_r = \#merged + \#unmerged\ clusters$.
(5) Repeat Step 1.

Step 2: Set $\mathbf{x_c} = \mathbf{x_0}$, $k = k+1$, $G_k = C(\mathbf{x_c}, r)$, $R = R - C(\mathbf{x_c}, r)$ and $U = U \cup \{\mathbf{x_c}\}$.

Step 3: Find $\mathbf{x}^* \in C(\mathbf{x_c}, r)$ such that $\mathbf{x}^* \notin U$ and $\max_{\mathbf{x}^*} density(C(\mathbf{x}^*, r))$.

Step 4: If $density(C(\mathbf{x}^*, r)) > \frac{1}{|S|}$, then:
Update current cluster G_k: $G_k = G_k \cup C(\mathbf{x}^*, r)$.
Update R: $R = R - C(\mathbf{x}^*, r)$.
Update U: $U = U \cup \{\mathbf{x}^*\}$.
Reset the new center: $\mathbf{x_c} = \mathbf{x}^*$.
Go to Step 3.
Otherwise, move to the next step.

Step 5: Set $k_r = k_r + 1$.
If $k_r > 1$, then execute lines (3) – (4).
If $r < m$ and $density(R) > 1\%$, then go to Step 1.

Step 6: While $r < m$, execute lines (1) – (4).

FIGURE 12.6: The HIERDENC algorithm.

cusing on both the quality of the clusters as well as the speed of the method and scalability to large datasets. MULIC is motivated by clustering of categorical datasets that have a *multilayered* structure. For instance, in networks a cluster (or module) may have a center of a few well-connected objects (nodes) surrounded by peripheries of sparser connectivity [27, 13, 8, 9, 11]. On such data, MULIC outperforms other algorithms that create a flat clustering. MULIC produces layered (or nested) clusters, which is different in several ways from traditional hierarchical clustering. MULIC requires no user-specified parameters and the resulting clusters have a clear separation. Layered clusters are useful in bioinformatics for finding protein modules and complexes, and for visualization purposes [7, 11, 13, 62, 68].

Input: a set S of objects.
Parameters: (1) $\delta\phi$: the increment for ϕ.
(2) *threshold* for ϕ : the maximum number of values that can differ between an object and the mode of its cluster.
Default parameter values: (1) $\delta\phi = 1$.
(2) *threshold* = the number of categorical attributes m.
Output: a set of clusters.
Method:

1. Order objects by decreasing aggregated frequency of their attribute values.
2. Insert the first object into a new cluster, use the object as the mode of the cluster, and remove the object from S.
3. Initialize ϕ to 1.
4. Loop through the following until S is empty or $\phi >$ *threshold*
   - **a.** For each object o in S
     - i. Find o's nearest cluster c by using the dissimilarity metric to compare o with the modes of all existing cluster(s).
     - ii. If the number of different values between o and c's mode is larger than ϕ, insert o into a new cluster
     - iii. Otherwise, insert o into c and update c's mode.
     - iv. Remove object o from S.
   - **b.** For each cluster c, if there is only one object in c, remove c and put the object back in S.
   - **c.** If in this iteration no objects were inserted in a cluster with *size* > 1, increment ϕ by $\delta\phi$.

FIGURE 12.7: The MULIC clustering algorithm.

Figure 12.7 shows the main part of the MULIC clustering algorithm. MULIC does not store the cube in memory and makes simplifications to decrease the runtime. A MULIC cluster starts from a dense area and expands outwards via a radius represented by the ϕ variable. When MULIC expands a cluster it does not search all member objects as HIERDENC does. Instead, it uses a mode that summarizes the content of a cluster. The mode of cluster c is a vector $\mu_c = \{\mu_{c1}, \cdots, \mu_{cm}\}$ where μ_{ci} is the most frequent value for the ith attribute in the given cluster c [47]. The MULIC clustering algorithm ensures that when an object o is clustered, it is inserted into the cluster c with the least dissimilar mode μ_c. The default dissimilarity metric between o and μ_c is the Hamming distance, although any metric could be used. A MULIC cluster consists of *layers* formed gradually, by relaxing the maximum dissimilarity criterion ϕ for inserting objects into existing clusters. MULIC does not require the user to specify the number of clusters and can identify outliers.

MULIC has several differences from traditional hierarchical clustering, which stores all distances in an upper square matrix and updates the distances after each merge [50, 61]. MULIC clusters have a clear separation. MULIC does not require a cut-off to extract the clusters, as in hierarchical clustering; this is of benefit for some MULIC applications, such as the one on protein interaction networks discussed in [7]. One of the drawbacks of hierarchical clustering is that the sequence of cluster mergings will affect the result and "bad" mergings cannot be undone later on in the process. Moreover, if several large clusters are merged, then interesting local cluster structure is likely to be lost. MULIC does not merge clusters during the object clustering process.

The best-case complexity of MULIC has a lower bound of $\Omega(mNk)$ and its worst-case complexity has an upper bound of $O(mN^2\frac{threshold}{\delta\phi})$. The MULIC complexity is comparable to that of k-Modes of $O(mNkt)$, where t is the number of iterations [47].

12.4.4 Model-Based Clustering

Model-based clustering assumes that objects match a model, which is often a statistical distribution. Then, the process aims to cluster objects such that they match the distribution. The model may be user-specified as a parameter and the model may change during the process. In bioinformatics, model-based clustering methods integrate background knowledge into gene expression, interactomes, and sequences [23, 66, 45, 51, 52, 69]. Building models is an oversimplification; user assumptions may be false and then results will be inaccurate. Another disadvantage of model-based clustering (especially neural networks) is slow processing time on large datasets.

12.4.4.1 BILCOM Empirical Bayesian (Mixed Categorical and Numerical)

BILCOM "BI-Level Clustering Of Mixed categorical and numerical data types" is a model-based method [10]. This algorithm uses categorical data clustering as an example to guide the clustering of numerical data. This process adapts an empirical Bayesian approach, with categorical data as the guide. In previous biological applications to genes, Gene Ontology (GO) annotations were the categorical data and gene expression data was numerical. Model-based clustering can find arbitrary shaped gene expression clusters by including background knowledge [1, 24, 45, 51, 52, 69, 20, 64].

The BILCOM algorithm performs clustering at two levels, where the first level clustering acts as a foundation for the second level, thus simulating a pseudo-Bayesian process. BILCOM was primarily applied to datasets from the biomedical domain. In these sets, categorical data represent semantic information on the objects, while numerical data represent experimental results. Categorical similarity is emphasized at the first level and numerical similarity at the second level. The first level result is the basis given as input to the second level and the second level result is the output of BILCOM. Figure 12.8 shows an example, where the first level and second level involve four clusters. The second level clusters consist of subclusters. Object A is assigned to different first and second level clusters, because the numerical similarity at the second level is stronger than the categorical similarity at the first level. On the other hand, object B is assigned to the same clusters in both levels, because both categorical and numerical similarities support this classification. Thus, BILCOM considers categorical and numerical similarities of an object to the clusters to which it may be assigned.

Different types of data are used at the first and second levels. The numerical data represent experimental results involving the objects. For example, the numerical data used at the second level might look as follows: BILIRUBIN : 0.39; ALBUMIN : 2.1; PROTIME : 10. The categorical data represent what was observed to be true about the objects before the experiment. For example, the categorical data used at the first level might be existing information on objects looking as follows: SEX : male; STEROID : yes; FATIGUE : no.

The main disadvantage of BILCOM is that the user has to specify the number of clusters.

12.4.4.2 AutoClass (Mixed Categorical and Numerical)

AutoClass is a clustering algorithm for mixed datatypes, which uses a Bayesian method for determining the optimal classes based on prior distributions [71]. AutoClass finds the most likely classifications of objects in clusters, given a prior distribution for each attribute, symbolizing prior beliefs of the user. In the first step the user selects a probabilistic distribution for each of the m attributes in the dataset. As an example, let the evidence on an object be $X = \{age = 28, blood\text{-}type = A, weight = 73kg\}$; blood-type is a categorical attribute that is modeled with a Bernoulli distribu-

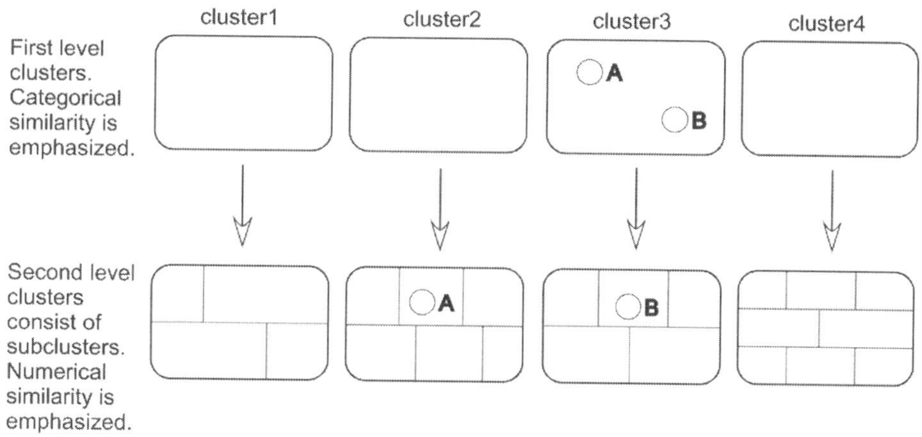

FIGURE 12.8: Overview of the BILCOM clustering process.

tion, while age and weight are numerical attributes modeled with a normal (Gaussian) distribution. At each loop, AutoClass changes the classifications of objects in clusters, such that each object is assigned to the cluster with the attribute probability distributions (considering the current means and variances) that give the highest probability of observing the object's values. Additionally, AutoClass iteratively investigates different numbers of clusters, which are not user-specified. Then, AutoClass changes the means and variances of the probability distributions for the attributes in each cluster. The iteration continues until the clusters and the attributes' probability distributions stabilize. The AutoClass output is a mixture of several likely clusterings, where a clustering classifies each object into the most likely cluster on the basis of the attributes' probability distributions.

Drawbacks of AutoClass include that users have to specify the model spaces to be searched in and wrong models may produce wrong results. Also, AutoClass can be slow to converge to a clustering result on some datasets.

12.4.4.3 SVM Clustering (Mixed Categorical and Numerical)

Support Vector Machines (SVMs) provide a method for supervised learning. SVMs construct a separating hyperplane using a set of training data, as shown in Figure 12.9. Despite their supervised nature, SVMs have been applied to categorical data to find clusters in an unsupervised manner. The approach involves randomly assigning objects to a pair of clusters and recomputing the separating hyperplane, until there is a convergence of object assignment and the hyperplane.

Previously, SVM-Internal Clustering (usually referred to as a one-class SVM) used internal aspects of Support Vector Machines to find a cluster as the smallest enclosing sphere in a dataset. The internal approach to SVM clustering lacked robustness and is biased towards clusters with a spherical shape in feature space. In the case of most real-world problems, the SVM-Internal Clustering algorithm could only detect the relatively small cluster cores.

To remedy this, an external-SVM clustering algorithm was introduced that clusters data vectors with no prior knowledge of each object's classification. Initially, every object in the dataset is randomly labeled and a binary SVM classifier is trained. After the initial convergence is achieved, the sensitivity + specificity will be low, likely near 1. The algorithm then improves this result by iteratively relabeling only the worst misclassified vectors, which have confidence factor values beyond some threshold, followed by rerunning the SVM on the newly relabeled dataset. The lowest confidence classifications, those objects with confidence factor values beyond some threshold, repeatedly

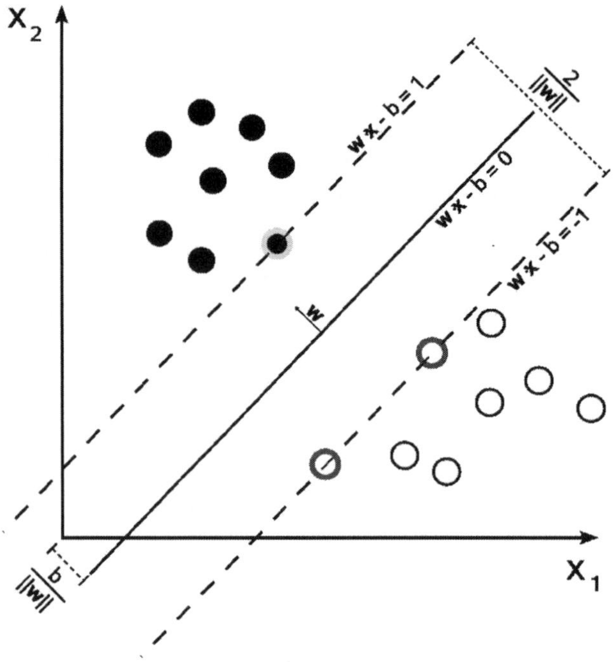

FIGURE 12.9: A Support Vector Machine tries to maximize the margin of separation between the hyperplane and the training data, which typically consists of samples from two classes. SVMs have been previously adapted for clustering. After the SVM is trained, new objects are predicted to belong in a class (as defined by the separating hyperplane) according to a measure of the distance between the testing data and the hyperplane.

have labels switched to the other class label. The SVM is retrained after each relabeling of the lowest confidence objects. This continues until no more progress can be made. Progress is determined by an increasing value of sensitivity+specificity, eventually nearly reaching 2. The repetition of the above process limits local minima traps [72].

The SVM clustering approach provides a way to cluster datasets without prior knowledge of the data's clustering characteristics or the number of clusters. This method is not biased toward the shape of the clusters, and unlike the SVM-Internal Clustering approach the formulation is not fixed to a single kernel class. Nevertheless, there are robustness and consistency issues that arise in realistic applications of SVM-External Clustering.

12.5 Conclusion

Desired features of categorical clustering algorithms for different applications include speed, minimal parameters, robustness to noise and outliers, redundancy handling, and object order independence. Desirable clustering features are used as evaluation criteria for clustering algorithms. Categorical clustering algorithms are separated into various approaches: partitioning (e.g., k-Modes), hierarchical (e.g., ROCK, Chameleon, LIMBO), density-based (e.g., MULIC, CACTUS, CLICKS, CLOPE), model-based (e.g., COBWEB, Autoclass). Some algorithms can handle mixed categorical and numerical data (*k*-Protoypes, BILCOM). Within an approach, inheritance relationships between clustering algorithms specify common features and improvements they make upon one another. Table 12.1 and Figure 12.3 summarize benefits and drawbacks of categorical clustering algorithms. For improved analysis of categorical datasets, it is important to match clustering algorithms to the requirements of an application.

The benefits and drawbacks of categorical clustering algorithms can be a basis for matching them to an application. Speed and accuracy are two competing goals in the design of categorical clustering algorithms. Making a fast algorithm usually involves trading off precision of the result. On the other hand, producing the most accurate result is not necessarily fast. Ideally, a set of probabilistically justified goals for categorical clustering would serve as a framework for approximation algorithms [54, 63]. This would allow designing and comparing categorical clustering algorithms on a more formal basis.

Bibliography

[1] B. Adryan and R. Schuh. Gene ontology-based clustering of gene expression data. *Bioinformatics* 20(16):2851–2852, 2004.

[2] C. Aggarwal, J. Han, J. Wang, and P.S. Yu. A framework for projected clustering of high dimensional data streams. In *Proceedings of the 30th VLDB Conference (VLDB'04), Toronto, Canada*, pages 852–863, 2004.

[3] C. C. Aggarwal and P. S. Yu. Finding generalized projected clusters in high dimensional spaces. In *Proceedings of the SIGMOD*, pages 70–81, 2000.

[4] R. Agrawal, J. Gehrke, D. Gunopulos, and P. Raghavan. Automatic subspace clustering of high dimensional data for data mining applications (1998). In *Proceedings ACM-SIGMOD '98 International Conference on Management on Data*, pages 94–105. Seattle, Washington, June 1998.

[5] H. Akaike. A new look at the statistical model identification. *IEEE TAC* 19:716–723, 1974.

[6] M. R. Anderberg. *Cluster Analysis for Applications*. Academic Press, New York, 1973.

[7] B. Andreopoulos. Clustering algorithms for categorical data. PhD Thesis, Dept of Computer Science & Engineering, York University, Toronto, Canada, 2006.

[8] B. Andreopoulos, A. An, V. Tzerpos, and X. Wang. Clustering large software systems at multiple layers. *Elsevier Information and Software Technology (IST)* 49(3):244–254, 2007.

[9] B. Andreopoulos, A. An, V. Tzerpos, and X. Wang. Multiple layer clustering of large software systems. In *Proceedings of the 12th IEEE Working Conference on Reverse Engineeering 2005 (WCRE 2005)*, pages 79-88, Pittsburgh, PA (Carnegie Mellon), USA, November 2005.

[10] B. Andreopoulos, A. An and X. Wang. Bi-level clustering of mixed categorical and numerical biomedical data. *International Journal of Data Mining and Bioinformatics (IJDMB)* 1(1):19-56, 2006.

[11] B. Andreopoulos, A. An, and X. Wang. Finding molecular complexes through multiple layer clustering of protein interaction networks. *International Journal of Bioinformatics Research and Applications (IJBRA)* 3(1):65–85, 2007.

[12] B. Andreopoulos, A. An and X. Wang. Hierarchical density-based clustering of categorical data and a simplification. In *Proceedings of the 11th Pacific-Asia Conference on Knowledge Discovery and Data Mining (PAKDD 2007)*, Springer LNCS 4426/2007, pages 11–22, Nanjing, China, May 22–25, 2007.

[13] B. Andreopoulos, A. An, X. Wang, M. Faloutsos, and M. Schroeder. Clustering by common friends finds locally significant proteins mediating modules. *Bioinformatics* 23(9):1124–1131, 2007.

[14] B. Andreopoulos, A. An, X. Wang, and D. Labudde. Efficient layered density-based clustering of categorical data. *Elsevier Journal of Biomedical Informatics* 42(2):365–376, Apr 2009.

[15] B. Andreopoulos, A. An, X. Wang, and M. Schroeder. A roadmap of clustering algorithms: finding a match for a biomedical application. *Briefings in Bioinformatics* 10(3):297–314, May 2009.

[16] P. Andritsos, P. Tsaparas, R. Miller, et al. LIMBO: Scalable clustering of categorical data. In *Proceedings of the 9th International Conference on Extending Database Technology EDBT'04*, pages 123–146, Heraklion, Greece. March 14–18, 2004.

[17] M. Ankerst, M. Breunig, H.P. Kriegel, and J. Sander. OPTICS: Ordering points to identify the clustering structure. *SIGMOD*, pages 49–60, 1999.

[18] G. Bader and C. Hogue. An automated method for finding molecular complexes in large protein interaction networks. *BMC Bioinformatics* 4(2), 2003.

[19] D. Barbara, Y. Li, and J. Couto. COOLCAT: An entropy-based algorithm for categorical clustering. In *Proceedings of CIKM 2002*, pages 582–589, 2002.

[20] R. Bellazzi and B. Zupan. Towards knowledge-based gene expression data mining. *Journal of Biomedical Informatics* 40(6):787-802, June 2007.

[21] P. Berkhin. Survey of Clustering Data Mining Techniques. Accrue Software, Inc. Technical report. San Jose, CA. 2002.

[22] S. Boriah, V. Chandola, and V. Kumar. Similarity measures for categorical data: A comparative evaluation. In *Proceedings of the Eighth SIAM International Conference on Data Mining*, pages 243–254, 2008.

[23] D. Brown. Efficient functional clustering of protein sequences using the Dirichlet process. *Bioinformatics* 24(16):1765–1771, 2008.

[24] M. Brown, W. Grundy, D. Lin D, et al. Knowledge-based analysis of microarray gene expression data by using support vector machines. *PNAS* 97(1): 262–267, 2000.

[25] T. Burnaby. On a method for character weighting a similarity coefficient, employing the concept of information. *Mathematical Geology* 2(1):25–38, 1970.

[26] G. Das and H. Mannila. Context-based similarity measures for categorical databases. In *Proceedings of PKDD 2000, 4th European Conference on Principles of Data Mining and Knowledge Discovery*, Pages 201– 210, Springer-Verlag, London, UK, 2000.

[27] Z. Dezso, Z.N. Oltvai, and A.L. Barabasi. Bioinformatics analysis of experimentally determined protein complexes in the yeast Saccharomyces cerevisiae. *Genome Research* 13, 2450–2454, 2003.

[28] C. Ding, X. He, H. Zha. Adaptive dimension reduction for clustering high dimensional data. In *Proceedings of ICDM 2002*, pages 107–114, 2002.

[29] M. Ester, H.P. Kriegel, J. Sander, X. Xu. A density-based algorithm for discovering clusters in large spatial databases with noise. *KDD* pages 226–231, 1996.

[30] G. Even, J. Naor, S. Rao, and B. Schieber. Fast approximate graph partitioning algorithms. *SIAM Journal on Computing* 28(6):2187–2214, 1999.

[31] Fisher D. Knowledge acquisition via incremental conceptual clustering. *Machine Learning* 2:139–172, 1987.

[32] V. Ganti, J. Gehrke, and R. Ramakrishnan. CACTUS-clustering categorical data using summaries. In *Proceedings of KDD'99* pages 73–83, San Diego, CA, USA, 1999.

[33] I. Gat-Viks, R. Sharan, and R. Shamir. Scoring clustering solutions by their biological relevance. *Bioinformatics* 19(18):2381–2389, 2003.

[34] D. Gibson, J. Kleiberg, and P. Raghavan. Clustering categorical data: An approach based on dynamical systems. In *Proceedings of 24th International Conference on Very Large Databases (VLDB'98)*, pages 311–323, New York City, USA, August 24–27, 1998.

[35] A. Gionis, A. Hinneburg, S. Papadimitriou, and P. Tsaparas. Dimension induced clustering. In *Proceedings of KDD'05*, pages 51–60, 2005.

[36] D. W. Goodall. A new similarity index based on probability. *Biometrics* 22(4):882–907, 1966.

[37] A. D. Gordon. *Classification*, 2nd Edition. London: Chapman and Hall CRC, 1999.

[38] J. Grambeier and A. Rudolph. Techniques of cluster algorithms in data mining. *Data Mining and Knowledge Discovery* 6:303–360, 2002.

[39] S. Guha, R. Rastogi, and K. Shim. ROCK: A Robust clustering algorithm for categorical attributes. *Information Systems* 25(5): 345–366, 2000.

[40] S. Guha, R. Rajeev, and S. Kyuseok. CURE: An efficient clustering algorithm for large databases. In *Proceedings of ACM SIGMOD'98*, pages 73–84, Seattle, USA, 1998.

[41] J. Han and M. Kamber. *Data Mining: Concepts and Techniques*, 2nd edition, Morgan Kaufmann, 2006.

[42] J. A. Hartigan. *Clustering Algorithms*. New York: Wiley. 1975.

[43] Z. He, X. Xu, S. Deng et al. Squeezer: An efficient algorithm for clustering categorical data. *Journal of Computer Science and Technology* 17(5):611–624, 2002.

[44] A. Hinneburg and D.A. Keim. An efficient approach to clustering in large multimedia databases with noise. *KDD*, pages 58–65, 1998.

[45] D. Huang and W. Pan. Incorporating biological knowledge into distance-based clustering analysis of microarray gene expression data. *Bioinformatics* 22(10):1259–1268. 2006.

[46] Z. Huang. Clustering large data sets with mixed numeric and categorical values. Knowledge discovery and data mining: Techniques and applications. *World Scientific*, 1997.

[47] Z. Huang. Extensions to the k-means algorithm for clustering large data sets with categorical values. *Data Mining and Knowledge Discovery* 2(3):283–304, 1998.

[48] Z. Huang and M. Ng. A fuzzy k-modes algorithm for clustering categorical data. *IEEE Transaction on Fuzzy Systems*, 7(4): 446–452. 1999.

[49] L. Hubert and P. Arabie. Comparing partitions. *Journal of Classification*, 2:193–218, 1985

[50] D. Jiang, J. Pei, and A. Zhang. DHC: A density-based hierarchical clustering method for time series gene expression data. *IEEE Symposium on Bioinformatics and Bioenginering*, pages 393–400, 2003.

[51] Y. Joo, J. Booth, Y. Namkoong, et al. Model-based Bayesian clustering (MBBC). *Bioinformatics* 24(6):874–875, 2008.

[52] A. Joshi, Y. Van de Peer, and T. Michoel. Analysis of a Gibbs sampler method for model-based clustering of gene expression data. *Bioinformatics* 24(2):176–183, 2008.

[53] G. Karypis, E. H. Han, and V. Kumar. Chameleon: A hierarchical clustering algorithm using dynamic modeling. *IEEE Computer Special Issue on Data Analysis and Mining* 32(8): 68–75, 1999.

[54] J. Kleinberg, C. Papadimitriou, and P. Raghavan. Segmentation problems. Internatinal Business Machines, Research Division, *STOC* 1998.

[55] R. Krauthgamer and J. R. Lee. The black-box complexity of nearest neighbor search. *ICALP* pages 262–276, 2004.

[56] P. Langfelder, B. Zhang, and S. Horvath. Defining clusters from a hierarchical cluster tree: The dynamic tree cut package for R. *Bioinformatics* 24(5):719–720, 2008.

[57] T. Li, S. Ma, and M. Ogihara. Entropy-based criterion in categorical clustering. *ICML*, 2004.

[58] D. Lin. An information-theoretic definition of similarity. In *Proc ICML 1998 (15th International Conference on Machine Learning)*, pages 296-304, San Francisco, CA, USA, 1998. Morgan Kaufmann Publishers Inc.

[59] L. McShane, M. Radmacher, B. Freidlin, et al. Methods for assessing reproducibility of clustering patterns observed in analyses of microarray data. *Bioinformatics* 18(11):1462–1469, 2002.

[60] C. J. Mertz and P. Murphy. UCI Repository of Machine Learning Databases, 1996.

[61] R. Mojena. Hierarchical grouping methods and stopped rules: An evaluation. *The Computer Journal* 20(4):359–63, 1977.

[62] A. Morgulis, G. Coulouris, Y. Raytselis, T.L. Madden, R. Agarwala, and A.A. Schaeffer. Database indexing for production megaBLAST searches. *Bioinformatics* 24(16):1757–64, August 2008.

[63] C. Papadimitriou. Algorithms, games, and the Internet. *STOC* pages 749–753, 2001.

[64] C. Pasquier, F. Girardot, K. Jevardat de Fombelle, et al. THEA: Ontology driven analysis of microarray data. *Bioinformatics* 20:2636-2643, 2004.

[65] I. Priness, O. Maimon, and I. Ben-Gal. Evaluation of gene-expression clustering by mutual information distance measures. *BMC Bioinformatics* 8(1):111, 2007.

[66] Y. Qi, F. Balem, C. Faloutsos, et al. Protein complex identification by supervised graph local clustering, *Bioinformatics* 24(13):i250–i258, 2008.

[67] L. Royer, M. Reimann, B. Andreopoulos, and M. Schroeder. Unravelling the modular structure of protein networks with power graph analysis. *PLoS Computational Biology* 4(7):e1000108, 1–17, July 2008.

[68] J. Sander, M. Ester, H.P. Kriegel, and X. Xu. Density-based clustering in spatial databases: the algorithm GDBSCAN and its applications. *Data Mining and Knowledge Discovery* 2(2):169–194, 1998.

[69] R. Schachtner, D. Lutter, P. Knollmueller, et al. Knowledge-based gene expression classification via matrix factorization. *Bioinformatics* 24(15):1688–1697, August 2008.

[70] E. S. Smirnov. On exact methods in systematics. *Systematic Zoology* 17(1):1–13, 1968.

[71] J. Stutz and P. Cheeseman. Bayesian classification (AutoClass): Theory and results. *Advances in Knowledge Discovery and Data Mining*, 153–180, Menlo Park, CA, AAAI Press, 1995.

[72] S. Winters-Hilt and S. Merat. SVM clustering. *BMC Bioinformatics* 8(7):S18, 2007.

[73] R. Xu. Survey of clustering algorithms. *IEEE Transactions on Neural Networks* 16(3):645–678, May 2005.

[74] Y. Yang, S. Guan, J. You. CLOPE: A fast and effective clustering algorithm for transactional data. In *Proceedings of KDD* 2002, pages 682–687, 2002.

[75] M. Zaki and M. Peters. CLICK: Clustering categorical data using K-partite maximal cliques. *TR04-11*, Rensselaer Polytechnic Institute, 2004

[76] M. J. Zaki and M. Peters. CLICKS: Mining subspace clusters in categorical data via K-partite maximal cliques. In *Proceedings of the 21st International Conference on Data Engineering (ICDE)* 2005, pages 355–356, Tokyo, Japan, 2005.

[77] Y. Zhang, A. W. Fu, C. H. Cai, and P. A. Heng. Clustering categorical data. In *Proceedings of ICDE* 2000.

Chapter 13

Document Clustering: The Next Frontier

David C. Anastasiu
University of Minnesota
Minneapolis, MN
anast021@umn.edu

Andrea Tagarelli
University of Calabria
Arcavacata di Rende, Italy
tagarelli@deis.unical.it

George Karypis
University of Minnesota
Minneapolis, MN
karypis@cs.umn.edu

13.1 Introduction

The proliferation of documents, on both the Web and in private systems, makes knowledge discovery in document collections arduous. Clustering has been long recognized as a useful tool for the task. It groups like-items together, maximizing intra-cluster similarity and inter-cluster distance. Clustering can provide insight into the make-up of a document collection and is often used as the initial step in data analysis.

While most document clustering research to date has focused on moderate length single topic documents, real-life collections are often made up of very short or long documents. Short documents do not contain enough text to accurately compute similarities. Long documents often span multiple topics that general document similarity measures do not take into account. In this chapter, we will first give an overview of *general purpose* document clustering, and then focus on recent advancements in the next frontier in document clustering: *long* and *short* documents.

13.2 Modeling a Document

Unlike the traditional clustering task, document clustering faces several additional challenges. Corpora are high-dimensional with respect to words, yet documents are sparse, are of varying length, and can contain correlated terms [2]. Finding a *document model*, a set of features that can be used to discriminate between documents, is key to the clustering task. The clustering algorithm and the measure used to compute similarity between documents is highly dependent on the chosen document model.

13.2.1 Preliminaries

For the problem of document clustering, we are given a collection of documents, or texts, $\mathcal{D} = \{d_1, \ldots, d_N\}$, called a corpus. The set of words $\mathcal{V} = \{w_1, \ldots, w_M\}$ represents the vocabulary of $\mathcal{D}$. Each document $d \in \mathcal{D}$ is a sequence of n_d words. We denote the term vector of a document by d. At times, we may consider d as being made up of contiguous, nonoverlapping chunks of text, called *segments*, which in turn are composed of sentences and words. A set of segments, $\mathcal{S}$, is called a *segment-set*. We denote with $\mathbf{S}_d$ the set of segment-sets from a document d and with $\mathbf{S} = \bigcup_{d \in \mathcal{D}} \mathbf{S}_d$ the set of segment-sets from all the documents in $\mathcal{D}$. The result of a document clustering is a set $C = \{C_1, \ldots, C_K\}$ of clusters. Table 13.1 lists main notations used throughout this chapter.

Probabilistic generative algorithms (cf. Section 13.3.4) learn a lower dimension (latent) feature space model that associates hidden topics (unobserved class variables) with word occurrences (observed data). The following notation applies to this class of algorithms. Documents are represented as sequences (rather than sets) of words, $d_i = (w_1, \ldots, w_{n_{d_i}})$. Words in a document are represented as unit-basis vectors, where w_l is a vector of size M with $w_l^u = 1$ and $w_l^v = 0$ for all indexes $u \neq v$. If the document is segmented, its segments are also considered sequences of words. However, a document can be either a sequence or a set of segments. Each z_k *topic* in $\mathcal{Z} = \{z_1, \ldots, z_K\}$, the set of latent topics, is a distribution over the vocabulary. *Topic proportions*, e.g., α and θ, are distributions over topics specifying the percentage of the document or segment that could be drawn from each topic. *Topic assignments* z and y tell which topic was selected as source for choosing a term or document, respectively.

We use plate notation, a standard representation for probabilistic generative models, to depict graphically the intricacies of some models. Plate notation should help the reader compare and con-

TABLE 13.1: Main Notations Used in This Chapter

Symbol	Description	Symbol	Description
$\mathcal{D}$	collection of documents	N	number of documents
$\mathbf{D}$	document-term matrix	S	number of segments
$d, \mathbf{d}$	document, document vector	M	number of terms
s	segment	α, θ	word-topic proportions
$\mathcal{S}$	segment-set	μ	document-topic proportions
$\mathbf{S}$	collection of segment-sets	β	word probabilities
$\mathbf{S}_d$	set of segment-sets in d	δ, η	distribution parameters
$\mathcal{C}$	document clustering solution	z	word-topic assignments
C	document cluster	y	document-topic assignments
K	number of clusters	w	observed words

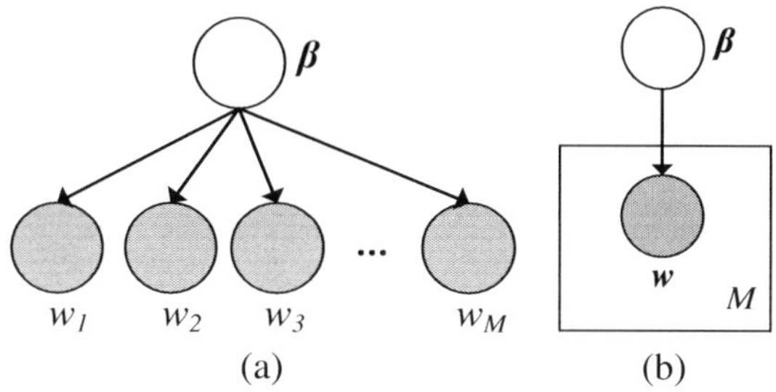

FIGURE 13.1: Example notations for a graphical model. The plate notation in (b) provides a more compact notation for the same model represented in (a).

trast the presented models. In this notation, rectangles (plates) represent repeated areas of the model. The number in the lower right corner of the plate denotes the number of times the included variables are repeated. Shaded and un-shaded variables indicate observed and unobserved (latent) variables respectively. In Figure 13.1, (a) and (b) both represent the same model in which M words are sampled from a distribution β. The depiction in (b) is more compact due to the use of plate notation.

13.2.2 The Vector Space Model

Most current document clustering methods choose to view text as a *bag of words*. Each document is considered to be a vector in the term-space, represented in its simplest form by the *term-frequency* (TF) vector

$$d_{tf} = (tf_1, tf_2, \ldots, tf_M),$$

where tf_i is the frequency of the ith term in the document. This gives the model its name, the *vector space model* (VSM).

A widely used refinement to the vector space model is to weight each term based on its *inverse document frequency* (IDF) in the document collection. The motivation behind this weighting is that terms appearing frequently in many documents have limited discrimination power and thus need to be de-emphasized. This is commonly done [91] by multiplying the frequency of the ith term by $\log(N/df_i)$, where df_i is the number of documents that contain the ith term (i.e., document frequency). This leads to the *tf-idf* representation of the document:

$$d_{tf\text{-}idf} = (tf\text{-}idf_1, tf\text{-}idf_2, \ldots, tf\text{-}idf_M).$$

Finally, to account for documents of different lengths, the length of each document vector is normalized to unit length ($||d_{tf\text{-}idf}|| = 1$); that is, each document is a vector in the unit hypersphere.

To maximize term co-occurrence in text, words can be reduced to a base form, through either stemming or lemmatization. Stemming [85] is a fast heuristic process that works on individual words, removing derivational affixes and in general cutting off the word ending in hopes of matching bases with other forms of the same word. Lemmatization uses dictionaries and morphological analysis, aiming to return the root of the word [73]. It analyzes words in context and is more computationally demanding than stemming. Synonyms of a word can also be replaced by a common form using lexical databases [50]. Some attempts have been made to capture word order and sentence structure in the VSM by encoding text as word or character n-grams (sequences of two or more items) [19, 75].

Similarity in vector space. The cosine similarity is the most used measure to compute similarity between two documents in the vector space. Given vectors d_1 and d_2, it is defined as

$$cos(d_1, d_2) = \frac{d_1 \cdot d_2}{||d_1|| \times ||d_2||},$$

where "·" represents the vector dot product operation. This formula can be simplified to $cos(d_1, d_2) = d_1 \cdot d_2$ for vectors of unit length.

Let $\mathbf{D}$ be the $N \times M$ document-term matrix, whose rows are the document-term frequency vectors. The pairwise similarities of all documents in the collection can be computed directly from $\mathbf{D}$ as

$$SIM = \mathbf{L}^{-1/2}\mathbf{X}\mathbf{L}^{-1/2},$$

where $\mathbf{X} = \mathbf{D}\mathbf{D}^T$ and $\mathbf{L}$ is an $N \times N$ diagonal matrix whose diagonal elements are the diagonal elements of $\mathbf{X}$. The left and right multiplication of $\mathbf{X}$ by $\mathbf{L}^{-1/2}$ scales the documents to unit length. The formula reduces to $SIM = \mathbf{D}\mathbf{D}^T$ if the document vectors are already unit length.

Other popular measures for comparing documents include the Euclidean, Manhattan, and Chebyshev distances, and the Jaccard coefficient similarity. The Euclidean distance, also known as the ℓ^2 norm, is simply the geometric distance in the M-dimensional space of the vectors, defined by

$$dist_2(d_1, d_2) = \sqrt{\sum_{i=1}^{M}(d_1^i - d_2^i)^2},$$

where d_1^i is the ith element in the d_1 document vector. The Manhattan (also known as the city-block distance or the ℓ^1 norm) and the Chebyshev (also known as the chessboard distance or the ℓ^∞ norm) distances are similarly defined as

$$dist_1(d_1, d_2) = \sum_{i=1}^{M} |d_1^i - d_2^i|$$

$$dist_\infty(d_1, d_2) = \max_{i=1}^{M} |d_1^i - d_2^i|$$

The Jaccard coefficient is a set similarity metric. It can be applied to a feature vector by considering its nonzero elements as set members. Using this logic, the Jaccard coefficient measures commonality, represented by the intersection of the two documents normalized by their union:

$$J(D_1, D_2) = \frac{|D_1 \cap D_2|}{|D_1 \cup D_2|},$$

where D_1 and D_2 are set representations of d_1 and d_2, respectively.

13.2.3 Alternate Document Models

Some document models have been proposed to overcome VSM limitations. Wang et al. [112] represent documents as word dependency graphs and compare them using graph similarity measures. The Matrix Space Model (MSM) [112] considers each document to be a set of segments, represented by a term-segment matrix. Concept-based models augment the original term vector by adding or replacing terms with some term category information, such as WordNet concepts [50], synsets [4], part-of-speech tags and hypernyms [94], or Wikipedia-based concepts [35].

Some models build corpus representations that allow computing semantic similarity between documents. The Generalized Vector Space Model (GVSM) [114] addresses the pairwise orthogonality assumption in the vector space model. It represents document vectors in terms of a suitably chosen set of orthonormal basic term vectors, allowing computation of term correlations. Latent Semantic Analysis (LSA) [24] finds a low-rank approximation of the document-term matrix, which effectively merges, in the latent space, dimensions associated with terms that have similar meanings.

Topic models describe a simple probabilistic process by which items can be generated in a collection. In this framework, documents are represented as mixtures of topics, in effect, *probability mass functions* (PMFs) defined over a lower-dimensional feature space representing topics. Topic models can describe words, segments, or documents, and are the basis for many generative algorithms discussed later in the chapter.

13.2.4 Dimensionality Reduction for Text

The number of unique terms in text corpora is often very high. *Dimensionality reduction* techniques aim to alleviate this problem by decreasing noise in the term space. This can be done by *feature selection*, which aims to choose an optimal subset of features given some objective function, or *feature transformation*, which seeks a lower-dimensional space mapping of the original feature space. The simplest selection technique prunes features with low or high document frequency. Frequently occurring terms are deemed uninformative, while rare terms constitute noise. *Stop words*, which are lexicon-specific frequent terms, are also removed. These simple selection techniques have been found in some cases to be as effective as more complicated supervised methods that select features based on information gain (IG), mutual information (MI), or χ^2 (Chi-Square) analysis [116].

Feature transformation algorithms project the data to some lower dimensional space. Principal Component Analysis (PCA) [49, 56] is the dominant unsupervised approach. It diagonalizes the covariance matrix $\mathbf{C}_D = \frac{1}{N-1}\mathbf{D}\mathbf{D}^T$ into $\frac{1}{N-1}(\mathbf{PD})(\mathbf{PD})^T$ and removes lesser principal components, i.e., reduces $\mathbf{P}$ to size $K \times N$, where $K < N$. Here, $\mathbf{P}$ is the matrix of principal components, whose rows are the eigenvectors of $\mathbf{D}\mathbf{D}^T$.

A related approach, Latent Semantic Analysis (LSA) [24], performs a singular-value decomposition of the document-term matrix, $\mathbf{D} = \mathbf{U}{\cdot}\mathbf{V}^T$ and keeps latent space representations of the document vectors associated with the first K singular values (largest eigenvalues). In the supervised domain, Linear Discriminant Analysis [34, 74] aims to find a latent space in which documents from different classes are well separated, by maximizing the Fisher criterion,

$$\mathbf{W} = \underset{\mathbf{W}}{\operatorname{argmax}} \frac{|\mathbf{W}^T\mathbf{S}_b\mathbf{W}|}{|\mathbf{W}^T\mathbf{S}_w\mathbf{W}|},$$

$$\mathbf{S}_b = \sum_{c \in C} n_c(\mu_c - \mu)(\mu_c - \mu)^T,$$

$$\mathbf{S}_w = \sum_{c \in C} \sum_{j:Y_j = c} (d_j - \mu_c)(d_j - \mu_c)^T,$$

where $\mathbf{S}_b$ and $\mathbf{S}_w$ are the between-class and within-class scatter matrices. Here, C is the set of class labels, μ is the collection mean, μ_c is the mean of documents in class c, n_c is the number of documents in class c, and Y_j is the label assigned to document j. The most discriminative projections are the eigenvectors associated with the largest eigenvalues of $\mathbf{S}_w^{-1}\mathbf{S}_b$.

A number of non-linear [9, 41, 101] and approximate [92, 68] extensions address the problems of non-linearly separable data and high computational complexity in the previous algorithms. While shown initially to be less effective than other methods [33], algorithms based on random projections are actively being investigated due to their lower computational complexity [3].

Feature transformation techniques have also been used for feature selection. Lu at al. choose a subset of features by analyzing principal components [71]. Hardin at al. compare SVM and Markov-Blanket based feature selection [43]. In the supervised domain, Yan at al. use the Orthogonal Centroid (OC) subspace learning algorithm to achieve optimal feature selection. As a way to bridge the gap between the two dimensionality reduction techniques, Yan at al. proposed TOFA [115], an optimization framework for both feature selection and feature transformation algorithms. Dy and Brodley [31], Vinay et al. [107], Aldo and Verleysen [66], and Cunningham [23] survey different aspects of dimensionality reduction. Additionally, Chapter 2 of this book includes a deeper discussion on *feature selection*.

13.2.5 Characterizing Extremes

As the two extremes of text data representations, long and short documents have additional characteristics that can impact how they are processed in information retrieval and data management tasks. For example, short texts often lack context, can have multiple interpretations, and use imprecise or incorrect language. Long documents are often domain specific and address multiple subjects. *Linguistic* characteristics include the size of the text and the type of language used to express ideas. *Topical* characteristics focus on the communicative function and targets of the documents. More specifically, we identify the following characterizing attributes:

- *Noise*, which is related to the use of informal language. Noisy texts are usually rich in contracted forms of words, colloquialisms, emotional punctuation and graphics, and frequently occurring typos.
- Amount of *context-shared information*, which is related to the sparseness of the text representation.
- *Community-focus*, which is regarded as the extent to which the contents of a document are of interest to a specific group of users (e.g., neighbors in a social network or a research community).
- *Domain-specificity*, which expresses the degree of alignment of the document vocabulary to a lexicon that is specific to a certain subject domain.

Note that the amount of noise and context-shared information are regarded as *linguistic* characteristics, whereas the remaining two fall into the *topical* category.

Figures 13.2 (a) and (b) graphically compare short and long documents under the above listed attributes. We have taken two of the most representative examples for each type of document: Web pages and scientific articles as long documents, and microblogs, such as tweets and search result snippets, as short documents. Increasing positions along each of the axes correspond to an increasing impact of a certain characteristic. As represented in the graphs, tweets generally feature high noise, low amount of context-shared information, high degree of community focus, low/mid domain-specificity; by contrast, scientific documents are usually less noisy and sparse, but more domain-specific.

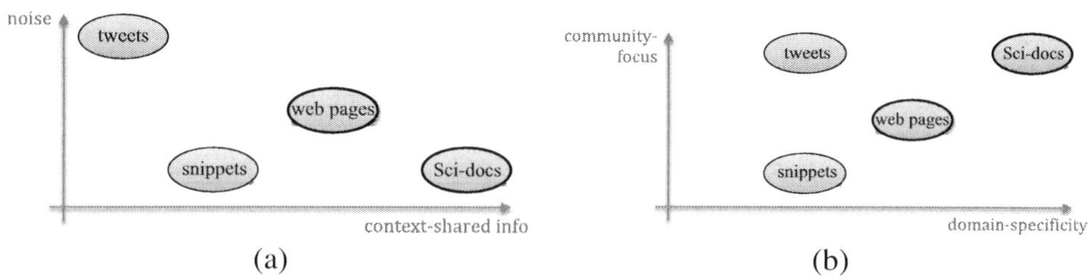

FIGURE 13.2: Comparison of example short and long documents: (a) linguistic and (b) topical characteristics. Thicker (as oppsed to thinner) ovals correspond to examples of long (as opposed to short) documents.

13.3 General Purpose Document Clustering

Most documents have moderate length, often address a single topic, and use nondescript language. Examples include Web pages, emails, encyclopedia articles, and newspaper articles. These documents have been the focus of the data mining community for many years. As a result, most document clustering algorithms to date pertain to clustering these standard document collections. In the following, we will give an overview of the most prominent of these algorithms.

13.3.1 Similarity/Dissimilarity-Based Algorithms

Traditionally, documents are grouped based on how similar they are to other documents. Similarity-based algorithms define a function for computing document similarity and use it as the basis for assigning documents to clusters. Each group (cluster) should have documents that are similar to each other and dissimilar to documents in other clusters.

Clustering algorithms fall into different categories based on the underlying methodology of the algorithm (*agglomerative* or *partitional*), the structure of the final solution (*flat* or *hierarchical*), or the multiplicity of cluster membership (*hard* or *soft*, *overlapping*, *fuzzy*). Agglomerative algorithms find the clusters by initially assigning each object to its own cluster and then repeatedly merging pairs of clusters until a certain stopping criterion is met. A number of different methods have been proposed for determining the next pair of clusters to be merged, such as group average (UPGMA) [53], single-link [97], complete link [62], CURE [39], ROCK [40], and Chameleon [59]. Hierarchical algorithms produce a clustering that forms a dendrogram, with a single all-inclusive cluster at the top and single-point clusters at the leaves. On the other hand, partitional algorithms, such as k-Means [72], k-Medoids [53, 60], graph partitioning based [117, 53, 100], and spectral partitioning based [14, 27], find the clusters by partitioning the entire dataset into either a predetermined or an automatically derived number of clusters. Depending on the particular algorithm, a k-way clustering solution can be obtained either directly or via a sequence of repeated bisections.

The Spherical k-Means algorithm (*Sk-Means*) [53] is used extensively for document clustering due to its low computational and memory requirements and its ability to find high-quality solutions. A spherical variant of the "fuzzy" version of k-Means, called Fuzzy Spherical k-Means (*FSk-Means*) [123, 64], produces an overlapping clustering by using a matrix of degrees of membership of objects with respect to clusters and a real value $f > 1$. The latter is usually called the "fuzzyfier," or fuzzyness coefficient, and controls the "softness" of the clustering solution. Higher f values lead to harder clustering solutions.

In recent years, various researchers have recognized that partitional clustering algorithms are well suited for clustering large document datasets due to their relatively low computational requirements [1, 99]. A key characteristic of many partitional clustering algorithms is that they use a global

TABLE 13.2: The Mathematical Definition of Various Clustering Criterion Functions

Criterion Function	Optimization Function	
$\mathcal{I}_\infty$	$\text{maximize} \sum_{i=1}^{K} \frac{1}{n_i} \left(\sum_{v,u \in S_i} \text{sim}(v,u) \right)$	(13.1)
$\mathcal{I}_\in$	$\text{maximize} \sum_{i=1}^{K} \sqrt{\sum_{v,u \in S_i} \text{sim}(v,u)}$	(13.2)
$\mathcal{E}_\infty$	$\text{minimize} \sum_{i=1}^{K} n_i \frac{\sum_{v \in S_i, u \in S} \text{sim}(v,u)}{\sqrt{\sum_{v,u \in S_i} \text{sim}(v,u)}}$	(13.3)
$\mathcal{G}_\infty$	$\text{minimize} \sum_{i=1}^{K} \frac{\sum_{v \in S_i, u \in S} \text{sim}(v,u)}{\sum_{v,u \in S_i} \text{sim}(v,u)}$	(13.4)
$\mathcal{G}_\in$	$\text{minimize} \sum_{r=1}^{K} \frac{\text{cut}(V_r, V - V_r)}{W(V_r)}$	(13.5)
$\mathcal{H}_\infty$	$\text{maximize} \frac{\mathcal{I}_\infty}{\mathcal{E}_\infty}$	(13.6)
$\mathcal{H}_\in$	$\text{maximize} \frac{\mathcal{I}_\in}{\mathcal{E}_\infty}$	(13.7)

Note: The notations in these equations are as follows: K is the total number of clusters, S is the total set of objects to be clustered, S_i is the set of objects assigned to the ith cluster, n_i is the number of objects in the ith cluster, v and u represent two objects, and $\text{sim}(v,u)$ is the similarity between two objects.

criterion function whose optimization drives the entire clustering process.[1] The criterion function is implicit for some of these algorithms (e.g., PDDP [14]), whereas for others (e.g., k-Means) the criterion function is explicit and can be easily stated. This later class of algorithms can be thought of as consisting of two key components. The first is the criterion function that needs to be optimized by the clustering solution, and the second is the actual algorithm that achieves this optimization. These two components are largely independent of each other.

Table 13.2 lists some of the most widely used criterion functions for document clustering. Zhao and Karypis analyze these criterion functions in both the hard and soft clustering scenarios and provide insights into their relative performance [121, 122, 123]. Various clustering algorithms and criterion functions described in this section are part of the CLUTO [58] clustering toolkit, which is available online at `http://www.cs.umn.edu/~cluto`.

13.3.2 Density-Based Algorithms

In contrast to similarity-based algorithms that often optimize a global clustering criterion function, density-based clustering algorithms focus on the local picture. DBSCAN [32] and OPTICS [7], typical density-based clustering algorithms, are designed to discover clusters of arbitrary shape in the presence of noise and have been shown effective for some text datasets. Users do not need to know the number of clusters in advance, but have to provide other parameters that are sometimes hard to identify, e.g., a density threshold and the radius of a neighborhood in the case of DBSCAN. Additionally, the indexing techniques the algorithms use for efficient neighborhood inquiry do not scale well to high-dimensional feature spaces.

[1] Global clustering criterion functions are an inherent feature of partitional clustering algorithms, but they can also be used in the context of agglomerative algorithms.

13.3.3 Adjacency-Based Algorithms

The document-term matrix naturally represents the adjacency between documents and words in a collection and can be interpreted as a graph. Spectral clustering finds cuts within the induced document-term matrix graph that produce optimal clusters. Zha et al. [120] partition the graph by minimizing a normalized sum of edge weights between unmatched vertex pairs in the graph. Their Spectral Recursive Embedding (SRE) algorithm provides an approximate solution to the problem by computing a partial singular value decomposition of a scaled document-term frequency matrix. Spectral clustering is covered in more detail in Chapter 8 of this book.

The optimal solution to the graph partitioning problem is NP-complete. Relaxations of this problem often lead to a generalized eigenvalue problem, which makes spectral clustering algorithms suitable for only small datasets with limited feature vectors. Ding et al. [28] introduce the Mcut algorithm, which solves a relaxed version of the optimization of the min-max cut objective function. They show that it produces more balanced partitions than other cuts, including the normalized cut.

Similar documents are often defined by a shared vocabulary. It stands to reason that finding word clusters in a collection can lead to identifying document clusters, and vice-versa. Co-clustering, also known as biclustering, tries to find blocks of related words and documents in the text domain, i.e. related rows and columns in the document-term matrix. Dhillon et al. [26] take an information-theoretic approach to solving the problem. In their solution, the optimal co-clustering maximizes the mutual information between document and term random variables, where the document-term matrix represents an empirical joint probability distribution of the two random variables. Equivalently, the optimal co-clustering minimizes the mutual information loss between the original random variables and the clustered random variables. Dhillon et al. formulate the problem as optimizing this loss function. At each iteration, the algorithm recomputes row cluster prototypes by using column clustering information and column cluster prototypes by using row clustering information. They show that the algorithm monotonically decreases the given objective function and is thus guaranteed to reach a local minimum in a finite number of steps.

Co-clustering can also be solved via graph-theoretic approaches. Rege et al. [89] propose the Isoperimetric Co-clustering Algorithm (ICA) which partitions the bipartite graph formed by documents and terms. It does so by heuristically minimizing the ratio of the partition perimeter and area, given an appropriate definition of graph-theoretic area. The advantage of ICA over classic spectral clustering approaches is that SVD is replaced with a solution to a system of linear equations, which is generally computationally less expensive. Gu and Zhou [38] propose a Dual Regularized Co-Clustering (DRCC) method based on semi-nonnegative matrix tri-factorization. Considering both documents and terms to be discrete samplings from separate manifolds, they construct two graphs that allow them to explore the geometric structure of the two manifolds. They ensure that both documents and terms are smooth with respect to their individual manifolds via regularizing the two graphs, enabling DRCC to utilize the encoded geometric information. The partitioning is then accomplished via semi-nonnegative matrix tri-factorization with two graph regularizers.

13.3.4 Generative Algorithms

While previous methods focus on the current picture of data, generative algorithms try to find how the documents arrived at their current state. Documents are made up of words that must be connected in certain patterns to form comprehensible language. If the generative models, the language factories, of documents could be identified, documents issued from the same models would use similar language and thus be considered similar. Generative algorithms assume documents can be represented as a mixture of probability distributions over the collection set of terms [47, 105, 15, 13, 124, 61]. For example, Probabilistic Latent Semantic Analysis (PLSA) [47, 46], a probabilistic extension of the dimensionality reduction approach based on LSA [24] (cf. Section 13.2.4), defines a statistical model in which the conditional probability be-

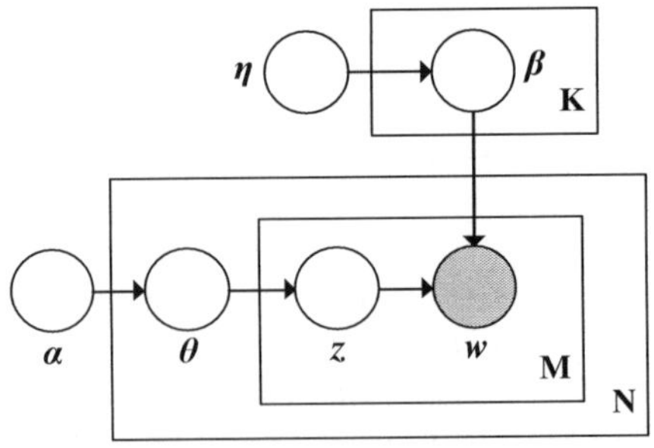

FIGURE 13.3: Plate notation for the LDA generative model.

tween documents and terms is modeled as a latent variable. An unobserved class variable is assigned to each observation (e.g., the occurrence of a term in a given document), since each document is created by a mixture of distributions.

Latent Dirichlet Allocation (LDA) [13] considers mixture models that express the so-called "exchangeability" of both terms and documents. In LDA, the generative process consists of three levels that involve the whole corpus, the documents, and the terms of each document. The algorithm first samples, for each document, a distribution over collection topics from a Dirichlet distribution. It then selects a single topic for each of a document's terms according to this distribution. Finally, each term is then sampled from a multinomial distribution over terms specific to the sampled topic. In this way, LDA defines a more sophisticated generative model for a document collection, whereas PLSA generates a model for each individual document. The complete LDA generation process, shown graphically through plate notation (cf. Section 13.2.1) in Figure 13.3, is detailed below.

1. For each topic, generate a multinomial distribution over terms, $\beta_k \sim Dir_M(\eta),\ k \in \{1,\ldots,K\}$
2. For each document $d_i,\ i \in \{1,\ldots,N\}$
   a. Generate a multinomial distribution over topics, $\theta_i \sim Dir_K(\alpha)$
   b. For each word w_{il} in document d_i
      i. Choose a topic z_{il} from the distribution in step a., e.g., $z_{il} \sim Multi(\theta_i)$
      ii. Choose word w_{il} from topic z_{il}, e.g., $w_{il} \sim Multi(\beta_{z_{il}})$

Many extensions to the initial probabilistic clustering algorithms have been developed. Chemudugunta et al. [20] propose a model that combines topic-level and word-level modeling of documents. To address the uncorrelated words assumption made by LDA, Wallach [108] generates a *bigram topic model* that incorporates a notion of word order. Bayesian nonparametric topic models [104, 12] find the number of topics exhibited in the collection as part of the inference, rather than requiring the user to provide it. Rosen-Zvi et al. [90] use a two-stage stochastic process to model the author–topic relationship. Blei [11] provides a general overview of and several future research directions for probabilistic topic models. Chapter 3 in this book provides a more in-depth look at probabilistic models for clustering.

Similarity in probabilistic space. Since generative models represent documents as probability distributions, a number of information theoretic distance metrics have been proposed for comparing two such documents. Let X be a discrete random variable defined on a sample space $X = \{x_1,\ldots,x_R\}, x_r \in \mathbb{R}, \forall r \in [1..R]$ and two PMFs $p = \{p_1,\ldots,p_R\}, q = \{q_1,\ldots,q_R\}$ for that variable. The Kullback-Leibler (KL) divergence quantifies in bits the proximity of p to q.

$$KL(p,q) = \sum_{i=1}^{R} p_i \log_2 \frac{p_i}{q_i}$$

Its value is nonnegative, is not symmetric, and will equal zero if the distributions match exactly. The Jensen-Shannon (JS) divergence is a symmetrized and smoothed version of the KL divergence, defined as

$$JS(p,q) = \frac{1}{2}KL(p, \frac{1}{2}(p+q)) + \frac{1}{2}KL(q, \frac{1}{2}(p+q)).$$

The Hellinger distance is a metric directly derived from the Bhattacharyya coefficient [57], which offers an important geometric interpretation in that it represents the cosine between any two vectors that are composed by the square root of the probabilities of their mixtures. Formally, the Hellinger distance is defined as $HL(p,q) = \sqrt{1 - BC(p,q)}$, where $BC(p,q) = \sum_{i=1}^{R} \sqrt{p(x_i)\, q(x_i)}$ is the Bhattacharyya coefficient for the two PMFs p and q.

13.4 Clustering Long Documents

Long documents often discuss multiple subjects. This presents added challenge to general purpose document clustering algorithms that tend to associate a document with a single topic. The key idea to solving this problem is to consider the document as being made up of smaller topically cohesive text blocks, named *segments*. Segments can be identified independent of or concurrent to the clustering procedure.

13.4.1 Document Segmentation

Text segmentation is concerned with the fragmentation of input text into smaller units (e.g., paragraphs) each possibly discussing a single main topic. Regardless of the presence of logical structure clues in the document, linguistic criteria and statistical similarity measures have been mainly used to identify thematically coherent, contiguous text blocks in unstructured documents [44, 10, 21].

The *TextTiling* algorithm [44] is the exemplary similarity block-based method for text segmentation. TextTiling is able to subdivide a text into multiparagraph, contiguous and disjoint blocks that represent passages, or subtopics. More precisely, TextTiling detects subtopic boundaries by analyzing patterns of lexical co-occurrence and distribution in the text. Terms that discuss a subtopic tend to co-occur locally. A switch to a new subtopic is detected when the co-occurrence of a given set of terms ends and the co-occurrence of another set of terms starts. All pairs of adjacent text blocks are compared using the cosine similarity measure and the resulting sequence of similarity values is examined in order to detect the boundaries between coherent segments.

Recent segmentation techniques have taken advantage of advances in generative topic modeling algorithms, which were specifically designed to identify topics within text. Brants et al. [15] use PLSA to compute word–topic distributions, fold in those distributions at the block level (in their case blocks are sentences), and then select segmentation points based on the similarity values of adjacent block pairs. Sun et al. [102] use LDA on a corpus of segments, compute intrasegment similarities via a Fisher kernel, and optimize segmentation via dynamic programming. Misra et al. [76] use a document-level LDA model, treat segments as new documents and predict their LDA models, and then perform segmentation via dynamic programming with probabilistic scores.

Modeling segmentation. The Segmented Topic Model (STM) [29] assumes that each segmented document has a certain mixture of latent topics and each segment within the document also has a mixture over the same latent topics as the documents. The shared latent topic pool provides a way to correlate documents and segments.

The basic idea of the LDA model is that documents can be represented as random mixtures over topics, depicted by word–topic proportions θ, where topics are distributions over words. The

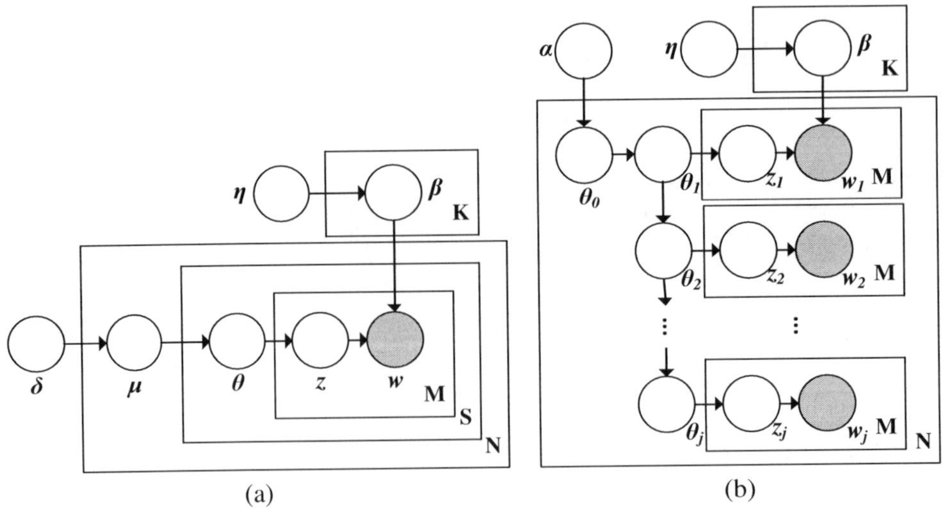

FIGURE 13.4: Plate notation for the STM (a) and LDSeq (b) generative models.

dimensionality K of the topic space (and thus of the Dirichlet distribution from which topics are drawn) is assumed known. The parameter β is treated as a $K \times M$ random matrix, where each row θ_k is drawn from an exchangeable Dirichlet distribution and is associated with one mixture component. This view of β is shared by STM and all other LDA-based models.

STM extends the LDA model by adding an additional layer in deriving word–topic proportions θ, which effectively correlates document topics with segment topics, modeling the topic structure within a segmented document. While LDA samples θ at the document level ($\theta_i \sim Dir_K(\alpha)$), STM extends document-level proportions (μ_i) to the segment level (θ_{ij}) with the aid of the two-parameter Poisson-Dirichlet Process (PDP). Du et al. [29] posit the following approximations on distributions, which enable this extension,

$$PDP(0, b, discrete(\theta)) \approx Dir(b\theta),$$
$$PDP(a, 0, discrete(\theta)) \approx Dir(a\theta),$$

where a and b are PDP *discount* and *strength* parameters and $a \to 0$. They justify the first approximation because the means and the first two central moments of the LHS and RHS are equal, and the second approximation based on an agreement up to $O(a^2)$ error in the means and first two central moments of the two sides. Since PDP is a prior conjugate to multinomial likelihoods, replacing the Dirichlet distribution with the PDP allows the authors to use collapsed Gibbs samplers in the STM inference, greatly reducing computational complexity. Figure 13.4 (a) depicts the plate notation representation of the STM model, whose generation process for each document is detailed below.

1. Generate document topic proportions, $\mu_i \sim Dir_K(\delta)$
2. For each segment s_{ij} in document d_i
   a. Generate segment topic proportions, $\theta_{ij} \sim PDP(a, b, \mu_i)$
   b. For each word w_{ijl} in segment s_{ij}
      i. Choose a topic $z_{ijl} \sim Multi_K(\theta_{ij})$
      ii. Choose word $w_{ijl} \sim Multi_M(\beta_{z_{ijl}})$

STM is also similar to LDCC, a four-level probabilistic model that also considers documents and segments as mixtures over latent topics. Unlike STM, LDCC considers documents and segments as random mixtures over different kinds of topics and associates a segment with a single topic. By using a single topic pool for both documents and segments, STM better models the structure of a normal document, in which document topics are a superset of the segment topics in the document. Similarly, segments can at times exhibit multiple topics, e.g., a paragraph about Ludwig van Beethoven's Violin Concerto in D major can draw from topics related to *violins*, *music*, *musical performance*, and *the life of Beethoven*. By assuming segments have a topic distribution, STM allows them to share multiple topics. In contrast, LDCC assigns a specific topic to each segment. LDCC will be presented in more detail in Section 13.4.3.

Consecutive segments. Du et al. also propose Sequential LDA (LDSeq) [30], an extension of STM that addresses the *bag of segments* document assumption. Considering segment order in a document, the topic distribution of a segment in LDSeq is dependent on that of the previous segment. The first segment, which does not have an antecedent, has a topic distribution dependent on the document topic distribution. Figure 13.4 (b) depicts the plate notation representation of the LDSeq model, whose generation process for each document is detailed below.

1. Generate document topic proportions, $\theta_{i,0} = \mu_i \sim Dir_K(\delta)$
2. For each segment s_{ij} in document d_i
   a. Generate segment topic proportions, $\theta_{ij} \sim PDP(a, b, \theta_{i,j-1})$
   b. For each word w_{ijl} in segment s_{ij}
      i. Choose a topic $z_{ijl} \sim Multi_K(\theta_{ij})$
      ii. Choose word $w_{ijl} \sim Multi_M(\beta_{z_{ijl}})$

While documents are segment sets in STM, LDSeq sees them as sequences of segments. Du et al. [30] take advantage of the fact that the PDP is self-conjugate, allowing them to model progressive topical dependency via a nested PDP, i.e., the PDP of the current segment uses the PDP of the previous segment as its base distribution ($\theta_{ij} \sim PDP(a, b, \theta_{i,j-1})$). This assumption may not be appropriate for all text domains, but it showcases, once again, the modularity and extensibility of the LDA model. LDSeq is also related to the LDSEG model, an extension of LDCC model that assumes a Markovian relationship between distributions of consecutive segments. LDSEG will be presented in more detail in Section 13.4.3.

13.4.2 Clustering Segmented Documents

Using techniques outlined above, a multi-topic document can be decomposed into segments that correspond to thematically coherent contiguous text passages in the original document. Segmentation can be used as a base step in *long document* clustering.

Segment-based document clustering. Tagarelli and Karypis [103] propose a framework for clustering of multi-topic documents that leverages the natural composition of documents into text segments in a "divide-et-impera" fashion. First, the documents are segmented using an existing document segmentation technique (e.g., TextTiling). Then, the segments in each document are clustered (potentially in an overlapping fashion) into groups, each referred to as a *segment-set*. Each segment-set contains the thematically coherent segments that may exist at different parts of the document. Thinking of them as mini-documents, the segment-sets across the different documents are clustered together into nonoverlapping thematically coherent groups. Finally, the segment-set clustering is used to derive a clustering solution of the original documents. The key assumption underlying this *segment-based document clustering* framework is that multi-topic documents can be decomposed into smaller single-topic text units (segment-sets) and that the clustering of these segment-sets can lead to an overlapping clustering solution of the original documents that accurately reflects the multiplicity of the topics that they contain.

Although parametric with respect to the clustering algorithm, the framework is designed to work with "hard" as well as "soft" clustering strategies; in particular, the authors test their framework using existing algorithms for clustering the segments within each document. For disjoint clustering solutions they use Spherical k-Means (*Sk-Means*), whereas for overlapping clustering solutions they use Fuzzy Spherical k-Means (*FSk-Means*) and *LDA* (cf. Section 13.3 for details). The authors also show that *overclustering* the segments, producing a relatively high degree of overlapping clustering of the segments, can circumvent the problem of identifying the correct number of segment clusters, which is necessary input for most partitioning clustering algorithms.

Once the within-document clustering has been performed on all the documents in the collection, the resulting set $\mathbf{S}$ of segment-sets becomes the input to the subsequent phase, which is designed to identify the document topics in the collection. The authors use a *bisecting* version of the Spherical k-Means algorithm to cluster the segments. The use of disjoint clustering is motivated by the fact that each of the segment-sets will describe a single topic from the original document.

Tagarelli and Karypis [103] devise a model akin to the *vector space model* (cf. Section 13.2.2) for representing a collection of segment-sets. Intuitively, they adapt the conventional *tf-idf* function to be *segment-set-oriented*, *segment-oriented*, or *document-oriented*. Similar to *tf-idf*, their weighting functions increase with the term frequency within the local text unit (segment), and with the term rarity across the whole collection of text objects (i.e., segments, segment-sets, or documents).

Let w be an index term and $\mathcal{S} \in \mathbf{S}$ be a segment-set. Let $tf(w, \mathcal{S})$ be the number of occurrences of w over all the segments in $\mathcal{S}$. The *segment-set-oriented* relevance weight of w with respect to $\mathcal{S}$ is computed by the *Segment-set Term Frequency–Inverse Segment-set Frequency* function:

$$stf\text{-}issf(w, \mathcal{S}) = tf(w, \mathcal{S}) \times \log\left(\frac{N_{\mathbf{S}}}{N_{\mathbf{S}}(w)}\right),$$

where $N_{\mathbf{S}}$ is the number of segment-sets in $\mathbf{S}$, and $N_{\mathbf{S}}(w)$ is the part of $N_{\mathbf{S}}$ that contains w.

At a higher level (i.e., at document level), the relevance weight of w with respect to $\mathcal{S}$ is computed by the *Segment-set Term Frequency–Inverse Document Frequency* function:

$$stf\text{-}idf(w, \mathcal{S}) = tf(w, \mathcal{S}) \times \log\left(\frac{N_{\mathbf{D}}}{N_{\mathbf{D}}(w)}\right),$$

where $N_{\mathbf{D}}$ is the number of documents in $\mathcal{D}$, and $N_{\mathbf{D}}(w)$ is the part of $N_{\mathbf{D}}$ that contains w.

Finally, at a lower level (i.e., at segment level), the relevance weight of w with respect to $\mathcal{S}$ is computed by the *Segment-set Term Frequency–Inverse Segment Frequency* function:

$$stf\text{-}isf(w, \mathcal{S}) = tf(w, \mathcal{S}) \times \exp\left(\frac{N_{\mathcal{S}}(w)}{N_{\mathcal{S}}}\right) \times \log\left(\frac{n_{\mathbf{S}}}{n_{\mathbf{S}}(w)}\right),$$

where $N_{\mathcal{S}}$ is the number of segments in $\mathcal{S}$, $n_{\mathbf{S}}$ is the number of segments in $\mathbf{S}$, and $N_{\mathcal{S}}(w)$ and $n_{\mathbf{S}}(w)$ are the portions of $N_{\mathcal{S}}$ and $n_{\mathbf{S}}$, respectively, that contain w. In the above formula, an exponential factor is used to emphasize the segment-frequency of the terms within the local segment-set. The rationale here is that terms occurring in many segments of a segment-set should be recognized as characteristic (discriminatory) of that segment-set, thus they should be weighted more than terms with low segment-frequency.

The final step in the framework is to use the disjoint clustering solution of the segment-sets in order to derive an overlapping solution of the initial document collection that correctly reflects the multiple topics that may exist in the collection's documents. Although alternative methods could be used to induce the final clustering, the authors take a simple assignment approach. Each cluster of segment-sets is considered to be a single topic, and each document is assigned to all the topics that contain at least one of its segment-sets.

The empirical evaluation that Tagarelli and Karypis [103] performed shows general improved clustering accuracy over nonsegmented document clustering techniques in both the soft and hard clustering strategies. The segment-based views over the documents allow for an effective identification of overlapping clustering solutions, and the authors' proposed segment-level overclustering improves the quality of both disjoint and overlapping clustering solutions. They also find that segment-based document clustering leads to cluster descriptions that are more "useful" according to a number of aspects, including higher coherence of terms within a description, higher presence of discriminating terms, and wider coverage of topics.

Clustering long legal documents. Lu et al. [70] apply a similar clustering strategy in the legal domain, where documents with multiple topics are very common. They develop a highly scalable soft clustering system centered around a topic segmentation-based clustering framework that also incorporates metadata information. The process of identifying highly refined issue-based clusters is broken down into three logical steps: (1) build a universe of legal issues (topics) to search in, (2) identify relevant documents for each issue in the topic universe, and (3) associate each document in the collection with one or more issues.

The document segmentation step leverages available metadata for the document collection. In particular, the algorithm represents a headnote, a brief summary of points of law within a document, as a compound vector with four different feature types: a term frequency vector for the *text* in the headnote, a frequency vector of *noun phrases* in the text, a vector of codes for applicable laws from a legal taxonomy (known as *key numbers*), and a *citation network* for the headnote. The similarity between two headnotes is then computed as the weighted sum of their respective feature type similarities, with heuristically determined weights. The usual cosine similarity with *tf-idf* weighting is used for comparing the first two feature types, and an analogous method is used for the third. Citation features are compared in terms of co-citations,

$$cite\_sim(h_i, h_j) = \frac{cite(h_i \cap h_j)}{cite(h_i \cup h_j)},$$

where $cite(h_i \cup h_j)$ is the number of documents citing at least one headnote, and $cite(h_i \cap h_j)$ is the number of documents in which both headnotes h_i and h_j are cited. The use of noun phrases as part of the feature set is motivated by an in-house study that found them to be closely related to legal concepts, which form the basis for topics in this domain.

Headnotes in each document are clustered using an agglomerative clustering algorithm employing an automatic stopping criteria. The algorithm merges two clusters by maximizing the ratio of intracluster and intercluster similarity, dubbed the *intratopic similarity threshold*, and thus does not require the number of clusters as input. The intratopic similarity threshold is determined heuristically. The resulting headnote clusters are considered *topics* within the document.

Given the large size of the topic set, the authors use dimensionality reduction on topics to reduce the computational complexity of the next step. To obtain a unique set of collection topics, document topics are clustered using a "canopy"-based soft clustering technique. A document classification engine and a ranker support vector machine (SVM) [22] are used to retrieve topics similar to some seed topics, the top ranked of which are merged with the seeds. Topic similarity is extended for this step in the framework to include classification engine scores and co-click similarity, a score based on users viewing (clicking on) the documents that the headnotes represent. The algorithm is executed recursively, using the output of each round after the first execution as the input of the next, until the intercluster similarity between any two clusters is lower than a threshold. The resulting clustering represents the set of *most important topics* within the collection.

The last step in the framework associates the collection documents with the discovered topics. For this step, the main document text is segmented and documents are assigned to clusters based on the similarity of their segments with the cluster. The quality of the resulting issue-based clustering was validated by human legal experts in multiple test categories.

A statistical model. Clustering segmented documents is not limited to VSM techniques. Ponti et al. [84] describe a statistical model for topically segmented documents and provide a clustering strategy for documents modeled this way. The key idea of their work is that a generative model that exploits the underlying composition of documents into segments is able to better capture dependencies among terms, alleviating some of the problems related to the bag-of-words assumption in

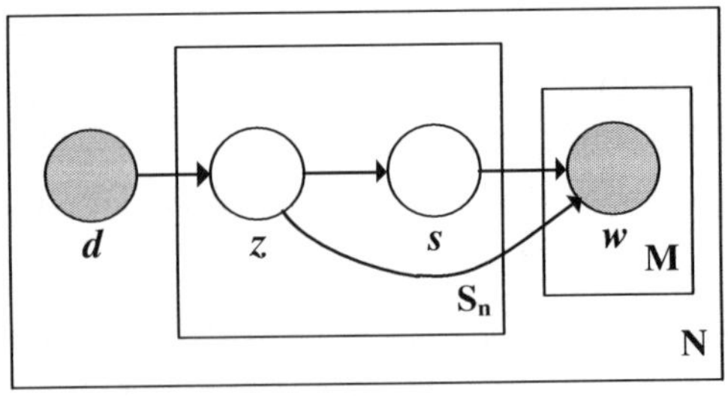

FIGURE 13.5: Plate notation for the SGM generative model.

large multi-topic documents. Term generation in such a model should be related not only to topics but also to segments. As a consequence, the latent variable that models topics should be directly associated to the within-document segments, rather than to the document as a whole. They propose *Segment-based Generative Model* (SGM), a model that explicitly considers segments within each document by introducing a segment model variable in the generative process.

SGM assumes that each document $d \in \mathcal{D}$ is a sequence of n_d words and, at the same time, a set S_d of contiguous, nonoverlapping text blocks, or *segments*. The segmentation strategy is decoupled from SGM, the authors using TextTiling in their implementation. SGM utilizes latent variable z to model topic distributions and the model variable s to represent document segments. Figure 13.5 illustrates the graphical model representation of SGM. The generative process performed by SGM on a corpus $\mathcal{D}$ of segmented documents can be summarized as follows.

1. Select a document d from $\mathcal{D} \Rightarrow \Pr(d)$
2. For each segment $s \in S_d$
   a. Choose a topic z for the document, $d \Rightarrow \Pr(z|d)$
   b. Associate topic-to-segment probability for segment s, $z \Rightarrow \Pr(s|z)$
   c. For each word w in the segment s
      i. Choose a word w from the current topic and segment, $w \Rightarrow \Pr(w|z,s)$

SGM provides a finer-grained document-to-topic modeling by taking into account text segments. Choosing a topic ($\Pr(z|d)$) in the generative process is based on the topic-to-segment association probability ($\Pr(s|z)$), intuitively providing a topical affinity for each segment given a selected topic. Words are then generated not only by topics, but also by segments ($\Pr(w|z,s)$). The above generative process can be translated into a joint probability model for triadic data, in which each observation is expressed by a triad defined on documents, segments, and words:

$$\Pr(d,s,w) = \Pr(d) \sum_{z \in \mathcal{Z}} \Pr(z|d) \Pr(s|z) \Pr(w|z,s).$$

Ponti et al. use Expectation-Maximization (EM) [25] to estimate model parameters and a centroid-based linkage agglomerative hierarchical method for clustering the resulting document PMFs. The prototype $\mathcal{P}_C$ of each cluster is represented as the mean of the PMFs of the documents within that cluster. The cluster merging criterion, which decides the pair of clusters to be merged at each step, utilizes the Hellinger distance (cf. Section 13.3.4) to compare the cluster prototypes. The merging score criterion computes the average distance between the prototypes of each pair of clusters ($\mathcal{P}_{C_i}$ and $\mathcal{P}_{C_j}$) and the prototype of the union cluster ($\mathcal{P}_{C_i \cup C_j}$). The pair of clusters with the minimum score is chosen to be merged. Intuitively, this criterion aims to choose the merged clustering that is closest to the original clustering. The algorithm stops when the cluster hierarchy is completed or the desired number of clusters is reached.

Extending the vector space. Wang et al. propose the *Matrix Space Model* (MSM) [110], in which each presegmented document is represented as a *tf-idf*-weighted term-segment frequency matrix instead of a term frequency vector. Segments are then cast as probabilistic distributions over a small set of l latent topics, which are used to realize a document clustering. Latent topic extraction is accomplished by approximating the document matrices $\mathbf{A}_i$ as $\mathbf{LM}_i\mathbf{R}^T$, where the nonnegative basis-matrices $\mathbf{L} \in \mathbb{R}^{m \times l_1}$ $(\mathbf{L} \geq 0)$ and $\mathbf{R} \in \mathbb{R}^{s \times l_2}$ $(\mathbf{R} \geq 0)$ jointly define the lower dimensional space, and matrices M_i are the low rank representation of the documents. l_1 and l_2 are user specified parameters defining the size of the latent space, m is the size of the term vocabulary, and s is the number of segments into which a document is split.

Given that matrices $\mathbf{L}$ and $\mathbf{R}$ are shared among the collection, the authors expect similar documents in the original space to also have a similar latent space representation. They formulate the latent space extraction as the constrained optimization problem,

$$\min_{\substack{\mathbf{L} \in \mathbb{R}^{m \times l_1} \; : \; \mathbf{L} \geq 0 \\ \mathbf{R} \in \mathbb{R}^{s \times l_2} \; : \; \mathbf{R} \geq 0 \\ \mathbf{M}_i \in \mathbb{R}^{l_1 \times l_2} \; : \; \mathbf{M}_i \geq 0}} \sum_{i=1}^{n} ||\mathbf{A}_i - \mathbf{LM}_i\mathbf{R}^T||_F^2,$$

where $||\cdot||_F$ is the standard Frobenius matrix norm and n is the number of collection documents. In the reconstruction, $\mathbf{LM}_i$ is associated with the posterior of each term belonging to the latent topics, while $\mathbf{M}_i\mathbf{R}^T$ is the posterior of each segment in the document belonging to the latent topics.

13.4.3 Simultaneous Segment Identification and Clustering

Assuming the previous definition of segments as topically coherent blocks of text in a document, segment identification boils down to finding the document topics. The document segments can then be extracted by considering the major topic shifts in the document word list. Considering cluster assignment, the result of topic modeling can be written as a document–topic probability matrix $\mathbf{P}$ where, $\mathbf{P}_{ik} = \Pr(z_k|d_i)$. A hard (or soft) k-way clustering on documents could be induced from $\mathbf{P}$ by assigning documents to the *topic cluster(s)* for which their respective probability values are highest (or above a threshold).

Co-clustering in the latent space. The above assignment strategy assumes an order-dependent word assignment in the topic model, which is not generally the case, as most models assume documents are orderless *bags of words*. One of the first models to address order placement within the document, Latent Dirichlet Co-Clustering (LDCC) [95], is an extension of LDA that simultaneously clusters words and documents. By focusing on meaningful segments of text, LDCC is more likely to assign adjacent words to coherent topics.

LDCC extends LDA by assuming each document is a random mixture of topics, which in turn are distributions over segments. Segments are then modeled as in LDA: for each segment, a distribution over collection topics is sampled from a Dirichlet distribution; a topic is then selected according to this distribution for each segment term; each term is finally sampled from a multinomial distribution over terms specific to the selected topic. Figure 13.6 (a) depicts the plate notation representation of the LDCC model, whose generation process for each document is detailed below.

1. Choose the number of segments for document d_i, $S_i \sim \mathit{Poisson}(\phi)$
2. Generate document topic proportions, $\mu_i \sim \mathit{Dir}_K(\delta)$
3. For each s_{ij} of the S_i segments in document d_i
    a. Choose a random topic for the segment, $y_{ij} \sim \mathit{Multi}(\mu_i)$
    b. Choose number of words for the segment, $N_{ij} \sim \mathit{Poisson}(\varepsilon)$
    c. Generate segment topic proportions, $\theta_{ij} \sim \mathit{Dir}_K(\alpha, y_{ij})$

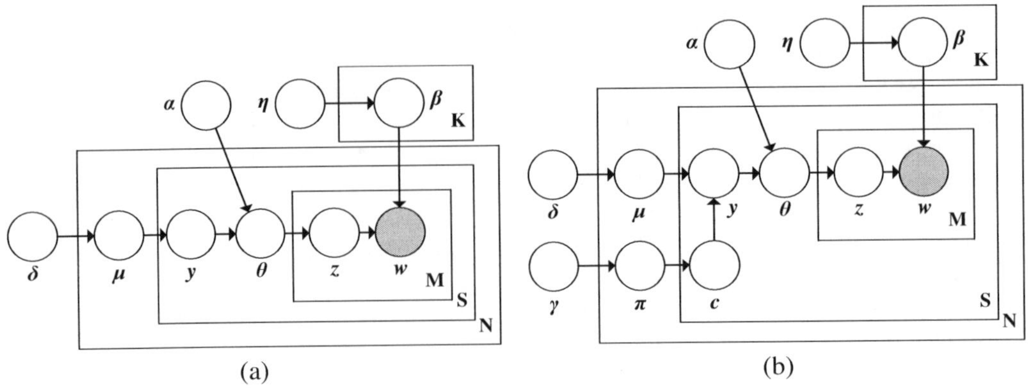

FIGURE 13.6: Plate notation for the LDCC (a) and LDSEG (b) generative models.

d. For each w_{ijl} in segment s_{ij}
   i. Choose a topic $z_{ijl} \sim Multi_K(\theta_{ij})$
   ii. Choose word w_{nsm} from $Pr(w_{ijl}|z_{ijl}, \beta)$

Accounting for segment correlation. LDCC intuitively assumes that documents are composed of single-topic segments. Yet consecutive segments often pertain to the same subject, in the same way that paragraphs in a chapter may cover different aspects of the same topic discussed therein. In their follow-up paper, Shafiei and Milios [96] extend LDCC to also identify topically coherent segments in text. The proposed model, LDSEG, assumes a high likelihood that a segment has the same distribution over words as the previous segment in the document and models this assumption through a Markov structure on the segment-topic distribution. A switching binary variable for the topic of each segment indicates whether its topic is the same as that of the previous segment. If it is not, a new topic is sampled for the current segment. The list of states for this switching variable also defines a segmentation in each document. Figure 13.6 (b) depicts the plate notation representation of the LDSEG model, whose generation process for each document is detailed below.

1. Choose the number of segments for document d_i, $S_i \sim Poisson(\phi)$
2. Generate document topic proportions, $\mu_i \sim Dir_K(\delta)$
3. For each segment s_{ij} in document d_i
   a. Choose $y_{ij} = y_{ij-1}$ with probability $Pr(c_{ij} = 1) = \pi$
   b. Otherwise, choose a random topic for the segment, $y_{ij} \sim Multi(\mu_i)$
   c. Choose number of words for the segment, $N_{ij} \sim Poisson(\varepsilon)$
   d. Generate segment topic proportions, $\theta_{ij} \sim Dir_K(\alpha, y_{ij})$
   e. For each w_{ijl} in segment s_{ij}
      i. Choose a topic $z_{ijl} \sim Multi_K(\theta_{ij})$
      ii. Choose word w_{nsm} from $Pr(w_{ijl}|z_{ijl}, \beta)$

A framework for generative clustering. Ponti and Tagarelli relax the segment topic coherence assumption and provide a topic-based framework for clustering multi-topic documents using generative models [83]. Instead of assigning documents to topic clusters, Ponti and Tagarelli cluster documents based on their topic distributions. The proposed framework executes three steps. First, the documents are processed using standard preprocessing techniques to obtain the document term matrix. Then, a generative model is applied to represent the documents in a topic latent space. The output of this step is a probability matrix expressing the topic mixture underlying the documents. In the final step, documents are clustered based on their topic mixtures, using an information theory

PMF distance metric to compare documents. The clustering algorithm is a centroid-based linkage agglomerative hierarchical algorithm, like the one used by Ponti et al. in [84], which was described earlier.

13.5 Clustering Short Documents

Clustering short documents faces additional challenges above those of general purpose document clustering. Short documents normally address a single topic, yet they may do so with completely orthogonal vocabulary. Noise, contracted forms of words, and slang are prevalent in short texts. In this section, we will first discuss general methods for clustering short documents and then focus on methods designed specifically for clustering Web documents and microblogs.

13.5.1 General Methods for Short Document Clustering

There has been a relatively large corpus of study on alternative approaches to the clustering of short texts. Wang et al. [111] propose a frequent-term-based parallel clustering algorithm specifically designed to handle large collections of short texts. The algorithm involves an information-inference mechanism to build a semantic text feature graph which is used by a k-NN-like classification method to control the degree of cluster overlapping. Pinto et al. [81] resort to the information-theory field and define a symmetric KL divergence to compare short documents for clustering purposes. Since the KL distance computation relies on the estimation of probabilities using term occurrence frequencies, a special type of back-off scheme is introduced to avoid the issue of zero probability due to the sparsity of text. Carullo et al. [18] describe an incremental online clustering algorithm that utilizes a generalized Dice coefficient as a document similarity measure. The algorithm requires two thresholds as input, one to control the minimum accepted similarity that any document must have to be assigned to a cluster, and the other to define the maximum similarity of a document that can still contribute to the definition of a cluster.

Particle-swarm optimization techniques and bio-inspired clustering algorithms have also been proposed for short text data. Ingaramo et al. [52] develop a partitional clustering algorithm to handle short texts of arbitrary size. The key aspect of that study is the adaptation of the AntTree algorithm [42], which integrates the "attraction of a cluster" and the Silhouette Coefficient concepts, to detecting clusters. Each ant represents a single data object as it moves in the clustering structure according to its similarity to other ants already connected to the tree under construction. Starting from an artificial support, all the ants are incrementally connected, either to that support or to other already connected ants. This process continues until all ants are connected to the structure, i.e., all objects are clustered.

Finding core terms. In [78], Ni et al. regard the short document clustering task of grouping short texts based on some selected "core" terms. The underlying idea is to recursively bisect one of the clusters according to the core term identified within that cluster. The core term of a cluster is the term that minimizes the value of the Ratio Min-Max Cut (RMcut) criterion over all possible bisections of that cluster. Following a strategy dubbed *TermCut*, a bisection of a specific cluster is obtained for each of the terms contained within the cluster. All documents containing the selected term are assigned to one subcluster and the rest of the documents (not containing the term) are assigned to the other.

To find its RMcut value, an input collection of short documents is modeled as a graph, where vertices represent documents and edges are weighted by the similarity between the adjoined documents. As is generally done for short documents, term frequencies are smoothed to be one or zero, and thus the document representation is simplified to be a vector of *idf* values. The RMcut value corresponding to a K-way clustering C of $\mathcal{D}$ is defined as

$$RMcut(C) = \sum_{k=1}^{K} \frac{cut(C_k, C - C_k)}{|C_k| \sum_{d_i, d_j \in C_k} sim(d_i, d_j)}.$$

The denominator part of the above formula takes into account both the intrasimilarity of each cluster and its size, where the latter is used to avoid producing very unbalanced clusters. Moreover, the edge-cut function $cut(\cdot,\cdot)$ acts as an intercluster similarity criterion; it is defined as the summation over the weights of all edges connecting vertices (documents) within a specified cluster to the vertices within the rest of the clusters.

Taking into account cluster frequencies for terms, we can observe that the sum of all pair-wise document similarities within a cluster is equal to the sum of the product of the squared inverse document frequency and cluster frequency over all terms in the cluster. Thus, the RMCut criterion can be efficiently computed as

$$RMcut(C) = \sum_{k=1}^{K} \frac{\sum_{l=1}^{M} (idf_l)^2 \, cf_{l,k} \, cf_{l,\neg k}}{|C_k| \sum_{l=1}^{M} (idf_l \, cf_{l,k})^2},$$

where $cf_{l,k}$ and $cf_{l,\neg k}$ denote the cluster frequency of the lth term within the kth cluster and within the rest of the clusters, respectively. Note that the overall complexity of the RMcut criterion is $O(N+M)$, since the inverse document frequency and the cluster frequency of the terms can be computed by a single scan of the documents in the collection, and the computation of the numerator and the denominator in the above formula is $O(M)$.

Following the *TermCut* strategy, Ni et al. [78] propose two algorithms. The first tries to bisect clusters until the desired number of clusters is reached. The second takes a *minimal RMcut decrease threshold* as input and, as the name suggests, continues the bisecting process until the decrease in the RMcut value falls below the given threshold. While the idea behind the RMcut criterion is very similar to that underlying the CLUTO criterion functions detailed in Equations 13.5 and 13.6 of Table 13.2, Ni et al. show that their bisecting strategy outperforms CLUTO for a number of short text datasets.

13.5.2 Clustering with Knowledge Infusion

Motivated by the lack of common vocabulary in short documents, many short document clustering algorithms first enrich or complement the statistical vector representation of short texts with external knowledge bases, such as WordNet or Wikipedia. Banerjee et al. [8] propose to enrich the original term-feature space of search results with the titles of the Wikipedia articles that are retrieved as relevant to two queries created for each result. The first query is based on the result title, while the other is based on the result description, or snippet. Scaiella et al. [93] propose a "graph-of-topics" model to represent each snippet, in which vertices correspond to Wikipedia pages that are identified by existing topic annotators, and the edges are weighted to determine the semantic relatedness between the linked topics. An online spectral clustering algorithm is used on an induced graph consisting of two types of vertices, topics and snippets, where the weighted edges express either topic similarities or topic-to-snippet memberships.

User actions have also been useful in identifying short document clusters. Wang and Zhai [109] exploit information contained in log data produced by a real search engine to cluster search result snippets. Carpineto and Romano [17] introduce a meta-clustering strategy that clusters snippets by integrating partitions separately obtained as a result of analyzing the *tf-idf* document–term matrix with SVD, NMF, and generalized suffix trees. They also define an evaluation measure that takes into account the behavior of Web users in terms of the time spent to satisfy their search needs.

Hu et al. [51] combine original text features with semantic features derived from external knowledge bases to support the clustering task. Applying standard NLP techniques, they model the short input text into a parsing-tree-like structure to support the extraction of nonredundant seed phrases, which they use in turn to generate external semantic features. More specifically, a naive punctuation-based segmentation of the text facilitates a subsequent shallow parsing step which identifies seed phrases. To avoid redundancy, each phrase is compared with all the other ones in the segment, and the phrase with the highest Wikipedia-based similarity is removed. The remaining seed phrases are used to retrieve external feature content, from either Wikipedia or WordNet, depending on the presence of stop words in the phrase. External features are extracted from titles and link text in the Wikipedia pages or similar term concepts in WordNet. Document features are finally selected based on *tf-idf* weights of original and external features. An additional parameter is introduced to control the influence of external features in the feature space. Hu et al. demonstrate the concurrent use of multiple types of external knowledge bases, along with internal semantics, to improve clustering of short texts.

13.5.3 Clustering Web Snippets

Document clustering research has traditionally focused on Web documents as a way to facilitate users' ability to quickly browse search results. Web documents could be clustered off-line, with a general purpose document clustering algorithm. However, this approach was shown ineffective [37, 16], because it is based on features that are frequent in the entire collection but irrelevant to the particular query. Instead, query-specific, online, postretrieval clustering, i.e., *clustering search results*, was shown to produce superior results [45]. A search result is generally composed of a title and a *snippet*, a short summary, often containing phrases from the document related to the search query. As such, clustering search results uses a subset of the collection vocabulary concentrated around the query terms.

Unlike the traditional clustering task, the primary focus of search result clustering is **not** to produce optimal clusters [109, 16, 5]. Rather, search result clustering is a highly *user-centric* task with several unique additional requirements. The algorithm must be fast, as users are unwilling to wait longer than a few seconds for search results. Clusters must exhibit interesting query subtopics or facets from the user's perspective. Finally, clusters must be assigned informative, expressive, meaningful and concise labels.

Scatter/Gather [82, 45] was an early cluster-based document browsing method that addressed the speed requirement by performing postretrieval clustering on top-ranked documents returned from a traditional information retrieval system. Zamir and Etzioni introduced the well-known Suffix Tree Clustering (STC) [118] algorithm, which creates interesting subtopic clusters based on phrases shared between documents. It follows the assumption that repeated phrases imply topics of interest within the result collection. STC treats a snippet as a string of words, builds a suffix tree over the collection of snippets, and traverses the suffix tree to extract base clusters. The algorithm then uses a binary similarity measure based on overlap of documents to create a base cluster graph. In this graph, each node corresponds to a group of snippets sharing a phrase. The final clustering solution is obtained by finding the connected components in the graph. Zamir and Etzioni also showed that using snippets for clustering is as effective as using whole documents.

Addressing the meaningful and concise label requirement of search result clustering, Anastasiu et al. [5] employ a strategy that generates labels before clusters. They first identify frequent phrases within a set of search results using a suffix tree built in linear time by Ukkonen's algorithm [106]. Then they select labels from the frequent phrases using a greedy set cover heuristic, where at each step a frequent phrase covering the most uncovered search results is selected until the whole cluster is covered or no frequent phrases remain. Results are then assigned to a label if they contain the terms in the label, uncovered results being placed in a special cluster named *Other*. Osiński et al. [79] also follow a label-before-cluster approach. They use dimensionality reduction techniques to induce cluster labels. Then, treating each label as a query over the snippet-set in the information retrieval sense, they populate the clusters with the retrieved results for the queries.

Common phrases can naturally describe clusters. This has inspired many other phrase-based hierarchical methods for clustering Web snippets. Kummamuru et al. [65] develop a monothetic clustering algorithm with the ultimate goal of automatically generating a concept hierarchy, where concepts are terms or phrases. At each level of the hierarchy being constructed, the algorithm progressively identifies topics such that the distinctiveness of the monothetic features describing the clusters is maximized, and at the same time document coverage in clusters is maximized. Li and Wu [67] first build a phrase-based document index by extracting salient phrases from snippets. The clustering method starts with all extracted phrases belonging to their individual clusters and combines the most similar clusters according to the constructed index. Each cluster is finally identified by a distinct phrase. The snippets whose indexing phrases belong to the same cluster are grouped together, while the remaining snippets are clustered based on their k-nearest neighbors. Zeng et al. [119] map the clustering problem to a phrase ranking problem, in which a regression model is first trained to rank the n-grams for a specified keyword. The model is then used to extract relevant phrases according to which the snippets are finally clustered.

13.5.4 Clustering Microblogs

The recent popularity of social networks has led to increasing demand for robust clustering algorithms for microblog data, or tweets. General purpose document clustering algorithms do not work well with these data due to the lack of co-occurring terms and context information in the short "documents." Researchers have tried to solve this problem by altering existing techniques or creating specialized document models. The most promising research direction relies on aligning or augmenting the short texts with external information.

Liu et al. [69] rely on an incremental similarity-threshold based clustering step to identify groups of similar tweets for the task of semantic role labeling. In a related problem of classifying tweets to a predefined set of generic classes, Sriram et al. [98] compare the *bag of words* model with other models based on short-text specific features such as use of shortened words or slang, time–event phrases, opinion phrases, or username mentions. In their evaluation, they find that non-*bag of words* models outperform the *bag of words* one. Park et al. [80] propose a hybrid approach that exploits external information from search result clustering to deal with the extraction of topics from blogs. A set of candidate terms with relatively high *tf-idf* values is initially extracted from all posts of a blog, and then used to feed a Web search engine. The resulting snippets for a specified candidate term are grouped into a hierarchy of clusters, and each of these clusters is compared and matched to the blog posts covering that term to finally determine how many subtopics are covered by the blog.

Topic modeling has recently also been shown effective in the microblog domain. Ramage et al. propose *Labeled LDA* [88], a version of LDA that incorporates available supervision, and use it on Twitter data [87] to characterize content, rank tweets, and recommend users to follow. Weng et al. [113] propose a PageRank-type algorithm for measuring topic-sensitive influence of microblog authors. They use LDA to discover latent topics and compute transition probabilities contingent on the topical similarity of users. Hong and Davison [48] study how to train topic models on microblog data to be used in standard text mining applications. They find that model based features can be very useful, but the length of the documents can greatly affect the effectiveness of trained topic models. Specifically, aggregating short messages leads to better models.

Aligning topics in short and long texts. Inspired by the idea of using external data sources, Jin et al. [55] train topic models on the short texts alongside a collection of auxiliary long texts. They realize that long texts cannot be perfectly aligned to the short. Thus, their Dual LDA (DLDA) algorithms distinguish between inconsistent topical structures across domains by correlating the

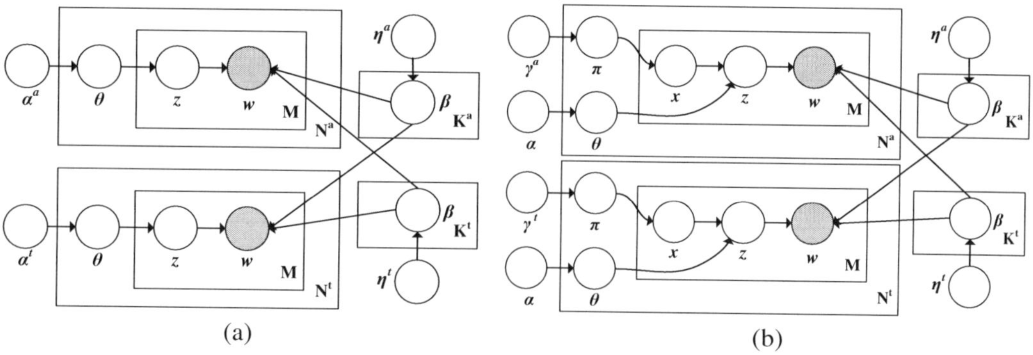

FIGURE 13.7: Plate notation for the α-DLDA (a) and γ-DLDA (b) generative models.

simultaneous training of two LDA models, the *target* model on the short texts and the *auxiliary* model on the long ones.

Depending on how the two models are related to each other, Jin et al. [55] propose two algorithms. α-DLDA models two separate sets of topics for auxiliary and target data and uses asymmetric Dirichlet priors to control the relative importance of the two when generating a document. α^t, the Dirichlet prior for generating topic mixing proportions for target documents, is given higher values for entries associated with target topics. Similarly, α^a is given higher values for entries associated with auxiliary topics. Figure 13.7 (a) illustrates the graphical model representation of α-DLDA.

γ-DLDA introduces a document-dependent binary switch that constrains each document to be generated either from the target model or from the auxiliary one. In addition to the multinomial distributions over topics, each document is also associated with a binomial distribution over target or auxiliary topics with a Beta prior γ. Similar to the α parameter in α-DLDA, γ^t is given higher values for entries associated with target topics. Figure 13.7 (b) depicts the plate notation representation of the γ-DLDA model, whose generation process for each document is detailed below.

1. For each target topic, generate a multinomial distribution over terms, $\beta^t_k \sim Dir_M(\eta^t)$, $k \in \{1,\ldots,K^t\}$
2. For each auxiliary topic, generate a multinomial distribution over terms, $\beta^a_k \sim Dir_M(\eta^a)$, $k \in \{1,\ldots,K^a\}$
3. For each corpus (auxiliary and target data), $c \in \{a,t\}$
   a. For each corpus document d_i, $i \in \{1,\ldots,N^c\}$
      i. Generate a multinomial distribution over target topics, $\theta^t_i \sim Dir_K(\alpha^t)$
      ii. Generate a multinomial distribution over auxiliary topics, $\theta^a_i \sim Dir_K(\alpha^a)$
      iii. Generate a binomial distribution over target vs. auxiliary topics, $\pi_i \sim Beta(\gamma^c)$
      iv. For each word w_{il} in document d_i
         1) Choose a value for $x_{il} \sim Binomial(\pi_i)$
         2) If $x_{il} = t$, choose a target topic $z_{il} \sim Multi(\theta^t_i)$
         3) If $x_{il} = a$, choose an auxiliary topic $z_{il} \sim Multi(\theta^a_i)$
         4) Choose word w_{il} from topic z_{il}, i.e. $w_{il} \sim Multi(\beta^{x_{il}}_{z_{il}})$

Jin et al. [55] compare their DLDA algorithms against direct clustering with CLUTO, topic model-based clustering on the individual collections and against several algorithms that transfer knowledge from the long texts when clustering the short. While α-DLDA and γ-DLDA outperformed the competition, the authors also note that methods utilizing long texts performed significantly better than the others, demonstrating the value of external information when clustering noisy short documents.

13.6 Conclusion

This chapter primarily focused on reviewing some recently developed text clustering methods that are specifically suited for long and for short document collections. These types of document collections introduce new sets of challenges. Long documents are by their nature multi-topic and as such the underlying document clustering methods must explicitly focus on modeling and/or accounting for these topics. On the other hand, short documents often contain domain-specific vocabulary, are very noisy, and proper modeling/understanding often requires the incorporation of external information. We strongly believe research in clustering long and short documents is in its early stages and many new methods will be developed in the years to come. Moreover, many real datasets are composed of not only standard, long, or short documents, but rather documents of mixed length. Current scholarship lacks studies on these types of data. Since different methods are often used for clustering standard, long, or short documents, new methods or frameworks should be investigated that address mixed collections.

Traditional document clustering is also faced with new challenges. Today's very large, high-dimensional document collections often lead to multiple valid clustering solutions. Subspace/projective clustering approaches [63, 77] have been used to cope with high dimensionality when performing the clustering task. Ensemble clustering [36] and multiview/alternative clustering approaches [54, 86], which aim to summarize or detect different clustering solutions, have been used to manage the availability of multiple, possibly alternative clusterings for a given dataset. Relatively little work has been done so far in document clustering research to take advantage of lessons learned from these methods. Integrating subspace/ensemble/multiview clustering with topic models or segmentation may lead to developing the next-generation clustering methods specialized for the document domain.

Some topics that we have only briefly touched on in this article are further detailed in other chapters of this book. Other topics related to clustering documents, such as semisupervised clustering, stream document clustering, parallel clustering algorithms, and kernel methods for dimensionality reduction or clustering, were left for further study. Interested readers may consult document clustering surveys by Aggarwal and Zhai [2], Andrews and Fox [6], and Steinbach et al. [99].

Bibliography

[1] Charu C. Aggarwal, Stephen C. Gates, and Philip S. Yu. On the merits of building categorization systems by supervised clustering. In *Proceedings of the Fifth ACM SIGKDD International Conference on Knowledge Discovery and Data Mining*, KDD '99, pages 352–356, New York, USA, 1999. ACM.

[2] Charu C. Aggarwal and ChengXiang Zhai. A survey of text clustering algorithms. In *Mining Text Data*, pages 77–128. Springer US, 2012.

[3] Nir Ailon and Bernard Chazelle. Faster dimension reduction. *Communications of the ACM*, 53(2):97–104, February 2010.

[4] Abdelmalek Amine, Zakaria Elberrichi, Michel SimoNet, and Mimoun Malki. Wordnet-based and n-grams-based document clustering: A comparative study. *International Conference on Broadband Communications, Information Technology & Biomedical Applications*, pages 394–401, 2008.

[5] David C. Anastasiu, Byron J. Gao, and David Buttler. A framework for personalized and collaborative clustering of search results. In *Proceedings of the 20th ACM International Conference on Information and Knowledge Management*, CIKM '11, pages 573–582, New York, USA, 2011. ACM.

[6] Nicholas O. Andrews and Edward A. Fox. Recent developments in document clustering. Technical report, Computer Science, Virginia Tech, 2007.

[7] Mihael Ankerst, Markus M. Breunig, Hans-Peter Kriegel, and Jörg Sander. OPTICS: Oordering points to identify the clustering structure. *SIGMOD Record*, 28(2):49–60, June 1999.

[8] Somnath Banerjee, Krishnan Ramanathan, and Ajay Gupta. Clustering short texts using wikipedia. In *Proceedings of the 30th ACM SIGIR Conference on Research and Development in Information Retrieval*, SIGIR '07, pages 787–788, 2007.

[9] G. Baudat and F. Anouar. Generalized discriminant analysis using a kernel approach. *Neural Computation*, 12(10):2385–2404, October 2000.

[10] Doug Beeferman, Adam Berger, and John Lafferty. Statistical models for text segmentation. *Journal of Machine Learning Research*, 34(1-3):177–210, 1999.

[11] David M. Blei. Probabilistic topic models. *Communications of the ACM*, 55(4):77–84, April 2012.

[12] David M. Blei, Thomas L. Griffiths, and Michael I. Jordan. The nested Chinese restaurant process and Bayesian nonparametric inference of topic hierarchies. *Journal of the ACM*, 57(2):7:1–7:30, February 2010.

[13] David M. Blei, Andrew Y. Ng, and Michael I. Jordan. Latent Dirichlet allocation. *Journal of Machine Learning Research*, 3:993–1022, 2003.

[14] Daniel Boley. Principal direction divisive partitioning. *Data Mining and Knowledge Discovery*, 2(4):325–344, December 1998.

[15] Thorsten Brants, Francine Chen, and Ioannis Tsochantaridis. Topic-based document segmentation with probabilistic latent semantic analysis. In *Proceedings of the 11th ACM CIKM International Conference on Information and Knowledge Management*, CIKM '02, pages 211–218, 2002.

[16] Claudio Carpineto, Stanislaw Osiński, Giovanni Romano, and Dawid Weiss. A survey of web clustering engines. *ACM Computing Surveys (CSUR)*, 41(3):1–38, 2009.

[17] Claudio Carpineto and Giovanni Romano. Optimal meta search results clustering. In *Proceedings of the 33rd International ACM SIGIR Conference on Research and Development in Information Retrieval*, SIGIR '10, pages 170–177, New York, USA, 2010. ACM.

[18] Moreno Carullo, Elisabetta Binaghi, and Ignazio Gallo. An online document clustering technique for short web contents. *Pattern Recognition Letters*, 30(10):870–876, 2009.

[19] William B. Cavnar. Using an n-gram-based document representation with a vector processing retrieval model. In *Third Text Retrieval Conference* (TREC-3), pages 269–278, 1994.

[20] Chaitanya Chemudugunta, Padhraic Smyth, and Mark Steyvers. Modeling general and specific aspects of documents with a probabilistic topic model. In *Advances in Neural Information Processing Systems 19*, NIPS '06, pages 241–248, 2006.

[21] Freddy Y. Y. Choi, Peter Wiemer-Hastings, and Johanna Moore. Latent semantic analysis for text segmentation. In *Proceedings International Conference on Empirical Methods in Natural Language Processing*, EMNLP '01, pages 109–117, 2001.

[22] Corinna Cortes and Vladimir Vapnik. Support-vector networks. *Machine Learning*, 20(3):273–297, September 1995.

[23] Pádraig Cunningham. Dimension reduction. Technical Report UCD-CSI-2007-7, University College Dublin, August 2007.

[24] Scott Deerwester, Susan T. Dumais, George W. Furnas, Thomas K. Landauer, and Richard Harshman. Indexing by latent semantic analysis. *Journal of the American Society for Information Science*, 41(6):391–407, 1990.

[25] Arthur P. Dempster, N. M. Laird, and D. B. Rubin. Maximum likelihood from incomplete data via the EM algorithm. *Journal of the Royal Statistical Society: Series B (Methodological)*, 39(1):1–38, 1977.

[26] Inderjit S. Dhillon, Subramanyam Mallela, and Dharmendra S. Modha. Information-theoretic co-clustering. In *Proceedings of the Ninth ACM SIGKDD International Conference on Knowledge Discovery and Data Mining*, KDD '03, pages 89–98, New York, USA, 2003. ACM.

[27] Chris Ding, Xiaofeng He, Hongyuan Zha, Ming Gu, and Horst Simon. Spectral min-max cut for graph partitioning and data clustering. In *Proceedings of the First IEEE International Conference on Data Mining*, pages 107–114, 2001.

[28] Chris H. Q. Ding, Xiaofeng He, Hongyuan Zha, Ming Gu, and Horst D. Simon. A min-max cut algorithm for graph partitioning and data clustering. In *Proceedings of the 2001 IEEE International Conference on Data Mining*, ICDM '01, pages 107–114, San Jose, CA, USA, 2001. IEEE Computer Society.

[29] Lan Du, Wray Buntine, and Huidong Jin. A segmented topic model based on the two-parameter Poisson-Dirichlet process. *Machine Learning*, 81(1):5–19, October 2010.

[30] Lan Du, Wray Lindsay Buntine, and Huidong Jin. Sequential latent Dirichlet allocation: Discover underlying topic structures within a document. In *Proceedings of the 2010 IEEE International Conference on Data Mining*, ICDM '10, pages 148–157, Washington, DC, USA, 2010. IEEE Computer Society.

[31] Jennifer G. Dy and Carla E. Brodley. Feature selection for unsupervised learning. *Journal of Machine Learning Research*, 5:845–889, December 2004.

[32] Martin Ester, Hans peter Kriegel, Jörg Sander, and Xiaowei Xu. A density-based algorithm for discovering clusters in large spatial databases with noise. In *Proceedings of the Second International Conference on Knowledge Discovery and Data Mining*, KDD '96, pages 226–231. AAAI Press, 1996.

[33] Xiaoli Zhang Fern and Carla E. Brodley. Random projection for high dimensional data clustering: A cluster ensemble approach. In *Proceedings of the Twentieth International Conference on Machine Learning*, pages 186–193, 2003.

[34] R. A. Fisher. The use of multiple measurements in taxonomic problems. *Annals of Eugenics*, 7(7):179–188, 1936.

[35] Evgeniy Gabrilovich and Shaul Markovitch. Computing semantic relatedness using Wikipedia-based explicit semantic analysis. In *Proceedings of the 20th International Joint Conference on Artifical Intelligence*, IJCAI '07, pages 1606–1611, San Francisco, CA, USA, 2007. Morgan Kaufmann Publishers Inc.

[36] Joydeep Ghosh and Ayan Acharya. Cluster ensembles. *Wiley Interdisciplinary Reviews: Data Mining and Knowledge Discovery*, 1(4):305–315, 2011.

[37] Alan Griffiths, H. Claire Luckhurst, and Peter Willett. Using interdocument similarity information in document retrieval systems. *Journal of the American Society for Information Sciences*, 37(1):3–11, 1986.

[38] Quanquan Gu and Jie Zhou. Co-clustering on manifolds. In *Proceedings of the 15th ACM SIGKDD International Conference on Knowledge Discovery and Data Mining*, KDD '09, pages 359–368, New York, USA, 2009. ACM.

[39] Sudipto Guha, Rajeev Rastogi, and Kyuseok Shim. CURE: an efficient clustering algorithm for large databases. In *Proceedings of the 1998 ACM SIGMOD International Conference on Management of Data*, SIGMOD '98, pages 73–84, New York, USA, 1998. ACM.

[40] Sudipto Guha, Rajeev Rastogi, and Kyuseok Shim. ROCK: A robust clustering algorithm for categorical attributes. In *Proceedings of the 15th International Conference on Data Engineering*, ICDE '99, pages 512–521, Washington, DC, USA, 1999. IEEE Computer Society.

[41] Jihun Hamm Daniel D. Lee, Sebastian Mika, and Bernhard Schölkopf. A kernel view of the dimensionality reduction of manifolds. In *Proceedings of the Twenty-First International Conference on Machine Learning*, ICML '04, page 47, doi:10.1145/1015330.1015417, New York, USA, 2004. ACM.

[42] Azzag Hanene, Christiane Guinot, and Gilles Venturini. AntTree: A web document clustering using artificial ants. In *Proceedings of the 16th European Conference on Artificial Intelligence*, ECAI, pages 480–484, 2004.

[43] Douglas Hardin, Ioannis Tsamardinos, and Constantin F. Aliferis. A theoretical characterization of linear SVM-based feature selection. In *Proceedings of the Twenty-First International Conference on Machine Learning*, ICML '04, page 48, doi:10.1145/1015330.1015421, New York, USA, 2004. ACM.

[44] Marti A. Hearst. TextTiling: segmenting text into multi-paragraph subtopic passages. *Computational Linguistics*, 23(1):33–64, March 1997.

[45] Marti A. Hearst and Jan O. Pedersen. Reexamining the cluster hypothesis: Scatter/gather on retrieval results. In *Proceedings of the 19th Annual International ACM SIGIR Conference on Research and Development in Information Retrieval*, SIGIR '96, pages 76–84, New York, USA, 1996. ACM.

[46] Thomas Hofmann. Probabilistic latent semantic indexing. In *Proceedings of the 22nd ACM SIGIR International Conference on Research and Development in Information Retrieval*, pages 50–57, 1999.

[47] Thomas Hofmann. Unsupervised learning by probabilistic latent semantic analysis. *Machine Learning*, 42(1–2):177–196, 2001.

[48] Liangjie Hong and Brian D. Davison. Empirical study of topic modeling in twitter. In *Proceedings of the First Workshop on Social Media Analytics*, SOMA '10, pages 80–88, New York, USA, 2010. ACM.

[49] Harold Hotelling. Analysis of a complex of statistical variables into principal components. *Journal of Educational Psychology*, 24:417–441, 1933.

[50] Andreas Hotho, Steffen Staab, and Gerd Stumme. WordNet improves text document clustering. In *Proceedings of the SIGIR 2003 Semantic Web Workshop*, pages 541–544, 2003.

[51] Xia Hu, Nan Sun, Chao Zhang, and Tat-Seng Chua. Exploiting internal and external semantics for the clustering of short texts using world knowledge. In *Proceedings of the 18th ACM Conference on Information and Knowledge Management*, CIKM '09, pages 919–928, 2009.

[52] Diego Ingaramo, Marcelo Errecalde, and Paolo Rosso. A general bio-inspired method to improve the short-text clustering task. In *Proceedings of the 11th International Conference on Computational Linguistics and Intelligent Text Processing*, CICLing '10, pages 661–672, 2010.

[53] Anil K. Jain and Richard C. Dubes. *Algorithms for Clustering Data.* Prentice-Hall, Inc., Upper Saddle River, NJ, USA, 1988.

[54] Prateek Jain, Raghu Meka, and Inderjit S. Dhillon. Simultaneous unsupervised learning of disparate clusterings. *Statistical Analysis and Data Mining*, 1(3):195–210, November 2008.

[55] O. Jin, N. N. Liu, K. Zhao, Y. Yu, and Q. Yang. Transferring topical knowledge from auxiliary long texts for short text clustering. In *Proceedings of the 20th ACM Conference on Information and Knowledge Management*, CIKM '11, pages 775–784, 2011.

[56] Ian T. Jolliffe. *Principal Component Analysis*, 2nd edition, Springer. October 2002.

[57] Thomas Kailath. The divergence and Bhattacharyya distance measures in signal selection. *IEEE Transactions on Communication Technology*, 15(1):52–60, 1967.

[58] George Karypis. CLUTO—a clustering toolkit. Technical Report #02-017, University of Minnesota, November 2003.

[59] George Karypis, Eui-Hong (Sam) Han, and Vipin Kumar. Chameleon: Hierarchical clustering using dynamic modeling. *Computer*, 32(8):68–75, August 1999.

[60] Leonard Kaufman and Peter J. Rousseeuw. *Finding Groups in Data: An Introduction to Cluster Analysis (Wiley Series in Probability and Statistics).* Wiley-Interscience, March 2005.

[61] Young-Min Kim, Jean-François Pessiot, Massih R. Amini, and Patrick Gallinari. An extension of PLSA for document clustering. In *Proceedings of the 17th ACM CIKM International Conference on Information and Knowledge Management*, CIKM '08, pages 1345–1346, 2008.

[62] Benjamin King. Step-wise clustering procedures. *Journal of the American Statistical Association*, 69:86–101, 1967.

[63] Hans-Peter Kriegel, Peer Kröger, and Arthur Zimek. Clustering high-dimensional data: A survey on subspace clustering, pattern-based clustering, and correlation clustering. *ACM Transactions on Knowledge Discovery from Data*, 3(1):1:1–1:58, March 2009.

[64] Krishna Kummamuru, Ajay Dhawale, and Raghu Krishnapuram. Fuzzy co-clustering of documents and keywords. In *Proceedings of the 12th IEEE International Conference on Fuzzy Systems*, pages 772–777, 2003.

[65] Krishna Kummamuru, Rohit Lotlikar, Shourya Roy, Karan Singal, and Raghu Krishnapuram. A hierarchical monothetic document clustering algorithm for summarization and browsing search results. In *Proceedings of the 13th International Conference on World Wide Web*, WWW '04, pages 658–665, New York, USA, 2004. ACM.

[66] John Aldo Lee and Michel Verleysen. Unsupervised dimensionality reduction: Overview and recent advances. In *International Joint Conference on Neural Networks*, IJCNN '10, pages 1–8, 2010.

[67] Zhao Li and Xindong Wu. A phrase-based method for hierarchical clustering of web snippets. In *Proceedings of the Twenty-Fourth AAAI Conference on Artificial Intelligence*, AAAI '10, pages 1947–1948, 2010.

[68] Li-Ping Liu, Yuan Jiang, and Zhi-Hua Zhou. Least square incremental linear discriminant analysis. In *Proceedings of the 2009 Ninth IEEE International Conference on Data Mining*, ICDM '09, pages 298–306, Washington, DC, USA, 2009. IEEE Computer Society.

[69] Xiaohua Liu, Kuan Li, Ming Zhou, and Zhongyang Xiong. Collective semantic role labeling for tweets with clustering. In *Proceedings of the Twenty-Second International Joint Conference on Artificial Intelligence*, volume 3 of *IJCAI'11*, pages 1832–1837. AAAI Press, 2011.

[70] Qiang Lu, Jack G. Conrad, Khalid Al-Kofahi, and William Keenan. Legal document clustering with built-in topic segmentation. In *Proceedings of the 20th ACM International Conference on Information and Knowledge Management*, CIKM '11, pages 383–392, New York, USA, 2011. ACM.

[71] Yijuan Lu, Ira Cohen, Xiang Sean Zhou, and Qi Tian. Feature selection using principal feature analysis. In *Proceedings of the 15th International Conference on Multimedia*, MULTIMEDIA '07, pages 301–304, New York, NY, USA, 2007. ACM.

[72] James B. MacQueen. Some methods for classification and analysis of multivariate observations. In *Berkeley Symposium on Mathematical Statistics and Probability*, volume 1, pages 281–297. University of Califoria Press, 1967.

[73] Christopher D. Manning, Prabhakar Raghavan, and Hinrich Schutze. *Introduction to Information Retrieval.* Cambridge University Press, 2008.

[74] Aleix M. Martínez and Avinash C. Kak. PCA versus LDA. *The IEEE Transactions on Pattern Analysis and Machine Intelligence*, 23(2):228–233, 2001.

[75] Yingbo Miao, Vlado Kešelj, and Evangelos Milios. Document clustering using character n-grams: A comparative evaluation with term-based and word-based clustering. In *Proceedings of the 14th ACM International Conference on Information and Knowledge Management*, CIKM '05, pages 357–358, New York, USA, 2005. ACM.

[76] Hemant Misra, François Yvon, Joemon M. Jose, and Olivier Cappe. Text segmentation via topic modeling: An analytical study. In *Proceedings of the 18th ACM Conference on Information and Knowledge Management*, CIKM '09, pages 1553–1556, New York, USA, 2009. ACM.

[77] Gabriela Moise, Arthur Zimek, Peer Kröger, Hans-Peter Kriegel, and Jörg Sander. Subspace and projected clustering: Experimental evaluation and analysis. *Knowledge and Information Systems*, 21(3):299–326, November 2009.

[78] Xingliang Ni, Xiaojun Quan, Zhi Lu, Liu Wenyin, and Bei Hua. Short text clustering by finding core terms. *Knowledge and Information Systems*, 27(3):345–365, June 2011.

[79] Stanislaw Osiński, Jerzy Stefanowski, and Dawid Weiss. LINGO: Search results clustering algorithm based on singular value decomposition. In *Intelligent Information Systems, Advances in Soft Computing*, pages 359–368. Springer, 2004.

[80] Jinhee Park, Sungwoo Lee, Hye-Wuk Jung, and Jee-Hyong Lee. Topic word selection for blogs by topic richness using web search result clustering. In *Proceedings of the 6th International Conference on Ubiquitous Information Management and Communication*, ICUIMC '12, pages 80:1–80:6, New York, USA, 2012. ACM.

[81] David Pinto, José-M. Benedí, and Paolo Rosso. Clustering narrow-domain short texts by using the Kullback-Leibler distance. In *Proceedings of the 8th International Conference on Computational Linguistics and Intelligent Text Processing*, CICLing '07, pages 611–622, 2007.

[82] Peter Pirolli, Patricia Schank, Marti Hearst, and Christine Diehl. Scatter/gather browsing communicates the topic structure of a very large text collection. In *Proceedings of the SIGCHI Conference on Human Factors in Computing Systems: Common Ground*, CHI '96, pages 213–220, New York, USA, 1996. ACM.

[83] Giovanni Ponti and Andrea Tagarelli. Topic-based hard clustering of documents using generative models. In *Proceedings of the 18th International Symposium on Foundations of Intelligent Systems*, ISMIS '09, pages 231–240, Berlin, Heidelberg, 2009. Springer-Verlag.

[84] Giovanni Ponti, Andrea Tagarelli, and George Karypis. A statistical model for topically segmented documents. In *Proceedings of the 14th International Conference on Discovery Science*, DS '11, pages 247–261, Berlin, Heidelberg, 2011. Springer-Verlag.

[85] Martin F. Porter. An algorithm for suffix stripping. In Karen Sparck Jones and Peter Willett, editors, *Readings in Information Retrieval*, pages 313–316. Morgan Kaufmann Publishers Inc., San Francisco, CA, USA, 1997.

[86] ZiJie Qi and Ian Davidson. A principled and flexible framework for finding alternative clusterings. In *Proceedings of the 15th ACM SIGKDD International Conference on Knowledge Discovery and Data Mining*, KDD '09, pages 717–726, New York, NY, USA, 2009. ACM.

[87] Daniel Ramage, Susan T. Dumais, and Daniel J. Liebling. Characterizing microblogs with topic models. In *Proceedings of the Fourth International Conference on Weblogs and Social Media*, ICWSM '10, pages 130–137, Washington, DC, USA, 2010. The AAAI Press.

[88] Daniel Ramage, David Hall, Ramesh Nallapati, and Christopher D. Manning. Labeled lDA: A supervised topic model for credit attribution in multi-labeled corpora. In *Proceedings of the 2009 Conference on Empirical Methods in Natural Language Processing*, volume 1 of *EMNLP '09*, pages 248–256, Stroudsburg, PA, USA, 2009. Association for Computational Linguistics.

[89] Manjeet Rege, Ming Dong, and Farshad Fotouhi. Bipartite isoperimetric graph partitioning for data co-clustering. *Data Mining and Knowledge Discovery*, 16(3):276–312, June 2008.

[90] Michal Rosen-Zvi, Chaitanya Chemudugunta, Thomas Griffiths, Padhraic Smyth, and Mark Steyvers. Learning author-topic models from text corpora. *ACM Transactions on Information and System Security*, 28(1):4:1–4:38, January 2010.

[91] Gerard Salton. *Automatic Text Processing: The Transformation, Analysis, and Retrieval of Information by Computer*. Addison-Wesley Longman Publishing Co., Inc., Boston, MA, USA, 1989.

[92] Badrul Sarwar, George Karypis, Joseph Konstan, and John Riedl. Incremental singular value decomposition algorithms for highly scalable recommender systems. In *Fifth International Conference on Computer and Information Science*, pages 27–28, 2002.

[93] Ugo Scaiella, Paolo Ferragina, Andrea Marino, and Massimiliano Ciaramita. Topical clustering of search results. In *Proceedings of the 5th International Conference on Web Search and Data Mining*, WSDM '12, pages 223–232, 2012.

[94] Julian Sedding and Dimitar Kazakov. WordNet-based text document clustering. In *Proceedings of the 3rd Workshop on RObust Methods in Analysis of Natural Language Data*, ROMAND '04, pages 104–113, Stroudsburg, PA, USA, 2004. Association for Computational Linguistics.

[95] M. Mahdi Shafiei and Evangelos E. Milios. Latent Dirichlet co-clustering. In *Proceedings of the Sixth International Conference on Data Mining*, ICDM '06, pages 542–551, Washington, DC, USA, 2006. IEEE Computer Society.

[96] M. Mahdi Shafiei and Evangelos E. Milios. A statistical model for topic segmentation and clustering. In *Proceedings of the Canadian Society for Computational Studies of Intelligence, 21st Conference on Advances in Artificial Intelligence*, Canadian AI '08, pages 283–295, Berlin, Heidelberg, 2008. Springer-Verlag.

[97] Peter H. A. Sneath and Robert R. Sokal. *Numerical Taxonomy. The Principles and Practice of Numerical Classification.* Freeman, 1973.

[98] Bharath Sriram, Dave Fuhry, Engin Demir, Hakan Ferhatosmanoglu, and Murat Demirbas. Short text classification in twitter to improve information filtering. In *Proceedings of the 33rd International ACM SIGIR Conference on Research and Development in Information Retrieval*, SIGIR '10, pages 841–842, New York, NY, USA, 2010. ACM.

[99] Michael Steinbach, George Karypis, and Vipin Kumar. A comparison of document clustering techniques. In *KDD Workshop on Text Mining*, 2000.

[100] Alexander Strehl and Joydeep Ghosh. A scalable approach to balanced, high-dimensional clustering of market-baskets. In *Proceedings of the 7th International Conference on High Performance Computing*, HiPC '00, pages 525–536, London, UK, UK, 2000. Springer-Verlag.

[101] Masashi Sugiyama. Local Fisher discriminant analysis for supervised dimensionality reduction. In *Proceedings of the 23rd International Conference on Machine Learning*, ICML '06, pages 905–912, New York, USA, 2006. ACM.

[102] Qi Sun, Runxin Li, Dingsheng Luo, and Xihong Wu. Text segmentation with LDA-based Fisher kernel. In *Proceedings of the 46th Annual Meeting of the Association for Computational Linguistics on Human Language Technologies: Short Papers*, HLT-Short '08, pages 269–272, Stroudsburg, PA, USA, 2008. Association for Computational Linguistics.

[103] Andrea Tagarelli and George Karypis. A segment-based approach to clustering multi-topic documents. In *Proceedings of SIAM Data Mining Conference Text Mining Workshop*, Atlanta, GA, USA, 2008.

[104] Yee Whye Teh, Michael I. Jordan, Matthew J. Beal, and David M. Blei. Hierarchical Dirichlet processes. *Journal of the American Statistical Association*, 101(476):1566–1581, 2006.

[105] Naonori Ueda and Kazumi Saito. Single-shot detection of multiple categories of text using parametric mixture models. In *Proceedings of the 8th ACM SIGKDD International Conference on Knowledge Discovery and Data Mining*, KDD '02, pages 626–631, 2002.

[106] Esko Ukkonen. On-line construction of suffix trees. *Algorithmica*, 14:249–260, 1995.

[107] Vishwa Vinay, Ingemar J. Cox, Ken Wood, and Natasa Milic-Frayling. A comparison of dimensionality reduction techniques for text retrieval. In *Proceedings of the Fourth International Conference on Machine Learning and Applications*, ICMLA '05, pages 293–298, Washington, DC, USA, 2005. IEEE Computer Society.

[108] Hanna M. Wallach. Topic modeling: Beyond bag-of-words. In *Proceedings of the 23rd International Conference on Machine Learning*, ICML '06, pages 977–984, New York, USA, 2006. ACM.

[109] Xuanhui Wang and ChengXiang Zhai. Learn from web search logs to organize search results. In *Proceedings of the 30th Annual International ACM SIGIR Conference on Research and Development in Information Retrieval*, SIGIR '07, pages 87–94, New York, USA, 2007. ACM.

[110] Xufei Wang, Jiliang Tang, and Huan Liu. Document clustering via matrix representation. In *Proceedings of the 2011 IEEE 11th International Conference on Data Mining*, ICDM '11, pages 804–813, Washington, DC, USA, 2011. IEEE Computer Society.

[111] Yongheng Wang, Yan Jia, and Shuqiang Yang. Short documents clustering in very large text databases. In *Proceedings of the WISE Workshops*, pages 83–93, 2006.

[112] Yujing Wang, Xiaochuan Ni, Jian-Tao Sun, Yunhai Tong, and Zheng Chen. Representing document as dependency graph for document clustering. In *Proceedings of the 20th ACM International Conference on Information and Knowledge Management*, CIKM '11, pages 2177–2180, New York, USA, 2011. ACM.

[113] Jianshu Weng, Ee-Peng Lim, Jing Jiang, and Qi He. TwitterRank: Finding topic-sensitive influential Twitterers. In *Proceedings of the Third ACM International Conference on Web Search and Data Mining*, WSDM '10, pages 261–270, New York, USA, 2010. ACM.

[114] Sunny K. M. Wong, Wojciech Ziarko, and Patrick C. N. Wong. Generalized Vector Space Model in information retrieval. In *Proceedings of the 8th Annual International ACM SIGIR Conference on Research and Development in Information Retrieval*, SIGIR '85, pages 18–25, New York, USA, 1985. ACM.

[115] Jun Yan, Ning Liu, Shuicheng Yan, Qiang Yang, Weiguo Fan, Wei Wei, and Zheng Chen. Trace-oriented feature analysis for large-scale text data dimension reduction. *The IEEE Transactions on Knowledge and Data Engineering*, 23(7):1103–1117, July 2011.

[116] Yiming Yang and Jan O. Pedersen. A comparative study on feature selection in text categorization. In *Proceedings of the Fourteenth International Conference on Machine Learning*, ICML '97, pages 412–420, San Francisco, CA, USA, 1997. Morgan Kaufmann Publishers Inc.

[117] Charles T. Zahn. Graph-theoretical methods for detecting and describing Gestalt clusters. *IEEE Transactions on Computers*, 20(1):68–86, January 1971.

[118] Oren Zamir and Oren Etzioni. Web document clustering: A feasibility demonstration. In *Proceedings of the 21st ACM SIGIR International Conference on Research and Development in Information Retrieval*, SIGIR '98, pages 46–54, 1998.

[119] Hua-Jun Zeng, Qi-Cai He, Zheng Chen, Wei-Ying Ma, and Jinwen Ma. Learning to cluster web search results. In *Proceedings of the 27th Annual International ACM SIGIR Conference on Research and Development in Information Retrieval*, SIGIR '04, pages 210–217, New York, USA, 2004. ACM.

[120] Hongyuan Zha, Xiaofeng He, Chris Ding, Horst Simon, and Ming Gu. Bipartite graph partitioning and data clustering. In *Proceedings of the Tenth International Conference on Information and Knowledge Management*, CIKM '01, pages 25–32, New York, USA, 2001. ACM.

[121] Ying Zhao and George Karypis. Criterion functions for document clustering: Experiments and analysis. Technical report, University of Minnesota, 2002.

[122] Ying Zhao and George Karypis. Empirical and theoretical comparisons of selected criterion functions for document clustering. *Machine Learning*, 55(3):311–331, 2004.

[123] Ying Zhao and George Karypis. Soft clustering criterion functions for partitional document clustering: A summary of results. In *Proceedings of the Thirteenth ACM International Conference on Information and Knowledge Management*, CIKM '04, pages 246–247, New York, USA, 2004. ACM.

[124] Shi Zhong and Joydeep Ghosh. Generative model-based document clustering: A comparative study. *Knowledge and Information Systems*, 8(3):374–384, 2005.

Chapter 14

Clustering Multimedia Data

Shen-Fu Tsai
Microsoft Inc.
Redmond, WA
stsai8@illinois.edu

Guo-Jun Qi
University of Illinois at Urbana-Champaign
Urbana, IL
qi4@illinois.edu

Shiyu Chang
University of Illinois at Urbana-Champaign
Urbana, IL
chang87@illinois.edu

Min-Hsuan Tsai
University of Illinois at Urbana-Champaign
Urbana, IL
mtsai2@illinois.edu

Thomas S. Huang
University of Illinois at Urbana-Champaign
Urbana, IL
huang@ifp.uiuc.edu

14.1 Introduction

Clustering multimedia is a broad and interesting area of research that typically deals with clusters of different spaces (such as feature space and semantic space) and with different levels of granularity. As opposed to the conventional clustering approaches which deal with simple generation and assignment of documents to clusters, clustering multimedia goes far beyond these approaches. In the multimedia domain, clusterings of different levels and different spaces can be meaningful and can complement each other to form a more consistent and robust interpretation and understanding of multimedia documents.

Hence, clustering multimedia data is one of the most widely used techniques in multimedia applications. These applications range from the construction of visual vocabulary as a preprocessing procedure of multimedia data, to automatic video structuring and image content summarization. Because of their vital role in many applications, the clustering methods have been widely studied across different levels of granularity—from low level features to high level concepts. It is a challenging problem since we often encounter large-scale, high-dimensional, multimodal, and even noisy data in many real-world multimedia applications. The solutions to the above challenges often highly depend on the modalities we process in real applications.

In this chapter, we organize the sections by the type of the modalities involved in the multimedia applications. We will provide several concrete examples, each of which shows the ideas and principles of developing effective clustering algorithms with multimedia data, including audio, text, image, video, and combinations of several types. We will review some of the popular clustering techniques in the multimedia domain, with the goal of covering as many multimedia applications as possible. For each application, we give a concise overview, summary of their motivation, and the methods that have been developed for that application. We believe that this chapter provides an introductory yet practical guide to help interested readers understand the nature of the multimedia clustering problems and their solutions in real world.

This chapter is organized as follows. In Section 14.2, we discuss the clustering techniques that have been applied to wide variety of image data, including the application to visual words learning (14.2.1), face clustering and annotation (14.2.2), photo album event recognition (14.2.3), image segmentation (14.2.4) as well as large-scale image classification (14.2.5). In Section 14.3, we explain different clustering algorithms used in the context of video and audio data, including video summarization (14.3.1), video event detection (14.3.2), video story clustering (14.3.3), and music summarization (14.3.4). Finally, in Section 14.4, we discuss the clustering with multimodal data primarily with respect to image and text data.

14.2 Clustering with Image Data

Clustering algorithms have been adapted to image data for various purposes. These algorithms are applied across different levels of the image data, from visual feature level to the image level and even to the image collection level. In this section, we discuss each of these three levels along with an application and describe how the clustering algorithms have been successfully applied to these applications. The applications include quantization of visual features for representation, facial image clustering for face annotation, and collection level clustering on photo album for event recognition.

14.2.1 Visual Words Learning

Visual words learning, which involves vector quantization, is among one of the earliest adaptions of clustering algorithms in multimedia applications. Inspired by the success of the bag-of-words (BoW) model used in text domain, bag-of-visual-words (BoVW) or bag-of-features (BoF) model is proposed to represent an image with a visual word histogram. Similar to the BoW model in the text domain, the BoVW model uses a number of visual vocabulary to map the low level image patches or visual features to obtain a higher level visual word representation. However, unlike the text domain where the text can be naturally tokenized by the punctuation and white spaces (for certain languages) to obtain the vocabulary, BoVW models involve an additional step, usually referred to as visual words learning or vocabulary learning. Visual words learning employs vector quantization (VQ) or more sophisticated methods (such as sparse coding) to learn a set of visual words as the basis functions or the vocabulary, which is then used in the second stage of the BoVW model to encode the low-level visual-feature vectors into a new feature space. There are a number of advantages for the bag-of-visual-words model such as the ability to handle partial occlusion.

As vector quantization is one of the key ingredients in the visual word learning step, clustering algorithms play an important role in the BoVW model. Among many existing clustering algorithms, the k-means clustering is the most widely used method due to its simplicity and efficiency. However, there are two issues with the k-means clustering method: the convergence of the algorithm and the determination of the parameter k. More specifically, the k-means clustering algorithm only converges to the local optimum and thus it is sensitive to the initialization of the cluster centroids. In addition, the parameter k of the algorithm has to be determined beforehand. Although there have been a few remedies proposed in the literature such as the gap statistics [30], X-means [23], G-means [35], and data spectroscopic clustering [26], the most popular way of choosing k is to empirically pick the parameter k with multiple random initializations of the centroids. This is due to the lack of prior knowledge on the density and the compactness of the data.

It is worth noting that the BoVW model is not the first approach to apply the idea of clustering image patches along with their visual features. For instance, in the texture classification problem, Leung and Malik [19] proposed to use the k-means clustering to quantize the densely sampled outputs of filter banks and form a dictionary of tiny surface patches called *prototype textons*. These textons are then used as the bases for describing the images of different materials in relatively low-dimensional histograms. Sivic and Zisserman [27] were among the first to adapt the BoVW model in the video frame retrieval problem. They constructed a visual vocabulary by applying k-means clustering with Mahalanobis distance function, which downweights the noisy dimensions and decorrelates the dimensions. With the built vocabulary they can first encode the local descriptors of the key frames as visual words. As a visual analogy of text retrieval, these visual words can then be used to build inverted index and term frequency–inverse document frequency (tf-idf) weighting scheme which will be used for the task of retrieval. Similarly, Csurka et al. [11] proposed a *bag of keypoints* method based on the low-level affine-invariant descriptors of image patches in the image categorization problem. They chose to use the k-means clustering algorithm as the vector quantization method for constructing the visual vocabulary and empirically decided the number of desired representative vectors (k) with multiple initialization.

Other than the k-means clustering algorithm, some other clustering algorithms were also used in the VQ stage for the construction of the vocabulary. For example, Raginsky and Lazebnik [25] used the agglomerative clustering algorithm that takes a matrix of pairwise distances between visual descriptors of an image to successively merge clusters to thus discover a small set of basic primitives for each image individually. With these basic primitives, the distribution of the descriptors of the image is then summarized in the form of the medoids (the most centrally located descriptor for each cluster) and the weights (proportional to the cluster size) of the clusters.

Although VQ methods with clustering algorithm have demonstrated their simplicity and efficiency, more sophisticated and powerful methods are proposed to learn better image representation.

One of these alternative methods is the sparse coding that enforces the sparsity constraint when learning the vocabulary or the codebook [34]. With the ability to nonlinearly encode the visual features and achieve less quantization error, these methods have achieved better performance compared to the VQ methods on several image recognition benchmarks. However, the sparsity constraint that involves solving L_1-norm optimization problems make these methods computationally expensive. Therefore, Wang et al. [32] proposed to exploit the k-means clustering in their locality-constraint linear coding to initialize the codebook and learn an approximate codebook. The strategy to use the VQ clustering method in their nonlinear coding process speeds up the algorithm while maintaining a good performance.

While the clustering algorithms have been widely used for the vector quantization step in the BoVW or BoF model in various multimedia applications, they are often unsupervised methods and thus the vocabulary learned will capture only the low-level similarity of the appearance. It is still an open research problem to study the ways in which one can allow the clustering algorithms to learn a vocabulary that is semantically meaningful.

14.2.2 Face Clustering and Annotation

Clustering algorithms have also been used as an effective tool for the problems of face recognition and identification. For example, Berg et al. [4] employed a clustering algorithm in the two-stage procedure for face recognition to break ambiguities in labeling and to identify incorrectly labeled faces. Specifically, they first extracted a list of candidate names for each image by analyzing the caption. With these candidate names they then used a modified k-means clustering algorithm to progressively refine the name selection from the candidates for each image and computed the mean of the visual features of the images associated with each name. They showed that this iterative procedure along with cluster pruning and merging would result in a reliable set of clusters with more face images of the same individual. Also, it was observed that the resulting set of images contained less noise.

Guillaumin et al. [15] approached the face image clustering problem with agglomerative hierarchical clustering. Specifically, they adopt complete-linkage clustering along with a logistic discriminant metric learning from labeled image pairs. In particular, the complete-linkage function, or the distance between a pair of clusters, is computed using Mahalanobis metric learned from a standard linear logistic discriminant model. The clustering process then yields a hierarchy of clusters by varying the maximum distance in terms of the learned Mahalanobis metric to merge the clusters.

To reduce the users' burden on large scale face annotation, Suh and Bederson [28] introduced a bulk annotation framework that incorporates hierarchical clustering techniques to cluster faces according to timestamps and torso information. With the clustered faces, the users can do annotation at the cluster level instead of single-image annotation. However, the clustering based on event and torso information has certain limitations and may produce inconsistent results, which would thus increase the workload of labeling the face images. To solve this problem, Tian et al. [29] proposed a partial clustering algorithm with the goal of providing small sized and high accuracy (evident) clusters to reduce the labeling effort instead of improving the overall clustering performance. They substitute the classic k-means clustering with their novel partial clustering algorithm after the spectral embedding onto the unit hypersphere. Their partial clustering algorithm focuses only on finding evident clusters and excludes all noisy samples into a litter-bin cluster by adding a uniform background noise distribution in their Gaussian Mixture Model (GMM) based clustering.

More recently, Zhu et al. [36] presented a novel iterative clustering algorithm with a rank-order dissimilarity distance for face annotation. A rank-order distance is proposed as a dissimilarity measure based on the observation that the top neighbors of two faces would highly overlap if the faces were from the same person, otherwise they would differ greatly. Using the neighborhood structure as the distance measure, the rank-order distance can handle nonuniformly distributed data as well as noise and outliers better than using the absolute distances such as L_1 or L_2 distance. Based on this rank-order distance, the authors used a clustering algorithm to group faces into a small number of clusters for face annotation. However, as observed by the authors, the faces of the same person usually form several "subclusters" due to the variations in illumination, pose, and expression. To improve the clustering performance, the authors further introduced a cluster-level rank-order distance in conjunction with a cluster-level normalized distance to merge these subclusters. Their algorithm shows superior performance on four face databases over several classical clustering methods such as affinity propagation, spectral clustering, and shared nearest neighbors in terms of both effectiveness (based on four evaluation measures—precision, recall, compression ratio, normalized mutual information) and efficiency (time taken for clustering).

14.2.3 Photo Album Event Recognition

Collection-level analysis of consumer photos has not drawn too much attention in the past. In [7] and [8], the authors explored the relationship between collection-level annotation and image-level annotation with the help of GPS and time information associated with the photos. Without using any metadata, the authors of [18] proposed to describe images and video frames using a set of scores of 21 visual concepts. Then, in order to account for typical and nontypical photos within each album, they (partially) matched up the photos from two albums when computing the Earth Mover's Distance (EMD) between them. The EMD is a more robust distance metric which computes the minimum weighted distance among all pairwise distances between two image sets subject to weight-normalization constraints. It allows partial match between data points (album) and can alleviate the influence of outlier images. Assume there are n_1 and n_2 images in albums x_1 and x_2, respectively. The EMD between them is a linear combination of ground distance $d(I_p^1, I_p^2)$ weighted by flow $f(I_p^1, I_p^2)$ between any two images $I_p^1 \in x_1$, $I_p^2 \in x_2$:

$$D(x_1, x_2) = \frac{\sum_{p=1}^{n_1} \sum_{q=1}^{n_2} f(I_p^1, I_q^2) d(I_p^1, I_q^2)}{\sum_{p=1}^{n_1} \sum_{q=1}^{n_2} f(I_p^1, I_q^2)} \tag{14.1}$$

where an optimal flow $f(I_p^1, I_p^2)$ is obtained from the following linear program:

$$\begin{aligned}
&\min_f \sum_{p=1}^{n_1} \sum_{q=1}^{n_2} d(I_p^1, I_q^2) f(I_p^1, I_q^2) \\
&s.t. \ \forall 1 \le p \le n_1, 1 \le q \le n_2, f(I_p^1, I_q^2) \ge 0, \\
&\sum_{p=1}^{n_1} f(I_p^1, I_q^2) \le w_q^2 \\
&\sum_{q=1}^{n_2} f(I_p^1, I_q^2) \le w_p^1 \\
&\sum_{p=1}^{n_1} \sum_{q=1}^{n_2} f(I_p^1, I_q^2) = \min\{\sum_{p=1}^{n_1} w_p^1, \sum_{q=1}^{n_2} w_q^2\}
\end{aligned} \tag{14.2}$$

where w_p^1 and w_q^2 are the weights of images I_p^1 and I_q^2 in albums x_1 and x_2, respectively. They are set $w_p^1 = 1/n_1, w_q^2 = 1/n_2$, and $d(I_p^1, I_q^2)$ takes the value of the Euclidean distance over concept scores. The interpretation of EMD is that a matching between images of two albums can be mapped to a unique flow described in Equation (14.2), and hence, even sets of $\{w_p^2\}_p$ and $\{w_q^1\}_q$ imply that all the images within the same album are "equally matched." The last set of constraints in Equation (14.2) deals with the imbalanced number of photos in two albums. The final similarity matrix at the album level is then $S(x_1, x_2) = \exp(-D(x_1, x_2)/r)$, where r is the mean of all pairwise distances between all training data points. The next step is the spectral clustering performed using similarity matrix S. Let the resulting clusters be $\{W_j\}_j$. The similarity between an album x to a cluster W_j is the maximum similarity between x and the members in W_j:

$$S(x, W_j) = \max_{x_k \in W_j} S(x_k, x) \tag{14.3}$$

The final album representation is a Bag of Features consisting of similarities to each cluster $(S(x,W_1),\ldots,S(x,W_n))^T$. The authors of [16] take a different approach. Instead of using mid level detections, they discovered frequent visual words within an event and then ranked them using the PageRank algorithm. A visual word is the concatenation of a 12-dimensional RGB color descriptor and a 32-dimensional SIFT descriptor. The ranking of the visual word is proportional to the number of times it appears and the number of times it matches with another visual word in another photo. Eventually, an album is represented by the mean of all visual word histograms within it.

There are a few issues in the approaches described above. Low-level visual words can be complemented by mid-level detection because the latter is semantically closer to higher-level events. When using mid-level detection, it is clear that not all detections are relevant to the target event class, and incorporating all of them can overfit and degrade the performance because the detection usually becomes noisy. Moreover, averaging all photos within an album mixes up typical and non-typical images. In [31], the authors address these issues with frequent object patterns in semantic events. Short object patterns are more robust to noise, compared to directly using all object detections as described in [18]. Ideally, an object pattern can be ("car," "street light," "person"), which means that the three objects "car," "street light," and "person" co-occur a lot in a certain event. In practice, object detection is a relative score and is not binary value. Hence, the actual proposed object pattern looks like ("car(6/7)," "street light(5/7)," "desk(2/7)"), which means that in a certain scene or event "car" has detection score at least 6 out of 7, "street light" has at least 5 out of 7, and "desk" has *at most* 2 out of 7. This discrete levels is due to the detection quantization, and such an object pattern approximates image-level decision tree classifier $F(m(I)) = 1$ if and only if

$$\begin{aligned} & m_{i_1}(I)s_1 \geq t_1 \wedge m_{i_2}(I)s_2 \geq t_2 \wedge m_{i_3}(I)s_3 \geq t_3 \ldots \\ & s_1, s_2, \ldots \in \{1, -1\} \end{aligned} \tag{14.4}$$

where $m_k(I)$ is the raw detection output of object k for image I. i_j is the object index, s_j controls the interval region (greater or less than), and t_j is the threshold of the jth object in the pattern. Finally, the album level feature is formed by aggregating the output of a large number of $F_k(m(I))$ for all photos in the albums.

14.2.4 Image Segmentation

In this section, we review a semiautomatic image segmentation algorithm via *graph cuts* to minimize the energy function [5]. Users can manually label the pixels of some object in an image, and then graph cuts automatically segment the image into different objects corresponding to these labels. The graph cuts simultaneously measure the smoothness of the labeling and the agreement between the labeled pixels and the labeling function. More precisely, we are given a graph $G = (\mathcal{P}, \mathcal{N})$, where $\mathcal{P}$ is the set of nodes corresponding to the pixels in an image, and $\mathcal{N}$ is the set of edges measuring the similarity between the pixels based on their colors and locations. Then a labeling function f assigns each pixel $p \in \mathcal{P}$ with a label f_p in a finite set $\mathcal{L}$ of object labels. The goal here is to obtain the labeling function f that minimizes the energy

$$E(f) = E_{smooth}(f) + E_{data}(f)$$

where, $E_{smooth}(f)$ measures the extent to which f is not piecewise smooth, while $E_{data}(f)$ measures the disagreement between f and the pixel labels.

In many general cases, minimizing the energy function is NP-hard, and it is impossible to rapidly compute the global minimum unless P = NP. Therefore, the two algorithms, α-expansion and $\alpha-\beta$-swap, focus on effectively minimizing it based on graph cuts to find a local minimum. Let $P = \{\mathcal{P}_l | l \in \mathcal{L}\}$ be a partition of nodes given by a labeling function f, where $\mathcal{P}_l = \{p \in \mathcal{P} | f_p = l\}$ is a subset of nodes assigned with label l. Then, we can define α-expansion $\alpha-\beta$-swap as follows:

α-expansion: Given a label α, a move from a partition P w.r.t. f to a new partition P' w.r.t. f' is called an α-expansion, if $\mathcal{P}_\alpha \subset \mathcal{P}_\alpha{}'$ and $\mathcal{P}_l{}' \subset \mathcal{P}_l$, for any label $l \neq \alpha$.
$\alpha-\beta$-swap: Given a pair of labels α and β, a move from a partition P w.r.t. f to a new partition P' w.r.t. f' is called an $\alpha-\beta$-swap, if $\mathcal{P}_l = \mathcal{P}_l{}'$, for any label $l \neq \alpha, \beta$.

Given a labeling f, [5] finds an optimal $\alpha-\beta$-swap or α-expansion to reduce the energy defined above. Then the energy can be minimized by a variant of the "fastest descent" techniques. Thus, the key is to develop an efficient algorithm to find the optimal swap and expansion given the current labeling f. Compared to other standard algorithms where there is only one label change at a time, this approach can change arbitrarily large sets of labels simultaneously. Moreover, this algorithm can guarantee that the obtained local minimum is within a known factor of the global one.

14.2.5 Large-Scale Image Classification

Before the recent 2000s, most research efforts on image classification and clustering were focused on medium-scale databases, where the image features could fit into a desktop's memory (typically 4GB–32GB). Two main reasons for such a limitation on large-scale computer vision task are first, during earlier days, large-scale image classification databases were not available primarily due to the expensive cost associated with class labeling, and second, effective and efficient learning algorithms are critical for large-scale classification problems. In the year 2009, a new ontological dataset *ImageNet* was introduced by Deng et al. [12], which can be used for many applications based on image clustering, including object recognition, image classification, and automatic object clustering.

After this database was made publicly available, a significant amount of research work has been conducted in these areas which has led to efficient and robust learning algorithms. Unlike the traditional multiclass classification problem, a few hundreds of categories were involved in large-scale tasks that require a high testing efficiency while producing good classification performance. In other words, traditional strategies—including one-vs-all ($O(k)$), one-vs-one ($O(k(k-1)/2)$), label ranking [10] ($O(k(k-1)/2)$) and Decision Directed Acyclic Graph [24] ($O(k(k-1)/2)$ in training, and $O(k)$ in testing—will not be suitable for large-scale problems. [3] and [13] proposed to explore structural relations between object classes to construct classification trees that could possibly achieve a sub-linear testing cost for multiclass classification problems at web-scale. [3] considers a two-step approach: learning a tree structure and classifier weights using a *Label Embedding Tree*. The basic idea behind this work is to construct a tree structure where each node of the tree is associated with a label set and an SVM-type classifier.

Using such a learned tree structure, the classification complexity could become sublinear. Splitting labels into respective label sets is similar to a clustering problem. However, unlike the standard clustering techniques such as k-means (and its variants), label clustering is performed on the label space. We will now briefly introduce the definitions and notations of the Label Tree and then explain the details of the label clustering process.

A Label Tree is a tree $T = (N, E, F, L)$ with $n+1$ indexed nodes $N = \{0, \ldots, n\}$, a set of edges $E = \{(p_1, c_1), (p_{|E|}, c_{|E|})\}$ that are ordered pairs of parent and child node indices, label predictors $F = \{f_1, \ldots, f_n\}$, and label sets $L = \{l_0, \ldots, l_n\}$ associated with each node. Each node is associated with a subset of class labels and a linear classifier that determines which branch to follow. The root of such a tree contains all possible labels, which are indexed to 0. Except for the root node, all other nodes only belong to one parent node, with a fixed number of children. The label sets L indicate the set of labels to which the given vertex of the tree belongs. [3] considers a disjoint classification path to construct its tree, which implies only K unique leaf nodes will be observed, and two nodes i and j at the same depth do not have any common labels ($l_i \cap l_j = \emptyset$).

To learn the structure of the tree, k one-vs-all classifiers are trained to obtain a "confusion matrix" which will then be used as the affinity matrix for spectral clustering. k one-vs-rest classifiers are denoted by $\{f_1, \ldots, f_k\}$ independently (no tree structure is used). And the confusion matrix can be formulated as $C_{ij} = |\{(x, y_i) \in \mathrm{v} : \arg\min_r f_r(x) = j\}|$ on validation set v. For each internal node l of the tree, from root to leaf, partition its label set l_l between its children's label set $L_l = \{l_c : c \in N_l\}$, where $N_l = \{c \in N : (l, c) \in E\}$ and $\bigcup_{c \in N_l} l_c = l_l$, by maximizing

$$R_l(L_l) = \sum_{c \in N_l} \sum_{y_p, y_q \in l_c} A_{pq} \tag{14.5}$$

where $A = \frac{1}{2}(C + C^T)$ is a symmetrized confusion matrix and subject to constraints preventing trivial solutions (all labels belongs to one set). To optimize this function, which is a graph cut problem, the standard spectral clustering is applied [21]. Furthermore, we define D to be the diagonal matrix whose (i, i)th element is the sum of A's ith row and construct

$$L = D^{-\frac{1}{2}} A D^{\frac{1}{2}} \tag{14.6}$$

Then b largest eigenvectors of L are computed to form the matrix $X = [x_1, x_2, \ldots, x_b]$ by stacking the eigenvectors column-wise. After that, the matrix Y is constructed from X by renormalizing each of X's rows to a unit length and trading each row of Y as a point in $\mathcal{R}^b$. Then, they are clustered into b clusters using K-means. Finally, samples in the original space are assigned to cluster j if and only if row i of the matrix Y was assigned to cluster j to obtain such a label spaced clustering at a specific node. Using such a top-to-down label clustering scheme, an overall Label Tree structure is learned. Moreover, once the tree structure is obtained, label embedded hierarchical SVM can be learned for each node on the tree by minimizing the empirical tree loss defined as follows.

$$R_{emp}(x, y|T, w) = \frac{1}{m} \sum_{i=1}^{m} I(f_T(x) \neq y_i) \tag{14.7}$$

$$= \frac{1}{m} \sum_{i=1}^{m} \max_{p \in B(x)} I(y_i \notin l_p) \tag{14.8}$$

where

$$B(x) = \{b_1(x), b_2(x), \ldots, b_{D(x)}(x)\} \tag{14.9}$$

$$b_j(x) = argmax_{\{c:(b_{j-1}(x), c) \in E\}} f_c(x) \tag{14.10}$$

Though such a label spaced clustering incorporated with hierarchical SVM learning can perform testing in sublinear time for multiclass classification with a large number of classes, the major drawback of this approach is that the spectral clustering produces only disjoint subsets. It is difficult to learn a classifier for disjoint subsets when a class cannot be clustered into a label set reliably. The clustering technique proposed in [3] requires k one-vs-all classifiers as a prior for the spectral clustering, which is expensive when the number of classes is large. Recently, Deng et al. [13] presented a novel approach that can simultaneously determine the structure of the tree (label cluster of each label set) and learn the classifiers for each node in the tree, which not only avoids initial k one-vs-all training problem, but also gains control over the efficiency vs. accuracy trade-off in designing such a classification tree.

In [13], a local greedy method is formulated, for any given node r, $c(r) = \{1, \ldots, K\}$, and Q represents a prespecified number of children. At each node, a partition matrix $P \in \{0, 1\}^{Q \times K}$ and classifier weights $w \in \mathcal{R}^{D\times Q}$ need to be determined. $P_{qk}=1$ if class label k appears in child q and $P_{qk}=0$ otherwise. To measure the accuracy, the loss function is defined as follows:

$$L(w,x,y,P) = 1 - P(\hat{q},y) \tag{14.11}$$

which implies that label clustering should keep a higher classification accuracy. In the meantime, the partition for each label set within the cluster will also contribute to the overall efficiency measure which is formulated as follows:

$$A(w,x,P) = \frac{1}{K}\sum_{k=1}^{K} P(\hat{q},k) \tag{14.12}$$

Therefore, the joint balance of efficiency and accuracy trade-off for clustering label sets and enhancing the classification performance leads to the following optimization problem:

$$\min_{w,P} \quad \frac{1}{m}\sum_{i=1}^{m} A(w,x_i,P) \tag{14.13}$$

$$s.t. \quad \frac{1}{m}\sum_{i=1}^{m} L(w,x_i,y_i,P) \leq \varepsilon \tag{14.14}$$

$$\varepsilon \geq 0 \tag{14.15}$$

$$P \in \{0,1\}^{Q\times K} \tag{14.16}$$

However, directly optimizing the above optimization problem is intractable. With proper relaxation, alternatively optimizing over w and P can be formulated as a convex programmming problem:

$$\hat{L}(w,x,y,P) = \max\{0, 1 + \max_{q\in A_i, r\in B_i} w_r^T x_i - w_q^T x_i\} \tag{14.17}$$

Here, A_i is the set of children that contains class y_i, and B_i is the rest of the children. Therefore, optimizing w with a fixed P can be done as follows:

$$\min_{W} \ \lambda \sum_{q=1}^{Q} ||w_q||_2^2 + \frac{1}{m}\sum_{i=1}^{m} \hat{L}(w,x,y,P) \tag{14.18}$$

Furthermore, if we fix w and optimize over P, the terms can be rearranged as follows:

$$\min_{P} \quad A(P) = \sum_{q,k} P_{qk} \frac{1}{mK}\sum_{i=1}^{m} I(\hat{q}_i = q) \tag{14.19}$$

$$s.t. \quad 1 - \sum_{q,k} P_{qk} \frac{1}{mK}\sum_{i=1}^{m} I(\hat{q}_i = q \wedge y_i = k) \leq \varepsilon \tag{14.20}$$

$$\varepsilon \geq 0 \tag{14.21}$$

$$P_{qk} \in \{0,1\} \ \ \forall q,k \tag{14.22}$$

In such a way, clustering for each label set and the associated classifier are iteratively learned, thus yielding a local optimal solution.

14.3 Clustering with Video and Audio Data

In the previous section, we showed several examples of clustering algorithms with image data. It is known that videos consist of image sequences. However, clustering algorithms with video data are far more sophisticated than simply applying image clustering algorithm to the image sequences. Videos have their own structure, from the key images to the snapshots with natural transitions between each other. Within these shots are the scenes with more definitive story lines, such as stories and events, which contain the relatively complete information the video makers would like to tell us. Therefore, in this section, we first present how the clustering algorithms can be applied to structure the videos and summarize their content in order to enable users to effectively browse the large video corpus. We also show two automatic detectors which automatically extract video events and stories with the help of cross-media clustering algorithms with the associated visual and textual contents. These approaches demonstrate the challenges of developing clustering algorithms with heterogeneous content in videos. In addition to the video content, we often encounter the users demand to automatically structure the audio contents. Therefore, in this section, we also present a clustering algorithm with music data which shows how the audio content can be summarized effectively.

14.3.1 Video Summarization

It is an intensive task to find the informative or desired video sequences from a large volume of video corpus. To efficiently organize the unstructured video content in more compact form, several video summarization methods [14, 22] have been developed to aid users to browse through voluminous videos efficiently and capture the video content of their interest.

One of the most straightforward ways for video summarization is to use a set of keyframes to summarize the video content, each of which can be extracted from a video shot or subshot. Several optimization criteria are proposed to construct the set of keyframes which maximize the video content information that is covered by the selected frames. Most of the video summarization methods rely on user interactions to generate the keyframes, or some heuristic criteria such as minimizing the keyframe similarity or maximizing the time intervals between keyframes. However, these simple methods usually suffer from poor summarization results due to the suboptimal or heuristic criteria that are adopted.

As opposed to these naive approaches, [14] provides a more sophisticated way to summarize the video sequences in a more principled manner. They perform Singular Value Decomposition (SVD) on the visual feature representations of video frames, where each frame is represented by a feature vector in a lower dimensional space. SVD maps these input feature vectors onto a refined space with reduced dimensions, where the video content value is quantified by the visual changes in a frame cluster $\mathcal{S}$:

$$\mathrm{CON}(\mathcal{S}) = \sum_{\psi_i \in \mathcal{S}} \|\psi_i\|^2 \tag{14.23}$$

where ψ_i is the mapped point in the refined space. It is observed that, in the refined space, there is a strong correlation between the degree of visual changes in a frame cluster and the distance of the constituent frames in the cluster to the origin. In other words, it was shown that the frames in a video segment with more visual changes are usually projected into the points farther from the origin in the refined space by SVD. Based on this observation, a video summarization system was developed by extracting the continuous video clusters containing the frames whose mapped points are farther from the origin.

In another approach, [22] advances the video summarization by an analysis of global video structures and video highlights. Specifically, a video is represented as a weighted undirected graph with a set of shots, where the vertices are the feature points of the shot and edges connecting each pair of vertices measure the similarity between the shots. A normalized cut algorithm is used to recursively bipartition into clusters of shots. A temporal graph is then constructed using the obtained temporally adjacent clusters as nodes and the transition probabilities among the clusters as edges. A directed edge is added to the temporal graph if a shot in one cluster node has a temporally successive shot in another cluster node. The temporal graph is a state transition diagram that models the evolution of a video between the states, where a state is equivalent to a cluster. Dijkstra's algorithm is applied to the temporal graph to find the shortest path from the cluster containing the first shot in the video to the cluster containing the last shot. Then, the two adjacent clusters in the obtained path are disconnected if there is no path between them. The videos are then partitioned into two different scenes and the clusters along the obtained shortest path can be used to summarize the video. The algorithm shows effective and efficient for video skimming and summarization [22].

14.3.2 Video Event Detection

Video clustering algorithms have also been applied in the analysis and detection of abnormal or suspicious events in many surveillance applications. This allows for automatically providing early warning alerts to the human operators about any potentialy harmful human actions or social events. The basic assumption here is that the normal events are usually similar to each other. These events typically form some dominant clusters with the common patterns. On the contrary, the abnormal events are usually distinctive compared to the normal events, and thus, they will not be part of a cluster. This suggests that by clustering the video segments, the events clustered into dominant clusters are more likely to be the normal ones, while the other events far away from the dominant clustered patterns are probably the abnormal ones.

[17] proposes an abnormal event detection system of such a paradigm. This system contains two components in its abnormal event detection pipeline. In the first step, all video segments in the training set are clustered by the Dynamic Hierarchical Clustering (DHC) algorithm. To cluster these video sequences, a set of Hidden Markov Models (HMMs) is constructed to represent the video sequences, and the video sequences with similar HMMs are hierarchically merged. To compare the similarity between HMMs in DHC, the authors proposed to use the difference of BIC (Bayesian Information Criterion) between the models before and after the merge of video sequences:

$$d(i,j,\cdots) = BIC(ij,\cdots) - BIC(i,j,\cdots) \tag{14.24}$$

where $BIC(ij,\cdots)$ is the BIC value after two video sequences i and j are merged to fit into HMM, and $BIC(i,j,\cdots)$ is the BIC value before these two video sequences are merged.

A large negative value of BIC difference $d(i,j,\cdots)$ means the HMMs are more similar to each other. This similarity can be used to compare the similarity between any number of models, rather than only pairwise similarity. To reduce the overfitting problem of model fitting with few samples, the clustering results are updated at each merging step, and a new HMM is trained after all the video sequences are merged and reclassified to their corresponding clusters. This will alleviate the overfitting problem and will correctly assign the possibly wrongly clustered videos. In addition, a 2-depth search strategy is also proposed within DHC where 3 clusters can be merged to form a new cluster rather than only merging a pair of clusters each time as done in a typical 1-depth search strategy. This will accelerate the hierarchical clustering process with fewer levels of clustering trees. To avoid the unaffordable search effort due to the increased search space in a 2-depth strategy, the authors [17] also developed a set of exclusion rules to stop the merging of cluster triplets. These exclusion rules can be efficiently tested as they involve computing only of pair-wise model similarity.

With the obtained clustering results and corresponding models for each cluster, a mixture model of the cluster HMMs is developed to explain each video sequence in the corpus as follows:

$$P(i) = \sum_{c=1}^{C} \pi_c L_c^i \tag{14.25}$$

where $P(i)$ is the probability that a video sequence i is generated by this mixture model, π_c is the mixing coefficient, and L_c^i is the likelihood of the ith video sequence being generated by the HMM for the cth cluster.

It is also asserted that the mixing coefficient π_c for each component in the above mixture model can be used to represent the dominance of each cluster. The videos generated from the dominant clusters are considered to be normal ones, while a video in a cluster with mixing coefficient less than a certain threshold is considered to be an abnormal one.

14.3.3 Video Story Clustering

Story-level clustering of videos has been an indispensable ingredient to discover the evolving stories based on different event themes. [14] develops a video clustering algorithm to merge the stories with the same theme based on the occurrences of concepts in the stories which are extracted from both video frames and the keywords from speech transcripts. All the video stories are represented by a bipartite graph, where each story is linked to the keyframe clusters and textual keywords associated with them. The association between a story and the associated keyframe clusters and keywords is computed based on two factors: (i) their occurrence frequency and (ii) the story frequency. A co-clustering algorithm based on the spectral clustering [33] was then applied to simultaneously group the stories into clusters with the same event theme as well as the keyframe and keyword clusters which describe the same story theme.

Let us denote the keyword-story matrix by A_1 and keyframe cluster–story matrix by A_2. These two matrices are then concatenated to obtain the affinity matrix $A = [A_1^T A_2^T]^T$ for the bipartite graph. The normalized affinity matrix is then constructed: $A_n = D_1^{-1/2} A D_2^{-1/2}$, where D_1 and D_2 are two diagonal matrices whose elements are the summation of each row vector of A_1 and A_2, respectively. To obtain k clusters from the bipartite graph, $l = \lceil \log_2 k \rceil$ singular vectors of A_n are computed corresponding to the second smallest singular value until the $l+1$th singular value. The k-means algorithm is then run on the obtained l-dimensional space using these singular values to yield the desired k-way partition of the bipartite graph.

The authors of [37] proposed a video clustering algorithm across multiple sources. It was noticed that one news event with the same topic is usually reported by multiple news channels, and it is interesting to cluster these news videos with the same theme together. To achieve this goal, a K-partite graph which constitutes a number of node sets is constructed. Each of the node sets contains the video segments within one news source, and a pair of nodes in different sets is connected by an edge whose weight is the similarity between the video segments associated with these two nodes. Also, the nodes in the same set cannot be connected to each other. Similar to the clustering algorithm in [14], a spectral clustering algorithm is developed to cluster the videos from different sources into the stories with the same topics. As the K-partite graph is constructed across the multiple sources, the obtained stories will cover the videos from different sources instead of the single news channel.

14.3.4 Music Summarization

In [9], the authors investigate segmentation and clustering for the task of music summarization. Both are achieved based on similarity matrix of windows and segments, respectively. In this approach, a similarity matrix for predefined windows based on spectrogram distance is constructed. Then, kernel matching on this matrix to find specific patterns associated with segment boundary is applied. Let us denote the spectral data computed for N windows of a digital audio file by vectors $\{v_i : i = 1, \cdots, N\}$. The similarity matrix $\mathbf{S}$ is computed as follows:

$$\mathbf{S}(i,j) = d_{cos}(v_i, v_j) = \frac{v_i^T v_j}{|v_i||v_j|} \tag{14.26}$$

When analyzing the structure of $\mathbf{S}$, it can be noticed that the boundary between two coherent audio segments produces a checkerboard pattern, because the two segments will exhibit high within-segment (self-) similarity thus producing adjacent square regions of high similarity along the main diagonal of $\mathbf{S}$. The two segments will also produce rectangular regions of low between-segment (cross-) similarity off the main diagonal. The boundary is the crux of this checkerboard pattern. After segments are found, a similarity matrix for the segments is computed according to the Kullback-Leibler (KL) divergence, followed by clustering via SVD. Let us assume that, in segments p_i and p_j, the spectral feature has normal densities $G(\mu_i, \Sigma_i)$ and $G(\mu_j, \Sigma_j)$, respectively. The similarity between the two segments is calculated as follows:

$$d_{seg}(p_i, p_j) = \exp(-d_{KL}(G(\mu_i, \Sigma_i) \| G(\mu_j, \Sigma_j)) - d_{KL}(G(\mu_j, \Sigma_j) \| G(\mu_i, \Sigma_i))) \quad (14.27)$$

And the *segment-indexed* similarity matrix $\mathbf{S}_S$ is

$$\mathbf{S}_S(i,j) = d_{seg}(p_i, p_j) \quad \forall\ i,j = 1, \cdots, P \quad (14.28)$$

The SVD of $\mathbf{S}_S = \mathbf{U}\Lambda V^T$ is computed to decompose $\mathbf{S}_S$:

$$\mathbf{S}_S(i,j) = \sum_{k=1}^{K} \lambda_k \mathbf{U}(i,k)\mathbf{V}(j,k) = \sum_{k=1}^{K} \mathbf{B}_k(i,j) \quad (14.29)$$

Then,

$$b_k(j) = \sum_{i=1}^{P} \mathbf{B}_k(i,j) \quad (14.30)$$

is evaluated and measures the similarity of segment p_j to the segments in the kth segment cluster. Accordingly, segment p_i is assigned to cluster

$$k^* = \arg \max_{k=1,\cdots,K} b_k(i) \quad (14.31)$$

The authors go further to designate segments with large singular values as summarization, because they are usually repeated and, therefore, serve as summaries of popular music.

[20] identifies key phrases in pop or rock songs according to the clustering result. First, for each overlapping frame, a *Mel-cepstral feature* is calculated. Each initial segment consists of a fixed number of frames, within which the Mel-cepstral features are assumed to follow a Gaussian distribution. A bottom-up clustering is then applied; i.e., the clustering algorithm iteratively finds the two clusters with lowest distortion between them (if it is less than a threshold, then they are combined). The authors use a modified KL distance and call it $KL2$. They use

$$KL2(A;B) = KL(A;B) + KL(B;A) \quad (14.32)$$

where A and B are Gaussian distributions:

$$KL(A;B) = E_A\{\log(pdf(A)) - \log(pdf(B))\} \quad (14.33)$$

Assuming A and B are Gaussian distributions:

$$KL2(A;B) = \frac{\Sigma_A}{\Sigma_B} + \frac{\Sigma_B}{\Sigma_A} + (\mu_A - \mu_B)^2 \times (\frac{1}{\Sigma_A} + \frac{1}{\Sigma_B}) \quad (14.34)$$

After applying hierarchical clustering or HMM-based clustering, the most frequent cluster (label) is considered and the longest section containing this label in the first half of the song is chosen as the key phrase.

14.4 Clustering with Multimodal Data

Multimodal data refers to data with multiple modalities. More specifically, data with different modalities are linked within the structure of multimodal data, and the goal is to cluster the portion of the data with specified modality(ies). For example, images and the surrounding text on the web pages are linked together, and the goal is to cluster images, text, or both. The first naive method would be to cluster based only on the data from the target modality and ignore the rest, and the second would be to concatenate all modalities to form the feature vector for each entity. The former apparently does not make use of all the data available, while the latter does not consider the particular linkage structure of multimodal data. As we shall see, there are better ways to model the multimodal data that lead to better clusterings. We will specifically focus on the multimodal scenario of combining images with text since it was heavily investigated in the literature.

In [1], Barnard et al. propose a generative model to model the joint distribution of image segment and words given a hierarchy of image clustering. Each image cluster corresponds to a leaf node in the hierarchy, and the probability of generating a segment or word is summed up along the path from root to the leaf, implying the high level nodes are associated with abstract concepts and the lower nodes are associated with more concrete concepts. Mathematically, let D denote image and word observation associated with a document d, c denote the cluster index, i denote the words or image segment, and l denote the level. Then, the generative model described above can be written as follows:

$$P(D|d) = \sum_{c} P(c) \prod_{i \in D} \left(\sum_{l} P(i|l,c)P(l|c,d) \right) \tag{14.35}$$

The probability of generating an observation D given cluster c is summed up across all clusters, where on each path from the root to a cluster c, the probability of observing D is the product of the probabilities of observing all segments of D (which is the summation along the path because each concept along the path may generate the segment). The authors also propose a simpler variant which is document-independent:

$$P(D) = \sum_{c} P(c) \prod_{i \in D} \left(\sum_{l} P(i|l,c)P(l|c) \right) \tag{14.36}$$

It was shown through experiments that this model improves the quality of image clusters obtained and the annotation performance. The resulting art photo clusters also show that visual and textual features are complementary to each other and when effectively used can improve the clustering of these complex and abstract images.

Cai et al. [6] use spectral clustering based on visual, textual, and link information to cluster web image search results. Link structure among webpage, block, and image are exploited and combined with text similarity for spectral clustering. An image graph W_I can be constructed from a block graph W_B as follows:

$$W_I(i,j) = \sum_{i \in \alpha, j \in \beta} W_B(\alpha,\beta) \tag{14.37}$$

where i, j represent index images, α, β represent index blocks, and $i \in \alpha$ indicates image i is contained in block α. Equation (14.37) sums up similarities between blocks that contain a particular pair of images and obtain the similarity between the images. As for block similarity $W_B(\alpha,\beta)$, they are derived from **page-to-block** and **block-to-page** relations. Page-to-block function $X(p,\alpha)$ dictates the importance of a block α in a webpage p. It is proportional to the size of the block divided by the distance between the center of the block and the page. Block-to-page function $Z(\alpha, p)$ is nonzeros if and only if block α links to page p. Thus, to evaluate $W_B(\alpha,\beta)$ conditional probability interpretation is used as follows:

$$\begin{aligned} W_B(\alpha,\beta) &= Prob(\beta|\alpha) \\ &= \sum_{p \in P} Prob(p|\alpha) Prob(\beta|p) \\ &= \sum_{p \in P} Z(\alpha,p) X(p,\beta) \end{aligned} \tag{14.38}$$

Spectral clustering is used to cluster on textual feature. To combine the two, the authors use

$$S_{combine}(i,j) = \begin{cases} S_{textual}(i,j) \leq 1 & \text{if } S_{link}(i,j) = 0 \\ 1 & \text{if } S_{link}(i,j) > 0 \end{cases} \tag{14.39}$$

Then, the images within clusters are further arranged by their visual features. The resulting organization is much more user-friendly for browsing, and the clusters obtained are more reasonable than those produced using only the visual content of the images.

Bekkerman and Jeon [2] proposes applying *combinatorial Markov random field* (Comraf) to multimodal clustering. A combinatorial Markov random field is an MRF in which at least one node is a combinatorial random variable, which can be partitionings, partial orderings, etc., of a given finite set. Thus, the result of inference on a Comraf can be directly converted to a clustering. Given a Comraf model over m combinatorial random variables $\mathbf{X^c} = \{X_0^c, \ldots, X_{m-1}^c\}$, where X_0^c corresponds to the target modality, the task is to find the most probable instantiation of $\mathbf{X^c} : \mathbf{x^c}_{MPE} = \arg\max_{\mathbf{x^c}} P(\mathbf{x^c})$. Because X_0^c is the only target modality, for all $i > 0$, X_i^c is considered known and will correspond to a clustering of all singleton clusters. Moreover, the structure of the Comraf is an asterisk with target modality in the center. The authors chose weighted mutual information to be the pairwise potential function between two modalities, i.e.,

$$(x_0^c)_{MPE} = \arg\max_{x_0^c} \sum_{j=1}^{m-1} f_j(x_0^c, x_j^c) = \arg\max_{x_0^c} \sum_{j=1}^{m-1} w_j I(\hat{X}_0; X_j) \tag{14.40}$$

where $I(\hat{X}_0; X_j)$ evaluates the mutual information between two clusterings and w_j is the weight. In practice, the observed modalities are words, rectangular blobs, color, and texture. Through experiments, this framework is shown to obtain 2.5–3 times higher accuracy compared with a unimodal k-means algorithm.

14.5 Summary and Future Directions

This chapter concisely reviews, using some concrete examples, different categories of classical multimedia applications which directly or indirectly involve clustering techniques. For each example, we discussed the application-specific technical challenges for developing effective and efficient clustering algorithms, for example, multimodality and high-dimensionality. One of the prominent emerging multimedia applications involves the interaction between multimedia data with the fast developed social networks. For example, the users or groups who own or favor the multimedia documents have attracted many research efforts. On one hand, this additional dimensionality of social relations might provide useful information and insights from user affinities and behavior. On the other hand, these relationships are often dynamic and incomplete, thus making the problem more complicated and challenging.

Bibliography

[1] Kobus Barnard, Pinar Duygulu, and David Forsyth. Clustering art. In *CVPR*, pages 434–441, 2001.

[2] Ron Bekkerman and Jiwoon Jeon. Multi-modal clustering for multimedia collections. In *CVPR*, 2007.

[3] Samy Bengio, Jason Weston, and David Grangier. Label embedding trees for large multi-class tasks. In *NIPS*, pages 163–171, 2010.

[4] Tamara Berg, Alexander Berg, Jaety Edwards, Michael Maire, Ryan White, Yee-Whye Teh, Erik Learned-Miller, and D. A. Forsyth. Names and faces in the news. In *Proceedings of the 2004 IEEE Computer Society Conference on Computer Vision and Pattern Recognition*, CVPR'04, pages 848–854, 2004.

[5] Yuri Boykov, Olga Veksler, and Ramin Zabih. Fast approximate energy minimization via graph cuts. *IEEE Transactions on Pattern Analysis and Machine Intelligence*, 29(11):1222–1239, November 2001.

[6] Deng Cai, Xiaofei He, Zhiwei Li, Wei-Ying Ma, and Ji-Rong Wen. Hierarchical clustering of www image search results using visual, textual and link information. In *Proceedings of the 12th Annual ACM International Conference on Multimedia*, MULTIMEDIA '04, pages 952–959, New York, NY, USA, 2004. ACM.

[7] Liangliang Cao, Jiebo Luo, Henry S. Kautz, and Thomas S. Huang. Annotating collections of photos using hierarchical event and scene models. In *CVPR*, pages 1–8, 2008.

[8] Liangliang Cao, Jie Yu, Jiebo Luo, and Thomas S. Huang. Enhancing semantic and geographic annotation of web images via logistic canonical correlation regression. In *ACM Multimedia*, pages 125–134, 2009.

[9] M. Cooper and J. Foote. Summarizing popular music via structural similarity analysis. *2003 IEEE Workshop on Applications of Signal Processing to Audio and Acoustics*, pages 127–130, October 19, 2003.

[10] Koby Crammer and Yoram Singer. On the learnability and design of output codes for multi-class problems. *Machine Learning*, 47(2-3):201–233, 2002.

[11] Gabriella Csurka, Christopher R. Dance, Lixin Fan, Jutta Willamowski, and Cedric Bray. Visual categorization with bags of keypoints. In *Workshop on Statistical Learning in Computer Vision, ECCV*, pages 1–22, 2004.

[12] Jia Deng, Wei Dong, Richard Socher, Li-Jia Li, Kai Li, and Fei-Fei Li. ImageNet: A large-scale hierarchical image database. In *CVPR*, pages 248–255, 2009.

[13] Jia Deng, Sanjeev Satheesh, Alexander C. Berg, and Fei-Fei Li. Fast and balanced: Efficient label tree learning for large scale object recognition. In *NIPS*, pages 567–575, 2011.

[14] Yihong Gong and Xin Liu. Video summarization using singular value decomposition. In *Proc. of CVPR*, pages 174–180, 2000.

[15] Matthieu Guillaumin, Jakob Verbeek, and Cordelia Schmid. Is that you? Metric learning approaches for face identification. In *ICCV*, pages 498–505, 2009.

[16] Naveed Imran, Jingen Liu, Jiebo Luo, and Mubarak Shah. Event recognition from photo collections via pagerank. In *Proceedings of the 17th ACM International Conference on Multimedia*, MM '09, pages 621–624, New York, NY, USA, 2009.

[17] Fan Jiang, Ying Wu, and Aggelos K. Katsaggelos. A dynamic hierarchical clustering method for trajectory-based unusual video event detection. *Transactions on Image Processing*, 18(4):907–913, April 2009.

[18] Wei Jiang and Alexander C. Loui. Semantic event detection for consumer photo and video collections. In *ICME*, pages 313–316, 2008.

[19] Thomas Leung and Jitendra Malik. Representing and recognizing the visual appearance of materials using three-dimensional textons. *International Journal of Computer Vision*, 43(1):29–44, June 2001.

[20] Beth Logan and Stephen Chu. Music summarization using key phrases. In *Proceedings of IEEE ICASSP*, pages 749–752, 2000.

[21] Andrew Y. Ng, Michael I. Jordan, and Yair Weiss. On spectral clustering: Analysis and an algorithm. In *NIPS*, pages 849–856, 2001.

[22] Chong Wah Ngo, Yu-Fei Ma, and Hong-Jiang Zhang. Video summarization and scene detection by graph modeling. *IEEE Transactions on Circuits and Systems for Video Technology*, 15:296–305, 2005.

[23] Dan Pelleg and Andrew W. Moore. X-means: Extending k-means with efficient estimation of the number of clusters. In *ICML*, pages 727–734, 2000.

[24] John C. Platt, Nello Cristianini, and John Shawe-Taylor. Large margin DAGs for multiclass classification. In *Advances in Neural Information Processing Systems 12*, pages 547–553, 2000.

[25] Maxim Raginsky and Svetlana Lazebnik. Locality sensitive binary codes from shift-invariant kernels. In *Advances in Neural Information Processing Systems*, 2009.

[26] Tao Shi, Mikhail Belkin, and Bin Yu. Data spectroscopy: Eigenspace of convolution operators and clustering. *The Annals of Statistics*, 37, 6B:3960–3984, 2009.

[27] J. Sivic and A. Zisserman. Video Google: A text retrieval approach to object matching in videos. In *Proceedings of the International Conference on Computer Vision*, volume 2, pages 1470–1477, October 2003.

[28] Bongwon Suh and Benjamin Bederson. Semi-automatic image annotation using event and torso identification. In *Tech Report HCIL-2004-15*, Computer Science Department, University of Maryland, College Park, 2004.

[29] Yuandong Tian, Wei Liu, Rong Xiao, Fang Wen, and Xiaoou Tang. A face annotation framework with partial clustering and interactive labeling. In *CVPR*, pages 1–8, 2007.

[30] Robert Tibshirani, Guenther Walther, and Trevor Hastie. Estimating the number of clusters in a dataset via the gap statistic. *Journal of the Royal Statistical Society: Series B (Statistical Methodology)*, 63:411–423, 2000.

[31] Shen-Fu Tsai, Liangliang Cao, Feng Tang, and Thomas S. Huang. Compositional object pattern: a new model for album event recognition. In *ACM Multimedia*, pages 1361–1364, 2011.

[32] Jinjun Wang, Jianchao Yang, Kai Yu, Fengjun Lv, Thomas Huang, and Yihong Gong. Locality-constrained linear coding for image classification. In *CVPR*, 2010.

[33] Xiao Wu, Chong-Wah Ngo, and Qing Li. Co-clustering of time-evolving news story with transcript and keyframe. In *Proceedings of the 2005 IEEE International Conference on Multimedia and Expo, ICME 2005*, July 6-9, 2005, Amsterdam, The Netherlands, pages 117–120. IEEE, 2005.

[34] Jianchao Yang, Kai Yu, Yihong Gong, and Thomas S. Huang. Linear spatial pyramid matching using sparse coding for image classification. In *2009 IEEE Computer Society Conference on Computer Vision and Pattern Recognition (CVPR 2009)*, 20-25 June 2009, Miami, Florida, USA, pages 1794–1801, 2009.

[35] Zhonghua Zhao, Shanqing Guo, Qiuliang Xu, and Tao Ban. G-means: A clustering algorithm for intrusion detection. In *Proceedings of the 15th International Conference on Advances in Neuro-Information Processing—Volume Part I*, pages 563–570, 2009.

[36] Chunhui Zhu, Fang Wen, and Jian Sun. A rank-order distance based clustering algorithm for face tagging. In *CVPR'11*, pages 481–488, 2011.

[37] Dong-Qing Zhang, Ching Yung Lin, Shi-Fu Chang, and John R. Smith. Semantic video clustering across sources using bipartite spectral clustering. In *Spectral Clustering, International Conference on Multimedia and Expo*, pages 117–120, 2004.

Chapter 15

Time-Series Data Clustering

Dimitrios Kotsakos
University of Athens
Athens, Greece
dimkots@di.uoa.gr

Goce Trajcevski
Evanston, IL
Northwestern University
goce@eecs.northwestern.edu

Dimitrios Gunopulos
University of Athens
Athens, Greece
dg@di.uoa.gr

Charu C. Aggarwal
IBM T. J. Watson Research Center
Yorktown Heights, NY
charu@us.ibm.com

15.1 Introduction

Time-series data is one of the most common forms of data encountered in a wide variety of scenarios such as the stock markets, sensor data, fault monitoring, machine state monitoring, environmental applications, or medical data. The problem of clustering finds numerous applications in the time-series domain, such as the determination of groups of entities with similar trends. Time-series clustering has numerous applications in diverse problem domains:

- *Financial Markets:* In financial markets, the values of the stocks represent time series which continually vary with time. The clustering of such time series can provide numerous insights into the trends in the underlying data.
- *Medical Data:* Different kinds of medical data such as EEG readings are in the form of time series. The clustering of such time-series can provide an understanding of the common shapes in the data. These common shapes can be related to different kinds of diseases.
- *Earth Science Applications:* Numerous applications in earth science, such as temperature or pressure, correspond to series, which can be mined in order to determine the frequent trends in the data. These can provide an idea of the common climactic trends in the data.
- *Machine State Monitoring:* Numerous forms of machines create sensor data, which provides a continuous idea of the states of these objects. These can be used in order to provide an idea of the underlying trends.
- *Spatio-temporal Data:* Trajectory data can be considered a form of multivariate time-series data, in which the X-coordinates and Y-coordinates of objects correspond to continuously varying series. The trends in these series can be used in order to determine the important trajectory clusters in the data.

Time-series data falls within the class of *contextual data* representations. Many kinds of data such as time-series data, discrete sequences, and spatial data fall in this class. Contextual data contains two kinds of attributes:

- *Contextual Attribute:* For the case of time-series data, this corresponds to the time dimension. These attributes provide the reference points at which the behavioral values are measured. The time-stamps could correspond to actual time values at which the data points are measured, or they could correspond to *indices* at which these values are measured.
- *Behavioral Attribute:* This could correspond to any kind of behavior which is measured at the reference point. Some examples include stock ticker values, sensor measurements such as temperature, and other medical time series.

The determination of clusters of time series is extremely challenging because of the difficulty in defining similarity across different time series which may be scaled and translated differently on both the temporal and behavioral dimensions. Therefore, the concept of similarity is a very important one for time-series data clustering, and this chapter will devote a section to the problem of time-series similarity measures. Note that once a similarity measure has been defined for time-series data, it can be treated as an abstract object on which a variety of similarity-based methods such as spectral methods or partitioning methods can be used.

Time-series data allow diverse formulations for the clustering process, depending upon whether the series are clustered on the basis of their online correlations, or whether they are clustered on the basis of their shapes. The former is usually performed with an online approach based on a small past window of history, whereas the latter is typically performed with an off-line approach on the entire series. Therefore, this chapter will also carefully study the diverse formulations which arise in the context of time-series data clustering, map them to different application-specific scenarios, and also discuss the techniques for the different formulations.

Clustering of time-series data, like clustering for all types of data, has the goal of producing clusters with high intracluster similarity and low intercluster similarity. Specifically, objects belonging to the same cluster must exhibit high similarity to each other, while objects belonging to different clusters must exhibit low similarity, thus high distance from each other. However, some specific properties that are part of the nature of the time-series data—such as the *high dimensionality*, the presence of *noise*, and the *high feature correlation* [53, 25]— pose unique challenges for designing effective and efficient clustering algorithms. In time-series clustering it is crucial to decide what kind of similarity is important for the clustering application. Accordingly, an appropriate clustering algorithm and an appropriate distance measure should be chosen. For example, Euclidean distance reflects similarity in time, while dynamic time warping (DTW) reflects similarity in shape. Other approaches, like model-based clustering methods such as Hidden Markov Models (HMM) or ARMA processes [56] are followed when similarity in change matters.

A significant difference between time-series data clustering and clustering of objects in Euclidean space is that the time series to be clustered may not be of equal length. When this is not the case, so all time series are of equal length, standard clustering techniques can be applied by representing each time series as a vector and using a traditional L_p-norm distance [18]. With such an approach, only similarity in time can be exploited, while similarity in shape and similarity in change are disregarded. In this study we split the clustering process into two basic steps. The first one is the choice of the similarity measure that will be employed, while the second one concerns the grouping algorithm that will be followed. In the section discussing the first step, we classify the proposed approaches into two categories: one for one-dimensional time series and one for multidimensional time series. In the section discussing the clustering algorithms, we present them following the traditional classification scheme that defines three main classes of clustering algorithms: (a) k-means and its variants, (b) hierarchical approaches, and (c) density-based clustering. In addition to these three main classes, in Subsection 15.5.4 we discuss methods for trajectory clustering, as we consider this field to be of individual interest.

15.2 The Diverse Formulations for Time-Series Clustering

Time-series clustering lends itself to diverse formulations, which are highly application-specific in nature. For example, do we wish to determine sets of time series with similar online trends, or do we wish to determine time series with similar shapes? The objective function criteria in these cases are clearly very different. Furthermore, the application-specific goals in these cases are also quite different, and the corresponding similarity functions used in the analytical process are therefore also quite different.

Broadly speaking, two main formulations exist for time series data clustering:

- *Correlation-based Online Clustering:* In this case, a bunch of time series are clustered in *real time* on the basis of the correlations among the different time series. Such applications are common in scenarios such as financial markets, where it is desirable to determine groups of stocks which move together over time. Such methods can also be used in order to find time series which have similar trends, or at least correlated trends, over time. In these cases, a short window of history is used for the clustering process, and the similarity functions across the different series are based on interattribute correlations. These methods are also intimately related to the problems of time-series forecasting and regression. In fact, a few such clustering methods such as *Selective Muscles* [58] use clustering methods for stream selection and efficient forecasting. Other methods in the literature use such techniques for sensor selection [1, 2]. An important aspect of such methods is that they often need to be performed in real time, as the streams are evolving over time. In many cases, it is also desirable to detect the significant changes in the clustering behavior. This provides insights into how the stream correlation trends have changed significantly.

- *Shape-based Off-line Clustering:* In these cases, time series of similar *shapes* are determined from the data. The main challenge here is to define an appropriate shape-based similarity function. This can be extremely challenging because it requires the definition of similarity functions which can capture the similarities in the shapes of the time series. Depending upon the domain, the time series may be translated, scaled, and time warped. Therefore, numerous similarity functions such as the Euclidean function or dynamic time warping are used for time-series clustering. Clearly, the appropriateness of a similarity function depends upon the underlying data domain. Furthermore, some of the similarity functions are computationally quite expensive, since they require the use of dynamic programming methods for computation. Depending upon the application, the exact values on the timestamps may sometimes not be important, as long as the shapes of the time series are similar. For example, consider the case, where the EEG readings of different patients are used to create a database of multiple time series. In such cases, the exact value of the timestamp at which the EEG reading was measured is not quite as important. This is very different from online correlation-based clustering, where all the series are evaluated over approximately the same timestamps. Some versions of shape-based clustering are also used for online-clustering, such as the methods which perform subsequence clustering.

Clearly, the precise choice of model and similarity function depends upon the specific problem domain. Therefore, this chapter will examine the different formulations in the context of different application domains.

The problem becomes more complex as one moves to the multivariate scenario. The simplest multivariate scenario is that of trajectory clustering, which can be considered a form of bivariate or trivariate time-series data, depending upon whether 2- or 3-dimensional trajectories are considered. The aforementioned cases also apply to the case of trajectories, where it may be desirable to either determine objects which move together (online correlation methods) or determine trajectories which have similar shape (off-line shape methods). In the latter case, the exact time at which the trajectories were created may be of less importance. This chapter will carefully address the case of multivariate time-series clustering and will use spatiotemporal data as specific application domain of interest. While spatiotemporal data is used as a specific case in this chapter, because it provides the only known set of multivariate clustering methods, the techniques for spatiotemporal data clustering can be generalized to any bivariate and possibly nonspatial time-series data, by "pretending" that the two series correspond to a trajectory in 2-dimensional space. Furthermore, the core techniques for trajectory-based methods can also be generalized to higher dimensional multivariate time series, by treating them as n-dimensional trajectories. Therefore, the existing research on trajectory clustering provides some of the most powerful results on how multivariate time-series clustering can be properly addressed.

15.3 Online Correlation-Based Clustering

Online correlation-based clustering methods are closely related to the problem of forecasting. Such methods are typically based on clustering the streams on the basis of their correlations with one another in their past window of history. Thus, the similarity function between the different series uses an intrastream regression function in order to capture the correlations across different streams.

These methods are typically based on windows of the immediate history. Specifically, a window of length p is used for regression analysis, and the different series are clustered on the basis of these trends. Some of the most common methods for segmenting such streams define *regression-based similarity functions* on the previous window of history. Note that two streams in the same cluster need not be positively correlated. In fact, two streams with perfect negative correlation may also be assumed to belong to the same cluster, as long as the *predictability* between the different streams is high. This is quite often the case in many real scenarios in which some streams can be predicted almost perfectly from others.

15.3.1 Selective Muscles and Related Methods

The *Selective Muscles* method [58] is designed to determine the k best representatives from the current time series which can be used in order to predict the other series. This approach can be considered a version of the k-medoid clustering for online predictability-based clustering of time series. One important aspect of the original *Selective Muscles* approach is that it was designed for finding the k best representatives which predict *one particular* stream in the data. On the other hand, in unsupervised correlation clustering, it is desirable to determine the best set of representatives which can predict all the streams in the data. However, the two problems are almost exactly the same in principle, since the same approach in *Selective Muscles* can be used with an aggregated function over the different time series. The problem can be posed as follows:

Problem 15.3.1 *Given a dependent variable $\overline{X_i}$ and $d-1$ independent variables $\{\overline{X_j} : j \neq i\}$, determine the top $b < d$ streams to pick so as to minimize the expected estimation error.*

Note that the estimation is performed with a standard model such as the *Autoregressive (AR)* or the *Autoregressive Integrated Moving Average (ARIMA)* models. A more general version of the problem does not treat any particular variable as special and may be defined as follows:

Problem 15.3.2 *Given d variables $\overline{X_1} \ldots \overline{X_d}$, determine the top $b < d$ streams to pick so as to minimize the average estimation error over all d streams.*

Note that while the second problem is slightly different from the first, and more relevant to online clustering, it is no different in principle.

The approach in [58] uses a greedy method in order to select the k representatives for optimizing the predictability of the other streams. The approach for selecting the representatives is as follows. In each iteration, a stream is included in the representative set, which optimizes the estimation error. Subsequently, the next stream which is picked is based on maximizing the aggregate impact on the estimation error, considering the streams which have already been picked. In each iteration, the stream is added to the set of representatives, for which the incremental impact on the estimation error is as large as possible.

However, the technique in [58] is optimized in order to pick the k streams which optimize a specific dependent variable. A method in [2] uses the greedy approach in order to pick the streams with the use of the formulation discussed in Problem 15.3.2. A graph-based representation is used to model the dependencies among the streams with the use of a linkage structure. A greedy algorithm is used in order to pick the optimal set of representatives. It has been shown in [2] that the approach has an approximation bound of $(e-1)/e$ over the optimal choice of representatives.

A method in [1] performs the clustering directly on the streams by defining a *regression-based similarity function*. In other words, two streams are deemed to be similar, if it is possible to predict one stream from the other. Otherwise the immediate window of history is used in order to cluster the streams. The work in [1] is a general method, which also associates a cost of selecting a particular

```
Algorithm CorrelationCluster(Time Series Streams: [1...n]
     NumberOfStreams: k;
begin
  J = Randomly sampled set of k time series streams;
  At the next time stamp do
  repeat
    Add a stream to J, which leads to
      maximum decrease in regression error;
    Drop the stream from J which leads to
      least increase of regression error;
  until(J did not change in last iteration)
end
```

FIGURE 15.1: Dynamically maintaining cluster representatives.

representative. In general, for the clustering problem, it is not necessary to model costs. A simplified version of the algorithm, in which all costs are set to the same value is provided in Figure 15.1. The similarity between two streams i and j is equal to the regression error in predicting stream j from stream i with the use of any linear model. A particular form of this model has been discussed in [1]. Note that the similarity function between streams i and j is not symmetric, since the error of predicting stream i from stream j is different from the error of predicting stream j from stream i.

These methods for online correlation based stream clustering are very useful in many applications, since it is possible to select small subsets of streams, from which all the other streams can be effectively predicted. A number of other methods, which are not necessarily directly related to muscles select representatives from the original data streams. Such methods are typically used for sensor selection.

15.3.2 Sensor Selection Algorithms for Correlation Clustering

Sensor selection algorithms are naturally related to correlation clustering. This is because such methods typically pick a representative set of streams which can be used in order to predict the other streams in the data. The representative streams are used as a proxy for collecting streams from all the different streams. Such an approach is used in order to save energy. This naturally creates clusters in which each stream belongs to the cluster containing a particular representative. A number of techniques have been designed in recent years in order to determine correlations between multiple streams in real time [43, 48, 58, 61]. The techniques in [43, 48] use statistical measures in order to find lag-correlations and forecast the behavior of the underlying data stream. The works in [2, 16] propose methods for sensor selection with the use of domain-specific linkage knowledge and utility-feedback, respectively. Methods for observation selection are proposed in [27], when the importance of a sensor-set can be *premodeled* as a submodular function. Methods for using adaptive models for time-series prediction were proposed in [57]. The method in [58] uses linear regression in order to determine the correlations in the underlying data and use them for forecasting. The technique in [61] proposes general monitoring techniques to determine statistical correlations in the data in real time.

Correlation clustering algorithms in time series can also be used in order to monitor key *changes* in the correlation trends in the underlying stream. This is because when the correlations between the different streams change over time, this leads to changing membership of streams in different clusters in the data. Such changes are relevant to determining key temporal anomalies in the data and are discussed in [4].

15.4 Similarity and Distance Measures

For more conventional *shape-based* clustering algorithms, the use of similarity and distance measures is crucial. In [34] Liao classifies time-series distance measures into three classes:

(a) feature-based

(b) model-based

(c) shape-based

While this classification makes sense and analyzes in-depth the proposed time-series distance measures and their differences, it lacks information about whether each distance measure is appropriate for comparing multivariate time series as well. Over the last few years, multivariate time-series clustering has attracted the interest of the time-series research community as it has a variety of interesting applications. In this sense, our study of similarity/distance measures that are used in time-series clustering is supplementary to that of Liao [34] and provides another interesting insight. One of the most important decisions when performing time-series clustering is the similarity/distance measure that will be chosen in order to compare the time series. Given a time-series similarity measure, time-series clustering can be performed with one of the several proposed techniques that will be reviewed in the next section. Time-series similarity has been a long and deeply studied problem for the past two decades. In the literature one can find many proposed distance measures, with each of them being appropriate for different applications and having specific advantages and disadvantages. Many classification schemes have been proposed regarding time-series similarity measures. A popular classification criterion is the time-series representation scheme that is assumed by the various measures, as the time series can be represented either by the raw data or by transformations of the original data such as the frequency transformation representations (Discrete Fourier Transformation (DFT) and Discrete Wavelet Transformation (DWT)) or other representation models (Landmark and Important Points Representation, ARIMA Model and LPC Cepstral Coefficient Representation, Symbolic Representation, Signature Representation [10]). In this section we employ the raw data representation, where a time series T of length n is represented as an ordered sequence of values $T = [t_1, t_2, \dots, t_n]$. T is called raw representation of the time-series data. Another important feature and classification criterion when classifying similarity/distance measures for time series is whether a measure is appropriate for univariate or multivariate time series. In this section we review the most commonly used similarity/distance measures when it comes to time-series clustering and distinguish them to univariate and multivariate measures.

15.4.1 Univariate Distance Measures

In this subsection, we briefly present some similarity/distance measures that are widely used by the majority of time-series clustering methods proposed in the literature and reviewed in this chapter. The distance/similarity measures presented here handle one-dimensional time series and most of them are extended by many multidimensional time series similarity/distance approaches that will be described in the next subsection.

15.4.1.1 L_p Distance

The L_p-norm is a distance metric, since it satisfies all of the nonnegativity, identity, symmetry, and triangle inequality conditions. An advantage of the L_p-norm is that it can be computed in linear time to the length of the trajectories under comparison; thus, its time complexity is O(n), n being the length of the time series. In order to use the L_p-norm, the two time series under comparison must be of the same length.

The Minkowski of order p or the L_p-norm distance, being the generalization of Euclidean distance, is defined as follows:

$$L_p - \text{norm}(T_1, T_2) = D_{M,p}(T_1, T_2) = \sqrt[p]{\sum_{i=1}^{n} (T_{1i} - T_{2i})^p} \quad (15.1)$$

The Euclidean distance between two one-dimensional time series T_1 and T_2 of length n is a special case of the L_p-norm for $p = 2$ and is defined as

$$D_E(T_1, T_2) = L_2 - \text{norm}(T_1, T_2) = \sqrt{\sum_{i=1}^{n} (T_{1i} - T_{2i})^2} \quad (15.2)$$

L_1-norm (p = 1) is named the Manhattan distance or city block distance.

15.4.1.2 Dynamic Time Warping Distance

Dynamic Time Warping (DTW) is a well-known and widely used shape-based distance measure. DTW computes the warping path $W = w_1, w_2, ..., w_K$ with $\max(m,n) \le K \le m+n-1$ of minimum distance for two time series of lengths m and n.

DTW stems from the speech processing community [45] and has been very popular in the literature of time-series distance measures [6]. Moreover, it has been extended by many approaches to handle the multidimensional case of trajectory matching. With use of dynamic programming DTW between two one-dimensional time series T_1 and T_2 of length m and n, respectively, can be computed as follows:

(a) $D_{\text{DTW}}(T_1, T_2) = 0$, if $m = n = 0$

(b) $D_{\text{DTW}}(T_1, T_2) = \infty$, if $m = n = 0$

(c) $D_{\text{DTW}}(T_1, T_2) = \text{dist}(T_{11}, T_{21}) + \mathit{minFactor}$, otherwise

where *minFactor* is computed as

$$\mathit{minFactor} = \min \begin{cases} D_{\text{DTW}}(\text{Rest}(T_1), \text{Rest}(T_2)) \\ D_{\text{DTW}}(\text{Rest}(T_1), T_2) \\ D_{\text{DTW}}(T_1, \text{Rest}(T_2)) \end{cases}$$

where $\text{dist}(T_{1,1}, T_{2,1})$ is typically the L_2-norm. The Euclidean distance, as long as all L_p-norms, described in 15.4.1.1, performs a one-to-one mapping between the data points of the time series under comparison. Thus, it can be seen as a special case of the DTW distance, which performs a one-to-many mapping. DTW is a more robust distance measure than L_p-norm because it allows time shifting and thus matches similar shapes even if they have a time-phase difference. An important point is that DTW does not satisfy the triangular inequality which could be a problem while indexing time series. However, in the literature there are several lower bounds that serve as solutions for indexing DTW offering faster performance [51]. Another advantage of DTW over L_p-norm is that DTW can handle different sampling intervals in the time series. This is a very important feature especially for long time series that span many years as the data sampling strategy may change over a long time.

15.4.1.3 EDIT Distance

EDIT distance (ED) comes from the field of string comparison and measures the number of insert, delete, and replace operations that are needed to make two strings of possibly different lengths identical to each other. More specifically, the EDIT distance between two strings S_1 and S_2 of length m and n, respectively, is computed as follows:

(a) $D_{ED}(S_1,S_2) = m$, if $n = 0$

(b) $D_{ED}(S_1,S_2) = n$, if $m = 0$

(c) $D_{ED}(S_1,S_2) = D_{ED}(\text{Rest}(S_1),\text{Rest}(S_2))$, if $S_{1_1} = S_{2_1}$

(d) $D_{ED}(S_1,S_2) = \min \begin{cases} D_{ED}(\text{Rest}(S_1),\text{Rest}(S_2))+1 \\ D_{ED}(\text{Rest}(S_1),S_2)+1 \\ D_{ED}(S_1,\text{Rest}(S_2))+1 \end{cases}$, otherwise

Although ED for strings is proven to be a metric distance, the two ED-related time-series distance measures DTW and Longest Common Subsequence (LCSS) that will be described in the next subsection are proven not to follow the triangle inequality. Lei Chen [11] proposed two extensions to EDIT distance, namely Edit distance with Real Penalty (ERP) to support local time shifting and Edit Distance on Real sequence (EDR) to handle both local time shifting and noise in time series and trajectories. Both extensions have a high computational cost, so Chen proposes various lower bounds, indexing and pruning techniques to retrieve similar time series more efficiently. Both ERP and DTW can handle local time shifting and measure the distance between two out-of-phase time series effectively. An advantage that ERP has over DTW is that the former is a metric distance function, whereas the latter is not. DTW does not obey triangle inequality, and therefore, traditional index methods cannot be used to improve efficiency in DTW-based applications. On the other hand, ERP is proved to be a metric distance function [11], and therefore, traditional access methods can be used. EDR, as a distance function, proves to be more robust than Euclidean distance; DTW and ERP are more accurate than LCSS, which will be described in the next subsection. EDR is not a metric, thus the author proposes three nonconstraining pruning techniques (mean value Q-grams, near triangle inequality, and histograms) to improve retrieval efficiency.

15.4.1.4 Longest Common Subsequence

The Longest Common Subsequence (LCSS) distance is a variation of EDIT distance described in 15.4.1.3 [52]. LCSS allows time series to stretch in the time axis and does not match all elements, thus being less sensitive to outliers than L_p-norms and DTW. Specifically, the LCSS distance between two real-valued sequences S_1 and S_2 of length m and n, respectively, is computed as follows:

(a) $D_{\text{LCSS},\delta,\varepsilon}(S_1,S_2) = 0$, if $n = 0$ or $m = 0$

(c) $D_{\text{LCSS},\delta,\varepsilon}(S_1,S_2) = 1 + D_{\text{LCSS},\delta,\varepsilon}(\text{HEAD}(S_1),\text{HEAD}(S_2))$
if $|S_{1,m} - S_{2,m}| < \varepsilon$ and $|m-n| \leq \delta$

(d) $\max \begin{cases} D_{\text{LCSS},\delta,\varepsilon}(\text{HEAD}(S_1),S_2) \\ D_{\text{LCSS},\delta,\varepsilon}(S_1,\text{HEAD}(S_2)) \end{cases}$, otherwise

where $\text{HEAD}(S_1)$ is the subsequence $[S_{1,1},S_{1,2},\ldots,S_{1,m-1}]$, δ is an integer that controls the maximum distance in the time axis between two matched elements, and ε is a real number $0 < \varepsilon < 1$ that controls the maximum distance that two elements are allowed to have to be considered matched, as depicted in Figure 15.2.

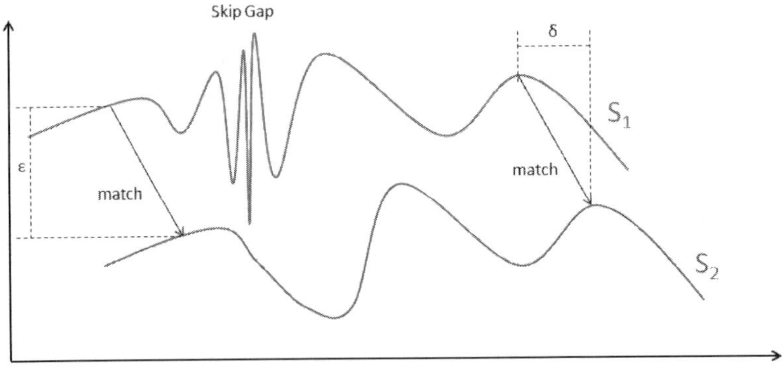

FIGURE 15.2 (See color insert): LCSS distance, thresholds δ and ε.

Apart from being used in time-series clustering, LCSS distance is often used in domains like speech recognition and text pattern mining. Its main drawback is that often it is needed to scale or transform one sequence to the other. A detailed study of LCSS variations and algorithms is presented in [5].

15.4.2 Multivariate Distance Measures

A metric distance function, ERP, is proposed that can support local time shifting in time series and trajectory data. A second distance function, EDR, is proposed to measure the similarity between time series or trajectories with local time shifting and noise [10].

15.4.2.1 Multidimensional L_p Distance

The L_p-norm between two d-dimensional time series T_1 and T_2 of length n extends the L_p-norm for the one-dimensional case and is defined as

$$L_p - \text{norm}(T_1, T_2) = D_{M,p}(T_1, T_2) = \sqrt[p]{\sum_{i=1}^{n}(T_{1i} - T_{2i})^p} = \sqrt[p]{\sum_{i=1}^{n}\sum_{j=1}^{d}(T_{1ij} - T_{2ij})^p} \tag{15.3}$$

As in the one-dimensional case, multidimensional L_p-norm has been proven to be very sensitive to noise and to local time shifting [31]. A wide variety of methods described in this section, either use the Euclidean distance or expand it, in order to define new distance measures. Lee et al. use the multidimensional Euclidean distance, namely, the L_2-norm, to compare multidimensional time series [19]. They define the distance between two multivariate sequences of equal length as the mean Euclidean distance among all corresponding points in the sequences. For the case where the two compared sequences are not of same length, their approach slides the shorter one over the longer one, and the overall distance is defined as the minimum of all mean distances, computed as described above. Lin and Su [35] use the Euclidean distance in order to define the distance from a point p to a trajectory T as follows: $D_{\text{point}}(p,T) = \min_{q \in T} ED(p,q)$ where $ED(p,q)$ represents the Euclidean distance between points p and q. D_{point} is used to define the one-way distance (OWD) from one trajectory T_1 to another trajectory T_2 as the integral of the distance from points of T_1 to trajectory T_2, divided by the length of T_1. OWD is not symmetric, so the distance between trajectories T_1 and T_2 is the average of their one-way distances: $D(T_1,T_2) = \frac{1}{2} \cdot (D_{\text{OWD}}(T_1,T_2) + D_{\text{OWD}}(T_2,T_1))$. Similarly, using an alternate grid representation for trajectories, Lin and Su define the distance between two grid cells as their Euclidean distance, and through it they define the distance between two trajectories [35]. Moreover, they provide a semiquadratic algorithm with complexity $O(mn)$ for grid trajectory distance computation, where n is the length of trajectories and m is the number of local min points. The grid OWD computation algorithm turns out to be faster than the quadratic complexity needed to compute the DTW between two trajectories. The experimental evaluation proves that OWD outperforms DTW in accuracy and performance. However, OWD does not take into account the time information in trajectories, so no discussion about trajectories with different sampling rates can be made. Frentzos et al. [14] focus on the problem of identifying spatiotemporally similar trajectories, by taking into account the time information in trajectories, except for their spatial shapes. They introduce a dissimilarity metric, DISSIM, between two trajectories Q and R by integrating their Euclidean distance over a definite time interval when both Q and R are valid. This way, DISSIM takes into account the time dimension in both trajectories. Moreover, DISSIM can be used for trajectories with different sampling rates, if the nonrecorded data points are approximated by linear interpolation, assuming that the objects follow linear movements. The linear-interpolation technique for missing values can be applied to LCSS and EDR measures too, as pointed out in [14]. Lee et al. solve the problem of searching for similar multidimensional sequences in a database by computing the distance between two sequences through their MBRs [31]. The database sequences as well as the query sequence are partitioned into optimal subsequences that are represented by their MBR. The query processing is based on these MBRs, so scanning and comparing the entire data sequences are avoided. The distance between two MBRs is defined as the minimum Euclidean distance between the two corresponding hyper-rectangles. Based on this distance, the authors introduce two lower-bounding distance metrics and propose a pruning algorithm to efficiently process similarity queries.

15.4.2.2 Multidimensional DTW

DTW (Dynamic Time Warping) between two d-dimensional time series T_1 and T_2 of length m and n, respectively, is defined as the one-dimensional case as described in 15.4.1.2.

Just like in the one-dimensional case, multidimensional DTW allows stretching in time axis, matches all elements, and is extensively used in the speech recognition domain. DTW, in contrast to Euclidean distance, does not require the two time series under comparison to be of the same length and is not sensitive to local time shifting. However, DTW is not a metric, since it doesn't follow triangle inequality and its time complexity is $O(mn)$, which means that it is computationally expensive for long time series and is useful for only short ones, comprising a few thousand points. Moreover, DTW, like Euclidean distance, has been proven to be sensitive to noise. [49] contains a very detailed description of the computation and the semantics of DTW. Vlachos et al. apply DTW on handwriting data [51]. Before comparing two trajectories, they are transformed into a rotation invariant Angle/Arc-Length space in order to remove the translation, rotation, and scaling components. In the new space, the technique of warped matching is used, in order to compensate for shape variations. Salvador and Chan propose FastDTW, an approximation of DTW in linear time and space [49]. Their approach creates multiple resolutions of the compared time series, coarsening them and representing them with fewer points. Then the standard DTW is run over the lowest resolution and the produced wrap path is passed over to the next higher resolution. Finally the path is refined and this process continues until the original resolution of the time series is reached. FastDTW, being a suboptimal approximation of DTW and producing errors up to 19.2%, is faster because the number of cells it evaluates scales linearly with the length of the time series. Through experimental evaluation, the authors prove the accuracy and efficiency improvements over Sakoe-Chiba bands and data abstraction, which are two other popular DTW approximations [46]. However, the authors do not report results on multidimensional time series, so the performance of FastDTW when applied on this type of data has to be examined.

15.4.2.3 Multidimensional LCSS

Vlachos et al. proposed two nonmetric distance functions as an extension of LCSS for multidimensional time series. The method proved to be robust to noise, especially when compared to DTW and ERP [52]. LCSS does not depend on continuous mapping of the time series; thus, the approaches using or extending it tend to focus on the similar parts between the examined sequences. LCSS, in contrast to the Euclidean distance, does not take into account unmatched elements and matches only the similar parts. It therefore allows trajectories to stretch in the time axis. DTW and Euclidean distance try to match every element, so they are more sensitive to outliers. However, when using LCSS the time series under comparison must have the same sampling rates. In [52] the two-dimensional LCSS between two two-dimensional trajectories T_1 and T_2 of length m and n, respectively, is computed as follows:

(a) $D_{\mathrm{LCSS},\delta,\varepsilon}(T_1,T_2) = 0$, if $n = 0$ or $m = 0$

(c) $D_{\mathrm{LCSS},\delta,\varepsilon}(T_1,T_2) = 1 + D_{\mathrm{LCSS},\delta,\varepsilon}(\mathrm{HEAD}(T_1),\mathrm{HEAD}(T_2))$
if $|T_{1,m,x} - T_{2,n,x}| < \varepsilon$ and $|T_{1,m,y} - T_{2,n,y}| < \varepsilon$ and $|i - j| \le \delta$

(d) $\max \begin{cases} D_{\mathrm{LCSS},\delta,\varepsilon}(\mathrm{HEAD}(T_1),T_2) \\ D_{\mathrm{LCSS},\delta,\varepsilon}(T_1,\mathrm{HEAD}(T_2)) \end{cases}$, otherwise

where $\mathrm{HEAD}(T_1)$, δ, and ε are defined as in 15.4.1.4. Two-dimensional LCSS can easily be extended to d dimensions.

15.4.2.4 Multidimensional Edit Distance

In 2005, Chen et al. proposed EDR, Edit Distance on Real sequence, in order to address the problem of comparing real noisy trajectories with accuracy and robustness, claiming that EDR is more robust and accurate than DTW, LCSS, ERP, and Euclidean distance [11]. EDR is defined as the number of insert, delete, or replace operations to convert a trajectory T_1 into another T_2. Specifically, applying the Edit Distance on sequences of real numbers, rather than strings as it was originally proposed in [10] by Levenshtein, the authors define EDR as follows:

(a) $\mathrm{EDR}(T_1,T_2) = m$, if $n = 0$

(b) $\mathrm{EDR}(T_1,T_2) = n$, if $m = 0$

(d) $\mathrm{EDR}(T_1,T_2) = \min \begin{cases} \mathrm{EDR}(\mathrm{Rest}(T_1),\mathrm{Rest}(T_2)) + sc \\ \mathrm{EDR}(\mathrm{Rest}(T_1),T_2) + sc \\ \mathrm{EDR}(T_1,\mathrm{Rest}(T_2)) + sc \end{cases}$, otherwise

where $sc = 0$ if T_{11} and T_{21} match, and $sc = 1$ otherwise. Elements T_{1i} and T_{2i} are supposed to match if the distance between them in all dimensions is below a threshold ε, similarly to the way LCSS distance described in 15.4.1.4 and 15.4.2.3 defines matching. This way, EDR manages to cope with noisy multivariate time series by not being affected by outliers and to handle shifting in the time axis like ERP distance. In their experimental evaluation, Chen et al. prove their claims about the improvements of EDR over DTW, ERP, and LCSS when applied on noisy sequences.

15.4.2.5 Multidimensional Subsequence Matching

There are a variety of methods proposed for multidimensional subsequence matching that can be used for data mining tasks such as clustering and classification. SPRING is a dynamic-programming based method that identifies the subsequences of evolving numerical streams that are closest to a query in constant space and linear time in the dataset size [47].

Kotsifakos et al. have introduced SMBGT, a subsequence matching method, that allows for gaps in both the query and the target sequences and constrains the maximum match length between the two [26]. In their study, they apply SMBGT to build a Query-by-Humming system that given a hummed query song, retrieves the top K most similar songs in a database. The proposed similarity measure, SMBGT, given a query Q and a target sequence X, finds the subsequence of X that best matches Q. The experimental evaluation of the proposed similarity measure was performed on 2-dimensional time series of notes of arbitrary length. Given sequences Q and X and their respective subesequences $Q[qs,qe]$ and $X[xs,xe]$ of equal length, SMBGT(Q,X) is defined as follows. Let $\mathcal{G}_Q$ and $\mathcal{G}_X$ be the indices of $Q[qs,qe]$ and $X[xs,xe]$ in Q and X. If $q_{\pi_i} \approx_{\varepsilon} x_{\gamma_i}, \forall \pi_i \in \mathcal{G}_Q, \forall \gamma_i \in \mathcal{G}_X, i = 1,...,|\mathcal{G}_Q|$, and

$$\pi_{i+1} - \pi_i - 1 \leq \beta, \gamma_{i+1} - \gamma_i - 1 \leq \alpha \tag{15.4}$$

then, the pair $\{Q[qs,qe], X[xs,xe]\}$ is a common bounded-gapped subsequence of Q and X. The longest such subsequence with $xs - xe \leq r$ is called SMBGT(Q,X). ε controls tolerance, α and β control allowed gaps in sequences X and Q, respectively, and r controls the maximum alignment length. A graphical example of SMBGT distance in two dimensions is illustrated in Figure 15.3.

In the experimental evaluation is shown that the main advantage of SMBGT over compared subsequence matching methods (SPRING, Edit distance, and DTW) is that it can handle high noise levels better. In some applications, such as the studied Query-by-Humming problem, this is extremely important. Other approaches that perform subsequence matching are described in [20, 5, 7].

15.5 Shape-Based Time-Series Clustering Techniques

In this section we review the major and most widely used techniques for shape-based time-series clustering. The two most popular approaches are k-means clustering and hierarchical clustering. Most techniques either extend or use one of these two clustering methods; thus, we classify the reviewed approaches accordingly.

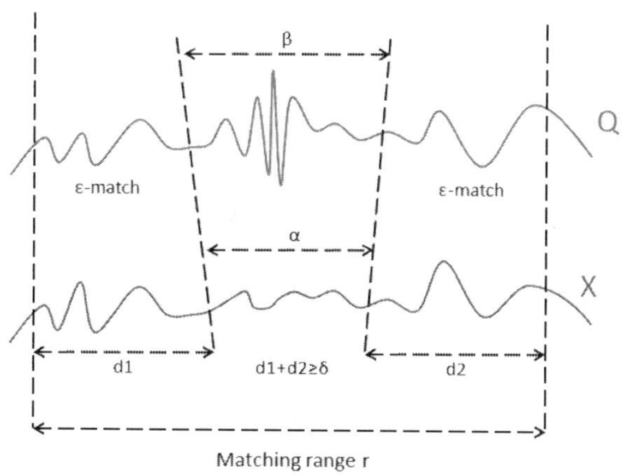

FIGURE 15.3 (See color insert): SMBGT distance between a query sequence Q and a database sequence X.

15.5.1 *k*-Means Clustering

One of the most widely used clustering techniques is *k*-means clustering, and this fact holds for time-series data clustering as well. *k*-means is a simple partitioning clustering algorithm, as it groups similar objects in the same cluster, and using an iterative refinement technique, it minimizes an objective error function. A general description of the algorithm is the following:

1. Find k initial cluster centers by selecting k random objects.
2. Assign each object to the most similar cluster. The most similar cluster is the cluster with the closest center, according to some distance function, e.g., Euclidean or DTW.
3. Recalculate the k cluster centers by averaging all the assigned objects for each cluster.
4. Repeat steps 2 and 3 until cluster centers no longer move. The objective error function, which is the sum of squared errors among each cluster center and its assigned objects has been minimized.

The complexity of k-means algorithm is $O(k \cdot N \cdot r \cdot D)$, where k is the number of desired clusters, N is the number of objects to be clustered (which equals the size of the dataset), r is the number of iterations until convergence is reached and D is the dimensionality of the object space [38]. In a slight modification of k-means algorithm, called k-medoids clustering, in Step 3, each cluster center is represented by the cluster object that is located nearest to the cluster center. For clustering large datasets of time series, k-means and k-medoids are preferred over other clustering methods, due to their computational complexity. However, both k-means and k-medoids require an initial cluster center selection which affects the clustering results, as both are hill-climbing algorithms, converging on a local and not a global optimum. The main disadvantage of k-means is that the number k of clusters must be specified a priori. This imposes the possibility that the optimal number of clusters for a specific dataset is not known before the clustering process, so k-means will produce a suboptimal clustering result.

Many k-means time-series clustering approaches use Euclidean distance as a distance metric and a corresponding averaging technique to compute new cluster centers at each step. However, DTW distance is considered a better distance for most time-series data mining applications. Until very recently and the work of Meesrikamolkul et al. [39] there was no DTW-based k-means clustering approach with satisfying performance. In Step 3 of the algorithm an averaging approach is needed in order to calculate the k new cluster centers. Unfortunately, DTW averaging produces sequences of equal or greater length than the original ones, thus decreasing a clustering system's accuracy, because the new cluster centers do not preserve the characteristics of the cluster objects. In [39], the authors propose a shape-based k-means clustering technique that uses DTW as distance measure and improves the time complexity of DTW averaging. In their approach, called Shape-based Clustering for Time Series (SCTS), they propose a DTW averaging method they call Ranking Shape-based Template Matching Framework (RSTMF), where a cluster center is calculated by averaging a pair of time series with Cubic-Spline Dynamic Time Warping (CSDTW) averaging. RSTMF computes an approximate ordering of the sequences before averaging, instead of calculating the DTW distance between all pairs of sequences within each cluster before selecting the most similar pair. DTW is a computationally expensive distance and therefore DTW-based k-means clustering using DTW averaging can become also computationally expensive.

Vlachos et al. proposed an anytime variation of the k-means algorithm that is based on an initial approximation of the raw data by wavelets [53]. As the process is repeated, the approximation becomes finer and the process stops when the approximation resembles the original data or the clustering results do not change. Their approach reduces the running time of original k-means and improves the quality of clustering results. Moreover, they demonstrate that time series can be effectively approximated by higher level representations while still preserving their shape characteristics useful to classification or clustering tasks. According to the Liao algorithm classification, the algorithm of Vlachos et al. is placed in the feature-based category, as it operates on a reduced-dimensionality approximation of the original time series using the Haar wavelet basis. In [36] the same researchers describe an extension to the work presented in [53], where the same approach is followed, in this work with an Expectation Maximization (EM) method serving the clustering process. EM is more a soft version of k-means than a fundamentally different approach, in the sense that each data object has a degree of membership in each cluster, whereas in k-means each object must belong to exactly one cluster. The major difference between EM and k-means is that EM produces a richer variety of cluster shapes than k-means, which favors spherical clusters.

15.5.2 Hierarchical Clustering

There are two types of hierarchical clustering, *agglomerative* and *divisive*. *Agglomerative* hierarchical clustering starts by regarding each data object as a different cluster and continues by searching the most similar pair of clusters. Then the most similar pair is merged into one cluster and the process continues until the desired number of clusters is reached.

Agglomerative hierarchical clustering has a variety of options for choosing which two clusters are the closest to each other and thus should be merged in the current step. Some of them are listed below:

- Single linkage: In single linkage selection, the distance between two clusters is defined as the shortest distance among all their member objects. Specifically, the single-link distance between clusters C_i and C_j is the following:

$$D_{SL}(C_i, C_j) = \min x \in C_i, y \in C_j(\text{dist}(x,y)) \tag{15.5}$$

  where *dist* is the chosen distance measure.

- Complete linkage: In complete linkage selection, the distance between two clusters is defined as the longest distance among all their member objects. Specifically, the complete-link distance between clusters C_i and C_j is the following:

$$D_{CL}(C_i, C_j) = \max_{x \in C_i, y \in C_j} (\text{dist}(x,y)) \tag{15.6}$$

  where *dist* is the chosen distance measure.

- Average linkage: In average linkage selection, the distance between two clusters is defined as the average distance among all their member objects. Specifically, the average-link distance between clusters C_i and C_j is the following:

$$D_{AV}(C_i, C_j) = \text{avg}\, x \in C_i, y \in C_j(\text{dist}(x,y)) \tag{15.7}$$

  where *dist* is the chosen distance measure.

Divisive hierarchical clustering is the inverse process of agglomerative hierarchical clustering in that it starts by regarding the whole dataset as a single cluster and continues by recursively dividing it into smaller ones.

Hierarchical clustering has better visualization capabilities than k-means, as the clustering process forms a dendrogram. In contrast to k-means, it takes no parameters and can be stopped and traced back at any point to the desired clustering level. Its major drawback is its quadratic computational complexity which makes hierarchical clustering practically useful only for small datasets. As opposed to k-means clustering, the hierarchical algorithm is a deterministic algorithm, which means that when applied on the same dataset it provides the same clustering results in every run.

In [19] the authors perform an evaluation of clustering methods applied on a long time series of medical data and implement agglomerative hierarchical clustering. Their experimental results show that complete-linkage cluster selection produces more reasonable clusters and better formed dendrograms, in that the input data sequences are more uniformly distributed in the output clusters. However, the superiority of complete-linkage selection over other methods is not proven nor believed to hold on all datasets and clustering applications.

In [18] the authors propose a hierarchical clustering method followed by a k-means fine-tuning process using DTW distance, where the objective function that is minimized is a sum of DTW distances from each object to a prototype of the cluster to which it belongs. The cluster prototype can be the DTW average of the objects belonging to the cluster, which has been proven to be inaccurate [42], the cluster medoid, or a locally optimal prototype that has been computed with a warping path based local search.

15.5.3 Density-Based Clustering

In [13] Ester et al. propose DBSCAN as a way to identify clusters of points utilizing the fact that intercluster density is higher than that among points that belong to different clusters. The intuition behind their approach is that objects belonging to a cluster must be surrounded by a minimum number of objects at a small distance, thus defining the notion of neighborhood density. In the DBSCAN algorithm, points that are located in a neighborhood of high density are defined as *core* points, whereas points that do not have a core point in their neighborhood are defined as *noise* points, and are discarded. Clusters are formed around *core* points, and clusters that are in the same neighborhood are merged.

Ertöz et al. argue that traditional DBSCAN cannot be used effectively in high-dimensional data such as time series, because the notion of Euclidean density is meaningless as the number of dimensions increases [12]. Instead, they propose the use of the k-nearest neighbor approach to multivariate density estimation, where a point is considered to be in a region with high probability density if it has a lot of highly similar neighbors. Using this notion, they eliminate noise and outliers, by identifying dense clusters in the data.

15.5.4 Trajectory Clustering

While multidimensional time series can be used to represent trajectories,[1] another commonly accepted interpretation is that a trajectory is a data type representing the movement of an object. In the past couple of decades, Moving Objects Databases have become a research trend of their own, and various representation methods, storage, and indexing techniques, along with spatiotemporal queries processing methodologies, have been introduced [17, 60]. Clustering and mining of spatiotemporal trajectories is of interest in various application domains such as traffic management, transportation optimizations, and ecological studies of animals motions/migrations.

Vlachos et al. define a trajectory as the set of positional information of a moving object ordered by time [51]. Lin and Su [35] disregard the time information in trajectories and focus on the shape information only. They point out that continuous representation of trajectories is usually costly and difficult to compute, so they represent trajectories in terms of line segment sequences. Specifically, each trajectory is defined as an ordered sequence of points $T = [p_1, p_2, ..., p_n]$ that are connected with straight line segments. Alternatively, Chen et al. propose to explicitly incorporate the time values in the representation; hence, a trajectory S in the two-dimensional plane is defined as a sequence of triples $S = [(t_1; s_{1_x}; s_{1_y}), ..., (t_n; s_{n_x}; s_{n_y})]$ [11].

A fair amount of work has been devoted to trajectory clustering, where not only shape but also speed and direction are important. Some methods [44] also discuss the online correlation aspect

[1]For that matter, a trajectory can be perceived as a special case of multidimensional time series.

of trajectory clustering, though these methods are not discussed in detail in this chapter. Globally, the efforts can be grouped in several categories: *relative motion patterns, flocks, convoys, moving clusters*, and *swarms* (cf. [21]). Going into great details about the peculiarities of each category of works is beyond the scope of this chapter, therefore, in the sequel we provide an overview of a few techniques in order to illustrate some specific issues (and solutions) arising in the domain of trajectories clustering.

Lee et al. [30] describe a trajectory clustering approach that belongs to the category of density-based clustering approaches. For a given collection of trajectories, the proposed method can operate on trajectories of different lengths and produce a set of clusters and a representative trajectory for each cluster. The authors argue on the meaningfulness of subtrajectory clustering; therefore, the clusters they generate are sets of trajectory partitions. A trajectory partition is a line segment p_ip_j, where p_i and p_j $(i < j)$ are points from the trajectory. The representative trajectory for each cluster is a trajectory partition that is common to all the trajectories that belong to that cluster. Trajectory partitions that belong to the same cluster have a relatively small distance to each other, according to the respective distance function which operates on line segments and corresponds to the weighted sum of the *perpendicular distance*, *parallel distance*, and *angle distance* (cf. [30]). After partitioning the trajectories, the method proceeds to the line segment clustering process based on DBSCAN (discussed in [13]). The main difference with DBSCAN is that in line segment clustering which is applied here, not all density connected groups of line segments can become clusters, because many line segments can belong to the same trajectory. Thus, each cluster that contains line segments from less than a desired number of trajectories is discarded. For the computation of the representative trajectory for each cluster, the authors use an average direction vector and sweep a vertical line across the line segments in its direction. The complexity of the proposed algorithm is $O(n^2)$ if no spatial index is used, and $O(n \log n)$ otherwise, with n being the total number of line segments that are produced by the partitioning process. In the subsequent work [29], trajectory-based and region-based clustering were combined to build a feature generation framework for trajectory classification. Region-based clustering disregards movement information and clusters regions of trajectories that are mostly of one class. The trajectory-based clustering extends the previous work [30] by using class label information in the clustering process. After trajectory partitioning, region-based clustering is performed and the partitions that cannot be represented by homogeneous regions are the input of the trajectory-based clustering module. The proposed framework generates a hierarchy of features in a top-down approach, namely features produced by region-based clustering do not include movement patterns and are thus of higher level whereas trajectory-clustering features are less general.

One observation regarding the above approaches is that they do not properly incorporate the temporal dimension of the trajectories (i.e., they work with *routes*). One of the first works that brought the temporal awareness in the realm of clustering spatiotemporal trajectories clustering was [24], which introduced the concept of a *moving cluster*—a set of objects that move close to each other for a given time duration. One can think of it as a temporal sequence of spatial clusters, preserving the property that the number of common objects among consecutive clusters is maintained above a certain threshold Θ. An example of a moving cluster with $\Theta \geq 75\%$ (i.e., 3/4 of the objects are within a cluster at any given moment) is shown in the left portion of Figure 15.4. A concept that uses different criteria for grouping the trajectories is the one of *convoys* [22] that corresponds to a group that has *at least m* objects which are density-connected with respect to *distance e* and *cardinality m* during *k consecutive time-instants*. For comparison, the right portion of Figure 15.4 shows the formation of a convoy of three trajectories over three consecutive time-instants, illustrating the difference with the moving clusters.

Many other criteria for grouping trajectories have been proposed, generating different corresponding models, for example, dynamic/evolving convoys, flocks, swarms, and we refer the reader to [21] for a recent survey. We close this section with *data-driven* observations regarding trajec-

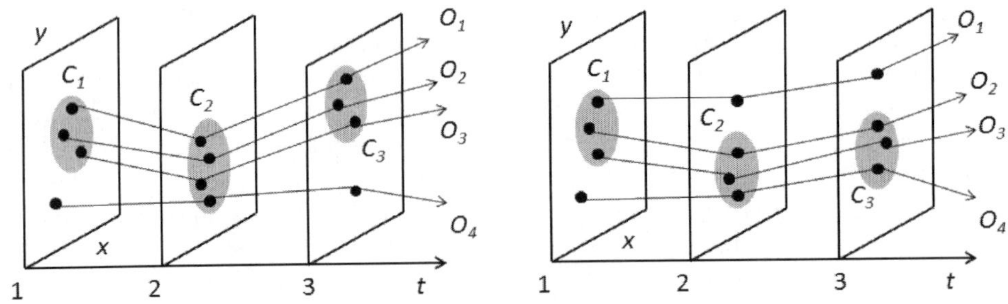

FIGURE 15.4: Trajectories grouping: Moving clusters and convoys.

tories clustering, given that in the recent years, the GPS traces and sensor-based location data are becoming widely available:

- The sheer volume of the *(location,time)* data may become large enough to incur high computation cost for clustering trajectories. Hence, oftentimes, some *simplification* techniques may be employed to reduce the size of the data [8], before proceeding with the clustering.
- The data streams are often subject to imprecision in the measurements as well as noise during communication/transmission. Hence, in order to improve the efficiency and effectiveness of the clustering techniques, it may be desirable to apply *trajectory smoothing* techniques as a preprocessing step to their clustering [9].

15.6 Time-Series Clustering Applications

Time-series clustering is a very interesting domain and has increasingly many applications. Widespread smartphone networks and mobile computing environments as long as the corresponding active communities present a field where multidimensional spatiotemporal trajectory clustering is important and necessary. Location-oriented applications and services can use trajectory clustering techniques to improve query evaluation performance. One example is route recommendations, as tourist guides can benefit from identifying similar user routes to recommend places or tours to users. Figure 15.5 illustrates how a route recommendation tourist application would utilize trajectory clustering information to classify a new user and recommend paths to follow.

The online scenario is particularly common in financial markets, machine monitoring, and anomaly detection. In fact, since outliers and clusters are connected by a complementary relationship, multivariate regression models are often used in order to identify broad trends in the data. Data points which do not match this broad trend are declared outliers [4].

In another example, mobile social networking applications can avoid controversial privacy concerns by using distributed techniques to identify and use similar trajectories in their network without disclosing the traces, as does the SmartTrace system [28]. Such systems provide the functionality of nearest neighbor search, where a user can determine other users that have exposed similar spatiotemporal behavior, such as visiting the same places, without knowing the exact trajectories or revealing their trajectory either. Both centralized and distributed approaches have been proposed to evaluate trajectory similarity queries, while the former serve applications where the transfer of data to the central site is inexpensive and the latter are appropriate for environments with expensive or not-always-connected mediums, such as wireless sensor networks [59] or smartphone networks.

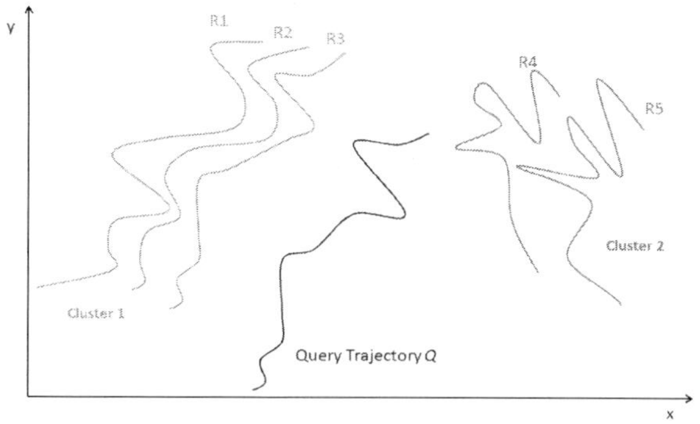

FIGURE 15.5: Query trajectory Q represents the route of a user of a mobile route-recommendation application. The application performs trajectory clustering and classifies the user to Cluster 1 (containing trajectories $R1$, $R2$, and $R3$); thus, it recommends paths similar to those users of Cluster 1 followed.

It becomes apparent that trajectory clustering is suitable for applications where privacy and anonymity are needed. A classic example of such cases is social sensing applications [44].

In another example, video surveillance and tracking systems can largely benefit from trajectory and time-series clustering, due both to the insight in scene monitoring that movement clustering provides and to privacy and security issues that have arisen. Abnormal events such as pedestrians crossing the street or dangerous vehicle movements can be represented as outliers to clusters of normal movement patterns [23]. In [23] the authors use a hierachical clustering method to overcome the overfitting in HMM trajectory similarity that is often used in surveillance video analysis, where video events are represented as object trajectories.

Automatic counting of pedestrians in detection and tracking systems is another application example. In [3] the authors propose a method to reduce the difference between the number of tracked pedestrians and the real number of individuals, as most detection and tracking systems overestimate the number of targets. The authors apply agglomerative hierarchical trajectory clustering, assuming that trajectories belonging to the same human body are more similar to each other than trajectories produced by the movement of different individuals. In this process they employ different trajectory representation schemes, including time series and independent component analysis representation, and different distance/similarity measures, including Hausdorff Distance and LCSS.

Time-series clustering is necessary in a variety of other domains, including music retrieval [26, 32, 40], speech recognition [55, 50], and financial and socioeconomic time-series applications [25, 15].

15.7 Conclusions

Time-series data have diverse formulations because of the variety of applications in which they can be used. The two primary formulations for time-series clustering use online and off-line analysis. The application domains for these cases are quite different. The online formulation is often used for real-time analysis and applications such as financial markets or sensor selection. The online scenario is also relevant to social sensing applications. The off-line scenario is more useful for applications in which the key shapes in the data need to be discovered for diagnostic purposes.

Bibliography

[1] C. Aggarwal, Y. Xie, and P. Yu. On dynamic data-driven selection of sensor streams, *KDD Conference*, 1226–1234, 2011.

[2] C. Aggarwal, A. Bar-Noy, and S. Shamoun. On sensor selection in linked information networks, *DCOSS Conference*, 1–8, 2011.

[3] G. Antonini and J.-P. Thiran. Counting pedestrians in video sequences using trajectory clustering. *IEEE Transactions on Circuits and Systems for Video Technology*, 16(8): 1008-1020, 2006.

[4] S. Bay, K. Saito, N. Ueda, and P. Langley. A framework for discovering anomalous regimes in multivariate time-series data with local models. Technical report, Center for the Study of Language and Information, Stanford University, 2004.

[5] L. Bergroth, H. Hakonen, and T. Raita. A survey of longest common subsequence algorithms. *SPIRE*, 39–48, 2000.

[6] D. Berndt and J. Clifford. Using dynamic time warping to find patterns in time series. *KDD Workshop*, 10(16):359-370, 1994.

[7] B. Bollobás, G. Das, D. Gunopulos and H. Mannila. Time-series similarity problems and well-separated geometric sets. *Nordic Journal of Computing*, 8(4):409–423, 2001.

[8] H. Cao, O. Wolfson, and G. Trajcevski. Spatio-temporal data reduction with deterministic error bounds. *VLDB J.*, 15(3):211–228, 2006.

[9] F. Chazal, D. Chen, L. Guibas, X. Jiang, and C. Sommer. Data-driven trajectory smoothing. *GIS*, 251–260, 2011.

[10] L. Chen. Similarity search over time series and trajectory data. PhD Thesis, 2005.

[11] L. Chen, M. Tamer Ozsu, and V. Oria. Robust and fast similarity search for moving object trajectories. *SIGMOD '05*, 491–502, 2005.

[12] L. Ertöz, M. Steinbach, and V. Kumar. Finding clusters of different sizes, shapes, and densities in noisy, high dimensional data. *SDM*, 2003.

[13] M. Ester, H.-P. Kriegel, J. Sander and X. Xu. A density-based algorithm for discovering clusters in large spatial databases with noise. *KDD*, 226-231, 1996.

[14] E. Frentzos, K. Gratsias, and Y. Theodoridis. Index-based most similar trajectory search. In *Proceedings of IEEE 23rd International Conference on Data Engineering (ICDE)*, 816–825, 2007.

[15] T.-C. Fu, C. Law, K. Chan, K. Chung, and C. Ng. Stock time series categorization and clustering via SB-tree optimization. *FSKD*, 1130–1139, 2006.

[16] D. Golovin, M. Faulkner, and A. Krause. Online distributed sensor selection. *IPSN Conference*, 220–231, 2010.

[17] R. Güting and M. Schneider. *Moving Objects Databases*. Morgan Kaufmann, 2005.

[18] V. Hautamaki, P. Nykanen, and P. Franti. Time-series clustering by approximate prototypes. *ICPR*, 1–4, 2008.

[19] S. Hirano and S. Tsumoto. Empirical comparison of clustering methods for long time-series databases. In *Proceedings of the Second International Conference on Active Mining*, 268-286, 2003.

[20] N. Hu, R. Dannenberg, and A. Lewis. A probabilistic model of melodic similarity. *Proceedings of the 2002 International Computer Music Conference*, 509–515, 2002.

[21] H. Jeung, M. Yiu, and C. Jensen. Trajectory pattern mining. In Yu Zheng and Xiaofang Zhou (Eds.) *Computing with Spatial Trajectories*, 143–177, 2011.

[22] H. Jeung, M. Yiu, X. Zhou, C. Jensen, and H. Shen. Discovery of convoys in trajectory databases. *VLDB*, 1068–1080, 2008.

[23] F. Jiang, Y. Wu, and A. Katsaggelos. Abnormal event detection from surveillance video by dynamic hierarchical clustering. *ICIP*, 145–148, 2007.

[24] P. Kalnis, N. Mamoulis, and S. Bakiras. On discovering moving clusters in spatio-temporal data. *SSTD*, pp. 364–381, 2005.

[25] K. Kalpakis, D. Gada, and V. Puttagunta. Distance measures for effective clustering of ARIMA time-series. *ICDM*, 273–280, 2001.

[26] A. Kotsifakos, P. Papapetrou, J. Hollmen, and D. Gunopulos. A subsequence matching with gaps-range-tolerances framework: A query-by-humming application. *PVLDB*, 4(11): 761–771, 2011.

[27] A. Krause, and C. Guestrin, Near-optimal observation selection using submodular functions. *AAAI Conference*, 1650–1654, 2007.

[28] C. Laoudias, M. Andreou, and D. Gunopulos. Disclosure-free GPS trace search in smartphone networks. *Proceedings of the 2011 IEEE 12th International Conference on Mobile Data Management*, 78–87, 2011.

[29] J.-G. Lee, J. Han, X. Li, and H. Gonzalez. TraClass: Trajectory classification using hierarchical region-based and trajectory-based clustering. *VLDB Conference*, 1081–1094, 2008.

[30] J.-G. Lee, J. Han, and K.-Y. Whang. Trajectory clustering: A partition-and-group framework. *SIGMOD Conference*, 593–604, 2007.

[31] S. Lee, S. Chun, D. Kim, J. Lee, and C. Chung. Similarity search for multidimensional data sequences. *Proceedings of 16th International Conference on IEEE Data Engineering*, 599–608, 2000.

[32] K. Lemström and E. Ukkonen. Including interval encoding into edit distance based music comparison and retrieval. *AISB*, 53–60, 2000.

[33] V. Levenshtein. Binary codes capable of correcting deletions, insertions, and reversals. *Soviet Physics Doklady*, 10:707–710, 1966.

[34] T. Warren Liao. Clustering of time series data—A survey. *Pattern Recognition* 38(11):1857–1874 (2005).

[35] B. Lin, and J. Su. Shapes based trajectory queries for moving objects. In *Proceedings of the 13th annual ACM International Workshop on Geographic Information Systems*, ACM, 21–30, 2005.

[36] J. Lin, M. Vlachos, E. Keogh, and D. Gunopulos. Iterative incremental clustering of time series. *EDBT*, 106–122, 2004.

[37] J. Lin, M. Vlachos, E. Keogh, and D. Gunopulos. Multi-resolution K-means clustering of time series and applications to images. *Workshop on Multimedia Data Mining (MDM)*, SIGKDD, Washington DC, 2003.

[38] J. MacQueen. Some methods for classification and analysis of multi-variate observations. *Proceedings of the Fifth Berkeley Symposium on Mathematical Statistics and Probability*, 281–297 1967.

[39] W. Meesrikamolkul, V. Niennattrakul, and C. Ratanamahatana. Shape-based clustering for time series data. *PAKDD*, 530–541, 2012.

[40] M. Mongeau and D. Sankoff. Comparison of musical sequences. *Computers and the Humanities*, 24(3):161–175, 1990.

[41] M. Munich and P. Perona. Continuous dynamic time warping for translation-invariant curve alignment with applications to signature verification. *ICCV*, 108–115, 1999.

[42] V. Niennattrakul and C. Ratanamahatana Inaccuracies of shape averaging method using dynamic time warping for time series data. *International Conference on Computational Science*, 513–520, 2007.

[43] S. Papadimitriou, J. Sun, and C. Faloutsos. Dimensionality reduction and forecasting of time-series data streams. *Data Streams: Models and Algorithms*, (Ed.) Charu Aggarwal, Springer, Chapter 12, 261–288, 2007.

[44] G. Qi, C. Aggarwal, and T. Huang. Online Community Detection in Social Sensing, *WSDM Conference*, 617–626, 2013.

[45] N. Roussopoulos, S. Kelley, and Fr. Vincent. Nearest neighbor queries. *SIGMOD Conference*, 71–79, 1995.

[46] H. Sakoe and S. Chiba. Dynamic programming algorithm optimization for spoken word recognition. *IEEE Transactions on Acoustics, Speech, and Signal Processing*, 26:143–165, 1978.

[47] Y. Sakurai, C. Faloutsos, and M. Yamamuro. Stream monitoring under the time warping distance. *ICDE 2007*, 1046–1055, 2007.

[48] Y. Sakurai, S. Papadimitriou, and C. Faloutsos. BRAID: Stream mining through group lag correlations, *ACM SIGMOD Conference*, 599–610, 2005.

[49] S. Salvador and P. Chan. FastDTW: Toward accurate dynamic time warping in linear time and space. In *KDD Workshop on Mining Temporal and Sequential Data*, 70–80, 2004.

[50] D. Tran and M. Wagner. Fuzzy c-means clustering-based speaker verification. *Proceedings of the 2002 AFSS International Conference on Fuzzy Systems*, 318–324, 2002.

[51] M. Vlachos, D. Gunopulos, and G. Das. Rotation invariant distance measures for trajectories. In *Proceedings of the Tenth ACM SIGKDD International Conference on Knowledge Discovery and Data Mining*, 707–712, 2004.

[52] M. Vlachos, D. Gunopulos, and G. Kollios. Robust similarity measures for mobile object trajectories. *13th International Workshop on Database and Expert Systems Applications*, 721–726, 2002.

[53] M. Vlachos, M. Hadjieleftheriou, D. Gunopulos, and E. Keogh. Indexing multi-dimensional time-series with support for multiple distance measures. In *Proceedings of the 9th International Conference on Knowledge Discovery & Data Mining (SIGKDD)*,Washington, DC, 216–225, 2003.

[54] M. Vlachos, J. Lin, E. Keogh, and D. Gunopulos. A wavelet-based anytime algorithm for k-means clustering of time-series. *Workshop on Clustering High-Dimensionality Data and Its Applications*, SIAM Datamining, San Francisco, CA, USA, 2003.

[55] J. Wilpon and L. Rabiner. A modified k-means clustering algorithm for use in isolated work recognition. *IEEE Transactions on Acoustics, Speech and Signal Processing*, ASSP, 33:587–594, 1985.

[56] Y. Xiong and D.-Y. Yeung. Mixtures of ARMA models for model-based time series clustering. *ICDM 2002*, 717–720, 2002.

[57] L. Yann-Ael, S. Santini, and G. Bontempi. Adaptive model selection for time series prediction in wireless sensor networks. *Signal Processing*, 87:3010–3020, 2007.

[58] B.-K. Yi, N. D. Sidiropoulos, T. Johnson, H. Jagadish, C. Faloutsos, and A. Biliris. Online data mining for co-evolving time sequences. *ICDE Conference*, 13–22, 2000.

[59] D. Zeinalipour-Yazti, S. Lin, and D. Gunopulos. Distributed spatio-temporal similarity search. *CIKM*, 14–23, 2006.

[60] Y. Zheng and X. Zhou. *Computing with Spatial Trajectories.* Springer, 2011.

[61] Y. Zhu and D. Shasha. StatStream: Statistical monitoring of thousands of data streams in real time. *VLDB Conference*, 358–369, 2002.

Chapter 16

Clustering Biological Data

Chandan K. Reddy
Wayne State University
Detroit, MI
reddy@cs.wayne.edu

Mohammad Al Hasan
Indiana University - Purdue University
Indianapolis, IN
alhasan@cs.iupui.edu

Mohammed J. Zaki
Rensselaer Polytechnic Institute
Troy, NY
zaki@cs.rpi.edu

16.1 Introduction

With the advancement of recent technologies, a vast amount of biological data is being generated. As data banks increase their size, one of the current challenges in biology is to be able to infer some of the critical functions from such complex data. To analyze complex biological systems, researchers usually aim to identify some patterns that co-occur in the form of groups. Clustering analysis is an exploratory technique that discovers rich patterns from vast data, and hence, it has become an indispensable tool for various knowledge discovery tasks in the field of computational biology. Clustering is a powerful and widely used technique that organizes and elucidates the structure of biological data. Clustering data from a wide variety of biological experiments has proven to be immensely useful at deriving a variety of insights, such as the shared regulation or function of genes.

In analyzing this complex biological data, one can observe that the activities of genes are not independent of each other. It has been shown that genes with the same function (or genes involved in the same biological process) are likely to be co-expressed [56]. Hence, it is important to study groups of genes rather than to perform a single gene analysis. In other words, it is crucial to identify subsets of genes that are relevant to the biological problem under study. Analyzing such subsets of data yields crucial information about the biological processes and the cellular functions. Thus, *clustering the gene expression profiles* can provide insights into gene function, gene regulation, and cellular processes.

It has also been shown that proteins of known functions tend to cluster together [92]. The network distance is correlated with functional distance, and the proteins that are closer to one another tend to have similar biological functions [96]. Hence, *clustering the protein–protein interaction networks* is crucial in discovering the functions of proteins and thus understanding the inner workings of cells [84]. The most important building blocks of living organisms, such as DNA, RNA, mRNA, polypeptides, and proteins have linear structure and can be represented as sequences. *Clustering biological sequence data* aims to group together the biological sequences that are related. The identified clusters can help in providing a better understanding of the genome.

This chapter comprehensively reviews different kinds of biological data where clustering has provided promising and biologically meaningful results. More specifically, we will discuss the role of clustering for gene expression data, biological networks, and sequence data. For each type of data, the challenges and the most prominent clustering algorithms that have been successfully studied will be described. The rest of this chapter is organized as follows. Section 16.2 describes various types of clustering and the corresponding state-of-the-art clustering techniques for each category in the context of microarray data analysis. Section 16.3 provides details about several protein interaction network clustering algorithms. Section 16.4 describes the state-of-the-art biological sequence clustering algorithms. Software packages that implement most of the popular biological clustering algorithms are discussed in Section 16.5. Finally, Section 16.6 concludes our discussion.

16.2 Clustering Microarray Data

The recent advances in DNA microarray technology allow genome-wide expression profiling. It has revolutionized the analysis of genes and proteins and has made it possible to simultaneously measure the expression levels of tens of thousands of genes. The expression level of a gene is a measurement for the frequency with which the gene is expressed, and it can be used to estimate the current amount of the protein in a cell for which the gene codes [58]. The availability of such massive data has transformed the field of gene expression analysis [18].

Gene expression data clustering provides a powerful tool for studying functional relationships of genes in a biological process. Identifying correlated expression patterns of genes represents the basic challenge in this clustering problem. The underlying hypothesis here is based on a popular phenomenon known as *guilt-by-association* principle which states that genes with similar functions exhibit similar expression patterns (they are co-expressed together) [109, 25]. Hence, it becomes critical to study the relationships between the genes among various biological conditions. Clustering methods allow the biologists to capture the relationships between genes and identify the co-expressed genes in a given microarray study. Clustering also plays a critical role in other related biological applications. In addition to clustering the genes, there is also some research work on clustering the conditions to identify phenotype subtypes [44]. Clustering can also be used to extract regulatory motifs from the gene expression data [28].

More formally, the gene expression data is typically organized in a two-dimensional matrix format where *the rows correspond to genes and the columns correspond to some biological conditions (or samples)*. The columns usually represent various possible phenotypes such as normal cells, cancerous cells, drug treated cells, or time-series points. Also, the number of genes is significantly larger than the number of conditions. Clustering has been successfully employed as one of the key steps in high-throughput expression data analysis [24]. Several clustering techniques have been successfully applied to cluster the genes, conditions, and/or samples [56].

In this section, we will first describe some of the popular proximity measures used and then categorize the clustering methods proposed in the literature in the context of gene expression data analysis. We will then briefly describe the most representative methods that are widely used for analyzing gene expression datasets. Finally, we will provide a discussion about biologically validating the results of the clustering methods.

16.2.1 Proximity Measures

Before explaining more details on the clustering methods, we will define the proximity measures that are used to quantify the similarity (or distance) between two genes across all the conditions. The most popular measures used in the context of gene expression clustering are the following.

- *Euclidean distance*: Given two genes g_i and g_j, the distance between the two genes can be measured as

$$Euclidean(g_i, g_j) = \sqrt{\sum_{k=1}^{N} (g_{ik} - g_{jk})^2}$$

where N is the total number of samples (columns or features). g_{ik} represents the kth column of vector g_i. One of the problems with Euclidean distance measure is its inability to capture shifting and scaling patterns that commonly occur in gene expression data [2]. To avoid this problem, typically, these gene vector representations are Z-score normalized by subtracting the mean of the gene vector from individual column values and then dividing them by the variance of the original gene vector [20]. This will make the mean value of the resultant vector zero and the variance value one for each gene vector.

- *Pearson's correlation coefficient*: It measures the similarity between the shapes of the expression profiles of two genes as follows:

$$Pearson(g_i, g_j) = \frac{\sum_{k=1}^{N}(g_{ik} - \mu_{g_i})(g_{jk} - \mu_{g_j})}{\sqrt{\sum_{k=1}^{N}(g_{ik} - \mu_{g_i})^2}\sqrt{\sum_{k=1}^{N}(g_{jk} - \mu_{g_j})^2}}$$

  where μ_{g_i} and μ_{g_j} represent the mean of the expression values for the genes g_i and g_j, respectively. This measure has been widely used in the analysis of gene expression data but it is not robust to outliers in the data.

- *Spearman correlation coefficient*: To make the similarity measure robust to the underlying distributions and outliers, Spearman correlation coefficient considers the rank ordering of the expression values. Rather than using the original expression values for each gene, the Spearman correlation coefficient uses the rank of each sample value for that particular gene [56]. It is defined to be the Pearson correlation coefficient between the ranked expression values. Since the original expression values are completely discarded, the results from this measure are almost always inferior to those obtained using Pearson's correlation coefficient in the context of standard gene expression clustering.

- *Mutual Information*: Mutual Information (MI) is an information-theoretic approach which uses a generalization of pairwise correlation coefficient to compare two gene expression profiles. *MI* can be used to measure the degree of independence between two genes [19]. The *MI* between a pair of genes g_i and g_j is computed as follows:

$$MI_{ij} = H_i + H_j - H_{ij}$$

  where H denotes the entropy which is given as follows:

$$H_i = -\sum_{k=1}^{N} p(g_{ik}) log(p(g_{ik}))$$

  It can be seen that the higher the entropy, the more randomly distributed are gene expression levels across the conditions. Also, *MI* becomes zero if the expression levels of genes i and j are statistically independent since their joint entropy $H_{ij} = H_i + H_j$. A higher value of *MI* indicates that the two genes are nonrandomly associated to each other. Even though *MI* has shown some promising results in the context of clustering [83], it is more widely used in the context of constructing gene co-expression networks.

16.2.2 Categorization of Algorithms

The existing clustering methods can be categorized into the following groups as shown below. Each of these categories will be explained more elaborately in this section.

1. **Standard (single-dimensional) clustering**: In this category, a standard clustering technique can be applied on the gene expression data to cluster the genes or to cluster the samples (or conditions). Such clustering methods can be used to capture the relationships between genes and identify the co-expressed genes based on the microarray data collected. It should be noted that these standard clustering techniques can be applied not only on the genes but also on biological conditions.

2. **Biclustering**: This is also referred to as co-clustering [71]. Methods in this category aim to discover local patterns in complex noisy gene expression data by simultaneously clustering both genes (rows) and conditions (columns). These methods are effective in identifying clusters of genes that are correlated only under a subset of conditions. [71].

3. **Triclustering:** The goal of triclustering is to find coherent clusters that are similar across three dimensions ($genes \times conditions \times time$) [123]. In the triclustering approach, the genes are clustered across a subset of conditions under a subset of time points.

4. **Time-Series clustering**: In some microarray studies, the gene expression values are collected over different time points that correspond to various experimental conditions [11]. One of the key characteristics of such time-series data is that it typically exhibits a strong autocorrelation between successive time points. In such scenarios, it is critical to capture the inherent relationships between genes over time in order to accurately perform clustering of the genes [6].

16.2.3 Standard Clustering Algorithms

In this section, we will briefly explain the most widely studied clustering methods for analyzing gene expression data.

16.2.3.1 Hierarchical Clustering

Hierarchical clustering algorithms (discussed in detail in Chapter 4) first create a dendrogram for the genes, where each node represents a gene cluster and is merged/split using a similarity measure. There are two categories of hierarchical clustering that are studied in the context of gene expression analysis.

- **Agglomerative clustering** *(bottom-up approach)*: This approach starts with each gene as an individual cluster, and at each step of the algorithm, the closest pair of clusters are merged until all the of the genes are grouped into one cluster. Eisen et al. [33] applied an agglomerative clustering algorithm called UPGMA (Unweighed Pair Group Method with Arithmetic Mean). Using this approach, each cell of the gene expression matrix is colored and the rows of the matrix are reordered based on the hierarchical dendrogram structure and a consistent node-ordering rule. An illustration of a simple dendrogram for gene expression data is shown in Figure 16.1.

- **Divisive clustering** *(top-down approach)*: This approach starts with a single cluster that contains all the genes; then repeatedly the clusters are split until each cluster contains one gene. Based on a popular deterministic annealing algorithm, authors in [3] proposed a divisive approach to obtain gene clusters. The algorithm first chooses two random initial centroids. An iterative Expectation-Maximization algorithm is then applied to probabilistically assign each gene to one of the clusters. The entire dataset is recursively split until each cluster contains only one gene.

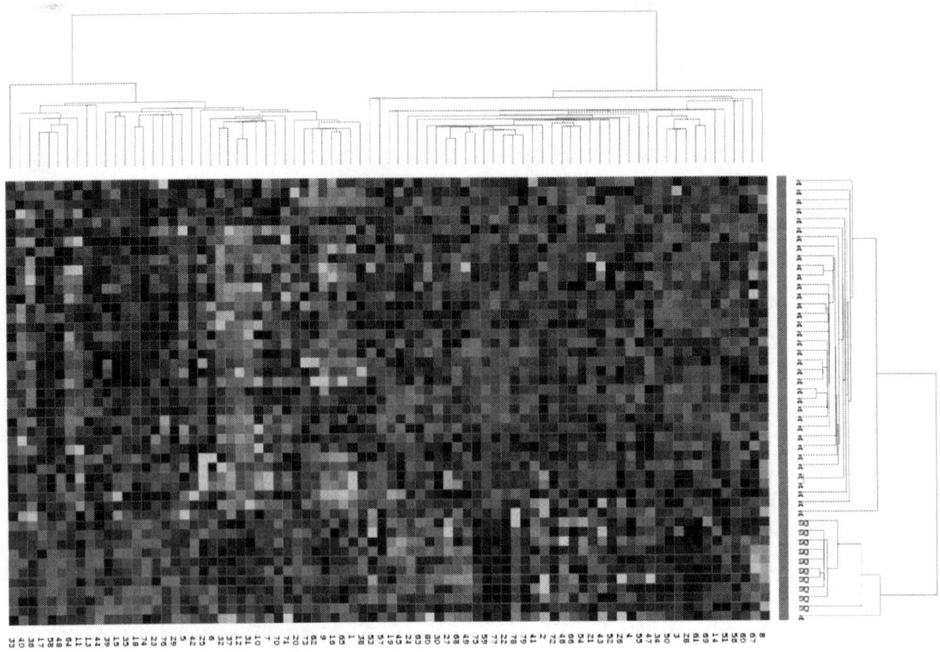

FIGURE 16.1 (See color insert): A simple dendrogram based on hierarchical clustering of rows and columns for gene expression data. The figure has been adapted from [39]. Here, rows correspond to the genes and columns correspond to the conditions. The color scale ranges from saturated green for log ratios -3.0 and below to saturated red for log ratios 3.0 and above.

16.2.3.2 Probabilistic Clustering

Since some of the genes are regulated by several biological pathways, it is important to obtain overlapping clusters; i.e., some genes might appear in multiple clusters. Probabilistic clustering provides an intuitive solution to this problem by implicitly modeling this overlapping nature through assigning probabilistic memberships. The most popular choice of probabilistic clustering of the data is by developing a model-based approach. *Model-based clustering* algorithms assume that the gene expression data is generated by a finite mixture of probability distributions [118]. The primary challenge here is to estimate the best probabilistic model that represents the patterns in the complex data. A popular Model-based CLUSTering algorithm, MCLUST, uses multivariate Gaussian distributions for clustering microarray data [118]. The basic idea here is that each cluster of genes is generated by an underlying Gaussian distribution. More details about clustering using Gaussian mixture models are given in Chapter 3. It should be noted that before applying any model-based approach, the raw gene expression data is first transformed or normalized. Several feature transformation methods have been shown to achieve good results when applying the model-based clustering. The other choice for probabilistic clustering is to apply a Fuzzy C-means (FCM) algorithm to cluster the gene expression data [26]. More details about this algorithm are available in Chapter 4.

16.2.3.3 Graph-Theoretic Clustering

In the graph-theoretical approach, a proximity graph is constructed where the nodes are the genes and the edges are the similarities between the nodes. After constructing the graph, the problem of clustering is transformed into finding minimum cut or maximal cliques in the proximity graph. CLuster Identification via Connectivity Kernels (CLICK) is a graph-theoretical algorithm that defines clusters as highly connected components in the proximity graph [95]. This algorithm does not make any prior assumptions on the number or the structure of the clusters.

Cluster Affinity Search Technique (CAST) is another graph-based clustering algorithm [14]. This algorithm alternates between adding high affinity elements to the current cluster and removing low affinity elements from this cluster. The affinity between gene i and cluster C is defined as the sum of the similarities between gene i and all genes in cluster C. The CAST algorithm then aims to find cluster assignments so that the affinity between genes and their clusters is maximized. It adds a gene to a cluster if the affinity is higher than some prespecified threshold. CAST repeats this operation over all genes and clusters, until all genes are assigned to at least one of the clusters. One of the advantages of this algorithm is that it does not require a predefined number of clusters and can efficiently handle outliers.

Another graph-theoretic algorithm that uses Minimum Spanning Tree (MST) for clustering gene expression data has been proposed in [114]. One of the key properties of using the MST representation is that each cluster of the expression data corresponds to one subtree of the MST, which will then transform a multidimensional clustering problem to a tree partitioning problem. The simple structure of an MST facilitates efficient implementations of rigorous clustering algorithms, and it does not depend on detailed geometric shape of a cluster. The implementation of this algorithm is available in a software package called EXpression data Clustering Analysis and VisualizATiOn Resource (EXCAVATOR).

16.2.3.4 Self-Organizing Maps

Self-Organizing Maps (SOMs) [64] is a clustering algorithm that is based on neural networks with a single layer. The clusters are identified by mapping all data points to the output neurons [103]. SOMs require the number of clusters and the grid layout of the neuron map as the user input. An unsupervised neural network algorithm called the Self-Organizing Tree Algorithm, (SOTA), was studied in [49]. SOTA is a top-to-bottom divisive hierarchical clustering method that is built using SOMs.

16.2.3.5 Other Clustering Methods

k-means clustering has been applied to the problem of clustering gene expression data [50]. In spite of its simplicity and efficiency, it is not well suited to the problem of gene expression clustering due to the following reasons:

- Gene expression data typically contains a lot of noise. The standard k-means algorithm is known to be sensitive to noise due to the objective function (root mean square) it optimizes.

- Since the number of gene clusters is unknown beforehand for gene expression data, there is a need to run this algorithm several times with different inputs. For large datasets, such an approach becomes impractical.

However, other variations of the k-means algorithm have been proposed to overcome some its drawbacks [104].

To improve the accuracy of clustering the tumor samples, resampling methods such as bagging, were proposed in [32]. In these ensemble methods, clustering is applied to bootstrap learning sets, and the resulting multiple partitions are combined. The main intuition of using bagging is to reduce variability in the partitioning results through averaging. Clustering gene expression data using Principal Components Analysis (PCA) was studied in [120]. However, it was shown that clustering with the principle components instead of the original variables often degrades the cluster quality. Therefore, PCA was not recommend before clustering except in specific cases [120].

16.2.4 Biclustering

Standard clustering techniques discussed in the previous section typically assume that closely related genes must have similar expression profiles across all the conditions [69]. However, this assumption does not hold in all of the biological experiments. From a practical point of view, not all the genes are involved in each biological pathway, and some of these pathways may be active under only a subset of the samples [75]. Hence, biclustering was proposed to overcome the limitations of the traditional clustering algorithms [71].

The objective of biclustering is to simultaneously cluster both rows and columns in a given matrix. Biclustering algorithms aim to discover local patterns that cannot be identified by the traditional one-way clustering algorithms. A bicluster can be defined as a subset of genes that are correlated under a subset of biological conditions (or samples). Biclustering has been used in several applications such as clustering microarray data [71], identifying protein interactions [68], and other data mining applications such as collaborative filtering [40] and text mining [18].

The concept of biclustering is illustrated using a simple example in Figure 16.2. In this example, the expression levels of three genes over 10 conditions are shown. Considering all of the ten samples, it is evident that there is no strong correlation between the three genes (Figure 16.2(a)). However, it can seen that there is a strong correlation between the three genes in a subset of the conditions, namely $\{2,3,5,8\}$ (Figure 16.2(b)). Hence, we will consider these three genes to be highly correlated though traditional proximity measures that consider all of the conditions will determine that these three genes are not correlated. Biclustering has emerged as a powerful tool to simultaneously cluster both dimensions of a data matrix by utilizing the relationship between the genes and the samples. It has been proven that the task of finding all the significant biclusters is an NP-hard problem [21].

There are *several challenges* that arise while searching for biclusters in gene expression data. A subset of genes can be correlated only in a small subset of conditions due to the *heterogeneity of the samples*. Such heterogeneity arises due to the complexities involved with different diseases, different patients, different timepoints, or different stages within a disease. In addition, since genes can be *positively or negatively correlated* [54], it is important to allow both types of correlations in the same bicluster. Moreover, there are *several types* of biclusters that can be biologically relevant [71]. A gene can be involved in more than one biological pathway; therefore, there is a need for a biclustering algorithm that *allows overlapping between the biclusters* [27, 75], i.e., the same gene

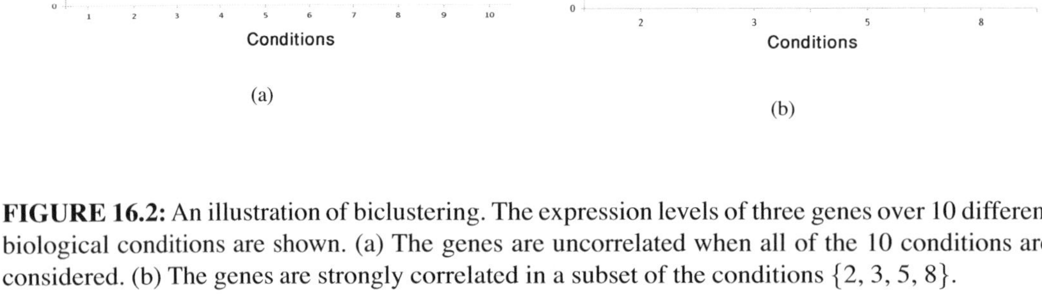

FIGURE 16.2: An illustration of biclustering. The expression levels of three genes over 10 different biological conditions are shown. (a) The genes are uncorrelated when all of the 10 conditions are considered. (b) The genes are strongly correlated in a subset of the conditions $\{2, 3, 5, 8\}$.

can be a member of more than one bicluster. Finally, a bicluster will have to capture the *positively and/or negatively co-expressed set of genes* since the genes in the same biological pathway can be positively and/or negatively correlated [54, 117, 76].

16.2.4.1 Types and Structures of Biclusters

We will now discuss the different types of biclusters that might appear in gene expression data [71]. Let μ be a typical value in the bicluster. α_i is the adjustment for row i and β_j is the adjustment for column j.

- Biclusters with *constant values.* All the elements in this type have the same value. $a_{ij} = \mu$ (Figure 16.3a).
- Biclusters with *constant values on rows.* $a_{ij} = \mu + \alpha_i$ (Figure 16.3b).
- Biclusters with *constant values on columns.* $a_{ij} = \mu + \beta_j$ (Figure 16.3c).
- Biclusters with *(additive) coherent values.* Each row and column is obtained by addition of the previous row and column by a constant value. $a_{ij} = \mu + \alpha_i + \beta_j$ (Figure 16.3d).
- Biclusters with *(multiplicative) coherent values.* Each row and column is obtained by multiplication of the previous row and column by a constant value. $a_{ij} = \mu \times \alpha_i \times \beta_j$ (Figure 16.3e).
- Biclusters with *coherent evolutions.* In this type, the coherence of the values is not considered. Only the direction of change of values is important (Figure 16.3(f)).

2.0	2.0	2.0	2.0
2.0	2.0	2.0	2.0
2.0	2.0	2.0	2.0
2.0	2.0	2.0	2.0

(a) Constant values.

1.0	1.0	1.0	1.0
2.0	2.0	2.0	2.0
3.0	3.0	3.0	3.0
4.0	4.0	4.0	4.0

(b) Constant rows.

1.0	2.0	3.0	4.0
1.0	2.0	3.0	4.0
1.0	2.0	3.0	4.0
1.0	2.0	3.0	4.0

(c) Constant columns.

1.0	4.0	5.0	0.0
4.0	7.0	8.0	3.0
3.0	6.0	7.0	2.0
5.0	8.0	9.0	4.0

(d) Coherent values (additive).

1.0	2.0	3.0	4.0
1.0	2.0	3.0	4.0
1.0	2.0	3.0	4.0
1.0	2.0	3.0	4.0

(e) Coherent values (multiplicative).

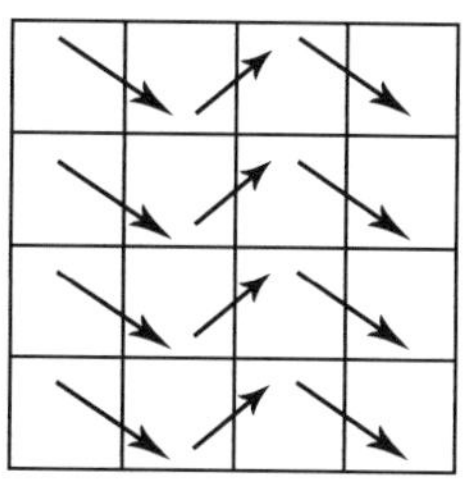

(f) Coherent evolutions.

FIGURE 16.3: Examples of different types of biclusters.

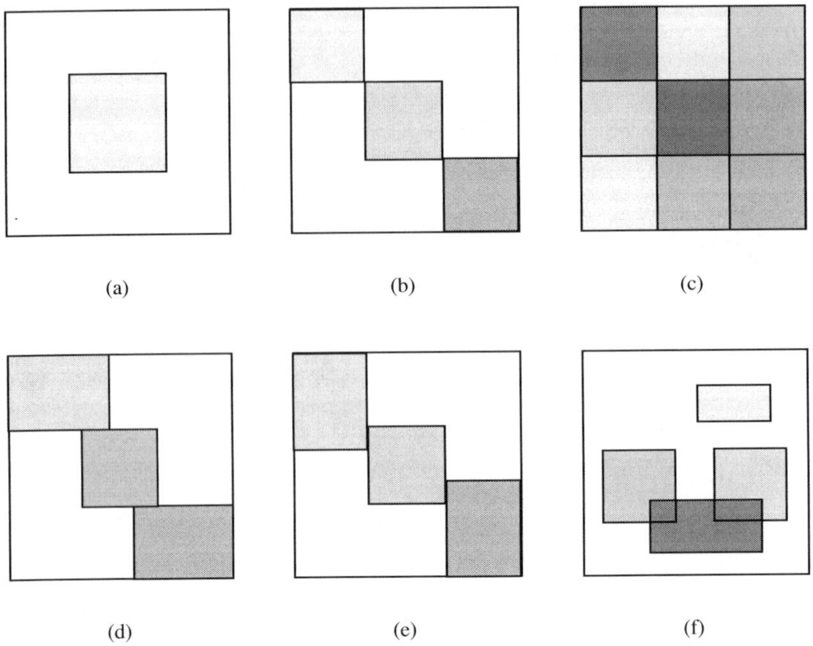

FIGURE 16.4 (See color insert): Examples of bicluster structures. (a) Single bicluster. (b) Exclusive row and column biclusters. (c) Checkerboard pattern biclusters. (d) Exclusive row biclusters. (e) Exclusive column biclusters. (f) Arbitrarily positioned overlapping biclusters.

In addition to the variations in the types of the biclusters, there are other important sources of variations, such as the variations in the size and the position of the biclusters in the gene expression data. Though the earlier biclustering algorithms used to find only a single bicluster at a time (Figure 16.4(a)), most of the recent approaches attempt to find several biclusters simultaneously. When there are several biclusters in the data, some of the standard bicluster structures are as follows.

- Exclusive row and column biclusters which form rectangular diagonal blocks after reordering rows and columns (Figure 16.4(b)).
- Checkerboard pattern biclusters that are completely nonoverlapping (Figure 16.4(c)).
- Exclusive row biclusters that might have overlapping columns (Figure 16.4(d)).
- Exclusive column biclusters that might have overlapping rows (Figure 16.4(e)).
- Arbitrarily positioned overlapping biclusters (Figure 16.4(f)).

16.2.4.2 Biclustering Algorithms

In the first biclustering algorithm proposed by Cheng and Church [21], the mean-squared residue (MSR) score was used as a measurement of the coherence between two genes. Given a gene expression submatrix X that has I genes and J conditions, the residue is computed as follows:

$$H(I,J) = \frac{1}{|I||J|} \sum_{i \in I, j \in J} (x_{ij} - x_{Ij} - x_{iJ} + x_{IJ})^2 \tag{16.1}$$

where $x_{iJ} = \frac{\sum_{j \in J} x_{ij}}{|J|}$ is the row mean, $x_{Ij} = \frac{\sum_{i \in I} x_{ij}}{|I|}$ is the column mean and $x_{IJ} = \frac{\sum_{i \in I, j \in J} x_{ij}}{|I| * |J|}$ is the overall mean of the matrix X. x_{ij} is a particular element (ith row and jth column) of the original matrix. A perfect bicluster will have MSR$=0$. The MSR function has been used in many biclustering algorithms [21, 115, 27, 75].

This algorithm starts with the original data matrix; then, a set of row/column deletions and additions are applied to produce one bicluster, which will be replaced with random numbers. This procedure is repeated until a certain number of biclusters is obtained. The algorithm has two main limitations: (i) It finds only one bicluster at a time, and (ii) random interference (masking the discovered biclusters with random numbers) reduces the quality of the biclusters and obstructs the discovery of other biclusters. After this algorithm was proposed, a plethora of new heuristic algorithms that aim to extract biclusters from noisy gene expression data have been developed. We will mention only a few here; for a detailed discussion on several existing biclustering algorithms, we refer the readers to an excellent survey on this topic [71].

Coupled two-way clustering (CTWC) technique was proposed in [41]. In this technique, a subset of genes (conditions) are used to cluster the conditions (genes), while the Order-Preserving Submatrices (OPSMs) [13] algorithm finds local patterns in which the expression levels of all genes induce the same linear ordering of the experiments. However, the OPSM algorithm finds only one bicluster at a time and captures only positively correlated genes. Iterative Signature Algorithm (ISA) [51] is a statistical biclustering algorithm which defines a transcription module (bicluster) as a coregulated set of genes under a set of experimental conditions. ISA starts from a set of randomly selected genes (or conditions) that are iteratively refined until they are mutually consistent. At each iteration, a threshold is used to remove noise and to maintain coregulated genes and the associated coregulating conditions.

16.2.4.3 Recent Developments

Recently, there have been many emerging trends in the field of biclustering: (i) Identifying overlapping biclusters and handling both positive and negative correlations within a bicluster has gained some attention due to their biological importance [117]. Some of the recent algorithms [27, 75] allow for overlapping biclusters and find k row clusters and l column clusters simultaneously. (ii) *Differential biclustering* [76, 77, 37] aims to find gene sets that are correlated under a subset of conditions in one class of conditions but not in the other class. Identifying such class-specific biclusters can provide valuable knowledge for understanding the roles of genes in several diseases [76]. The classes could represent different tissue types (normal vs cancerous), different subject types (e.g., male vs female), different group types (African-American vs Caucasian American) [61], different stages of cancer (early stage vs developed stage) [76], or different time points [42]. (iii) *Query-based biclustering* algorithms [30, 122, 4] allow for identifying biclusters that are centered around a set of seed genes of interest. In these algorithms, new search strategies are developed to exploit the expression profiles of certain genes of interest that are used to guide the bicluster searching mechanism. These approaches are extremely handy when one wants to compare the expression profiles of a certain set of genes with that of the existing knowledge which can be obtained by querying for similar profile genes from a large-scale expression compendia.

16.2.5 Triclustering

The goal of triclustering is to find coherent subspace clusters that are similar across three dimensions. The motivation of this task comes from the biological domain where finding coherent clusters along the gene-sample-time (temporal) or gene-sample-region (spatial) dimensions is of great value. Although the problems that are suitable for triclustering had been addressed earlier by ad-hoc methods [55], Zhao and Zaki [123] proposed the first formal algorithm for triclustering. After that, a few more algorithms have also been proposed in recent years [98, 57].

Given G, the set of genes; S, the set of samples; and T, the set of time points; a tricluster C is a submatrix of the dataset $D = G \times S \times T$, where $C = X \times Y \times Z = \{c_{ijk}\}$, with $X \subseteq G, Y \subseteq S$, and $Z \subseteq T$ provided that certain conditions of homogeneity are satisfied. For example, a simple condition might be that all values $\{c_{ijk}\}$ are identical or approximately equal. If we are interested in finding common gene co-expression patterns across different samples and times, we can find clusters that have similar values in the G dimension, but possibly different values in the S and T dimensions. Other homogeneity conditions can also be defined, such as similar values in S dimension and order preserving submatrix. [71]. Let $\mathcal{B}$ be the set of all triclusters that satisfy the given homogeneity conditions, then $C \in \mathcal{B}$ is called a maximal tricluster iff there does not exist another cluster $C' \in \mathcal{B}$ such that $C \subset C'$. In most of the cases, we are interested in only the maximal triclusters.

Zhao and Zaki's method for triclustering is known as TRICLUSTER. It accepts the 3-dimensional dataset D; the minimum size thresholds, mx, my, and mz, that define the size of the clusters in three dimensions; and a maximum ratio threshold, ε, which represents the maximum allowed deviation among values in different cells of a cluster. It then constructs a range multigraph data structure for the data matrix at each of the timestamps. It uses the range multigraphs of a timestamp to obtain a set of robust biclusters (involving dimensions G and S) that are observed for that time value. As a final step, it merges similar biclusters across different timestamps to obtain maximal triclusters. For this step, it represents each of the biclusters that it has found in the earlier step as a node in a graph, and defines the node–node (bicluster–bicluster) relationships based of the similarity on the time dimension; then the maximal triclusters are simply the maximal cliques in this graph. As an optional step, it also merges and deletes clusters based on the degree of overlap among various dimensions. Interested readers can read more details on the TRICLUSTER algorithm from the original paper by the authors [123]. Key features of TRICLUSTER are that it is flexible and can accept various homogeneity criteria. Also, it is robust and generates only maximal triclusters. The downside of this method is that it requires a large number of parameters and domain knowledge is necessary to set the parameter values optimally.

16.2.6 Time-Series Gene Expression Data Clustering

In order to determine the complete set of genes that are expressed under a set of new conditions and to determine the interaction between these genes in these new conditions, it is important to measure a time course of expression experiments [11]. In such microarray studies, the gene expression values are collected over different time points which correspond to the experimental conditions. One of the key characteristics of such time-series data is that while static data from a sample population are assumed to be i.i.d. (independent and identically distributed), time series gene expression data typically exhibit a strong autocorrelation between successive timepoint values. In such scenarios, it is critical to capture the inherent relationships between genes over time in order to perform clustering [6]. One of the first models to cluster time series gene expression data was developed in [70]. This clustering algorithm was based on the mixed effects model using B-splines and was applied on the yeast cell cycle gene expression data. The estimated gene expression trajectory was also used to fill in the missing gene expression levels for any time point using the available data in the same cluster.

Hidden Markov Models (HMMs) have also been used to cluster time-series gene expression data [91]. The primary advantage of using HMMs is that they explicitly take into account the temporal nature of the expression patterns which will produce high quality clusters. Given gene expression data, the goal is to find a partition of the data into K HMMs which will maximize the likelihood of the data given the learned HMM model. For gene expression data, the emission probabilities are assumed to be Gaussians with fixed variance. The authors of [91] developed a new algorithm for clustering genes based on a mixture of HMMs. The parameters of this model are learned using an iterative Expectation-Maximization style algorithm. The two iterative steps in this algorithm are (i) genes are associated with the HMM that would have most likely generated their time courses and (ii) the parameters of each HMM are estimated using the genes assigned to it. This algorithm requires the number of time points to be much larger than the number of states. Though this algorithm is suitable for clustering long time-series data, it does not work well on short time-series data.

To tackle the challenges with short time-series data, the authors of [36] proposed an algorithm specifically designed for clustering short time-series expression data. The algorithm works by assigning genes to a predefined set of model profiles that capture the potential distinct patterns. After determining the significance of each of these profiles, the most significant ones are retained for further analysis and can be combined to form clusters. Using immune response data, the authors have shown that their algorithm can correctly detect the temporal profile of relevant functional categories. Using Gene Ontology-based evaluation, the algorithm outperformed both general clustering algorithms and algorithms designed specifically for clustering time-series gene expression data. STEM is a toolkit that is available based on this work for the analysis and clustering of short time series gene expression data [35]

In certain scenarios, a gene might not instantaneously be correlated with other genes at that time but a gene might regulate another gene after a certain time. In order to identify time-lagged coregulated gene clusters, [53] proposes a novel clustering algorithm for effectively extracting the time-lagged clusters. This algorithm first generates complete time-lagged information for gene clusters by processing several genes simultaneously. Instead of considering the lags for the entire sequence, it considers only small interesting parts (subsequences) of the genes that are coregulated while there is no distinct relationship between the remaining part. It identifies localized time-lagged co-regulations between genes and/or gene clusters. It builds a novel mechanism that aims to extract clusters (which are referred to as q-clusters) of (time-lagged) coregulated genes over a subset of consecutive conditions. Each such cluster essentially contains information of genes that have similar expression patterns over a set of consecutive conditions. More recently, authors in [113] have extended the concept of time-lagged clustering to three-dimensional clustering.

16.2.7 Cluster Validation

All the clustering algorithms that are described in the previous section will yield either groups of co-expressed genes or groups of samples with a common phenotype [22]. Reliability of the clusters, which measures the probability that the clusters are not formed by chance, is a commonly used metric for validating the results of these clustering algorithms. To compute the p-values of a cluster, typically the genes from a given cluster are mapped to the functional categories defined in annotated databases such as Martinsried Institute of Protein Sciences (MIPS) or Gene Ontology (GO) [87]. Typically, a hypergeometric distribution is used to calculate the probability of having at least k genes from a cluster of size n genes by chance in a biological process containing f genes from a total size of N genes as follows:

$$P = 1 - \sum_{i=0}^{k} \frac{\binom{f}{i}\binom{N-f}{n-i}}{\binom{N}{n}}$$

This test measures if a gene cluster is enriched with genes from a particular functional category to a greater extent than what would be expected by chance. The range of the p-values is from 0 to 1. Lower p-values indicate biological significance of the clusters.

Another popular metric for evaluating the goodness of the clusters is the Figure of Merit (FOM) which was originally proposed in [119] to estimate the predictive power of clustering algorithms. The FOM measure computes the mean deviation of the expression levels of genes in a particular condition relative to their corresponding cluster means. Thus, a small value of FOM indicates high predictive ability of the resulting clusters.

16.3 Clustering Biological Networks

Proteins control the functions of the cell [52]. Understanding these functions requires not only studying the proteins and but also studying their interactions [8]. Protein interactions are essential elements of all the biological processes [16]. Pairwise protein interactions have been identified and validated using recent technologies. These interactions have been obtained by different methods such as mass spectrometry, two-hybrid methods, and genetic studies. The whole network of protein–protein interactions for a given organism describes the interactome of that organism. The protein–protein interaction (PPI) network is represented as a graph in which the nodes represent the protein and the edges represent the interactions between the corresponding proteins [84].

It has been shown that proteins of known functions tend to cluster together [92]. The network distance is correlated with functional distance, and the proteins that are closer to one another tend to have similar biological function [96]. Therefore, studying the PPI networks is crucial in discovering the functions of proteins and thus understanding inner workings of cells [84]. Clustering the PPI network can be used to predict the unknown functional categories of proteins.

16.3.1 Characteristics of PPI Network Data

We will first describe some of the key characteristics of PPI networks [8]:

1. *Scale-free structures* [52]: PPI networks contain hub proteins which are typically involved with many interactions. In other words, most of the proteins in the network have few interactions, and only a few proteins will have a lot of interactions with other proteins. Applying existing clustering techniques on these networks would produce a few giant clusters (containing the hub nodes) and the remaining clusters would be very small. Hence, the clustering process should be adapted to produce better results in terms of the size of the clusters.

2. *Disassortativity:* In many forms of scale-free networks (such as social networks), highly connected nodes usually are strongly connected with each other. This property is known as *assortativity*. However, in protein interaction networks this is not the case. Hubs are not directly linked to each other thus causing the disassortativity in such networks though the average path lengths are relatively shorter compared to other networks.

3. *Multifunctionality:* The same protein can be involved in several biological processes [105]. Due to these overlapping structures, it becomes difficult to extract groups of proteins in such a way that a single protein is present in multiple groups.

16.3.2 Network Clustering Algorithms

Several network clustering algorithms have been proposed in the PPI literature. Some of the most popular ones will be discussed in this section.

16.3.2.1 Molecular Complex Detection

Molecular Complex Detection (MCODE) algorithm was one of the first computational methods that was proposed to detect protein complexes based on the protein connectivity values in the PPI networks. The MCODE algorithm aims to detect the densely connected nodes in complex networks [9]. The algorithm first assigns a weight to each node based on each protein's local network density using the highest k-core of the node neighborhood. A k-core is a graph of minimal degree k. As an alternative to the clustering coefficient, the MCODE algorithm defines the core-clustering coefficient of a node u as the density of the highest k-core of the immediate neighborhood of u. The final weight for any node is the product of the node core-clustering coefficient and the highest k-core level.

In the next step, the MCODE algorithm selects the highest weighted node and recursively moves outward from this seed node including the nodes whose weights are above a certain threshold. If a node is included, its neighbors are recursively checked in the same manner to see if they are part of the complex until no more nodes can be added to the complex, and this process is repeated for the next highest unseen weighted node. The experimental results from this algorithm indicate that the protein complexes obtained are generally small in number and each of the results is a much larger complex.

16.3.2.2 Markov Clustering

The Markov clustering (MCL) algorithm is one of the widely studied graph clustering algorithms [106]. Markov clustering has been applied to PPI networks such as yeast and human PPI networks, and the generated clusters accurately map to known protein complexes [65, 85]. MCL is an iterative algorithm that has two main alternating steps: expansion and inflation. Initially, the graph is translated into a stochastic Markov matrix so that the sum of the values in each of the columns is 1. This matrix represents the transition probabilities between all pairs of nodes. The resulting stochastic matrix is then clustered using the following steps [38, 106]. *(i) Expansion:* The expansion step spreads the flow out of a node to potentially new nodes and enhances flow in dense regions. This step leads to strengthening the strong edges and further weakening the weaker edges. *(ii) Inflation:* The inflation step aims to strengthen the intracluster flow and weaken the intercluster flow to obtain the natural clusters. The above two steps are repeated until convergence. The algorithm is said to be converged when we obtain a doubly idempotent matrix (a nonnegative column-homogenous matrix idempotent under matrix multiplication). Also, it has been shown that MCL is more robust to noise and identifies meaningful clusters [107].

16.3.2.3 Neighborhood Search Methods

The Restricted Neighborhood Search Clustering (RNSC) algorithm was proposed in [63] to cluster PPI data of Saccharomyces cerevisiae, Drosophila melanogaster, and Caenorhabditis elegans and predict protein complexes. This algorithm optimizes a cost-based local search algorithm based loosely on the tabu search metaheuristic [43]. Clustering of a graph begins with an initial random clustering and a cost function. Nodes are then randomly added or removed from clusters to find a partition that minimizes the cost function value. This cost function is based on the number of invalid connections (absence of intracluster connections or presence of intercluster connections) incident from a particular node. To achieve high accuracy in predicting the protein complexes, some post-processing is performed based on the functional homogeneity of the complex, density thresholding, and minimum size thresholding.

16.3.2.4 Clique Percolation Method

In order to extract the overlapping structures from complex networks, Palla et al. [79] proposed the Clique Percolation Method (CPM). CPM has the ability to generate the overlapping clustering by identifying the k-clique percolation communities. A k-clique corresponds to a complete subgraph of size k. A cluster is defined to be the maximal union of k-cliques that can be reached from others through a series of adjacent k-cliques. A software package named CFinder [1] implements this method and is currently being used even in other application domains such as social networks [78].

16.3.2.5 Ensemble Clustering

An ensemble clustering approach applies different clustering methods on the PPI network data and then combines the clustering results of all these methods (the base clustering algorithms) into a single comprehensive clustering result of the PPI networks [8]. The following three base clustering algorithms were used in this work.

(i) Repeated bisections [100]: This method starts with having the complete dataset as one cluster. Then the following steps is repeated until the desired number of clusters is obtained: (1) select a cluster to bisect into two clusters and (2) compute the similarity score for each cluster. This algorithm optimizes the $I2$ criterion defined as follows:

$$I2 = max \sum_{i=1}^{k} \sqrt{\sum_{v,u \in C_i} S(u,v)}$$

where k is the number of clusters, C_i is the set of objects in cluster i and $S(u,v)$ is the similarity between the two objects u and v.

(ii) Direct k-way partitioning [59]: This method works as follows: (1) select a set of k objects (seeds), and (2) compute the similarity between each object and the seed, and (3) assign each object to the most similar cluster. This procedure is repeated to optimize the $I2$ criterion.

(iii) Multilevel k-way partitioning [59]: This algorithm has three main steps: coarsening, initial partitioning, and refinement. In the coarsening step, k-way partitioning is used to cluster the graph into a set of smaller graphs. In the second step, the partitions are projected back onto the original graph by iterating over intermediate partitions. Finally, the refinement step reduces the edge-cut while conserving the balance constraints.

The clustering results of the above mentioned base clustering methods are combined by applying the following three phases:

1. *Cluster purification:* The similarity between the objects within each cluster is computed, and the weak clusters are removed such that each protein is a member of at least one third of the remaining clusters. The result of this step is represented using a binary cluster membership matrix where each column is a cluster produced by the base clustering algorithms and each row is a protein.

2. *Dimensionality reduction:* A dimensionality reduction method, such as PCA, can be used to reduce the number of dimensions in the cluster membership matrix. This will avoid the problem of the curse of dimensionality.

3. *Consensus clustering*: Two different consensus clustering algorithms are applied: (i) *the recursive bisection* algorithm where the best of the three base clustering algorithms is chosen and (ii) *the agglomerative hierarchical clustering* where the desired k-way clustering solution is computed using the agglomerative method.

Weighted consensus: This method considers the weight of the edges between proteins within each cluster [8]. In **soft consensus** clustering, the same protein can belong to more than one cluster. The cluster membership can be computed as a factor of the distance from the nodes in the cluster. The soft clustering solves the problem of multifunctional proteins. This ensemble method produced better results on yeast PPI networks compared to several other methods.

16.3.2.6 Other Clustering Methods

In [48], a clustering algorithm was developed that combines information from expression data and biological networks and computes a joint clustering of genes and nodes of the biological network. This method was validated using expression data of the yeast, and the results were explained in terms of the biochemical network and the gene expression data. In [82], a biclustering-based approach was proposed to cluster the PPI data and generate both overlapping and nonoverlapping clusters. This algorithm was applied on human and yeast networks.

16.3.3 Cluster Validation and Challenges

Using different clustering methods, various results are obtained from a given PPI network. Hence, it is important to compare and evaluate the performance of these clustering algorithms. Being an unsupervised approach, it is often quite difficult to evaluate and compare the performance of various clustering algorithms used in the context of PPI network analysis. Validating the clusters is primarily done by calculating the p-values for the clusters [110]. Statistically significant p-values indicate that the set of proteins in the same clusters are involved in the same biological functions. The GO database provides the annotation of known molecular functions and biological processes [7]. Similar to the evaluation of the gene clusters from the expression data, a hypergeometric distribution is used to calculate the probability of having at least k genes from a cluster of size n genes by chance in a biological process containing f genes from a total size of N genes. Lower p-values indicate biological significance of the clusters.

If there are known protein complexes, they can be used as a gold-standard to evaluate the performance of the clustering algorithm by comparing the predicted cluster to the known ones. Clustering the PPI data is still a challenging problem due to various reasons [110]. The false positive and false negative interactions in the protein network makes it difficult to evaluate the quality of the resulting clusters. Also, there is a need for clustering methods that allow overlapping between the clusters. Similar to other applications, identifying the optimal number of clusters in the PPI data is a challenging task. In addition, the large amount of available data makes it computationally expensive to analyze the PPI networks.

16.4 Biological Sequence Clustering

Sequence clustering is an essential task in biological data analysis, because the most important building blocks of living organisms, such as DNA, RNA, mRNA, polypeptides, and proteins, have a linear structure and can be represented as sequences. For DNA, RNA, and mRNA, the sequences are made of nucleic acids, also called bases and for polypeptides and proteins, the sequences are made of amino acids. Earlier studies on biological sequences were mostly limited to pairwise sequence alignment and multiple sequence alignment. However, in recent years, scientists are amassing an enormous amount of sequence data as a result of improvement on the high-throughput sequencing. Similarity search in such a large sequence database using alignment algorithms is extremely costly. So, alignment-based similarity search is performed on a small cluster of highly similar sequences. Sequence clustering plays a significant role in finding those small clusters.

16.4.1 Sequence Similarity Metrics

Many of the existing clustering methods that we have discussed in the earlier sections, such as bi-clustering, graph-theoretic clustering, and Markov clustering can be used for clustering sequences if a suitable distance (or similarity) metric is available. Formally speaking, a distance metric takes a pair of sequences and returns a real number which denotes the distance between the given sequences. Once all pairwise distances among a set of sequences are found, the similarity information can be encoded in a matrix or in a graph. In the case of a matrix, the rows and the columns correspond to the sequences, and for the graph, the nodes represent the sequences and an edge represents a pair of *similar* sequences for a chosen similarity threshold. Any traditional clustering algorithm can find clusters once a similarity matrix or a similarity graph is available.

For clustering biological sequences, finding a suitable distance metric is challenging due to the complexity associated with these sequences. For instance, a protein sequence may be composed of various functional domains, and two protein sequences may share only a few of those domains; in that case the overall similarity between these two proteins will be weak. However, if the matched functional domains are highly significant, these proteins should belong to the same cluster. Also, when clustering genome sequences, a similarity metric should consider only the coding part of the DNA and discard a significant part of the genome sequences, such as junk DNA and tandem repeats. Below, we discuss a collection of distance metrics that can be used for measuring the distance between a pair of sequences. It is important to note that though we use the term *metric*, many of the similarities measurement may not be a "metric" using its mathematical definition.

16.4.1.1 Alignment-Based Similarity

The most popular distance metric for sequence data is the Levenshtein distance, or edit distance [46]. It denotes the number of edits needed to transform one string into the other with the allowable edit operations being insertion, deletion, or substitution of a single character. For a pair of sequences, this distance can be computed by performing the global alignment between the sequences; for two sequences of length l_1 and l_2, the global alignment cost is $O(l_1 l_2)$. This is costly considering that biological sequences (particularly, DNA sequences) are typically long. So, Levenshtein distance is not an ideal metric for biological sequences for the task of sequence clustering. Another limitation of this distance metric is that it captures the optimal global alignment between two sequences, whereas for clustering biological sequences, local similarities between two sequences should be considered. Finally, the dependency of the edit distance metric on the length of the sequence makes it a poor metric for the cases where the length of the sequences in the dataset varies significantly.

To capture the local similarity between two sequences, Smith-Waterman's local alignment score [46] can be used. Instead of aligning the entire sequence the local alignment algorithm aligns similar regions between two sequences. The algorithm compares segments of all possible lengths and optimizes the similarity measure. Though it overcomes some of the problems of global alignment, it is as costly as the global alignment algorithm and is simply impractical for clustering thousands of biological sequences.

During the eighties and the nineties, two popular tools, known as FASTA [80] and BLAST (Basic Local Alignment Search Tool) [5], were developed to improve the scalability of similarity search from biological sequence databases. Both tools accept a query sequence and return a set of statistically significant similar sequences from a sequence database. The main benefit of these tools over an alignment-based method is that they are highly scalable, as they adopt smart heuristics for finding similar segments after sacrificing the strict optimality. Also, specifically BLAST has various versions (such as PSI-BLAST, PHI-BLAST, BLAST-tn) that are customized based on the kind of sequences and the kind of scoring matrix used. For a given set of sequences, FASTA and BLAST can also be used to find a set of similar sequences that are pairwise similar. For a pair of similar sequences, BLAST also returns bit score (also known as BLAST-score) which represents the similarity strength that can be used for clustering sequences.

16.4.1.2 Keyword-Based Similarity

To model the effect of local alignment explicitly, some sequence similarity metrics consider the q-gram-based method, where a sequence is simply considered as a bag of short segments of fixed length (say, q); thus, a sequence can be represented as a vector, in which each component corresponds to the frequency of one of the q-length segments. Then the similarity between two sequences is measured using any metric that measures the similarity between two vectors, such as dot product, Euclidean distance, Jaccard coefficient, or even *tf-idf* [89]. This approach is also known as a keyword-based method, as one can consider each sequence as a document and each q-gram as a keyword in the document. The biggest advantage of similarity computation using a keyword-based method is that it is fast. Another advantage is that this method represents a sequence using an $\mathbb{R}^n$ vector, which can accommodate some of the clustering algorithms (such as, k-means) that work only on vector-based data.

16.4.1.3 Kernel-Based Similarity

In recent years, kernel-based sequence similarity metrics also got popular. In [67, 66], the authors present several families of k-gram-based string kernels, such as restricted gappy kernels, substitution kernels, and wildcard kernels, all of which are based on feature spaces indexed by k-length subsequences (k-mers) from the string alphabet. Typically, kernels are used for supervised classification with support vector machines (SVM), however, they can also be used as a similarity metric for unsupervised clustering.

16.4.1.4 Model-Based Similarity

Probabilistic models, such as HMM are also used for finding similarity metrics. For a given set of sequences to be clustered, such a method trains one HMM for each of the sequences. Then, the similarity between two sequences can be obtained from the similarity (or distance) between the corresponding HMMs. In the past, few authors have proposed approaches for computing the distance between two HMMs [86]; early approaches were based on the Euclidean distance of the discrete observation probability, others on entropy, or on co-emission probability of two HMM models, or on the Bayes probability of error [10].

16.4.2 Sequence Clustering Algorithms

In an earlier section, we discussed that given a similarity metric, we can use any clustering method for clustering biological sequences. Nevertheless, there have been many clustering methods that are explicitly proposed for clustering biological sequences. We discuss them under the following groups.

16.4.2.1 Subsequence-Based Clustering

A subsequence-based clustering method mines a set of frequent subsequences from each of the sequences and uses them as features for clustering the sequences. The idea of such clustering is similar to the task of document clustering using the "bag-of-words" representation of a document. A traditional sequence mining method returns a large number of subsequences that are frequent, but all such subsequences are not good features for clustering. So, a good clustering method needs to choose a subset of these subsequences as features so that when projected on the feature-set the similarity between a pair of sequences is computed correctly. Different algorithms of sequence clustering vary in the way they choose the subsequence feature set. The advantages of subsequence-based clustering methods is that they are typically fast compared to other sequence clustering methods. However, this approach ignores the relative position of various subsequences, so they cannot model some of the sequential relations, such as ordering and sequential dependency.

One of the first among subsequence-based sequence clustering methods was proposed by Guralnik and Karypis [45]. They used traditional frequent sequence mining methods [121] to mine subsequences, but to control the number of subsequences, they imposed minimum and maximum length constraint on the mined subsequences; then they used a subset of the mined subsequences as the feature set for clustering. To select the feature set, they followed two approaches: global and local. The global approach prunes the feature space by selecting a set of independent subsequences, where dependency is defined as the overlap between the symbols of the sequences or as the overlap between their support list. On the other hand, the local approach finds independent features locally from each of the sequences, where two features are independent if they are supported by a nonoverlapping segment of the corresponding sequences. Once the feature set is defined, they used a k-means algorithm to cluster the sequences.

A more sophisticated variation of the subsequence-based method was proposed in [111]. In this paper, the authors introduce the notion of frequent summarization subsequence (FSS) and represent each sequence by a collection of those FSSs. Intuitively, an FSS is a keyword that can be viewed as a discriminating feature for clustering the input data sequences. However in this work, the authors use a *tf-idf* kind of weighting on each symbol to assign weight on each of these FSSs. They also present an effective method that directly mines all the FSSs from the sequence data. The final clustering method has two stages. The first stage generates microclusters, which are obtained by simply grouping the sequences with a shared FSS in a cluster. Typically, the number of microclusters is larger than the desired number of clusters, so a second stage is used to merge the microclusters using a hierarchical agglomerative clustering method, until the desired number of clusters is obtained.

16.4.2.2 Graph-Based Clustering

A graph-based sequence clustering method represents the sequences in a similarity graph, in which a vertex represents a sequence and an edge represents the similarity relation between the corresponding pair of sequences. In such a representation, a partition of the similarity graph represents a clustering of the input sequences. The crucial requirement in a graph-based sequence clustering method is to obtain the similarity graph in an efficient manner. A brute-force approach to obtain a similarity graph computes the similarity values between all $\binom{n}{2}$ pairs of sequences (here, n is the number of sequences) and then uses a user-defined threshold to add edges between sequences in a similarity graph. Clearly this is inefficient, as it requires to compute $O(n^2)$ similarity scores explicitly, so many graph-based sequence clustering algorithms use an efficient method for similarity graph construction. Once a similarity graph is obtained, one of the many available graph-clustering [90] methods can be used to obtain the desired clustering. In Algorithm 35, we present a pseudocode for a graph-based sequence clustering algorithm; based on the specific method for the similarity routine (Line 4) and the graph clustering (Line 9), various graph-based sequence clustering methods can be obtained.

The graph-based sequence clustering method that is shown in Algorithm 35 is sometimes not scalable as the computation cost grows quadratically with the number of sequences. Since the similarity computation of each pair of sequences is independent, a straightforward remedy to the lack of scalability is to use distributed or parallel computing to perform the tasks in Lines 2–8. There also exist algorithmic solutions; instead of finding the similarity between all the sequence-pairs explicitly, these solutions adopt methods that find similar segment-pairs across all the input sequences simultaneously. Then, they obtain the similarity between input sequences by scanning the frequency of highly significant similar segment-pairs. For instance, if two sequences share a large number of similar segment-pairs, the pair obtains a high similarity score, and hence the method adds an edge between those two sequences in the similarity graph. The advantage of such methods is that they avoid the all-pair similarity computation with has quadratic complexity and replace it with a method that has linear or sublinear complexity.

Line 9 of Algorithm 35 calls a graph-clustering algorithm. Many of the existing graph-clustering methods, such as spectral clustering [97, 29] or Markov clustering [31], have quadratic ($O(|V|^2)$) complexity and, hence, are not efficient for this task. Also, unlike the earlier task of similarity graph

Algorithm 35 Generic Graph-Based Sequence Clustering

Require: Sequence database ($\mathcal{S}$)
Similarity threshold (σ)
Cluster count (k)

$G(V,E) = \textit{build-graph}(V = \mathcal{S}, E = \emptyset)$
for each edge $s_1 \in \mathcal{S}$ **do**
  for each edge $s_2 \in \mathcal{S}$ **do**
    if $\textit{sim}(s_1, s_2) \geq \sigma$ and $s_1 \neq s_2$ **then**
      $E = E \cup (s_1, s_2)$
    end if
  end for
end for
$C = \textit{graph-clustering}(G, k)$
return C

construction, obtaining a parallel or distributed method for clustering graph is nontrivial. So, the majority of the methods choose a simple graph clustering algorithm. One such method is single-link clustering (SLC), which is an agglomerative clustering method that merges two of the existing clusters based on the largest similarity between a pair of sequences that are taken from those two clusters. The process continues until the desired number of clusters is obtained. The complexity of SLC is $O(|E|)$, which is much cheaper than $O(|V|^2)$ for sparse graphs. Since similarity graphs are very sparse, SLC method on such graph is highly efficient. However, the clustering quality of SLC can be poor; for instance, it can happen that the distance of an object from another object belonging to a different cluster can be smaller than the distance of the first object to another object belonging to the same cluster.

One of the earliest sequence clustering methods that uses a graph-based technique is GeneRAGE [34]. It performs an all-against-all sequence similarity search using BLAST; if the similarity between two proteins is higher than a given threshold, an edge is added between those two proteins in the similarity graph. Along this process, GeneRAGE also performs some preprocessing on the similarity graphs. For example, it identifies whether a protein is multidomain by considering the transitivity of similarity among other proteins that are similar to the said protein; if multidomain proteins are found, they are allowed to be part of multiple clusters. For the clustering task, GeneRAGE uses SLC. Another earlier clustering method, called d2_cluster [17] also uses SLC for clustering EST (expressed sequence tags) and full-length cDNA sequences. In a recent work [73], Miele et. al. propose a memory-efficient implementation of SLC by following the well-known Union-Rank data structure for disjoint sets.

Kawaji et al. [60] mention the limitations of single-link clustering and propose to use recursive graph partitioning to find clusters from the similarity graph. However, the balanced bipartition technique that they propose uses a one-change optimization scheme, which is very costly. In [81], the authors use spectral clustering with the multiway normalized cut criteria for clustering the similarity graph; however, as we mentioned earlier, spectral clustering-based methods are not scalable because they require finding eigenvectors and eigenvalues of similarity matrix—a computationally intensive task. Another clustering method called BAG [62] also uses partitioning of the similarity graph, but its partitioning method is targeted to find biconnected components of the similarity graph. For building the similarity graph, BAG uses the FASTA algorithm in an efficient manner; for every sequence $i : 1 \leq i \leq n$, it calls $\text{FASTA}(s_i, S)$ and finds the pair of similar sequences with only $O(n)$ number of FASTA calls, instead of $O(n^2)$ such calls. Very recently, Voevodski et al. [108] proposed a method that finds the similarity graph with only $k(< n)$ number of BLAST calls, where the value of k is decided by following an active learning paradigm.

16.4.2.3 Probabilistic Models

The probabilistic approach is popular for sequence modeling. For instance, the earliest approaches for modeling protein families use profile-HMM, a hidden Markov model-based probabilistic approach. However, the main objective of profile-HMM is to perform multiple sequence alignment. Since then, several methods have been proposed to use HMM for sequence clustering. We discuss a few of them below.

Smyth's work [99] is one of the first that uses HMM for clustering biological sequences. His method has two steps: the first step devises a pairwise distance between observed sequences by computing a symmetrized similarity. This similarity is obtained by training an HMM for each sequence, so that the log-likelihood (LL) of each model, given each sequence, can be computed. This information is used to build an LL matrix which is then used to cluster the sequences into k groups, using a hierarchical algorithm. In the second step, one HMM is trained for each cluster; the resulting k models are then merged into a composite global HMM, where each HMM is used to design a disjoint part of this composite model. This initial estimate is then refined using the standard Baum-Welch procedure. As a result, a global HMM modeling of all the sequences is obtained. The number of clusters (k) is selected using a cross-validation method.

CLUSEQ [116] uses conditional probability distribution (CPD) for characterizing a cluster, i.e., for different clusters the CPD of the next symbol given a preceding segment is different, and hence the CPD can be used for clustering biological sequences. To store and retrieve the CPD of various segments, CLUSEQ uses a novel data structure, called a probabilistic suffix tree (PST); every node in PST stores a probability vector to store the probability distribution of the next symbol given the label of the node as the preceding segment. Usages of PST make CLUSEQ very efficient. In the following paragraphs, we describe CLUSEQ in more detail.

The CLUSEQ algorithm accepts a sequence database along with user-defined values for various threshold parameters. At the beginning, all sequences in the database are unclustered. Then, an iterative process is employed to continuously improve the quality of the clustering until no further improvement can be made. In each iteration, CLUSEQ starts with a set of new clusters; a few random sequences from the unclustered set are chosen to be clusters themselves. Then, it examines each sequence to evaluate its similarity to each of the clusters and updates the cluster membership of that sequence, if necessary. At the end of the iteration, CLUSEQ also merges multiple clusters, if their membership overlaps significantly. The key part of CLUSEQ is the step which computes the similarity of a sequence to a cluster. For this, CLUSEQ uses CPD of symbols given a segment. For efficient computation of CPD, the sequences in a cluster are stored in a probabilistic suffix tree. If S is a cluster and $\sigma = s_1 s_2 \cdots s_l$ is a sequence, its likelihood to be a member of the cluster S is given by $P_S(\sigma) = P_S(s_1) \times P_S(s_2|s_1) \times \ldots \times P_S(S_l|S_1 \ldots S_{l-1})$, where $P_S(s_i|s_1 \cdots s_{i-1})$ is the conditional probability that the symbol s_i is the next symbol right after the segment $s_1 \ldots s_{i-1}$ in the sequence cluster S. For each cluster, CLUSEQ maintains a PST, which stores the above conditional probabilities effectively. The strengths of CLUSEQ are that is is specifically adapted for biological sequence clustering and it is very efficient. The weakness is that is has several user-defined parameters that need to be selected appropriately for good clustering results. Also, due to the randomness, the clustering is nondeterministic, and its quality depends on the random choice of the initial sequences that constitute the cluster seeds.

In a recent work [112], the authors propose another CPD-based model, called DHCS, which overcomes some of the difficulties of the CLUSEQ; specifically, DHCS does not choose seed sequences randomly, rather it employs a two-tier Markov Model, where the first tier provides a good initialization for the CPD model in the second tier. The two-tier structure guarantees that the statistical models in the DHCS algorithm are constructed in a statistically significant way without resorting to pairwise comparison of sequences.

16.4.2.4 Suffix Tree and Suffix Array-Based Method

For fast processing of sequence similarity, efficient data structures that index small sequence-segments are also proposed. The main objective of using a data structure is to avoid the full pairwise similarity computation, which is infeasible for many clustering tasks because of its quadratic complexity. For example, databases of ESTs sometimes contain millions of ESTs and efficient data structure is critical for clustering those sequences.

In [72], the authors use suffix arrays to cluster ESTs; suffix arrays are particularly suitable because ETSs are short subsequences of cDNA sequences. To identify all matching blocks of length k, this method first adds all suffixes into a suffix array. Then it sorts the suffixes and groups the suffixes that share a k-length prefix. Then, for each pair of suffixes sharing at least one matching block, it finds the largest consistent matching blocks and the corresponding matching score. Finally, starting with the highest scoring sequence pair, it builds hierarchical clusters using single-link clustering. In [47], the authors propose a clustering tool called CLAGen, that uses a suffix tree for sequence clustering. CLAGen constructs a suffix tree from the input sequences; this construction is highly efficient with only linear time complexity. Then, CLAGen uses the suffix tree for searching and overlapping common subsequences so that it can obtain sequence pairs that are highly similar. From the pairwise similarity information, CLAGen finds sequence clusters. A nice feature of CLAGen is that it annotates the gene clusters by using information from the BLAST search. The CLUSEQ algorithm that we discussed earlier also uses a suffix tree, but it uses it to find the conditional probability of a symbol given a sequence in a cluster.

16.5 Software Packages

Implementations of the well-known clustering algorithms for microarray, protein interaction and sequence data are available online. We summarize some of the most popular ones that are currently being used.

The most popular tool for cluster analysis and visualization of microarray data is **Cluster** [33] which was developed by Michael Eisen. It contains some of the basic clustering algorithms such as k-means, hierarchical and self-organizing maps. **TreeView** is a software (associated with Cluster software) that provides a graphical user interface to browse through the clustering results. Both of these work only in the Windows environment and can be downloaded from `http://rana.lbl.gov/EisenSoftware.htm`.

Michiel de Hoon of the University of Tokyo has created a version of Cluster (called **Cluster 3.0**) [23] that implements the same algorithms for different platforms. Routines for hierarchical (pairwise simple, complete, average, and centroid linkage) clustering, k-means and k-medians clustering, and 2D self-organizing maps are included. This software can be downloaded from `http://bonsai.hgc.jp/~mdehoon/software/cluster/`. Extensions of these modules to Python and Perl languages are also available.

Java TreeView is a Java-based visualization software that can be used to view the original microarray data and also the corresponding clustering results [88]. This is an open source software which can be downloaded from `http://sourceforge.net/projects/jtreeview/`.

EXCAVATOR (EXpression data Clustering Analysis and VisualizATiOn Resource) was developed by the Protein Informatics Group of Oak Ridge National Laboratory. It uses concepts from graph theory (such as minimum spanning trees) to represent gene expression data. In addition to providing different distance measures for computing clusters, it contains many other additional features such as automatic selection of the number of clusters, background noise removal, and identifying gene expression profiles that are similar to a specified set of seed genes. This software is available at http://digbio.missouri.edu/software/Excavator/index.html. **EMMIX-GENE** (available at http://www.maths.uq.edu.au/~gjm/emmix-gene/) is a software for mixture model-based clustering for gene expression data.

EXPANDER (EXPression ANalyzer and DisplayER) is a gene expression analysis and visualization tool that contains several clustering methods such as *k*-means, hierarchical clustering, and self-organizing maps, and it enables visualizing the gene expression data and their clusters [94]. It is a Java-based tool which can be downloaded from http://acgt.cs.tau.ac.il/expander/. The **Pvclust** is an R package that can be used to assess the uncertainty in hierarchical cluster analysis. It calculates p-values for each cluster using bootstrap resampling technique [102]. It can be downloaded from http://www.is.titech.ac.jp/~shimo/prog/pvclust/. **Genesis** is a Java-based, platform-independet package for simultaneously analyzing and visualizing gene expression datasets. It contains several popular clustering algorithm implementations such as hierarchical clustering, self-organizing maps, and *k*-means clustering [101]. It can be downloaded from http://genome.tugraz.at/genesisclient/genesisclient_download.shtml. **wCLUTO** (available at http://glaros.dtc.umn.edu/gkhome/cluto/wcluto/overview) is a web-enabled data-clustering application that is designed for the clustering and data-analysis requirements of gene-expression analysis. wCLUTO is built on top of the CLUTO clustering library. Users can upload their datasets, select from a number of clustering methods, perform the analysis on the server, and visualize the final results. A **synthetic generator** for gene expression data is available at http://www.che.udel.edu/eXPatGen/. Such a simulator can generate hypothetical expression patterns that can be used to evaluate and compare different clustering methods.

A popular tool that implements many popular biclustering algorithms (such as CC, ISA, and OPSM) is the **Biclustering Analysis Toolbox (BicAT)** which is available at http://www.tik.ee.ethz.ch/sop/bicat/. BicAT is a Java-based software that provides a nice graphical user interface for analyzing gene expression data [12]. **TriCluster** [123] is the first triclustering algorithm for microarray expression clustering. Tricluster first mines all the biclusters across the gene-sample slices, and then it extends these into triclusters across time dimensions. This software can be downloaded from http://www.cs.rpi.edu/~zaki/software/TriCluster.tar.gz. **STEM** (Short Time-series Expression Miner) is a Java program for clustering, comparing, and visualizing short time-series gene-expression data from microarray experiments. It can be downloaded from http://www.cs.cmu.edu/~jernst/stem/.

clusterMaker [74] provides a unified platform for various traditional clustering and network clustering techniques. It is available at http://www.cgl.ucsf.edu/cytoscape/cluster/clusterMaker.html. All of the network partitioning cluster algorithms create collapsible "meta nodes" to allow interactive exploration. These clustering algorithms have been developed as a plugin to the Cytoscape software [93]. **Cytoscape** is an open source bioinformatics software platform for visualizing complex molecular interaction networks. It is available at http://www.cytoscape.org/.

CFinder is a free software for finding and visualizing overlapping dense groups of nodes in networks, based on the Clique Percolation Method (CPM) [1]. It is a fast program for locating and visualizing overlapping, densely interconnected groups of nodes in undirected graphs, and allowing the user to easily navigate between the original graph and the web of these groups. This software is available at http://www.cfinder.org/. The code for **Markov clustering** is available at http://www.micans.org/mcl/. It provides a collection of network analysis tools focused on analysis of very large networks, scaling up to millions of nodes and hundreds of millions of edges. The software for **MCODE** algorithm [9] is made available in the form of a Cytoscape plugin at http://baderlab.org/Software/MCODE. The **network analysis tools (NeAT)** [15] provides a user-friendly web access to a collection of modular tools for the analysis of networks (graphs) and clusters. It includes a set of tools that support basic graph operations and clustering algorithms. NeAT is designed to cope with large datasets and provides a flexible toolbox for analyzing biological networks stored in various databases or obtained from high-throughput experiments. This software is available at `http://rsat.ulb.ac.be/neat/`.

CD-HIT is a very widely used program for clustering protein and DNA sequence. It is available from `http://weizhong-lab.ucsd.edu/cd-hit/`. The website has a server that provides CD-HIT services online. Alternatively, the user can also download the program for off-line use. The CD-HIT package has various sequence clustering tools that are customized for different biological sequences. The standard BLAST package includes a program called BLASTclust that can be used to cluster either protein or DNA sequences. BLASTclust accepts a number of parameters that can be used to control the stringency of clustering including thresholds for score density, percent identity, and alignment length. Besides clustering, BLASTclust can also be used to create a nonredundant set of sequences from a source database. The following website `http://toolkit.tuebingen.mpg.de/blastclust` provides access to a blastClust server.

16.6 Discussion and Summary

This chapter provides a survey of different data clustering techniques that have been successfully applied in the context of biological data. We have discussed different types of biological data where data clustering has produced promising and biologically meaningful results. For clustering gene expression data, different categories of clustering, namely, biclustering, triclustering, and time-series clustering, have been discussed along with some of the standard clustering algorithms. Different properties of biological networks along with the most widely studied network clustering algorithms have also been discussed. Various biological sequence similarity measures and clustering algorithms have been studied. Along with the clustering algorithms, we also have discussed the biological validation of the clusters obtained.

As more and more biologists become familiar with the advancements in data clustering algorithms, we can envision a data-driven biological science instead of a hypothesis-driven science. In addition, we also hope that the new insights that these clustering algorithms provide can drive a new set of biologically constrained experiments that can capture better insights about cellular functions.

Bibliography

[1] B. Adamcsek, G. Palla, I. J. Farkas, I. Derenyi, and T. Vicsek. CFinder: Locating cliques and overlapping modules in biological networks. *Bioinformatics*, 22(8):1021–1023, 2006.

[2] J. S. Aguilar-Ruiz. Shifting and scaling patterns from gene expression data. *Bioinformatics*, 21(20):3840–3845, October 2005.

[3] U. Alon, N. Barkai, D. A. Notterman, K. Gishdagger, S. Ybarradagger, D. Mackdagger, and A. J. Levine. Broad patterns of gene expression revealed by clustering analysis of tumor and normal colon tissues probed by oligonucleotide arrays. *Proceedings of the National Academy of Sciences of the United States of America*, 96(12):6745–6750, June 1999.

[4] F. Alqadah, J. S. Bader, R. Anand, and C. K. Reddy. Query-based biclustering using formal concept analysis. In *Proceedings of SIAM International Conference on Data Mining*, pages 648–659, 2012.

[5] S. Altschul, W. Gish, W. Miller, E. Myers, and D. Lipman. Basic local alignment search tool. *Journal of Molecular Biology*, 215:403–410, 1990.

[6] I. P. Androulakis, E. Yang, and R. R. Almon. Analysis of time-series gene expression data: Methods, challenges, and opportunities. *Annual Review of Biomedical Engineering*, 9:205–228, 2007.

[7] M. Ashburner, C. A. Ball, J. A. Blake, D. Botstein, H. Butler, J. M. Cherry, A. P. Davis, K. Dolinski, S. S. Dwight, J. T. Eppig, M. A. Harris, D. P. Hill, L. Issel-Tarver, A. Kasarskis, S. Lewis, J. C. Matese, J. E. Richardson, M. Ringwald, G. M. Rubin, and G. Sherlock. Gene-Ontology: Tool for the unification of biology. The Gene Ontology Consortium. *Nature Genetics*, 25(1):25–29, 2000.

[8] S. Asur, D. Ucar, and S. Parthasarathy. An ensemble framework for clustering protein-protein interaction networks. *Bioinformatics*, 23(13):i29–i40, 2007.

[9] G. Bader and C. Hogue. An automated method for finding molecular complexes in large protein interaction networks. *BMC Bioinformatics*, 4(1):2, 2003.

[10] C. Bahlmann and H. Burkhardt. Measuring HMM similarity with the Bayes probability of error and its application to online handwriting recognition. In *Proceedings of the Sixth International Conference on Document Analysis and Recognition*, ICDAR '01, pages 406–411, 2001.

[11] Z. Bar-Joseph. Analyzing time series gene expression data. *Bioinformatics*, 20(16):2493–2503, 2004.

[12] S. Barkow, S. Bleuler, A. Prelic, P. Zimmermann, and E. Zitzler. BicAT: A biclustering analysis toolbox. *Bioinformatics*, 22(10):1282–1283, 2006.

[13] A. Ben-Dor, B. Chor, R. Karp, and Z. Yakhini. Discovering local structure in gene expression data: The order-preserving submatrix problem. *Journal of Computational Biology*, 10(3–4):373–384, 2003.

[14] A. Ben-Dor and Z. Yakhini. Clustering gene expression patterns. In *Proceedings of the Third Annual International Conference on Computational Molecular Biology*, RECOMB '99, pages 33–42, New York, USA, 1999.

[15] S. Brohée, K. Faust, G. Lima-Mendez, O. Sand, R. Janky, G. Vanderstocken, Y. Deville, and J. van Helden. NeAT: A toolbox for the analysis of biological networks, clusters, classes and pathways. *Nucleic Acids Research*, 36(Web Server issue), July 2008.

[16] S. Brohee and J. van Helden. Evaluation of clustering algorithms for protein-protein interaction networks. *BMC Bioinformatics*, 7(1):488, 2006.

[17] J. Burke, D. Davison, and W. Hide. d2_cluster: A validated method for clustering EST and full-length cDNAsequences. *Genome Research*, 9(11):1135–1142, November 1999.

[18] S. Busygin, O. Prokopyev, and P. M. Pardalos. Biclustering in data mining. *Computers and Operations Research*, 35(9):2964–2987, 2008.

[19] A.J. Butte and I.S. Kohane. Mutual information relevance networks: functional genomic clustering using pairwise entropy measurements. In *Pacific Symposium on Biocomputing*, volume 5, pages 418–429, 2000.

[20] C. Cheadle, M. P. Vawter, W. J. Freed, and K. G. Becker. Analysis of microarray data using Z score transformation. *Journal of Molecular Diagnostics*, 5(2):73–81, May 2003.

[21] Y. Cheng and G. M. Church. Biclustering of expression data. In *Proceedings of the Eighth International Conference on Intelligent Systems for Molecular Biology*, pages 93–103, 2000.

[22] S. Datta and S. Datta. Comparisons and validation of statistical clustering techniques for microarray gene expression data. *Bioinformatics*, 19(4):459–466, 2003.

[23] M. J. L. de Hoon, S. Imoto, J. Nolan, and S. Miyano. Open source clustering software. *Bioinformatics*, 20(9):1453–1454, 2004.

[24] F. De Smet, J. Mathys, K. Marchal, G. Thijs, B. De Moor, and Y. Moreau. Adaptive quality-based clustering of gene expression profiles. *Bioinformatics*, 18(5):735–746, 2002.

[25] R. De Smet and K. Marchal. Advantages and limitations of current network inference methods. *Nature Reviews Microbiology*, 8(10):717–729, October 2010.

[26] D. Dembele and P. Kastner. Fuzzy c-means method for clustering micrroarray data. *Bioinformatics*, 19(8):973–980, 2003.

[27] M. Deodhar, G. Gupta, J. Ghosh, H. Cho, and I. Dhillon. A scalable framework for discovering coherent co-clusters in noisy data. In *ICML '09: Proceedings of the 26th Annual International Conference on Machine Learning*, pages 241–248, 2009.

[28] P. D'haeseleer, S. Liang, and R. Somogyi. Genetic network inference: From co-expression clustering to reverse engineering. *Bioinformatics*, 16(8):707–726, 2000.

[29] I. S. Dhillon, Y. Guan, and B. Kulis. Kernel *k*-means: Spectral clustering and normalized cuts. In *Proceedings of the Tenth ACM SIGKDD International Conference on Knowledge Discovery and Data Mining*, KDD '04, pages 551–556, 2004.

[30] T. Dhollander, Q. Sheng, K. Lemmens, B. De Moor, K. Marchal, and Y. Moreau. Query-driven module discovery in microarray data. *Bioinformatics*, 23(19):2573–2580, 2007.

[31] S. Dongen. A cluster algorithm for graphs. Technical report, CWI (Centre for Mathematics and Computer Science), Amsterdam, The Netherlands, 2000.

[32] S. Dudoit and J. Fridlyand. Bagging to improve the accuracy of a clustering procedure. *Bioinformatics*, 19(9):1090–1099, 2003.

[33] M. B. Eisen, P. T. Spellman, P. O. Brown, and D. Botstein. Cluster analysis and display of genome-wide expression patterns. *Proceedings of the National Academy of Sciences*, 95(25):14863–14868, December 1998.

[34] A. J. Enright and C. A. Ouzounis. GenerAGE: A robust algorithm for sequence clustering and domain detection. *Bioinformatics*, 16(5):451–457, 2000.

[35] J. Ernst and Z. Bar-Joseph. STEM: a tool for the analysis of short time series gene expression data. *BMC Bioinformatics*, 7(1):191, 2006.

[36] J. Ernst, G.J. Nau, and Z. Bar-Joseph. Clustering short time series gene expression data. *Bioinformatics*, 21(Suppl 1):i159–i168, 2005.

[37] G. Fang, R. Kuang, G. Pandey, M. Steinbach, C. L. Myers, and V. Kumar. Subspace differential coexpression analysis: Problem definition and a general approach. *Pacific Symposium on Biocomputing*, pages 145–156, 2010.

[38] B. J. Frey and D. Dueck. Clustering by passing messages between data points. *Science*, 315(5814):972–976, 2007.

[39] M. E. Garber, O. G. Troyanskaya, K. Schluens, S. Petersen, Z. Thaesler, Pacyna M. Gengelbach, M. van de Rijn, G. D. Rosen, C. M. Perou, R. I. Whyte, and others. Diversity of gene expression in adenocarcinoma of the lung. *Proceedings of the National Academy of Sciences*, 98(24):13784–13789, 2001.

[40] T. George and S. Merugu. A scalable collaborative filtering framework based on co-clustering. In *Proceedings of the Fifth IEEE International Conference on Data Mining*, ICDM '05, pages 625–628, Washington, DC, USA, 2005.

[41] G. Getz, E. Levine, and E. Domany. Coupled two-way clustering analysis of gene microarray data. *Proceedings of National Academy of Sciences of the United States of America*, 97:12079–12084, 2000.

[42] R. Gill, S. Datta, and S. Datta. A statistical framework for differential network analysis from microarray data. *BMC Bioinformatics*, 11(1):95, 2010.

[43] F. Glover. Tabu search—Part I. *ORSA Journal on Computing*, 1(3):190–206, 1989.

[44] T. R. Golub, D. K. Slonim, P. Tamayo, C. Huard, M. Gaasenbeek, J. P. Mesirov, H. Coller, M. L. Loh, J. R. Downing, M. A. Caligiuri, C. D. Bloomfield, and E. S. Lander. Molecular classification of cancer: Class discovery and class prediction by gene expression monitoring. *Science*, 286(5439):531–537, October 1999.

[45] V. Guralnik and G. Karypis. A scalable algorithm for clustering sequential data. In *Proceedings of the 2001 IEEE International Conference on Data Mining*, ICDM '01, pages 179–186, 2001.

[46] D. Gusfield. *Algorithms on strings, trees, and sequences*. Cambridge University Press, 1997.

[47] S. Han, S. G. Lee, K. H. Kim, C. J. Choi, Y. H. Kim, and K. S. Hwang. CLAGen: A tool for clustering and annotating gene sequences using a suffix tree algorithm. *BioSystems*, 84(3):175–182, June 2006.

[48] D. Hanisch, A. Zien, R. Zimmer, and T. Lengauer. Co-clustering of biological networks and gene expression data. *Bioinformatics*, 18(suppl 1):S145–S154, 2002.

[49] J. Herrero, A. Valencia, and J. Dopazo. A hierarchical unsupervised growing neural network for clustering gene expression patterns. *Bioinformatics*, 17(2):126–136, 2001.

[50] L. J. Heyer, S. Kruglyak, and S. Yooseph. Exploring expression data: Identification and analysis of coexpressed genes. *Genome Research*, 9(11):1106–1115, November 1999.

[51] J. Ihmels, S. Bergmann, and N. Barkai. Defining transcription modules using large-scale gene expression data. *Bioinformatics*, 20(13):1993–2003, 2004.

[52] H. Jeong, S. P. Mason, A. L. Barabasi, and Z. N. Oltvai. Lethality and centrality in protein networks. *Nature*, 411(6833):41–42, 2001.

[53] L. Ji and K. Tan. Identifying time-lagged gene clusters using gene expression data. *Bioinformatics*, 21(4):509–516, 2005.

[54] L. Ji and K. Tan. Mining gene expression data for positive and negative co-regulated gene clusters. *Bioinformatics*, 20(16):2711–2718, 2004.

[55] D. Jiang, J. Pei, M. Ramanathan, C. Tang, and A. Zhang. Mining coherent gene clusters from gene-sample-time microarray data. In *Proceedings of the Tenth ACM SIGKDD International Conference on Knowledge Discovery and Data Mining*, KDD '04, pages 430–439, 2004.

[56] D. Jiang, C. Tang, and A. Zhang. Cluster analysis for gene expression data: A survey. *IEEE Transactions on Knowledge and Data Engineering*, 16(11):1370–1386, November 2004.

[57] H. Jiang, S. Zhou, J. Guan, and Y. Zheng. gTRICLUSTER: A more general and effective 3d clustering algorithm for gene-sample-time microarray data. In *Proceedings of the 2006 International Conference on Data Mining for Biomedical Applications*, BioDM'06, pages 48–59, 2006.

[58] K. Kailing, H. Kriegel, and P. Kroger. Density-connected subspace clustering for high-dimensional data. In *SDM*, pages 256–257, 2004.

[59] G. Karypis and V. Kumar. Multilevel k-way partitioning scheme for irregular graphs. *Journal of Parallel and Distributed Computing*, Vol. 48, 96–129, 1998.

[60] H. Kawaji, Y. Yamaguchi, H. Matsuda, and A. Hashimoto. A graph-based clustering method for a large set of sequences using a graph partitioning algorithm. *Genome Informatics*, 12:93–102, 2001.

[61] G. C. Kennedy, H. Matsuzaki, S. Dong, W. M. Liu, J. Huang, G. Liu, X. Su, M. Cao, W. Chen, J. Zhang, W. Liu, G. Yang, X. Di, T. Ryder, Z. He, U. Surti, M. S. Phillips, Boyce M. T. Jacino, S. P. Fodor, and K. W. Jones. Large-scale genotyping of complex DNA. *Nature Biotechnology*, 21(10):1233–1237, 2003.

[62] S. Kim and J. Lee. BAG: A graph theoretic sequence clustering algorithm. *International Journal of Data Mining and Bioinformatics*, 1(2):178–200, 2006.

[63] A. D. King, N. Przulj, and I. Jurisica. Protein complex prediction via cost-based clustering. *Bioinformatics*, 20(17):3013–3020, 2004.

[64] T. Kohonen. *Self-organization and associative memory*. 3rd edition. Springer-Verlag, New York, NY, USA, 1989.

[65] N. J. Krogan, G. Cagney, H. Yu, G. Zhong, X. Guo, A. Ignatchenko, J. Li, S. Pu, N. Datta, et al. Global landscape of protein complexes in the yeast Saccharomyces cerevisiae. *Nature*, 440(7084):637–643, March 2006.

[66] C. Leslie and R. Kuang. Fast string kernels using inexact matching for protein sequences. *Journal of Machine Learning Research*, 5:1435–1455, 2004.

[67] C. S. Leslie, E. Eskin, A. Cohen, J. Weston, and W. S. Noble. Mismatch string kernels for discriminative protein classification. *Bioinformatics*, 20(4):467–476, 2004.

[68] J. Li, K. Sim, G. Liu, and L. Wong. Maximal quasi-bicliques with balanced noise tolerance: Concepts and co-clustering applications. In *Proceedings of the SIAM International Conference on Data Mining SDM'08*, pages 72–83, April 2008.

[69] J. Liu, Z. Li, X. Hu, and Y. Chen. Biclustering of microarray data with mospo based on crowding distance. *BMC Bioinformatics*, 10(Suppl 4):S9, 2009.

[70] Y. Luan and H. Li. Clustering of time-course gene expression data using a mixed-effects model with B-splines. *Bioinformatics*, 19(4):474–482, 2003.

[71] S. C. Madeira and A. L. Oliveira. Biclustering algorithms for biological data analysis: A survey. *IEEE/ACM Transactions on Computational Biology and Bioinformatics*, 1(1):24–45, 2004.

[72] K. Malde, E. Coward, and I. Jonassen. Fast sequence clustering using a suffix array algorithm. *Bioinformatics*, 19(10):1221–1226, 2003.

[73] V. Miele, S. Penel, and L. Duret. Ultra-fast sequence clustering from similarity networks with silix. *BMC Bioinformatics*, 12(1):116, 2011.

[74] J. H. Morris, L. Apeltsin, A. M. Newman, J. Baumbach, T. Wittkop, G. Su, G. D. Bader, and T. E. Ferrin. clusterMaker: A multi-algorithm clustering plugin for Cytoscape. *BMC Bioinformatics*, 12(1):436+, November 2011.

[75] O. Odibat and C. K. Reddy. A generalized framework for mining arbitrarily positioned overlapping co-clusters. In *Proceedings of the SIAM International Conference on Data Mining (SDM)*, pages 343–354, 2011.

[76] O. Odibat, C. K. Reddy, and C. N. Giroux. Differential biclustering for gene expression analysis. In *Proceedings of the ACM Conference on Bioinformatics and Computational Biology (BCB)*, pages 275–284, 2010.

[77] Y. Okada and T. Inoue. Identification of differentially expressed gene modules between two-class DNA microarray data. *Bioinformation*, 4(4):134–137, 2009.

[78] G. Palla, A. L. Barabási, and T. Vicsek. Quantifying social group evolution. *Nature*, 446(7136):664–667, 2007.

[79] G. Palla, I. Derényi, I. Farkas, and T. Vicsek. Uncovering the overlapping community structure of complex networks in nature and society. *Nature*, 435(7043):814–818, 2005.

[80] W. R. Pearson and D. J. Lipman. Improved tools for biological sequence comparison. *Proceedings of the National Academy of Sciences*, 85(8):2444–2448, 1988.

[81] W. Pentney and M. Meila. Spectral clustering of biological sequence data. In *Proceedings of the 20th National Conference on Artificial Intelligence—Volume 2*, AAAI'05, pages 845–850, 2005.

[82] C. Pizzuti and S. E. Rombo. A coclustering approach for mining large protein-protein interaction networks. *IEEE/ACM Transactions on Computational Biology and Bioinformatics*, 9:717–730, 2012.

[83] I. Priness, O. Maimon, and I. Ben-Gal. Evaluation of gene-expression clustering via mutual information distance measure. *BMC Bioinformatics*, 8(1):111, 2007.

[84] N. Przulj, D. A. Wigle, and I. Jurisica. Functional topology in a network of protein interactions. *Bioinformatics*, 20(3):340–348, 2004.

[85] S. Pu, J. Vlasblom, A. Emili, J. Greenblatt, and S. J. Wodak. Identifying functional modules in the physical interactome of Saccharomyces cerevisiae. *Proteomics*, 7(6):944–960, March 2007.

[86] L. R. Rabiner. A tutorial on hidden Markov models and selected applications in speech recognition, In Alex Waibel and Kai-Fulee (Eds.) *Readings in Speech Recognition*, pages 267–296. Morgan Kaufmann, San Francisco, CA, 1990.

[87] M. D. Robinson, J. Grigull, N. Mohammad, and T. R. Hughes. FunSpec: A web-based cluster interpreter for yeast. *BMC Bioinformatics*, 3:35+, 2002.

[88] A. J. Saldanha. Java Treeview—Extensible visualization of microarray data. *Bioinformatics*, 20(17):3246–3248, November 2004.

[89] G. Salton and C. Buckley. Term-weighting approaches in automatic text retrieval. *Information Processing and Management*, 24(5): 513–523, 1988.

[90] S. E. Schaeffer. Graph clustering. *Computer Science Review*, 1(1):27–64, 2007.

[91] A. Schliep, A. Schonhuth, and C. Steinhoff. Using hidden Markov models to analyze gene expression time course data. *Bioinformatics*, 19(Suppl 1):i255–i263, 2003.

[92] B. Schwikowski, P. Uetz, and S. Fields. A network of protein-protein interactions in yeast. *Nature Biotechnology*, 18(12):1257–1261, December 2000.

[93] P. Shannon, A. Markiel, O. Ozier, N. S. Baliga, J. T. Wang, D. Ramage, N. Amin, B. Schwikowski, and T. Ideker. Cytoscape: A software environment for integrated models of biomolecular interaction networks. *Genome Research*, 13(11):2498–2504, November 2003.

[94] R. Sharan, A. Maron-Katz, and R. Shamir. CLICK and EXPANDER: A system for clustering and visualizing gene expression data. *Bioinformatics*, 19(14):1787–1799, 2003.

[95] R. Sharan and R. Shamir. Click: A clustering algorithm with applications to gene expression analysis. In *ISMB*, pages 307–316, 2000.

[96] R. Sharan, I. Ulitsky, and R. Shamir. Network-based prediction of protein function. *Molecular Systems Biology*, 3(1), March 2007.

[97] J. Shi and J. Malik. Normalized cuts and image segmentation. *IEEE Transactions on Pattern Analysis and Machine Intelligence*, 22(8):888–905, 2000.

[98] K. Sim, Z. Aung, and V. Gopalkrishnan. Discovering correlated subspace clusters in 3d continuous-valued data. In *IEEE 10th International Conference on Data Mining (ICDM), 2010*, pages 471–480. IEEE, 2010.

[99] P. Smyth. Clustering sequences with hidden Markov models. In *Advances in Neural Information Processing Systems*, volume 9, pages 648–654, MIT Press, Cambridge, MA, 1997.

[100] M. Steinbach, G. Karypis, and V. Kumar. A comparison of document clustering techniques. *KDD Workshop on Text Mining*, Vol. 400, pages 109–111, 2000.

[101] A. Sturn, J. Quackenbush, and Z. Trajanoski. Genesis: Cluster analysis of microarray data. *Bioinformatics*, 18(1):207–208, 2002.

[102] R. Suzuki and H. Shimodaira. Pvclust: An R package for assessing the uncertainty in hierarchical clustering. *Bioinformatics*, 22(12):1540–1542, 2006.

[103] P. Tamayo, D. Slonim, J. Mesirov, Q. Zhu, S. Kitareewan, E. Dmitrovsky, E. S. Lander, and T. R. Golub. Interpreting patterns of gene expression with self-organizing maps: Methods and application to hematopoietic differentiation. *Proceedings of the National Academy of Sciences of the United States of America*, 96(6):2907–2912, March 1999.

[104] S. Tavazoie, J. D. Hughes, M. J. Campbell, R. J. Cho, and G. M. Church. Systematic determination of genetic network architecture. *Nature Genetics*, 22(3):281–285, July 1999.

[105] D. Ucar, S. Asur, U. Catalyurek, and S. Parthasarathy. Improving functional modularity in protein-protein interactions graphs using hub-induced subgraphs. *Lecture Notes in Computer Science*, 4213:371, 2006.

[106] S. Van Dongen. Graph clustering by flow simulation. PhD Thesis. University of Utrecht, 2000.

[107] J. Vlasblom and S. Wodak. Markov clustering versus affinity propagation for the partitioning of protein interaction graphs. *BMC Bioinformatics*, 10(1):99, 2009.

[108] K. Voevodski, M. Balcan, H. Röglin, S. Teng, and Y. Xia. Active clustering of biological sequences. *Journal of Machine Learning Research*, 13:203–225, 2012.

[109] B. H. Voy, J. A. Scharff, A. D. Perkins, A. M. Saxton, B. Borate, E. J. Chesler, L. K. Branstetter, and M. A. Langston. Extracting gene networks for low-dose radiation using graph theoretical algorithms. *PLoS Computational Biology*, 2(7):e89, 2006.

[110] J. Wang, M. Li, Y. Deng, and Y. Pan. Recent advances in clustering methods for protein interaction networks. *BMC Genomics*, 11(Suppl 3):S10, 2010.

[111] J. Wang, Y. Zhang, L. Zhou, G. Karypis, and C. C. Aggarwal. Contour: An efficient algorithm for discovering discriminating subsequences. *Data Mining and Knowledge Discovery*, 18(1):1–29, 2009.

[112] T. Xiong, S. Wang, Q. Jiang, and J. Z. Huang. A new Markov model for clustering categorical sequences. In *Proceedings of the IEEE 11th International Conference on Data Mining*, ICDM '11, pages 854–863, 2011.

[113] X. Xu, Y. Lu, K. Tan, and A. K. H. Tung. Finding time-lagged 3d clusters. In *Proceedings of the 2009 IEEE International Conference on Data Engineering*, ICDE '09, pages 445–456. IEEE Computer Society, 2009.

[114] Y. Xu, V. Olman, and D. Xu. Clustering gene expression data using a graph-theoretic approach: An application of minimum spanning trees. *Bioinformatics*, 18(4):536–545, 2002.

[115] J. Yang, H. Wang, W. Wang, and P. Yu. Enhanced biclustering on expression data. In *Proc. of 3rd IEEE Symposium on BioInformatics and BioEngineering*, pages 321–327, 2003.

[116] J. Yang and W. Wang. CLUSEQ: Efficient and effective sequence clustering. In *Proceedings of the 19th IEEE International Conference on Data Engineering (ICDE)*, pages 101–112, 2003.

[117] W. Yang, D. Dai, and H. Yan. Finding correlated biclusters from gene expression data. *IEEE Transactions on Knowledge and Data Engineering*, 23(4):568–584, April 2011.

[118] K. Y. Yeung, C. Fraley, A. Murua, A. E. Raftery, and W. L. Ruzzo. Model-based clustering and data transformations for gene expression data. *Bioinformatics*, 17(10):977–987, 2001.

[119] K. Y. Yeung, D. R. Haynor, and W. L. Ruzzo. Validating clustering for gene expression data. *Bioinformatics*, 17(4):309–318, 2001.

[120] K. Y. Yeung and W. L. Ruzzo. Principal component analysis for clustering gene expression data. *Bioinformatics*, 17(9):763–774, 2001.

[121] M. J. Zaki. Spade: An efficient algorithm for mining frequent sequences. *Machine Learning*, 42(1):31–60, 2001.

[122] H. Zhao, L. Cloots, T. Van den Bulcke, Y. Wu, R. De Smet, V. Storms, P. Meysman, K. Engelen, and K. Marchal. Query-based biclustering of gene expression data using probabilistic relational models. *BMC Bioinformatics*, 12(Suppl 1):S37, 2011.

[123] L. Zhao and M. J. Zaki. TRICLUSTER: An effective algorithm for mining coherent clusters in 3d microarray data. In *Proceedings of the 2005 ACM SIGMOD International Conference on Management of Data*, pages 694–705. ACM Press, New York, 2005.

Chapter 17

Network Clustering

Srinivasan Parthasarathy
The Ohio State University
Columbus, OH
`srini@cse.ohio-state.edu`

S M Faisal
The Ohio State University
Columbus, OH
`faisal@cse.ohio-state.edu`

17.1 Introduction

Networks are ubiquitous—ranging from network of computers in the *World Wide Web*, to connections between users on social networks, from Protein–Protein Interaction (PPI) networks to citation network among authors, from follower–followee network (on Twitter and similar networks) to dependency structure between constituent tasks of a large program. Recent advances in technology have resulted in a diverse range of domains generating network data. Some example domains range from social [120, 169] to biological [81] from scientific [37, 119] to ecological [14, 153] and the like. Study of such networks can convey important information about community structures among the nodes, connection patterns, influence of nodes, etc. [11, 12, 68, 124].

Clustering is one of the most important operations to apply on a network for mining valuable information from the network. Many other clustering algorithms on other data domains can be solved as a special case of network clustering. For instance relational data can be expressed as entity-relation graphs, general object-driven data sets can be represented as graphs with a (sparse)-similarity matrix between nodes representing the different objects (spectral clustering of relational data is based on this concept) and so on. Some of the early algorithms (e.g., Chameleon [85]) on clustering generic data sets are designed as a special case of graph clustering. Graph clustering is a very powerful abstraction for all kinds of data clustering. Over decades graph clustering has been studied intensively because of its practical applications and importance [90, 57, 78]. More recent algorithms have been developed keeping the structure and size of modern networks in mind. Examples include multilevel graph partitioning algorithms such as Metis [86], Graclus [42], and MLR-MCL [148]. These algorithms are scalable and can deal with some of the biggest graphs [148]; some with millions of vertices and billions of edges Spectral clustering methods form an important class of graph partitioning algorithms. They use weighted cuts [151] and are very effective in terms of quality in the context of social networks. Another important class of graph clustering algorithms involve Markov Clustering (MCL) [45] that is based on stochastic flow simulation. One drawback of the original MCL, however, is the lack of scalability. But improvements to the algorithm have been proposed and the limitations can be effectively redressed while retaining the advantageous features [148, 150]. There has also been the use of *hybrid* algorithms such as Metis+MQI.

Some of the practical applications of graph clustering range from analysis of social networks [125] to analysis of Protein–Protein Interaction (PPI) networks [19], from VLSI design [46] to load balancing in distributed computing environment [135]. Partitioning and grouping the vertices of a graph based on objective functions such as similarity and distance have become important aspects of large network analytics. Some additional applications include community discovery for proxy caches in World Wide Web context and detection of link farms [20, 65], for personalized recommendation systems [141], for efficient routing in mobile ad-hoc networks [157], and so on. Other important applications are summarization of activities within network for a better understanding of the interaction between groups and how the network evolves [10], identification of influential nodes for marketing purposes [43, 100], and prediction of user rating for given items [93].

In recent years, a number of scalable algorithms have been proposed that efficiently produce high quality clusters from graphs. Networks can be perfectly represented with graph data structures and clustering networks, in general, similar to the problems of graph clustering. In this chapter, we will discuss various techniques proposed in the literature for clustering networks/graphs.

17.2 Background and Nomenclature

Before discussing various approaches in detail we would like to clarify the terms *clustering*, *graph partitioning*, and *community discovery*, which are often used interchangeably in the literature. The term *clustering* is often described as a method by which one assigns or organizes records, objects, or entities into groups (clusters) wherein members of each group are more similar to one another (by some definition) and less similar with elements that lie outside of said group. Clustering may further be categorized as *hard* or *soft*—the former refers to the case where each element strictly belongs to at most one group while the latter refers to the case where each element may belong to multiple groups; *complete* or *partial*—the former refers to the case where all elements are placed in one or more groups and the latter refers to the case where most elements are placed into groups; *balanced* or *skewed*—the former refers to the case where the cardinalities of groups are roughly similar, the latter when there is high skew among the cardinalities.

Graph partitioning is a term that is often associated with a hard form of clustering on graphs, that is, complete and typically balanced. This concept is natural in many applications such as VLSI design [46], and the partitioning of physical nodes within a cluster environment [171].

Community discovery is a term that refers to a clustering on graphs that yields tight knit clusters, typically from a topological standpoint. Community discovery algorithms may yield skewed or balanced, partial or complete, and hard or soft groupings depending on the needs of the domain. Although very similar to graph clustering, the term is frequently used in domains where one is interested only in finding the most densely connected components of the graph and not in the cluster assignment of each of the vertices. In such scenarios, a complete clustering of the whole network is not necessary. Rather, finding the most interesting groups often suffices. Lots of computations can be saved by clustering only the densely connected parts of the network [65].

17.3 Problem Definition

Networks are represented using graph data structures that contain sets of vertices and edges between vertices. In formal notations, a graph $\mathcal{G} = (\mathcal{V}, \mathcal{E})$ consists of a set of n vertices $\mathcal{V} = \{v_1, v_2, ..., v_n\}$ and a set of edges $\mathcal{E}$. An edge $e_{i,j}$ represents a connection between vertex v_i and v_j. Both vertices and edges can have weights associated with them. A *weighted* edge represents not only a connection between the nodes, but also a measure of similarity, distance, etc., between the two nodes. Graphs can also be *directed* or *undirected*. Edges are symmetric ($e_{i,j} = e_{j,i}$) in undirected graphs while they are not in directed graphs ($e_{i,j} \neq e_{j,i}$).

Let k be a positive integer. The graph partitioning problem can be posed as a *k-way* partition of G into a set of nonempty subsets $C_1, ...C_k$ of $\mathcal{G}$ such that $\bigcup_{i=1}^{k} C_i = \mathcal{G}$. In general, the partitioning problem has two objectives: the graph should be clustered in such a way that each cluster contains roughly the same number of vertices and the number of edges between clusters is minimized though other objectives can be defined based on application requirements.

Given the formulation of the general graph clustering problem, we first discuss commonly used evaluation criteria followed by a detailed discussion and analysis of different graph clustering algorithms proposed in the literature.

17.4 Common Evaluation Criteria

In this section, we give an overview of commonly used performance evaluation criteria for clustering algorithms. A variety of measures have been proposed in the literature that capture the goodness of a partition of the graph [131].

The simplest function to measure the quality of a partition is "cut" and the most direct way to construct the partition is to solve the mincut problem [105]. Given the number k of partitions, the mincut approach chooses a partition $C_1,...,C_k$ that minimizes

$$cut(C_1,C_2,...,C_k) = \frac{1}{2}\sum_{i=1}^{k} W(C_i,\bar{C_i})$$

For $k = 2$ mincut is relatively simple and an easy problem to solve [158]. But in practice, the partitions are not often satisfactory. More specifically, the solution tends to separate one individual vertex from the rest of the graph which is not expected [105]. In partitioning, each group should be large enough. There have been two most popular objective functions to include this aspect of a partition.

RatioCut, proposed in [74], of a group of vertices C is the sum of the edge weights connecting C to the rest of the graph normalized by size of C. It is denoted as

$$RatioCut(C_1,...,C_k) = \frac{1}{2}\sum_{i=1}^{k} \frac{W(C_i,\bar{C_i})}{|C_i|}$$

Probably the most popular quality function for graph partitioning is *normalized cut* [151, 114]. The *normalized cut* of a group of vertices C is the sum of the weights of the edges connecting C to the rest of the graph normalized by total edge weight of C and total edge weight of the rest of the vertices in the graph. Mathematically, the normalized cut of a partition of the graph $C \subset \mathcal{V}$ is denoted as

$$Ncut(C_1,...,C_k) = \frac{1}{2}\sum_{i=1}^{k} \frac{W(C_i,\bar{C_i})}{d(C_i)}$$

where $d(C_i)$ is the total degree of cluster C_i

The intuition in both RatioCut and Normalized Cut is to make sure partitions are "balanced" [105]. Groups with low normalized cut represent good communities because they are well connected among themselves and sparsely connected to the rest of the graph [131].

The *conductance* is defined in [83] and is closely related. It is defined as

$$Conductance(C) = \frac{\Sigma_{i\in C, j\in \bar{C}} W(i,j)}{min(\Sigma_{i\in C} d(i), \Sigma_{i\in \bar{C}} d(i))}$$

The normalized cut or conductance of a partition of a graph into k clusters $C_1,...,C_k$ is the sum of the normalized cut or conductance of the individual partitions C_i for $i = 1,...,k$ [42].

Another popular measure of partition goodness of a graph is *Modularity* [125]. This measure is the sum of the differences, for each cluster, between fraction of internal edges and fraction of edges that are expected to be inside a random cluster with the same total degree. Mathematically,

$$Q = \sum_{i=1}^{k} \left[\frac{W(C_i,C_i)}{e} - \left(\frac{d(C_i)}{2e}\right)^2 \right]$$

where C_is are clusters, e is number of edges, and $d(C_i)$ represents the total degree of cluster C_i. Unfortunately, optimizing any of these "normalized" objective functions is NP-hard [64, 151].

In following sections, we review some of the partitioning algorithms proposed in the literature for solving the problem of graph clustering. Some of these methods date back to the early 1970s and have played important roles in laying out the direction for solving graph clustering problems. Primarily the problem was viewed as a graph bisection problem and the goal was to partition the graph into two roughly equal sized partitions so that the interpartition edges are minimized; either in number or in weight. Many of such bisection algorithms were later extended to handle the problem of *k-way* partitioning.

Different approaches have been taken to solve the generic problem of graph clustering. Some of them rely on geometric properties while others need only the graph itself and no additional information. Some approaches are deterministic in the sense that they always produce the same result while some employ randomization. In this section, we will present some of the representative algorithms to cover the whole spectrum.

17.5 Partitioning with Geometric Information

Sometimes a geometric layout of the graph can be known from additional information about the graph. For instance, a structure in a d-dimensional space can be easily represented as a graph where the geometric coordinates are attached to the vertices. Such graphs are also referred to as *meshes*. Algorithms that deal with *meshes* usually use only the geometric information and are therefore limited to graphs that have such geometric information available. In fact, these algorithms do not use any edge information and, hence, cannot be extended very easily to handle graphs with weighted edges [51]. But they often produce acceptable results for mesh clustering. The overall approach is to divide the underlying space into two parts in such a way that the points are divided into two roughly equal groups. Let us look at some of the common methods.

17.5.1 Coordinate Bisection

This is the simplest coordinate-based method that involves finding a hyperplane that is orthogonal to a chosen coordinate axis and that divides the points into two equal parts. This can be done simply by looking at the corresponding coordinate values and finding a value so that half of the coordinate values are smaller and half are greater. For instance, if y-axis is chosen, a value $\bar{y}$ is chosen that divides the y-coordinate values of the points into two groups. There are two ways in which this technique can be applied recursively:

- *Repeatedly bisecting the same axis:* This leads to thin partitions leading to long boundaries which typically results in high cut-size
- *Alternately bisecting the coordinate axes:* This usually leads to a better partitioning and lower cut-size

The obvious drawback of coordinate bisection is the dependency on the coordinate system. That is, the same graph might be partitioned differently in a different coordinate system.

17.5.2 Inertial Bisection

The basic idea of Inertial Bisection [77, 135] is to choose an axis that runs through the "middle" of the points. Mathematically this corresponds to choosing a line L such that the sum of squared distances from the points to the line L is minimized. The line L is the axis of minimal rotational inertia. In a convex domain, this axis aligns itself with the shape of the mesh and the spatial extend in the directions orthogonal to the axis is minimized. This usually results in a minimized the cut-size.

17.5.3 Geometric Partitioning

Miller et al. introduced the method in [115] that, for certain graphs, finds a vertex separator that divides the graph in two roughly equal sizes with high probability. The following definitions are necessary in order to state the theorem given in [115].

Definition 17.5.1 *A k-ply neighborhood system in d dimensions is a set* $D_1, D_2, \ldots, D_n$ *of closed disks in* $\mathbb{R}^d$ *such that no point of* $\mathbb{R}^d$ *is strictly interior to more than k disks.*

Definition 17.5.2 *An* (α, k) *overlap graph is a graph defined in terms of a k-ply neighborhood system* $D_1, D_2, \ldots, D_n$ *and a constant* $\alpha \geq 1$. *Each disk* D_i *is represented by a vertex and there is an edge between two vertices if expanding the radius of the smaller of the disks by a factor of* α *causes the disks to overlap. Mathematically,*

$$\mathcal{E} = \{e_{i,j} | D_i \cap \alpha D_j \neq 0 \text{ and } \alpha D_i \cap D_j \neq 0\}$$

A regular n-by-n mesh is a (1,1)-overlap graph. The theorem given in [115] is as follows:

Theorem 1 *Given* $\mathcal{G} = (\mathcal{V}, \mathcal{E})$, *an* (α, k)*-overlap graph in d dimensions with n nodes, there is a vertex separator* $\mathcal{V}_S$ *so that* $\mathcal{V} = \mathcal{V}_1 \dot{\cup} \mathcal{V}_S \dot{\cup} \mathcal{V}_2$ *with the following properties:*

1. $\mathcal{V}_1$ *and* $\mathcal{V}_2$ *each have at most* $n(d+1)/(d+2)$ *vertices, and*
2. $\mathcal{V}_S$ *has at most* $O(\alpha k^{1/d} n^{(d-1/d)})$ *vertices.*

Here $\dot{\cup}$ represents *disjoint* union, i.e., $\mathcal{U} \dot{\cup} \mathcal{V}$ is $\mathcal{U} \cup \mathcal{V}$ with $\mathcal{U} \cap \mathcal{V} = \emptyset$. The separator size given by the theorem is "asymptotically optimal" for regular meshes in simple geometric shapes.

The proof of the theorem leads to a randomized algorithm running in linear time that will find a separator of the size given in the theorem with high probability. A d-dimensional sphere defines the separator and the algorithm randomly chooses the separating circle from a distribution that, with high probability, satisfies the conclusions of the theorem. We will now define some terminologies before describing the algorithm.

A ***stereo-graphic projection*** is a mapping of points in $\mathbb{R}^d$ to the unit-sphere centered at the origin in $\mathbb{R}^{d+1}$.

The ***center-point*** of a given set of points in $\mathbb{R}^d$ is such that every hyperplane through the center-point divides the set into two subsets so that the sizes differ by at most a ratio of 1:d.

The algorithm is as follows.

Algorithm 36 Geometric Bisection

Project up: Stereographically project points from $\mathbb{R}^d$ to unit-sphere in $\mathbb{R}^{d+1}$
Find Centerpoint: Find centerpoint z of the projected points
Conformal Map: Map the points back on the sphere so that the centerpoint lies at the origin; that is, to rotate the sphere around the origin and then dilate points so that the new centerpoint lies in the origin
Find great circle: Intersect the $\mathbb{R}^{d+1}$-sphere with a random d-dimensional hyperplane through the origin
Unmap: Unmap the great circle to a circle C in $\mathbb{R}^d$ by inverting the dilation, rotation, and stereographic projection
Find Separator: A *point* separator is found by choosing all points whose corresponding disk, magnified by a factor of α, intersects C

Some simplifications to the algorithm were suggested in [66] by Gilbert et al. It is easier to find the edge separator by the edges cut by the circle C than by finding the point separator because to find the edge separator, it is not necessary to know the neighborhood system for the graph. The theorem guarantees a ratio of $1 : d+1$ between the two partitions whereas the goal of bisection is to find two partitions of the same size. This is done by moving the hyperplane along the normal vector until it divides the points evenly.

Finding the centerpoint is a polynomial time process and, hence, is slow. Heuristics can be used to find an approximate centerpoint as opposed to the *real* one in linear time using a randomized algorithm. The computation can be sped up even more by using only a randomly chosen subset of points. Although a random great circle has a good possibility to induce good partitions, experiments in [66] show that it is worthwhile to generate different circles and choose the one that delivers the result.

The methods discussed in this section work quite well on meshes but they have some drawbacks. First, these methods assume geometric adjacency among the connected vertices. Second, they require geometric information about the points. The second condition is critical because even if the first condition is satisfied, the mesh layout may not be known. In the following section, we describe methods that do not use geometric information. The algorithms in the following sections rely only on the connectivity information of a graph and, hence, are applicable to a wider range of practical problems. They do not assume the availability of geometric information and can be applied to more generic classes of problems.

17.6 Graph Growing and Greedy Algorithms

One simple idea for graph partitioning is to select a starting vertex and keep adding vertices to it, based on some criteria, until the partition is of the desired size. These algorithms are often classified as "greedy" and "graph-growing" algorithms [51]. In "greedy" algorithms (e.g., [30]) the next vertex is chosen greedily, one that appears to be the "best" in some sense (e.g., minimum increment in cut-size). In "graph-growing" algorithms, on the other hand, the resultant subgraphs are grown following a certain order (e.g., breadth-first).

A simple greedy partitioning algorithm is given in [51] where the process starts by selecting a *pseudo-peripheral* vertex (one of a pair of maximum distant vertices in the graph [146]) and marking it the first partition. The algorithm then iteratively keeps adding vertices to the current partition by selecting one of the unmarked vertices that has the least number of unmarked neighbors until the partition is big enough. After that, a new starting point for the next partition is chosen by selecting the unmarked vertex with least number of unmarked neighbors. By choosing new vertices adjacent to the current partition, it tries to keep the subpartitions connected. Note that the neighborhood information of each adjacent vertex needs to be available in order to select the next vertex.

Another "greedy" algorithm is Farhat-algorithm [54] which is a graph-growing algorithm but chooses the starting vertices of each partition in a greedy way. The original algorithm works on nodes and elements of a FEM mesh. The adapted version for graphs works by growing partitions by selecting unmarked neighbors of currently marked vertices and adding them to the current partition. One can choose a pseudo-peripheral vertex as the first vertex. The algorithm adds all unmarked neighbors of the vertices in the current partition to the current partition until the partition is big enough. It then proceeds to the next partition in the following way. Among all vertices chosen for last partition, the one with least nonzero number of unmarked neighbors is chosen and all these neighbors are marked and added to the current partition. It then follows the iterative method where all unmarked neighbors of vertices of the current partition are added to the current partition until it is big enough and so on.

The graph-growing algorithms are very fast and are able to divide the graph into the desired number of partitions directly, avoiding recursive bisection. As a result, their running time is typically independent of the number of desired subpartitions. Although fast, the quality of the partitions is not always as good and the last subpartitions tend to be disconnected when partitioning complicated graphs into several subpartitions [51]. Given a predefined size of "big enough," clusters often tend to be of the same size which may not result in the best partitions of a graph from a qualitative point of view. They are also sensitive to the choice of the starting vertex. But one can run the process with different starting vertices and choose the best result since these algorithms are fast.

17.6.1 Kernighan-Lin Algorithm

One of the earliest graph-partitioning algorithms is the Kernighan-Lin algorithm [90], often abbreviated as K/L. It was originally developed to optimize the placement of electronic circuits onto printed circuit cards to minimize the number of connections between cards. The K/L algorithm does not create partitions, rather *improves* them iteratively. The original idea was to take random partitions and apply K/L to them.

Let us introduce some notations before explaining the algorithm. It is more convenient to describe the algorithm on a graph with weighted edges. Let $(\mathcal{V}, \mathcal{E}, \mathcal{W}_{\mathcal{E}})$ be a graph with given subpartitions A and B such that $\mathcal{V}_{\mathcal{A}} \dot{\cup} \mathcal{V}_{\mathcal{B}} = \mathcal{V}$. The *diff-value* of a vertex v is defined to be the amount the cut-size will decrease when it is moved to the other partition. That is, for $v \in \mathcal{V}_{\mathcal{A}}$:

$$diff(\mathcal{V}) = diff(v, \mathcal{V}_{\mathcal{A}}, \mathcal{V}_{\mathcal{B}}) := \sum_{b \in \mathcal{V}_{\mathcal{B}}} w_e(e_{v,b}) - \sum_{a \in \mathcal{V}_{\mathcal{A}}} w_e(e_{v,a})$$

It is obvious that moving a vertex from one partition to the other changes only the vertex's diff-value and the diff-values of its the neighbors. The *gain-value* of a pair of vertices is the change in cut-size if the pair is swapped.

Lemma 17.6.1 *For two partitions $\mathcal{A}$ and $\mathcal{B}$ and $a \in \mathcal{V}_{\mathcal{A}}$ and $b \in \mathcal{V}_{\mathcal{B}}$, if a and b are interchanged, the gain is*

$$gain(a,b) = gain(a,b,\mathcal{V}_{\mathcal{A}}, \mathcal{V}_{\mathcal{B}}) := diff(a) + diff(b) - 2w_e(e_{a,b})$$

Proof 3 *Let c be the cost due to all connections between $\mathcal{A}$ and $\mathcal{B}$ that do not involve a or b. Then,*

$$T = c + \sum_{y \in \mathcal{V}_{\mathcal{B}}} w_e(e_{a,y}) + \sum_{x \in \mathcal{V}_{\mathcal{A}}} w_e(e_{b,x}) - w_e(e_{a,b})$$

Let T' be the new cost after swapping a and b. Then, we have:

$$T' = c + \sum_{x \in \mathcal{V}_{\mathcal{A}}} w_e(e_{a,x}) + \sum_{y \in \mathcal{V}_{\mathcal{B}}} w_e(e_{b,y}) + w_e(e_{a,b})$$

Now, gain is the change in cost:

$$\begin{aligned} T - T' &= \sum_{y \in \mathcal{V}_{\mathcal{B}}} w_e(e_{a,y}) - \sum_{x \in \mathcal{V}_{\mathcal{A}}} w_e(e_{a,x}) \\ &\quad + \sum_{x \in \mathcal{V}_{\mathcal{A}}} w_e(e_{b,x}) - \sum_{y \in \mathcal{V}_{\mathcal{B}}} w_e(e_{b,y}) \\ &\quad - w_e(e_{a,b}) - w_e(e_{a,b}) \\ &= diff(a) + diff(b) - 2w_e(e_{a,b}) \end{aligned}$$

■

Let us describe the algorithm:

Algorithm 37 Kernighan-Lin

Given two partitions $\mathcal{V}_{\mathcal{A}}, \mathcal{V}_{\mathcal{B}}$
  Compute diff-value for all vertices
  Unmark all vertices
  Let, k_0 = cut-size
  for $i = 1$ to $\min(|\mathcal{V}_{\mathcal{A}}|, |\mathcal{V}_{\mathcal{B}}|)$ **do**
    Find the pair $((a_i, b_i) \mid a_i \in \mathcal{V}_{\mathcal{A}}$ & $b_i \in \mathcal{V}_{\mathcal{B}})$ with biggest gain among all unmarked vertices
    Mark a_i and b_i
    for each neighbor v of a_i or b_i **do**
      Update diff(v) assuming a_i and b_i has been swapped, i.e.,

$$diff(v) := diff(v) + \{ \begin{array}{ll} 2w_e(e_v, a_i) - 2w_e(e_v, b_i) & \text{for} \quad v \in \mathcal{V}_{\mathcal{A}} \\ 2w_e(e_v, b_i) - 2w_e(e_v, a_i) & \text{for} \quad v \in \mathcal{V}_{\mathcal{B}} \end{array}$$

    end for
    $k_i = k_{i-1}$ - gain(a_i, b_i), that is, k_i would be the cut-size if $a_1, a_2, ..., a_i$ and $b_1, b_2, ..., b_i$ had been swapped
  end for
  Select the smallest j such that $k_j = min_i(k_i)$
  Swap the first j pairs. That is,

$$\mathcal{V}_{\mathcal{A}} = \mathcal{V}_{\mathcal{A}} - \{a_1, a_2, ..., a_j\} \cup \{b_1, b_2, ..., b_j\}$$
$$\mathcal{V}_{\mathcal{B}} = \mathcal{V}_{\mathcal{B}} - \{b_1, b_2, ..., b_j\} \cup \{a_1, a_2, ..., a_j\}$$

  Repeat until no further cut-size improvement is achieved

In each iteration, the algorithm swaps pairs of vertices to maximize the gain. This process is continued until all vertices of the smaller partition are swapped. One important fact is, the algorithm does not stop as soon as there are no more improvements to be made; rather it continues, even accepting negative gains in order to climb out of the local-minima.

[57] implemented the algorithm that takes $O(|\mathcal{E}|)$ time for one iteration. The reduction in time is achieved by choosing single nodes to be swapped as opposed to pairs. It also applies a bunch of other optimizations. There are many variations of the Kernighan-Lin algorithm [78, 86, 104] often trading execution time against quality or generalizations. Some examples are

- Limiting the number of swapping to a fixed number based on the observation that largest gains are achieved early on.
- Limiting the number of iterations.
- Evaluating diff-values of vertices near the boundary.

17.7 Agglomerative and Divisive Clustering

Agglomerative approaches start with considering each node as a separate cluster. In subsequent steps "similar" clusters are merged together until the intended number of clusters are left or until no two clusters are similar enough to be merged. Divisive approaches, on the other hand, work in the opposite direction. That is, they start with the whole graph as a single cluster and then subsequently split into smaller clusters [131]. Both approaches essentially produce a binary tree, known as a *dendrogram*, where leaves represent individual vertices and internal nodes are clusters.

The idea of *edge betweenness* was proposed to be used in a divisive algorithm for community detection by Newman and Girvan in [125]. *Edge betweenness* is defined in such a way that edges connecting different communities are more likely to receive a higher betweenness score than intracommunity edges. By removing edges with high betweenness scores, communities in the graph can be found. One simple edge betweenness measure is the *shortest path betweenness* that relies on the fact that shortest paths between vertices in different clusters are bound to pass through the few intercluster edges. As a result, those few intercluster edges will receive a high betweenness score. This has been proposed and used in the Girvan–Newman (GN) [68] algorithm. This algorithm removes edges with high *betweenness* scores to split the network. There are also variations such as *random-walk betweenness* and *current-flow betweenness* that assign the betweenness score based on different principles. The high level structure of the algorithm is as follows [131]:

Algorithm 38 Divisive Clustering Algorithm [105]

Input: Graph G, number k of clusters

repeat
  Calculate *betweenness* score of each edge using any suitable *betweenness* measure
  Remove the edge with maximum *betweenness* score
  Recalculate *betweenness* score of all remaining edges
until Desired number of clusters are obtained

Experiments showed that the choice of a particular *betweenness* measure is not very crucial as long as the *recalculation* step is executed [131]. The results for different *betweenness* measures differ only slightly. Although very intuitive, the algorithm is very *costly* in terms of computation. Computation of the *betweenness* score for all edges is $O(|\mathcal{V}||\mathcal{E}|)$ time operation. The whole algorithm takes $O(|\mathcal{V}|^3)$ time.

A greedy agglomerative clustering technique for optimizing *modularity* was proposed by Newman in [123]. Similar to agglomerative algorithms, this approach starts with each node as a different community. In subsequent steps, smaller communities are merged to form larger communities so that the *modularity* of the graph increases after the merge. *Modularity* is defined as the difference between the fraction of edges that fall within communities and the expected number of edges that fall within communities if they fall at random without regard for the community structures. [123] suggests that a value greater than 0.3 often indicates good community structure. More examples are given in [125]. The algorithm merges only communities that share at least one edge because otherwise it does not improve the *modularity*. Hence, this step has complexity $O(|\mathcal{E}|)$. Maintaining and updating an additional data structure that stores the fraction of shared edges between each pair of communities in the current partition has worst-case complexity of $O(|\mathcal{V}|)$. Since there are $|\mathcal{V}|-1$ mergers, the total complexity is $O|\mathcal{V}|^2)$. The complexity was improved by the use of data structures such as max-heaps to $O(|\mathcal{E}|d\log|\mathcal{V}|)$ by Clauset et. al. in [37]. Here d denotes the depth of the dendrogram describing the partitions found during the execution of the algorithm.

17.8 Spectral Clustering

Spectral clustering has become one of the most popular clustering methods in recent years. It is simple to implement and can be solved efficiently with standard linear algebra software. Spectral clustering does not use geometric information of the graph. It does not operate on the graph itself but on a mathematical representation of the graph. One advantageous feature of spectral clustering is that the decision about each vertex is taken based on a more global view of the problem. In the following sections we discuss spectral clustering in detail.

17.8.1 Similarity Graphs

Given a set of data points $x_1, x_2, ..., x_n$ and some notion of similarity $s_{i,j} \geq 0$ between all pairs x_i and x_j, a simple and nice way of representing data is in the form of *similarity graph* $\mathcal{G} = (\mathcal{V}, \mathcal{E})$. Each data point x_i is represented by a vertex v_i in the graph and two vertices v_i and v_j are connected if the similarity $s_{i,j}$ between corresponding data points x_i and x_j is greater than some threshold value. The edge is weighted by $s_{i,j}$. The clustering problem can be reformulated as the problem of partitioning the similarity graph so that the edges between different groups have very low weights.

Let $\mathcal{G} = (\mathcal{V}, \mathcal{E})$ be a weighted undirected graph, the weighted *adjacency matrix* of the graph is matrix $W = w_{i,j}$ for $i, j = 1, ..., n$. The *degree* of a vertex v_i is defined to be

$$d_i = \sum_{j=1}^{n} w_{i,j}$$

Because $w_{i,j} = 0$ if v_i and v_j are not connected, this sum runs over vertices adjacent to v_i. The *degree matrix* D is defined to be diagonal matrix with $d_1, ..., d_n$ on the diagonal. Let $\mathbb{F} = (f_1, ..., f_n)' \in \mathbb{R}^n$ as the vector with $f_i = 1$ if $v_i \in A$ and $f_i = 0$ otherwise. A few more notions are as follows:

$$W(A,B) := \sum_{i \in A, j \in B} w_{ij}$$

Given a subset $A \subset V$, the "size" of A is as follows:

$$\begin{aligned} |A| &:= Number\ of\ vertices\ in\ A \\ vol(A) &:= \sum_{i \in A} d_i \end{aligned}$$

Subset $A \subset \mathcal{V}$ of a graph is *connected* if there is a path between each pair of vertices in A such that no intermediate vertex is from $\mathcal{V} \backslash A$ *or* $\bar{A}$. Subset A is called a *connected component* if it is connected and there are no connections between vertices in A and $\bar{A}$. The *nonempty* sets $A_1, ..., A_k$ form a partition of the graph if $A_i \cap A_j = \emptyset$ and $\cup_{i=1}^{k} A_i = \mathcal{V}$.

17.8.2 Types of Similarity Graphs

Given a set of data points $x_1, x_2, ..., x_n$ and a pairwise similarity $s_{i,j}$ (or dissimilarity $d_{i,j}$) measure, there are several popular methods to convert these points into a similarity graph [105]. In this section we discuss several well-known methods for constructing a similarity graph from a set of points and the pairwise similarity/dissimilarity function.

ε-neighborhood graph: In this method, all points with pairwise distance smaller than ε are connected. Since the distance is smaller than ε for all connected pairs, weighting the edges usually does not convey much more information. Hence, these graphs are typically unweighted.

***k*-nearest neighbor graph:** In this approach, vertex v_i is connected with v_j if v_j is among k-nearest neighbors of v_i. However, since the neighborhood relationship is not symmetric, this definition leads to a directed graph. There are two standard ways to deal with this issue:

- The simplest solution is to just ignore the directions of the edges. In this method v_i and v_j are connected if v_j is among k-nearest neighbors of v_i **or** if v_i is among k-nearest neighbors of v_j. Resulting graph is called *k-nearest neighbor graph.*
- The second way is to connect v_i and v_j only if both v_j is in k-nearest neighbors of v_i **and** v_i is in k-nearest neighbors of v_j. The resulting graph is called *mutual k-nearest neighbor graph.*

After connecting the vertices, the edges are weighted by the similarity of the endpoints.

Fully connected graph: In this case all points are connected with each other and the edges are weighted by the similarity function $s_{i,j}$. An important issue is that since all pairs are connected, it is important that the similarity function reflects the local neighborhoods. A popular similarity function is the Gaussian similarity function $s(x_i,x_j) = exp(-||x_i - x_j||^2/(2\sigma^2))$. The parameter σ controls the width of the neighborhood and plays a role similar to that of ε in the ε-neighborhood graph. The paper [105] provides a discussion on the choice of similarity graph construction methods and the resultant cluster quality based on a toy graph.

17.8.3 Graph Laplacians

Graph Laplacian matrices are the principal tool for spectral clustering and there is a dedicated area of study for those matrices (see [36]). There are variants of graph Laplacians, and unfortunately, there is no unique definition of a matrix that exactly is "graph Laplacian" [105]. Let $\mathcal{G} = (\mathcal{V}, \mathcal{E}, \mathcal{W})$ be an undirected, weighted graph with weight matrix $\mathcal{W}$ with $w_{i,j} \geq 0$. While using eigenvectors of a matrix, we do not assume them to be normalized, i.e., constant vector $\mathbb{F}$ and $a\mathbb{F}$ for some $a \neq 0$ are the same eigenvectors. Eigenvalues will be ordered in ascending order respecting multiplicities.

17.8.3.1 Unnormalized Graph Laplacian

The unnormalized graph Laplacian matrix L is defined as

$$L = D - W$$

Some of the important properties of graph Laplacian matrices that are critical for spectral clustering are given below. Mohar gives an overview of the properties of the matrices in more detail in [116, 117].

Properties of graph Laplacian: Matrix L has the following properties:

- for every vector $f \in \mathbb{R}^n$,

$$f'Lf = \frac{1}{2} \sum_{i,j=1}^{n} w_{ij}(f_i - f_j)^2$$

- L is *symmetric* and *positive semi-definite.*
- The smallest eigenvalue of L is 0 with the corresponding eigenvector $\mathbb{F}$ (constant one vector).
- L has n nonnegative, real-valued eigenvalues $0 = \lambda_1 \leq \lambda_2 \leq ... \leq \lambda_n$.

Proof of these properties is available in [105]. It is noticeable that self-edges in a graph do not change the corresponding graphs Laplacian. The unnormalized graph Laplacian, its eigenvectors and eigenvalues can describe many properties of graphs [116, 117]. One such property that is important for spectral clustering is as follows:

Number of connected components and spectrum of L: For an undirected graph $\mathcal{G}$ with nonnegative weights, the *multiplicity* k of the *eigenvalue 0 of* L equals the number of connected components in the graph. Proof of this proposition can be found in [105].

17.8.3.2 Normalized Graph Laplacians

The following two matrices are known as normalized graph Laplacians [36]:

$$\begin{aligned} L_s &:= D^{-1/2}LD^{-1/2} = I - D^{-1/2}WD^{-1/2} \\ L_{rw} &:= D^{-1}L = I - D^{-1}W \end{aligned}$$

L_s is a symmetric matrix while L_{rw} is related to *random walk*. Some of the properties of normalized graph Laplacians are as follows:

- for every vector $f \in \mathbb{R}^n$,

$$f'L_s f = \frac{1}{2}\sum_{i,j=1}^{n} w_{ij}\left(\frac{f_i}{\sqrt{d_i}} - \frac{f_j}{\sqrt{d_j}}\right)^2$$

- λ is an eigenvalue of L_{rw} with eigenvector u iff λ is an eigenvalue of L_s with eigenvector $w = D^{1/2}u$.

- λ is an eigenvalue of L_{rw} with eigenvector u iff the generalized eigenproblem $Lu = \lambda Du$ is solved by λ and u.

- 0 is an eigenvalue of L_{rw} with constant one eigenvector $\mathbb{F}$ and an eigenvalue of L_s with eigenvector $D^{1/2}\mathbb{F}$.

- Both L_s and L_{rw} are positive semidefinite with n nonnegative real-valued eigenvalues $0 = \lambda_1 \leq \lambda_2 \leq \ldots \leq \lambda_n$.

For proofs of these properties, please see [105].

Number of connected components and spectra of L_s and L_{rw}: Similar to unnormalized graph Laplacians, for (non-negative) weighted undirected graph $\mathcal{G}$ the multiplicity k of the eigenvalue 0 of both L_s and L_{rw} equals the number of connected components in the graph.

17.8.4 Spectral Clustering Algorithms

Let $x_1, x_2, \ldots, x_n$ be n data points and $s_{ij} = s(x_i, x_j)$ be a pairwise similarity function between points. We assume s_{ij} is symmetric and nonnegative. Let $S = (s_{ij})_{i,j=1\ldots n}$ be the similarity matrix. The following algorithm describes the unnormalized spectral clustering.

Algorithm 39 Unnormalized Spectral Clustering [105]

Input: Similarity Matrix $S \in \mathcal{R}^{n\times n}$, number k of clusters
  Construct Similarity Graph (using a method from 17.8.2)
  Let $\mathcal{W}$ be the weighted adjacency matrix
  Calculate unnormalized Laplacian L
  Compute first k eigenvectors $l_1, l_2, \ldots, l_k$ of L
  Let $U \in \mathbb{R}^{n\times k}$ be the matrix containing vectors $l_1, l_2, \ldots, l_k$ as columns
  Let $y_i \in \mathbb{R}^k$ be the ith row of U for $i = 1, 2, \ldots, n$
  Cluster points y_i for $i = 1, 2, \ldots, n$ into k clusters $C_1, C_2, \ldots, C_k$ using k-means algorithm
Output clusters $A_1, A_2, \ldots, A_k$ with $A_i = \{j | y_j \in C_i\}$

For normalized graph Laplacians, there are two versions of normalized spectral clustering. Shi and Malik [151] proposed the following normalized spectral clustering algorithm [105]. Since the algorithm uses *generalized eigenvectors* of L that correspond to eigenvectors of L_{rw}, it is called normalized spectral clustering.

Algorithm 40 Normalized Spectral Clustering [151]

Input: Similarity Matrix $S \in \mathcal{R}^{n\times n}$, number k of clusters
Construct Similarity Graph (using a method from 17.8.2)
Let $\mathcal{W}$ be the weighted adjacency matrix
Calculate unnormalized Laplacian L
Compute first k generalized eigenvectors $l_1, l_2, ..., l_k$ of the generalized eigenproblem $Lu = \lambda Du$
Let $U \in \mathbb{R}^{n\times k}$ be the matrix having vectors $l_1, l_2, ..., l_k$ as columns
Let $y_i \in \mathbb{R}^k$ be the ith row of U for $i = 1, 2, ..., n$
Cluster points y_i for $i = 1, 2, ..., n$ into k clusters $C_1, C_2, ..., C_k$ using k-means algorithm
Output clusters $A_1, A_2, ..., A_k$ with $A_i = \{j|y_j \in C_i\}$

The following algorithm by Ng et al. [126] uses matrix L_s instead of L_{rw}.

Algorithm 41 Normalized Spectral Clustering [126]

Input: Similarity Matrix $S \in \mathcal{R}^{n\times n}$, number k of clusters
Construct Similarity Graph (using a method from 17.8.2)
Let $\mathcal{W}$ be the weighted adjacency matrix
Calculate normalized Laplacian L_s
Compute first k eigenvectors $l_1, l_2, ..., l_k$ of L_s
Let $U \in \mathbb{R}^{n\times k}$ be the matrix having vectors $l_1, l_2, ..., l_k$ as columns
Normalize rows of U to norm 1 to obtain matrix $T \in \mathbb{R}^{n\times k}$, i.e., set $t_{ij} = u_{ij}/(\sum_k u_{ik}^2)^{1/2}$
Let $y_i \in \mathbb{R}^k$ be the ith row of U for $i = 1, 2, ..., n$
Cluster points y_i for $i = 1, 2, ..., n$ into k clusters $C_1, C_2, ..., C_k$ using k-means algorithm
Output clusters $A_1, A_2, ..., A_k$ with $A_i = \{j|y_j \in C_i\}$

The three spectral clustering algorithms given here are rather similar, differing only in their use of the graph Laplacians. However, the critical part is the conversion of data points x_i to points $y_i \in \mathbb{R}^k$, and because of the properties of graph Laplacians this change is useful.

17.9 Markov Clustering

Markov Clustering (MCL), proposed by Stijn van Dongen, clusters graph by simulation of stochastic flows on a graph [45]. MCL is based on iterative application of two operations on the transition probability matrix or stochastic flow matrix of the graph: *Expand* and *Inflate*. *Expand(M)* is simply a matrix–matrix multiplication as follows:

$$M_{Expand} = M \times M$$

while *Inflate(M,r)* corresponds to raising each element of matrix M to its rth power and normalizing the columns to sum to 1. r is the *inflation* parameter ($r > 1$) and is typically set to 2.

$$M_{Inflate}(i,j) = \frac{M(i,j)^r}{\sum_k M(k,j)^r}$$

These two steps are followed by a *Prune* step that prunes away smaller values in each column (smaller with respect to the values in respective columns, of course). Remaining values are renormalized to make sure each of the columns of the matrix sums to 1, and hence, the matrix is a stochastic flow matrix. Starting with the initial flow matrix, the whole process is iterated until convergence. *Expand* step spreads flow out of a vertex to new vertices. This enhances the intracluster flows as there are more paths between two nodes within the same cluster than between two nodes in two different clusters [150]. Nonlinearity is brought into the process through the *Inflate* operation that strengthens the intracluster flow and weakens the intercluster flow. The process sets up a positive feedback loop forcing all nodes within a tightly linked group to flow to one *attractor* node within the group [131]. MCL is particularly popular within the bioinformatics community because of its effectiveness in clustering protein–protein interaction networks [19, 102]. The pseudocode for the MCL algorithm is as follows.

Algorithm 42 Markov Clustering, MCL

```
A := A + I // Add self loop to the vertices
M := AD^{-1} // M is the canonical flow matrix
repeat
    M := M_Expand := Expand(M)
    M := M_Inflate := Inflate(M, r)
    M := Prune(M)
until M converges
Interpret M as the resulting clustering
```

MCL has two major shortcomings [148]. First, it is slow because it involves a matrix–matrix multiplication. Especially, during the first few iterations when the flow matrix is *dense*, the *Expand* step becomes very time consuming. Second, it has a tendency to produce *imbalanced* clusters, e.g., *singleton* clusters or clusters with few nodes, or to produce one very big cluster.

17.9.1 Regularized MCL (RMCL): Improvement over MCL

Recently there have been variants of MCL in an effort to address major drawbacks of MCL with respect to *scalability* and *imbalanced clustering*. Regularized MCL (RMCL) [148, 150] has succeeded in solving the problem associated with imbalanced clustering, especially the problem related to producing singleton clusters. Satuluri and Parthasarathy [148] observed that the reason behind MCL's producing too many clusters is the fact that it allows columns of pairs of neighboring nodes in flow matrix M to diverge significantly. This happens because MCL uses the adjacency matrix of the input graph only at the start of the algorithm to initialize the flow matrix and, in the iterative step, uses only the current flow matrix, which allows MCL to "overfit." RMCL addresses the problem by *regularizing* (or *smoothing*) the flow distributions with respect to neighbors, i.e., by taking into account the neighborhood structure in each *Expand* step. Essentially it changes the *Expand* ($M := M \times M$) step of MCL into a *Regularize* ($M := M \times M_G$) step where M_G is the canonical transition matrix of the graph. That is,

$$M_{Regularize} = M \times M_G$$

The pseudocode of the algorithm is given in Algorithm 43.

Algorithm 43 Regularized Markov Clustering, RMCL [148]

$A := A + I$ // Add self-loop to the vertices
$M := AD^{-1}$ // M is the canonical flow matrix
repeat
  $M := M_{Regularize} := M \times M_G$
  $M := M_{Inflate} := Inflate(M, r)$
  $M := Prune(M)$
until M converges
Interpret M as the resulting clustering

17.10 Multilevel Partitioning

The multilevel approach has been successfully applied to a variety of problems [165] and has resulted in fast and high quality results. This powerful framework can also be applied to graph partitioning and often produces fast, accurate, and high quality partitions. The basic idea is to coarsen the graph successively to get a small enough graph, partition the small graph, and use the result to successively project the partition back to the original graph [131]. A schematic of the multilevel framework is given in Figure 17.1. Many well-known graph partitioning algorithms use multi-level graph partitioning techniques. Some of them are multilevel spectral clustering [13], **Metis** (optimizing K/L objective function) [86], **Graclus** (optimizing normalized and other weighted cuts) [42] and **MLR-MCL** [148].

There are three main components of any multilevel graph partitioning approach. They are briefly discussed below:

1. **Coarsening:** The goal of coarsening is to get a smaller graph that retains the most important

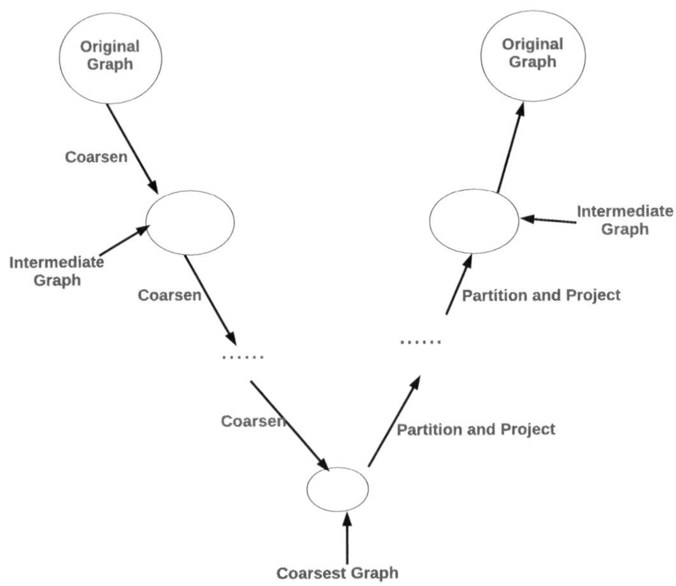

FIGURE 17.1: Multilevel graph partitioning.

characteristics of the original graph. Often this step is applied repeatedly until the resultant graph reduces to a manageable "small enough" size. As a consequence, partitioning can be applied on the small graph very efficiently. One popular method is to construct a *maximal matching* on the graph [105]. A *matching* is a subset of edges ($\mathcal{M} \subset \mathcal{E}$) such that no two edges share the same endpoint. A *matching* is *maximal* if no edge can be added to the matching. Once a *matching* has been constructed, vertices at the ends of each edge are collapsed into a super node in the coarsened graph. Specifically, for edge $e_{i,j} \in \mathcal{M}$, vertices v_i and v_j are collapsed into a single vertex. The weight of the new vertex is $w(v_i) + w(v_j)$ and the neighbors are combined neighbors of v_i and v_j [105].

Different techniques exist for construction of a *maximal matching*. One can use randomized methods for coarsening graphs fairly quickly. A simple *randomized matching* (RM) is given in [78] which randomly selects eligible edges until the matching is maximal. A slight modification of RM which is known as *heavy edge matching* (HEM) has been proposed in [86]. The basic idea is to try to reduce the total edge weight of the coarser graph which results in reduced cut-size in coarser as well as the original graph. However, one problem associated with this approach is that it can miss some heavy weight edges as pointed out in [105]. Another modification was proposed in [72] by Gupta that suggests sorting the edges by weight and then choosing the heaviest permissible edge. This is called *heaviest edge matching*. Since sorting is an additional cost for this approach, it is often applied in the later stages of the coarsening process.

2. **Partitioning the Coarsest Graph:** Once the graph is small enough, it is simple to apply any partitioning algorithm on the coarsest graph. One may even try several methods or the same randomized method several times. Some of the popular choices include graph-growing, spectral clustering or simple Kernighan-Lin algorithm with random starting partitions. Note that since the size of the graph is small, it is feasible to apply a slow method like spectral clustering in order to get high quality output [105].

3. **Uncoarsening and Projecting Up:** In this step the partition of the coarsest graph is used to initialize partitions on finer (bigger) graph. This is called "projecting up". Since each vertex in finer graph can be traced to a vertex in the coarser graph, assigning the vertices of finer graph to appropriate partitions is simple once we have the partition on the coarser graph. The finer connectivity given by uncoarsening is used to refine the partition. This is done using some local search, or some variants of Kernighan-Lin algorithm (Metis [86]) or weighted kernel k-means (Graclus [42]) [131, 105]. This step is continued until the original graph has been reached.

Metis [86] supports different matching schemes for the coarsening step including RM and HEM. For partitioning the coarsest graph, Metis implements four different schemes: three based on graph-growing heuristics and one based on spectral bisection. During the uncoarsening phase, Metis uses the K/L algorithm [57] as a partition refinement algorithm. Since the K/L algorithm can take many iterations before converging, Metis also implements *Greedy Refinement (GR)* that performs only a single iteration and a *Boundary Kernighan-Lin Refinement (BKLR)* that allows swapping only vertices that are along the boundary of the bisection. More detailed discussion about Metis can be found in [86].

Graclus [42] generalizes the coarsening process of Metis to a *max-cut coarsening* procedure to make it effective for a wider range of objectives. For clustering the coarsest graph, three base clustering approaches are explored—the extremely efficient region-growing algorithm of Metis [86], the spectral clustering algorithm, and the bisection method. In the Refinement step, Graclus uses weighted kernel k-means. A more detailed discussion about the algorithm can be found in [42].

Multilevel Regularized MCL (MLR-MCL) [148] combines RMCL (Section 17.9.1) with the powerful multilevel framework to solve the *scalability* issue associated with MCL. The coarsening step involves finding a *matching* on the graph and collapsing incident vertices on each edge of the matching into a super node. RMCL is run on the resulting coarsest graph for a few (4 or 5) iterations. The projection step involves plotting the flows into vertices on a coarser graph to the vertices in a refined graph and running RMCL for a few iterations on the refined graph. This continues until the original graph has been reached. Finally, RMCL is run on the original graph until convergence. A detailed explanation of the algorithm can be found in [148].

17.11 Local Partitioning Algorithms

Local algorithms solve the partitioning problem for a given vertex or a set of vertices without looking at the whole graph. These algorithms are important in the context of large graphs when one is interested in a few vertices as opposed to all the vertices in the graph. They are interesting because the complexity of the algorithms, to a large extent, no longer depends on the size of the graph, but rather on the size of the *solution*. The intuition being that random walks from inside a well-connected group of nodes will not mix well enough since the cluster boundary acts as a bottleneck that prevents the probability of easily going out of the cluster [131].

Such local clustering using random walks has been described in [156, 154]. Let $p_{t,u}$ be the probability distribution of a t-step random walk starting at vertex u. Let Γ be the permutation of the vertices of the graph according to descending order of degree-normalized probability for each t. That is,

$$\frac{p_t(\Gamma_i)}{d(\Gamma_i)} \geq \frac{p_t(\Gamma_{i+1})}{d(\Gamma_{i+1})}$$

Let us define sweep set $S^t_j = \Gamma_1, ..., \Gamma_j$. Given all random walks within a component converge to the same stationary distribution [131], let Ψ_u be the final stationary distribution of the random walk. The main theoretical result exploited says that either the difference between $p^t(S^t_j)$ and $\Psi_u(S^t_j)$ is small or there exists a cut with low conductance among sweep sets. Thus, by checking conductance of sweep sets S^t_j at each step t, one can discover clusters of low conductance. This work was extended for seed sets as opposed to seed vertex by Andersen and Lang [6]. It has been shown that the local clustering approach can recover the original community for real datasets with a random subset of vertices belonging to a known community as seeds.

Spielman and Teng's algorithm [156] was improved by Andersen et al. [5] by the simulation of *random walks with restarts* (Personalized PageRank). Another local graph clustering algorithm is Nibble [155].

Improving partitions by Flow-Based Postprocessing: Partitions of a graph can be improved using algorithms for computing maximum flow in flow networks. Flake et al. [58] use a focused crawler to obtain an approximate community and then set up a max-flow/min-cut problem that produces the actual community in order to discover web communities. A strategy for improving conductance of any arbitrary cut of the graph was discussed by Lang and Rao in [97]. For a given cut $(C, \bar{C})$ of a graph, their algorithm finds the best improvement among all cuts $(C', \bar{C'})$ such that C' is a strict subset of C. They formulate a new instance of the max-flow problem such that the polynomial time solution to this problem can be used to find C' with lowest conductance among all subsets of C. This method is referred to as Max-flow Quotient-cut Improvement (MQI). They use Metis+MQI recursively to bipartition the input graph. In a step, Metis is used first to bipartition the graph and then MQI is used to improve the partition. MQI can improve partitions given by local clustering as well [6].

Community Discovery via Shingling: Web documents can be clustered using *shingles* and *fingerprints* (also known as *sketches*) as proposed in [18]. A length-s shingle of an object is s of all parts of the object. For instance, a length-s shingle of a document is a contiguous subsequence of length s contained in the document. A *sketch* is a fixed-size subset of all shingles with a specific length. The property that makes *sketch* an object's *fingerprint* is the fact that the similarity between two objects' sketches is approximately equivalent to the similarity between the objects themselves. Here, the similarity refers to *Jaccard similarity*, i.e.,

$$Sim(A,B) = \frac{|A \cap B|}{|A \cup B|}$$

Shingling has been applied for extracting dense communities from large graphs [65]. In this work, each node's outgoing links are used to get the first-level shingling where each vertex v is associated with a certain number of shingles c, each of which includes n nodes selected from nodes to which v points. After this an inverted index is built that contains each first-level shingle and a list of vertices associated with the shingle. Second-level shingles (also known as meta-shingles) and sketches are built from first-level shingles. Two first-level shingles are considered as relevant if they share at least one meta-shingle. A graph is constructed where nodes stand for first-level shingles and edges indicate the meta-shingles relationship. Clusters of first-level shingles refer to connected components in this new graph. Communities are extracted by mapping first-level shingles clusters back to original nodes and including associated common meta-shingles. The algorithm is applicable to bipartite and directed as well as undirected graphs and is very efficient and scalable [131].

Different Definitions of Communities: *Community* in a graph is most commonly defined as a subset of vertices well connected internally and loosely connected to the rest of the graph. But there have been alternative characterization and definitions of *community* in the literature.

One recent community structure has been proposed by Asur and Parthasarathy in [8]. They define *viewpoint neighborhoods* as a group of influential and salient nodes from the viewpoint of a single node or a subset of nodes in the graph. *Viewpoint neighborhood* basically refers to the clusters of nodes local to the node or the subset of nodes. The authors use activation spread models in their algorithm to extract viewpoint neighborhoods. These models are general enough to incorporate various notions of influence and salience. Viewpoint neighborhood provides a new and exciting tool for analysis of large graphs.

Another class of communities, as found by Leskovec et al. [101], is *whiskers* which is a group of nodes that are connected to the rest of the graph by only one edge (similarly, groups of nodes connected to the rest of the graph with 2 edges are called 2-*whiskers*). The authors find that according to the measure of *conductance*, some of the best communities in a wide variety of real-world networks are simply *whiskers*. They propose a core-and-whiskers model for structures of networks where most networks have a core part surrounded by whiskers connected to the rest of the graph by only one or two edges.

17.12 Hypergraph Partitioning

A *hypergraph* is a generalization of a graph where an edge, called *hyperedge*, can connect more than two vertices. Hypergraphs appear naturally in many problems such as Boolean SATisfiability problem and circuit layout [129]. Hypergraphs are also popular in the areas of workload partitioning in parallel processing, restructuring sparse matrices [27], load balancing and scientific computing [28], scheduling of batch-shared I/O tasks in Grid [91] as well as in automatic management of memory hierarchies in global address space programming [94].

Formally, a hypergraph is defined as $\mathcal{H} = (\mathcal{V}, \mathcal{N})$ where $\mathcal{V}$ is the set of vertices and $\mathcal{N}$ is the set of nets (hyperedges) among those vertices. Weights can be easily associated with the vertices of the graph. Every net $n_j \in \mathcal{N}$ is a subset of vertices in the graph. That is, $n_j \subseteq \mathcal{V}$. Vertices in a net are called its *pins*.

A k-way partitioning of a hypergraph partitions the vertices into k disjoint nonempty partitions. Typically a solution to the problem tries to minimize a cost function. One standard cost function is *net cut* which is simply the sum of weights of the edges that span more than one partition. Constraints are usually added to the partitioning problem. Some such constraints are *fixed constraint* where certain vertices are fixed in their partitions and *balance constraint* where the total vertex weight in each partition is balanced [129]. Optimal hypergraph partitioning with *balanced constraint* is known to be NP-hard [63]. However, heuristic algorithms with near-linear runtime have been developed. Some move-based heuristics for k-way partitioning are given in [22, 57, 90] and some refinements are proposed in [104, 47, 4, 84, 48].

Clustering a hypergraph, in general, refers to the process of finding a coarser hypergraph from the input hypergraph by merging vertices into larger groups (clusters). Weight of a cluster is simply the sum of weights of the vertices within the cluster [129].

Performance is always a big concern for hypergraph partitioning algorithms because in many of their applications the size of the input graph increases considerably every year. For instance, in VLSI design the number of transistors grows exponentially following Moore's law, and thus, the algorithms applicable to them must scale for the larger inputs to be used effectively. In the following section, we give a brief review of different partitioning themes.

Exhaustive Search: These methods produce optimal partitions, but have an exponential asymptotic complexity. Evaluation of each solution is the bottleneck and can be sped up by incrementally evaluating the cost objective. Iteratively updating the cut of a solution when a vertex is moved is quite straightforward [25]. However, the complexity grows *exponentially* in the number of vertices and makes it infeasible for large, practical cases.

Branch and Bound: Intelligent pruning of the search space can be applied to exhaustive methods in order to improve their scalability. This technique is known as Branch and Bound (B&B). It performs a recursive *depth first search* on the tree of partial assignment (of vertices to partitions) and finds the best partition for the next unassigned vertex. In the worst case it can search the entire solution space resulting in an exponential time complexity, but maintains optimality by bounding away suboptimal results. Different constraints can be used for pruning and bounding [129].

Fiduccia–Mattheyses Heuristic: Even B&B methods take an impractical amount of time to produce the optimal output for any realistic input. The Fiduccia–Mattheyses (FM) heuristic is an amortized near-linear time heuristic for iterative improvement of hypergraph partitions. The FM algorithms prioritize *moves* by *gain*. Here *move* is the change of assignment of a vertex to a partition and *gain* is the corresponding change to the cost function. The algorithm runs in passes where each vertex is moved exactly once. The initial solution is usually produced by some randomized algorithm and then passes are continued until convergence.

Multilevel FM Framework: The multilevel framework provides the best known results for large scale hypergraphs [129]. It involves three major steps as does any multilevel algorithm, namely, coarsening, partitioning, and uncoarsening. MLFM is one of the best techniques for partitioning practical sized hypergraphs.

Other techniques: Several other techniques have been proposed in the literature for partitioning hypergraphs [44, 50, 174]; most have some drawbacks that make it impractical to use them for real applications. Meta-heuristics such as simulated annealing and tabu search often produce better quality results at the cost of an impractical increase in runtime [50]. Other techniques have some constraints associated with them that limit their applicability in some practical cases. One such example is the spectral techniques [44]. Yang and Wong's [174] method relies on min-cut max-flow algorithms and efficient network flow algorithms. These are polynomial-time algorithms that cannot take the balance constraint into account which results in costly trial-and-error in flow-based partitioning [129].

17.13 Emerging Methods for Partitioning Special Graphs

The abundance of data these days is spawning newer challenges in the context of network/graph clustering. Graphs with millions of nodes and billions of edges are commonplace and even larger graphs with a billion nodes are emerging. These novel issues have led to the formulation of special types of graphs and newer challenges. In addition to having large number of vertices and edges, modern graphs come with other interesting properties. For example, some of the graphs are also *dynamic*, in the sense that they are evolving continuously. Events such as addition and removal of nodes, edges, and communities take place continuously in these graphs. Other issues include the directionality of edges, availability of rich content information, etc.

Given these huge graphs and their special characteristics, we are now facing unprecedented challenges of processing and analyzing them. With that in mind, some of the newer and emerging challenges, existing approaches and methods related to processing of modern, special graphs are briefly outlined below. It should be noted that this is by no means a comprehensive list of emerging challenges, techniques, and methods. We briefly touch upon some of the major challenges and existing approaches to dealing with these problems. Note that, these techniques are still under active research and lots of innovations have yet to be made.

17.13.1 Bipartite Graphs

Bipartite graphs are widely used in many graph applications. A bipartite graph (or bigraph) is a special type of graph whose vertices can be divided into two disjoint sets such that no edge connects two vertices from the same set. That is, for graph $\mathcal{G} = (\mathcal{V}, \mathcal{E})$ the set of vertices $\mathcal{V}$ is divided into two sets $\mathcal{V}_1$ and $\mathcal{V}_2$ such that $\mathcal{V}_1 \cap \mathcal{V}_2 = \emptyset$ and each edge $e \in \mathcal{E}$ connects vertices (v_1, v_2) where $v_1 \in \mathcal{V}_1$ and $v_2 \in \mathcal{V}_2$. Many practical graphs can be represented as a bipartite graph. For example, in a text corpus terms and documents can be represented as a bipartite graph where one set of vertices represents the documents while the other represents terms that appear in the corpus. The edges in the graph represents the co-occurrence of the term and the document in the corpus. Other examples include graphs representing the relationship of buyers and items in a departmental store or reviewers and movies in a movie recommendation system. Bipartite graphs, having been used in the literature to solve clustering problems, are extremely important in *Matching Systems* partitioning.

Zha et al. [176] uses bipartite graph partitioning for data clustering by minimizing the normalized sum of edge weights between unmatched pair of vertices in the underlying bipartite graph. The formulation of the bipartite graph naturally leads to partial Singular Value Decomposition (SVD) problems for an underlying edge weight matrix. This work computes a partial SVD of the associated edge weight matrix of the graph to obtain an approximate solution to the minimization problem.

Dhillon [41] uses bipartite graph partitioning for clustering documents and words *simultaneously*. He models the document collection as bipartite graph between documents and words and use spectral co-clustering for bipartitioning. Using the theoretical property of the proposed algorithm, he demonstrates that the algorithm provides an optimal solution to a real relaxation of the NP-complete co-clustering objective.

Qiu [138] uses bipartite graphs to model images and their content descriptors in image databases. A partitioning algorithm is developed for co-clustering images and their content description so that each image cluster is automatically associated with the set of features that best describe the image content. This work develops a Hopfield Network [79]-based solution for partitioning a bipartite graph.

Fern and Brodley [56] uses bipartite graph partitioning to solve the cluster ensemble problems of how to combine multiple clusterings to yield superior clustering result. This work solves the problem by proposing a *lossless* reduction that constructs a bipartite graph from a given cluster ensemble that models both instances and clusters of the ensemble as vertices of the graph. Consequently, this graph is partitioned to solve the problem.

17.13.2 Dynamic Graphs

So far we have discussed clustering algorithms that assume that the underlying network is stable and static. But in many real networks this assumption cannot be made because the networks are changing continuously. For instance, social networks keep changing, and hence, the assumption of a static structure is not very practical. As a result, several questions arise such as: How do the communities evolve over time? How are communities formed? How should existing algorithms be modified to accommodate these dynamic networks? How persistent are the communities? What is the temporal pattern in the network? Some of the recent work has started looking at these problems. We give an overview of some of the representative works that have looked into these emerging questions.

An event-based approach that provides a structured way to reason about how communities and individuals in these networks evolve over time and what characterizes their behavior was proposed by Asur et al. [10] where typical events involving communities are *form, dissolve, merge, split, continue, k-merge, k-split*, etc., and events involving individuals are *join, appear, disappear*, etc. The authors demonstrate that their framework can effectively detect behavioral indices such as *stability*, *influence*, and a *diffusion model* and use that to analyze real-life evolving networks "incrementally." This model can predict future behaviors such as collaboration between groups as well as identify influential nodes. One important capability of this model is that semantic content can be integrated into it very naturally.

A simple approach to dealing with dynamic networks is to treat each snapshot of the network independently and apply conventional clustering algorithms on each. But this may result in unwanted fluctuations in community structures between snapshots. One extreme example is given in [29] where an "optimal" clustering would cause the resultant clusters to change radically and using a consistent feature would provide consistent clustering while providing arbitrarily close to optimal answers. This problem was handled by constructing temporal slices of the network and discovering community in individual slices to detect temporal change in community structure between slices. Berger-Wolf and Saia [16] took partitions of nodes at each timestamp as input in order to find a *metagroup* of sequence of similar groups. They defined three extreme metagroups as *most persistent, most stable*, and *largest* and discussed algorithms for extracting them. Given the affiliation of each individual in each timeslice, Tantipathananandh et al. [164] tried to identify "true"community affiliations of them in a dynamic setting by formulating this as a combinatorial optimization problem and proving it to be NP-hard. They solve the problem using approximate greedy heuristics and dynamic programming [131].

Another interesting approach to dynamic analysis is referred to as *evolutionary clustering* as proposed by Chakrabarti et al. [29]. This approach takes a holistic view of the community discovery across time-slices by constraining division in a time-slice from diverging too much from previous time-slices. The critical contribution is, as opposed to first extracting communities on snapshots and finding connections among them across snapshots, that it considers *quality* of snapshot and *history cost* as a whole [131]. It allows the study of community structure and its evolution at the same time as well as the compromise between these two by linear combination of snapshot *quality* and *history cost*. Different clustering techniques have also been adapted for the framework.

Use of the *Minimum Description Length (MDL)* principle was proposed by Sun et al. [160] as an alternative approach to clustering dynamic graphs. Graphs of consecutive timestamps are grouped into graph stream segments divided by change-points that represent drastic change in network structure. Minimizing the cost of this solution, however, was proved to be NP-hard, leading to a greedy algorithm called GraphScope based upon alternating minimizations. It decides when to start a new stream segment and looks for ways of finding communities among the snapshots in a single segment. GraphScope does not require any input parameter.

In order to extend *spectral* clustering to dynamic networks, Chi et al. [33] propose two frameworks, *Preserving Cluster Quality* (PCQ) and *Preserving Cluster Membership* (PCM), to measure history cost. PCQ defines how well the partition, at a particular time, performs on the data at the previous time. PCM, on the other hand, measures similarity between the two consecutive partitions. This framework allows insertion and removal of nodes in the graph.

FacetNet uses probabilistic community membership models for dynamic community discovery and was proposed by Lin et al. [103]. The authors use K/L-divergence to measure quality and history cost. One key benefit of such probabilistic models is that they allow membership to *multiple* communities for individuals by assigning weights indicating the degree of membership. When certain conditions hold, optimization of total cost is equivalent to maximizing log-likelihood function $L(U_t) = \log P(W_t|U_t) + \log P(U_t|U_{t-1})$ where W_t is data at time t and U_t is the cover at t [131].

Kim and Han [92] revisited the cost function and found that smoothing at the clustering level can degrade the performance. What they suggest as a remedy is to *push down* the cost to each pair of nodes to get a temporal-smoothed version of pairwise node distance and then apply density-based clustering. Greedy local clustering mapping based on mutual information was used to make the model capable of dealing with arbitrary creation/dissolution and growing/shrinking of a community over time [131].

17.13.3 Heterogeneous Networks

Traditional clustering algorithms assume a homogeneous underlying network where the nodes and edges are of uniform types. In the real world, however, the nodes are often different from each other and so are the edges; for instance, relationships based on communication methods as shown in [73] or both nodes and edges may differ at the same time [162]. Another example is the IMDB network where nodes can represent movies, actors, directors, etc., and consequently, the edges can represent different relationships. We often have to deal with heterogeneous social networks as well. These diverse networks bring great opportunities as well as harder challenges—opportunity in the sense that we may gain valuable information from such diverse information rich networks and challenges because we do not know of any obvious methods yet for appropriately clustering such heterogeneous networks.

SONAR API aims at aggregating social network information from emails, instant messages, charts, blogs, and so on for the purpose of user recommendations based on an aggregated network. It was designed by Guy et al. [73] who showed that the recommendations based on an aggregated network performed better than any of the input networks. However, methods to find the best combination scheme were not discussed.

Cai et al. [24] focused on finding the best linear combination of different source networks aimed at building a target network with adjacency matrix $\tilde{M}$ and regressing it on source networks M_i.

$$\mathbf{a}^{opt} = \arg\min_{\mathbf{a}} ||\tilde{M} - \sum_{i=1}^{n} a_i M_i||^2 \tag{17.1}$$

Here a_is are coefficients of corresponding source networks. Since the target network is rarely known in full, it depends on the users provision of a few example target relationships and finds a linear programming formulation that efficiently solves the regression problem [131].

Sun et al. [162] proposed the NetClus algorithm that clusters *star* network schema. In these networks each record is a compound of a single *target type* and several *attribute types*. A generative model is used to iteratively rank the posterior probabilities for cluster assignment until convergence. Using ranking distribution for each type of object, typewise influence ranking can be retrieved. But as mentioned above, this algorithm is limited to only networks with *star* network schema [131]. The

RankClus algorithm of Sun et al. [161] similarly deals with only bi-type networks where the vertex set has only two types of vertices.

Ensemble clustering [9, 159] refers to a class of clustering techniques where results of multiple clustering are combined to get the final result. We envision that this class of clustering can also be a potential solution for dealing with heterogeneous networks.

Some recent approaches use a combination of content and link structure for clustering purposes [142]. There has also been a focus on clustering web images with the use of associated text from the web page [23] but these approaches do not use linkage structure for clustering. Qi et al. [137] focused on jointly clustering media objects, textual context objects, and users in social media networks. The authors propose a Heterogeneous Random Field (HRF) model to model structure and content of social media networks and determine clusters. This work introduces an energy function on edges and shows that the most probable clusters on the graph are found by minimizing the energy function. One advantage of this algorithm is that it can detect noisy links which are fairly commonplace in real networks.

Aggarwal et al. [1] focused on community detection from the perspective of heterogeneity of link density in social networks. Their method uses local methods that adapt well to local variations in density to extract interesting, coherent, and balanced clusters from all parts of a network. They use a min-hash based approach to find a small number of local communities specific to each node and then merge them into a (concise) set of global communities.

17.13.4 Directed Networks

Network clustering and community discovery algorithms, in general, work with undirected networks. But in the real world, many interesting and important networks are essentially *directed.* Examples include graphs representing Web Pages, Twitter users, or citations between research papers. Many studies have just ignored the directionality of the edges and very few have worked on community detection in directed networks [60]. But simply ignoring the edge directionalities can lead to false results [131].

Recently there has been some work on this matter. Researchers have extended their algorithms to take the directionality information for edges into account. Directed versions of Normalized Cuts have been defined using random-walks interpretation of Normalized Cuts [114] by multiple researchers. Let P be the transition matrix of a random walk on directed graph and π be its stationary distribution vector (i.e., the PageRank vector) satisfying the condition that $\pi P = \pi$. The directed Normalized Cut for a group $S \subset V$ is [35, 80, 113]

$$Ncut_{dir}(S) = \frac{\sum_{i \in S, j \in \bar{S}} \pi(i) P(i,j)}{\sum_{i \in S} \pi(i)} + \frac{\sum_{j \in \bar{S}, i \in S} \pi(j) P(j,i)}{\sum_{j \in \bar{S}} \pi(j)}$$

Spectral clustering can be used to minimize the objective function above by postprocessing the top eigenvectors of the directed Laplacian defined as [35, 80, 113]:

$$\mathcal{L} = I - \frac{\pi^{1/2} P \pi^{-1/2} + \pi^{-1/2} P' \pi^{1/2}}{2}$$

Here P and π are defined as above. The directed version of modularity [125] was introduced by Leicht and Newman [99] as follows:

$$Q = \frac{1}{e} \sum_{ij} [A_{ij} - \frac{d_i^{in} d_j^{out}}{e}] \delta_{c_i, c_j}$$

where e is number of edges, d_i^{in} and d_i^{out} refer to the *indegree* and *outdegree* of node i, respectively,

$\delta_{i,j}$ is the Kronecker delta symbol, and c_i is the label of the community to which vertex i is assigned. The definition of modularity matrix **B** is modified as $B_{ij} = A_{ij} - \frac{d_i^{in} d_j^{out}}{e}$ to fit the new metric into spectral optimization proposed in [124]. Since **B** alone may not be symmetric, the modularity function is rewritten as

$$Q = \frac{1}{4e}\mathbf{s}^T(\mathbf{B}+\mathbf{B}^T)s$$

Here s is the vector whose elements are s_i such that s_i is $+1$ if vertex i is assigned to community 1 and -1 if vertex i is assigned to community 2. But as Fortuanto and Barthéleney [61] point out, the algorithm may still suffer from resolution problem.

Satuluri and Parthasarathy [149] claim that such objective functions still favor clusters with high interconnectivity structures, and hence, clustering with low-directed normalized cut or high-directed modularity are often not the most meaningful way to cluster directed graphs. They argue that high interconnectivity is not necessarily required for a group of nodes to form a meaningful group in a directed graph. They propose a more general framework that converts the input directed graph into a weighted undirected graph using a symmetric similarity measure for the vertices of the directed graph. They demonstrate that similarity measures using in-link and out-link similarity while discounting common links perform better than existing approaches [131].

Macropol and Singh [106] proposed *Top Graph Clusters (TopGC)* that probabilistically finds "best" (top scoring) clusters on directed edge weighted graphs in *linear* time using Locality Sensitive Hashing (LSH) [67] for similarity search. The key idea is to create an LSH signature based on the node neighborhoods and calculate the Jaccard Index [18]. Next, they create a signature of a certain length for each node and use that for matching two nodes. But given some limitations regarding weights and other issues, they modify their algorithm to overcome these based on weighted neighborhood and pruning of search space.

17.13.5 Combining Content and Relationship Information

One of the major issues associated with the analysis of modern network is the incorporation of *content* information with the *relationship* information. Even though the relationship part of social networks has been studied extensively in the literature, there have been only a few efforts at coupling content information with the structural data. Without content information, a network's relationship information is simply a plain graph with vertices and edges. It gets interesting when we consider the content information at the vertices and edges. With the advent of content-rich nodes and edges in modern networks, it is now necessary to include this information into the analysis of these networks. Contents are often in the form of text, images, events, tags, etc. With the availability of such information, communities are expected to be not only topologically well connected but also semantically coherent and meaningful with respect to their content. For example, consider an email network where an edge between two nodes represents an email communication. Now if we just look at the edge between nodes, a *spammer* is most likely to become the hub of the network and be the center of clusters which is hardly what we want. Taking into account the content we can filter out the spam emails and extract more meaningful clusters. Although many previous studies have worked on datasets with content information, in most cases the content has been used just to extract links between nodes based on similarity or some other measure but not for community extraction [131].

Content information may be available in various formats. Content associated with vertices may come from user profile in social networks or material created by the user while content with edges my come from interactions. The goal is to combine content and structural information for finding more meaningful communities.

Here we review some approaches that use Bayesian generative models for incorporating textual contents. Wang et al. [168] propose a Group-Topic model which is an extension of stochastic block structures models [127] where both relations and their attributes are considered. An entity is related to another if they behave the same way to an event and if texts associated with the event are this relationship's attributes. Moreover, each event corresponds to one of T latent topics, and hence, group membership of an entity is no longer constant; rather it changes with respect to different topics. The discovery of groups by topics and vice-versa is guided by this framework of directed probabilistic model [131].

The notion of *semantic community* and two corresponding Community-User-Topic (CUT) models were introduced by Zhou et al. [178] where the objective is to extract semantic community from communication documents. The CUT_1 model is more similar to conventional community discovery algorithms because a community is still nothing more than a group of users. Here, the distribution of topics is conditioned on users, who are, in turn, conditioned on communities. The CUT_2 model, on the other hand, assumes a tighter connection between community and topic and lets communities decide topics and topics decide users. The experiments report that the CUT_2 model finds higher quality semantic communities and is computationally more efficient as well.

Another model called Community-Author-Recipient-Topic (CART) model in an email communication network setting was presented by Pathak et al. [133]. This model assumes that discussion among users within a community is relevant to the users as well as the community and constrains all users involved and topics discussed in the email conversation to belong to a single community while same users and topics in a different conversation can be assigned to different communities [131]. It is claimed that this model emphasizes the *joint* effect of topic and relationships on community structures more than the other models discussed above. However, a common concern with all three methods discussed above is that inference of the generative model using Gibbs sampling may converge slowly leading to a longer running time for large-scale datasets.

Moser et al. [118] introduce the problem of Connected X Clusters (CXC) inspired by traditional graph clustering. The algorithm requires each cluster to be *internally connected* using relationship information and assumes each cluster to be compact and distinctive from neighboring ones (by content information). The proposed algorithm is called *JointClust* and is essentially an agglomerative clustering method. After determining cluster atoms based on number of initial centroids, it merges cluster atoms in a bottom-up fashion based on an extension of the traditional Silhouette coefficient known as *Joint Silhouette Coefficient* [88]. The advantage is that it does not require a prespecified cluster number. However, it still takes the minimum size of each cluster as a parameter.

Negoescu et al. [121] proposed an algorithm for identifying groups on the Flickr image-sharing website. This algorithm defines groups to be a set of self-organized users who are elements of the final communities. A community is also called a *hypergroup*. The algorithm first extracts bag-of-tags from the groups' images which are treated as the content generated by the group. Then Latent Dirichlet Allocation (LDA) is applied to get the distribution of latent topics over each group. Different similarity measures can be used to build a similarity matrix for groups, and the original problem is converted into a clustering problem on a similarity matrix. This algorithm is applicable to finding communities of users as well [131]. The concern, again, is the efficiency that is associated with all latent-topic-based approaches.

Sun et al. [163] propose a topic modeling framework, called *iTopicModel*, on arbitrary document networks. This framework builds a generative topic model that considers both text and structure information for documents. Their model provides a joint distribution function for both text and structure of documents. Estimation of the topic model is done by maximizing the log-likelihood of the joint probability using an EM-based iterative solution.

17.13.6 Networks with Overlapping Communities

In modern networks, especially in social networks, each node in the network can belong to multiple groups (i.e., communities). Another example is the biological network where each node (protein) can have multiple functions. Kelley et al. [89] show that *overlap* is a significant feature of many social networks. Most of the partitioning algorithms explained in this chapter target a disjoint clustering of the network. Recently, there has been a growing interest in algorithms that can detect overlapping community structures.

Given a graph $\mathcal{G} = (V, E)$, a *cover*, $C = (c_1, c_2, ..., c_k)$, is a set of clusters found by overlapping community detection [96], and each node is associated with a community by a *belonging factor* $b_1, b_2, ..., b_k$ [122]. Generally it is assumed that $0 \leq b_i \leq 1$ and $\sum_i b_i = 1$. Node assignment to clusters can be *crisp* or *fuzzy* [71]. In *crisp* assignment, a node either belongs to a community or does not; but in *fuzzy* assignment the membership of a node in a cluster is expressed as a *belonging factor* [172].

Xie et al. [172] categorize algorithms for overlapping community detection into following five categories based on how communities are identified.

- *Clique Percolation Algorithm (CPM)* assumes that a community consists of overlapping fully connected subgraphs and, hence, searches for adjacent cliques in order to detect communities. CFinder is the implementation of CPM which has a polynomial time complexity in many applications [128] and does not terminate in many large social networks [172]. Related works under this category include subgraph intensity threshold for weighted networks by Farkas et al. [55] and SCP by Kumpula et al. [95] that finds clique communities of given size. SCP is faster than CPM and allows multiple weight threshold in a single run.

- *Line Graph and Link Partitioning* explores the idea of *link partitioning* and calls a node in the graph *overlapping* if links connected to it are partitioned into more than one clusters. Some works in this context are presented in [3, 53, 52]. Despite being intuitively natural, there is no guarantee that it gives better detection than node-based methods [59] because it depends on an ambiguous definition of community.

- *Local Expansion and Optimization* are based on growing a natural [96] or partial community that relies on local benefit function characterizing the quality of a densely connected group of nodes. Baumes et al. [15] propose a two-step process: RankRemoval finds the seed communities for the second step Iterative Scan (IS). LFM, proposed in [96], expands a community from random seed with respect to a *fitness* function. Havemann et al. proposed MONC [75] that uses a modified fitness function of LFM. A thorough study of local expansion and optimization algorithms can be found in [172].

- *Fuzzy Detection* algorithms quantify a soft *membership vector* or *belonging factor* [70] for each node which represents the strength of association between each pairs of nodes and communities. Nepusz et al. [122] model the problem as nonlinear constrained optimization that can be solved using simulated annealing methods. Zhang et al. [177] propose a spectral clustering-based algorithm that uses fuzzy c-means (FCM) to obtain a soft assignment. Use of mixture models has also been explored in this context and some examples are SPAEM [143] and FOG [39]. OSBM [98] and MOSES [110] are based on the Stochastic Block Model (SBM) [127] generative model. More detailed descriptions of above mentioned and related algorithms can be found in [172].

- *Agent Based and Dynamical Algorithms* extend the label propagation algorithm [139] to overlapping community detection. In COPRA [70] each node averages the *belonging* coefficients of neighbors in a synchronous fashion to update its coefficients in each time step. SLPA [173] follows pairwise interaction rules to spread labels between nodes based on a general speaker-listener-based information propagation process. Chen et al. [32] propose a game theoretic framework where a community is associated with a Nash local equilibrium and each agent has a *gain* and *loss* function. The assumption is agents are *selfish* and form communities based on respective utilities. Breve et al. [17] propose a process where particles walk and compete with each other to occupy nodes, and particles represent different communities. A comprehensive study on this topic can be found in [172].

Many other algorithms have been proposed in the literature to address the problem. CONGO [69] allows a node to split into multiple copies, extends Girvan-Newman's (GN) divisive clustering algorithm [68], and optimizes for associated high computational complexity of it for speed. Xie et al. present an exhaustive discussion on available algorithms in [172].

- *Overlapping Markov Clustering* has recently been proposed by Shih and Parthasarathy [152]. This approach extends Regularized Marcov Clustering (RMCL) [148] and allows soft clustering. This method, called SR-MCL, executes RMCL algorithm iteratively but penalizes flows going to previous attractor nodes. As a consequence, RMCL produces slightly different clustering each time. SR-MCL then combines all this clustering to generate overlapping clusters. It also performs some postprocessing to remove unqualified and redundant clusters [152].

17.13.7 Probabilistic Methods

In this section we briefly talk about some of the probabilistic methods in the area of network clustering.

- *Generative Model-Based Methods* assume that the data is part of some unobserved probability distribution. Generative models describe probabilistically how a dataset may be formed. In other words, they are the hypothesis of the underlying distribution that created the data. Assuming the hypothesis to be true, algorithms like Expectation Maximization can be used to find clustering that best agrees with the underlying model [7]. Fu and Banerjee [62] show that mixture models can be constructed as generative model as well. Magdon-Ismail and Purnell [107] use spectral clustering to map the network into a d-dimensional space and train a Gaussian Mixture Model (GMM) using the Expectation Maximization (EM) algorithm. Increase in log-likelihood of adding a cluster is used to determine the number of communities [172].

  For groups in the networks, SBM is another type of generative model [127]. However, fitting an empirical network to SBM requires inferring model parameters. In OSBM [98] each node is associated with a latent vector with K independent Boolean variables where K is the number of communities. As in [144], the latent vector is inferred by maximizing the posterior probability conditioned on the presence of edges. Because the factorization in the observed condition distribution for edges given the latent vector is in general intractable, OSBM requires more effort than mixture models [172]. McDaid and Hurley combine OSBM with local optimization scheme in MOSES [110] where the fitness function is defined based on the observed condition distribution. MOSES greedily expands community from edges and has a worst-case time complexity of $O(en^2)$ where e is the number of edges expanded. Recently McDaid et al. [111] extended the SBM of Nowicki and Snijders [127] exploiting parameter collapsing to integrate out block parameters. This model defines a posterior over the number of clusters and cluster memberships and allows the number of clusters to be directly estimated. Another SBM based model was proposed by McDaid et al. in [112] for finding communities in networks.

- *Exponential Random Graph (p^*) Models* [170] for social networks have enjoyed a growing interest from the research community in recent years. These models regard possible ties among nodes of a network as random variables. The general form of the exponential random graph model for the network is determined by assumptions about dependencies among these random tie variables. An observed network is regarded as one realization from a set of possible networks with similar characteristics. In other words, any observed network is an outcome of some unknown stochastic process. The goal is to formulate a model to propose a theoretically principled hypothesis for the unknown process [145].

17.14 Conclusion

Clustering networks with efficiency and scalability is a hard challenge in general. In this chapter, we present a comprehensive survey on different approaches to network clustering and describe a wide range of clustering algorithms based on their working principles. We discuss different formulations of the problem in the literature and most commonly used evaluation criteria for measuring the quality of the resulting clusters. Next, we describe core methods and representative algorithms in each category starting from early geometric partitioning through to modern spectral methods and Markov clustering. We then discuss major emerging challenges and state-of-the-art approaches to solving them in the area of network clustering. In this concluding section, we briefly portray some of the modern research themes that, as we envision, will guide the advancement of future research in the area of network clustering in a broader sense.

- **Summarization and Ranking:** One of the biggest challenges with modern networks is the fact that they are large. With increasing sizes of the networks, it becomes extremely challenging to mine them and convey meaningful information. One intuitive and effective approach to dealing with large networks is to summarize and rank information for efficient analytic operations. Ranking patterns in order of importance helps experts focus on the most significant part of the large network and explore from there. It also facilitates *interactive browsing* of different parts of the network based on the analysts' interest. Given a region of interest, the analyst may want to apply operations similar to *rollup* or *drill down* for further analysis. Although there have been significant research efforts in the generic area of summarization and ranking, application and adaption of these techniques in the context of dynamic large networks needs to be explored by more researchers.

- **Visualization and Interactive Browsing:** Visualizing modern billion-node networks is a hard challenge because one often runs out of pixels on a commodity display device. It gets harder when one targets visualizing topological characteristics and behavior of such networks. Efficient interactive visualization and browsing is key to efficient analysis of modern networks. But providing these capabilities entails a number of big challenges from the domain of visualization and graph miming. Unfortunately, this area has seen limited work in the context of network partitioning thus far [20, 167, 175]. Visualization and Browsing has multiple important roles in the context of network analytics: First of all, as a front end *interactive* tool for visualizing dynamic networks at different levels of abstractions on demand. Next, as a tool for understanding and analyzing the network. Also, as a means to validate and guide the clustering process interactively. All these operations are challenging and, hence, more research in this area are necessary.

- **Scalable Algorithms for Distributed Systems:** Due to the complex structure, increasing size, and the tremendous amount of information associated with modern networks, *scalability* of the clustering algorithms is going to be a critical factor in determining the usability of a particular algorithm in real scenarios. Algorithms that can scale well for billion-node networks on a parallel and distributed systems framework are going to be necessary to process the ever growing networks. Recently, substantial research efforts have been made toward

TABLE 17.1: Different Clustering Methods

Clustering Method	Category	Tunable Parameters	Free Software
Partitioning with Geometric Information	Balanced Bipartitioning		Chaco [77]
Coordinate Bisection	Balanced Bipartitioning		Party [136]
Inertial Bisection	Balanced Bipartitioning		Chaco [77]
Geometric Bisection	Bipartitioning	Overlap Graph Parameters α and k	Chaco [77], Zoltan [179]
Kernighan-Lin Algorithm	Hard Partitioning	Number of partitions k	Party [136], Chaco [77]
Agglomerative and Divisive Algorithms	Hierarchical Clustering	Number of partitions k	Orange [38], hcluster [49]
Spectral Clustering	Hard Clustering	Number of partitions k	Chaco [77], MATLAB package [31]
Markov Clustering	Hard Clustering	Inflation Param. r controls Granularity of clusters	MCL software [166]
Multilevel Partitioning	Hard Clustering	Coarsening & Uncoarsening Overhead	Metis [86], Graclus [42], Chaco [77], SCOTCH [134]
MLR-MCL	Hard & Soft Clustering	Inflation Parameter r, Size of Coarsest Graph c	MLR-MCL Software [147]
Hypergraph Partitioning	Hard Clustering	Number of Partitions k, Number of Cells in Coarsest Hypergraph	PaToH [26], Zoltan [179], hMetis [87]
Bipartite Graph Partitioning	Hard Clustering		Co-clustering software [34], Metis [86]
Dynamic Graph Partitioning	Hard & Soft Clustering		
Directed Graphs Partitioning	Hard & Soft Clustering	Number of Clusters, Definition of Substructures	GraphClust [140]
Networks with Overlapping Communities	Soft Clustering		MOSES Software [109]
Probabilistic Methods	Hard & Soft Clustering		SBM & SCF Software [108]

scalable solutions to network clustering problems. These include scalable, parallel, and distributed algorithms for processing graphs and networks that exceed the storage capacity of a single machine. At the algorithmic level, multilevel algorithms that depend on graph coarsening offer advantages over other conventional methods that work directly on the input graph [42, 86, 148]. Streaming algorithms [2] as well as architecture-aware algorithms on GPU and multicores offer an orthogonal approach [21, 132]. Algorithms on platforms such as Hadoop [82, 130] are also gaining popularity because of the recent trend toward commodity clusters of shared-nothing architectures and cloud computing.

- **Use of Domain Knowledge:** Underutilization of domain knowledge during the model building process has been a common tendency in data mining research. Researchers in the area of data mining often intentionally discard important domain knowledge during the training phase that allows them to independently validate the effectiveness of the proposed methods during testing [131]. Such tendency often limits the robustness of the solutions and scientific advances within the domain. It is necessary to reconsider how the domain knowledge can be incorporated into the approaches. We believe domain knowledge is too valuable a resource to simply ignore during the discovery process as it can effectively guide the whole process. Especially modern graphs are well known to have a tremendous amount of valuable information and metadata associated with them. Hence, robust algorithms are necessary that make use of the available rich information to produce better results during the clustering process.

Even though the problem of network partitioning has been studied for decades, large modern networks with a substantial amount of information associated with the nodes and edges pose novel, complicated and hard-to-overcome challenges, demanding a fresh investigation, novel techniques, and robust innovative approaches for dealing with them. Generic scalable algorithms for parallel and distributed systems, architecture-aware solutions on GPUs and multicore systems, and algorithms using Map-Reduce [40] framework and running on cloud systems need to be developed to handle these ever growing graphs. Given that this is an exciting area of research with lots of promises, we expect to see many more novel algorithms, principled approaches, and exciting results on this topic in future.

Acknowledgments

We are thankful to the editors and anonymous reviewers for their valuable comments, insightful suggestions, and constructive feedback that greatly helped improving this article.

This work is supported by **NSF Grant IIS-0917070** and **NSF Grant CCF-1217353**. Any opinions, findings, and conclusions or recommendations expressed in this material are those of the author(s) and do not necessarily reflect the views of the National Science Foundation.

Bibliography

[1] C. C. Aggarwal, Y. Xie, and P. S. Yu. Towards community detection in locally heterogeneous networks. In *SDM*, pages 391–402. SIAM / Omnipress, 2011.

[2] C. C. Aggarwal, Y. Zhao, and P. S. Yu. On clustering graph streams. In *SDM*, pages 478–489, 2010.

[3] Y-Y. Ahn, J. P. Bagrow, and S. Lehmann. Link communities reveal multiscale complexity in networks. *Nature*, 466(7307):761–764, 2010.

[4] C. J. Alpert, J. Huang, and A. B. Kahng. Multilevel circuit partitioning. In *Proceedings of the 34th ACM/IEEE Design Automation Conference*, pages 530–533, 1998.

[5] R. Andersen, F. Chung, and K. Lang. Local graph partitioning using PageRank vectors. In *FOCS '06: Proceedings of the 47th Annual IEEE Symposium on Foundations of Computer Science*, pages 475–486, Washington, DC, USA, 2006. IEEE Computer Society.

[6] R. Andersen and K. J. Lang. Communities from seed sets. In *WWW '06: Proceedings of the 15th International Conference on World Wide Web*, page 232. ACM, 2006.

[7] A. Anthony and M. desJardins. Generative models for clustering: The next generation. In *AAAI Spring Symposium: Social Information Processing*, pages 7–10. AAAI, 2008.

[8] S. Asur and S. Parthasarathy. A viewpoint-based approach for interaction graph analysis. In *KDD '09: Proceedings of the 15th ACM SIGKDD International Conference on Knowledge Discovery and Data Mining*, pages 79–88, New York, NY, USA, 2009. ACM.

[9] S. Asur, S. Parthasarathy, and D. Ucar. An ensemble approach for clustering scalefree graphs. In *LinkKDD Workshop*, 2006.

[10] S. Asur, S. Parthasarathy, and D. Ucar. An event-based framework for characterizing the evolutionary behavior of interaction graphs. In *KDD '07: Proceedings of the 13th ACM SIGKDD International Conference on Knowledge Discovery and Data Mining*, pages 913–921, New York, NY, USA, 2007. ACM.

[11] A. L. Barabási and E. Bonabeau. Scale-free networks. *Scientific American*, 288(5):60, 2003.

[12] A. L. Barabási and R. E. Crandall. Linked: The new science of networks. *American journal of Physics*, 71:409, 2003.

[13] S. T. Barnard and H. D. Simon. Fast multilevel implementation of recursive spectral bisection for partitioning unstructured problems. *Concurrency Practice and Experience*, 6(2):101–118, 1994.

[14] J. Bascompte, P. Jordano, C. J. Melián, and J. M. Olesen. The nested assembly of plant–animal mutualistic networks. *Proceedings of the National Academy of Sciences of the United States of America*, 100(16):9383, 2003.

[15] J. Baumes, M. K. Goldberg, M. S. Krishnamoorthy, M. M. Ismail, and N. Preston. Finding communities by clustering a graph into overlapping subgraphs. In Nuno Guimaraes and Pedro T. Isaias, editors, *IADIS AC*, pages 97–104. IADIS, 2005.

[16] T. Y. Berger-Wolf and J. Saia. A framework for analysis of dynamic social networks. In *KDD '06: Proceedings of the 12th ACM SIGKDD International Conference on Knowledge Discovery and Data Mining*, page 528. ACM, 2006.

[17] F. Breve, L. Zhao, and M. Quiles. Uncovering overlap community structure in complex networks using particle competition. In *Proceedings of the International Conference on Artificial Intelligence and Computational Intelligence*, AICI '09, pages 619–628, Berlin, Heidelberg, 2009. Springer-Verlag.

[18] A. Z. Broder, S. C. Glassman, M. S. Manasse, and G. Zweig. Syntactic clustering of the web. *Computer Networks and ISDN Systems*, 29(8-13):1157–1166, 1997.

[19] S. Brohee and J. Van Helden. Evaluation of clustering algorithms for protein-protein interaction networks. *BMC Bioinformatics*, 7(1):488, 2006.

[20] G. Buehrer and K. Chellapilla. A scalable pattern mining approach to web graph compression with communities. In *WSDM '08: Proceedings of the International Conference on Web Search and Web Data Mining*, pages 95–106, New York, NY, USA, 2008. ACM.

[21] G. Buehrer, S. Parthasarathy, and M. Goyder. Data mining on the cell broadband engine. In *Proceedings of the 22nd Annual International Conference on Supercomputing*, pages 26–35. ACM, 2008.

[22] T. N. Bui, F. T. Leighton, S. Chaudhuri, and M. Sipser. Graph bisection algorithms with good average case behavior. *Combinatorica*, 7(2):171–191, June 1987.

[23] D. Cai, X. He, Z Li, W. Ma, and J Wen. Hierarchical clustering of www image search results using visual, textual and link information. In *Proceedings of the 12th annual ACM International Conference on Multimedia*, pages 952–959. ACM Press, 2004.

[24] D. Cai, Z. Shao, X. He, X. Yan, and J. Han. Mining hidden community in heterogeneous social networks. In *Proceedings of the 3rd International Workshop on Link Discovery*, page 65. ACM, 2005.

[25] A. E. Caldwell, A. B. Kahng, and I. L. Markov. Optimal partitioners and end-case placers for standard-cell layout. In *Proceedings of the 1999 International Symposium on Physical Design*, ISPD '99, pages 90–96, New York, NY, USA, 1999. ACM.

[26] U. V. Çatalyürek. Partitioning Tools for Hypergraph (PaToH). `http://bmi.osu.edu/~umit/software.html`.

[27] U. V. Catalyurek, C. Aykanat, and B. Ucar. On two-dimensional sparse matrix partitioning: Models, methods, and a recipe. *SIAM J. Sci. Comput.*, 32(2):656–683, February 2010.

[28] U. V. Catalyurek, E. G. Boman, K. D. Devine, R. Heaphy, L. Ann, and R. Ohio. Hypergraph-based dynamic load balancing for adaptive scientific computations. *Parallel and Distributed Processing Symposium, IPDPS 2007. IEEE International.* IEEE, pages 711–724, 2007.

[29] D. Chakrabarti, R. Kumar, and A. Tomkins. Evolutionary clustering. In *KDD '06: Proceedings of the 12th ACM SIGKDD International Conference on Knowledge Discovery and Data Mining*, pages 554–560. ACM New York, NY, USA, 2006.

[30] G. Chartrand and O. Ollermann. *Applied and Algorithmic Graph Theory.* McGraw Hill, 1993.

[31] W. Chen, C. Lin, Y. Song, and H. Bai. MATLAB spectral clustering package 1.1, 2010. `http://mloss.org/software/view/133/`.

[32] W. Chen, Z. Liu, X. Sun, and Y. Wang. A game-theoretic framework to identify overlapping communities in social networks. *Data Mining and Knowledge Discovery*, 21(2):224–240, September 2010.

[33] Y. Chi, X. Song, K. Hino, and B. L. Tseng. Evolutionary spectral clustering by incorporating temporal smoothness, October 18 2007. US Patent App. 11/874,395.

[34] H. Cho, Y. Guan, and S. Sra. Co-clustering software (version 1.1). `http://www.cs.utexas.edu/users/dml/Software/cocluster.html`.

[35] F. Chung. Laplacians and the Cheeger inequality for directed graphs. *Annals of Combinatorics*, 9(1):1–19, 2005.

[36] F. R. K. Chung. *Spectral Graph Theory*, volume 92. American Mathematical Society, 1997.

[37] A. Clauset, M. E. J. Newman, and C. Moore. Finding community structure in very large networks. *Physical Review E*, 70(6):66111, 2004.

[38] T. Curk, J. Demsar, Q. Xu, G. Leban, U. Petrovic, I. Bratko, G. Shaulsky, and B. Zupan. Microarray data mining with visual programming. *Bioinformatics*, 21:396–398, February 2005.

[39] G. Davis and K. Carley. Clearing the FOG: Fuzzy, overlapping groups for social networks. *Social Networks*, 30(3):201–212, 2008.

[40] J. Dean and S. Ghemawat. MapReduce: Simplified data processing on large clusters. In *OSDI*, pages 137–150, 2004.

[41] I. S. Dhillon. Co-clustering documents and words using bipartite spectral graph partitioning. In *Proceedings of the Seventh ACM SIGKDD International Conference on Knowledge Discovery and Data Mining*, KDD '01, pages 269–274, New York, 2001.

[42] I. S. Dhillon, Y. Guan, and B. Kulis. Weighted graph cuts without eigenvectors: A multilevel approach. *IEEE Trans. Pattern Anal. Mach. Intell.*, 29(11):1944–1957, November 2007.

[43] P. Domingos and M. Richardson. Mining the network value of customers. In *KDD '01: Proceedings of the Seventh ACM SIGKDD International Conference on Knowledge Discovery and Data Mining*, pages 57–66. ACM, 2001.

[44] W. E. Donath and A. J. Hoffman. Algorithms for partitioning graphs and computer logic based on eigenvectors of connection matrices. *IBM Technical Disclosure Bulletin*, 15(3):938–944, 1972.

[45] S. V. Dongen. *Graph Clustering by Flow Simulation*. PhD thesis, University of Utrecht, 2000.

[46] S. Dutt and W. Deng. Cluster-aware iterative improvement techniques for partitioning large VLSI circuits. *ACM Transactions on Design Automation of Electronic Systems*, 7(1):91–121, January 2002.

[47] S. Dutt and W. Deng. VLSI circuit partitioning by cluster-removal using iterative improvement techniques. In *Proceedings of the 1996 IEEE/ACM International Conference on Computer-Aided Design*, ICCAD '96, pages 194–200, Washington, DC, USA, 1996. IEEE Computer Society.

[48] S. Dutt and H. Theny. Partitioning using second-order information and stochastic-gain functions. In *Proceedings of the 1998 International Symposium on Physical Design*, ISPD '98, pages 112–117, New York, NY, USA, 1998. ACM.

[49] D. Eads. *hcluster: Hierarchical Clustering for Scipy*, 2008.

[50] P. Eles, Z. Peng, K. Kuchcinski, and A. Doboli. System level hardware/software partitioning based on simulated annealing and tabu search. *Design Automation for Embedded Systems*, 2(1): 5–32, 1997.

[51] U. Elsner. *Graph Partitioning—A Survey*, Technische Universitat Chemnitz, 1997.

[52] T. S. Evans. Clique graphs and overlapping communities. *Journal of Statistical Mechanics: Theory and Experiment*, P12037, 2010.

[53] T. S. Evans and R. Lambiotte. Line graphs, link partitions, and overlapping communities. *Physical Review E*, 80(1):016105+, July 2009.

[54] C. Farhat. A simple and efficient automatic FEM domain decomposer. *Computers and Structures*, 28(5):579–602, 1988.

[55] I. Farkas, D. Abel, G. Palla, and T. Vicsek. Weighted network modules. *New Journal of Physics*, 9(6):180, 2007.

[56] X. Z. Fern and C. E. Brodley. Solving cluster ensemble problems by bipartite graph partitioning. In *In Proceedings of the International Conference on Machine Learning*, Article 36, 2004.

[57] C. M. Fiduccia and R. M. Mattheyses. A linear-time heuristic for improving network partitions. In *19th Conference on Design Automation, 1982*, pages 175–181, 1982.

[58] G. W. Flake, S. Lawrence, and C. L. Giles. Efficient identification of web communities. In *KDD '00: Proceedings of the sixth ACM SIGKDD International Conference on Knowledge Discovery and Data Mining*, page 160. ACM, 2000.

[59] S. Fortunato. Community detection in graphs. *CoRR*, abs/0906.0612, 2009.

[60] S. Fortunato. Community detection in graphs. *Physics Reports*, 486:75–174, 2010.

[61] S. Fortunato and M. Barthélemy. Resolution limit in community detection. *Proceedings of the National Academy of Sciences*, 104(1):36, 2007.

[62] Q. Fu and A. Banerjee. Multiplicative mixture models for overlapping clustering. In *ICDM*, pages 791–796. IEEE Computer Society, 2008.

[63] M. R. Garey and D. S. Johnson. *Computers and Intractability; A Guide to the Theory of NP-Completeness*. W. H. Freeman & Co., New York, USA, 1990.

[64] M. R. Garey and L. Johnson. Some simplified NP-complete graph problems. *Theoretical Computer Science*, 1(3):237–267, 1976.

[65] D. Gibson, R. Kumar, and A. Tomkins. Discovering large dense subgraphs in massive graphs. In *VLDB '05: Proceedings of the 31st International Conference on Very Large Data Bases*, Trondheim, Norway, August 30-September 2, 2005, page 721. ACM, 2005.

[66] J. Gilbert, G. L. Miller, and S. Teng. Geometric mesh partitioning: Implementation and experiments. *SIAM J. Scientific Computing*, 19(6):2091–2110, 1998.

[67] A. Gionis, P. Indyk, and R. Motwani. Similarity search in high dimensions via hashing. In *Proceedings of the 25th International Conference on Very Large Data Bases*, VLDB '99, pages 518–529, San Francisco, CA, USA, 1999. Morgan Kaufmann Publishers Inc.

[68] M. Girvan and M. E. J. Newman. Community structure in social and biological networks. *Proceedings of the National Academy of Sciences*, 99(12):7821, 2002.

[69] S. Gregory. A fast algorithm to find overlapping communities in networks. In *Proceedings of the 2008 European Conference on Machine Learning and Knowledge Discovery in Databases—Part I*, ECML PKDD '08, pages 408–423, Berlin, Heidelberg, 2008. Springer-Verlag.

[70] S. Gregory. Finding overlapping communities in networks by label propagation. *New Journal of Physics*, 12(10):103018, 2010.

[71] S. Gregory. Fuzzy overlapping communities in networks. *CoRR*, abs/1010. 1523, February 2011.

[72] A. Gupta. Fast and effective algorithms for graph partitioning and sparse matrix ordering. *IBM Journal of Research and Development*, 41:171–183, 1996.

[73] I. Guy, M. Jacovi, E. Shahar, N. Meshulam, V. Soroka, and S. Farrell. Harvesting with SONAR: The value of aggregating social network information. In *Proceeding of the 26th Annual SIGCHI Conference on Human Factors in Computing Systems*, pages 1017–1026. ACM, 2008.

[74] L. W. Hagen and A. B. Kahng. New spectral methods for ratio cut partitioning and clustering. *IEEE Trans. on CAD of Integrated Circuits and Systems*, 11(9):1074–1085, 1992.

[75] F. Havemann, M. Heinz, A. Struck, and J. Gläser. Identification of overlapping communities and their hierarchy by locally calculating community-changing resolution levels. *CoRR*, abs/1012.1269, 2010.

[76] B. Hendrickson and R. Leland. The Chaco User's Guide: Version 2.0. Technical Report SAND94–2692, Sandia National Laboratories, Albuquerque, NM, USA, 1994.

[77] B. Hendrickson and R. Leland. An empirical study of static load balancing algorithms. *Proceedings of IEEE Scalable High Performance Computing Conference*, pages 682–685, 1994.

[78] B. Hendrickson and R. Leland. A multilevel algorithm for partitioning graphs. In *Proceedings of the 1995 ACM/IEEE Conference on Supercomputing (CDROM)*, Supercomputing '95, New York, Article 28, 1995.

[79] J. Hertz, A. Krogh, and R. G. Palmer. *Introduction to the Theory of Neural Computation*, volume 1. Addison-Wesley, 1991.

[80] J. Huang, T. Zhu, and D. Schuurmans. Web communities identification from random walks. *Lecture Notes in Computer Science*, 4213:187, 2006.

[81] H. Jeong, S. P. Mason, A. L. Barabási, and Z. N. Oltvai. Lethality and centrality in protein networks. *Nature*, 411(6833):41–42, 2001.

[82] U Kang, C. E. Tsourakakis, and C. Faloutsos. Pegasus: Mining peta-scale graphs. *Knowledge and Information Systems*, 27(2):303–325, 2010.

[83] R. Kannan, S. Vempala, and A. Veta. On clusterings—Good, bad and spectral. In *FOCS '00*, pages 367–377. IEEE Computer Society, 2000.

[84] G. Karypis, R. Aggarwal, V. Kumar, and S. Shekhar. Multilevel hypergraph partitioning: Applications in VLSI domain. *IEEE Transactions on Very Large Scale Integration (VLSI) Systems*, 7(1):69–79, 1999.

[85] G. Karypis, E-H. (Sam) Han, and V. Kumar. Chameleon: A hierarchical clustering algorithm using dynamic modeling. *Computer*, 32(8): 68-75, 1999.

[86] G. Karypis and V. Kumar. A fast and high quality multilevel scheme for partitioning irregular graphs. *SIAM Journal on Scientific Computing*, 20(1): 359-392, 1998.

[87] G. Karypis and V. Kumar. hmetis—hypergraph & Circuit Partitioning. `http://glaros.dtc.umn.edu/gkhome/metis/hmetis/overview`.

[88] L. Kaufman and P. J. Rousseeuw. Finding groups in data; An introduction to cluster analysis. *Wiley Series in Probability and Mathematical Statistics*. John Wiley & Son Hoboken, NJ, 1990.

[89] S. Kelley, M. Goldberg, M. Magdon-Ismail, K. Mertsalov, and A. Wallace. Defining and discovering communities in social networks. In M. Thai and P. Pardalos, editors, *Handbook of Optimization in Complex Networks*, pages 139–168. Springer Science+Business Media New York, 2011.

[90] B. W. Kernighan and S. Lin. An efficient heuristic procedure for partitioning graphs. *The Bell System Technical Journal*, 49:291–307, 1970.

[91] G. Khanna, N. Vydyanathan, T. Kurc, U. Catalyurek, P. Wyckoff, and J. Saltz. A hypergraph partitioning based approach for scheduling of tasks with batch-shared I/O. In *Proceedings of the 5th IEEE/ACM International Symposium on Cluster Computing and the Grid (CCGrid)*, pages 792–799. 2005.

[92] M. S. Kim and J. Han. A particle-and-density based evolutionary clustering method for dynamic networks. *Proceedings of the VLDB Endowment*, 2(1):622–633, 2009.

[93] Y. Koren. The BellKor Solution to the Netflix Grand Prize. *KorBell Team's Report to Netflix*, 2009.

[94] S. Krishnamoorthy, U. Catalyurek, J. Nieplocha, A. Rountev, and P. Sadayappan. Hypergraph partitioning for automatic memory hierarchy management. In *Proceedings of the 2006 ACM/IEEE Conference on Supercomputing SC '06*, Article 96, New York, USA, 2006.

[95] J. M. Kumpula, M. Kivela, K. Kaski, and J. Saramaki. A sequential algorithm for fast clique percolation. *Physical Review E*, 78(2):8, 2008.

[96] A. Lancichinetti, S. Fortunato, and J. Kertesz. Detecting the overlapping and hierarchical community structure in complex networks. *New Journal of Physics*, 11(3):033015, 2009.

[97] K. Lang and S. Rao. A flow-based method for improving the expansion or conductance of graph cuts. *Lecture Notes in Computer Science*, 3064:325–337, 2004.

[98] P. Latouche, E. Birmele, and C. Ambroise. Overlapping stochastic block models with application to the French political blogosphere. *The Annals of Applied Statistics*, 5(1):309–336, 2009.

[99] E. A. Leicht and M. E. J. Newman. Community structure in directed networks. *Physical Review Letters*, 100(11):118703, 2008.

[100] J. Leskovec, L. A. Adamic, and B. A. Huberman. The dynamics of viral marketing. *ACM Transactions on the Web (TWEB)*, 1(1):5, 2007.

[101] J. Leskovec, K. J. Lang, A. Dasgupta, and M. W. Mahoney. Community structure in large networks: Natural cluster sizes and the absence of large well-defined clusters. *CoRR*, abs/0810.1355, 2008.

[102] L. Li, C. J. Stoeckert, and D. S. Roos. OrthoMCL: Identification of ortholog groups for eukaryotic genomes. *Genome Research*, 13(9):2178–2189, September 2003.

[103] Y. R. Lin, Y. Chi, S. Zhu, H. Sundaram, and B. L. Tseng. FacetNet: A framework for analyzing communities and their evolutions in dynamic networks. In *WWW '08: Proceeding of the 17th International Conference on World Wide Web*, pages 685–694, New York, 2008. ACM.

[104] L. Liu, M. Kuo, S. Huang, and C. Cheng. A gradient method on the initial partition of Fiduccia-Mattheyses algorithm. In *Proceedings of the 1995 IEEE/ACM International Conference on Computer-Aided Design*, ICCAD '95, pages 229–234, Washington, DC, 1995.

[105] U. Luxburg. A tutorial on spectral clustering. *Statistics and Computing*, 17(4):395–416, December 2007.

[106] K. Macropol and A. Singh. Scalable discovery of best clusters on large graphs. *Proceedings of VLDB Endowment*, 3(1-2):693–702, September 2010.

[107] M. Magdon-Ismail and J. T. Purnell. SSDE-cluster: Fast overlapping clustering of networks using sampled spectral distance embedding and gmms. In *SocialCom/PASSAT*, pages 756–759. IEEE, 2011.

[108] A. McDaid. Collapsed SBM software. `https://github.com/aaronmcdaid/collapsedSBM`.

[109] A. McDaid and N. Hurley. Detecting highly overlapping communities with model-based overlapping seed expansion (MOSES). In *Advances in Social Networks Analysis and Mining (ASONAM), 2010 International Conference on*, pp. 112-119. IEEE, 2010.

[110] A. McDaid and N. Hurley. Detecting highly overlapping communities with model-based overlapping seed expansion. In *Proceedings of the 2010 International Conference on Advances in Social Networks Analysis and Mining*, ASONAM '10, pages 112–119, Washington, DC, USA, 2010. IEEE Computer Society.

[111] A. F. McDaid, B. Murphy, N. Friel, and N. Hurley. Clustering in networks with the collapsed Stochastic Block Model. *CoRR*, abs/1203. 3083, March 2012.

[112] A. F. McDaid, T. B. Murphy, N. Friel, and N. J. Hurley. Model-based clustering in networks with stochastic community finding. *CoRR*, abs/1205.1997, 2012.

[113] M. Meila and W. Pentney. Clustering by weighted cuts in directed graphs. In *Proceedings of the 7th SIAM International Conference on Data Mining*, pages 135–144. 2007.

[114] M. Meila and J. Shi. A random walks view of spectral segmentation. *AI and Statistics (AISTATS)*, 2001.

[115] G. L. Miller, S. Teng, W. Thurston, and S. A. Vavasis. Geometric separators for finite-element meshes. *SIAM Journal of Scientific Computing*, 19(2):364–386, March 1998.

[116] B. Mohar. The Laplacian spectrum of graphs. In Y. Alavi et al. (Eds.) *Graph Theory, Combinatorics, and Applications*, volume 2, pages 871–898, John Wiley & Sons, 1991.

[117] B. Mohar. Some applications of Laplace eigenvalues of graphs. In Geňa Hahn and Gert Sabidussi (Eds.) *Graph Symmetry: Algebraic Methods and Applications, Vol 497 of NATO ASI Series C*, pages 227–275. Springer, The Netherlands, 1997.

[118] F. Moser, R. Ge, and M. Ester. Joint cluster analysis of attribute and relationship data without a-priori specification of the number of clusters. In *KDD '07: Proceedings of the 13th ACM SIGKDD International Conference on Knowledge Discovery and Data Mining*, pages 510–519. ACM New York, NY, 2007.

[119] F. Murray. Innovation as co-evolution of scientific and technological networks: exploring tissue engineering. *Research Policy*, 31(8-9):1389–1403, 2002.

[120] S. F. Nadel. *The Theory of Social Structure*. Cohen and West, London, 1957.

[121] R. A. Negoescu, B. Adams, D. Phung, S. Venkatesh, and D. Gatica-Perez. Flickr hypergroups. In *Proceedings of the Seventeen ACM International Conference on Multimedia*, pages 813–816, 2009.

[122] T. Nepusz, A. Petróczi, L. Négyessy, and F. Bazsó. Fuzzy communities and the concept of bridgeness in complex networks. *Physical Review E*, 77(1):016107, January 2008.

[123] M. E. J. Newman. Fast algorithm for detecting community structure in networks. *Physical Review E*, 69(6):066133, 2004.

[124] M. E. J. Newman. Modularity and community structure in networks. *Proceedings of the National Academy of Sciences*, 103(23):8577, 2006.

[125] M. E. J. Newman and M. Girvan. Finding and evaluating community structure in networks. *Physical Review E*, 69:026113, 2004.

[126] A. Y. Ng, M. I. Jordan, and Y. Weiss. On spectral clustering: Analysis and an algorithm. In T. Ditterich, S. Becker, and Z. Ghahramani (Eds.) *Advances in Neural Information Processing Systems*, pages 849–856. MIT Press, 2001.

[127] K. Nowicki and T. A. B. Snijders. Estimation and prediction for stochastic blockstructures. *Journal of the American Statistical Association*, 96(455):1077–1087, 2001.

[128] G. Palla, I. Derenyi, I. Farkas, and T. Vicsek. Uncovering the overlapping community structure of complex networks in nature and society. *Nature*, 435:814, 2005.

[129] D. A. Papa and I. L. Markov. Hypergraph partitioning and clustering. In Teofilo F. Gonzalez (Ed.) *Approximation Algorithms and Metaheuristics*, Chapman & Hall/CRC Press, Boca Raton, FL, 2007.

[130] S. Papadimitriou and J. Sun. Disco: Distributed co-clustering with Map-Reduce: A case study towards petabyte-scale end-to-end mining. In *Eighth IEEE International Conference on Data Mining, 2008. ICDM'08*, pages 512–521, 2008.

[131] S. Parthasarathy, Y. Ruan, and V. Satuluri. Community discovery in social networks: Applications, methods and emerging trends. In Charu C. Aggarwal, editor, *Social Network Data Analytics*, chapter 4, pages 79–113. Springer, Boston, MA, 2011.

[132] S. Parthasarathy, S. Tatikonda, G. Buehrer, and A. Ghoting. Architecture conscious data mining: Current directions and future outlook, In H. Kargupta, J. Han, P.S. Yu, R. Motwari, and V. Kumar (Eds.), *Next Generation of Data Mining*, pages 261–280. Chapman and Hall/CRC, Boca Raton, FL, USA, 2008.

[133] N. Pathak, C. DeLong, A. Banerjee, and K. Erickson. Social topic models for community extraction. In *The 2nd SNA-KDD Workshop*, volume 8, 2008.

[134] F. Pellegrini and J. Roman. Scotch: A software package for static mapping by dual recursive bipartitioning of process and architecture graphs. In *Proceedings of the International Conference and Exhibition on High-Performance Computing and Networking*, HPCN Europe 1996, pages 493–498, Springer-Verlag, London, UK, 1996.

[135] A. Pothen. Graph partitioning algorithms with applications to scientific computing. In David E. Keyes, Ahmed Sameh, and V. Venkatakrishnan (Eds.) *Parallel Numerical Algorithms*, pages 323–368. Kluwer Academic Press, 1997.

[136] R. Preis and R. Diekmann. Party—A software library for graph partitioning. In B.H.V. Topping (Ed.) *Advances in Computational Mechanics with Parallel and Distributed Processing*, pages 63–71. Civil-Comp Press, 1997.

[137] G. Qi, C. C. Aggarwal, and T. S. Huang. On clustering heterogeneous social media objects with outlier links. In *Proceedings of the Fifth ACM International Conference on Web Search and Data Mining*, WSDM '12, pages 553–562, New York, 2012.

[138] G. Qiu. Bipartite graph partitioning and content-based image clustering. *Visual Media Production 2004CVMP 1st European*, pages 87–94, 2004.

[139] U. N. Raghavan, R. Albert, and S. Kumara. Near linear time algorithm to detect community structures in large-scale networks. *Physical Review E—Statistical, Nonlinear and Soft Matter Physics*, 76(3 Pt 2):036106, 2007.

[140] D. R. Recupero and D. Shasha. Graphclust. `http://www.cs.nyu.edu/cs/faculty/shasha/papers/GraphClust.html`.

[141] P. K. Reddy, M. Kitsuregawa, P. Sreekanth, and S. S. Rao. A graph based approach to extract a neighborhood customer community for collaborative filtering. In *Proceedings of Databases in Networked Information Systems: Second International Workshop, DNIS 2002*, Aizu, Japan, December 16-18, 2002: page 188. Springer-Verlag New York, 2002.

[142] M. Rege, M. Dong, and J. Hua. Graph theoretical framework for simultaneously integrating visual and textual features for efficient web image clustering. In *Proceedings of the 17th International Conference on World Wide Web*, WWW '08, pages 317–326, New York, 2008.

[143] W. Ren, G. Yan, and X. Liao. A simple probabilistic algorithm for detecting community structure in social networks. *Network*, page 7, 2007.

[144] W. Ren, G. Yan, X. Liao, and L. Xiao. Simple probabilistic algorithm for detecting community structure. *Physical Review E*, 79:036111, March 2009.

[145] G. Robins, P. Pattison, Y. Kalish, and D. Lusher. An introduction to exponential random graph (p*) models for social networks. *Social Networks*, 29(2):173–191, May 2007.

[146] Y. Saad. *Iterative Methods for Sparse Linear Systems*, 2nd edition. Society for Industrial and Applied Mathematics, Philadelphia, PA, 2003.

[147] V. Satuluri. MLR-MCL graph clustering software. `http://www.cse.ohio-state.edu/~satuluri/research.html`.

[148] V. Satuluri and S. Parthasarathy. Scalable graph clustering using stochastic flows: Applications to community discovery. In *KDD '09*, pages 737–746, New York, NY, USA, 2009. ACM.

[149] V. Satuluri and S. Parthasarathy. Symmetrizations for clustering directed graphs. In *Workshop on Mining and Learning with Graphs, MLG 2010*, 2010.

[150] V. Satuluri, S. Parthasarathy, and D. Ucar. Markov clustering of protein interaction networks with improved balance and scalability. In *Proceedings of the ACM Conference on Bioinformatics and Computational Biology*, pages 247–256, 2010.

[151] J. Shi and J. Malik. Normalized cuts and image segmentation. *IEEE Transactions on Pattern Analysis and Machine Intelligence*, 22:888–905, 1997.

[152] Y. K. Shih and S. Parthasarathy. Identifying functional modules in interaction networks via overlapping Markov clustering. In *Bioinformatics*, 28(18):i473–i479, 2012.

[153] R. V. Solé and M. Montoya. Complexity and fragility in ecological networks. *Proceedings of the Royal Society of London. Series B: Biological Sciences*, 268(1480):2039, 2001.

[154] D. A. Spielman and N. Srivastava. Graph sparsification by effective resistances. In *STOC '08: Proceedings of the 40th Annual ACM Symposium on Theory of Computing*, pages 563–568, New York, 2008. ACM.

[155] D. A. Spielman and S. Teng. A local clustering algorithm for massive graphs and its application to nearly-linear time graph partitioning, *CoRR*, abs/0809. 3232, 2008.

[156] D. A. Spielman and S. H. Teng. Nearly-linear time algorithms for graph partitioning, graph sparsification, and solving linear systems. In *Proceedings of the 36th Annual ACM Symposium on Theory of Computing*, pages 81–90, New York, 2004.

[157] M. Steenstrup. Cluster-based networks. In C. Perkins (Ed.) *Ad Hoc Networking*, page 138. Addison-Wesley Longman Publishing Co., Reading, PA, USA, 2001.

[158] M. Stoer and F. Wagner. A simple min-cut algorithm. *J. ACM*, 44(4):585–591, July 1997.

[159] A. Strehl, J. Ghosh, and C. Cardie. Cluster ensembles—A knowledge reuse framework for combining multiple partitions. *Journal of Machine Learning Research*, 3:583–617, 2002.

[160] J. Sun, C. Faloutsos, S. Papadimitriou, and P. S. Yu. Graphscope: Parameter-free mining of large time-evolving graphs. In *KDD '07: Proceedings of the 13th ACM SIGKDD International Conference on Knowledge Discovery and Data Mining*, pages 687–696. New York, 2007.

[161] Y. Sun, J. Han, P. Zhao, Z. Yin, H. Cheng, and T. Wu. RankClus: Integrating clustering with ranking for heterogeneous information network analysis. In *Proceedings of the 12th International Conference on Extending Database Technology: Advances in Database Technology*, pages 565–576, 2009.

[162] Y. Sun, Y. Yu, and J. Han. Ranking-based clustering of heterogeneous information networks with star network schema. In *KDD '09: Proceedings of the 15th ACM SIGKDD International Conference on Knowledge Discovery and Data Mining*, pages 797–806, 2009.

[163] Y. Sun, J. Han, Ji. Gao, and Yi. Yu. iTopicModel: Information network-integrated topic modeling. In *ICDM*, pages 493–502. IEEE Computer Society, 2009.

[164] C. Tantipathananandh, T. Berger-Wolf, and D. Kempe. A framework for community identification in dynamic social networks. In *KDD '07: Proceedings of the 13th ACM SIGKDD International Conference on Knowledge Discovery and Data Mining*, pages 717–726, 2007.

[165] S. H. Teng. Coarsening, sampling, and smoothing: Elements of the multilevel method. *Algorithms for Parallel Processing*, 105:247–276, 1999.

[166] S. Van Dongen. Markov cluster algorithm for graphs. `http://micans.org/mcl/`.

[167] N. Wang, S. Parthasarathy, K. L. Tan, and A. K. H. Tung. CSV: Visualizing and mining cohesive subgraphs. In *Proceedings of the 2008 ACM SIGMOD International Conference on Management of Data*, pages 445–458, 2008.

[168] X. Wang, N. Mohanty, and A. McCallum. Group and topic discovery from relations and their attributes. *Advances in Neural Information Processing Systems*, 18:1449, 2006.

[169] S. Wasserman and K. Faust. *Social Network Analysis: Methods and Applications*. Cambridge University Press, 1994.

[170] S. Wasserman and P. Pattison. Logit models and logistic regressions for social networks: I. An introduction to Markov graphs and p*. *Psychometrika*, 61:401–425, 1996.

[171] J. B. Weissman and A. S. Grimshaw. Network partitioning of data parallel computations. In *HPDC*, pages 149–156, 1994.

[172] J. Xie, S. Kelley, and B. K. Szymanski. Overlapping community detection in networks: The state of the art and comparative study. *CoRR*, abs/1110.5813, 2011.

[173] J. Xie, B. K. Szymanski, and X. Liu. SLPA: Uncovering overlapping communities in social networks via a speaker-listener interaction dynamic process. *CoRR*, abs/1109.5720, 2011.

[174] H. Yang and D. F. Wong. Efficient network flow based min-cut balanced partitioning. In *Proceedings of the 1994 IEEE/ACM International Conference on Computer-Aided Design*, ICCAD '94, pages 50–55, Los Alamitos, CA, 1994.

[175] X. Yang, S. Asur, S. Parthasarathy, and S. Mehta. A visual-analytic toolkit for dynamic interaction graphs. In *Proceeding of the 14th ACM SIGKDD International Conference on Knowledge Discovery and Data Mining*, pages 1016–1024, 2008.

[176] H. Zha, X. He, C. Ding, H. Simon, and M. Gu. Bipartite graph partitioning and data clustering. In *Proceedings of the 10th International Conference on Information and Knowledge Management*, CIKM '01, pages 25–32, New York, 2001.

[177] S. Zhang, R-S. Wang, and X-S. Zhang. Uncovering fuzzy community structure in complex networks. *Physical Review E—Statistical, Nonlinear and Soft Matter Physics*, 76(4.2):046103, 2007.

[178] D. Zhou, E. Manavoglu, J. Li, C. L. Giles, and H. Zha. Probabilistic models for discovering e-communities. In *WWW '06: Proceedings of the 15th International Conference on World Wide Web*, pages 173–182, 2006.

[179] Zoltan home page. `http://www.cs.sandia.gov/Zoltan`, 1999.

Chapter 18

A Survey of Uncertain Data Clustering Algorithms

Charu C. Aggarwal
IBM T. J. Watson Research Center
Yorktown Heights, NY
charu@us.ibm.com

18.1 Introduction

Many data sets which are collected often have uncertainty built into them. In many cases, the underlying uncertainty can be easily measured and collected. When this is the case, it is possible to use the uncertainty in order to improve the results of data mining algorithms. This is because the uncertainty provides a probabilistic measure of the relative importance of different attributes in data mining algorithms. The use of such information can enhance the effectiveness of data mining algorithms, because the uncertainty provides a guidance in the use of different attributes during the mining process. Some examples of real applications in which uncertainty may be used are as follows:

- Imprecise instruments and hardware are sometimes used in order to collect the data. In such cases, the level of uncertainty can be measured by prior experimentation. A classic example of such hardware is sensors, in which the measurements are often imprecise.

- The data may be input by statistical methods, such as forecasting. In such cases, the uncertainty may be inferred from the methodology used in order to perform the function.
- Many privacy-preserving data mining techniques use probabilistic perturbations [11] in order to reduce the fidelity of the underlying data. In such cases, the uncertainty may be available as an end result of the privacy-preservation process. Recent work [5] has explicitly connected the problem of privacy-preservation with that of uncertain data mining and has proposed a method which generates data, which is friendly to the use of uncertain data mining methods.

The problem of uncertain data has been studied in the traditional database literature [14, 43], though the issue has seen a revival in recent years [3, 5, 15, 19, 21, 22, 40, 49, 51, 52]. The driving force behind this revival has been the evolution of new hardware technologies such as sensors which cannot collect the data in a completely accurate way. In many cases, it has become increasingly possible to collect the uncertainty along with the underlying data values. Many data mining and management techniques need to be carefully redesigned in order to work effectively with uncertain data. This is because the uncertainty in the data can change the results in a subtle way, so that deterministic algorithms may often create misleading results [3]. While the raw values of the data can always be used in conjunction with data mining algorithms, the uncertainty provides additional insights which are not otherwise available. A survey of recent techniques for uncertain data mining may be found in [10].

The problem of clustering is a well-known and important one in the data mining and management communities. The problem has been widely explored in the context of deterministic data. Details of a variety of clustering algorithms may be found in [38, 34]. The clustering problem has been widely studied in the traditional database literature [28, 47, 56] because of its applications to a variety of customer segmentation and data mining problems.

Uncertainty modeling is very relevant in the context of a number of different clustering applications. An example is illustrated in [42] in which uncertainty was incorporated into the clustering process in the context of a sales merchandising application. Since the problem of data clustering is closely related to that of classification, the methods for uncertain data clustering can also be used to enable algorithms for other closely related data mining problems such as outlier detection [9] and classification [3]. This is because clustering serves as a general-purpose summarization tool, which can be used in the context of a wide variety of problems.

The presence of uncertainty significantly affects the behavior of the underlying clusters because the presence of uncertainty along a particular attribute may affect the expected distance between the data point and that particular attribute. In most real applications, there is considerable skew in the uncertainty behavior across different attributes. The incorporation of uncertainty into the clustering behavior can significantly affect the quality of the underlying results.

The problem of uncertain data clustering is often confused with that of *fuzzy clustering* [50]. In the case of uncertain data clustering, the uncertainty belongs to the *representation of the source objects which are being clusters*, and the actual clustering model may be either probabilistic or deterministic. In the case of fuzzy clustering [50], the source objects are typically deterministic, and the membership of objects to clusters is probabilistic. In other words, each object has a degree of belongingness to the different clusters, which is "fuzzy" or probabilistic in nature.

In this chapter, we will provide a survey of clustering algorithms for uncertain data. The main classes of clustering algorithms for uncertain data are as follows:

- **Mixture-Modeling Algorithms:** Mixture modeling techniques use probabilistic models for clustering uncertain data. A classic example of such an approach is given in [33], which uses an EM-approach [23] for the clustering process.

- **Density-Based Methods:** A density-based method for uncertain data was proposed in [40]. This is referred to as the FDBSCAN algorithm. This approach modifies the DBSCAN algorithm to the case of uncertain data. An alternative method modifies the OPTICS algorithm to the case of uncertain data [41]. This is referred to as the FOPTICS algorithm.

- **Partitional Methods:** The K-means algorithm has been modified for the case of uncertain data [16, 48, 44, 20, 27]. Typically, the main challenge in these methods is that the uncertain distance computations for the k-means algorithms are too slow. Therefore, the focus is on improving efficiency by using pruning methods [48], speeding up distance computations [44], or by using fast approximation algorithms, which provide worst-case bounds [20, 27].

- **Streaming Algorithms:** The problem of clustering uncertain data has been extended to the case of data streams [8]. For this purpose, we extend the microclustering approach [6] to the case of data streams.

- **High-Dimensional Algorithms:** High-dimensional data poses a special challenge in the uncertain data, because the data is distributed in a very sparse way to begin with. The addition of uncertainty and noise further adds to the sparsity. Therefore, effective methods need to be designed for approximately determining clusters in such applications.

In this chapter, we will provide a detailed discussion of each of the above algorithms for uncertain data. This chapter is organized as follows. In the next section, we will discuss mixture model clustering of uncertain data. In Section 18.3, we will describe density-based clustering algorithms for uncertain data. These include extensions of popular deterministic algorithms such as the DBSCAN and OPTICS algorithms. In Section 18.4, we will discuss partitional algorithms for clustering uncertain data. Most of these methods are extensions of the k-means and k-median algorithms. This includes methods such as the UK-means, CK-means, and a number of approximation algorithms for clustering uncertain data. Section 18.5 discusses streaming algorithms for clustering uncertain data. Section 18.6 discusses high-dimensional algorithms for clustering uncertain data. The uncertain data clustering problem has also been explored in the context of the possible worlds model in Section 18.7. Section 18.9 contains the conclusions and summary.

18.2 Mixture Model Clustering of Uncertain Data

Mixture model clustering [23] is a popular method for clustering deterministic data, and it models the clusters in the underlying data in terms of a number of probabilistic parameters. For example, the data can be modeled as a mixture of Gaussian clusters, and then the parameters of this mixture can be learned from the underlying data. The core idea [23] is to determine model parameters, which ensure a maximum likelihood fit of the *observed instantiations* of the data with the proposed model. A popular method in order to determine these model parameters is the EM algorithm, which uses an Expectation-Maximization approach to iteratively update the parameters with the observed data instances.

The work in [33] generalizes this approach to the case of uncertain data, where each data value may be drawn from an interval. The main difference between the uncertain version of the algorithm and the deterministic version is that each instantiation is now an uncertain value of the record, rather than a deterministic value. Correspondingly, the EM algorithm is also changed in order to evaluate the expressions in the E-step and M-step as an expectation over the uncertain range of the data value. We note that the approach can be used fairly easily for any uncertain distribution, which is represented in the form of a probability histogram of values.

Another algorithm known as the *MMVar* algorithm has been proposed in [29], in which the centroid of a cluster C is defined as an uncertain object CMM, that represents the mixture model of C. The cluster compactness criterion used by the *MMVar* algorithm is the minimization of the variance of the cluster centroid.

18.3 Density-Based Clustering Algorithms

Density-based methods are very popular in the deterministic clustering literature, because of their ability to determine clusters of arbitrary shapes in the underlying data. The core idea in these methods is to create a density profile of the data set with the use of kernel density estimation methods. This density profile is then used in order to characterize the underlying clusters. In this section, we will discuss two variations of such density-based methods, which are the FDBSCAN and FOPTICS methods.

18.3.1 FDBSCAN Algorithm

The presence of uncertainty changes the nature of the underlying clusters, since it affects the distance function computations between different data points. A technique has been proposed in [40] in order to find density-based clusters from uncertain data. The key idea in this approach is to compute uncertain distances effectively between objects which are probabilistically specified. The fuzzy distance is defined in terms of the distance distribution function. This distance distribution function encodes the probability that the distances between two uncertain objects lie within a certain user-defined range. Let $d(\overline{X},\overline{Y})$ be the random variable representing the distance between $\overline{X}$ and $\overline{Y}$. The distance distribution function is formally defined as follows.

Definition 18.3.1 *Let $\overline{X}$ and $\overline{Y}$ be two uncertain records, and let $p(\overline{X},\overline{Y})$ represent the distance density function between these objects. Then, the probability that the distance lies within the range (a,b) is given by the following relationship:*

$$P(a \leq d(\overline{X},\overline{Y}) \leq b) = \int_a^b p(\overline{X},\overline{Y})(z)dz \tag{18.1}$$

Based on this technique and the distance density function, the method in [40] defines a *reachability probability* between two data points. This defines the probability that one data point is directly reachable from another with the use of a path, such that each point on it has density greater than a particular threshold. We note that this is a direct probabilistic extension of the deterministic reachability concept which is defined in the DBSCAN algorithm [24]. In the deterministic version of the algorithm [24], data points are grouped into clusters when they are reachable from one another by a path which is such that every point on this path has a minimum threshold data density. To this effect, the algorithm uses the condition that the ε-neighborhood of a data point should contain at least *MinPts* data points. The algorithm starts off at a given data point and checks if the ε neighborhood contains *MinPts* data points. If this is the case, the algorithm repeats the process for each point in this cluster and keeps adding points until no more points can be added. One can plot the density profile of a data set by plotting the number of data points in the ε-neighborhood of various regions, and plotting a smoothed version of the curve. This is similar to the concept of probabilistic density estimation. Intuitively, this approach corresponds to the continuous contours of intersection between the density thresholds of Figures 18.1 and 18.2 with the corresponding density profiles. The density threshold depends upon the value of *MinPts*. Note that the data points in any

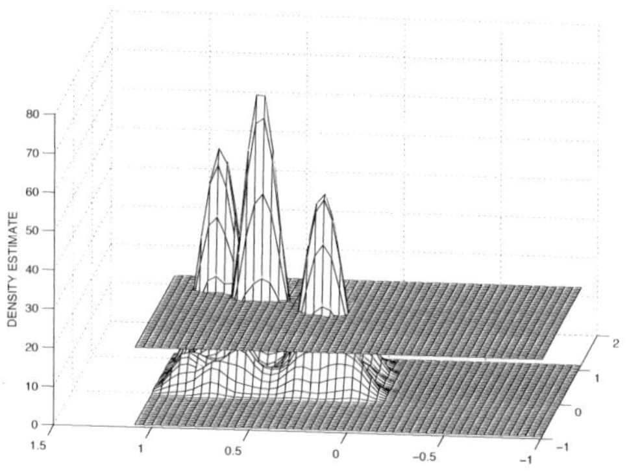

FIGURE 18.1: Density-based profile with lower density threshold.

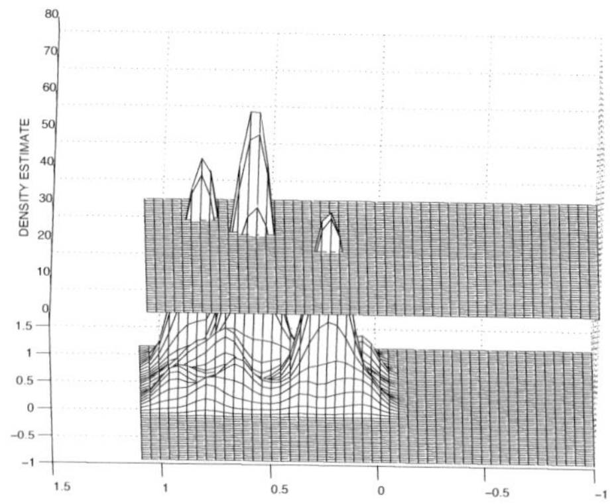

FIGURE 18.2: Density-based profile with higher density threshold.

contiguous region will have density greater than the threshold. Note that the use of a higher density threshold (Figure 18.2) results in 3 clusters, whereas the use of a lower density threshold results in 2 clusters. The fuzzy version of the DBSCAN algorithm (referred to as FDBSCAN) works in a similar way as the DBSCAN algorithm, except that the density at a given point is uncertain because of the underling uncertainty of the data points. This corresponds to the fact that the number of data points within the ε-neighborhood of a given data point can be estimated only probabilistically and is essentially an uncertain variable. Correspondingly, the reachability from one point to another is no longer deterministic, since other data points may lie within the ε-neighborhood of a given point with a certain probability, which may be less than 1. Therefore, the additional constraint that the computed reachability probability must be greater than 0.5 is added. Thus, this is a generalization of the deterministic version of the algorithm in which the reachability probability is always set to 1.

18.3.2 FOPTICS Algorithm

Another related technique discussed in [41] is that of hierarchical density-based clustering. An effective (deterministic) density-based hierarchical clustering algorithm is OPTICS [12]. We note that the core idea in OPTICS is quite similar to DBSCAN and is based on the concept of *reachability distance* between data points. While the method in DBSCAN defines a *global density parameter* which is used as a threshold in order to define reachability, the work in [41] points out that different regions in the data may have different data density, as a result of which it may not be possible to define the clusters effectively with a single density parameter. Rather, many different values of the density parameter define different (hierarchical) insights about the underlying clusters. The goal is to define an implicit output in terms of ordering data points, so that when the DBSCAN is applied with this ordering, one can obtain the hierarchical clustering at any level for different values of the density parameter. The key is to ensure that the clusters at different levels of the hierarchy are consistent with one another. One observation is that clusters defined over a lower value of ε are completely contained in clusters defined over a higher value of ε, if the value of *MinPts* is not varied. Therefore, the data points are ordered based on the value of ε required in order to obtain *MinPts* in the ε-neighborhood. If the data points with smaller values of ε are processed first, then it is assured that higher density regions are always processed before lower density regions. This ensures that if the DBSCAN algorithm is used for different values of ε with this ordering, then a consistent result is obtained. Thus, the output of the OPTICS algorithm is not the cluster membership, but it is the order in which the data points are processed. We note that since the OPTICS algorithm shares so many characteristics with the DBSCAN algorithm, it is fairly easy to extend the OPTICS algorithm to the uncertain case using the same approach as was used for extending the DBSCAN algorithm. This is referred to as the FOPTICS algorithm. Note that one of the core concepts needed to order data points is to determine the value of ε which is needed in order to obtain *MinPts* in the corresponding neighborhood. In the uncertain case, this value is defined probabilistically, and the corresponding expected values are used to order the data points. A different hierarachical clustering algorithm with the use of an information-theoretic approach was proposed in [30].

18.4 Partitional Clustering Algorithms

Partitional clustering methods are algorithms which extend the k-means and k-medoid principles to the case of uncertain data. In this section, we will discuss these methods. The advantage of using partitional clustering methods is their relative simplicity and quick execution.

18.4.1 The UK-Means Algorithm

A common approach to clustering is the k-means algorithm. In the k-means algorithm, we construct clusters around a predefined number of cluster centers. A variety of distance functions may be used in order to map the points to the different clusters. A k-means approach to clustering uncertain data was studied in the context of moving object data [16, 48]. In the case of moving objects, the actual locations of the objects may change over time as the data is reported intermittently. Thus, the position of a vehicle could be an arbitrary or circle region which uses the reported location as its center and has a size which is dependent upon the speed and direction of the vehicle. A probability density function could be used to model the probability of the presence of the vehicle at a given location at a particular time.

One possibility is to simply replace each uncertain data point by a representative point such as its centroid, and apply the (deterministic) k-means clustering method directly to it. The UK-means clustering approach is very similar to the K-means clustering approach, except that we use the *expected distance* from the data's uncertainty region to the representative of the candidate cluster to which it is assigned. It was shown in [16] that the use of expected distances has clear advantages over an approach which uses deterministic clustering algorithms over representative data points. This approach is referred to as the UK-means algorithm.

A key challenge is the computation of the expected distances between the data points and the centroids for the k-means algorithm. A natural technique for computing these expected distances is to use Monte-Carlo sampling, in which samples for the data points are used in order to compute the uncertain distances. Another technique is to create discrete buckets from both distributions and compute the expected distances by a pairwise weighted average from different pairs of buckets. Thus, if one probability density function (pdf) is discretized into m_1 buckets, and another pdf is discretized into m_2 buckets, such an approach would require $m_1 \cdot m_2$ distance computations. The Monte-Carlo approach can be very expensive because a large number of samples may be required in order to compute the distances accurately. Similarly, a large number of discrete buckets may be required in order to compute the pairwise distances accurately. The work in [16] uses a purely brute-force version of the UK-means algorithm in which no optimization or pruning of the distance computations is performed. This version can be impractical, especially if a high level of accuracy is required in the clustering process. Clearly, some kind of pruning is required in order to improve the efficiency of the approach.

The work in [48] improves on the work of [16] and designs a pruned version of the UK-means algorithm. The idea here is to use branch-and-bound techniques in order to minimize the number of expected distance computations between data points and cluster representatives. The broad idea is that once an upper bound on the minimum distance of a particular data point to some cluster representative has been quantified, it is necessary to perform the computation between this point and another cluster representative, if it can be proved that the corresponding distance is greater than this bound. In order to compute the bounds, the minimum bounding rectangle for the representative point for a cluster region is computed. The uncertain data point also represents a region over which the object may be distributed. For each representative cluster, its minimum bounding rectangle (MBR) is used to compute the following two quantities with respect to the uncertain data point:

- The minimum limit on the expected distance between the MBR of the representative point and the uncertain region for the data point itself.
- The maximum limit on the expected distance between the MBR of the representative point and the uncertain region for the data point itself.

These upper and lower bound computations are facilitated by the use of the minimum bounding rectangles in conjunction with the triangle inequality. We note that a cluster representative can be pruned, if its maximum limit is less than the minimum limit for some other representative. The approach in [48] constructs a k-d tree on the cluster representatives in order to promote an orderly pruning strategy and minimize the number of representatives which need to be accessed. It was shown in [48] that such an approach significantly improves the pruning efficiency over the brute-force algorithm.

18.4.2 The CK-Means Algorithm

While the work in [16] claims that UK-means provides qualitatively superior results to deterministic clustering, the work in [44] shows that the model utilized by the UK-means is actually equivalent to deterministic clustering, by replacing each uncertain data point by its expected value. Thus, the UK-means approach actually turns out to be equivalent to deterministic clustering. This contradicts the claim in [16] that the UK-means algorithm provides superior results to a deterministic clustering method which replaces uncertain data points with their centroids. We further note that most of the computational complexity is created by the running time required for expected distance calculations. On the other hand, deterministic distance computations are extremely efficient and are almost always superior to any method which is based on expected distance computations, whether or not pruning is used.

The UK-means algorithm aims to optimize the mean square expected distance about each cluster centroid. A key step is the computation of the expected square distance of an uncertain data point $\overline{X_i}$ with a cluster centroid $\overline{Y}$, where the latter is approximated as a deterministic entity. Let $\overline{Y}$ be the centroid of a cluster, and $\overline{X_1} \ldots \overline{X_r}$ be the set of data points in the cluster. Then, the expected mean square distance of data point $\overline{X_i}$ about $\overline{Y}$ is given by $E[||\overline{X_i} - \overline{Y}||^2]$. Then, if $\overline{Y}$ is approximated as a deterministic entity, we can show the following.

Lemma 18.4.1 *Let $\overline{X_i}$ be an uncertain data point, and $\overline{Y}$ be a deterministic point. Let $\overline{c_i} = E[\overline{X_i}]$ and $var(\overline{X_i})$ represent the sum of the variances of the pdfs in $\overline{X_i}$ over all dimensions. Then, we have*

$$E[||\overline{X_i} - \overline{Y}||^2] = E[||\overline{X_i}||^2] - ||\overline{c_i}||^2 + ||\overline{c_i} - \overline{Y}||^2$$
$$= var(\overline{X_i}) + ||\overline{c_i} - \overline{Y}||^2$$

We will provide a proof of a generalized version of this lemma slightly later (Lemma 18.4.2). We further note that the value of $E[||\overline{X_i}||^2] - ||\overline{c_i}||^2$ is equal to the variance of the uncertain data point $\overline{X_i}$ (summed over all dimensions). The term $||\overline{c_i} - \overline{Y}||^2$ is equal to the *deterministic* distance of the $\overline{Y}$ to the centroid of the uncertain data point $\overline{X_i}$. Therefore, the expected square distance of an uncertain data point $\overline{X_i}$ to the centroid $\overline{Y}$ is given by the square sum of its deterministic distance and the variance of the data point $\overline{X_i}$. The variance of the data point is not dependent on the value of $\overline{Y}$. Therefore, while computing the expected square distance to the different centroids for the UK-means algorithm, it suffices to compute the deterministic distance to the centroid $\overline{c_i}$ of $\overline{X_i}$ instead of computing the expected square distance. This means that by replacing each uncertain data point $\overline{X_i}$ with its centroid, the UK-means can be replicated exactly with an efficient deterministic algorithm.

It is important to note that the equivalence of the UK-means method to a deterministic algorithm is based on the approximation of treating each cluster centroid (in intermediate steps) as a deterministic entity. In practice, some of the dimensions may be much more uncertain than others in the clustering process over most of the data points. This is especially the case when different dimensions are collected using collection techniques with different fidelity. In such cases, the cluster centroids should not be treated as deterministic entities. Some of the streaming methods for uncertain data clustering such as those discussed in [8] also treat the cluster centroids as uncertain entities in order to enable more accurate computations. In those case, such deterministic approximations are not possible. Another method, which treats cluster centroids as uncertain entities was later proposed independently in [31]. The work on clustering streams, while treating centroids as uncertain entities will be discussed in a later section of this chapter.

18.4.3 Clustering Uncertain Data with Voronoi Diagrams

The work in [48] uses minimum bounding boxes of the uncertain objects in order to compute distance bounds for effective pruning. However, the use of minimax pruning can sometimes be quite restrictive in efficiently characterizing the uncertain object, which may have arbitrary shape. An approach which is based on voronoi diagrams, also improves the UK-means algorithms by computing the voronoi diagrams of the current set of cluster representatives [37]. Each cell in this voronoi diagram is associated with a cluster representative. We note that each cell in this voronoi diagram has the property that any point in this cell is closer to the cluster representative for that cell than any other representative. Therefore, if the MBR of an uncertain object lies completely inside a cell, then it is not necessary to compute its distance to any other cluster representatives. Similarly, for any pair of cluster representatives, the perpendicular bisector between the two is a hyperplane which is equidistant from the two representatives and is easily derivable from the voronoi diagram. In the event that the MBR of an uncertain object lies completely on one side of the bisector, we can deduce that one is the cluster representatives is closer to the uncertain object than the other. This allows us to prune one of the representatives.

As in [48], this work is focused on pruning the number of expected distance computations. It has been shown in [37] that the pruning power of the voronoi method is greater than the minimax method proposed in [48]. However, the work in [37] does not compare its efficiency results to those in [44], which are based on the equivalence of UK-means to a deterministic algorithm and does not require any expected distance computations at all. It would seem that any deterministic method for k-means clustering (as proposed in the reduction of [44]) should be much more efficient than a method based on pruning the number of expected distance computations, no matter how effective the pruning methodology might be.

18.4.4 Approximation Algorithms for Clustering Uncertain Data

Recently, techniques have been designed for approximation algorithms for uncertain clustering in [20]. The work in [20] discusses extensions of the k-mean and k-median version of the problems. Bicriteria algorithms are designed for each of these cases. One algorithm achieves a $(1+\varepsilon)$-approximation to the best uncertain k-centers with the use of $O(k \cdot \varepsilon^{-1} \cdot \log^2(n))$ centers. The second algorithm picks $2k$ centers and achieves a constant-factor approximation.

A key approach proposed in [20] is the use of a transformation from the uncertain case to a weighted version of the deterministic case. We note that solutions to the weighted version of the deterministic clustering problem are well known and require only a polynomial blow-up in the problem size. The key assumption in solving the weighted deterministic case is that the ratio of the largest to smallest weights is polynomial. This assumption is assumed to be maintained in the transformation. This approach can be used in order to solve both the uncertain k-means and k-median version of the problem with the aforementioned approximation guarantees. We refer the reader to [20, 27] for details of these algorithms.

18.4.5 Speeding Up Distance Computations

We note that there are two main ways in which the complexity of distance computations in a k-means algorithm can be reduced. The first is by using a variety of pruning tricks, which cuts down on the *number* of distance computations between data points and cluster representatives. The second is by speeding up the expected distance computation itself. This kind of approach can be especially useful where the pruning effectiveness of a technique such as that proposed in [48] is not guaranteed. Therefore, a natural question arises as to whether one can speed up the uncertain distance computations, which cause the performance bottleneck in these methods.

The work in [54] designs methods for speeding up distance computations for the clustering process. We note that such fast distance computations can be very effective not only for the UK-means algorithm, but for any clustering technique which is dependent on expected distance computations. The work in [54] proposes a number of methods for performing distance computations between uncertain objects, which provide different tradeoffs between effectiveness and efficiency. Specifically, for a pair of uncertain objects $\overline{X}$ and $\overline{Y}$, the following methods can be used in order to compute the distances between them:

- **Certain Representation:** Each uncertain object can be replaced by a certain object, corresponding to the expected values of its attributes. The distances between these objects can be computed in a straightforward way. While this approach is very efficient, it provides very poor accuracy.

- **Sampling:** It is possible to repeatedly sample both objects for pairs of instantiations and compute the distances between them. The average of these computed distances can be reported as the expected value. However, such an approach may require a large number of samples in order to provide a high quality approximation.

- **Probability Histograms:** Each uncertain object can be approximated by a set of bins, which corresponds to its probability histogram. Then, for every pair of bins between the two objects, the probability of that instantiation and the distance between the average values of those bins is computed. The weighted average over all pairs of bins is reported. Such an approach can still be quite inefficient in many scenarios, where a large number of bins is required to represent the probability histogram effectively.

- **Gaussian Mixture Modeling with Sample Clustering:** Each uncertain object can be approximated with a mixture of Gaussians. Specifically, we sample each uncertain object with the use of its pdf, and then cluster these samples with deterministic k-means clustering. Each of these clusters can be fit into a Gaussian model. Then, the pairwise weighted average distances between each of the components of the mixture can be computed.

- **Single Gaussian Modeling:** It turns out that it is not necessary to use multiple components in the mixture model for the approximation process. In fact, it suffices to use a single component for the mixture.

The last result is actually not very surprising in light of Lemma 18.4.1. In fact, the Gaussian assumption is not required at all, and it can be shown that the distance between a pair of uncertain objects (for which the pdfs are independent of one another) can be expressed purely as a function of their means and variances. Therefore, we propose the following (slight) generalization of Lemma 18.4.1.

Lemma 18.4.2 *Let $\overline{X_i}$ and $\overline{Y_i}$ be two uncertain data points, with means $\overline{c_i}$ and $\overline{d_i}$ respectively. Let the sum of the variances across all dimensions of these points be $var(\overline{X_i})$ and $var(\overline{Y_i})$, respectively. Then, we have*

$$E[||\overline{X_i} - \overline{Y_i}||^2] = ||\overline{c_i} - \overline{d_i}||^2 + var(\overline{X_i}) + var(\overline{Y_i}) \tag{18.2}$$

Proof: We can expand the term within the expectation on the left-hand side as follows:

$$E[||\overline{X_i} - \overline{Y_i}||^2] = E[||(\overline{X_i} - \overline{c_i}) + (\overline{c_i} - \overline{d_i}) + (\overline{d_i} - \overline{Y_i})||^2] \tag{18.3}$$

We further note that the three expressions within the round brackets on the right-hand side are statistically independent of one another. This means that their covariances are zero. Furthermore, the expected values of $(\overline{X_i} - \overline{c_i})$ and $(\overline{d_i} - \overline{Y_i})$ are both 0. This can be used to show that the expectation of the product of any pair of terms within the round brackets on the right-hand side of Equation 18.3 is 0. This implies that we can rewrite the right-hand side (RHS) as follows:

$$E[||\overline{X_i} - \overline{Y_i}||^2] = E[||\overline{X_i} - \overline{c_i}||^2] + (\overline{c_i} - \overline{d_i})^2 + E[||\overline{d_i} - \overline{Y_i}||^2] \tag{18.4}$$

The first term on the RHS of the above expression is $var(\overline{X_i})$ and the last term is $var(\overline{Y_i})$. The result follows.

The aforementioned results suggest that it is possible to compute the distances between pairs of uncertain objects very efficiently, as long as the uncertainties in different objects are statistically independent. Another observation is that these computations *do not require knowledge of the full probability density function of the probabilistic records*, but can be made to work with the more modest assumption about the standard error $var(\cdot)$ of the underlying uncertainty. This is a more reasonable assumption for many applications. Such standard errors are included as a natural part of the measurement process, though the full probability density functions are rarely available. This also suggests that a lot of work on pruning the number of expected distance computations may not be quite as critical to efficient clustering as has been suggested in the literature.

18.5 Clustering Uncertain Data Streams

In many applications such as sensor data, the data may have uncertainty, due to errors in the readings of the underlying sensors. This may result in uncertain *streams* of data. Uncertain streams pose of special challenge because of the dual complexity of high volume and data uncertainty. As we have seen in earlier sections, efficiency is a primary concern in the computation of expected distances, when working with probability density functions of data points. Therefore, it is desirable to work with simpler descriptions of the underlying uncertainty. This will reduce both the underlying data volume and complexity of stream computations. In recent years, a number of methods have specifically been proposed for clustering uncertain data streams.

18.5.1 The UMicro Algorithm

In this section, we will introduce *UMicro*, the Uncertain MICROclustering algorithm for data streams. We assume that we have a data stream which contains d dimensions. The actual records in the data are denoted by $\overline{X_1}$, $\overline{X_2}$, ... $\overline{X_N}$.... We assume that the estimated error associated with the

jth dimension for data point $\overline{X_i}$ is denoted by $\psi_j(\overline{X_i})$. This error is defined in terms of the standard deviation of the error associated with the value of the jth dimension of $\overline{X_i}$. The corresponding d-dimensional error vector is denoted by $\overline{\psi(X_i)}$. Thus, the input to the algorithm is a data stream in which the ith pair is denoted by $(\overline{X_i}, \overline{\psi(X_i)})$.

We note that most of the uncertain clustering techniques work with the assumption that the entire probability density function is available. In many real applications, a more realistic assumption is that only the standard deviations of the errors are available. This is because complete probability distributions are rarely available and are usually inserted only as a modeling assumption. An overly ambitious modeling assumption can also introduce modeling errors. It is also often quite natural to be able to estimate the standard error in many modeling scenarios. For example, in a scientific application in which the measurements can vary from one observation to another, the error value is the standard deviation of the observations over a large number of measurements. In a k-anonymity-based data (or incomplete data) mining application, this is the standard deviation of the partially specified (or imputed) fields in the data. This is also more practical from a stream perspective, because it reduces the volume of the incoming stream and reduces the complexity of stream computations.

The microclustering model was first proposed in [56] for large data sets and subsequently adapted in [6] for the case of deterministic data streams. The *UMicro* algorithm extends the microclustering approach of [6] to the case of uncertain data. In order to incorporate the uncertainty into the clustering process, we need a method to incorporate and leverage the error information into the microclustering statistics and algorithms. As discussed earlier, it is assumed that the data stream consists of a set of multidimensional records $\overline{X}_1 \ldots \overline{X}_k \ldots$ arriving at time stamps $T_1 \ldots T_k \ldots$. Each $\overline{X_i}$ is a multidimensional record containing d dimensions which are denoted by $\overline{X_i} = (x_i^1 \ldots x_i^d)$. In order to apply the microclustering method to the uncertain data mining problem, we also need to define the concept of error-based microclusters. We define such microclusters as follows.

Definition 18.5.1 *An uncertain microcluster for a set of d-dimensional points $X_{i_1} \ldots X_{i_n}$ with timestamps $T_{i_1} \ldots T_{i_n}$ and error vectors $\overline{\psi(X_{i_1})} \ldots \overline{\psi(X_{i_n})}$ is defined as the $(3 \cdot d + 2)$tuple $(\overline{CF2^x}(C)$, $\overline{EF2^x}(C), \overline{CF1^x}(C), t(C), n(C))$, wherein $\overline{CF2^x}(C)$, $\overline{EF2^x}(C)$, and $\overline{CF1^x}(C)$ each correspond to a vector of d entries. The entries in $\overline{EF2^x}(C)$ correspond to the error-based entries. The definition of each of these entries is as follows:*

- *For each dimension, the sum of the squares of the data values is maintained in $\overline{CF2^x}(C)$. Thus, $\overline{CF2^x}(C)$ contains d values. The* p*th entry of $\overline{CF2^x}(C)$ is equal to $\sum_{j=1}^{n} (x_{i_j}^p)^2$. This corresponds to the second moment of the data values along the* p*th dimension.*
- *For each dimension, the sum of the squares of the errors in the data values is maintained in $\overline{EF2^x}(C)$. Thus, $\overline{EF2^x}(C)$ contains d values. The* p*th entry of $\overline{EF2^x}(C)$ is equal to $\sum_{j=1}^{n} \psi_p(X_{i_j})^2$. This corresponds to the sum of squares of the errors in the records along the* p*th dimension.*
- *For each dimension, the sum of the data values is maintained in $\overline{CF1^x}(C)$. Thus, $\overline{CF1^x}(C)$ contains d values. The* p*th entry of $\overline{CF1^x}(C)$ is equal to $\sum_{j=1}^{n} x_{i_j}^p$. This corresponds to the first moment of the values along the pth dimension.*
- *The number of points in the data is maintained in $n(C)$.*
- *The timestamp of the last update to the microcluster is maintained in $t(C)$.*

We note that the uncertain definition of microclusters differs from the deterministic definition, since we have added additional d values corresponding to the error information in the records. We will refer to the uncertain microcluster for a set of points C by $\overline{ECF}(C)$. We note that error-based microclusters maintain the important *additive property* [6] which is critical to its use in the clustering process. We restate the additive property as follows.

Property 18.5.1 *Let C_1 and C_2 be two sets of points. Then all nontemporal components of the error-based cluster feature vector $\overline{ECF}(C_1 \cup C_2)$ are given by the sum of $\overline{ECF}(C_1)$ and $\overline{ECF}(C_2)$.*

The additive property follows from the fact that the statistics in the individual microclusters are expressed as a separable additive sum of the statistics over individual data points. We note that the single temporal component $t(C_1 \cup C_2)$ is given by $\max\{t(C_1), t(C_2)\}$. We note that the additive property is an important one, since it ensures that it is easy to keep track of the cluster statistics as new data points arrive. Next, we will discuss the process of uncertain microclustering. The *UMicro* algorithm works using an iterative approach which maintains a number of microcluster centroids around which the clusters are built. It is assumed that one of the inputs to the algorithm is n_{micro}, which is the number of microclusters to be constructed. The algorithm starts off with a number of null clusters and initially creates new singleton clusters, to which new points are added subsequently. For any incoming data point, the closest cluster centroid is determined by using the *expected distance* of the uncertain data point to the *uncertain microclusters*. The process of expected distance computation for the closest centroid is tricky and will be subsequently discussed. Furthermore, for the incoming data point, it is determined whether it lies within a *critical uncertainty boundary* of the microcluster. If it lies within this critical uncertainty boundary, then the data point is added to the microcluster, otherwise a new microcluster needs to be created containing the singleton data point. In order to create a new microcluster, either it must be added to the current set of microclusters, or it needs to replace one of the older microclusters. In the initial stages of the algorithm, the current number of microclusters is less than n_{micro}. If this is the case, then the new data point is added to the current set of microclusters as a separate microcluster with a singleton point in it. Otherwise, the new data point needs to replace one of the older microclusters. For this purpose, we always replace the least recently updated microcluster from the data set. This information is available from the temporal timestamp in the different microclusters. The overall framework for the uncertain stream clustering algorithm is illustrated in Figure 18.3. Next, we will discuss the process of computation of individual subroutines such as the expected distance or the uncertain boundary.

```
Algorithm UMicro(Number of Clusters: n_micro)
begin
S = {};
{ Set of micro-clusters }
repeat
   Receive the next stream point X̄;
   { Initially, when S is null, the computations below
    cannot be performed, and X̄ is simply
    added as a singleton micro-cluster to S }
   Compute the expected similarity of X̄ to the closest
    micro-cluster 𝓜 in S;
   Compute critical uncertainty boundary of 𝓜;
   if X̄ lies inside uncertainty boundary
   add X̄ to statistics of 𝓜
   else
   add a new micro-cluster to S containing singleton
      point X̄;
   if |S| = n_micro + 1 remove the least recently
        updated micro-cluster from S;
until data stream ends;
end
```

FIGURE 18.3: The UMicro algorithm.

In order to compute the expected similarity of the data point $\overline{X}$ to the centroid of the cluster C, we need to determine a closed form expression which is expressed only in terms of $\overline{X}$ and $ECF(C)$. We note that just as the individual data points are essential random variables with a given error, the centroid $\overline{Z}$ of a cluster C is also a random variable. We make the following observation about the centroid of a cluster:

Lemma 18.5.1 *Let $\overline{Z}$ be the random variable representing the centroid of cluster C. Then, the following result holds true:*

$$E[||Z||^2] = \sum_{j=1}^{d} CF1(C)_j^2/n(C)^2 + \sum_{j=1}^{d} EF2(C)_j/n(C)^2 \tag{18.5}$$

Proof: We note that the random variable Z_j is given by the current instantiation of the centroid and the mean of $n(C)$ different error terms for the points in cluster C. Therefore, we have

$$Z_j = CF1(C)_j/n(C) + \sum_{\overline{X}\in C} e_j(\overline{X})/n(C) \tag{18.6}$$

Then, by squaring $\overline{Z_j}$ and taking the expected value, we obtain the following:

$$E[Z_j^2] = CF1(C)_j^2/n(C)^2 + 2\cdot \sum_{\overline{X}\in C} E[e_j(\overline{X})]\cdot CF1(C)_j/n(C)^2 + E[(\sum_{\overline{X}\in C} e_j(\overline{X}))^2]/n(C)^2 \tag{18.7}$$

Now, we note that the error term is a random variable with standard deviation $\psi_j(\cdot)$ and zero mean. Therefore, $E[e_j] = 0$. Further, since it is assumed that the random variables corresponding to the errors of different records are independent of one another, we have $E[e_j(\overline{X})\cdot e_j(\overline{Y})] = E[e_j(\overline{X})]\cdot E[e_j(\overline{Y})] = 0$. By using these relationships in the expansion of the above equation, we get

$$\begin{aligned} E[Z_j^2] &= CF1(C)_j^2/n(C)^2 + \sum_{\overline{X}\in C} E[e_j(\overline{X})^2]/n(C)^2 = CF1(C)_j^2/n(C)^2 + \sum_{\overline{X}\in C} \psi_j(\overline{X})^2/n(C)^2 \\ &= CF1(C)_j^2/n(C)^2 + EF2(C)_j/n(C)^2 \end{aligned}$$

By adding the value of $E[Z_j^2]$ over different values of j, we get

$$E[||Z||^2] = \sum_{j=1}^{d} CF1(C)_j^2/n(C)^2 + \sum_{j=1}^{d} EF2(C)_j/n(C)^2 \tag{18.8}$$

This proves the desired result.

Next, we will use the above result to directly estimate the expected distance between the centroid of cluster C and the data point $\overline{X}$. We will prove the following result:

Lemma 18.5.2 *Let v denote the expected value of the square of the distance between the uncertain data point $\overline{X} = (x_1 \dots x_d)$ (with instantiation $(x_1 \dots x_d)$ and error vector $(\psi_1(\overline{X}) \dots \psi_d(\overline{X}))$ and the centroid of cluster C. Then, v is given by the following expression:*

$$v = \sum_{j=1}^{d} CF1(C)_j^2/n(C)^2 + \sum_{j=1}^{d} EF2(C)_j/n(C)^2 + \sum_{j=1}^{d} x_j^2 + \sum_{j=1}^{d} (\psi_j(\overline{X}))^2 - 2\sum_{j=1}^{d} x_j\cdot CF1(C)_j/n(C) \tag{18.9}$$

Proof: Let $\overline{Z}$ represent the centroid of cluster C. Then, we have

$$v = E[||\overline{X} - \overline{Z}||^2] = E[||\overline{X}||^2] + E[||\overline{Z}||^2] - 2E[\overline{X} \cdot \overline{Z}] = E[||\overline{X}||^2] + E[||\overline{Z}||^2] - 2E[\overline{X}] \cdot E[\overline{Z}]$$

Next, we will analyze the individual terms in the above expression. We note that the value of X is a random variable, whose expected value is equal to its current instantiation, and it has an error along the jth dimension which is equal to $\psi_j(\overline{X})$. Therefore, the expected value of $E[||\overline{X}||^2]$ is given by

$$E[||\overline{X}||^2] = (E[X])^2 + \sum_{j=1}^{d} (\psi_j(\overline{X}))^2 = \sum_{j=1}^{d} x_j^2 + \sum_{j=1}^{d} (\psi_j(\overline{X}))^2$$

Now, we note that the jth term of $E[Z]$ is equal to the jth dimension of the centroid of cluster C. This is given by the expression $CF1(C)_j/n(C)$, where $CF1_j(C)$ is the jth term of the first order cluster component $CF1(C)$. Therefore, the value of $E[X] \cdot E[Z]$ is given by the following expression:

$$E[X] \cdot E[Z] = \sum_{j=1}^{d} x_j \cdot CF1(C)_j/n(C) \tag{18.10}$$

The results above and Lemma 18.5.1 define the values of $E[||X||^2]$, $E[||Z||^2]$, and $E[X \cdot Z]$. Note that all of these values occur in the right-hand side of the following relationship:

$$v = E[||\overline{X}||^2] + E[||\overline{Z}||^2] - 2E[\overline{X}] \cdot E[\overline{Z}] \tag{18.11}$$

By substituting the corresponding values in the right-hand side of the above relationship, we get

$$v = \sum_{j=1}^{d} CF1(C)_j^2/n(C)^2 + \sum_{j=1}^{d} EF2(C)_j/n(C)^2 + \sum_{j=1}^{d} x_j^2 + \sum_{j=1}^{d} (\psi_j(\overline{X}))^2 - 2\sum_{j=1}^{d} x_j \cdot CF1(C)_j/n(C) \tag{18.12}$$

The result follows.

The result of Lemma 18.5.2 establishes how the square of the distance may be computed (in expected value) using the error information in the data point $\overline{X}$ and the microcluster statistics of C. Note that this is an efficient computation which requires $O(d)$ operations, which is asymptotically the same as the deterministic case. This is important since distance function computation is the most repetitive of all operations in the clustering algorithm, and we would want it to be as efficient as possible.

While the expected distances can be directly used as a distance function, the uncertainty adds a lot of noise to the computation. We would like to remove as much noise as possible in order to determine the most accurate clusters. Therefore, we design a dimension-counting similarity function which prunes the uncertain dimensions during the similarity calculations. This is done by computing the variance σ_j^2 along each dimension j. The computation of the variance can be done by using the cluster feature statistics of the different microclusters. The cluster feature statistics of all microclusters are added to create one global cluster feature vector. The variance of the data points along each dimension can then be computed from this vector by using the method discussed in [56]. For each dimension j and threshold value *thresh*, we add the *similarity value* $\max\{0, 1 - E[||X - Z||_j^2]/(thresh * \sigma_j^2)\}$ to the computation. We note that this is a similarity value rather than a distance value, since larger values imply greater similarity. Furthermore, dimensions which have a large amount of uncertainty are also likely to have greater values of $E[||X - Z||_j^2]$ and are often pruned from the computation. This improves the quality of the similarity computation.

Next, we describe the process of computing the uncertain boundary of a microcluster. Once the closest microcluster for an incoming point has been determined, we need to decide whether it should be added to the corresponding microclustering statistics, or whether a new microcluster containing a singleton point should be created. We create a new microcluster, if the incoming point lies outside the uncertainty boundary of the microcluster. The uncertainty boundary of a microcluster is defined in terms of the standard deviation of the distances of the data points about the centroid of the microcluster. Specifically, we use t standard deviations from the centroid of the cluster as a boundary for the decision of whether to include that particular point in the microcluster. A choice of $t = 3$ ensures a high level of certainty that the point does not belong to that cluster with the use of the normal distribution assumption. Let $\overline{W}$ be the centroid of the cluster C, and let the set of points in it be denoted by $\overline{Y_1} \ldots \overline{Y_r}$. Then, the uncertain radius U is denoted as follows:

$$U = \sum_{i=1}^{r} \sum_{j=1}^{d} E[||Y_i - W||_j^2] \tag{18.13}$$

The expression on the right-hand side of the above equation can be evaluated by using the relationship of Lemma 18.5.2.

18.5.2 The LuMicro Algorithm

A variation of the *UMicro* algorithm has been discussed in [55], which incorporates the concept of tuple uncertainty into the clustering process. The primary idea in this approach is that the *instance uncertainty* of a cluster is quite important, in addition to the expected distances of assignment. If T is the set of possible probabilistic instances of a tuple, then the instance uncertainty $U(T)$ is defined as follows:

$$U(T) = -\sum_{x_i \in T} p(x_i) \cdot \log(p(x_i)) \tag{18.14}$$

We note that the value of $U(T)$ is somewhat akin to the concept of entropy, is always at least 0, and takes on the least value of 0 for deterministic data. This concept can also be generalized to a cluster (rather than a single tuple) by integrating all possible probabilistic instances into the computation. As more data points are added to the cluster, the tuple uncertainty decreases, because the data in the cluster tends to be biased toward a few common tuple values. Intuitively, this is also equivalent to a reduction in entropy. The *LuMicro* algorithm implements a very similar approach as the *UMicro* method in terms of assigning data points to their closest clusters (based on expected distance), except that the distance computation is only used to narrow down to a smaller set of candidate centroids. The final decision on centroid assignment is performed by determining the cluster to which the addition of the data point would result in the greatest reduction in uncertainty (or entropy). Intuitively, this can be considered an algorithm which incorporates distance-based and probabilistic entropy-based concepts into the clustering process. Unlike the *UMicro* algorithm, the *LuMicro* method works with the full probability distribution functions of the underlying records, rather than only the error values because the computation of the uncertainty values requires knowledge of the full probability distribution of the tuples.

18.5.3 Enhancements to Stream Clustering

The method for clustering uncertain data streams can be further enhanced in several ways:

- In many applications, it is desirable to examine the clusters over a specific time horizon rather than the entire history of the data stream. In order to achieve this goal, a pyramidal time frame [6] can be used for stream classification. In this time frame, snapshots are stored in different orders depending upon the level of recency. This can be used in order to retrieve clusters over a particular horizon with very high accuracy.
- In some cases, the behavior of the data stream may evolve over time. In such cases, it is useful to apply a *decay-weighted* approach. In the decay-weighted approach, each point in the stream

is a weighted by a factor which decays over time. Such an approach can be useful in a number of scenarios in which the behavior of the data stream changes considerably over time. In order to use the decay-weighted approach, the key modification is to define the microclusters with a weighted sum of the data points, as opposed to the explicit sums. It can be shown that such an approach can be combined with a lazy-update method in order to effectively maintain the microclusters.

18.6 Clustering Uncertain Data in High Dimensionality

Recently, this method has also been extended to the case of projected and subspace clustering of high-dimensional uncertain data [32, 4]. The high-dimensional scenario suffers from data sparsity, which makes it particularly susceptible to noise. The addition of uncertainty typically increases the noise and reduces the correlations among different dimensions. This tends to magnify the high dimensional sparsity issue and makes the problem even more challenging.

In the case of the standard clustering problem, the main effect of uncertainty is the impact on the distance computations. However, in the uncertain case, the uncertainty also affects the choice of dimensions to be picked. The reason for this is that different dimensions in the data can have very different levels of uncertainty. Clearly, the level of uncertainty in a given dimension is critical information in characterizing the clustering behavior along a particular dimension. This is particularly important for the high dimensional case in which a very large number of dimensions may be available with varying clustering behavior and uncertainty. The interplay between the clustering of the values and the level of uncertainty may affect the subspaces which are most optimal for the clustering process. In some cases, if the uncertainty data is not used in the mining process, this may result in a clustering which does not truly reflect the underlying behavior.

For example, consider the case illustrated in Figure 18.4. In this case, we have illustrated two clusters which are denoted by "Point Set A" and "Point Set B." In each case, we have also illustrated the uncertainty behavior with elliptical contours. The two data sets are identical, except that the uncertainty contours are very different. In the case of point set A, it is better to pick the projection along the X-axis, because of lower uncertainty along that axis. On the other hand, in the case of point set B, it is better to pick the projection along the Y-axis because of lower uncertainty in that direction. This problem is further magnified when the dimensionality increases, and the different dimensions have different patterns of data and uncertainty distributions. We will examine the interplay between data uncertainty and projections along different dimensionalities for the clustering process. We will

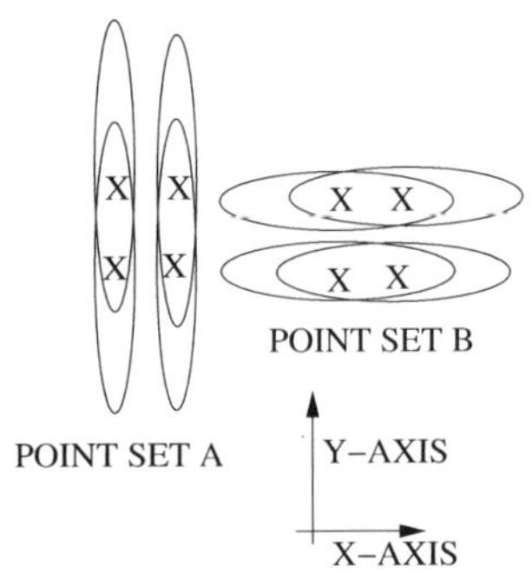

FIGURE 18.4: Effect of uncertainty in picking projections.

show that the incorporation of uncertainty information into critical algorithmic decisions leads to much better quality of the clustering.

In this section, we will discuss two different algorithms, one of which allows overlap among the different clusters, and the other designs a method for strict partitioning of the data in the uncertain streaming scenario. The first case creates a *soft partitioning* of the data, in which data points belong to clusters with a probability. This is also referred to as *membership degree*, a concept which we will discuss in the next subsection.

18.6.1 Subspace Clustering of Uncertain Data

A subspace clustering algorithm for uncertain data was proposed in [32]. The algorithm uses a grid-based method, which attempts to search on the space of medoids and subspaces for the clustering process. In the grid-based approach, the support is counted with a width w on the relevant subset of dimensions. Thus, for a given medoid m, we examine a distance w from the medoid along each of the relevant dimensions. The support of the hypercubes of this grid provide us with an idea of the dense subspaces in the data. For the other dimensions, unlimited width is considered. A Monte-Carlo sampling approach is used in order to search on the space of possible medoids. The core unit of the algorithm is a Monte-Carlo sampling approach, which generates a single good subspace cluster from the database.

In order to achieve this goal, a total of *numMedoids* are sampled from the underlying data. For each such medoid, its best possible local subspace is constructed in order to generate the grid-based subspace cluster. The quality of this local subspace is identified, and the best medoid (and associated subspace cluster) among all the *numMedoids* different possibilities is identified. In order to generate the local subspaces around the medoid, a support parameter called *minSup* is used. For all local subspaces (corresponding to grid width w), which have support of at least *minSup*, the *quality* of the corresponding subspace is determined. The quality of a local subspace cluster is different from the support in order to account for the different number of dimensions in the different subspaces. If this quality is the best encountered so far, then we update the best medoid (and corresponding subspace) encountered so far. The quality function for a medoid m and subspace S is related to the support as follows:

$$quality(m,S) = support(m,S) * 1/\beta^{|S|} \tag{18.15}$$

Here $\beta \in (0,1)$ normalizes for the different number of dimensions in the different subspaces S. The idea is that a subspace with a larger number of dimensions, but with the same support, is considered to be of better quality. A number of different methods can be used in order to compute the support of the subset S of dimensions:

- **Expectation-Based Support:** In this case, the support is defined as the number of data points, whose centroids lie within a given distance w of the medoid along the relevant dimensions. Essentially, this method for support computation is similar to the deterministic case of replacing uncertain objects with their centroids.

- **Minimal Probability-Based Support:** In this case, the support is defined as the number of data points that have a minimum probability of being within a distance of w from the medoid along each of the relevant dimensions.

- **Exact Probability-Based Support:** This computes the sum of the probabilities that the different objects lie within a distance of w along each of the relevant dimensions. This value is actually equal to the expected number of objects which lie within a width of w along the relevant dimensions.

We note that the last two measurements require the computation of a probability that an uncertain object lies within a specific width w of a medoid. This probability also reflects the *membership degree* of the data point to the cluster.

We note that the aforementioned technique only generates a single subspace cluster with the use of sampling. A question arises as to how we can generalize this in order to generate the *overall* clustering. We note that repeated samplings may generate the same set of clusters, a scenario which we wish to avoid. In order to reduce repeated clusters, two approaches can be used:

- Objects which have a minimal probability of belonging to any of the previously generated clusters are excluded from consideration for being medoids.

- The probability of an object being selected as a medoid depends upon its membership degree to the previously generated clusters. Objects which have very low membership degrees to previously generated clusters have a higher probability of being selected as medoids.

The work in [32] explores the different variations of the subspace clustering algorithms, and shows that the methods are superior to methods such as UK-means and deterministic projected clustering algorithms such as PROCLUS [7].

18.6.2 UPStream: Projected Clustering of Uncertain Data Streams

The *UPStream* algorithm is designed for the high dimensional uncertain *stream scenario*. This algorithm can be considered an extension of the *UMicro* algorithm. The error model of the *UPStream* algorithm is quite different from the algorithm of [32] and uses a model of error standard deviations rather than the entire probability distribution. This model is more similar to the *UMicro* algorithm.

The data stream consists of a set of incoming records which are denoted by $\overline{X_1} \ldots \overline{X_i} \ldots$. It is assumed that the data point $\overline{X_i}$ is received at the timestamp T_i. It is assumed that the dimensionality of the data set is d. The d dimensions of the record $\overline{X_i}$ are denoted by $(x_i^1 \ldots x_i^d)$. In addition, each data point has an error associated with the different dimensions. The error (standard deviation) associated with the jth dimension for data point $\overline{X_i}$ is denoted by $\psi_j(\overline{X_i})$.

In order to incorporate the greater importance of recent data points in an evolving stream, we use the concept of a *fading function* $f(t)$, which quantifies the relative importance of the different data points over time. The fading function is drawn from the range $(0,1)$ and serves as a multiplicative factor for the relative importance of a given data point. This function is a monotonically decreasing function and represents the gradual fading of importance of a data point over time. A commonly used decay function is the exponential decay function. The exponential decay function $f(t)$ with parameter λ is defined as follows as a function of the time t:

$$f(t) = 2^{-\lambda \cdot t} \tag{18.16}$$

We note that the value of $f(t)$ reduces by a factor of 2 every $1/\lambda$ time units. This corresponds to the half-life of the function $f(t)$. We define the half-life as follows.

Definition 18.6.1 *The half-life of the function $f(\cdot)$ is defined as the time t at which $f(t) = (1/2) \cdot f(0)$. For the exponential decay function, the half-life is $1/\lambda$.*

In order to keep track of the statistics for cluster creation, two sets of statistics are maintained:

- Global data statistics which keep track of the variance along different dimensions of the data. This data is necessary in order to maintain information about the scaling behavior of the underlying data.

- Fading microcluster statistics which keep track of the cluster behavior, the projection dimensions as well as the underlying uncertainty.

Let us assume that the data points that have arrived so far are $\overline{X_1} \ldots \overline{X_N} \ldots$. Let t_c be the current time.

- For each dimension, the weighted sums of the squares of the individual dimensions of $\overline{X_1} \ldots \overline{X_N} \ldots$ over the entire data stream are maintained. There are a total of d such entries. The ith component of the global second-order statistics is denoted by $gs(i)$ and is equal to $\sum_{j=1}^{N} f(t_c - T_j) \cdot (x_j^i)^2$.
- For each dimension, the sums of the individual dimensions of $\overline{X_1} \ldots \overline{X_N} \ldots$ over the entire data stream are maintained. There are a total of d such entries. The ith component of the global first-order statistics is denoted by $gf(i)$ and is equal to $\sum_{j=1}^{N} f(t_c - T_j) \cdot (x_j^i)$.
- The sum of the weights of the different values of $f(T_j)$ are maintained. This value is equal to $\sum_{j=1}^{N} f(t_c - T_j)$. This value is denoted by gW.

The above statistics can be easily maintained over a data stream since the values are computed additively over arriving data points. At first sight, it would seem that the statistics need to be updated at each clock tick. In reality, because of the multiplicative nature of the exponential distribution, we only need to update the statistics on the arrival of each new data point. Whenever a new data point arrives at time T_i, we multiply each of the statistics by $e^{-\lambda \cdot T_i - T_{i-1}}$ and then add the statistics for the incoming data point $\overline{X_i}$. We note that the global variance along a given dimension can be computed from the above values. Therefore, the global variance can be maintained continuously over the entire data stream.

Observation 18.6.1 *The variance along the ith dimension is given by* $\frac{gs(i)}{gW} - \frac{gf(i)^2}{gW^2}$.

The above fact can be easily proved by using the fact that for any random variable Y the variance $var(Y)$ is given by $E[Y^2] - E[Y]^2$. We will denote the global standard deviation along dimension i at time t_c by $\sigma(i, t_c)$. As suggested by the observation above, the value of $\sigma(i, t_c)$ is easy to maintain by using the global statistics discussed above.

An uncertain microcluster $C = \{X_{i_1} \ldots X_{i_N}\}$ is represented as follows.

Definition 18.6.2 *The uncertain microcluster for a set of d-dimensional points* $\overline{X_{i_1}} \ldots \overline{X_{i_n}}$ *with time-stamps given by* $T_{i_1} \ldots T_{i_n}$, *and error vectors* $\overline{\psi(X_{i_1})} \ldots \overline{\psi(X_{i_n})}$ *is defined as the* $(3 \cdot d + 3)$ *tuple* $\overline{ECF(C)} = (\overline{CF2(C)}, \overline{EF2(C)}, \overline{CF1(C)}, t(C), W(C), n(C))$, *and a d-dimensional bit vector* $\mathcal{B}(C)$, *wherein the corresponding entries are defined as follows:*

- *For each of the d dimensions, we maintain the weighted sum of the squares of the data values in* $\overline{CF2(C)}$. *The pth entry is given by* $\sum_{j=1}^{n} f(t - T_{i_j}) \cdot (x_{i_j}^p)^2$.
- *For each of the d dimensions, we maintain the weighted sum of the squares of the errors (along the corresponding dimension) in* $\overline{EF2(C)}$. *The pth entry is given by* $\sum_{j=1}^{n} f(t - T_{i_j}) \cdot \psi_p(X_{i_j})^2$.
- *For each of the d dimensions, we maintain the weighted sum of the data values in* $\overline{CF1(C)}$. *The pth entry is given by* $\sum_{j=1}^{n} f(t - T_{i_j}) \cdot x_{i_j}^p$.
- *The sum of the weights is maintained in* $W(C)$. *This value is equal to* $\sum_{j=1}^{n} f(t - T_{i_j})$.
- *The number of data points is maintained in* $n(C)$.
- *The last time at which a data point was added to the cluster is maintained in* $t(C)$.
- *We also maintain a d-dimensional bit-vector* $\mathcal{B}(C)$. *Each bit in this vector corresponds to a dimension. A bit in this vector takes on the value of 1, if that dimension is included in the projected cluster. Otherwise, the value of the bit is zero.*

This definition is quite similar to the case of the *UMicro* algorithm, except that there is also a focus on maintaining dimension-specific information and the time-decay information. We note that the microcluster definition discussed above satisfies two properties: the *additive property* and the *multiplicative property*. The additive property is common to all microclustering techniques:

Observation 18.6.2 Additive Property *Let C_1 and C_2 be two sets of points. Then the components of the error-based cluster feature vector (other than the timestamp) $\overline{ECF(C_1 \cup C_2)}$ are given by the sum of $\overline{ECF(C_1)}$ and $\overline{ECF(C_2)}$.*

The additive property is helpful in streaming applications, since the statistics for the microclusters can be modified by simply adding the statistics for the incoming data points to the microcluster statistics. However, the microcluster statistics also include time-decay information of the underlying data points, which can potentially change at each timestamp. Therefore, we need an effective way to update the microcluster statistics without having to explicitly do so at each timestamp. For this purpose, the *multiplicative property* is useful.

Observation 18.6.3 Multiplicative Property *The decaying components of $\overline{ECF(C)}$ at time t_c can be obtained from the component values at time $t_s < t_c$ by multiplying each component by $2^{-\lambda \cdot (t_c - t_s)}$ provided that no new points have been added to a microcluster.*

The multiplicative property follows from the fact the statistics decay at the multiplicative rate of $2^{-\lambda}$ at each tick. We note that the multiplicative property is important in ensuring that a *lazy-update process* can be used for updating the decaying microclusters, rather than at each clock-tick. In the lazy-update process, we update a microcluster only when a new data point is added to it. In order to do so, we first use the multiplicative property to adjust for time decay, and then we use the additive property to add the incoming point to the microcluster statistics.

The *UPStream* algorithm uses a continuous partitioning and projection strategy in which the different microclusters in the stream are associated with a particular projection, and this projection is used in order to define the assignment of data points to clusters. The input to the algorithm is the number of microclusters k which are to be determined by the algorithm. The algorithm starts off with a empty set of clusters. The initial set of k data points is assigned to singleton clusters in order to create the initial set of seed microclusters. This initial set of microcluster statistics provides a starting point which is rapidly modified by further updates to the microclusters. For each incoming data point, probabilistic measures are computed over the projected dimensions in order to determine the assignment of data points to clusters. These assignments are used to update the statistics of the underlying clusters. These updates are combined with a probabilistic approach for determining the expected distances and spread along the projected dimensions. In each update iteration, the details of the steps performed are as follows:

- We compute the global moment statistics associated with the data stream by using the multiplicative and additive properties. If t_s is the last time of arrival of a data stream point, and t_c is the current time of arrival, then we multiply the moment statistics by $2^{-\lambda \cdot (t_c - t_s)}$ and add the current data point.

- For each microcluster, we compute and update the set of dimensions associated with it. This computation process uses both the uncertainty information of data points within the different microclusters. A critical point here is that the original data points which have already been received from the stream are not available, but only the summary microcluster information is available. The results in [4] show that the summary information encoded in the microclusters is sufficient to determine the projected dimensions effectively.

- We use the projected dimensions in order to compute the expected distances of the data points to the various microclusters. The closest microcluster is picked based on the expected projected distance. As in the previous case, the original data points which have already been received from the stream are not available. The information encoded in the microclusters is sufficient to compute the expected distances.

- We update the statistics of the microclusters based on the incoming data points. The additive and the multiplicative properties are useful for updating the microclusters effectively for each incoming data point. Since the microcluster statistics contains information about the last time the microcluster was updated, the multiplicative property can be used in order to update the decay behavior of that microcluster. Subsequently, the data point can be added to the corresponding microcluster statistics with the use of the additive property.

The steps discussed above are repeated for each incoming data point. The entire clustering algorithm is executed by repeating this process over different data stream points.

18.7 Clustering with the Possible Worlds Model

The "possible worlds model" is the most generic representation of uncertain databases in which no assumptions are made about the independence of different tuples in the database or across different dimensions [1]. All the algorithms discussed so far in this chapter make the assumption of independence between tuples and also among different dimensions. In practice, many uncertain databases, in which the records are generated by mutually exclusive or correlated events, may be highly dependent in nature. Such databases are drawn from the possible worlds model, and a particular instantiation of the database may have a high level of dependence among the different tuples. Such a method for possible worlds-based clustering has been proposed in [53].

A general-purpose method for performing data analytics in such scenarios is to use Monte-Carlo sampling to generate different instantiations of the database and then apply the algorithms to each sample [35]. Subsequently, the output of the algorithms on the different samples is merged in order to provide a single global result. We note that the key to the success of this method is the design of an effective sample generator for the uncertain data. In this case, an effective methodology is the use of the value generator functions [35] for VG+ function.

For the case of the clustering application, a total of M possible worlds is generated with the use of the VG+ function. Each of these local samples is then clustered with the use of the deterministic DBSCAN algorithm [24]. In practice, any clustering methodology can be used, but we work with DBSCAN because it was used in the case of the possible world clustering proposed in [53]. Since the different samples are completely independent of one another, it is possible to use a high level of parallelism in the clustering process. This results in a total of M possible clusterings of the different samples.

The final step is to merge these M different clusterings into a single clustering. For this purpose, a *clustering aggregation* method which is similar to that proposed in [26] is leveraged. A similarity graph is generated for each of the clusterings. Each uncertain tuple in the database is treated as a node, and an edge is placed between two tuples if they appear in the same cluster in that particular sample. Thus, a total of M possible similarity graphs can be generated. These M different similarity graphs are merged into a single global similarity graph with the use of techniques discussed in [26]. The final set of clusters is determined by determining the clustered regions of the global similarity graph.

18.8 Clustering Uncertain Graphs

In recent years, uncertain graphs have been studied extensively, because of numerous applications in which uncertainty is present on the edges. Many forms of graphs in biological networks are derived through statistical analysis. Therefore, the links are uncertain in nature. Thus, an uncertain graph is defined as a network $G = (N, A, P)$, where N is the set of nodes, A is the set of edges, and P is a set of probabilities such that each edge in A is associated with a probability in P.

Many techniques can be used in order to perform the clustering:

- It is possible to use the probabilities as the weights on the edges. However, such an approach does not explicitly account for the connectivity of the underlying network and its interaction with the combinatorial nature of the underlying graph. Intuitively, a good cluster in the network is one which is hard to disconnect.

- The possible worlds model has been used in [39] in order to perform the clustering. The edit distance is used on the underlying network in order to perform the clustering. A connection is established with the problem of correlation clustering [13] in order to provide an approximation algorithm for the problem.

- The problem of graph clustering is explicitly connected to the problem of subgraph reliability in [36, 45]. The work in [36] determines methods for finding "reliable" subgraphs in uncertain graphs. These subgraphs are those which are hard to disconnect, based *on a combination of* the combinatorial structure of the graph and the edge uncertainty probabilities. Thus, such an approach is analogous to the deterministic problem of finding dense subgraphs in deterministic graphs. However, it is not specifically focussed on the problem of *partitioning* the graph. A solution which finds reliable partitions from uncertain graphs is proposed in [45].

18.9 Conclusions and Summary

In this chapter, we discussed recent techniques for clustering uncertain data. The uncertainty in the data may be specified either in the form of a probability density function or in the form of variances of the attributes. The specification of the variance requires less modeling effort, but is more challenging from a clustering point of view. The problem of clustering is significantly affected by the uncertainty, because different attributes may have different levels of uncertainty embedded in them. Therefore, treating all attributes evenly may not provide the best clustering results. This chapter provides a survey of the different algorithms for clustering uncertain data. Most of the conventional classes of deterministic algorithms such as mixture modeling, density-based algorithms, partitioning algorithms, streaming algorithms, and high-dimensional algorithms have been extended to the case of uncertain data. For the streaming and high-dimensional scenarios, uncertain data also creates additional challenges because of the following reasons:

- In the streaming scenario, the uncertain data has additional volume. The distance calculations are also much slower in such cases.

- In the high-dimensional scenario, the sparsity problem is exacerbated by uncertainty. This is because the uncertainty and noise reduce the correlations among the dimensions. Reduction of correlation between dimensions also results in an increase in sparsity.

We discussed several algorithms for the high dimensional and streaming case, which can be used for effective clustering of uncertain data.

Bibliography

[1] S. Abiteboul, P. Kanellakis, and G. Grahne. On the representation and querying of sets of possible worlds. In *ACM SIGMOD Conference*, 1987.

[2] C. C. Aggarwal. *Managing and Mining Uncertain Data*, Springer, 2009.

[3] C. C. Aggarwal. On density based transforms for uncertain data mining. In *ICDE Conference Proceedings*, pages 866–875, 2007.

[4] C. C. Aggarwal. On high-dimensional projected clustering of uncertain data streams. In *ICDE Conference*, pages 1152–1154, 2009.

[5] C. C. Aggarwal. On unifying privacy and uncertain data models. In *ICDE Conference Proceedings*, pages 386–395, 2008.

[6] C. C. Aggarwal, J. Han, J. Wang, and P. Yu. A framework for clustering evolving data streams. In *VLDB Conference*, pages 81–92, 2003.

[7] C. C. Aggarwal, C. Procopiuc, J. Wolf, P. Yu, and J.-S. Park. Fast algorithms for projected clustering. In *ACM SIGMOD Conference*, pages 61–72, 1999.

[8] C. C. Aggarwal and P. S. Yu. A framework for clustering uncertain data streams. In *ICDE Conference*, pages 150–159, 2008.

[9] C. C. Aggarwal and P. S. Yu. Outlier detection with uncertain data. In *SDM Conference*, pages 483–493, 2008.

[10] C.C. Aggarwal, and P. S. Yu. A survey of uncertain data algorithms and applications. *IEEE Transactions on Knowledge and Data Engineering*, 21(5):609–623, 2009.

[11] R. Agrawal and R. Srikant. Privacy-preserving data mining. In *ACM SIGMOD Conference*, pages 439–450, 2000.

[12] M. Ankerst, M. M. Breunig, H.-P. Kriegel, and J. Sander. OPTICS: Ordering points to identify the clustering structure. In *ACM SIGMOD Conference*, pages 49–60, 1999.

[13] N. Bansal, A. Blum, and S. Chawla. Correlation clustering, *Machine Learning*, 56(1–3):89–113, 2004.

[14] D. Barbara, H. Garcia-Molina, and D. Porter. The management of probabilistic data. *IEEE Transactions on Knowledge and Data Engineering*, 4(5):487–502, 1992.

[15] D. Burdick, P. Deshpande, T. Jayram, R. Ramakrishnan, and S. Vaithyanathan. OLAP over uncertain and imprecise data. In *The VLDB Journal—The International Journal on Very Large Data Bases*, 16(1):123–144, 2007.

[16] M. Chau, R. Cheng, B. Kao, and J. Ng. Uncertain data mining: An example in clustering location data. In *PAKDD Conference*, pages 199–204, 2006.

[17] A. L. P. Chen, J.-S. Chiu, and F. S.-C. Tseng. Evaluating aggregate operations over imprecise data. *IEEE Transactions on Knowledge and Data Engineering*, 8(2):273–294, 1996.

[18] R. Cheng, Y. Xia, S. Prabhakar, R. Shah, and J. Vitter. Efficient indexing methods for probabilistic threshold queries over uncertain data. In *VLDB Conference Proceedings*, pages 876–887, 2004.

[19] R. Cheng, D. Kalashnikov, and S. Prabhakar. Evaluating probabilistic queries over imprecise data. In *SIGMOD Conference*, pages 551–562, 2003.

[20] G. Cormode and A. McGregor. Approximation algorithms for clustering uncertain data. In *PODS Conference*, pages 191–200, 2008.

[21] N. Dalvi and D. Suciu. Efficient query evaluation on probabilistic databases. In *VLDB Conference Proceedings*, pages 523–544, 2004.

[22] A. Das Sarma, O. Benjelloun, A. Halevy, and J. Widom. Working models for uncertain data. In *ICDE Conference Proceedings*, 2006.

[23] A. P. Dempster, N. M. Laird, and D. B. Rubin. Maximum likelihood for incomplete data via the EM algorithm. *Journal of the Royal Statistical Society, Series B*, 39:1–38, 1977.

[24] M. Ester, H.-P. Kriegel, J. Sander, and X. Xu. A density based algorithm for discovcering clusters in large spatial databases with noise. In *KDD Conference*, pages 226–231, 1996.

[25] H. Garcia-Molina and D. Porter. The management of probabilistic data. *IEEE Transactions on Knowledge and Data Engineering*, 4(5):487–501, 1992.

[26] A. Gionis, H. Mannila, and P. Tsaparas. Clustering aggregation. *ACM TKDD Journal*, 1(1): 4, 2007.

[27] S. Guha and K. Munagala. Exceeding expectations and clustering uncertain data. In *ACM PODS Conference*, pages 269–278, 2009.

[28] S. Guha, R. Rastogi, and K. Shim. CURE: An efficient clustering algorithm for large databases. *ACM SIGMOD Conference*, pages 73–84, 1998.

[29] F. Gullo, G. Ponti, and A. Tagarelli. Minimizing the variance of cluster mixture models for clustering uncertain objects. *IEEE ICDM Conference*, pages 839–844, 2010.

[30] F. Gullo, G. Ponti, A. Tagarelli, and S. Greco. A hierarchical algorithm for clustering uncertain data via an information-theoretic approach. *IEEE ICDM Conference*, pages 821–826, 2008.

[31] F. Gullo and A. Tagarelli. Uncertain centroid-based partitional clustering of uncertain data, *VLDB Conference*, pages 610–621, 2012.

[32] S. Gunnemann, H. Kremer, and T. Seidl. Subspace clustering for uncertain data. In *SIAM Conference on Data Mining*, pages 385–396, 2010.

[33] H. Hamdan and G. Govaert. Mixture model clustering of uncertain data. In *Proceedings of IEEE ICFS Conference*, pages 879–884, 2005.

[34] A. Jain and R. Dubes. *Algorithms for Clustering Data*. Prentice Hall, New Jersey, 1998.

[35] R. Jampani, F. Xu, M. Wu, L. L. Perez, C. M. Jermaine, and P. J. Haas. MCDB: A Monte Carlo approach to managing uncertain data. In *ACM SIGMOD Conference*, pages 687–700, 2008.

[36] R. Jin, L. Liu, and C. Aggarwal. Finding highly reliable subgraphs in uncertain graphs, *ACM KDD Conference*, pages 992–1000, 2011.

[37] B. Kao, S. D. Lee, D. W. Cheung, W. S. Ho, K. F. Chan. Clustering uncertain data using Voronoi diagrams. In *IEEE ICDM Conference*, pages 333–342, 2008.

[38] L. Kaufman and P. Rousseeuw. *Finding Groups in Data: An Intrduction to Cluster Analysis.* Wiley Interscience, 1990.

[39] G. Kollios, M. Potamias, and E. Terzi. Clustering large probabilistic graphs. *IEEE TKDE Journal*, pages 325–333, 2013.

[40] H.-P. Kriegel and M. Pfeifle. Density-Based Clustering of Uncertain Data. In *ACM KDD Conference Proceedings*, pages 672–677, 2005.

[41] H.-P. Kriegel and M. Pfeifle. Hierarchical density based clustering of uncertain data. In *ICDM Conference*, pages 672–689, 2005.

[42] M. Kumar, N. Patel, and J. Woo. Clustering seasonality patterns in the presence of errors. In *ACM KDD Conference Proceedings*, pages 557–563, 2002.

[43] L. V. S. Lakshmanan, N. Leone, R. Ross, and V. S. Subrahmanian. ProbView: A flexible probabilistic database system. *ACM Transactions on Database Systems*, 22(3):419–469, 1997.

[44] S. D. Lee, B. Kao, and R. Cheng. Reducing UK-means to K-means. In *ICDM Workshops*, pages 483–488, 2006.

[45] L. Liu, R. Jin, C. Aggarwal, and Y. Shen. Reliable clustering on uncertain graphs, *ICDM Conference*, pages 459–468, 2012.

[46] S. I. McClean, B. W. Scotney, and M. Shapcott. Aggregation of imprecise and uncertain Information in databases. *IEEE Transactions on Knowledge and Data Engineering*, 13(6):902–912, 2001.

[47] R. Ng and J. Han. Efficient and effective clustering algorithms for spatial data mining. In *VLDB Conference*, pages 144-155, 1994.

[48] W. Ngai, B. Kao, C. Chui, R. Cheng, M. Chau, and K. Y. Yip. Efficient clustering of uncertain data. In *ICDM Conference Proceedings*, pages 436–445, 2006.

[49] D. Pfozer and C. Jensen. Capturing the uncertainty of moving-object representations. In *SSDM Conference*, pages 111–132, 1999.

[50] M. Sato, Y. Sato, and L. Jain. *Fuzzy Clustering Models and Applications.* Physica–Verlag, Heidelberg, 1997.

[51] S. Singh, C. Mayfield, S. Prabhakar, R. Shah, and S. Hambrusch. Indexing uncertain categorical data. In *ICDE Conference*, pages 616–625, 2007.

[52] Y. Tao, R. Cheng, X. Xiao, W. Ngai, B. Kao, and S. Prabhakar. Indexing multi-dimensional uncertain data with arbitrary probabality density functions. In *VLDB Conference*, pages 922–933, 2005.

[53] P. Volk, F. Rosenthal, M. Hahmann, D. Habich, and W. Lehner. Clustering uncertain data with possible worlds. *ICDE Conference*, pages 1625–1632, 2009.

[54] L. Xiao and E. Hung. An efficient distance calculation method for uncertain objects, *CIDM Conference*, pages 10–17, 2007.

[55] C. Zhang, M. Gao, and A. Zhou. Tracking high quality clusters over uncertain data streams, *ICDE Conference*, pages 1641–1648, 2009.

[56] T. Zhang, R. Ramakrishnan, and M. Livny. BIRCH: An efficient data clustering method for very large databases. In *ACM SIGMOD Conference Proceedings,* pages 103–114, 1996.

Chapter 19

Concepts of Visual and Interactive Clustering

Alexander Hinneburg
Martin-Luther University
Halle/Saale, Germany
hinneburg@informatik.uni-halle.de

19.1 Introduction

Clustering algorithms group data objects together based on some notion of distance or similarity. This resembles visual tasks that are easy for humans: spotting a cluster of stars in the night sky or identifying a cluster of old houses within a modern city. The human visual system "has evolved to facilitate quick and considered detection of the visually like and unlike through a wide variety of cues – e.g. location and relative proximity, movement, shape, colour, texture, and matching against predetermined patterns. Consequently, visualization is a natural and powerful resource for cluster analysis; it is especially valuable in identifying unanticipated structure"[33].

Many algorithms have been proposed to formalize the concept of a cluster and to automatically detect such clusters in large sets of data objects. Many algorithms use Euclidean distance. When the notion of Euclidean distance is extended to vector spaces with more than three dimensions, clustering by proximity becomes formally possible for more complex data objects. Complex data objects are often described by a large but fixed set of features. Therefore, they are coded as high-dimensional vectors. Thus, the intuition of visually perceptible clusters is carried over to high-dimensional cases. While technically and mathematically possible, the three-dimensional intuitive understanding of distances and clusters can be misleading in high-dimensional cases due to several counterintuitive phenomena in such spaces [1].

Clusters of data objects that are computed by automatic algorithms are influenced by two major factors. First, complex data objects are described by *high-dimensional feature vectors*. In nearly all cases these are lossy descriptions of the original objects that neglect some aspects. Second, the *definitions of clusters* used by different algorithms involve complex algorithmic operations rendering the algorithm into a black box for end users. Therefore, there is in most cases no easy answer to simple questions such as: Why are two specific points put together into the same cluster? Furthermore, nearly all definitions of clusters include pathological cluster configurations that are unwanted but sometimes difficult to detect automatically.

The consequences of the interplay of the two factors—high-dimensional feature vectors and definitions of clusters—are in different degrees not directly accessible to humans. Thus, visual aids will enhance the understandability of cluster analysis. Visualization techniques have been devised that use four main concepts to enhance the understanding of clusters. The first idea is to cluster the data directly using visual and interactive tools. Thus, no automated clustering algorithm is used. Instead, all decisions about how to group the data objects into clusters are made by the user. Therefore, the produced results should be well understandable. This approach avoids problems caused by the second factor—the definitions of clusters by a black box algorithm. However, the first factor—understanding data sets of high-dimensional feature vectors—becomes more difficult to be visually communicated to the user.

The second approach is to use an automated clustering algorithm. However, the black box is opened and intermediate states of the algorithm are visually communicated to the user. This implies the hope that the user better understands the clusters, when the cluster construction itself is somehow documented. Further, the user might interact with the algorithm to steer the search toward well-interpretable clusterings.

The third approach puts the user into the position of a model selector. Many different clustering algorithms with different parameters and distance measures are run on the same data set. The results constitute the space of clusterings that is presented to the user. The visual representation of the space of clusterings helps the users to navigate through the clustering results. Alternative clusterings can be compared and possible interpretations can be explored. The structured space helps the user to reduce the number of such pairwise comparisons of clusterings.

The last major concept is to use visualization techniques that help to inspect the found clusters in the context of the particular application at hand. Those approaches often break with the visual metaphor of a cluster of stars. Instead, data objects themselves, the cluster information, and the semantic context are visualized.

In the remainder of the chapter we introduce all of these concepts and discuss the applicability to real world problems. Section 19.2 discusses visualization techniques to derive data clusters in a completely interactive and visual way. In detail techniques bases on scatterplots as well as parallel coordinates are discussed. Section 19.3 shows approaches to visually steer automatic, unsupervised clustering algorithms and Section 19.4 discusses the interactive comparison of several readily derived clusterings. Visual inspection of clustering results and sensemaking is discussed in Section 19.5 using the example of clustering document streams. Section 19.6 summarizes the chapter.

19.2 Direct Visual and Interactive Clustering

Automated clustering algorithms are difficult to understand for nonexperts. Therefore, it seems natural to ask whether the black box with the clustering algorithm can be replaced by an interactive and visual procedure to find clusters. Such procedure could be operated by a domain expert, who understands both, data and context. This would lift the burden of understanding the impact of choosing a clustering algorithm and finding parameters for it that produce understandable and meaningful clusters.

Tools toward this end have been developed in the contexts of statistics and information visualization. A major challenge is to present multi- and high-dimensional data to the user in a meaningful way that allows visual cluster detection by hand. High-dimensional vectors are not easily mapped to visual attributes in a direct way. We discuss methods based on scatterplots as well as parallel coordinates in the next two subsections.

19.2.1 Scatterplots

Multidimensional vectors cannot be shown directly as points in a space when the dimensionality is larger two or three. Therefore, several low-dimensional projections are shown as multiple two-dimensional scatterplots instead. Those low-dimensional projections are linked either spatially or temporally to visualize high-dimensional phenomena.

A typical example for spatially linked projections is a matrix of scatter plots [19, 10]. While presenting all two-dimensional projections of the high-dimensional data to the data analyst in one view, the large mass of such two-dimensional projections makes it difficult to visually recognize interesting structures in the data. Further, even when well-separated clusters are spotted in some of the two-dimensional projections, it is difficult or impossible to tell whether clusters in different projections are aligned. This problem can be alleviated by using an interaction technique called brushing and linking. By coloring a cluster in one projection, the selected points are also colored in the other projections. The pattern of the colored points in the other projections indicates whether two clusters in different projections are aligned.

An alternative to brushing and linking is to link projections temporally. A well-known example is Grand Tour [6] that cycles continually through all two-dimensional projections. Between two axis-parallel projections, the shown projections are linearly interpolated. When the two axis-parallel projections share one variable, the transition from one projection to the other corresponds to a rotation around the shared variable axis. This is well understandable for users. Through the continually changing positions of all points in the projections, the alignment of clusters in both projections is visually recognizable without brushing and linking. However, this property is lost, when the two projections do not share a common variable or the two projections are not consecutive in the Grand Tour.

Both visualization techniques, matrix of scatterplots and Grand Tour, lack effective means to navigate through the space of two-dimensional projections. Hurley and Oldford [26] introduced to this end *navigational graphs*. Each node in a navigational graph represents a two-dimensional axis parallel projection onto two variables. An edge between two nodes is drawn if the two nodes share a variable. The `RnavGraph` tool [45] uses navigational graphs together with scatter plots to facilitate the exploration of high-dimensional data. It allows one to interactively drag a button from one node to another along an edge. This translates to a rotation in a three-dimensional space spanned by the variables in nodes adjacent to the edge. The rotation of the data is shown during the transition in the scatterplot window of `RnavGraph`. As the user can control the speed and direction of the rotation by dragging the button, a smooth transition of points and clusters becomes recognizable.

Figure 19.1 shows an example of such a transition using the four-dimensional Iris data with dimensions petal width, petal length, sepal width, and sepal length. Figure 19.1(a) shows the projection onto (sepal width, petal width), which is indicated by the large empty node in the upper left corner of the navigational graph. Figures 19.1(b) and 19.1(c) show intermediate projections toward the projection (sepal length, petal width), where large empty node is in middle of the edge and Figure 19.1(d) shows the state after the transition is complete. The smoothly changing scatterplots on the right show that changing from (sepal width, petal width) to (sepal length, petal width) reduces

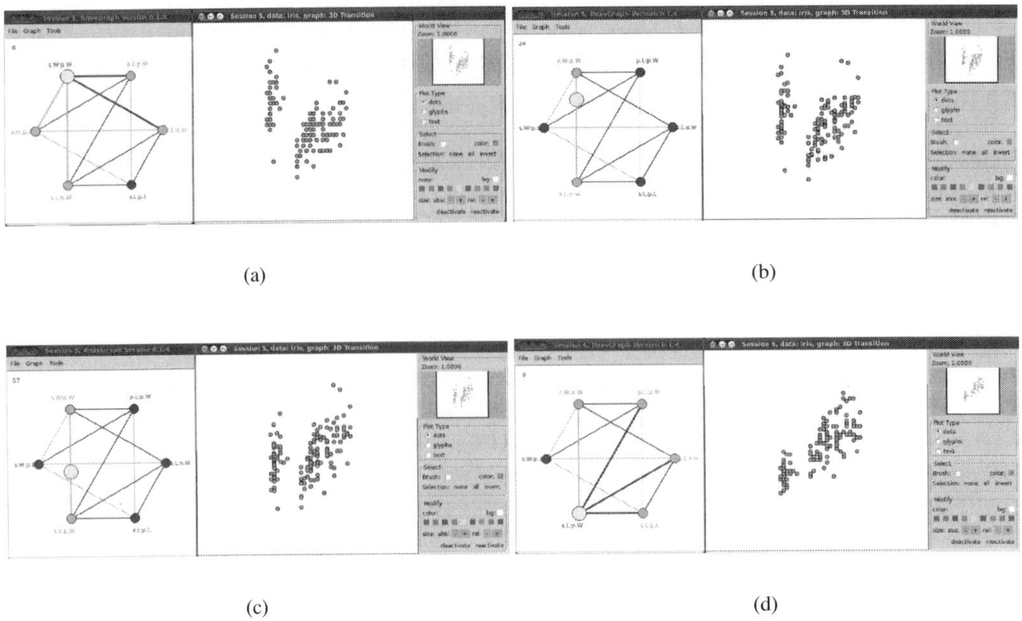

FIGURE 19.1: Usage of `RnavGraph` with Iris data. The left part of each subfigure shows the navigational graph. The large empty node in each graph indicates the current two-dimensional projection. The subfigures show the transition between the projections on (sepal width, petal width) and (sepal length, petal width).

the variance of the left cluster. The user can drag large empty node back and forth and observe simultaneously during this simple interaction the related changes in the scatterplot. The navigational graph on the right side acts as a simple map that represents symbolically the space of projections.

Using interactive brushing the user can color different points to mark visually detected clusters. Points can be deactivated in the plot as well to focus on specific data subsets. In case simple two-dimensional projections do not help to differentiate between visually close points, the full multi-dimensional data of the objects can be visually shown as star glyphs as well. More details can be found in the manual of the `RnavGraph` package [45].

The navigational graph does not necessarily include all two-dimensional projections as nodes. Data with larger dimensionality d would require $\binom{d}{2}$ nodes, which would render such navigational graph useless as an effective map for the space of projections. In [33] the use of scagnostic measures [49] is suggested to pick two-dimensional projections with interesting data distributions. Scagnostics are briefly introduced by Tukey and Tukey [44]. The ideas are further developed and described in more detail in [49]. In a nutshell, a number of k measures are computed for each of the distributions in the $\binom{d}{2}$ axis-parallel projections of a d-dimensional data set to guide the selection of interesting projections. The measures are designed to quantify a wide range of characteristics that appear in two-dimensional distributions. To illustrate the measures, the two-dimensional projections of the mtcars-dataset [22] were quantified by the proposed $k = 9$ measures [49]. A subset of produced measurement vectors, each with a dimensionality of nine, is shown with the respective distributions in a biplot in Figure 19.2. A biplot [15] reduces the multidimensional vectors to first two principal components and plots the reduced vectors together with the projected unit vectors of the original data space.

`RnavGraph` computes all of those scagnostic measures. A particular projection is included in

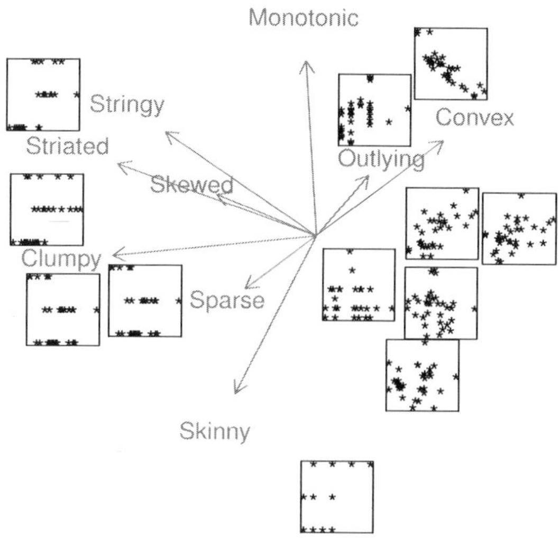

FIGURE 19.2: Biplot of nine scagnostic measures with a subset of the two-dimensional projections of the mtcars-dataset.

the navigational graph, if it has top scores in any of the measures with respect to the total set of projections. This strategy helps to cope with multidimensional data. However, as the number of projections to be evaluated with scagnostic measures grows quadratically with dimensionality of the data, this approach is not suitable for very high-dimensional data such as images or documents. Oldford and Wadell [33] suggest in this case to use some dimensionality reduction method first.

The `IPCLUS`-system [2] combines dimensionality reduction and cluster specification. It repetitively presents two-dimensional projections to the user and asks for manual cluster specification within those projections. Projections are selected by an iterative process that starts with the full-dimensional space and reduces the dimensionality by cycling through the following steps: (i) sampling a small subset of points called polarization anchors, (ii) computing neighbor sets for each polarization anchor, (iii) centering each neighbor set with the respective polarization anchor, and (iv) retain the principal components with least variance preservation of the union of the centered neighbor sets. The number of principal components retained is reduced by a constant factor in each iteration until a two-dimensional subspace is reached. The assignments of the data objects to clusters that the user might have specified within a projection are stored. The cluster specifications from all projections are combined to final clusters.

A related approach is taken by the `HD-Eye` system [23]. It evaluates several one- and two-dimensional projections to see whether they yield a potential separation of clusters. The potential separation is visually encoded into a row of icons, one icon for each potential cluster. The colors and the shapes of the icons show cluster sizes and degrees of separation, respectively. Thus, the user can choose, which projections are most interesting to examine further in more detail. However, in contrast to `RnavGraph`, `IPCLUS` and `HD-Eye` offer the user rather limited options to navigate and explore the space of projections.

In all cases, clusters are specified in scatterplots by selecting and coloring points. As not all dimensions are shown in a two-dimensional projection, `RnavGraph` can show star-plot glyphs of individual data objects that visualize the full data vectors. The idea of glyphs and icons for cluster visualization is further developed in [9]. The new icon technique–called Dicon–overcomes the problem of cluster visualization once a set of data objects has been interactively grouped together. Star-plot glyphs could show only something like a centroid or a medoid of the newly created cluster.

However, cluster size and the contributions of the individual data objects to the cluster are hidden. The icons proposed in [9] for cluster visualization can grow from individual data objects to larger clusters. During the process of cluster growing, the icons get visually larger. Therefore, the user gets the impression of interactively building clusters.

The principle of growing icons, first assigns all data objects a certain amount of visual space. When a data object is not yet part of any cluster, that visual area is a connected and convex region. The area of a data object is subdivided into as many subareas as there are dimensions. The sizes of the subareas are proportional to the data values in the feature vector describing the respective data object. This assumes that the feature vectors consist of quantitative attributes only. The color of each subarea indicates the respective dimension. When data objects are grouped together, their visual spaces add up to form the icon of the newly built cluster. The subareas of the individual data objects are rearranged such that subareas of same color form a connected larger subarea. Thus, adding a data object to a clusters corresponds visually to adjoin the subareas to the visual space of the cluster. Once the subareas are fit into the cluster, they are still discernible, thus the user could click on the cluster and the other subareas of the respective data objects are highlighted. Beside the relative contributions of the dimensions, the visualization of a cluster icon can also show some information about the data distribution within the cluster, e.g., the Kurtosis. Limitations of Dicon are (i) the number of dimensions is bound to the number of discernible colors and (ii) the number of data objects is limited by the screen resolution.

19.2.2 Parallel Coordinates

The difficulty of displaying multidimensional points with more than three dimensions in a single view instead of multiple linked views can be alleviated by using parallel coordinates [28, 27]. Parallel coordinates display a multidimensional data vector as a set of connected line segments (polyline) drawn in a two-dimensional space. The active domain of each dimension of the multidimensional data space is represented by an axis. All axes are scaled to the same size and drawn in parallel. Each end point of a polyline segment is placed at one of the parallel axes. The position of it there is computed by mapping the numerical value of the data vectors' respective dimensions onto the line segment of the respective axis. Thus, the polyline corresponding to a multidimensional data vector connects the numerical values of that vector mapped onto the parallel axes. An example of a single data vector taken from the Iris data set is shown in Figure 19.3(a).

Clusters would show up in parallel coordinates as bands that become dense at one or several dimensions, e.g., see the petal length and petal width at the bottom of Figure 19.3(b). As lines in the original geometric space show up as common intersections of many polyline segments in parallel coordinates [27], this visualization technique has been considered to explore correlations in the data [30]. However, parallel coordinates suffer from several types of visual clutter as shown in Figure 19.3(b). The problems include the many crossings of polyline segments and the ambiguity of the graphical vector representation in case two different vectors share a (nearly) common data value in one of the dimensions that is not displayed at the border of the visualization.

Several approaches have been suggested that address these problems and especially enhance the visual recognition of clusters in parallel coordinate displays. A general idea is to combine density with parallel coordinates in order to emphasize high density regions in the plot. Clusters in data are closely related to the density in the data space, as cluster centers are often regions of high density. Data density is a function defined on the whole data space that measures how many data vectors are close to a given location in the data space [40, 42]. A simple example of a density function is a histogram. One-dimensional histograms showing marginal distributions of single dimensions have been added to parallel coordinates as bars of different thickness that are centered around the axes [20]. Two-dimensional histograms defined by pairs of consecutive dimensions have been used in [5]. The density of a data vector in the projection to a particular pair of consecutive dimensions is mapped to the brightness of the line segment that connects the respective data values. The drawing

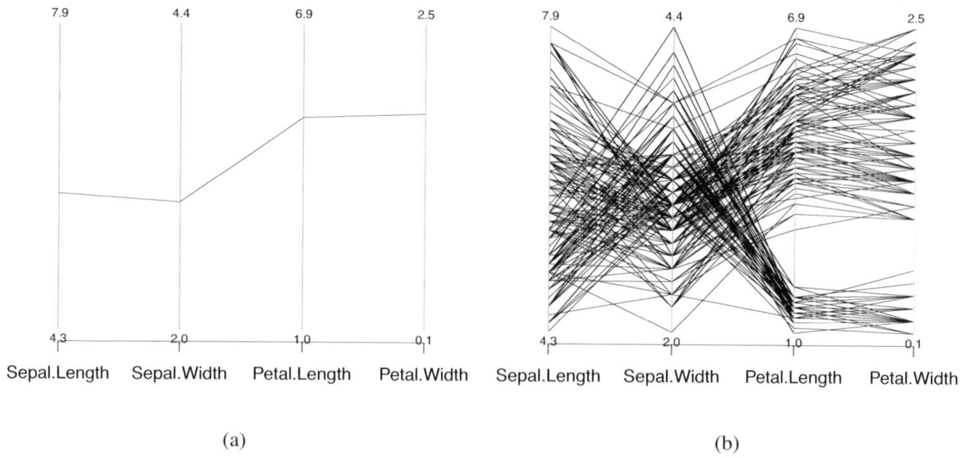

FIGURE 19.3: Parallel coordinates with Iris data: (a) a single data vector $(5.9, 3, 5.1, 1.8)$ and (b) the full Iris data set with a total of 150 data vectors.

algorithm orders the line segments of each pair of consecutive dimensions such that brighter line segments are always drawn on top of darker ones. Dense clusters appear as bright ribbons. Note that this visualizations assumes a black background. A similar effect is obtained when pixels of lines segments are drawn brighter when they are hit more often by other line segments [47]. Using modern SVG-rendering devices such as Firefox or Inkscape, this feature can be easily implemented using lines with opacity less than 100%. See Figure 19.4 for an example.

A further technique in that direction is proposed in [52]. One-dimensional histograms that are parallel to axes are introduced in between the axes that count how many line segments pass through a specific histogram bin. The average of the normalized bin counts that a particular polyline passes assigns a numerical value called average density to each data vector. Interactively specified transfer functions for different color and brightness channels are used to translate average density into color. The polylines of the data vectors are colored with respect to that transfer function. During the interactive specification, the user can see the effects of the coloring in the parallel coordinates plot and can adjust the transfer function as necessary.

A second kind of enhancement is to replace the straight line segments by curves. The proposed approaches are differently motivated. In order to alleviate ambiguity, curved segments are suggested in [16]. Incoming segments meeting at the same position at an axis can now be uniquely associated

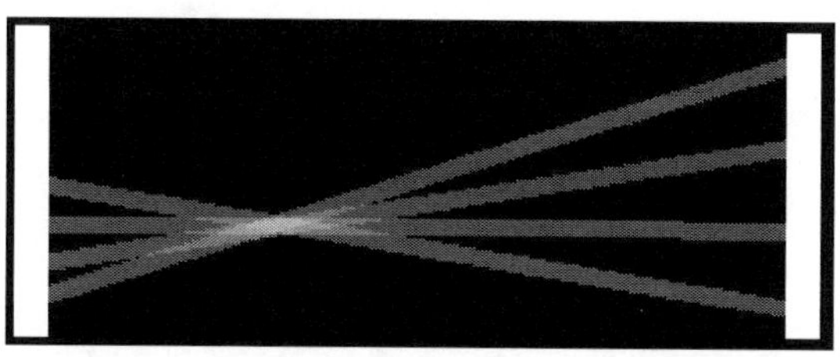

FIGURE 19.4: Parallel coordinates with brighter intersections of line segments. The graphic is produced using SVG where the line segments have opacity of 37%.

to the respective outgoing segments by having similar slopes. The Visual-Clustering-approach [52] use curves in the same spirit as bundled edges [24]. Original line segments that go in the same direction are replaced by curves that are drawn close together. This reduces the visual space formerly needed by the line segments, and consequently visual clutter is reduced in the plot. The curves are defined by introducing a few new middle points at each segment between two consecutive axes. The positions of all middle points are computed by minimizing a heuristically defined energy function. This energy function balances two terms, one forcing the middle points to keep the original position and a counterpart that accounts for attraction forces between the curves. Both terms involve parameters that need to be heuristically set by the user. The energy minimization involves solving a large linear program using a simplex-based lp-solver. Therefore, the layout computation is between minutes and hours, which does not allow for interactive search of suitable parameter settings for the energy function.

A third use of curved segments is motivated by integrating parallel coordinates and scatterplots. The idea in [50] is to plot middle points between consecutive axes that define the curves as splines in such a way that the positions of middle points resemble the distances between data vectors in some projections of the high-dimensional data space. Thus, a curve connects two data values of the same data vector, and the middle point at this curve represents the data vectors relative distances in some projected space. This integrates scatterplots represented by middle points and parallel coordinates represented by curves into a single plot. Given d dimensions the technique allows us to integrate at most $d-1$ of such scatterplots. The positions of the middle points are derived from distances of the data vectors in some low-dimensional projection using multidimensional scaling (MDS) [36]. However, the visualization technique does not restrict this low-dimensional projected space to be the axis-parallel projection onto the two consecutive dimensions framing the integrated scatterplot.

A much simpler embedding of scatterplots into parallel coordinates has been proposed in [25]. This integration shows the two-dimensional scatter plots between consecutive axes. Axes labels are effectively shared by rotating the scatterplots 45°. Figure 19.5 shows an example of *Iris* data.

The user study presented in [25] also shows that the embedding of scatterplots into parallel coordinates is most helpful for users to detect clusters. This visualization performed significantly better with respect to response time and accuracy of the number of detected clusters than standard parallel coordinates, parallel coordinates with cluster enhancement by color and brightness as well as parallel coordinates with curves discussed above. The user study also shows that animation techniques such as Grand Tour [6, 47] for parallel coordinates are less effective for cluster identification.

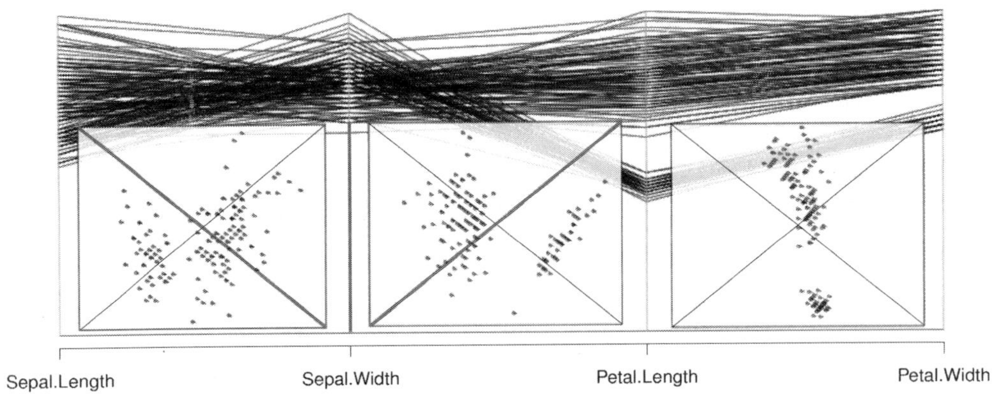

FIGURE 19.5: Scatterplots embedded into Parallel Coordinates showing the Iris data. The bold lines in the scatter plots show the directions of the sepal-width axis. Note, the intersections of the axes in the scatterplots are not the origin but the data mean.

19.2.3 Discussion

Visualization of multidimensional data is an effective tool to explore and identify clusters, when the numbers of data vectors, clusters, and dimensions of the data are not large. Depending on the visualization technique a few hundred up to a few thousand data vectors can be processed. The number of clusters shown in the applications in the cited papers seldom exceeds ten. The dimensionality of the data is in almost all cases is below 15.

Given such setting, visual interactive clustering without using an automated algorithm has the potential to allow effective exploration of the data and the identification of meaningful groupings that are useful in certain applications. This mainly complements automated algorithms that have their strengths on scalability to large data sets of high dimensionality possibly comprising a large number of clusters.

19.3 Visual Interactive Steering of Clustering

A few approaches have explored the possibility to combine an automated clustering algorithm with an interactive visualization technique. The proposed combination can be quite specific, which means that special properties of algorithm and/or the data visualization are required. The general design pattern of such combination is to replace some part of the algorithm with an interactive visual procedure that allows the user to influence the automated clustering algorithm during the computation of the clustering. A simple example is to replace the stopping criterion of an iterative clustering algorithm by visualizing the current state of the algorithm with respect to the given data and let the user visually decide whether the algorithm has reached a satisfying solution. Another example is to use interactive visualization techniques to build a hierarchical clustering. Depending on the underlying algorithm, the user decides, based on visualizations, which cluster found so far could be split further.

The last example is to influence iterative learning algorithms like self-organizing maps (SOM) during the initialization or during the learning procedure itself. The user can influence the algorithm by visually changing parts of the data structure, e.g. some neurons in the SOM, that are otherwise automatically initialized or updated. That kind of interaction biases the algorithm into a some direction that the algorithm would possibly have not explored by following the internal learning procedure. In case of machine learning by optimization of some criterion, such user interaction could help the algorithm to break free of some local optima that would correspond to nonmeaningful clusterings.

19.3.1 Visual Assessment of Convergence of Clustering Algorithm

Some clustering algorithms work in an iterative way by updating internal data structures. However, not all such algorithms are guarantied to converge toward a nontrivial solution. An example is topology learning neural gas [31]. Given a data set of multi-dimensional vectors and a predefined number of neurons (aka cluster centers), the topology learning neural gas algorithm cycles after random initialization of the neurons positions through the following steps:

- Randomly select a data vector.
- Move the neurons that are close to the selected data vector toward it.
- Add an edge with zero age between the two neurons that are the closest and second closest ones. If such an edge already exists, zero the age.
- Increase the ages of all existing edges by one.

- Remove all edges with ages above some predefined threshold.

Depending on the data and the parameters for measuring closeness of neurons, adaption toward selected data points, and age threshold, the algorithms finds arbitrarily shaped clusters as connected components of the graph defined by neurons and the edges between them. However, it is obvious that trivial solutions—no edges except the most recently inserted ones—could appear, e.g., in case of too low age threshold.

An algorithm that is similar to the topology learning neural gas algorithm is combined with a visualization technique based on parallel coordinates [51]. In that algorithm the static data vectors take the role of neurons which is a special case of topology learning neural gas. The visual algorithm starts by drawing all data vectors in same color and same opacity into a parallel coordinates plot. Then it cycles through the following steps:

- Randomly select a data vector.
- Increase the opacity of complete polylines of the close neighbors of the selected data vector
- Reduce the opacity of polylines of all data vectors by a given ratio.

Depending on parameters, the algorithm produces a series of visualizations starting with a crowded parallel coordinates plot that fades during the run of the algorithm. Outliers, noise, and data vectors at the borders of clusters fade faster than data vectors close to a cluster center because the former are less often selected as neighbors, and therefore, their opacity is less often increased than the opacity of the latter ones. The user can visually decide to stop the fading process when cluster centers become clearly visible. Alternatively, the full series of visualizations can be computed and afterwards interesting frames are visually selected. In case of hierarchically nested clusters, larger clusters would show up first and then subclusters within larger clusters would appear. Once a suitable visualization is found, parameters like the number of clusters needed for further automatic processing can be determined. In the case that no clustering turns up in the visualization and the parallel coordinates plot just fades until all polylines have almost zero opacity, different parameters of the fading process could be tried.

19.3.2 Interactive Hierarchical Clustering

Three-dimensional projections of high-dimensional data are much less explored in information visualization than two-dimensional ones. One visual difficulty of three-dimensional projection is that points with different coordinates are plotted at the same position on the two-dimensional screen. This ambiguity makes it difficult for the user to judge distances between points, and therefore, clusters are less reliably detected. The possibility to interactively rotate the view helps the user to alleviate the visual ambiguity by identifying point groups with similar motion.

Another visual technique to turn three-dimensional visualizations into a useful tool for clustering is to use surfaces that engulf already computed clusters [43]. The surfaces transform clusters from point clouds to solid three-dimensional objects. The use of partial transparency, shading, and light reflection helps the user to resolve visual ambiguities in the visualization and to recognize the particular spatial extensions of clusters. The goal of this approach is to produce distinguishable three-dimensional objects. The simplest approach that encloses each cluster with a sphere or an ellipsoid would produce visualizations with clusters that look quite uniform. Furthermore, details of the data distribution within a cluster could be deemphasized using overly simple convex shapes only. Therefore, a nonconvex surface—called BLOB—is constructed for each cluster by (1) placing spheres at border points of the cluster and (2) finding minimal radius of the spheres such that the cluster is still enclosed by the union of the spheres and the volume defined by that union does not break into disconnected regions. The method used in [43] does not guarantee finding the global

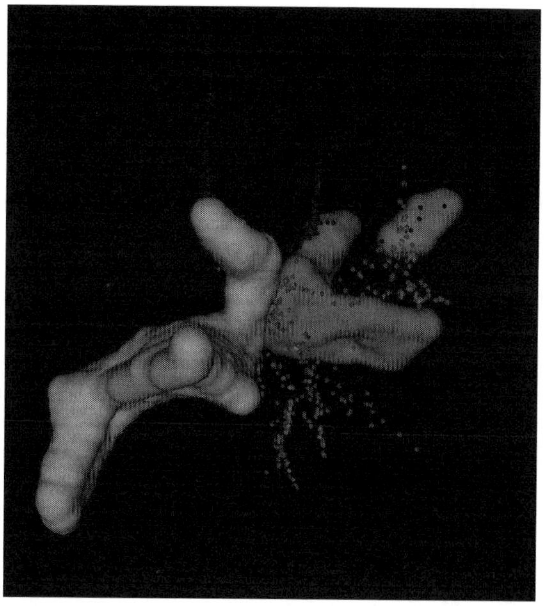

FIGURE 19.6 (See color insert): Two selected three-dimensional clusters shown with their hull. Note that the red cluster consists of two separate regions.

minimum of that optimization; however, it produces visually suitable results for nonpathological cases.

The algorithm and the visualization proposed in [43] integrates a hierarchical graph clustering algorithm based on edge collapsing with a three-dimensional visualization. The graph clustering algorithm first places the data objects in the three-dimensional view space using a spring-embedder that models data vectors as points in the three-dimensional space, and the distances between original data vectors are represented by the stiffness of springs placed between the points. The embedding of the original data in the three-dimensional space is computed by finding a low-energy state of the spring-system. This is similar to MDS [36]. The cluster hierarchy is computed using centroid linkage in the three-dimensional space based on the embedded three-dimensional points. Hierarchical clusters are shown as BLOBs within BLOBs, a technique that is called H-BLOBs.

The concept of using three-dimensional projections for clustering is further developed in [34]. Instead of relying on a readily computed cluster hierarchy, the user can interactively construct the hierarchy top-down by selecting a cluster and compute subclusters with some automated algorithm. The results are then shown as three-dimensional objects within larger transparent three-dimensional objects. The interactive system is demonstrated in videos that are available from `http://infoserver.lcad.icmc.usp.br/infovis2/3Dproj`. An example is shown in Figure 19.6.

The Hierarchical Clustering Explorer (`http://www.cs.umd.edu/hcil/hce/`) uses an alternative approach described in [41]. This tool relies on standard visualizations like scatterplots and parallel coordinates combined with dendrogram plots to interactively analyze high-dimensional data. Lessons learned from that project are that it is important to select the most relevant features first using interactive visual techniques and automated feature ranking methods. Then hierarchical clustering can be performed in the subspaces spanned by the selected features.

19.3.3 Visual Clustering with SOMs

A Self-Organizing Map (SOM) is a clustering-like algorithm originally invented by Kohonen [29] that is inspired by the architecture of the brain [35]. Given a set of high-dimensional data vectors, the learning algorithm folds a two-dimensional grid consisting of nodes that act as cluster centers into the high-dimensional data space. Each inner node of the grid is statically connected to four neighbor nodes via an edge. The learning algorithm cycles through the data vectors and determines the nearest node for each of them. The nearest node and to a lesser degree the neighbors of the node with respect to the grid topology are moved a certain amount toward the current data vector in the high-dimensional space. That amount decreases during training at a given learning rate. Thus, the map reaches a stable configuration overall after some iterations.

The learning algorithm tries to preserve the neighborhood with respect to the two-dimensional grid in the high-dimensional data space. That means, neighboring nodes tend to be mapped to close positions in the high-dimensional space. Therefore, a path along the edges of the two-dimensional map is often also a path with small local jumps in the high-dimensional space. The map visualization that shows the grid is therefore a nonlinear embedding of the two-dimensional plane in the high-dimensional space. Simple visualizations show class attributes or other features at the nodes of such a map.

An interactive visualization technique is to plot some small object visualization—e.g., some glyphs—at the node positions [38]. In principle, every type of icon or glyph such as Chernoff-faces or star-plots works for this combination with a SOM. The only restriction is the effectiveness of the glyph type for high-dimensional data, which is usually limited to visualize vectors with at most 10 or 15 dimensions. The two-dimensional map combined with glyph visualizations then shows smooth or rough transitions between neighboring nodes.

An interesting feature of the technique proposed in [38] is that the user can specify some glyphs interactively before the initialization and the learning of the SOM. The glyphs translate directly to high-dimensional vectors. The node positions of the interactively specified glyphs in the high-dimensional space are kept fixed during the SOM learning. Thus, the layout of the SOM can be directly specified by the user in an interactive and visual way. The gaps between the specified nodes in the map are automatically filled by the SOM learning algorithm. The technique of visual interactive editing SOMs can also be used during the training procedure to steer the SOM learning algorithm toward configurations the user prefers. The technique has also been used with small geographic maps instead of glyphs [3].

19.3.4 Discussion

The few examples that combine automated algorithms with interactive visualizations are rather simple with respect to both the algorithm and the visualization technique. Both ingredients are proven techniques in all cases. Whether the combination of automated algorithm and visualizations is really superior over fully automated algorithms on one side and purely interactive approaches on the other is yet to be studied. To the best of our knowledge, there exist no user studies that compare automated approaches with interactive visual clustering. Performance metrics to evaluate visual analytic systems in general have been proposed in [37]. However, it is not just the application of standard usability testing as done in human-computer-interface (HCI) research to visual analytics systems. The complex nature of the dialog between the analyst and the computer software using visualizations involves cognitive processes that are yet to be explored [4]. Thus, an open research question still exists regarding how to evaluate visual analytics systems in general and those for interactive visual clustering in particular.

19.4 Interactive Comparison and Combination of Clusterings

A novel concept that combines capabilities of humans with those of computers is computer-assisted clustering [17]. The premise is that all automatic clustering approaches are driven by their own well-justified optimization function to partition the given data. However, it is difficult tell beforehand which of the existing clustering algorithms will produce a partition the user recognizes as "insightful" or "useful" with respect to the application at hand. Therefore, computer-assisted clustering computes a large collection of clusterings using proven methods with different settings for parameters and distance functions. This is called the collective wisdom of the statistical community. All pairs of these clusterings are compared using variational information [32] and a distance matrix of all clusterings is computed. The metric space of clusterings defined by the distance matrix is embedded into the two-dimensional plane using Sammon's multidimensional scaling algorithm [36]. This two-dimensional map produces a visualization that allows the user to recognize the impact of choosing one of the different clustering methods given the data at hand. Furthermore, the map is used to interpolate between similar clusterings. Next, we introduce details of the construction of the space of clusterings, the visualization of it and discuss the overall approach.

19.4.1 Space of Clusterings

Given some data with N objects, the idea of computer-assisted clustering is to apply a large number M of automated clustering algorithms to it and select suitable clusterings afterwards. A clustering is here defined as a partition of the data instances. A first prerequisite of the approach is to compile a list of automated clustering algorithms that should be included. Further, several cluster algorithms are able to work with a set of different distance/similarity measures. Thus, suitable distance/similarity measures have to be chosen for each clustering algorithm as well. While picking algorithms and distance/similarity measures can be done independently of given data, parameters of the algorithms cannot be chosen in such a data-independent way. Despite no details about the parameter setting problem being given in [17] nor the respective supplementary material, general parameters like numbers of clusters could be specified once for all methods that need such an input. If in doubt as to what would be the right number of clusters, several different choices could be specified. This parameter is needed to specify the number of components of finite mixture models as well as to convert hierarchical clusterings into flat ones. The result of this step is a set of M clusterings, called the wisdom of the statistical community.

The space of flat clusterings (partitions) is build on a metric that compares two different clusterings. All such comparison functions are built on the confusion matrix, also called association matrix and contingency table. For completeness, we briefly describe the basics of variational information [32], on which metric computer-assisted clustering is based. Given two clusterings C and C' of N objects, the confusion matrix is a $K \times K'$ matrix where each entry is the size of the intersection between a cluster $C_k \in C$ and $C'_{k'} \in C'$. The normalized variant of the confusion matrix specifies the discrete joint distribution of the cluster labels of the two clusterings:

$$P(k,k') = \frac{|C_k \cap C'_{k'}|}{N} \tag{19.1}$$

This quantity says how likely it is that a randomly drawn data object is a member of both clusters C_k and $C'_{k'}$. The marginal distributions quantify the probabilities that a data object is member of a cluster in the respective clustering:

$$P(k) = \frac{|C_k|}{N} \qquad P'(k') = \frac{|C'_{k'}|}{N} \tag{19.2}$$

All these distributions are needed to define the mutual information $I(C,C')$ between two clusterings:

$$I(C,C') = \sum_{k=1}^{K} \sum_{k'=1}^{K'} P(k,k') \log \frac{P(k,k')}{P(k) \cdot P'(k')} \tag{19.3}$$

as well as for the definitions of the entropies of both clusterings:

$$H(C) = -\sum_{k=1}^{K} P(k) \log P(k) \qquad H(C') = -\sum_{k'=1}^{K'} P'(k') \log P'(k') \tag{19.4}$$

Following [32], variational information between two clusterings is defined as

$$VI(C,C') = H(C) + H(C') - 2I(C,C') \tag{19.5}$$

Most important for computer-assisted clustering, variational information is a metric over the space of possible clusterings [32], which means it is always positive, equals zero if and only if the two clusterings are equal, is symmetric, and obeys the triangle inequality. The derivation of variational information, proofs, and discussions of the properties as well as relations to other clustering comparison measures can be found in [32].

Variational information is used to compute distances between all of the M clusterings produced by the different algorithms with their different distance/similarity measures and parameters that were selected to represent the wisdom of the statistical community. The compiled distance matrix has dimensions $M \times M$ and defines a metric space of clusterings for the given data set. This space is a discrete space that means distances are defined only for those clusterings that are included in the wisdom of the statistical community. For purpose of visualization, the metric space of the clusterings is embedded into the two-dimensional plane where distance is measured by Euclidean distance between points. Each of the M two-dimensional points $\{x_1, \ldots, x_M\}$ represents a clustering in this space. The embedding is computed using Sammon's multidimensional scaling [36]. As the two-dimensional plane is a continuous space, [17] propose an interpolation scheme that assigns clusterings to the points in the plane that do not correspond to one of the clusterings computed so far.

The interpolation scheme takes a particular point x of the two-dimensional plane as input. Then, all precomputed clusterings are weighed using kernel density estimation in the two-dimensional plane, where each clustering is a two-dimensional point. Kernel density estimation places a kernel at each of the two-dimensional points $\{x_m\}_{m=1\ldots M}$ that represent the M clusterings. The kernels used in [17] are Gaussians $\mathcal{N}(x|x_m, \sigma^2)$. The weight of the mth clustering with respect to the particular point x is the normalized contribution of the corresponding kernel to the density estimate at x:

$$w_m(x) = \frac{\mathcal{N}(x|x_m, \sigma^2)}{\sum_{m'=1}^{M} \mathcal{N}(x|x_{m'}, \sigma^2)} \tag{19.6}$$

The smoothing parameter σ^2 needs to be adjusted by the user. The M weights together with corresponding clusterings are used to construct a voting matrix V. The mth clustering C_m is represented as an $N \times K_m$ matrix. A row in that matrix assigns a data object to the clusters. In case of hard cluster assignment, a row contains only a single one and zeros in all other entries, and in case of soft cluster assignment, a row consists of nonnegative numbers that sum up to one. The weighed clusterings are concatenated to form the voting matrix $V = [w_1 C_1, \ldots, w_M C_M]$, which has dimensions $N \times \sum_{m=1}^{M} K_m$. The voting matrix V defines a similarity matrix $S = V \cdot V'$, which in turn defines the interpolated clustering for the selected point x. It is argued in [17] that any clustering method can be used to compute the interpolated clustering assigned to x based on the similarity matrix S. The number of clusters is set to the weighed average of those of the neighboring clusterings. The interpolation scheme extends the discrete set of clusterings to a larger set of clusterings, such that every point on the two-dimensional plane corresponds to a clustering. This two-dimensional plane is called the space of clusterings.

19.4.2 Visualization

The space of clusterings is visualized as a scatter plot. The names of the clustering methods are placed at the locations $\{x_m\}_{m=1...M}$ computed by the embedding algorithm. The user can continuously move a point like a cursor over the scatter plot and the respective clustering is shown in a separate window. In principle, every cluster visualization method can be used to show the clustering. When the user examines example clusterings from different parts of the space of clusterings, he or she learns about the structure of the space. As the space of clusterings is visualized like a map with meaningful distances not all clusterings have to be examined. Looking at examples should help to understand the nearby clusterings as well.

A crucial assumption of the visualization of the space of clusterings is that the visualization of an individual clustering can be intuitively consumed by the user and facilitates a rapid interpretation of the single clusters as well as the total clustering. The example application shown in [17] meets that precondition. The data objects in the example application are biographies of US presidents. Thus, each document has a short and meaningful title, namely, just the president's name. Therefore, a clustering consists of groups of president names that often can be easily interpreted. Furthermore, the number of biographies is rather small, which allows nearly interactive exploration of the space of clusterings. In general, such ideal circumstances are rarely met in practice, even in the case of document clustering. Document titles may become less expressive in the case of large inhomogeneous corpora. Even just showing groups of many document titles on a limited display in way that helps the user to deduce an interpretation is a nontrivial task. Thus, many visualization problems with respect to scaling to large data as well as transfering computer-assisted clustering to other application domains remain unresolved. Furthermore, there is plenty room to study the use of brushing-and-linking techniques in combination with the visualization of the space of clusterings.

19.4.3 Discussion

The idea of visualizing the space of clusterings corresponds to the model selection task, which is heavily studied in machine learning and statistics [39, 18, 46]. Instead of relying on some statistical or information-theoretic criteria that could be computed in some way, computer-assisted clustering delegates this task to the user. The success of this approach depends heavily on the effectiveness of the used visualization of the individual clusterings.

19.5 Visualization of Clusters for Sense-Making

Clusters are computed using quite abstract concepts such as distance between data objects living in some feature spaces. Users do not easily understand these concepts. Therefore, clustering results are not easily accessible for them, and it is difficult to find out whether they can trust the clustering results with respect to the application at hand. Once a clustering is computed, the ultimate goal of a user is to interpret the clusters and make sense of them. Visualization can support such a sense-making process. We discuss temporal document clustering as an example.

Document clustering [12] as well as topic models [8] represent clusters/topics as distributions or feature vectors over the vocabulary of unique words. When looking at the words ordered by decreasing probability with respect to a cluster/topic, users often recognize a semantic topic, even if general terms describing the topic are not mentioned among the top-words. However, naming a cluster/topic is very difficult to do automatically. Even humans, who know the document corpus well, have problems with that task. Thus, visualizations of clusters/topics are demanded to support the sense-making. The task becomes even more challenging in the case of temporal document clus-

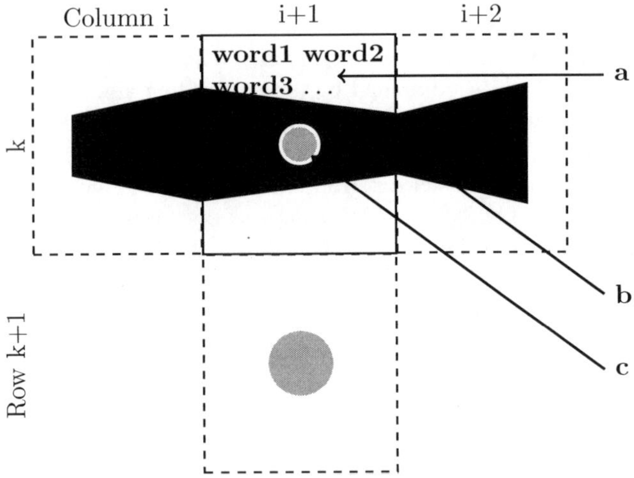

FIGURE 19.7: The cell in row k and column $(i+1)$ of Topic Table corresponds to the kth cluster/topic of the ith batch of documents. Features of Topic Table are (a) top-words of cluster/topic (bold face words are new), (b) width of the background river of each row is proportional to the similarity between the respective clusters/topics of the consecutive batches, (c) the radii of the background circles indicate the relative sizes/strengths of the clusters/topics with respect to the same batch, and the color of the background river indicates similarity among all clusters/topics across all time points: similar clusters/topics have a similar color.

tering, where each time point is associated with a separate batch of documents. Applying clustering independently to the individual document batches produces a set of clusters for each time point. The artificial cluster labels assigned by the algorithms often depend on the initialization of the algorithms; e.g., k-means or other types of models fitted with some sort of inference like expectation maximization or Gibbs sampling could produce the same clustering but with permuted cluster labels when run several times on the same data. Thus, clusters from consecutive time points with the same artificial cluster labels do not necessarily match semantically.

The dynamic topic model [7] and adaptive probabilistic semantic indexing [13] solve the problem of label switching between clusters/topic of consecutive time points. This allows to us present the clusterings/topics of all document batches as a large table where rows and columns indicate clusters and time points, respectively. A cell in this table shows the top-words of a particular cluster/topic at a certain time point. The Topic-Table visualization [14] shows additional information in such a table by visually stacking graphics behind the words to help the user to visually structure the table. The added data are newness of top-words, similarity between consecutive clusters/topics with same index, relative strengths of clusters/topics, and global similarities between all clusters/topics. Details are shown in Figure 19.7.

The color mapping is computed in two steps. First, all pairwise similarities/distances between the clusters/topics represented by the individual table cells are computed. Second, the similarity/distance matrix is used to compute an embedding of the clusters/topics into a three-dimensional color space. Thus, each cluster/topic becomes a point in the color space. Using Sammon's multidimensional scaling algorithm [36] the similarities/distances are preserved in the color space as much as possible. By drawing the river in each cell of the topic table with the respective color, global connections between similar clusters/topics can be easily spotted.

An alternative metaphor is the ThemeRiver [21]. It shows the relative sizes of a series of clusters with the same label as stacked layers of different widths. See Figure 19.8 for an example. The TIARA System [48] uses this visualization technique to visualize dynamic documents clus-

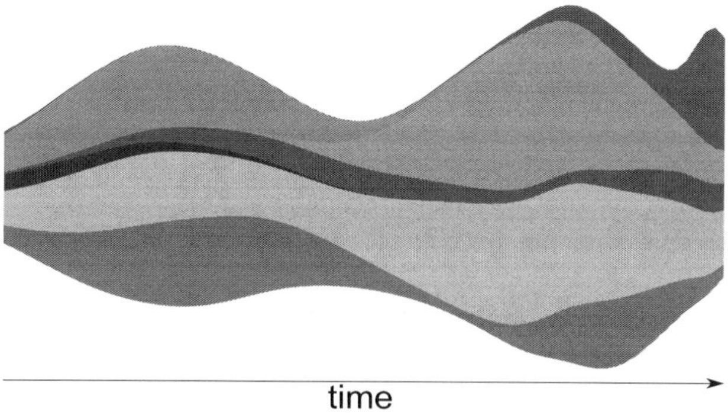

FIGURE 19.8: ThemeRiver visualization of stacked layers. Each layer represents the strength of a series of document clusters at different times.

ters/topics. The clusters/topics are indicated by different colors. In contrast to TopicTable, color here has the function to discriminate between distinct clusters/topics. The top words of the clusters/topics are inserted into the layers at time points when a cluster/topic is strongly present, and therefore, the respective time series is represented by a wide layer. The user can interactively explore the visualization. Less dominant clusters/topics that have no space to insert top-words could be zoomed and annotated with words after a certain minimum width is reached. Search functionality is included to easily retrieve documents related to clusters/topics.

A different extension of the ThemeRiver metaphor for temporal document clustering is TextFlow [11]. Instead of squeezing as many words as possible into the layers representing the clusters, the user can interactively pick a bunch of keywords. The temporal relations between a keyword and the clusters/topics are shown as topic threads (polygon lines) that are drawn on top of the layers. In case two keywords do co-occur in a certain cluster/topic, a wave bundle is drawn that weaves the threads of the keywords together. The amplitude of the wave bundle encodes to occurrences of the involved keywords.

Furthermore, so-called critical events are shown by special icons. Such icons show sources, sinks, splits, and merges. A source indicates the appearance of a new cluster/topic and is shown as filled circle. A sink marks the death of a cluster/topic and is shown as a donut. Splits and merges are shown as the letter Y rotated to the right and left, respectively.

Last, TextFlow expands the layout of the stacked graphs to include secondary branches as well. The original layout as proposed in the ThemeRiver approach connects a cluster/topic present at a certain time point only with the best matching cluster/topic of the next time point. The TextFlow approach also computes and shows connections to the second best matching clusters/topics. This extension works only for dynamic cluster/topic models that construct temporal relations as a post-inference process that means links between clusters back and forth in time are computed after the clusters have been already found at the individual time points. The layout with secondary branches is computed using force-directed simulations to reduce crossing of branches and to smooth cluster/topic layers.

In summary, all three approaches for visualizing temporal document clusters/topics emphasize different aspects of the sense-making process. All visualizations can be combined with interactive methods to dig deeper into the data. Thus, visualization is not the only component in the sense-making process. Other interactive features like semantic zooming, searching relevant documents and so on are important as well. There are no user studies, or stories of successful applications available yet that show which visualizations combined with interactive features are really helpful for end users. New evaluation measures are needed that capture usefulness originating in the interplay of visualization search and analytics.

19.6 Summary

Visual and interactive clustering is an active research field that combines information visualization techniques with concepts and algorithms for cluster analysis. The main challenge is to bridge the chasm between the intuitive concept of clustering that users understand and the reality of data distributions in abstract feature spaces. The consequence of this chasm is that users have difficulties understanding the results of clustering algorithms. Current approaches of visual clustering use different ways to make clusters understandable, namely, (a) completely replacing the automated algorithm by visual interactive procedures, (b) visual interactive steering of an automated clustering algorithm, (c) visual interactive selection of readily computed clusterings, and (d) visual representation of clusters to aid sense-making of the results. A general problem with all these approaches is scalability with respect to large data sets, with respect to high dimensionality of feature vectors, and with respect to a large number of clusters. Overview visualizations showing all clusters, data objects or dimensions in a single view become less effective in the case of very large data. Therefore, interactive search interfaces need to be integrated into the concepts of visual clustering. First steps into this direction have been demonstrated with systems such as TIARA [48] that have a search index as an integral part of the visualization system. Furthermore, new evaluation measures for interactive clustering procedures are needed to guide the development of such systems.

Bibliography

[1] Charu Aggarwal, Alexander Hinneburg, and Daniel Keim. On the surprising behavior of distance metrics in high dimensional space. In Jan Van den Bussche and Victor Vianu, editors, *Database Theory—ICDT 2001*, volume 1973 of *Lecture Notes in Computer Science*, pages 420–434. Springer Berlin / Heidelberg, 2001. doi:10.1007/3-540-44503-X27.

[2] Charu C. Aggarwal. A human-computer cooperative system for effective high dimensional clustering. In *Proceedings of the Seventh ACM SIGKDD International Conference on Knowledge Discovery and Data Mining*, KDD '01, pages 221–226, New York, USA, 2001. ACM.

[3] Gennady Andrienko, Natalia Andrienko, Sebastian Bremm, Tobias Schreck, Tatiana von Landesberger, Peter Bak, and Daniel A. Keim. Space-in-time and time-in-space self-organizing maps for exploring spatiotemporal patterns. *Computer Graphics Forum*, 29(3):913–922, 2010.

[4] Richard Arias-Hernández, John Dill, Brian Fisher, and Tera Marie Green. Visual analytics and human-computer Interaction. *Interactions*, 18(1):51–55, January 2011.

[5] Almir Olivette Artero, Maria Cristina Ferreira de Oliveira, and Haim Levkowitz. Uncovering clusters in crowded parallel coordinates visualizations. In *Proceedings of the IEEE Symposium on Information Visualization*, InfoVis '04, pages 81–88, Washington, DC, USA, 2004. IEEE Computer Society.

[6] Daniel Asimov. The grand tour: A tool for viewing multidimensional data. *SIAM Journal on Scientific and Statistical Computing*, 6(1):128–143, January 1985.

[7] David M. Blei and John D. Lafferty. Dynamic topic models. In *Proceedings of the 23rd International Conference on Machine Learning*, ICML '06, pages 113–120, New York, USA, 2006. ACM.

[8] David M. Blei, Andrew Y. Ng, and Michael I. Jordan. Latent Dirichlet allocation. *Journal of Machine Learning Research*, 3:993–1022, March 2003.

[9] Nan Cao, David Gotz, Jimeng Sun, and Huamin Qu. Dicon: Interactive visual analysis of multidimensional clusters. *IEEE Transactions on Visualization and Computer Graphics*, 17(12):2581–2590, December 2011.

[10] William S. Cleveland. *The Elements of Graphing Data*. Wadsworth, Belmont, CA, USA, 1985.

[11] Weiwei Cui, Shixia Liu, Li Tan, Conglei Shi, Yangqiu Song, Zekai Gao, Huamin Qu, and Xin Tong. TextFlow: Towards better understanding of evolving topics in text. *IEEE Transactions on Visualization and Computer Graphics*, 17(12):2412–2421, December 2011.

[12] Inderjit S. Dhillon and Dharmendra S. Modha. Concept decompositions for large sparse text data using clustering. *Machine Learning*, 42:143–175, 2001. doi:10.1023/A:1007612920971.

[13] André Gohr, Alexander Hinneburg, Rene Schult, and Myra Spiliopoulou. Topic evolution in a stream of documents. In *Proceedings of the SIAM International Conference on Data Mining, SDM 2009*, pages 859–872, 2009.

[14] André Gohr, Myra Spiliopoulou, and Alexander Hinneburg . Visually summarizing semantic evolution in document streams with Topic Table.

In A. Fred, J.L.G. Dietz, K. Liu, and J. Filipe, editors, *Knowledge Discovery, Knowledge Engineering and Knowledge Management*, volume 272 of *Communications in Computer and Information Science*. Springer, pages 136–150. 2012.

[15] J.C. Gower and D. J. Hand. *Biplots*. London, UK, Chapman & Hall, 1996.

[16] Martin Graham and Jessie Kennedy. Using curves to enhance parallel coordinate visualisations. *International Conference on Information Visualisation*, pages 10–16, 2003.

[17] Justin Grimmer and Gary King. General purpose computer-assisted clustering and conceptualization. *Proceedings of the National Academy of Sciences*, 108(7):2643–2650, 2011.

[18] Mark H Hansen and Bin Yu. Model selection and the principle of minimum description length. *Journal of the American Statistical Association*, 96(454):746–774, 2001.

[19] J.A. Hartigan. Printer graphics for clustering. *Journal of Statistical Computation and Simulation*, 4(3):187–213, 1975.

[20] Helwig Hauser, Florian Ledermann, and Helmut Doleisch. Angular brushing of extended parallel coordinates. In *Proceedings of the IEEE Symposium on Information Visualization (InfoVis'02)*, INFOVIS '02, pages 127–130, Washington, DC, USA, 2002. IEEE Computer Society.

[21] Susan Havre, Elizabeth Hetzler, Paul Whitney, and Lucy Nowell. ThemeRiver: Visualizing thematic changes in large document collections. *IEEE Transactions on Visualization and Computer Graphics*, 8(1):9–20, January 2002.

[22] Harold V. Henderson and Paul F. Velleman. Building multiple regression models interactively. *Biometrics*, 37(2):391–411, 1981.

[23] Alexander Hinneburg, Daniel A. Keim, and Markus Wawryniuk. HD-Eye: Visual mining of high-dimensional data. *IEEE Computer Graphics and Applications*, 19:22–31, 1999.

[24] Danny Holten. Hierarchical edge bundles: Visualization of adjacency relations in hierarchical data. *IEEE Transactions on Visualization and Computer Graphics*, 12(5):741–748, September 2006.

[25] Danny Holten and Jarke J. Van Wijk. Evaluation of cluster identification performance for different PCP variants. *Computer Graphics Forum*, 29(3):793–802, 2010.

[26] C. Hurley and R. Oldford. Graphs as navigational infrastructure for high dimensional data spaces. *Computational Statistics*, 26:585–612, 2011. doi:10.1007/s00180-011-0228-6.

[27] Alfred Inselberg. *Parallel Coordinates. Visual Multidimensional Geometry and Its Applications*. Springer, 2009.

[28] Alfred Inselberg and B. Dimsdale. Parallel coordinates: A tool for visualizing multidimensional geometry. In *IEEE Visualization*, pages 361–378, 1990.

[29] Teuvo Kohonen. Self-organized formation of topologically correct feature maps. *Biological Cybernetics*, 43:59–69, 1982. doi:10.1007/BF00337288.

[30] Jing Li, Jean-Bernard Martens, and Jarke J. van Wijk. Judging correlation from scatterplots and parallel coordinate plots. *Information Visualization*, 9(1):13–30, March 2010.

[31] T. M. Martinetz and K. J. Schulten. A "neural-gas" network learns topologies. In T. Kohonen, K. Mäkisara, O. Simula, and J. Kangas, editors, *Artificial Neural Networks*, pages 397–402. North-Holland, Amsterdam, 1991.

[32] Marina Meila. Comparing clusterings—an information based distance. *Journal of Multivariate Analysis*, 98(5):873–895, 2007.

[33] Wayne Oldford and Adrian Waddell. Visual clustering of high-dimensional data by navigating low-dimensional spaces. In *Proceedings of the 58th World Statistics Congress, ISI 2011*, 2011.

[34] J. Poco, R. Etemadpour, F.V. Paulovich, T.V. Long, P. Rosenthal, M.C.F. Oliveira, L. Linsen, and R. Minghim. A framework for exploring multidimensional data with 3d projections. *Computer Graphics Forum*, 30(3):1111–1120, 2011.

[35] Raúl Rojas. *Neural Networks: A Systematic Introduction*. Springer-Verlag, New York, USA, 1996.

[36] J. W. Sammon. A nonlinear mapping for data structure analysis. *IEEE Transactions on Computers*, 18:401–409, 1969.

[37] J. Scholtz. Beyond usability: Evaluation aspects of visual analytic environments. *Symposium on Visual Analytics Science and Technology*, pages 145–150, 2006.

[38] Tobias Schreck, Jürgen Bernard, Tatiana Tekušová, and Jörn Kohlhammer. Visual cluster analysis in trajectory data using editable Kohonen maps. In *IEEE Symposium on Visual Analytics Science and Technology*, pages 3–10. IEEE Computer Society, 2008.

[39] Stanley Sclove. Application of model-selection criteria to some problems in multivariate analysis. *Psychometrika*, 52:333–343, 1987. doi:10.1007/BF02294360.

[40] D.W. Scott. *Multivariate Density Estimation*. Wiley, 1992.

[41] Jinwook Seo and Ben Shneiderman. Interactively exploring hierarchical clustering results. *Computer*, 35:80–86, 2002.

[42] B. W. Silverman. *Density Estimation for Statistics and Data Analysis*. Chapman & Hall, 1986.

[43] T. C. Sprenger, R. Brunella, and M. H. Gross. H-BLOB: A hierarchical visual clustering method using implicit surfaces. In *Proceedings of the Conference on Visualization '00*, VIS '00, pages 61–68, Los Alamitos, CA, USA, 2000. IEEE Computer Society Press.

[44] John W. Tukey and Paul A. Tukey. Computer graphics and exploratory data analysis: An introduction. In *Proceedings of the Sixth Annual Conference and Exposition: Computer Graphics'85*, 3:773–785. National Computer Graphics Association, 1985.

[45] Adrian Waddell and Wayne Oldford. RnavGraph: A visualization tool for navigating through high-dimensional data. In *Proceedings of the 58th World Statistics Congress, ISI 2011*, 2011.

[46] Larry Wasserman. Bayesian model selection and model averaging. *Journal of Mathematical Psychology*, 44(1):92–107, 2000.

[47] Edward J. Wegman and Qiang Luo. High dimensional clustering using parallel coordinates and the Grand Tour. *Computing Science and Statistics*, 28:361–368, 1996.

[48] Furu Wei, Shixia Liu, Yangqiu Song, Shimei Pan, Michelle X. Zhou, Weihong Qian, Lei Shi, Li Tan, and Qiang Zhang. TIARA: A visual exploratory text analytic system. In *Proceedings of the 16th ACM SIGKDD International Conference on Knowledge Discovery and Data Mining*, KDD '10, pages 153–162, New York, NY, USA, 2010. ACM.

[49] Leland Wilkinson, Anushka Anand, and Robert Grossman. Graph-theoretic scagnostics. *IEEE Symposium on Information Visualization*, pages 167–164, 2005.

[50] Xiaoru Yuan, Peihong Guo, He Xiao, Hong Zhou, and Huamin Qu. Scattering points in parallel coordinates. *IEEE Transactions on Visualization and Computer Graphics*, 15(6):1001–1008, November 2009.

[51] Hong Zhou, Weiwei Cui, Huamin Qu, Yingcai Wu, Xiaoru Yuan, and Wei Zhuo. Splatting the lines in parallel coordinates. *Computer Graphics Forum*, 28(3):759–766, 2009.

[52] Hong Zhou, Xiaoru Yuan, Huamin Qu, Weiwei Cui, and Baoquan Chen. Visual clustering in parallel coordinates. *Computer Graphics Forum*, 27(3):1047–1054, 2008.

Chapter 20

Semisupervised Clustering

Amrudin Agovic
Reliancy, LLC
Saint Louis Park, MN
aagovic@cs.umn.edu

Arindam Banerjee
University of Minnesota at Twin Cities,
Minneapolis, MN
banerjee@cs.umn.edu

20.1 Introduction

Semisupervised clustering (SSC) has become an important part of data mining. With an ever increasing volume of data in several problem domains, it is more important than ever to leverage known information and observed relationships among data points to guide clustering.

Clustering methods are broadly divided into two groups depending on the data representation they use: *feature-based*, where each data point has a representation in terms of a feature vector or a structured representation such as sequence, time series, or graphs, and *graph-based*, where a similarity graph among the data points is given. Methods such as k-means and mixture of Gaussians work with feature-based representations, whereas spectral clustering methods work with graph-based representations.

Existing works on SSC can be broadly divided into two groups, depending on whether one adds semisupervision to a feature-based or a graph-based clustering algorithm. Much of early work in SSC focussed on extending feature-based clustering methods to the semisupervised setting. In particular, the literature has focussed on two types of semisupervision: *pointwise* [2], where the cluster labels of a small number of points are available to guide clustering, and *pairwise* [37, 3, 4, 20], where "must-link" and "cannot-link" constraints between some pairs of points are available. Such methods have been extensively studied over the past decade [6], with emphasis on suitably generalizing feature based clustering algorithms such as k-means and its variants to leverage the semisupervision. These methods have been generalized to incorporate metric learning in the context of SSC and also as parameter estimation and inference in suitable graphical models [9, 3, 5, 4]. We discuss this family of SSC methods in Section 20.2.

There are several SSC approaches based on the graph-based representation and graph-based clustering methods. The literature on graph-based SSC has primarily focused on pointwise semisupervision. Spectral clustering methods are widely used for unsupervised clustering with graph-based representations [36, 17] and can be viewed as solving a relaxation of suitable graph-cut problems. In the semisupervised setting, one can approach the problem as one of semisupervised graph-cuts, where the labeled points with the same cluster label are expected to be in the same cut. We illustrate that the relaxed versions of the semisupervised graph-cut problems can be solved by suitable semisupervised spectral clustering methods, which in turn are intimately related to label propagation methods [15].

We also present a generalized label propagation (GLP) framework which includes a variety of graph-based semisupervised learning methods developed over the past decade, and includes the semisupervised spectral clustering methods as special cases. Further, the same framework also includes semisupervised nonlinear embedding methods as special cases. Based on the unified treatment, we provide a generic recipe for converting nonlinear embedding methods to semisupervised label propagation methods. We illustrate the generic recipe by deriving semisupervised label propagation methods based on three well-known nonlinear embedding methods, viz locally linear embedding (LLE) [32], Laplacian eigenmaps (LE) [8], and local tangent space alignment (LTSA) [44]. We also present an empirical performance evaluation of some of the existing semisupervised label propagation methods as well as the ones which can be derived from nonlinear embedding. We illustrate that there are no clear winners across all datasets, although a few methods consistently show up among the top performing methods.

It is important to note that for pointwise semisupervision, where labels on certain individual points are available, the semisupervised clustering and classification problems are closely related. For feature-based data representation, the methods for semisupervised clustering [2] can in principle be compared against transductive classification methods [26, 27], although different loss functions are usually meaningful in the two settings. For graph-based representation, the methods are indeed related especially if the effectiveness of the learning is measured by accuracy of predicted cluster/class labels in the unlabeled set.

The rest of the chapter is organized as follows. In Section 20.2, we discuss approaches for SSC based on pointwise and pairwise semisupervision for feature-based representations and related clustering algorithms. In Section 20.3, we describe semisupervised graph-cuts and spectral clustering. In Section 20.4, we introduce the GLP formulation, present a unified view of existing label propagation methods, and illustrate their relationship to semisupervised graph-based clustering. We discuss semisupervised manifold embedding and a set of embedding-based label propagation methods in Section 20.5. We present empirical results in Section 20.6 and conclude in Section 20.7.

20.2 Clustering with Pointwise and Pairwise Semisupervision

Semisupervised clustering with pointwise and pairwise semisupervision has been widely studied for feature-based clustering methods. In this section, we review some of the approaches from the literature [6].

20.2.1 Semisupervised Clustering Based on Seeding

One of the earliest ideas on SSC focussed on pointwise label supervision, where the cluster ids of a small number of points are made available. SSC based on seeding focuses on centroid-based clustering algorithms, such as k-means [2, 11]. In an unsupervised setting, such algorithms usually start from a random initialization and perform expectation-maximization (EM)-style iterative updates of cluster memberships and cluster parameters. The main idea in SSC based on seeding is to improve the initialization based on the available cluster ids. In particular, [2] looks at two variants applied to k-means: *seeded k-means* and *constrained k-means*. Seeded k-means initializes the cluster centroids using the available cluster ids and then runs the iterative updates for k-means. In seeded k-means, the cluster id assignment of the labeled set can change during the iterative updates if the objective function improves as a result. Let $\mathcal{X} = \{x_1, \ldots, x_N\}, x_i \in \mathbb{R}^d$ denote a set of data points, let K be the number of clusters and let set $S = \cup_{l=1}^{K} S_l$ denote the initial seeds. The seeded k-means algorithm can be summarized in the following steps.

- Initialize cluster centers $\mu_h \leftarrow \frac{1}{|S_h|} \sum_{x \in S_h} x$, for $h = 1, \ldots, K$
- Assign each data point to the cluster $h^* = \underset{h}{\operatorname{argmin}} ||x - \mu_h||^2$
- Estimate cluster centers $\mu_h \leftarrow \frac{1}{|X_h|} \sum_{x \in X_h} x$, for $h = 1, \ldots, K$
- Repeat last two steps until convergence

Constrained k-means also initializes the cluster centroids with the available cluster ids, but the cluster id assignments of these labeled points are not allowed to change during the iterative updates. Seeded constrained k-means is thus more appropriate when the initial seed labeling is noise free.

SSC based on seeding can be generalized to other clustering methods beyond k-means, and [2] empirically evaluate the spherical-k-means method. The seeding approach can be viewed in the probabilistic clustering setting where one uses the EM algorithm to learn mixture models [21], and each mixture component corresponds to a cluster. In particular, the semisupervision is used to set the posterior probability $p(z_i|\mathbf{x}_i, \Theta)$ of the labeled points to be 1 for the true cluster, and 0 otherwise. Empirical results in [2] illustrate the advantages of the seeding approach, with clustering performance sharply increasing with a small number of seeds. The method has also been applied to the scenario where the labeled data covers only a fraction of the clusters, and the centroids of the other clusters have to be randomly initialized.

20.2.2 Semisupervised Clustering Based on Pairwise Constraints

One of the most popular frameworks for SSC is based on pairwise constraints. Given a dataset $\mathcal{X}$, pairwise semisupervision is typically given in the form of two types of constraints: *must-link* and *cannot-link* constraints. If $(\mathbf{x}_i, \mathbf{x}_j) \in \mathcal{M}$, the set of must-link constraints, then the clustering algorithm is encouraged to keep $(\mathbf{x}_i, \mathbf{x}_j)$ in the same cluster; if $(\mathbf{x}_i, \mathbf{x}_j) \in \mathcal{C}$, the set of cannot-link constraints, then the clustering algorithm is encouraged to keep $(\mathbf{x}_i, \mathbf{x}_j)$ in different clusters. Several approaches use a suitable constraint violation penalty, whereas some approaches even consider the constraints to be binding [37, 3, 5, 20, 38].

One of the early approaches to SSC with pairwise constraints focused on modifying the k-means algorithm [38] to incorporate must-link and cannot-link constraints. In each iteration of the algorithm, a sorted list of suitable clusters is considered for every point. The algorithm moves down the list until a cluster assignment is found that does not violate any constraints. If no such cluster is found, the algorithm terminates.

The approach presented in [38] posed a considerable improvement over the unsupervised version of k-means and allowed semisupervised background knowledge to be incorporated. However, one obvious drawback from a practical standpoint is the requirement for all constraints to be satisfied. One has to ensure that no contradicting constraints are specified.

Alternative approaches focused on modifying the k-means objective to take into account pairwise constraint violations [3] in terms of penalties. In particular, the pairwise constrained clustering with k-means (PCKMeans) objective function can be written as:

$$J_{PCC}(M,Y) = \frac{1}{2}\sum_{\mathbf{x}_i \in \mathcal{X}} \|\mathbf{x}_i - \mu_{y_i}\|^2 + \sum_{(\mathbf{x}_i,\mathbf{x}_j)\in\mathcal{M}} w_{ij}\mathbf{1}(y_i \neq y_j) + \sum_{(\mathbf{x}_i,\mathbf{x}_j)\in\mathcal{C}} \bar{w}_{ij}\mathbf{1}(y_i = y_j)\,, \tag{20.1}$$

where $w_{ij}, \bar{w}_{ij}$ are appropriate constants which serve as penalties for constraint violation, $M = \{\mu_1, \ldots, \mu_k\}$ is the set of cluster means, and $Y = \{y_1, \ldots, y_n\}$, where $y_i \in \{1, \ldots, k\}$ are the cluster ids of the data points. Several generalizations of the above formulation have been considered in the literature. One of the prominent threads of development involves incorporating metric learning which suitably modifies how distances are computed between data points and cluster centroids.

Local Metric Learning: Metric learning approaches in the context of clustering focus on a parameterized divergence function, where the parameters can be suitably chosen based on semisupervision. One can consider a semisupervised metric learning approach, where the following objective is optimized based on pointwise or pairwise semisupervision

$$J_{LML}(M,Y,\mathcal{A}) = \frac{1}{2}\sum_{\mathbf{x}_i \in \mathcal{X}} \left\{(\mathbf{x}_i - \mu_{y_i})^T A_{y_i}(\mathbf{x}_i - \mu_{y_i}) + \log(\det(A_{y_i}))\right\}\,, \tag{20.2}$$

where A_{y_i} is a positive definite matrix for cluster y_i, and $\mathcal{A} = \{A_1, \ldots, A_k\}$. One can consider $A_{y_i}^{-1}$ as the covariance of the multivariate Gaussian distribution corresponding to cluster y_i. Since each cluster $h = 1, \ldots, k$ has its own parameter A_h, this can be considered a local metric learning approach.

In [9], a hybrid approach is proposed which is capable of both incorporating constraints in terms of must-link and cannot link constraints, and also adapting the underlying distance measure. The idea in [9] is to combine the strengths of the two approaches. The method works by combining the two clustering objectives in Equation 20.1 and Equation 20.2. The combined objective is given by

$$J_{PCCM}(M,Y,\mathcal{A}) = \sum_{\mathbf{x}_i \in \mathcal{X}} \left\{ (\mathbf{x}_i - \mu_{y_i})^T A_{y_i} (\mathbf{x}_i - \mu_{y_i}) + \log(\det(A_{y_i})) \right\} \quad (20.3)$$
$$+ \sum_{(\mathbf{x}_i,\mathbf{x}_j) \in \mathcal{M}} w_{ij} f_M(\mathbf{x}_i,\mathbf{x}_j) \mathbf{1}[y_i \neq y_j] + \sum_{(\mathbf{x}_i,\mathbf{x}_j) \in \mathcal{C}} \bar{w}_{ij} f_C(\mathbf{x}_i,\mathbf{x}_j) \mathbf{1}[y_i = y_j] ,$$

where f_M and f_C are the constraint violation functions which depend on $(\mathbf{x}_i, \mathbf{x}_j)$. The function f_M is defined such that the penalty for violation of must-link constraints between points that are distant should be higher. Such a construction makes sense since if two distant points need to be must-linked, then the metric currently in use needs a major update, and a high penalty will be able to accomplish that. Since a violated must-link constraint involves two clusters, the penalty function is defined in terms of both metrics:

$$f_M(\mathbf{x}_i,\mathbf{x}_j) = \frac{1}{2}(\mathbf{x}_i - \mathbf{x}_j)^T A_{y_i} (\mathbf{x}_i - \mathbf{x}_j) + \frac{1}{2}(\mathbf{x}_i - \mathbf{x}_j)^T A_{y_j} (\mathbf{x}_i - \mathbf{x}_j) . \quad (20.4)$$

In the case of cannot-link violations, the penalty corresponding to nearby points should be higher. The following penalty term is considered:

$$f_C(\mathbf{x}_i,\mathbf{x}_j) = \frac{1}{2}(\mathbf{x}'_{y_i} - \mathbf{x}''_{y_i})^T A_{y_i} (\mathbf{x}'_{y_i} - \mathbf{x}''_{y_i}) - \frac{1}{2}(\mathbf{x}_i - \mathbf{x}_j)^T A_{y_i} (\mathbf{x}_i - \mathbf{x}_j) , \quad (20.5)$$

where $(\mathbf{x}'_{y_i}, \mathbf{x}''_{y_i})$ are the maximally separated set of points in the data set according to A_{y_i}. These weight functions are introduced in order to produce a more relevant adherence to the constraints. The objective function is optimized by using the EM algorithm, which alternates between updating the cluster assignments, and updating the cluster means μ_h and metrics A_h until convergence. For scalability and numerical stability, [9] suggest using metrics with simpler structures, such as diagonal matrices, for high-dimensional problems.

Global Metric Learning: A generalization of the PCKMeans setting was considered in the literature [9, 5, 4], where the distance metric used to measure the within-cluster distortion is globally parameterized, and the parameters are learned as part of the SSC process. In particular, if A is a positive definite matrix, one considers the following problem:

$$J_{GML}(M,Y,A) = \frac{1}{2} \sum_{\mathbf{x}_i \in \mathcal{X}} (\mathbf{x}_i - \mu_{y_i})^T A (\mathbf{x}_i - \mu_{y_i}) + \sum_{(\mathbf{x}_i,\mathbf{x}_j) \in \mathcal{M}} w_{ij} \mathbf{1}(y_i \neq y_j) + \sum_{(\mathbf{x}_i,\mathbf{x}_j) \in \mathcal{C}} \bar{w}_{ij} \mathbf{1}(y_i = y_j) . \quad (20.6)$$

Although the above exposition focuses on the k-means clustering setting, the formulation can be generalized to several other settings, such as information theoretic clustering and spherical clustering. The two main choices in each setting are (i) a parameterized function $d_A(\mathbf{x}_i, \mu_{y_i})$, which measures the divergence of each data point from its corresponding cluster mean, and (ii) suitable penalty functions $f_ML(i,j)$ and $f_{CL}(i,j)$ which respectively measure the cost of violating a must-link and a cannot-link constraint. In Equation 20.6, we have $d_A(\mathbf{x}_i, \mu_{y_i}) = (\mathbf{x}_i - \mu_{y_i})^T A (\mathbf{x}_i - \mu_{y_i}), f_{ML}(i,j) = 1$, and $f_{CL}(i,j) = 1$.

Parameterized divergence functions: One can consider more general divergence functions, including parameterized versions of information theoretic and directional divergences. In particular, information theoretic clustering uses the following divergence function [22, 4]: $d_I(\mathbf{x}_i, \mu_h) = \sum_{j=1}^d x_{ij} \log \frac{x_{ij}}{\mu_{hj}} - \sum_{j=1}^d (x_{ij} - \mu_{hj})$, where $\mathbf{x}_i, \mu_h$ are assumed to be positive vectors or probability distributions. In the latter case, the divergence reduces to the KL-divergence [19]. The parameterized version of I-divergence considers a diagonal matrix $A = \text{diag}(a_j)$ with nonnegative weight a_j corresponding to each dimension j leading to $d_{I,A}(\mathbf{x}_i, \mu_h) = \sum_{j=1}^d a_j x_{ij} \log \frac{x_{ij}}{\mu_{hj}} - \sum_{j=1}^d a_j (x_{ij} - \mu_{hj})$. One can similarly define a parameterized version of directional divergence based on the cosine similarity as $d_{cos,A}(\mathbf{x}_i, \mu_h) = 1 - \frac{\mathbf{x}_i^T A \mu_h}{\|\mathbf{x}_i\|_A \|\mu_h\|_A}$, where $\|\mathbf{x}\|_A = \sqrt{\mathbf{x}^T A \mathbf{x}}$.

While parameterized divergences give the additional flexibility of rotating and/or scaling the feature space appropriately in order to match the given pairwise semisupervision, without proper regularization, such formulations can yield degenerate solutions. For example, setting $A = 0$, the zero matrix makes all distances zero and, hence, minimizes the objective function, but it is not meaningful. To avoid such degenerate solutions, one considers suitable regularizers $R(A)$ on A or putting constraints on A, e.g., making sure A is positive semidefinite, as is commonly done in metric learning [41].

Constraint violation functions: The second major choice involves the constraint violation functions $f_{ML}(i,j)$ and $f_{CL}(i,j)$. For the must-link constraints, one often considers penalty functions which are proportional to the distance, i.e., $f_{ML}(i,j) = d_A(\mathbf{x}_i,\mathbf{x}_j)$. For points which are far apart according to the distance function, a must-link constraint provides valuable information regarding the structure of the clustering. As a result, the penalty is more if such a constraint is violated. For cannot-link constraints, one considers penalty functions which decrease with the distance, i.e., $f_{CL}(i,j) = d_A^{\max} - d_A(\mathbf{x}_i,\mathbf{x}_j)$, where $d_A^{\max}$ is a suitable upper bound on the pairwise distances between points in the dataset. Thus, cannot-link constraints between points which are nearby are given more weight as they suggest that the cluster structure may be substantially different from the one suggested by the distance measure. One can consider alternative definitions of the functions as appropriate for a given problem domain.

With the above two generalizations, the objective function for PCC can be written as

$$\begin{aligned} J_{PCCM}(M,Y,A) = & \frac{1}{2}\sum_{\mathbf{x}_i \in X} d_A(\mathbf{x}_i,\mu_{y_i}) + R(A) \\ & + \sum_{(\mathbf{x}_i,\mathbf{x}_j)\in \mathcal{M}} w_{ij} f_M(i,j)\mathbf{1}(y_i \neq y_j) + \sum_{(\mathbf{x}_i,\mathbf{x}_j)\in C} \bar{w}_{ij} f_C(i,j)\mathbf{1}(y_i \neq y_j) \, . \end{aligned} \tag{20.7}$$

Such a general formulation can be interpreted as the log-likelihood of a probabilistic graphical model known as the hidden Markov random field (HMRF) [5, 4]. The HMRF model for SSC consists of the following sets of variables and parameters: (i) an observable set $X = (x_1,\ldots,x_n)$ corresponding to the given data points X; (ii) an unobservable (hidden) set $Y = (y_1,...,y_n)$ corresponding to cluster assignments of points in X, where each hidden variable y_i encodes the cluster label of the point x_i and takes values from the set of cluster indices $(1,\ldots,K)$; (iii) a set of generative model parameters $\Theta = \{A,M\}$, which consists of divergence parameters A and cluster representatives $M = (\mu_1,\ldots,\mu_K)$; and (iv) an observable set of constraint variables $C = (c_{12},c_{13},\ldots,c_{n-1,n})$, where each c_{ij} is a tertiary variable taking on a value from the set $(-1,0,1)$, where $c_{ij} = 1$ indicates that $(x_i,x_j) \in \mathcal{M}$, the set of must-link constraints, $c_{ij} = -1$ indicates that $(x_i,x_j) \in C$, the set of cannot-link constraints, and $c_{ij} = 0$ corresponds to pairs (x_i,x_j) that are not constrained.

Consider the posterior distribution $P(Y,\Theta|X,C)$, which by Bayes rule is proportional to

$$P(X,Y,\Theta|C) = P(\Theta|C)P(Y|\Theta,C)P(X|Y,\Theta,C) \, . \tag{20.8}$$

We assume the model parameters $\Theta = (M,A)$ to be independent of the constraints C, so that $P(\Theta|C) = P(\Theta) = P(M,A) = P(M)P(A)$, where we further assume M and A to be independent. The latent variables Y are assumed to form a pairwise Markov random field (MRF) with only pairwise potentials [39, 4], so that

$$P(Y|\Theta,C) = \frac{1}{Z}\exp\left\{-\sum_{i,j} f(i,j)\right\}, \tag{20.9}$$

where Z is the partition function, and the pairwise potentials are given by

$$f(i,j) = \begin{cases} w_{ij} f_{ML}(i,j) \, , & \text{if } c_{ij} = 1 \text{and } y_i \neq y_j \\ \bar{w}_{ij} f_{CL}(i,j) \, , & \text{if } c_{ij} = -1 \text{and } y_i = y_j \\ 0 \, , & \text{otherwise} \, . \end{cases} \tag{20.10}$$

The conditional distribution $P(X|Y,\Theta,C) = P(X|Y,\Theta) = \prod_i p(\mathbf{x}_i|y_i,\Theta)$ is determined by a mixture model, where the mixture component is determined by y_i, and Θ contains the parameters of the mixture model. The literature has considered a variety of components, including multivariate Gaussians distributions, von Mises-Fisher distributions, and multinomial distributions [4]. More generally, one can consider a functional form of the conditional distribution determined by a parameterized divergence function: $p(\mathbf{x}_i|y_i,\Theta) = \frac{1}{Z_\Theta}\exp(-d_A(\mathbf{x}_i,\mu_{y_i}))$, where Z_Θ is the partition function. The negative log-likelihood of the posterior distribution, is given by

$$-\log P(Y,\Theta|X,C) = \sum_i d_A(\mathbf{x}_i,\mu_{y_i}) + \sum_{i,j} f(i,j) - \log P(\Theta) + n\log Z_\Theta + \log Z + \log P(X|C)\,. \quad (20.11)$$

Since $\log P(X|C)$ is a constant, with $R(\Theta) = n\log Z_\Theta - \log P(\Theta)$, Equation 20.11 is equivalent to the objective function in Equation 20.8.

Given a dataset, such models are learned using a suitable version of the EM algorithm. In particular, for the objective function in Equation 20.11, one has to estimate the parameters $\Theta = (M,A)$ and the cluster memberships $Y = \{y_1,\ldots,y_n\}$ for every data point. The EM algorithm is initialized with some value $\Theta^{(0)}$ for the parameters and then proceeds with the following alternating updates until convergence:

E-step Given $\Theta^t = (M^t,A^t)$, update cluster memberships Y^{t+1} to minimize the objective

M-step: *M*-update Given (A^t,Y^{t+1}), update cluster means M^{t+1} to minimize the objective

M-step: *A*-update Given (M^{t+1},Y^{t+1}), update distortion parameter A^{t+1} to minimize the objective

For a variety of semisupervised clustering problems, the above SSC approach has been shown to perform well and has, hence, evolved as one of the standard approaches to SSC with pairwise constraints [5, 4, 6].

20.2.3 Active Learning for Semisupervised Clustering

Since SSC works with a few pairwise (or pointwise) labels, active learning is a natural framework to consider in this setting. Given a fixed number of allowed queries on pairwise labels, the goal is to decide which pairwise relationships to query in order to get must-link and cannot-link information. In [3], a two-stage method is outlined for active learning for SSC with pairwise labels. In the first stage, called *explore*, the focus is on getting at least one point from each cluster with a small number of queries. The first point is chosen at random and assigned to a cluster. All subsequent points are chosen by farthest first traversal, i.e., by picking the point which is farthest from all existing points, where farthest point to a set is measured by the distance to the nearest point in the set. Once a point is selected, pairwise queries are made with any one point from each of the existing clusters. If a must-link constraint is found with any of the existing clusters, then the point is assigned to that cluster, and the method picks the next farthest point. If no must-link constraint is found with any of the existing clusters, then a new cluster is initialized with this point as the member. The explore process continues until at least one point from each cluster is found or the budget of queries is exhausted. In the second stage, called *consolidate*, additional data points are selected at random and assigned to the correct clusters by pairwise querying. Given a data point, all clusters are first sorted in increasing order of distances to the corresponding cluster centroids. Pairwise queries are made with any one point from each of the clusters in sorted order until a point/cluster with a must-link constraint is found. In that case, the new point is assigned to that cluster, and the process continues by picking another point at random. The querying is expected to be efficient, i.e., for a k-clustering problem, much fewer than k queries will usually be needed. Empirically, [3] show the active learning strategy is shown to be effective in practice, leading to better test-set performance with fewer queries as compared to choosing the queries at random.

20.2.4 Semisupervised Clustering Based on User Feedback

A set of ideas for SSC was pursued by [18], where the semisupervision is done with the user in the loop. The development acknowledges the fact that there can be multiple ways semisupervision can be provided and advocates updating the clustering based on input received from the user. In particular, [18] consider a scenario where a user is iteratively providing feedback about the quality of clusters. Typically a divergence measure d is defined up front and used to produce a clustering. Based on the user's feedback on the initial clustering, the method attempts to adjust what it means to be similar. In other words the divergence measure d in this case is not fixed, but rather learned as the user provides feedback. It is instructive to note that such ideas in [18] were also considered by others [4], as discussed in Section 20.2.2, but [18] explicitly consider the user to be a part of the process.

The problem of interest in [18] is the clustering of documents. Once an initial clustering has been done, the user can provide the following types of feedback: (i) an indication that a given document is in the wrong cluster; (ii) given a more appropriate cluster, an indication that the document in question should be moved to it; (iii) whether two given documents should be within the same cluster; and (iv) whether two documents should be in different clusters. Indeed, the type of feedback is closely related to pointwise and pairwise semisupervision discussed earlier.

The user is assumed to have no knowledge about how many clusters there really are. After the feedback is incorporated into the distance measure, the clustering algorithm is rerun. These steps are repeated as long as user feedback is available or until convergence. In practice, only very few iterations of user feedback are considered.

Following a bag of words model, a document d is assumed to be generated from a multinomial mixture model θ over words w [31, 18]. Words are defined by a vocabulary V. Within this probabilistic setting, KL divergence is selected as a natural choice of distance measure to be used in clustering. In order to enable adjustments based on user feedback the authors propose a weighted version of KL divergence:

$$d_A(d_1||d_2) = \prod_{w_j \in V} a_j p(w_j|\theta_1) \log \frac{p(w_j|\theta_1)}{p(w_j|\theta_2)}, \tag{20.12}$$

where a_j are constrained to be positive. The divergence used in [18] is in fact a symmetrized version of the parametrized KL-divergence where one computes the weighted average KL-divergence of individual document frequencies to a discrete distribution with mean of the word frequencies of the documents, and the weighting is determined by the number of words in each document. Once user feedback is provided for a given pair of documents, d_A is optimized by suitably updating a_j. The iterative updates are similar to the ideas discussed in Section 20.2.2 based on a parametrized KL-divergence.

In terms of results, the approach appears to provide a clear boost in performance when compared to unsupervised clustering. In particular, it seems that the algorithm performs well even with very little user feedback. Approaches such as the one proposed in [18] are referred to as metric-based, since they incorporate semisupervision only by adapting a distance metric. Several approaches have been proposed in literature with different distance measures such as [28, 10, 42].

The algorithm proposed in [28] is a variant of the complete-link hierarchical agglomerative clustering, whereby the Euclidean distance metric is altered according to user feedback. For pairs of points which are considered must-link, a distance value of zero is used. For pairs of points which have cannot-link constraints associated with them, the maximum distance plus one is used. The motivation behind the approach is to combine pairwise constraints with spatial constraints, so that local neighborhoods are affected by the pairwise constraints. As a result performance is improved significantly compared to using pairwise constraints alone. Furthermore, outliers in the output are reduced since spatial relationships are taken into consideration.

20.2.5 Semisupervised Clustering Based on Nonnegative Matrix Factorization

In [29], a Nonnegative Matrix Factorization (NMF) framework is proposed for both concensus clustering and semisupervised clustering. For the purposes of this chapter, we will focus on the semisupervised clustering part. The main contribution of [29] is to show how both semisupervised k-means and semisupervised kernel k-means can be approached as an NMF problem.

The authors consider an indicator matrix $H = [h_1 \ldots h_k]$ with

$$h_k = [0,\ldots,0,\overbrace{1,\ldots,1}^{n_k},0,\ldots,0]^T / n_k^{1/2}, \tag{20.13}$$

where the nonzero entries in h_k represent the points belonging to cluster k, and n_k denotes the number of those points. [29] make the observation that the k-means clustering objective can be expressed as

$$\max_{H^T H = I, H \geq 0} \quad J_k = Tr(H^T W H), \tag{20.14}$$

where $w_{ij} = x_i^T x_j$ for k-means and $w_{ij} = \phi(x_i)^T \phi(x_j)$ for kernel k-means. By encoding must-link and cannot-link constraints with indicator matrices A and B, respectively, the authors show that the semisupervised k-means algorithm can be posed as an optimization problem of the form:

$$\max_{H^T H = I, H \geq 0} \quad Tr[H^T W H + \alpha H^T A H - \beta H^T B H]. \tag{20.15}$$

Letting $W^+ = W + \alpha \geq 0$, $W^- = \beta B$, and removing the orthogonality constraint, a relaxed version of the optimization problem can be expressed as

$$\max_{H \geq 0} ||(W^+ - W^-) - HH^T||^2. \tag{20.16}$$

This is an objective function where $(W^+ - W^-)$ is approximated by HH^T with nonnegativity constraints on H. Related problems are widely studied in the context of NMF, and the above problem can be solved with updates of the form

$$H_{ik} = H_{ik} \sqrt{\frac{(W^+ H)_{ik}}{(W^- H)_{ik} (HH^T H)_{ik}}} \tag{20.17}$$

The authors provide a proof of convergence as well as correctness. The NMF approach performs favorably in comparison to certain existing algorithms in the literature [29]. In [29], the SSC problem is posed as a semidefinite optimization problem which incorporates constraints. Conceptually this approach is not too different from [9], where the constraints are considered and co-variance matrices are estimated in an iterative fashion. In [29] a decomposition of $(W^+ - W^-)$ is estimated with W denoting a distance measure. As such [29] can also be seen combining both constrained-based and metric-based aspects.

20.3 Semisupervised Graph Cuts

In this section, we consider SSC with graph-based representations, with focus on semi-supervised graph-cuts. The key difference between the methods in Section 20.2 and the ones considered here is that the dataset is in the form of a similarity graph G among data points, rather than feature vectors corresponding to each data point.

We start by briefly reviewing some problems and concepts relevant to unsupervised graph cuts. Let $G=(V,E)$ be a weighted undirected graph with weight matrix W. If V_1,V_2 is a partitioning of V, i.e., $V_1 \cap V_2 = \emptyset, V_1 \cup V_2 = V$, then the value of the cut implied by the partitioning (V_1,V_2) is given by

$$cut(V_1,V_2) = \frac{1}{2} \sum_{v_i \in V_1, v_j \in V_2} w_{ij} \, . \tag{20.18}$$

The minimum cut problem is to find a partitioning (V_1,V_2) such that $cut(V_1,V_2)$ is minimized. Due to practical reasons, one often works with a normalized cut objective, such as the ratio-cut [25] or normalized-cut [34], which encourage the partitions V_1,V_2 to be more balanced. The objective for ratio-cut is as follows:

$$Rcut(V_1,V_2) = \frac{cut(V_1,V_2)}{|V_1|} + \frac{cut(V_2,V_1)}{|V_2|} \, . \tag{20.19}$$

The objective for normalized-cut is similar; however, it normalizes cuts by the weight of the edges in each partition. For any subset $V_h \subseteq V$, letting $Vol(V_h) = \sum_{i \in V_h} D_{ii}$ where $D_{ii} = \sum_j w_{ij}$, we have

$$Ncut(V_1,V_2) = \frac{cut(V_1,V_2)}{Vol(V_1)} + \frac{cut(V_2,V_1)}{Vol(V_2)} \, . \tag{20.20}$$

There are extensions of the above formulations for k-cuts [36], but we focus on the 2-cut setting in this chapter for ease of exposition.

Graph Laplacians: Next, we briefly review graph Laplacians, which will play an important role in our exposition. For an undirected weighted graph $G=(V,E)$ with weights $w_{ij} \geq 0$, let D be a diagonal matrix with $D_{ii} = \sum_j w_{ij}$. In the existing literature, there are three related matrices that are called the graph Laplacian, and there does not appear to be a consensus on the nomenclature [36]. These three matrices are intimately related, and we will use all of them in our analysis. The *unnormalized graph Laplacian* L_u is defined as

$$L_u = D - W \, . \tag{20.21}$$

The following property of the unnormalized graph Laplacian is important for our analysis: For any $f \in R^n$, we have

$$f^T L_u f = \frac{1}{2} \sum_{i,j} w_{ij} (f_i - f_j)^2 \, . \tag{20.22}$$

The matrix L_u is a symmetric and positive semidefinite. There are also two normalized graph Laplacians in the literature [17] given by

$$L_r = D^{-1} L_u = I - D^{-1} W \, , \tag{20.23}$$

$$L_s = D^{-1/2} L_u D^{-1/2} = I - D^{-1/2} W D^{-1/2} \, . \tag{20.24}$$

For the symmetrically normalized graph Laplacian, the following property holds. For any $f \in \mathbb{R}^n$, we have

$$f^T L_s f = \frac{1}{2} \sum_{i,j} w_{ij} \left(\frac{f_i}{\sqrt{D_{ii}}} - \frac{f_j}{\sqrt{D_{jj}}} \right)^2 \, . \tag{20.25}$$

We refer the reader to [30, 17, 36] for further details on Laplacians and their properties.

Semisupervised Graph Cuts: While the graph-cut problems outlined above are unsupervised, given pointwise semisupervision on some nodes, one can construct semisupervised graph-cut problems that respects the given information. In particular, the labels may indicate that two nodes with the same label should remain within the same subgraph after the cut [12, 13]. The same information can also be conveyed by specifying a must-link relationship between pairs (or subsets) of nodes, so that the must-linked nodes should ideally belong to the same subgraph. For ease of exposition, we consider the 2-cut setting. The extension to k-cuts is straightforward, and we report empirical results on such problems in Section 20.6.

Let A_1 be the subset of vertices with cluster id 1, and A_2 be the subset with cluster id 2. Clearly, A_1 and A_2 are disjoint subsets of V. The *semisupervised unnormalized cut* problem can be posed as follows: Find a partitioning (V_1, V_2) such that $cut(V_1, V_2)$ is minimized subject to the constraint $A_1 \subseteq V_1, A_2 \subseteq V_2$. In order to achieve balanced cuts, we also consider semisupervised versions of the ratio-cut (or normalized-cut) problem. In particular, the *semisupervised ratio-cut* problem can be posed as follows: Find a partitioning (V_1, V_2) such that $Rcut(V_1, V_2)$ is minimized subject to the constraint $A_1 \subseteq V_1, A_2 \subseteq V_2$. Similarly, one can pose the semisupervised normalized-cut problem using $Ncut(V_1, V_2)$ instead of $Rcut(V_1, V_2)$ above. The problems outlined above are NP-hard, and there has been some work on developing polynomial-time approximation schemes (PTASs) for related problems [12, 13].

20.3.1 Semisupervised Unnormalized Cut

Consider a graph partitioning given by V_1 and V_2. Let f be defined as follows:

$$f_i = \begin{cases} 1 & \text{if } v_i \in V_1 \\ -1 & \text{if } v_i \in V_2 \ . \end{cases} \tag{20.26}$$

From Equation 20.22, we now have

$$f^T L_u f = \frac{1}{2} \sum_{i,j=1}^{n} \mathbf{w}_{ij}(f_i - f_j)^2 = 4cut(V_1, V_2) \ . \tag{20.27}$$

For any given disjoint sets A_1, A_2 which constitute the semisupervision, we construct constraints on the labels as $y_i = +1$ if $v_i \in A_1$ and $y_i = -1$ if $v_i \in A_2$. Then, for all nodes in the labeled set, i.e., $v_i \in A_1 \cup A_2 = \mathcal{L}$, we have the constraint that $f_i = y_i$. Then, the semisupervised unnormalized cut problem can be written as

$$\min_{V_1, V_2} f^T L_u f, \quad s.t. \ f_i \text{ is as in Equation 20.26}, \ \ \forall v_i \in \mathcal{L}, f_i = y_i \ . \tag{20.28}$$

By relaxing the problem such that $f \in \mathbb{R}^n$ and noting that the constraint above is equivalent to $\Sigma_{i=1}^{\ell}(f_i - y_i)^2 \leq 0$, we obtain the following formulation:

$$\min_{f \in \mathbb{R}^n} f^T L_u f, \quad s.t. \ \sum_{i=1}^{\ell}(f_i - y_i)^2 \leq 0. \tag{20.29}$$

20.3.2 Semisupervised Ratio Cut

In the context of the ratio-cut problem, consider again a graph partitioning given by V_1 and V_2. Let f be defined as

$$f_i = \begin{cases} +\sqrt{|V_2|/|V_1|} & \text{if } v_i \in V_1 \\ -\sqrt{|V_1|/|V_2|} & \text{if } v_i \in V_2 \ . \end{cases} \tag{20.30}$$

Now, following Equation 20.22, we can express

$$f^T L_u f = \frac{1}{2} \sum_{i,j} w_{ij}(f_i - f_j)^2, = |V| Rcut(V_1, V_2), \tag{20.31}$$

where $|V|$ is a constant. From the predefined values of f we can see that $f^T\mathbf{1} = 0$, and $||f||^2 = n$. The objective function for the semisupervised ratio-cut problem can therefore be expressed as

$$\min_{V_1,V_2} f^T L_u f, \ \ s.t.\ f \perp \mathbf{1},\ ||f||^2 = n,\ f \text{ as in Equation 20.30},\ \forall v_i \in \mathcal{L},\ f_i = y_i\ . \tag{20.32}$$

We relax the problem and perform the optimization over $f \in \mathbb{R}^n$ such that $f \perp \mathbf{1}$. Note that in the unsupervised case, i.e., $\mathcal{L} = \emptyset$, the empty set, the solution to the problem is simply the second eigenvector of L corresponding to the second smallest eigenvalue. Now, relaxing the constraint[1] on $||f||$ and allowing f_i to mildly deviate from y_i on $v_i \in \mathcal{L}$, we get the following problem:

$$\min_{f\in\mathbb{R}^n} f^T L_u f, \ \ s.t.\ \ f \perp \mathbf{1}\ ,\ \ \sum_{i=1}^{\ell}(f_i - y_i)^2 \leq \varepsilon\ , \tag{20.33}$$

The key difference between the relaxed unnormalized formulation in Equation 20.29 and the normalized formulation in Equation 20.33 is the constraint $f \perp \mathbf{1} \Rightarrow \sum_i f_i = 0$, which ensures f lies in the subspace of $\mathbb{R}^n$ orthogonal to $\mathbf{1}$. The balancing constraint ensures the total score on positive predictions is the same as that on the negative predictions.

20.3.3 Semisupervised Normalized Cut

In the context of normalized cut, for a graph partitioning given by V_1 and V_2, let f be defined as follows:

$$f_i = \begin{cases} \sqrt{vol(V_2)/vol(V_1)} & \text{if } v_i \in V_1 \\ -\sqrt{vol(V_1)/vol(V_2)} & \text{if } v_i \in V_2\ . \end{cases} \tag{20.34}$$

Following an analysis similar to that of ratio-cut, a semisupervised normalized cut can be posed as the following optimization problem:

$$\min_{V_1,V_2} f^T L_u f, \ \ s.t.\ Df \perp \mathbf{1},\ f^T Df = vol(V)\ ,\ \ f \text{ as in Equation 20.34}, \forall v_i \in \mathcal{L}, f_i = y_i\ . \tag{20.35}$$

First, we relax the problem and perform the optimization over $f \in \mathbb{R}^n$ such that $f \perp \mathbf{1}$. With $g = D^{1/2}f$, the relaxed problem is

$$\min_{g\in\mathbb{R}^n} g^T D^{-1/2} L_u D^{-1/2} g\ \ s.t.\ \ g \perp D^{1/2}\mathbf{1},\ \ ||g||^2 = vol(V),\ \forall v_i \in \mathcal{L}, g_i = D^{1/2} y_i. \tag{20.36}$$

Note that if $\mathcal{L} = \emptyset$, then the solution to the problem is simply the second eigenvector of the symmetrically normalized Laplacian $L_s = D^{-1/2} L_u D^{-1/2}$ corresponding to the second smallest eigenvalue. Now, relaxing the constraint on $||g||$ and allowing g_i to mildly deviate from $D^{1/2}y_i$ on $v_i \in \mathcal{L}$, we get the following problem:

$$\min_{g\in\mathbb{R}^n} g^T L_s g\ s.t.\ g \perp D^{1/2}\mathbf{1},\ \sum_{i=1}^{\ell} ||g_i - D^{1/2} y_i||^2 \leq \varepsilon\ . \tag{20.37}$$

There have been notable attempts in the literature to directly solve some of the semisupervised graph-cut problems [12, 13]. Among such methods, the spectral graph transducer (SGT) [27] solves a problem closely related to the semisupervised ratio-cut problem, and reduces to the algorithm described in [7] under certain assumptions.

[1] Since the clustering depends on sign(f_i), the norm constraint $||f||^2 = n$ does not have an effect on the accuracy.

20.4 A Unified View of Label Propagation

The semisupervised graph-cut problems discussed in Section 20.3 belong to a widely studied family of problems which can be solved using label propagation methods [15]. In this section, we present a Generalized Label Propagation (GLP) framework, discuss various specific instantiations of the framework from the literature, and illustrate its connections to the semisupervised graph-cut problems.

20.4.1 Generalized Label Propagation

GLP formulation considers a graph-based semisupervised learning setting. Let W be the symmetric weight matrix and L be a corresponding graph Laplacian. Note that L may be any of the Laplacians discussed in Section 20.3, and we will see how different label propagation formulations result from specific choices of the Laplacian. Let $f \in \mathbb{R}^n$, where $n = \ell + u$, be the predicted score on each data point x_i, $i = 1, \ldots, n$; the predicted cluster label on x_i can be obtained as $\text{sign}(f_i)$. The GLP problem can be formulated as follows:

$$\min_{f \in \mathcal{S}} f^T L f, \qquad s.t. \sum_{i=1}^{\ell} (f_i - y_i)^2 \leq \varepsilon\,, \tag{20.38}$$

where $\varepsilon \geq 0$ is a constant and $\mathcal{S} \subseteq \mathbb{R}^n$. For most existing formulations $\mathcal{S} = \mathbb{R}^n$, whereas for a few $\mathcal{S} = \{f | f \in \mathbb{R}^n, f \perp \mathbf{1}\}$ where $\mathbf{1}$ is the all ones vector. The Lagrangian for the GLP problem is given by $L(f,\mu) = f^T L f + \mu \sum_{i=1}^{\ell} (f_i - y_i)^2$, where $\mu \geq 0$ is the Lagrangian multiplier. Some variants assume $y_i = 0$ for $i = (\ell+1), \ldots, n$, so the constraint will be of the form $\sum_{i=1}^{n} (f_i - y_i)^2 \leq \varepsilon$. Assuming the Laplacian to be symmetric, which is true for L_u and L_s, the first order necessary conditions are given by $(L + \mu I) f = \mu y$, where I is the identity matrix. Several existing methods work with the special case $\varepsilon = 0$, which makes the constraints binding so that $\sum_{i=1}^{\ell} (f_i - y_i)^2 = 0$ and $f_i = y_i$ on the labeled points. The first order conditions for the special case are given by $Lf = 0$. In the next several sections, we show how most of the existing label propagation methods for semisupervised learning can be derived directly as a special case of the GLP formulation or alternatively with special case choices of the Laplacian L, the constant ε, and the subspace $\mathcal{S}$.

20.4.2 Gaussian Fields

Motivated by the assumption that neighboring points in a graph will have similar labels in Gaussian fields (GFs), the following energy function is considered [46]:

$$E(f) = \frac{1}{2} \sum_{i,j=1}^{n} w_{ij} (f_i - f_j)^2\,. \tag{20.39}$$

The GF method computes labels by minimizing the energy function $E(f)$ with respect to f under the contraint that $f_i = y_i$ for all labeled points. As observed in [46], the energy function is harmonic, i.e., it is twice continuously differentiable and it satisfies Laplace's equation [24]. From the harmonic property of the energy function it follows that the predicted labels will satisfy: $f = D^{-1} W f$. In terms of block matrices corresponding to labeled and unlabeled points we have:

$$\begin{bmatrix} D_{\ell\ell} & 0 \\ 0 & D_{uu} \end{bmatrix} \begin{bmatrix} f_\ell \\ f_u \end{bmatrix} = \begin{bmatrix} W_{\ell\ell} & W_{\ell u} \\ W_{u\ell} & W_{uu} \end{bmatrix} \begin{bmatrix} f_\ell \\ f_u \end{bmatrix}.$$

Since $f_\ell = y_\ell$ due to the constraints,[2] the above system can be simplified to get a closed form for f_u given by

$$f_u = (D_{uu} - W_{uu})^{-1} W_{ul} y_l \,. \tag{20.40}$$

We can interpret the objective function in GF as a special case of the GLP problem in (20.38). In particular, using the identity in Equation 20.22 and noting that the constraints on the labeled points are binding, GF can be seen as a special case of GLP with $L = L_u$ and $\varepsilon = 0$, i.e.,

$$\min_{f \in \mathbb{R}^n} f^T L_u f \,, \quad s.t. \; \sum_{i=1}^{\ell} (f_i - y_i)^2 \leq 0 \,. \tag{20.41}$$

20.4.3 Tikhonov Regularization (TIKREG)

Given a partially labeled data set, TIKREG [7] is an algorithm for regularized regression on graphs, where the objective is to infer a function f over the graph. The objective function for TIKREG is given by

$$\min_{f \in \mathbb{R}^n} \frac{1}{2} \sum_{i,j=1}^{n} w_{ij} (f_i - f_j)^2 + \frac{1}{\gamma \ell} \sum_{i=1}^{\ell} (f_i - y_i)^2 \tag{20.42}$$

with the constraint that $f \perp \mathbf{1}$, i.e., f lies in the orthogonal subspace of $\mathbf{1}$, the all ones vector. The parameter γ is a suitably chosen (positive) real number. A closed form solution for the above problem is obtained [7] as

$$f = (\ell \gamma L_u + I_k)^{-1} (\hat{y} + \mu \mathbf{1}), \tag{20.43}$$

where $\hat{y} = (y_1, y_2, \ldots, y_l, 0, \ldots, 0)$, $I_k = diag(1, \ldots, 1, 0, \ldots, 0)$ with the number of ones equal to the number of labeled points. The orthogonality constraint on f is enforced through the Lagrange multiplier μ, which is optimally computed as

$$\mu = -\frac{\mathbf{1}^T (\ell \gamma L_u + I_k)^{-1} \hat{y}}{\mathbf{1}^T (\ell \gamma L_u + I_k)^{-1} \mathbf{1}} \,. \tag{20.44}$$

The objective function in Equation 20.42 can be viewed as a special case of the GLP objective in (20.38). As previously, the first term is $f^T L_u f$, where L_u is the unnormalized Laplacian. The second term corresponds to the constraint $\sum_i (f_i - y_i)^2 \leq \varepsilon$, in (20.38) where $1/\gamma\ell$ is the optimal Lagrange multiplier corresponding to the constraint. In other words, if $\varepsilon(1/\gamma\ell)$ is the constraint value that leads to the optimal Lagrange multiplier of $1/\gamma\ell$, the TIKREG problem can be seen as a special case of GLP:

$$\min_{f \in \mathbb{R}^n, f \perp \mathbf{1}} f^T L_u f \,, \quad s.t. \; \sum_{i=1}^{\ell} (f_i - y_i)^2 \leq \varepsilon(1/\ell\gamma) \,. \tag{20.45}$$

20.4.4 Local and Global Consistency

The Local and Global Consistency (LGC) approach [45] gives an alternative graph-based regularization framework for semisupervised learning. In particular, the LGC is formulated based on the following objective function [45]:

$$\min_{f \in \mathbb{R}^n} \frac{1}{2} \left(\sum_{i,j=1}^{n} w_{ij} \left(\frac{1}{\sqrt{D_{ii}}} f_i - \frac{1}{\sqrt{D_{jj}}} f_j \right)^2 + \mu \sum_{i=1}^{n} (f_i - y_i)^2 \right) \tag{20.46}$$

[2] We abuse notation and denote $[f_1, \ldots, f_\ell]^T$ by f_ℓ (similarly for y_ℓ) and $[f_{(\ell+1)}, \ldots, f_n]^T$ by f_u in the sequel.

with $\mu > 0$ as the regularization parameter. Note that LGC assumes that there is a valid y_i for all points; operationally, the $y_i, i = 1, \ldots, \ell$ is set to the true given label, whereas $y_i, i = \ell+1, \ldots, n$ is set to 0. The problem is solved using an iterative label propagation algorithm. Given a weight matrix W among the points, the weights are normalized to obtain $S = D^{-1/2}WD^{-1/2}$ where D is a diagonal matrix with $D_{ii} = \sum_j w_{ij}$. Starting from an initial guess $f^{(0)}$, the iterative algorithm proceeds with the following updates:

$$f^{(t+1)} = \alpha S f^{(t)} + (1-\alpha) y \,, \tag{20.47}$$

where $\alpha \in (0,1)$. As shown in [45], this update equation converges to $f^* = (1-\alpha)(I-\alpha S)^{-1}y$, which can be shown to optimize the objective function in (20.46) when $\alpha = 1/(1+\mu)$. We now show that the LGC formulation is a special case of the GLP formulation in (20.38). From the identity involving the normalized Laplacian in Equation 20.25, LGC can be seen as a special case of GLP as follows:

$$\min_{f \in \mathbb{R}^n} f^T L_s f \,, \quad s.t. \ \sum_{i=1}^{n} (f_i - y_i)^2 \leq \varepsilon(\mu) \,, \tag{20.48}$$

where $\varepsilon(\mu)$ is the constant corresponding to the optimal Lagrange multiplier μ. Note that since in LGC one starts with an initial label $y_i, i = 1, \ldots, n$, the constraint involves terms corresponding to all the points.

20.4.5 Related Methods

We review three other methods from the literature, viz, cluster kernels, Gaussian random walks, and local neighborhood propagation for graph-based semisupervised learning which are closely related to the GLP framework.

20.4.5.1 Cluster Kernels

The main idea in cluster kernels (CK) [16] is to embed the data into a lower dimensional space based on its cluster structure and then subsequently build a semisupervised learner on the low-dimensional data. If K denotes a suitable kernel on the data space, the embedding method focuses on the k primary eigenvectors of the symmetrized matrix $D^{-1/2}KD^{-1/2}$. If K corresponds to the edge weights on the graph $G = (V,E)$ between the points, i.e., $K = W$, then the embedding corresponds to the k eigenvectors of the symmetrized Laplacian $L_s = I - D^{-1/2}WD^{-1/2}$ corresponding to the smallest k eigenvalues. In particular, for $k = 1$, the embedding is given by the eigenvector corresponding to the smallest eigenvalue of L_s which is the solution to LGC in absence of any semisupervision. CK trains a suitable semisupervised learner on the low-dimensional embedding to obtain the final label assignments.

20.4.5.2 Gaussian Random Walks EM (GWEM)

Consider a random walk on the graph with transition probability $P = D^{-1}W$. The GWEM method [35] works with the m-step transition probability matrix P^m so that the probability of going from x_i to x_j is given by $p_{m|0}(x_j|x_i) = (P^m)_{ij}$. The random walk is assumed to start with uniform probability from any one of the nodes, so $P(x_i) = 1/n$. Using Bayes rule, one can obtain the posterior probabilities $P_{0|m}(x_i|x_j)$. Now, each point is assumed to have a (possibly unknown) distribution $p(y|x_i)$ over the cluster labels. For any point x_j, the posterior probability of cluster label y is given by $P(y_j = c|x_j) = \sum_i P(y_i = c|x_i) p_{0|m} p(x_i|x_j)$. The cluster label is based on $y_j = \text{argmax}_c \ P(y_j = c|x_j)$. Now, since $P(y_i|x_i)$ is unknown for the unlabeled points, an EM algorithm can be used to alternately maximize the log-posterior probability of known labels on the labeled points

$$\sum_{k=1}^{\ell} \log P(y_k|x_k) = \sum_{k=1}^{\ell} \log \sum_{i=1}^{N} P(y_i|x_i) P_{0|m}(x_i|x_k). \tag{20.49}$$

As shown in [35], the EM algorithm alternates between the E-step which estimates $P(x_i|x_k,y_k) \propto P(y_k|x_i)P_{0|m}(x_i|x_k)$, where k denotes an index over labeled points, and the M-step, which computes

$$P(y=c|x_i) = \frac{\sum_{k:y_k=c}^{\ell} P(x_i|x_k,y_k)}{\sum_{h=1}^{\ell} P(x_i|x_h,y_h)} . \tag{20.50}$$

We now show that GWEM can be interpreted in terms of spectral decomposition of a suitable asymmetrically normalized Laplacian L_r as in Equation 20.23. For a fixed number of steps m for the random walk, let $Z^T = P^m = (D^{-1}W)^m$. Note that Z^T itself is a transition probability matrix, and $Z_{ij} = P_{m|0}(x_i|x_j)$. Let D_Z be a diagonal matrix such that $D_{Z,ii} = \sum_j Z_{ij}$. Since the prior probability $P(x_i) = 1/n$, by Bayes rule we have

$$P_{0|m}(x_j|x_i) = \frac{P_{m|0}(x_i|x_j)}{\sum_{i'} P_{m|0}(x_{i'}|j)} = (D_Z^{-1}Z)_{ij} . \tag{20.51}$$

Let $f_j = P(y_j|x_j)$. When the EM algorithm converges we will have

$$f = D_Z^{-1}Zf \Rightarrow (I - D_Z^{-1}Z)f = 0 , \tag{20.52}$$

where $f_i = y_i$ for the labeled points. Since $D_Z^{-1}Z$ is a transition probability matrix, from Equation 20.23 we note that $(I - D_Z^{-1}Z)$ can be viewed as a asymmetrically normalized Laplacian L_r so that $L_r f = 0$. Finally, since $D_Z f = Zf$ resembles the fixed point equation for GFs, a block decomposition as in Equation 20.40 yields $f_u = (D_{z,uu} - Z_{uu})^{-1} Z_{u\ell} y_\ell$.

20.4.5.3 Linear Neighborhood Propagation

Linear Neighborhood Propagation (LNP) [40] is another recent approach, which differs from the other methods as LNP computes a stochastic transition matrix U directly from the data. In particular, one computes a probability distribution over neighboring points so that their expectation best approximates the point under consideration: $\min_{\mathbf{u}_i} \left\| x_i - X_i^N \mathbf{u}_i \right\|^2$, where $\mathbf{u}_i$ is probability distribution over the neighbors of x_i and X_i^N is a matrix each of whose columns is a neighbor of x_i. Once the transition probability matrix U is computed, the semisupervised learning problem is posed as follows:

$$\min_{f \in \mathbb{R}^n} \sum_{i,j=1}^{n} u_{ij}(f_i - f_j)^2 + \mu \sum_{i=1}^{n} (f_i - y_i)^2 , \tag{20.53}$$

where, similar to LGC [45], the labels $y_i, i = 1, \ldots, \ell$ are set to their true values, and the unknown labels $y_i, i = \ell+1, \ldots, n$ are set to 0. Similar to LGC, the LNP problem is solved by an iterative label propagation algorithm. Starting from an initial guess $f^{(0)}$, the iterative algorithm proceeds with the following updates:

$$f^{(t+1)} = \alpha U f^{(t)} + (1-\alpha)y , \tag{20.54}$$

where $\alpha = 1/(1+\mu) \in (0,1)$. The updates are the same as in Equation 20.47 for LGC [45] with the difference that U is not normalized symmetrically, but is a transition probability matrix of a random walk. In spite of the similarities, a careful consideration of the analysis in [40] reveals that update equation in Equation 20.54 does not solve the problem in Equation 20.53. On convergence, the iterative updates in Equation 20.54 leads to $f = (I - \alpha U)^{-1}(1-\alpha)y$. On the other hand, setting derivatives of Equation 20.53 to zero leads to $f = \left(I - \alpha(U+U^T)/2\right)^{-1}(1-\alpha)y$. The issue arises in the analysis [40] when one assumes $[(I-U)+(I-U)^T]f \approx 2(I-U)f$, which is not true unless U is symmetric.

20.4.6 Label Propagation and Green's Function

We briefly describe an interesting relationship between label propagation and the discrete Green's function [23]. Green's functions are typically used to convert nonhomogenous partial differential equations with boundary conditions into an integral problem. In particular, the inverse Laplace operator with the zero mode removed can be interpreted as a Green's function for the discrete Laplace operator [23]. Let $\mathcal{G} = L^{\dagger}$ be the generalized inverse of the Laplacian L. The solutions for both GF and GWEM can be expressed as $f_u = (D_{uu} - W_{uu})^{-1} W_{ul} y_l = L^{\dagger}_{uu} z_{ul}$ where $z_{ul} = W_{ul} y_l$. Discarding the zero mode of L_{uu}, we have $f_u \approx \mathcal{G}_{uu} z_{ul}$. As argued in [23], discarding the zero mode is important to ensure that the Green's function exists; further, it does not affect the final result. Then f_u can be viewed as a solution to a partial differential equation with boundary value constraints. The interpretation is intuitive if the labeled points are treated as electric charges. In particular one assumes labeled points to be postive and negative charges. Using the Green's function one then computes the influence of these charges on unlabeled points [23]. For methods such as LGC and LNP the solution has the form $f = (I - A/(1+\mu))^{-1} \mu y/(1+\mu)$, with $A = D^{-1/2} W D^{-1/2}$ for LGC and $A = U$ for LNP. Considering the strong regularization limit as $\mu \to 0$ and removing the zero mode in L, we obtain: $f = L^{\dagger} y \approx \mathcal{G} y$.

20.4.7 Label Propagation and Semisupervised Graph Cuts

We now describe how relaxed versions of semisupervised graph cuts, as discussed in Section 20.3, lead to special cases of the GLP formulation for a suitable choice of the Laplacian L and the constraint ε and, hence, can be solved using label propagation methods. For semisupervised unnormalized cut, the objective function in (20.29) is a special case of our GLP formulation using an unnormalized graph Laplacian and $\varepsilon = 0$. In particular (20.29) is *exactly the same* as the formulation for Gaussian Fields [46]. For semisupervised ratio cut, the objective described in (20.33) is equivalent to the problem TIKREG solves [7]. For semisupervised normalized cut, the formulation in 20.37 is nearest to that of CK, but not the same since CK is a two-step method which uses the normalized Laplacian for embedding, and then applies a classification algorithm on the embedding. It is also similar to LGC [45], although the constraint in LGC includes all points with $y_i = 0$ for $i = (\ell+1), \ldots, n$ and does not involve the $D^{1/2}$ scaling on y_i in the constraint.

20.5 Semisupervised Embedding

In this section, we show how SSC methods based on graph-based representations, as discussed in Sections 20.3 and 20.4, can be viewed as doing semisupervised embedding. The geometric perspective helps in identifying relationships between existing embedding and label propagation methods, e.g., between Laplacian Eigenmaps [8] and Gaussian Fields [46]. Further, we illustrate that it is possible to derive novel SSC methods based on existing embedding methods, including Locally Linear Embedding (LLE) [33], Local Tangent Space Alignment (LTSA) [44] and Laplacian Eigenmaps (LE) [8]. While all such methods can be seen as a special case of the GLP formulation, they differ in the details—in particular, in the choice of the postive semidefinite matrix L and nature of constraints. Since our exposition is focussed on two clusters, the embedding will always be on $\mathbb{R}$, a one-dimensional space. For number of clusters $k > 2$, the embedding space may be of dimensionality k or $\log k$, depending on the representation used.

20.5.1 Nonlinear Manifold Embedding

Manifold embedding methods obtain a lower dimensional representation of a given dataset such that some suitable neighborhood structures are preserved. In this section we briefly review three popular embedding methods and demonstrate that their semisupervised generalizations solve a variant of the GLP formulation.

Locally Linear Embedding (LLE): In LLE [32], the assumption is that each point in the high-dimensional space can be accurately approximated by a locally linear region. In particular, the neighborhood dependencies are estimated by solving $\min_W \sum_i ||x_i - \sum_{j \in \mathcal{N}_i} w_{ij} x_j||^2$, such that $\sum_{j \in \mathcal{N}_i} w_{ij} = 1$, where $\mathcal{N}_i$ is the set of neighboring points of x_i. Then W is used to reconstruct the points in a lower dimensional space by solving:

$$\min_{f \in \mathbb{R}^n} \sum_i ||f_i - \sum_j w_{ij} f_j||^2 \ , \quad s.t. \ f \perp \mathbf{1}, \ ||f||^2 = n \ . \tag{20.55}$$

Letting $M = (I - W)^T (I - W)$, which is positive semidefinite and can be viewed as an iterated Laplace operator [8], we can rewrite the objective function as

$$\min_{f \in \mathbb{R}^n} f^T M f \ , \quad s.t. \ f \perp \mathbf{1} \ , \ ||f||^2 = n \ . \tag{20.56}$$

Laplacian Eigenmaps (LE): LE is based on the correspondence between the graph Laplacian and the Laplace Beltrami operator [8]. The symmetric weights between neighboring points are typically computed using the RBF kernel as $w_{ij} = \exp(-||x_i - x_j||^2/\sigma^2)$. Then W is used to reconstruct the points in a lower dimensional space by solving:

$$\min_{f \in \mathbb{R}^n} \frac{1}{2} \sum_{i,j} w_{ij} (f_i - f_j)^2 \ , \quad s.t. \ f \perp D\mathbf{1} \ , \ f^T D f = I \ . \tag{20.57}$$

Using (20.22), the objective function is $f^T L_u f$. Letting $g = D^{1/2} f$, with $M = L_s = D^{-1/2} L_u D^{-1/2}$ we can express the objective function as

$$\min_{g} g^T M g \ , \quad s.t. \ g \perp D^{1/2}\mathbf{1} \ , \ ||g||^2 = 1 \ . \tag{20.58}$$

Local Tangent Space Alignment (LTSA): In LTSA, the tangent space at each point is approximated using local neighborhoods, and a global embedding is obtained by aligning the local tangent spaces. If X_i^N denotes the matrix of neighbors of x_i, then it can be shown [44] that the principal components of X_i^N give an approximation to the tangent space of the embedding f_i. Let $g_{i1}, \ldots, g_{ik}$ be the top k principal components for X_i^N. Let $G_i = [e/\sqrt{k}, g_{i1}, \ldots, g_{id}]^T$. If $\mathcal{N}_i$ are the indices of the neighbors of x_i, submatrices of the alignment matrix M are computed as $M(\mathcal{N}_i, \mathcal{N}_i) \leftarrow M(\mathcal{N}_i, \mathcal{N}_i) + I - G_i G_i^T$ for $i = 1, \ldots, n$. Finally, using M, which is guaranteed to be positive semidefinite, an embedding is subsequently obtained by minimizing the alignment cost:

$$\min_{f} f^T M f \ , \quad s.t. \ f \perp \mathbf{1} \ , \ ||f||^2 = n. \tag{20.59}$$

We refer the reader to [44] for a detailed analysis of LTSA.

20.5.2 Semisupervised Embedding

In this section, we consider two variants of semisupervised embedding and its relationship to graph-based SSC. The variants differ in whether they consider the constraints associated with the corresponding unsupervised embedding problem. As discussed in Section 20.5.1, there are typically two types of constraints: $f \perp A\mathbf{1}$, where $A = I$ or $D^{1/2}$, and $||f||^2 = c$, a constant. Since the clustering is based on $\text{sign}(f_i)$, the norm constraint does not play any role and will be ignored for our analysis. The two variants we consider are based on whether $f \perp A\mathbf{1}$ is enforced or not, in addition to the constraints coming from the partially labeled data.

20.5.2.1 Unconstrained Semisupervised Embedding

Following [43], we want to obtain an embedding $f = [f_\ell \ \ f_u]^T$, where the exact embeddings of the first ℓ points are known and given by y_ℓ.[3] The objective for semisupervised embedding is given by

$$\min_f \ f^T M f \ , \quad s.t \ f_\ell = y_\ell \ , \tag{20.60}$$

where M is a suitable positive semidefinite matrix. Since f_ℓ is fixed, the problem can be cast in terms of block matrices as

$$\min_{f_u} \begin{bmatrix} f_\ell^T & f_u^T \end{bmatrix} \begin{bmatrix} M_{\ell\ell} & M_{\ell u} \\ M_{u\ell} & M_{uu} \end{bmatrix} \begin{bmatrix} f_\ell \\ f_u \end{bmatrix} . \tag{20.61}$$

Setting the first derivative to zero, one obtains

$$f_u = -M_{uu}^{-1} M_{u\ell} y_\ell \ . \tag{20.62}$$

In the context of label propagation for a 2-clustering setting, we will have $y_i = +1$ or $y_i = -1$ for $i = 1, \ldots, \ell$. In other words, the labeled points are being embedded to their true cluster label, and the rest will be embedded while trying to maintain the neighborhood structure. For LLE, $M = (I - W)^T (I - W)$ and we call the corresponding label propagation algorithm LLELP. Similarly, for LTSA, M is as discussed in Section 20.5.1, and the corresponding algorithm will be called LTSALP. For unconstrained LE from Equation 20.57, $M = L_u$, and the corresponding algorithm will be called LELP. For LELP, since $M = L_u$, the unnormalized Laplacian, from Equation 20.62 we have

$$\begin{aligned} f_u &= -L_{uu}^{-1} L_{u\ell} y_\ell = -(D_{uu} - W_{uu})^{-1} (D_{u\ell} - W_{u\ell}) y_\ell \\ &= (D_{u\ell} - W_{u\ell})^{-1} W_{u\ell} y_\ell \ , \end{aligned}$$

since $D_{u\ell} = 0$ as D is a diagonal matrix. We note that the solution is exactly the same as that for GF as in Equation 20.40 implying the *equivalence* of GF and LELP.

20.5.2.2 Constrained Semisupervised Embedding

In this section, we consider embedding problems when the orthogonality constraint of the form $f \perp A\mathbf{1}$ is enforced. In particular, we consider the following problem:

$$\min_f \ f^T M f \ , \quad s.t. \ \sum_i^\ell (f_i - y_i)^2 \le \varepsilon \ , \ f \perp A\mathbf{1} \ , \tag{20.63}$$

where $A = I$ for LLE and LTSA, and $A = D^{1/2}$ for LE. Let α and μ be the Lagrange multipliers for the two constraints, respectively. The first order necessary conditions obtained from the Lagrangian corresponding to (20.63) yield

$$f = (M + \alpha I_k)^{-1} (\alpha y + \mu A\mathbf{1}/2). \tag{20.64}$$

Since $\mathbf{1}^T A^T f = 0$, a direct calculation gives the optimal Lagrange multiplier as

$$\mu = -2\alpha \frac{\mathbf{1}^T A^T (M + \alpha I_k)^{-1} y}{\mathbf{1}^T A^T (M + \alpha I_k)^{-1} A\mathbf{1}} \ . \tag{20.65}$$

The multiplier α can also be computed by using existing results on solving quadratically constrained quadratic programs (QCQPs) with a single quadratic constraint [14]. For LLE, $M = (I - W)^T (I - W)$ and $A = I$, and we call the corresponding algorithm LLELPC. For LTSA, M is as discussed in Section 20.5.1 and $A = I$, and we call the corresponding algorithm LTSALPC. For LE as in Equation 20.58, $M = L_{sym} = I - D^{1/2} W D^{1/2}$ and $A = D^{1/2}$ and we call the corresponding algorithm LELPC.

[3]While the constraints can be relaxed to consider $\sum_{i=1}^{\ell} (f_i - y_i)^2 \le \varepsilon$, we do not focus on the general case here.

20.6 Comparative Experimental Analysis

In this section, we provide an empirical evaluation of SSC with graph-based representations. Our experiments are divided into two parts: First, we compare seven methods on 14 benchmark data sets in terms of their accuracy; later, we take a closer look at the performance of the label propagation methods obtained from the perspective of semisupervised manifold embedding. For the first set of experiments, the methods we consider include 6 standard methods: GF, LNP, CK, GWEM, TIKREG, and LGC, as discussed in Section 20.4. In addition, we include LTSALP, an approach based on semi-supervised LTSA embedding. For the second set of experiments, the methods we consider are the 6 embedding based methods introduced in Section 20.5.

Methodology: We conducted our experiments on 14 well-known benchmark data sets. They include the following 8 UCI data sets: Hepatitis, Cancer, Pima, Wine, Iris, Glass, USPS (1-4 only), and Letter (E and F only). In addition we also ran experiments on 6 text datasets, which are all subsets of the 20Newsgroup data set: Different100, Similar100, Same100, Different1000 and Same1000 [1]. Each dataset contains a subset of 3 newsgroups with varying degrees of difficulty for clustering. Different100 (1000) includes alt.atheism, rec.sport.baseball, and sci.space and are, hence, easy to cluster; Same100 (1000) includes comp.graphics, comp.os.ms-windows, comp.windows.x and are difficult to cluster; whereas Similar100 includes talk.politics.guns, talk.politics.mideast, and talk.politics.misc and are moderately difficult to cluster. We also used the well-known Classic-3 benchmark data set containing 3893 documents, whereby 1033 are from medical journals, 1400 are aeronautical system papers, and 1460 are information retrieval papers. For each method and each data set, we ran five-fold semisupervised cross-validation. In particular, the training points were chosen from four folds with increasing number of labeled points, and the test error was measured on the fifth fold. All points were used to construct the neighborhood graph. Further, for each approach involving parameters, e.g., CK using SVMs with RBF kernel, parameter values were selected by cross-validation. Performance is evaluated based on error rate of the true cluster label on the test set, unless noted otherwise.

20.6.1 Experimental Results

The performance of the 7 methods is shown in Figures 20.1, 20.2 and 20.3. All results reported are on the test set. To avoid clutter, we display the results for the top five methods based on average test-set error in Figures 20.1 and 20.2. The performance comparison results of 7 semisupervised learning methods on 14 datasets are shown in Tables 20.1–20.3 with different numbers of labeled points (30, 40, and 50). We make the following observations based on the results:

- There is no dominating method across all datasets. However, there is a set of methods that seems to be fairly consistently among the top few, and the performance of the top few methods is typically close. While across the top methods, the differences typically do not appear significant when compared to the worst methods, the improvements do tend to be significant. For instance, if we examine the results in Figure 20.1(e) for the USPS data set, we can see that the error rate for the top methods is around 1.5–1.8%, while the performance for the worst methods is between 3.0% and 7.6%. The improvements of the best methods are clearly significant in comparison to the worst methods judging by the standard deviation in error.
- Our results also indicate that no given method is always among the best methods for each data set. For any method, we could find at least one data set where its performance is among the worst. It appears that the assumptions made by various approaches do not work well across all cluster structures encountered in various data sets.

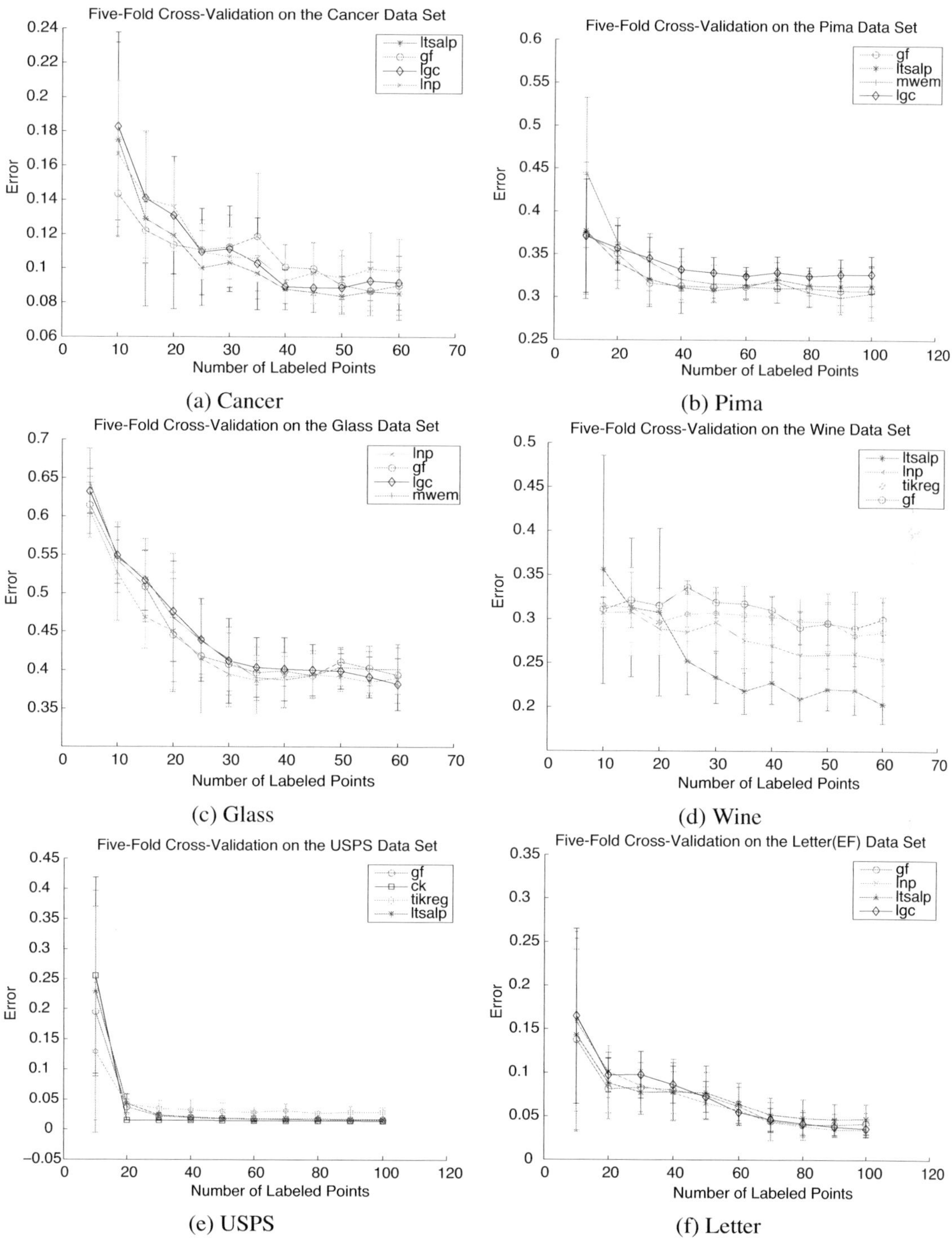

FIGURE 20.1: Five-fold cross-validation results on UCI datasets as increasingly many points are labeled.

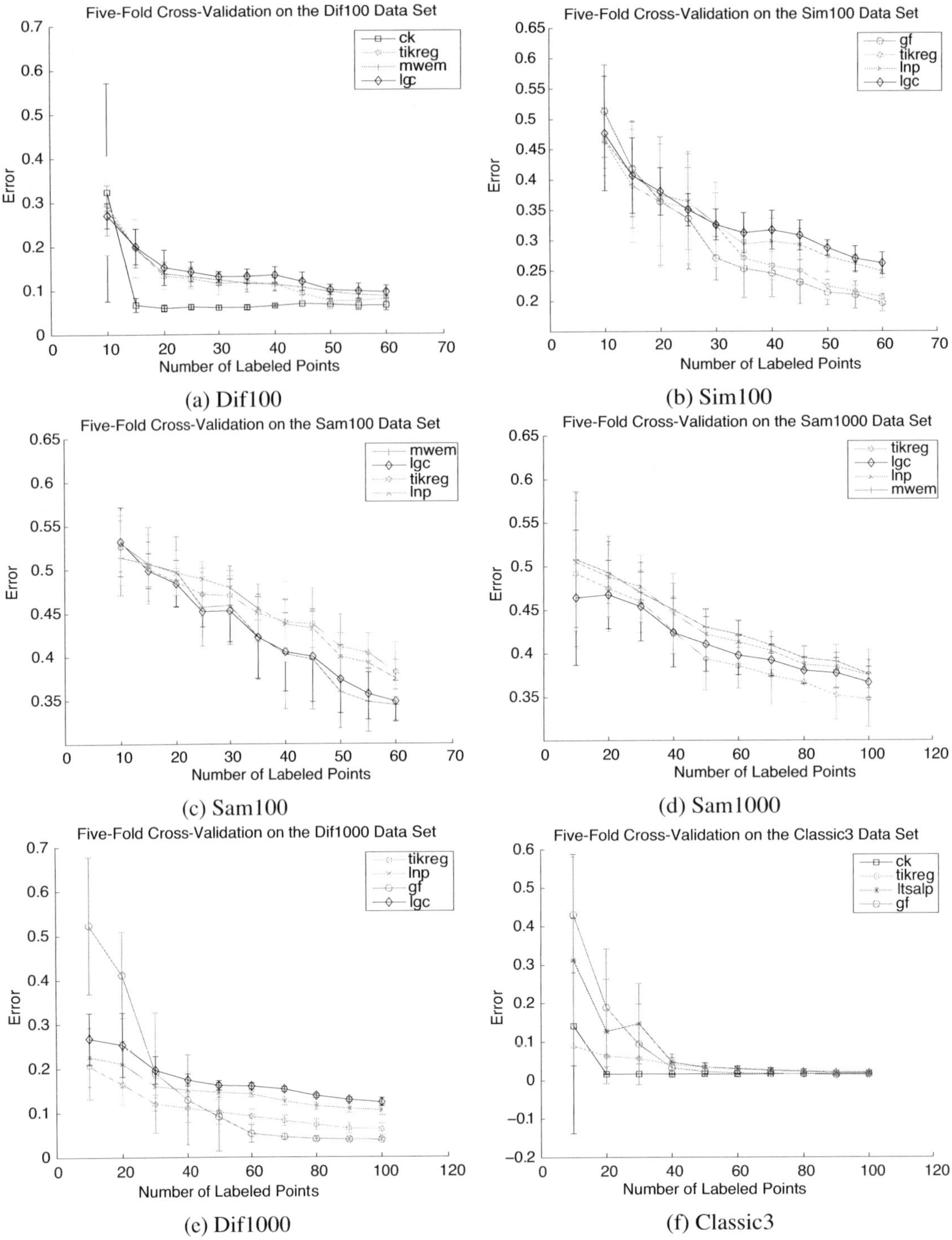

(a) Dif100 (b) Sim100

(c) Sam100 (d) Sam1000

(e) Dif1000 (f) Classic3

FIGURE 20.2: Five-fold cross-validation results on text datasets as increasingly many points are labeled.

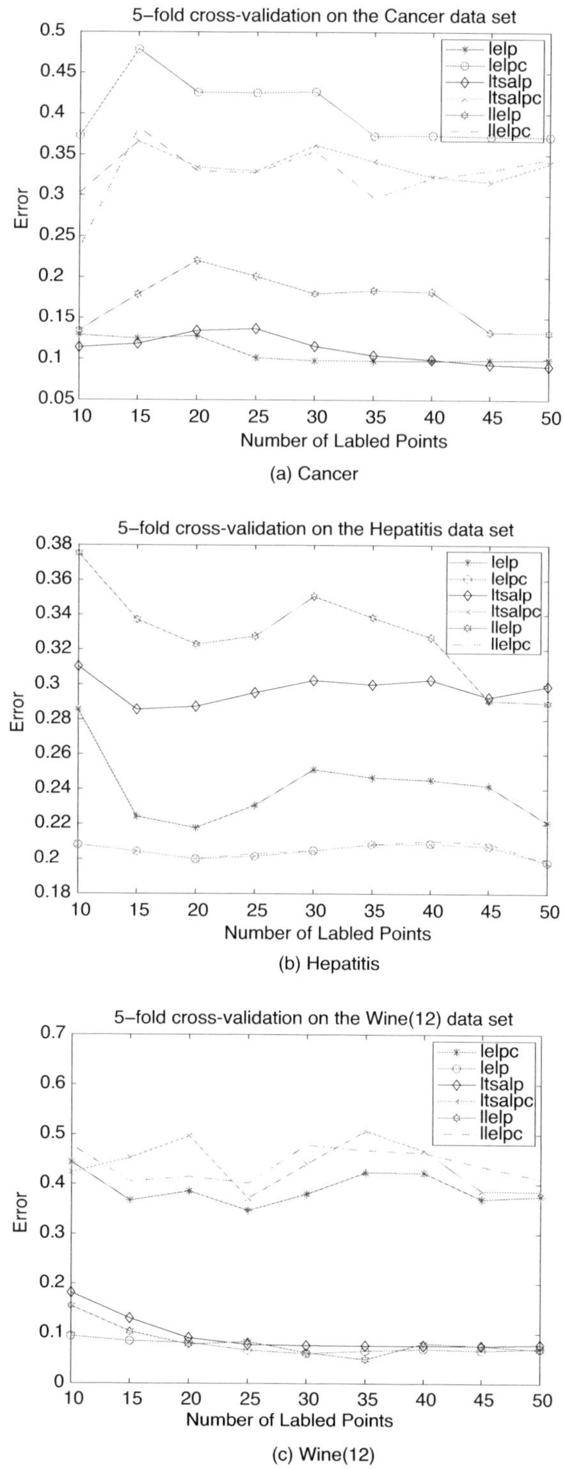

FIGURE 20.3: Comparison of (constrained) embedding based label propagation methods: LELP(C), LTSALP(C), and LLELP(C) on UCI datasets.

TABLE 20.1: Semisupervised Learning Performance Comparisons with 30 Labeled Points on 14 Datasets and 6 Methods

	gf	mwem	lnp	lgc	tikreg	ck	ltsalp
Hepatitis	21.1 ± 0.4	21.4 ± 0.7	25.8 ± 3.0	22.7 ± 1.2	**20.8 ± 0.0**	20.8 ± 0.0	24.3 ± 2.5
Cancer	11.2 ± 1.9	12.7 ± 2.5	10.6 ± 1.7	11.1 ± 2.5	10.8 ± 1.9	11.3 ± 1.3	**10.3 ± 1.5**
Wine	31.9 ± 1.7	32.2 ± 0.9	29.6 ± 3.5	32.4 ± 2.1	30.7 ± 2.2	30.1 ± 2.2	**23.4 ± 3.0**
Iris	5.5 ± 2.5	**4.8 ± 4.1**	6.0 ± 2.5	5.3 ± 3.8	5.7 ± 3.4	9.2 ± 1.0	5.3 ± 1.9
Glass	40.8 ± 4.0	41.4 ± 4.1	**39.5 ± 4.2**	41.2 ± 5.5	42.8 ± 5.5	49.6 ± 4.5	45.3 ± 4.9
Pima	**31.6 ± 2.5**	34.1 ± 3.3	36.0 ± 3.0	34.5 ± 2.5	33.2 ± 2.1	34.8 ± 0.2	32.1 ± 3.2
USPS	2.2 ± 0.4	13.0 ± 2.1	6.3 ± 0.9	4.9 ± 0.7	3.4 ± 1.4	**1.5 ± 0.1**	2.4 ± 0.4
Letter(EF)	8.3 ± 2.8	15.4 ± 2.8	8.5 ± 1.0	9.8 ± 2.7	8.3 ± 1.7	48.5 ± 0.5	**7.8 ± 2.6**
Dif100	7.3 ± 1.4	12.4 ± 1.3	13.6 ± 2.4	13.0 ± 1.2	11.5 ± 2.6	**6.1 ± 0.6**	12.7 ± 2.2
Sim100	**27.0 ± 3.6**	34.3 ± 5.3	32.7 ± 5.1	32.6 ± 2.5	32.5 ± 7.0	63.4 ± 2.1	41.0 ± 5.8
Sam100	48.8 ± 8.5	45.9 ± 4.5	47.9 ± 1.9	**45.3 ± 3.6**	47.0 ± 2.3	65.7 ± 2.0	51.7 ± 7.6
Dif1000	19.0 ± 13.6	23.3 ± 2.6	16.0 ± 1.9	19.7 ± 3.1	**12.0 ± 1.5**	65.8 ± 2.5	28.2 ± 14.6
Sam1000	54.9 ± 8.9	47.0 ± 3.4	47.7 ± 3.6	**45.4 ± 4.0**	46.0 ± 3.6	66.4 ± 1.4	61.0 ± 4.7
Classic3	9.3 ± 10.5	12.1 ± 1.7	29.2 ± 1.1	28.8 ± 0.7	5.7 ± 4.1	**1.6 ± 0.1**	14.7 ± 10.4

- TIKREG is among the most consistent methods on the text datasets. However, its performance is not consistent on the UCI data sets.
- In spite of an issue with its formulation (see Section 20.4.5.3), LNP is found to be quite competitive across several UCI and text datasets. Its parameter insensitivity makes it rather easy to use.
- When CK does well, it outperforms all the other methods. However, it does not have a consistent performance, which probably can be addressed by more thorough cross-validation over its parameter choices.
- GF seems to be slow starter, not performing well for a small number of labeled points, but

TABLE 20.2: Semisupervised Learning Performance Comparisons with 40 Labeled Points on 14 Datasets and 6 Methods

	gf	mwem	lnp	lgc	tikreg	ck	ltsalp
Hepatitis	21.4 ± 1.7	21.0 ± 1.0	23.8 ± 3.0	22.3 ± 1.8	**20.9 ± 0.6**	20.9 ± 0.6	25.4 ± 1.4
Cancer	10.1 ± 1.3	9.5 ± 0.9	9.3 ± 1.3	8.9 ± 1.0	10.6 ± 2.3	12.1 ± 2.3	**8.8 ± 1.2**
Wine	31.0 ± 1.6	31.4 ± 1.0	27.0 ± 3.0	31.2 ± 1.5	30.3 ± 2.4	30.1 ± 1.8	**22.8 ± 2.4**
Iris	**2.5 ± 2.0**	3.1 ± 0.5	6.4 ± 2.7	3.6 ± 1.9	4.0 ± 2.1	9.6 ± 1.4	4.4 ± 2.3
Glass	39.9 ± 1.9	**38.7 ± 3.6**	39.1 ± 4.0	40.1 ± 4.1	42.4 ± 2.1	44.4 ± 3.7	41.3 ± 3.8
Pima	31.3 ± 1.6	32.0 ± 2.7	35.0 ± 1.4	33.2 ± 2.5	33.7 ± 2.9	34.8 ± 0.2	**31.1 ± 3.0**
USPS	1.9 ± 0.3	9.0 ± 2.1	5.4 ± 1.5	4.0 ± 0.8	3.2 ± 1.8	**1.5 ± 0.1**	2.1 ± 0.7
Letter(EF)	8.0 ± 3.5	12.7 ± 2.6	**7.7 ± 2.4**	8.6 ± 2.1	7.9 ± 1.5	48.8 ± 0.6	7.8 ± 3.3
Dif100	6.8 ± 1.1	11.3 ± 1.6	13.5 ± 1.7	13.4 ± 1.8	11.4 ± 1.4	**6.5 ± 0.3**	11.8 ± 2.3
Sim100	**24.5 ± 3.9**	31.6 ± 4.5	29.8 ± 3.7	31.7 ± 3.2	25.8 ± 2.3	60.2 ± 3.0	37.8 ± 6.6
Sam100	41.6 ± 6.0	**40.3 ± 6.3**	43.7 ± 2.8	40.5 ± 4.5	44.0 ± 4.6	66.0 ± 2.9	44.3 ± 5.4
Dif1000	13.0 ± 10.2	20.8 ± 1.8	15.2 ± 1.4	17.5 ± 1.3	**11.1 ± 3.2**	65.7 ± 2.6	19.6 ± 8.1
Sam1000	48.7 ± 10.9	45.0 ± 3.0	44.6 ± 4.6	**42.4 ± 4.0**	42.5 ± 4.1	65.9 ± 1.0	55.4 ± 9.0
Classic3	3.2 ± 2.5	9.2 ± 0.5	28.5 ± 0.2	28.5 ± 0.3	4.1 ± 1.8	**1.6 ± 0.1**	4.7 ± 2.1

TABLE 20.3: Semisupervised Learning Performance Comparisons with 50 Labeled Points on 14 Datasets and 6 Methods

	gf	mwem	lnp	lgc	tikreg	ck	ltsalp
Hepatitis	22.1 ± 1.0	**21.7 ± 1.2**	25.0 ± 2.2	22.1 ± 1.6	21.7 ± 1.2	21.7 ± 1.2	24.8 ± 2.1
Cancer	9.0 ± 1.7	9.0 ± 0.7	9.4 ± 1.6	8.9 ± 0.7	11.3 ± 2.7	10.6 ± 1.8	**8.4 ± 1.0**
Wine	29.5 ± 2.4	29.1 ± 1.4	25.9 ± 5.8	30.2 ± 2.6	29.7 ± 3.3	28.0 ± 1.9	**22.0 ± 2.4**
Iris	**3.2 ± 1.1**	3.8 ± 0.4	6.6 ± 2.9	4.4 ± 1.1	4.2 ± 1.3	9.2 ± 2.3	3.2 ± 1.1
Glass	41.1 ± 1.9	40.4 ± 2.4	**39.1 ± 1.8**	39.9 ± 2.3	42.0 ± 2.8	42.9 ± 2.8	42.1 ± 4.4
Pima	31.1 ± 1.6	31.5 ± 2.2	35.3 ± 2.4	32.8 ± 1.9	33.7 ± 2.7	34.8 ± 0.3	**30.7 ± 0.8**
USPS	1.8 ± 0.2	7.6 ± 1.6	4.6 ± 1.2	3.6 ± 0.4	3.0 ± 1.4	**1.5 ± 0.1**	1.8 ± 0.4
Letter(EF)	7.4 ± 2.7	10.8 ± 3.3	6.5 ± 1.9	7.2 ± 1.8	**6.2 ± 2.8**	48.8 ± 0.6	7.7 ± 3.1
Dif100	6.9 ± 1.1	9.6 ± 1.6	10.7 ± 2.3	10.0 ± 1.2	7.5 ± 1.9	**6.7 ± 1.0**	10.2 ± 1.3
Sim100	**21.4 ± 2.1**	28.5 ± 3.1	27.3 ± 2.5	28.7 ± 1.3	22.4 ± 2.2	62.6 ± 3.0	30.6 ± 2.6
Sam100	**35.6 ± 3.6**	36.0 ± 4.2	40.0 ± 2.7	37.4 ± 3.9	41.1 ± 3.7	65.4 ± 2.0	39.6 ± 5.4
Dif1000	**9.1 ± 7.7**	18.9 ± 1.8	14.7 ± 1.2	16.2 ± 1.1	10.2 ± 2.8	63.5 ± 2.5	18.0 ± 6.6
Sam1000	43.0 ± 9.9	43.0 ± 2.1	42.2 ± 2.9	41.1 ± 3.1	**39.4 ± 3.6**	65.3 ± 1.4	55.6 ± 5.8
Classic3	2.2 ± 1.1	7.6 ± 1.5	28.5 ± 0.1	28.5 ± 0.4	3.4 ± 1.0	**1.6 ± 0.1**	3.4 ± 1.2

improving significantly as the number of labeled points increases; this can particularly be seen on the text data sets. As more labeled points are considered, GF becomes the best performing method on several text data sets. Generally, GF is among the consistently well-performing methods.

- GWEM does not demonstrate a consistent performance and seems quite sensitive to the predefined number of steps in the random walk.
- The semisupervised embedding method LTSALP is among the top performing methods in most of the UCI datasets. However, it performs poorly on the text datasets possibly indicating that the idea of aligning the tangent spaces may have to be suitably modified for sparse high-dimensional datasets.

20.6.2 Semisupervised Embedding Methods

We now compare the six label propagation methods based on semisupervised manifold embedding (see Section 20.5). In particular we examine both variants of Laplacian Eigenmaps-based Label Propagation (LELP, LELPC), Locally Linear Embedding-based Label Propagation (LLELP, LLELPC), and Local Tangent Space Alignment-based Label Propagation (LTSALP, LTSALPC). We compare the methods on the UCI datasets and show representative plots in Figure 20.3.[4] Based on our experiments, we observe that LELP and LTSALP performed well most consistently, while LLELP did well only on specific data sets such as Wine. The performance of LLELP seems to be more sensitive to the geometry of a data set. The effect of the orthogonality constraint on these methods can be understood when comparing the results for Cancer and Hepatitis in Figure 20.3. While the performance is clearly affected by the constraints, it does not necessarily result in improved performance. For example, the constraints lead to better performance in Hepatitis but worse performance in Cancer.

[4]We report results on 2-clustering problems in Figure 20.3. Since Wine is a 3-cluster dataset, we constructed a 2-cluster subset Wine(2) for these experiments.

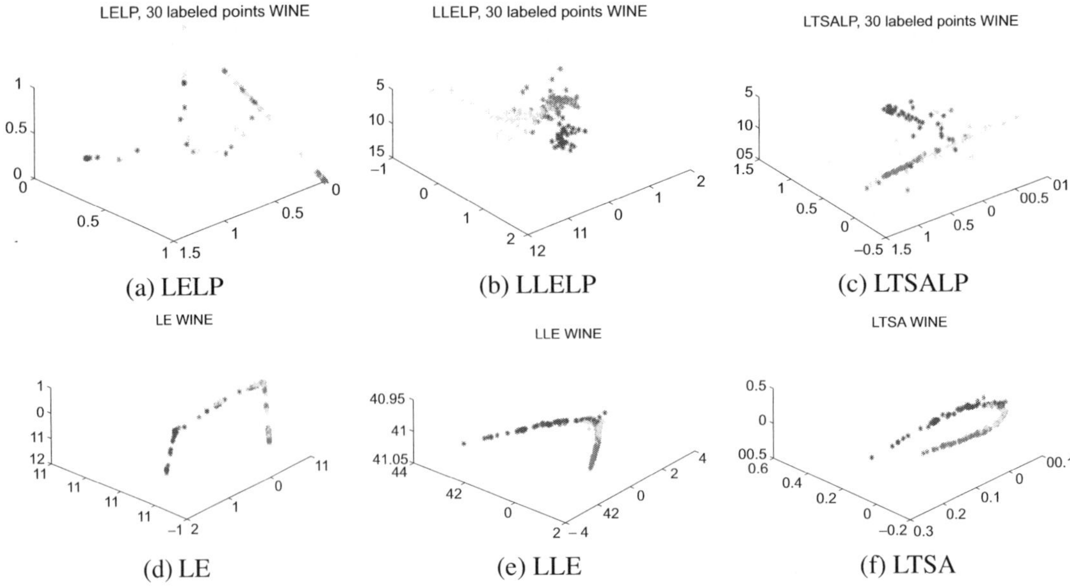

(a) LELP (b) LLELP (c) LTSALP

(d) LE (e) LLE (f) LTSA

FIGURE 20.4: Unsupervised and unconstrained semisupervised embedding on Wine. The prediction performance (top) is better if the unsupervised embedding (bottom) keeps the clusters separate.

For the embedding methods, the quality of semisupervised label propagation seems to depend on how well the unsupervised embedding preserves the cluster structure. Figure 20.4 illustrates the difference between unsupervised embedding and semisupervised embedding on Wine. Note that the unsupervised embedding obtained from LLE and LTSA maintains the cluster separation better than LE for this particular dataset. When semisupervision is added, the embedding obtained from all the methods changes suitably. The cluster separation is most clear in LLELP followed by LTSALP and LELP, a fact reflected in the test-set error rates in Figure 20.5. In general, SSC in graph-based representations with a semisupervised embedding method works well if the geometric structure of the cluster labels is well aligned with the biases of the embedding method.

20.7 Conclusions

The literature on semisupervised clustering can be broadly divided into two families, based on whether one considers feature-based or graph-based representation of the input data. The methods for feature-based representations usually generalize corresponding centroid-based clustering methods, such as k-means and variants, in order to incorporate constraints or penalty functions associated with the semisupervision. Such methods also consider metric learning under semisupervision and can be viewed as inference and parameter estimation in certain probabilistic graphical models. For graph-based representations, semisupervised clustering can be posed as a suitable semisupervised graph-cut problem. We developed a unified perspective to a variety of seemingly disparate methods for graph-based semisupervised learning and illustrate that semisupervised graph-cut problems can be viewed as a special case of such problems, which are often solved using label propagation. Interestingly, semisupervised nonlinear embedding methods also belong to the same broad family, and hence can be used for semisupervised clustering. Our empirical evaluation reveals that while there

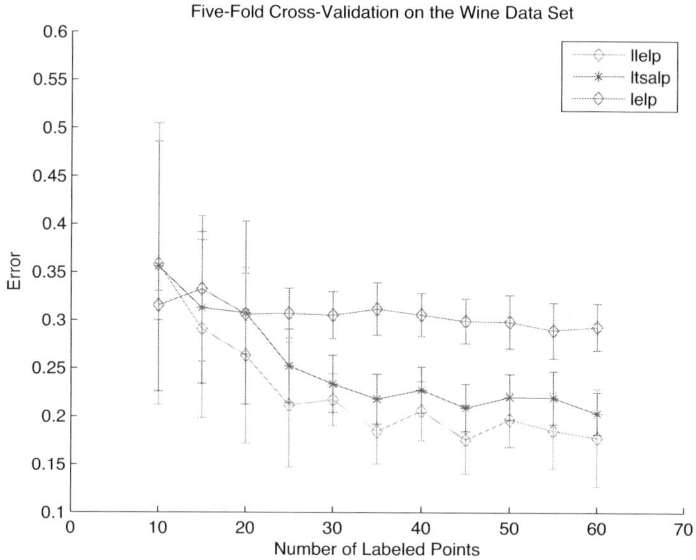

FIGURE 20.5: Comparison of embedding-based label propagation methods on Wine (corresponding to Figure 20.4).

is no clear winner in terms of performance, certain methods seem consistent across several datasets including some of the semisupervised embedding-based methods.

Bibliography

[1] A. Banerjee, I. Dhillon, J. Ghosh, and S. Sra. Clustering on the unit hypersphere using von Mises-Fisher distributions. *Journal of Machine Learning Research*, 6:1345–1382, 2005.

[2] S. Basu, A. Banerjee, and R. Mooney. Semi-supervised clustering by seeding. In *Proceedings of the 19th International Conference on Machine Learning (ICML)*, pages 27–34, 2002.

[3] S. Basu, A. Banerjee, and R. Mooney. Active semi-supervision for pairwise constrained clustering. In *Proceedings of the SIAM International Conference on Data Mining (SDM)*, 2004.

[4] S. Basu, M. Bilenko, A. Banerjee, and R. Mooney. Probabilistic semi-supervised clustering with constraints. In O. Chapalle, B. Schölkpof, and A. Zien (Eds.), *Semi-Supervised Learning*, MIT Press, 2006.

[5] S. Basu, M. Bilenko, and R. J. Mooney. A probabilistic framework for semi-supervised clustering. In *Proceedings of the 10th ACM International Conference on Knowledge Discovery and Data Mining (KDD)*, pages 59–68, 2004.

[6] S. Basu, I. Davidson, and K. Wagstaff, editors. *Constrained Clustering: Advances in Algorithms, Theory, and Applications*. Chapman & Hall/CRC Press, 2008.

[7] M. Belkin, I. Matveeva, and P. Niyogi. Regularization and semi-supervised learning on large graphs. In *Proceedings of the 18th Annual Conference on Learning Theory*, pages 624–638, 2004.

[8] M. Belkin and P. Niyogi. Laplacian eigenmaps for dimensionality reduction and data representation. *Neural Computation*, 15(6):1373–1396, June 2003.

[9] M. Bilenko, S. Basu, and R. Mooney. Integrating constraints and metric learning in semi-supervised clustering. In *Proceedings of the 21st International Conference on Machine Learning (ICML)*, New York, pages 81–88, 2004.

[10] M. Bilenko and R. J. Mooney. Adaptive duplicate detection using learnable string similarity measures. In *Proceedings of the Ninth ACM SIGKDD International Conference on Knowledge Discovery and Data Mining*, KDD '03, pages 39–48, New York, USA, 2003. ACM.

[11] C. M. Bishop. *Pattern Recognition and Machine Learning*. Springer, 2007.

[12] A. Blum and S. Chawla. Learning from labeled and unlabeled data using graph mincuts. In *Proceedings of the 18th International Conference on Machine Learning*, pages 19–26. Morgan Kaufmann, San Francisco, CA, 2001.

[13] A. Blum, J. Lafferty, M. R. Rwebangira, and R. Reddy. Semi-supervised learning using randomized mincuts. In *ICML '04: Proceedings of the Twenty-First International Conference on Machine Learning*, page 97–104, ACM, 2004.

[14] S. Boyd and L. Vandenberghe. *Convex Optimization*. Cambridge University Press, 2004.

[15] O. Chapelle, B. Scholkopf, and A. Zien, editors. *Semi-Supervised Learning*. MIT Press, 2006.

[16] O. Chapelle, J. Weston, and B. Schölkopf. Cluster kernels for semi-supervised learning. In *NIPS*, volume 15 of *NIPS*, pages 585–592, 2003.

[17] F. R. K. Chung. *Spectral Graph Theory*. American Mathematical Society, 1997.

[18] D. Cohn, R. Caruana, and A. Mccallum. Semi-supervised clustering with user feedback. Technical report, 2003.

[19] T. M. Cover and J. A. Thomas. *Elements of Information Theory*. Wiley-Interscience, 1991.

[20] I. Davidson and S. S. Ravi. Clustering with constraints: Feasibility issues and the k-means algorithm. In *SIAM International Conference on Data Mining (SDM)*, 2005.

[21] A. P. Dempster, N. M. Laird, and D. B. Rubin. Maximum likelihood from incomplete data via the EM algorithm. *Journal of the Royal Statistical Society, Series B*, 39:1–38, 1977.

[22] I. Dhillon, S. Mallela, and R. Kumar. A divisive information-theoretic feature clustering algorithm for text classification. *Journal of Machine Learning Research*, 3(4):1265–1287, 2003.

[23] C. Ding, H. D. Simon, R. Jin, and T. Li. A learning framework using Green's function and kernel regularization with application to recommender system. In *KDD '07: Proceedings of the 13th ACM SIGKDD International Conference on Knowledge Discovery and Data Mining*, pages 260–269. ACM, 2007.

[24] P. G. Doyle and L. J. Snell. Random walks and electric networks, Jan 2000.

[25] L. Hagen and A.B. Kahng. New spectral methods for ratio cut partitioning and clustering. In *IEEE Transactions on Computer-Aided Design of Integrated Circuits and Systems*, pages 1074–1085, 1992.

[26] T. Joachims. Transductive inference for text classification using support vector machines. In *Proceedings of the International Conference on Machine Learning (ICML)*, pages 200–209, 1999.

[27] T. Joachims. Transductive learning via spectral graph partitioning. In *Proceedings of the 20th International Conference on Machine Learning*, pages 290–297, 2003.

[28] D. Klein, S. D. Kamvar, and C. D. Manning. From instance-level constraints to space-level constraints: Making the most of prior knowledge in data clustering. In *Proceedings of the Nineteenth International Conference on Machine Learning*, ICML '02, pages 307–314, San Francisco, CA, USA, 2002. Morgan Kaufmann Publishers Inc.

[29] T. Li, C. Ding, and M. I. Jordan. Solving consensus and semi-supervised clustering problems using nonnegative matrix factorization. In *Proceedings of the 2007 Seventh IEEE International Conference on Data Mining*, ICDM '07, pages 577–582, Washington, DC, USA, 2007. IEEE Computer Society.

[30] B. Mohar. The Laplacian spectrum of graphs. In *Graph Theory, Combinatorics, and Applications*, pages 871–898. Wiley, 1991.

[31] K. Nigam, A. K. McCallum, S. Thrun, and T. M. Mitchell. Text classification from labeled and unlabeled documents using EM. *Machine Learning*, 39(2/3):103–134, 2000.

[32] S. Roweis and L. Saul. Nonlinear dimensionality reduction by locally linear embedding. *Science*, 290:2323–2326, 2000.

[33] S. T. Roweis and L. K. Saul. Nonlinear dimensionality reduction by locally linear embedding. *Science*, 290(5500):2323–2326, December 2000.

[34] J. Shi and J. Malik. Normalized cuts and image segmentation. In *CVPR '97: Proceedings of the 1997 Conference on Computer Vision and Pattern Recognition (CVPR '97)*, Washington, DC, USA, 1997. IEEE Computer Society.

[35] M. Szummer and T. Jaakkola. Partially labeled classification with Markov random walks. In *Proceedings of the 14th Annual Conference on Neural Information Processing Systems*, pages 945-952, 2001.

[36] U. von Luxburg. A tutorial on spectral clustering. *Statistics and Computing*, 17(4):395–416, 2007.

[37] K. Wagstaff and C. Cardie. Clustering with instance-level constraints. In *International Conference on Machine Learning (ICML)*, pages 1103–1100, 2000.

[38] K. Wagstaff, C. Cardie, S. Rogers, and S. Schrödl. Constrained *k*-means clustering with background knowledge. In *Proceedings of the Eighteenth International Conference on Machine Learning*, ICML '01, pages 577–584, San Francisco, CA, USA, 2001. Morgan Kaufmann Publishers Inc.

[39] M. J. Wainwright and M. I. Jordan. Graphical models, exponential families, and variational inference. *Foundations and Trends in Machine Learning*, 1(1–2):1–305, 2008.

[40] F. Wang and C. Zhang. Label propagation through linear neighborhoods. In *Proceedings of the 23rd International Conference on Machine Learning*, pages 55–67, 2006.

[41] E. P. Xing, A. Y. Ng, M. I. Jordan, and S. Russell. Distance metric learning, with application to clustering with side-information. In *Proceedings of the 15th Annual Conference on Neural Information Processing Systems (NIPS)*, pages 505–512, 2002.

[42] E. P. Xing, A. Y. Ng, M. I. Jordan, and S. Russell. Distance metric learning, with application to clustering with side-information. In *Advances in Neural Information Processing Systems 15*, pages 505–512. MIT Press, 2002.

[43] X. Yang, H. Fu, H. Zha, and J. Barlow. Semi-supervised nonlinear dimensionality reduction. In *ICML '06: Proceedings of the 23rd International Conference on Machine Learning*, pages 1065–1072. ACM Press, 2006.

[44] Z. Zhang and H. Zha. Principal manifolds and nonlinear dimensionality reduction via tangent space alignment. *SIAM Journal on Scientific Computing*, 26(1):313–338, 2005.

[45] D. Zhou, O. Bousquet, T. Lal, J. Weston, and B. Schölkopf. Learning with local and global consistency. In *Proceedings of the 16th Annual Conference on Neural Information Processing Systems*, pages 321–328, 2003.

[46] X. Zhu, Z. Ghahramani, and J. Lafferty. Semi-supervised learning using Gaussian fields and harmonic functions. In *Proceedings of the 20th International Conference on Machine Learning (ICML)*, pages 912–919, 2003.

Chapter 21

Alternative Clustering Analysis: A Review

James Bailey
The University of Melbourne
Melbourne, Australia
baileyj@unimelb.edu.au

21.1 Introduction

Clustering is one of the most fundamental and important techniques in knowledge discovery and is used in a wide variety of fields, ranging from biomedicine and information retrieval to financial analysis and Web mining. Clustering analysis provides a way to automatically identify patterns and relationships in complex data, to form hypotheses about their structure, and to make predictions for subclasses of objects.

There exists a large variety of knowledge discovery workflows that rely on clustering. In its basic form, clustering analysis is used to explore a complex dataset, by automatically identifying object groupings. Given an input dataset for analysis, a clustering algorithm, such as k-means, can be executed, producing a clustering as output. This output clustering consists of a set of clusters which partition the objects in the dataset.

It is well known that the process of clustering is subjective and the clustering that is output is strongly dependent on the nature of the specific clustering algorithm chosen. Indeed, vastly different outputs may be possible if one changes the clustering algorithm or varies the parameters input to a fixed algorithm. Alternative outputs are also possible according to different preprocessing methods for the input, such as when applying a feature selection step.

This inherent subjectiveness and instability of clustering is widely recognized and has provided impetus to the emerging area of multiple clustering analysis. The philosophy here is that making the assumption that only a single clustering exists for a dataset is too strict. Instead, one should expect that multiple *alternative* clusterings are reasonable for a dataset. Each one of these alternatives corresponds to a different grouping of the objects and reflects a different perspective, view, or hypothesis about the nature of the data.

Why might multiple clusterings be reasonable for the same dataset? First, the data being analyzed could be very complex, containing many features, which may be of different types. Different combinations of these features (or subspaces) may provide natural alternative perspectives of the data. The data might also consist of many instances, meaning a diversity of possible subpopulations, resulting in many possible views. Second, the data could be temporal in nature and evolving over time. As the data evolves, concept drift can occur, meaning that different groupings of the data become stronger or weaker. Third, the data may be spatial in nature, meaning that the different perspectives have a spatial origin. Fourth, the data objects may be diverse, due to datasets being merged or information having been integrated from multiple sources (e.g., a large clinical cohort study that pools data from multiple sites). Again, this may mean that multiple perspectives of the data are necessary and natural for knowledge discovery, rather than relying on just a single perspective.

Given that multiple clusterings or views of the data are possible, it is therefore also important to consider why they may be important for a user. First, clustering analysis is frequently exploratory in nature. A user often does not know what behavior he is looking for. What he needs is to navigate through and assess multiple alternatives, so different options can be evaluated. Conversely, the user may have a strong hypothesis (clustering) in mind and desire to verify that no other strong hypotheses are supported by the data. Second, users themselves can differ widely in their requirements and expectations. It is therefore unlikely that a single clustering will be appropriate for all users. Third, a common scenario in data mining studies is that the investigation focuses on a new or novel clustering algorithm and it is necessary to test the flexibility and limits of this proposed algorithm, to assess how many alternatives it is able to identify.

Due to these reasons, the area of multiple clustering analysis has been attracting considerable attention. Indeed, several recent workshops have been devoted to the topic [29, 30, 31, 37, 25]. The literature in the area is also growing fast, with a number of algorithms proposed, which particularly focus on the problem of generating alternative clusterings, that are each of high quality and also dissimilar to one another [15, 16, 12, 5, 27, 7, 2, 3, 9, 8, 28, 32, 22, 11, 10].

The focus of this chapter is to review algorithms for generating alternative clusterings, which is one of the prime tasks in the field of multiple clustering analysis. We also highlight connections to the areas of multiview clustering and subspace clustering, which are distinct, yet closely related. In multiview clustering, the aim is to learn a single clustering using multiple sources (representations) of the data [4, 20, 35, 6, 23, 17]. These sources usually contain the same set of objects, but with different features. In subspace clustering, the aim is to discover different subspaces, where each subspace contains a good cluster (as opposed to clustering). (See Chapter 9 for more on subspace clustering methods).

An outline of the rest of this chapter is as follows. In Section 21.2, we present necessary terminology and definitions. In Section 21.3.1, we present a taxonomy of alternative clustering techniques and discuss different dimensions of evaluation. In Sections 21.3.2, 21.3.3, 21.3.4 and 21.3.5, we review specific approaches for alternative clustering. In Section 21.4, we compare the areas of multiview clustering and subspace clustering to alternative clustering and identify similarities and dissimilarities. In Sections 21.5 and 21.6 we outline future directions and conclude.

21.2 Technical Preliminaries

Let D be a dataset containing N objects $o_1, \ldots, o_N$ and using n features $F_1, \ldots, F_n$. A (hard) clustering[1] C is a partition of the objects in D into k clusters $\{c_1, \ldots, c_k\}$, where each cluster is a set of objects and $c_i \cap c_j = \emptyset$. Let the universe of all clusterings of D be denoted as $C_{\mathcal{D}}$. We will also use the notation C_i to refer to cluster c_i of clustering C.

The quality of a clustering may be measured using a function $Qual : C_D \rightarrow [0,1]$ where higher values indicate higher quality. A large range of quality measures have been defined, with some well-known examples being the Dunn Index [13], the Davies-Bouldin Index [12], and the Silhouette Width [34].

Let $C_1 = \{c_1, \ldots, c_k\}$ and $C_2 = \{c'_1, \ldots, c'_{k'}\}$ be two clusterings of D. The similarity between C_1 and C_2 may be measured using a function $Sim : C_D \times C_D \rightarrow [0,1]$ where higher values indicate higher similarity. For measuring similarity, there are a number of possible measures, including the Rand Index [33], Adjusted Rand Index [21], Jaccard Index [18], Normalized Mutual Information [24], and Adjusted Mutual Information [36]. Measurement of similarity between clusterings is important, since it provides insight for the user into the relationship between them. When managing multiple clusterings, assessment of similarity may allow removal of redundant clusterings, selection of interesting clusterings, or increased understanding about clustering evolution. It is also a key step when exploring the convergence properties of a clustering algorithm or assessing its output compared to an expert generated clustering.

Given the large range of measures that can be "plugged in" for measuring quality and similarity, appropriate choices are often be made in an application dependent way. We will shortly describe the issues involved in generating alternative clusterings and the different dimensions along which the existing algorithms may be compared. A general description of the task is as follows.

Definition 21.2.1 *Generalized Alternative Clustering: Given a (possibly empty) collection of clusterings* $K = \{C_1, \ldots, C_m\}$ *provided as background knowledge (either* $K = \emptyset$*, or* $K \neq \emptyset$ *and* $m \geq 1$*), generate j alternative clustering(s)* $O = \{C_{m+1}, \ldots, C_{m+j}\}$*, such that i)* $\sum_{i=m+1}^{m+j} Qual(C_i)$ *is maximized and* $\sum_{i,j \in [1,m+j]} sim(C_i, C_j)$ *is minimized.*

The task here corresponds to generating a set of new (alternative) clusterings, where each individually is of high quality and also the pairwise similarity between the clusterings is low (the clusterings are distinctive). Three common cases are

- $|K| = 1$ and $|O| = 1$: *singular alternative clustering*
- $K = 0$ and $|O| = 2$: clusterings in O are generated in parallel: *simultaneous alternative clustering*
- $|K| > 1$ and $|O| = 1$: *sequential alternative clustering*

[1] It is also possible to use fuzzy clusterings as the basis for development, but the literature on alternative fuzzy clustering is less mature and we concentrate on the hard case.

In the next section, we consider the ways in which the behaviour of alternative clustering algorithms can be explained and specified.

21.3 Multiple Clustering Analysis Using Alternative Clusterings

In this section, we first review the different dimensions that may be used for assessing the behaviour of alternative clustering algorithms. We then describe in detail the different approaches, broken down according to style of technique.

21.3.1 Alternative Clustering Algorithms: A Taxonomy

Alternative clustering algorithms (ACAs) may be characterized in a range of different ways. We review the different options in turn.

Format of the input: The input to an ACA consists of the dataset to be clustered, which may be represented as feature valued instances or by a similarity matrix for all pairs of objects. A technique might additionally require features to be either continuous or discrete. In addition to these, the input may optionally include *background knowledge*, which is a single existing clustering or a collection of two or more existing clusterings that are already available. It is not specified where the background knowledge comes from; it might come from the application of a standard clustering algorithm or from user insights.

Format of the output: The output can consist of a single alternative clustering or two or more alternative clusterings. Some algorithms may place constraints on the number of clusters in each clustering (e.g., they must be equal), or may require that the number of clusters in the output matches with the number of clusters in the clustering that is input as background knowledge. Also, the output may consist of either an entire (alternative) clustering or a *partial* alternative clustering. The latter is useful if the user only wishes to change some characteristics (clusters) of the clustering(s) being used as background knowledge, while keeping other characteristics the same.

Style of output generation: If more than one alternative clustering can be output, is each alternative generated one at a time in a greedy fashion (*sequential generation*) or are all alternatives generated in parallel (*simultaneous generation*)? The latter may produce a more globally optimal solution. However, the former may be more realistic when one or more existing clusterings exist. It might also identify some strong clusterings which would be missed by a simultaneous generation technique.

Style of Technique: This describes the overall process of alternative clustering generation. One class of techniques is *unguided generation*, where no background knowledge is used for generation of alternatives. The other class of techniques is *guided generation*, where background knowledge is used as input and explicit effort is made to ensure dissimilarity between new clusterings and existing ones specified in the background knowledge. Guided generation techniques can be further broken down, according to whether the technique (i) relies on the use of inferred constraints to generate alternatives (*constraint based*), (ii) operates by generating feature spaces which are orthogonal to the existing feature space and to each other (*orthogonal feature space transformation*—such a transformation means the feature space of the alternative clustering(s) is different from that of the clustering(s) in the background knowledge), or (iii) uses an objective function based on information theoretic criteria, in order to optimize quality and dissimilarity characteristics (*information theoretic*).

Style of clustering algorithm: Is the method tied to a specific technique (e.g., *k*-means, hierarchical, or expectation maximization)? Or can any clustering algorithm be plugged in? The latter is typically possible for the orthogonal feature space transformation style, where once an orthogonal feature space is discovered, any clustering algorithm can be used to generate the alternative. This increases flexibility and means an appropriate clustering algorithm can be chosen according to the dataset and desired cluster characteristics.

Parameter requirements: Apart from the number of clusterings output and the number of clusters in each, are there any other parameters that must be specified to use the technique? Many techniques rely on the use of a regularization/tradeoff parameter, which is used to tune the relative weightings of quality and dissimilarity criteria in the objective function.

We now proceed with a description of the different approaches, grouped according to the style of the technique.

21.3.2 Unguided Generation

Unguided generation techniques do not employ any background knowledge for generating the alternative clustering(s). There are a variety of approaches in this category, ranging from the simple to the sophisticated.

21.3.2.1 Naive

The most basic technique is the naive method. Using this technique, alternative clusterings are obtained by (i) running a clustering algorithm multiple times, using different parameters each time, (ii) running different clustering algorithms, or (iii) a combination of (i) and (ii). The advantage of such an approach is that it is straightforward to implement and any clustering algorithm(s) may be used The principal disadvantage is that its behavior can be quite random and there is a risk of generating alternative clusterings that are very similar to one another. Background knowledge is also not taken into account. Due to the issue of redundancy, postprocessing is required to filter out clusterings having a high amount of overlap. Naive generation is a very common technique employed by users who are not familiar with alternative clustering. It is also often used in conjunction with consensus clustering, where the alternatives are combined into a single clustering using a voting strategy.

21.3.2.2 Meta Clustering

An extension of the naive technique is the approach of meta clustering [5]. Similar to naive, meta clustering does not make use of any background knowledge to generate alternatives. Instead, it adopts a more principled approach to achieve dissimilarity. In particular, *k*-means is repeatedly used with (i) random choices of initial centroids and (ii) different attribute weightings in the distance functions, according to a Zipf distribution. After generation of alternative clusterings in this fashion, meta clustering then treats the output clusterings as objects themselves and clusters them using a cluster difference distance function. This yields a meta level perspective for the clusterings, which can be explored by the user. While the generation strategy here is more sophisticated than the naive one, it still does not explicitly ensure that the output clusterings will be dissimilar, but only that they will have a reasonable chance to be dissimilar. Also, the random use of centroids in *k*-means again may result in duplicates. Furthermore, the use of differently weighted distance functions, while it increases the chance of dissimilarity between alternatives, may have an impact on the quality of the clusterings, since some weightings may produce unnatural output clusterings. For these reasons, as with the naive technique, the results are likely to need postprocessing to reduce redundancy. Again though, as does naive, meta clustering has the advantage of being simple and clean to implement.

21.3.2.3 Eigenvectors of the Laplacian Matrix

Dasgupta and Ng [10] show that alternative clusterings can be found by looking at different eigenvectors of the Laplacian matrix. The input is a similarity matrix S and no background knowledge is used. There is the strong requirement that each alternative clustering output is constrained to have two clusters. The approach is a spectral one, where the objects are represented as a graph with edges between nodes indicating pairwise similarities, and a partition is generated using a normalized cut criterion. Let $D_{i,i} = \sum_j S_{i,j}$ and the Laplacian matrix L is $L = D^{-1/2}(D-S)D^{-1/2}$. The first alternative clustering is found by applying 2-means to the objects represented by e_2, the eigenvector corresponding to the second smallest eigenvalue of L. The mth alternative clustering is produced by applying 2-means to the objects represented by the $(m+1)$th eigenvector of L. The dissimilarity objective is achieved by the orthogonality of the different eigenvectors. Quality is achieved by the 2-means algorithm, but the second and later alternatives will be "suboptimal," compared to the first, since the optimality decreases as m increases. The approach has the advantage of being simple to implement, but the limitation of two clusters in each clustering is restrictive.

21.3.2.4 Decorrelated k-Means and Convolutional EM

The approach by Jain et al [22] is a simultaneous one for generating two clusterings C_1 and C_2, without using any background knowledge. Supposing each of the output clusterings has k_1 and k_2 clusters, respectively; then, they are generated in a decorrelated fashion using k-means style. The objective function has the form:

$$\sum_{i=1}^{k_1} \sum_{x \in C_1^i} ||x - \mu^i||^2 + \sum_{j=1}^{k_2} \sum_{x \in C_2^j} ||x - \nu^j||^2 + \lambda \sum_{i,j} (\beta_j^T \mu^i)^2 + \lambda \sum_{i,j} (\alpha_i^T \nu^j)^2$$

where λ is a parameter used for regularization, μ^i and ν^j are the representative vectors of clusters C_1^i and C_2^j, respectively, and α_i and β_j are the mean vectors of C_1^i and C_2^j, respectively. The initial two terms correspond to k-means-type error terms, while the second two terms ensure dissimilarity (decorrelation) between the two clusterings. The objective function can be extended to generate more than 2 clusterings, by including an extra k-means-type error term for each new clustering and including a pairwise dissimilarity term for each possible pair of clusterings. An iterative approach is used to minimize the objective function. The regularization parameter λ is set empirically, and it is also possible to extend the objective to a kernelized version to handle nonlinearities. A disadvantage is that the representative vectors in the decorrelated k-means algorithm do not have a natural interpretation for the user.

In a companion proposal to decorrelated k-means, the work in [22] also outlines a convolutional expectation maximization (EM) algorithm, where it is assumed that the data can be modeled as the sum of two mixtures of distributions, each of which is associated with a clustering. One clustering has k_1 clusters; the other has k_2 clusters. Then, since the distribution of the sum of two independent random variables is the convolution of the distributions, the data is modeled as being sampled from a convolution of two mixtures. This then leads to the problem of learning a convolution of mixture distributions, using an EM method to determine the distributions' parameters. The technique is again simultaneous and can be kernelized.

21.3.2.5 CAMI

The CAMI [8] algorithm is designed to discover two alternative clusterings at the same time using the original data space. Formulating the clustering problem under mixture models, CAMI optimizes a dual-objective function in which the log-likelihood (accounting for clustering quality) is maximized, while the mutual information between two mixture models (accounting for the distinction between two clusterings) is minimized. The objective function of CAMI can be written as

$$\bar{L}(\Theta,D) = L(\Theta^1;D) + L(\Theta^2;D) - \eta \sum_{i,j} p(C_1^i, C_2^j) \log \frac{p(C_1^i, C_2^j)}{p(C_1^i)p(C_2^j)}$$

The first two terms correspond to the likelihood of each of the two clusterings that will be simultaneously discovered and Θ^1 and Θ^2 are their parameters. The third term corresponds to the dissimilarity between the clusterings C_1 and C_2 as measured by mutual information. The η is a regularization parameter used to trade off dissimilarity and quality (and which can be specified by the user). Using Gaussian mixture models, an EM approach can be used to optimize the objective function.

21.3.3 Guided Generation with Constraints

The next class of techniques uses constraints to guide the generation of one or more alternative clusterings. The type of constraints and the way they are used distinguishes each of the methods.

21.3.3.1 COALA

The COALA method takes as input a similarity matrix and a single existing clustering as background knowledge. It uses hierarchical algorithm. Using the existing clustering, a set of "cannot-link" constraints is generated, one for each pair of objects in the same cluster. Intuitively, it is less desirable for objects in these pairs to again be together in the same cluster of the alternative clustering. A hierarchical clustering approach is then used. At each iteration, COALA finds two candidate pairs of clusters for a possible merge, one denoted as $(q1,q2)$, called a qualitative pair, and the other denoted as $(o1,o2)$, called a dissimilar pair. The qualitative pair is the one with the minimum distance over all the pairs of clusters (ensuring the highest quality clusters when merged). The dissimilar pair has the minimum distance over all the pairs of clusters that also satisfies the cannot-link constraints (these pairs may be the same). COALA will select just one of these pairs to merge. Given a tradeoff factor parameter ω, if $\frac{d(q_1,q_2)}{d(o_1,o_2)} \geq \omega$, then the pair (o_1,o_2) is merged. Otherwise, the pair (q_1,q_2) is merged. By varying the value of ω, different behaviours can be achieved.

COALA is a simple and intuitive technique and has been used as a baseline method for comparison in a range of papers. A limitation of COALA is that it is specifically tied to a hierarchical clustering algorithm. It also was not formulated for the case of generating multiple alternative clusterings. However, it is easy to conceive generalizations in which multiple clusterings are used as background knowledge, yielding a larger set of cannot-link constraints for generating the alternative clustering.

21.3.3.2 Constrained Optimization Approach

The approach of Qi and Davidson in [32] uses constraints in a different way. It takes the original dataset $X = \{x_1, \ldots, x_n\}$ and transforms it to a new dataset $Y = \{y_1, \ldots, y_n\}$, where $Y = DX$ and D is a transformation matrix representing a distance metric. Any clustering algorithm can then be applied to the new dataset to generate an alternative to the original clustering.

The objective function is formulated as a constrained optimization task

$$\min_{B \succeq 0} D_{KL}(p_y(y)||p_x(x)) \;\; s.t. \;\; \frac{1}{n} \sum_{i=1}^{n} \sum_{j=1, x_i \notin C_j} ||(x_i - \mu_j)||_B^a \leq \beta$$

where $B = D^T D$, $||.||_B$ is the Mahalanobis distance using matrix B, and $B \succeq 0$ signifies that B is required to be positive semidefinite.

The original dataset X follows probability density function $p_x(x)$, and dataset Y follows probability density function $p_y(y)$. D_{KL} signifies the KL divergence between two distributions and $a \geq 1$ is a tradeoff parameter, with larger values ensuring higher dissimilarity of the alternative clustering.

The first part of the objective aims to ensure that the transformed data preserves the characteristics of the original data (with the KL distance being zero when they are identical). The second part of the objective ensures dissimilarity and discourages the original clusters being found by requiring each object in the new data space to be closer to the cluster centers of the cluster of which it was not originally part. To achieve a closed form solution to the objective function, a mixture model of multivariate Gaussian distributions can be assumed, having the same covariance matrix.

Some advantages of this approach are that (i) any clustering algorithm may be used to generate the alternative clustering, once the new dataset is obtained, and (ii) the approach extends naturally for discovering a partial alternative clustering. Users may specify properties of the original clustering they wish to keep (i.e., some original clusters or groups of objects should remain the same) and then solve the objective function with the intention of finding only some alternative clusters to add to the desired original clusters. A limitation of the approach is that it is somewhat unclear what kind of properties of the original dataset X get preserved in the new dataset Y, due to the generality of the KL-distance function.

21.3.3.3 MAXIMUS

Work by Bae, Bailey and Dong in [3] describes an algorithm known as MAXIMUS for discovering multiple alternative clusterings in a sequential manner. The MAXIMUS algorithm calculates the maximum dissimilarity between any currently available clusterings and a potential target alternative solution, by forming an integer programming model. The objective of this integer programming model is to maximize the distance between the density profiles of the known clusterings, versus the unknown target alternative clustering. It then uses the output of the model to generate an alternative clustering.

MAXIMUS is based on the use of a clustering similarity function known as *ADCO*, which can compare clusterings according to their spatial characteristics. At a high level, the *ADCO* measure constructs a spatial histogram for each cluster and represents a clustering as a vector containing the spatial histogram counts for the clusters. The two clusterings can then be compared using vector operations. Intuitively, the output of *ADCO* is a containment judgment between a clustering C_1 and a clustering C_2, expressed as "How much of clustering C_2 is contained in clustering C_1?" or "What percentage of clustering C_2 is contained in clustering C_1?"

Using the *ADCO* measure, one may generate a spatial template to ensure that a single alternative clustering has maximal (average) dissimilarity from the input background clusterings. This template describes how many objects must be present in bins within one-dimensional projections of the feature space. Using the template, a constrained k-means algorithm is used to derive a clustering for each bin. Next, consensus clustering is then used to combine the clusterings from all the bins into a single clustering. Thus, the quality of the alternative clustering is achieved by the use of k-means and consensus clustering. The dissimilarity objective is achieved by using the integer programming model and the *ADCO* measure to obtain a spatial template which can be expected to have very high dissimilarity from the background knowledge clusterings.

In order to use MAXIMUS, it is necessary to specify the binning strategy for representing the density profile (10 bins equidensity is recommended as a default). Unlike some other algorithms, MAXIMUS does not require the user to specify a regularization parameter to trade off between the quality and dissimilarity objectives.

21.3.4 Orthogonal Transformation Approaches

Our next class of approaches consider approach the task of alternative clustering from a feature space perspective. Using an existing clustering as background knowledge, this style of approach constructs a new feature space which is "orthogonal" to the data space that is characterised by the existing clustering. Once this orthogonal feature space is generated, any clustering algorithm can be used in this space to generate an alternative clustering. Thus, the objectives of quality and dissimilarity are decoupled, with the former being tied to the use of the chosen clustering algorithm and the latter being tied to the characteristics of the orthogonal space that gets generated. Overall, these approaches have an appealing mathematical formulation based on linear algebra. They are also relatively efficient. A limitation is that the orthogonality requirement may be too strict for some datasets and it is not always clear how it trades off against the quality of the clustering.

21.3.4.1 Orthogonal Views

Work by Cui et al in [7] presents two approaches that can generate multiple alternative clusterings, in a sequential manner. Each alternative clustering is determined by subsets of features of the data set, which are best described by the clustering. Given a clustering C_1, a subset of features that are well represented in C_1 is found and then another set of features, which are orthogonal to the first subset, is found. Their first approach carries out a transformation as follows: Each data object x_i from cluster j is projected onto its cluster center μ_j and then a residue is found by projection onto an orthogonal subspace:

$$x_i^{new} = (I - \frac{\mu_j \mu_j^T}{\mu_j^T \mu_j}) x_i$$

One then clusters the data in this orthogonal subspace to obtain an alternative clustering. The method may be executed iteratively to generate multiple alternative clusterings. A version where the input is a soft (fuzzy) clustering is also outlined.

In the second approach, a feature subspace F_2 that is a good representation for the clustering C_1 is first found using principal component analysis on the mean vectors of C_1. The data X is then projected to a subspace that is orthogonal to F_2, and a clustering algorithm is applied to the new data X^{new} to generate an alternative clustering C_2. Specifically,

$$X^{new} = ((I - F_2(F_2^T F_2)^{-1} F_2^T) X$$

Again, the method can be applied iteratively to generate further alternative clusterings.

21.3.4.2 ADFT

Work By Davidson and Qi [11] describes the ADFT approach to finding an alternative clustering, using a set of instance level constraints. This approach is also a transformation approach like that of [7]. However, instead of characterizing the background knowledge clustering C_1 according to mean vectors or a feature subset, it is characterized using instance must-link and cannot-link constraints and then a distance function D_{C_1} is learned using these constraints. This distance function can be decomposed using singular value decomposition into $D_{C_1} = HSA$, where H is the hanger matrix, S is the stretcher matrix, and A is aligner matrix.

Once the characteristic distance function D_{C_1} has been learned, an alternative distance function can be computed that is equal to $HS^{-1}A$. This alternative distance function is then employed to generate a new dataset $X^{new} = (HS^{-1}A)X$. The alternative clustering is then found by applying any clustering algorithm on X^{new}.

This method has an advantage over the approach of [7], since it can be applied in situations where the dimensionality of the dataset is smaller than the number of clusters (as is common for spatial data).

21.3.5 Information Theoretic

Another approach to the generation of alternative clusterings is based on the use of objective functions using information theoretic principles. Such approaches are mathematically attractive and incorporate the use of mutual (or similar) information to measure the strength of correlations between clustering. Several algorithms fall into this category, beginning with the approach of Gondek and Hofmann [16], which was the first (to our knowledge) alternative clustering algorithm to be proposed.

21.3.5.1 Conditional Information Bottleneck (CIB)

The conditional information bottleneck (CIB) approach for alternative clustering is described in [15, 16]. This algorithm takes as input an existing clustering C_1 as background knowledge and sequentially generates a single alternative clustering C_2 by optimizing the objective function

$$\max_{C_2} \left(I(C_2;F|C_1) - \lambda_1 I(C_2;X) + \lambda_2 I(C_2;F) \right)$$

where F is the features, X is the objects, and the existing clustering is C_1. The term $I(C_2;F|C_1)$ corresponds to the mutual information between the new alternative clustering being discovered and the features, given the predefined clustering. The term $I(C_2;X)$ corresponds to the mutual information between the desired alternative clustering and the objects (this is desired to be small, to avoid being overly confident about the groupings), and $I(C_2;F)$ corresponds to the mutual information between the desired alternative clustering and the features (we want this to be high). The symbols $\lambda_1 > 0$ and $\lambda_2 > 0$ are regularization parameters, used to trade off the different components of the objective function. The approach of [16] describes an alternating optimization scheme with deterministic annealing, which can be used for generating C_2 with this objective function. In practice, this style of approach has been found to behave particularly strongly for document datasets.

21.3.5.2 Conditional Ensemble Clustering

The CIB approach of [15, 16] was further extended in [14], which introduced the CondEns (Conditional Ensemble) alternative clustering algorithm.

CondEns operates in three stages. (1) Given the clustering $C_1 = \{c_1, \ldots, c_k\}$ as background knowledge, for each cluster c_i, a local clustering is generated using any clustering algorithm. This yields k local clusterings. (2) Each of the k local clusterings is extended into a global clustering, by assigning instances not already part of a local clustering, to one of its clusters. (3) The k global clusterings are then combined using a consensus technique based on the conditional information bottleneck, to yield a single alternative clustering.

As with the approach of [16], CondEns also performs well for text datasets. A limitation of CondEns is its guarantees about the dissimilarity of the alternative clustering are somewhat unclear, since the clusters in the original clustering c_1 may be quite similar among themselves. This means that the alternative clustering may in turn be similar to the background knowledge clustering.

21.3.5.3 NACI

The NACI algorithm was proposed by Dang and Bailey in [9] and targets scenarios where the borders between clusters in the alternative clustering may not be linearly separable.

At a high level, its objective function can be expressed as finding an alternative clustering C_2, given a clustering C_1 as background knowledge, according to

$$C_2 = \underset{C_2}{argmax}\{I(C_2;X) - \eta I(C_1;C_2)\}$$

where

$$I(C_1;C_2) = \sum_{C_1^i}\sum_{C_2^j}(p(C_1^i,C_2^k) - p(C_1^i)p(C_2^j))^2$$

where η is a regularization parameter, $p(\cdot,\cdot)$ is the probability density, and the mutual information $I(C_1;C_2)$ is in fact a quadratic form of the mutual information, which has the advantage of being amenable to density estimation using a Parzen window technique with Gaussian Kernel. This objective can then be used as a component within a hierarchical clustering framework to generate an alternative clustering. To use NACI, choices must be made for both the regularization parameter and the kernel parameter.

Another approach in the same spirit as NACI is that of minCEntropy [27], which instead of using a hierarchical algorithm with the quadratic mutual information, uses a k-means style algorithm with quadratic mutual information. The style is again sequential and requires specification of a kernel width parameter and a tradeoff parameter.

21.3.5.4 mSC

The final method we mention in this section is the mSC alternative clustering approach outlined in [28]. This is a spectral approach which can simultaneously generate multiple alternative clusterings.

Rather than being based on mutual information, it uses the Hilbert-Schmidt Independence Criterion (HSIC) to assess the correlation between clusterings. As with mutual information, the HSIC is also able to recognize nonlinear dependencies. Specifically, the mSC technique embeds the HSIC measure within a spectral clustering framework. The objective is a dual function, where at each iteration, one term is fixed and the other term is optimized. The user is able to specify the number of alternative clusterings that are desired and the number of clusters in each.

21.4 Connections to Multiview Clustering and Subspace Clustering

We have thus far reviewed a range of techniques that can be used for generating alternative clusterings, which is a core component for multiple clustering analysis. We now mention the connections that exist between alternative clustering analysis and two other directions: multiview clustering and subspace clustering.

Multiview clustering is also concerned with multiple clusterings, but from a different angle. In multiview clustering, one is provided with multiple sources or representations of data (multiple views) and wishes to learn a single clustering which is both consistent with and a good reflection of the multiple views. A prototypical example is Web pages, which may be modeled using features which describe the frequencies of words occurring in the page (View 1), or modeled using features which describe the links into the page (View 2), or modeled using features which describe the anchor text in the links going out from the page (View 3). It has been found that using the information in all views simultaneously, one can generate a better quality clustering than using only a single view obtained by merging the feature spaces. A particular benefit of multiview clustering is that using multiple views to produce a clustering can reduce the effect of noise within individual views. If one were to use a single view to derive a clustering, the presence of noise might corrupt the clusters and make the detection of cluster structure more difficult. Using multiple views to cluster, however, lessens the likelihood of noise within a view being dominant and instead emphasizes the commonalities between views and their contribution toward the overall cluster structure.

Broadly speaking, there are two kinds of approaches for multiview clustering [23]. In the first approach (centralized), the multiple views are used in parallel to cluster the dataset [4, 38, 6]. In the second approach (distributed), a clustering is generated for each view independently and the clusterings are later merged to produce a single clustering [23, 17].

For the centralized approach, Bickel and Scheffer [4] consider the setting of a dataset which has been generated by a mixture model and the objective of which is to determine the parameters for each of the components of the mixture. They develop both a multiview EM algorithm and a multiview k-means algorithm, which is based on an assumption of independence between views. They find that the multiview EM algorithm is able to optimize the agreement between the views and that it can achieve a significant improvement in performance compared to a single view version. They also evaluate an agglomerative multiview approach, but find that its results are not improved compared to a single view version. Zhou and Burges [38] consider a spectral clustering approach and propose an algorithm that generalizes the (single view) normalized cut to incorporate information from multiple views (graphs). The approach uses a random walk technique, that traverses the vertices of both graphs, to derive a multiple graph cut which is good on average for both graphs. They find their approach consistently performs better than just using a single view. Chaudhuri et al [6] address the problem of clustering in high dimensions and how to discover a lower dimensional subspace, in which a standard clustering algorithm can then be applied. In their work, this lower dimensional subspace is found using the information from multiple views, where each view is composed of a mixture of distributions. A canonical correlation technique is used for subspace learning.

For the distributed approach, Long et al [23] propose a pattern-based technique based on the use of a mapping function. After independently clustering each view, these clusterings are then combined into a single view using the mapping function. The objective function minimizes the averaged mapped distance of the views to the overall clustering, using an iterative algorithm. Greene and Cunningham [17] tackle the problem by proposing a matrix factorization approach. Specifically, a matrix is constructed that summarizes all the clusterings (one clustering per view). This matrix is then factorized (possibly with some approximation error) into the product of two nonnegative matrices. The first contains information about the contribution of each cluster from the views to the overall, final clustering. The second matrix describes the membership of objects in the final clustering.

A key issue for multiview clustering is how to balance the relative contributions of the views and ensure that noisy views do not degrade the final result [35]. Another key issue for multiview clustering is how to handle application-specific multiview integration. For example, the techniques needed to combine multiple views in a document domain may be quite different from what is appropriate for combining multiple views in a protein (bioinformatics) domain. The multiview paradigm has also been extended to the discovery of subspaces, rather than aiming to produce only one, overall clustering [20].

Like alternative clustering, subspace clustering is also concerned with discovering multiple solutions. Here though, the principal aim is to discover multiple clusters, each hidden in a lower dimensional subspace, rather than discovering multiple clusterings. The motivation is that the dataset can contain features which are irrelevant to and confusing for clustering structure. Removing these features can make the clustering structure clearer and of better quality. Some well-known examples include CLIQUE [1], MAFIA [26], and DENCLUE [19]. Issues for subspace clustering analysis include ensuring the dissimilarity between subspace clusters (which may otherwise have large overlap) and controlling the number of subspace clusters (which may be exponential in the number of features).

21.5 Future Research Issues

There are a number of issues that still remain to be explored in alternative clustering analysis.

1. Many good approaches have been proposed for generating alternative clusterings. These have tended to be evaluated on synthetic data, or small real-world data. Their degree of scalability is thus often untested. In order to extend the reach and applicability of alternative clustering, more serious evaluation will be needed that is based on the use of very large datasets.
2. Discovery of alternative clusterings is intuitively reasonable. However, it will be important to identify application scenarios and compelling case studies where alternative clusterings have influence for a real application area. Good visualization tools for alternative clusterings could have potential impact here.
3. A number of alternative clustering methods are capable of generating more than one alternative. This raises the issue of how many alternatives is sufficient. Is this an issue which is user dependent, much like choosing the number of clusters, or are there more principled ways to evaluate the viability of alternatives? Coupled with this issue is the companion question of how many clusters should be included in each alternative clustering.
4. The traditional notion of a "complete" alternative may sometimes be too strict. Instead, a user may sometimes desire partial alternatives, where the new clustering is similar in some respects, but different in other respects to the existing clustering(s). Work by Qi and Davidson [32] is a promising basis here.

21.6 Summary

We have reviewed the area of alternative clustering analysis. The impetus for the field has come from the complexity and heterogeneity of today's datasets. Users wish to obtain not only a single view or hypothesis of their data, but also be presented with several alternatives.

We have seen that a number of approaches for alternative clustering exist, possessing considerable diversity in their technical details. At the core of each, though, is the capability to generate new clusterings which achieve a balance between being novel and being different from clusterings that are already known.

The area has grown rapidly in the last few years and we believe it has a bright future. As the techniques become more widely known, the generation of alternative clusterings may become common place. This invites the following speculation: "In the future, might every clustering be accompanied by an alternative clustering?"

Bibliography

[1] R. Agrawal, J. Gehrke, D. Gunopulos, and P. Raghavan. Automatic subspace clustering of high dimensional data for data mining applications. In *International Conference on Management of Data*, pages 94–105, 1998.

[2] Eric Bae and James Bailey. COALA: A novel approach for the extraction of an alternate clustering of high quality and high dissimilarity. In *Proceedings of the International Conference on Data Mining (ICDM)*, pages 53–62, 2006.

[3] Eric Bae, James Bailey, and Guozhu Dong. A clustering comparison measure using density profiles and its application to the discovery of alternate clusterings. *Data Mining and Knowledge Discovery*, 21(3):427–471, 2010.

[4] Steffen Bickel and Tobias Scheffer. Multi-view clustering. In *Proceedings of the 4th IEEE International Conference on Data Mining (ICDM 2004)*, pages 19–26, 2004.

[5] R. Caruana, M. Elhawary, N. Nguyen, and C. Smith. Meta clustering. In *Proceedings of the International Conference on Data Mining (ICDM)*, pages 107–118, 2006.

[6] Kamalika Chaudhuri, Sham M. Kakade, Karen Livescu, and Karthik Sridharan. Multi-view clustering via canonical correlation analysis. In *Proceedings of the 26th Annual International Conference on Machine Learning (ICML)*, page 17, 2009.

[7] Ying Cui, Xiaoli Fern, and Jennifer Dy. Non-redundant multi-view clustering via orthogonalization. In *International Conference on Data Mining*, pages 133–142, 2007.

[8] Xuan Hong Dang and James Bailey. Generation of alternative clusterings using the CAMI approach. In *Proceedings of the SIAM International Conference on Data Mining, SDM 2010*, April 29–May 1, 2010, Columbus, Ohio, pages 118–129, 2010.

[9] Xuan Hong Dang and James Bailey. A hierarchical information theoretic technique for the discovery of non linear alternative clusterings. In *Proceedings of the 16th ACM SIGKDD International Conference on Knowledge Discovery and Data Mining*, Washington, DC, USA, July 25-28, 2010, pages 573–582, 2010.

[10] Sajib Dasgupta and Vincent Ng. Mining clustering dimensions. In *Proceedings of the 27th International Conference on Machine Learning (ICML-10)*, June 21-24, 2010, Haifa, Israel, pages 263–270, 2010.

[11] Ian Davidson and Zijie Qi. Finding alternative clusterings using constraints. In *Proceedings of the 8th IEEE International Conference on Data Mining (ICDM 2008)*, December 15-19, 2008, Pisa, Italy, pages 773–778, 2008.

[12] David L. Davies and Donald W. Bouldin. A cluster separation measure. *IEEE Transactions on Pattern Analysis and Machine Intelligence*, 1:224–227, 1979.

[13] J. Dunn. A fuzzy relative of the isodata process and its use in detecting compact well-separated clusters. *Journal of Cybernetics*, 3(3):32–57, 1974.

[14] D. Gondek. Non-redundant clustering with conditional ensembles. In *Proceedings of the International Conference on Knowledge Discovery and Data Mining (KDD)*, pages 70–77, 2005.

[15] D. Gondek and T. Hofmann. Conditional information bottleneck clustering. In *3rd International Conference on Data Mining, Workshop on Clustering Large Data Sets*, pages 36–42, 2003.

[16] D. Gondek and T. Hofmann. Non-redundant data clustering. In *Proceedings of International Conference on Data Mining (ICDM)*, pages 75–82, 2004.

[17] Derek Greene and Padraig Cunningham. A matrix factorization approach for integrating multiple data views. In *Machine Learning and Knowledge Discovery in Databases, European Conference, ECML PKDD*, pages 423–438, 2009.

[18] L. Hamers, Y. Hemeryck, G. Herweyers, M. Janssen, H. Keters, R. Rousseau, and A. Vanhoutte. Similarity measures in scientometric research: The Jaccard index versus Salton's cosine formula. *Information Processing and Management*, 25(3):315–318, 1989.

[19] A. Hinneburg and D. Keim. An efficient approach to clustering in large multimedia databases with noise. In *International Conference on Knowledge Discovery and Data Mining*, pages 58–65, 1998.

[20] Ming Hua and Jian Pei. Clustering in applications with multiple data sources—A mutual subspace clustering approach. *Neurocomputing*, 92:133–144, 2012.

[21] L Hubert and Phipps Arabie. Comparing partitions. *Journal of Classification*, 2(1):193–218, 1985.

[22] Prateek Jain, Raghu Meka, and Inderjit S. Dhillon. Simultaneous unsupervised learning of disparate clusterings. *Statistical Analysis and Data Mining*, 1(3):195–210, 2008.

[23] Bo Long, Philip S. Yu, and Zhongfei (Mark) Zhang. A general model for multiple view unsupervised learning. In *Proceedings of the SIAM International Conference on Data Mining, SDM*, pages 822–833, 2008.

[24] M. Meila. Comparing clusterings—An information based distance. *Journal of Multivariate Analysis*, 98(5):873–895, 2007.

[25] Emmanuel Muller, Stephan Gunnemann, Thomas Seidl, and Ines Farber. *Tutorial: Discovering Multiple Clustering Solutions Grouping Objects in Different Views of the Data.* ICDM 2010, SDM2011, ICDE 2012.

[26] H. Nagesh, S. Goil, and A. Choudhary. Mafia: Efficient and scalable subspace clustering for very large data sets. Technical report 9906-010, Northwestern University, 1999.

[27] Xuan Vinh Nguyen and Julien Epps. minCEntropy: A novel information theoretic approach for the generation of alternative clusterings. In *ICDM 2010, The 10th IEEE International Conference on Data Mining*, Sydney, Australia, 14-17 December 2010, pages 521–530, 2010.

[28] Donglin Niu, Jennifer G. Dy, and Michael I. Jordan. Multiple non-redundant spectral clustering views. In *Proceedings of the 27th International Conference on Machine Learning (ICML-10)*, June 21-24, 2010, Haifa, Israel, pages 831–838, 2010.

[29] *Proceedings of the 1st International Workshop on Discovering, Summarizing and Using Multiple Clusterings (MultiClust).* Held in conjunction with the *16th ACM SIGKDD International Conference on Knowledge Discovery and Data Mining (KDD).* July 25, 2010.

[30] *Proceedings of the 2nd International Workshop on Discovering, Summarizing and Using Multiple Clusterings (MultiClust).* Held in conjunction with the *European Conference on Machine Learning and Principles and Practice of Knowledge Discovery in Databases (ECML PKDD).* 5 September, 2011.

[31] *Proceedings of the 3rd International Workshop on Discovering, Summarizing and Using Multiple Clusterings (MultiClust).* Held in conjunction with the 2012 *SIAM International Conference on Data Mining (SDM).* April 26, 2012.

[32] Zijie Qi and Ian Davidson. A principled and flexible framework for finding alternative clusterings. In *Proceedings of the 15th ACM SIGKDD International Conference on Knowledge Discovery and Data Mining*, Paris, France, June 28–July 1, 2009, pages 717–726, 2009.

[33] W. Rand. Objective criteria for the evaluation of clustering methods. *Journal of the American Statistical Association*, 66(336):846–850, 1971.

[34] P. J. Rousseeuw. Silhouettes: A graphical aid to the interpretation and validation of cluster analysis. *Journal of Computational and Applied Mathematics*, 20:53–65, 1987.

[35] Grigorios Tzortzis and C. L. Likas. Multiple view clustering using a weighted combination of exemplar-based mixture models. *IEEE Transactions on Neural Networks*, 21(12):1925–1938, 2010.

[36] Nguyen Xuan Vinh, Julien Epps, and James Bailey. Information theoretic measures for clusterings comparison: Variants, properties, normalization and correction for chance. *Journal of Machine Learning Research*, 9999:2837–2854, 2010.

[37] *Workshop on Multi-view Data, High-dimensionality, External Knowledge: Striving for a Unified Approach to Clustering*. Held in conjunction with the 16th *Pacific-Asia Conference on Knowledge Discovery and Data Mining (PAKDD 2012)*. May 29, 2012.

[38] Dengyong Zhou and Christopher J. C. Burges. Spectral clustering and transductive learning with multiple views. In *Machine Learning, Proceedings of the Twenty-Fourth International Conference (ICML 2007)*, pages 1159–1166, 2007.

Chapter 22

Cluster Ensembles: Theory and Applications

Joydeep Ghosh
University of Texas at Austin
Austin, TX
ghosh@ece.utexas.edu

Ayan Acharya
University of Texas at Austin
Austin, TX
aacharya@utexas.edu

22.1 Introduction

The design of multiple classifier systems to solve difficult classification problems, using techniques such as bagging, boosting, and output combining [54, 62, 38, 36], has resulted in some of the most notable advances in classifier design over the past two decades. A popular approach is to train multiple "base" classifiers, whose outputs are combined to form a classifier ensemble. A survey of such ensemble techniques—including applications of them to many difficult real-world problems such as remote sensing, person recognition, one vs. all recognition, and medicine — can be found in [51]. Concurrently, analytical frameworks have been developed that quantify the improvements in classification results due to combining multiple models [61]. The extensive literature on the subject has shown that from independent, diversified classifiers, the ensemble created is usually more accurate as well as more reliable than its individual components, i.e., the base classifiers.

The demonstrated success of classifier ensembles provides a direct motivation to study effective ways of combining multiple clustering solutions as well. This chapter covers the theory, design, and application of *cluster ensembles*, which address the problem of combining multiple *"base clusterings"* of the same set of objects into a single consolidated clustering. Each base clustering refers to a *grouping* of the same set of objects or its transformed (or perturbed) version using a suitable clustering algorithm. The consolidated clustering is often referred to as the *consensus* solution. At first glance, this problem sounds similar to the problem of designing classifier ensembles. However, combining multiple clusterings poses additional challenges. First, the number of clusters produced may differ across the different *base* solutions [6]. The appropriate number of clusters in the consensus is also not known in advance and may depend on the scale at which the data is inspected. Moreover, cluster labels are symbolic and thus aligning cluster labels across different solutions requires solving a potentially difficult correspondence problem. Also, in the typical formulation,[1] the original data used to yield the base solutions are not available to the consensus mechanism, which has only access to the sets of cluster labels. In some schemes, one does have control on how the base clusterings are produced [21], while in others even this is not granted in order to allow applications involving knowledge reuse [56], as described later. Despite these added complications, cluster ensembles are inviting since typically the variations in quality across a variety of clustering algorithms applied to a specific dataset tends to be more than the typical variation in accuracies returned by a collection of reasonable classifiers. This suggests that cluster ensembles may achieve greater improvements over the base solutions, when compared with ensembles of classifiers [24].

In fact, the potential motivations for using cluster ensembles are much broader than those for using classification or regression ensembles, where one is primarily interested in improving predictive accuracy. These reasons include the following:

1. **Improved Quality of Solution**

   Just as ensemble learning has been proved to be more useful compared to single-model solutions for classification and regression problems, one may expect that cluster ensembles will improve the quality of results as compared to a single clustering solution. It has been shown that using cluster ensembles leads to more accurate results on average as the ensemble approach takes into account the biases of individual solutions [39, 31].

2. **Robust Clustering**

   It is well known that the popular clustering algorithms often fail spectacularly for certain datasets that do not match well with the modeling assumptions [33]. A cluster ensemble approach can provide a "meta" clustering model that is much more robust in the sense of being able to provide good results across a very wide range of datasets. As an example, by using an ensemble that includes approaches such as k-means, SOM, and DBSCAN that are typically better suited to low-dimensional metric spaces, as well as base clusterers designed for high-dimensional sparse spaces (spherical k-means, Jaccard-based graph clustering, etc.), one can perform well across a wide range of data dimensionality [56]. Authors in [53] present several empirical results on the robustness of the results in document clustering by using feature diversity and consensus clustering.

[1] In this chapter, we shall not consider approaches where the feature values of the original data or of the cluster representatives are available to the consensus mechanism, e.g. [30].

3. **Model Selection**

   Cluster ensembles provide a novel approach to the model selection problem by considering the match across the base solutions to determine the final number of clusters to be obtained [25].

4. **Knowledge Reuse**

   In certain applications, domain knowledge in the form of a variety of clusterings of the objects under consideration may already exist due to past projects. A consensus solution can integrate such information to get a more consolidated clustering. Several examples are provided in [56], where such scenarios formed the main motivation for developing a consensus clustering methodology. As another example, a categorization of web pages based on text analysis can be enhanced by using the knowledge of topical document hierarchies available from Yahoo! or DMOZ.

5. **Multiview Clustering**

   Often the objects to be clustered have multiple aspects or "views," and base clusterings may be built on distinct views that involve nonidentical sets of features or subsets of data points. In marketing applications for example, customers may be segmented based on their needs, psychographic or demographic profiles, attitudes, etc. Different views can also be obtained by considering qualitatively different distance measures, an aspect that was exploited in clustering multifaceted proteins to multiple functional groups in [5]. Consensus clustering can be effectively used to combine all such clusterings into a single consolidated partition. Strehl and Ghosh [56] illustrate empirically the utility of cluster ensembles in two orthogonal scenarios:

   - Feature Distributed Clustering (FDC): different base clusterings are built by selecting different subsets of the features but utilizing all the data points.
   - Object Distributed Clustering (ODC): base clusterings are constructed by selecting different subsets of the data points but utilizing all the features.

   Fern and Brodley [17] also show that clustering in high dimensions is much more effective compared to clustering with PCA when the data points are randomly projected onto a subspace, clustered in that subspace, and consensus clustering is performed with this ensemble.

6. **Distributed Computing**

   In certain situations, data is inherently distributed and it is not possible to first collect the entire data at a central site due to privacy/ownership issues or computational, bandwidth, and storage costs [46]. An ensemble can be used in situations where each clusterer has access to only a subset of the features of each object, as well as where each clusterer has access to only a subset of the objects [25, 56].

The problem of combining multiple clusterings can be viewed as a special case of the more general problem of comparison and consensus of data "classifications," studied in the pattern recognition and related application communities in the 70s and 80s. In this literature, *classification* was used in a broad sense to include clusterings, unrooted trees, graphs, etc, and problem-specific formulations were made (see [47] for a broad, more conceptual coverage). For example, in the building of phylogenetic trees, it is important to get a strict consensus solution, wherein two objects occur in the same consensus partition if and only if they occur together in all individual clusterings [13], typically resulting in a consensus solution at a much coarser resolution than the individual solutions. A quick overview with pointers to such literature is given by Ayad and Kamel [6]. Moreover, based on the success of classifier and cluster ensembles, efforts have been made to combine both types of ensembles [1, 23]. Unsupervised models can provide a variety of supplementary constraints which can be useful for improving the generalization capability of the resulting classifier, especially when labeled data is scarce. Also, they might be useful for designing learning methods that are aware of the possible differences between training and target distributions, thus being particularly interesting for applications in which concept drift might take place [2, 22]. A reasonable coverage of these problems is not feasible here, instead this article focuses on the cluster ensemble formulations and associated algorithms that have been proposed in the past decade.

This chapter is organized as follows. In Section 22.2, we formulate the cluster ensemble problem. In Section 22.3, different measures for comparing a pair of clustering solutions are introduced. Details of different cluster ensembles algorithms are presented in Section 22.4 followed by the applications of cluster ensembles in Section 22.5.

22.2 The Cluster Ensemble Problem

We denote a vector by a bold faced letter and a scalar variable or a set in normal font. We start by considering r base clusterings of a data set $\mathcal{X} = \{\mathbf{x}_i\}_{i=1}^n$ with the qth clustering containing $k^{(q)}$ clusters. The most straightforward representation of the qth clustering is $\lambda^{(q)} = \{C_\ell | \ell = 1,2,...,k^{(q)} \text{ and } C_\ell \subseteq \mathcal{X}\}$. Here, each clustering is denoted by a collection of subsets (not necessarily disjoint) of the original dataset. For hard partitional clustering (clustering where each object is assigned to a single cluster only), the qth clustering can alternatively be represented by a label vector $\lambda^{(q)} \in \mathbb{Z}_+^n$. In this representation, each object is assigned some cluster label, and 0 is used if the corresponding object is not available to that clusterer. The third possible way of representation of an individual clustering is by the binary membership indicator matrix $\mathbf{H}^q \in \{0,1\}^{1\times k^{(q)}}$ which is defined as $\mathbf{H}^q = \{h_{i\ell}^q | h_{i\ell}^q \in \{0,1\} \ \forall \mathbf{x}_i, C_\ell, \lambda^{(q)}\}$. For partitional clustering, we additionally have $\sum_{\ell=1}^{k^{(q)}} h_{i\ell}^q = 1 \ \forall \mathbf{x}_i \in \mathcal{X}$.

A *consensus function* Γ is defined as a function $\mathbb{Z}_+^{n\times r} \to \mathbb{Z}_{++}^n$ mapping a set of clusterings to an integrated clustering $\Gamma : \lambda^{(q)} | q \in \{1,2,.....,r\} \to \hat{\lambda}$. For conciseness, we shall denote the set of clusterings $\{\lambda^{(q)}\}_{q=1}^r$ that is available to the consensus mechanism by Λ. Moreover, the results of any hard clustering[2] of n objects can be represented as a binary, symmetric $n \times n$ *co-association matrix*, with an entry being 1 if the corresponding objects are in the same cluster and 0 otherwise. For the qth base clustering, this matrix is denoted by $S^{(q)}$ and is given by

$$S_{ij}^{(q)} = \begin{cases} 1 & (i,j) \in C_\ell(\lambda^{(q)}) \text{ for some } \ell \in \{1,2,...,k^{(q)}\} \\ 0 & \text{otherwise} \end{cases} \tag{22.1}$$

Broadly speaking, there are two main approaches to obtaining a consensus solution and determining its quality. One can postulate a probability model that determines the labeling of the individual solutions, given the true consensus labels, and then solve a maximum likelihood formulation to return the consensus [60, 64]. Alternately, one can directly seek a consensus clustering that agrees the most with the original clusterings. The second approach requires a way of measuring the similarity between two clusterings, for example, to evaluate how close the consensus solution is to each base solution. These measuring indices will be discussed in more details in Section 22.3. For now, let $\phi(\lambda^{(a)}, \lambda^{(b)})$ represent a similarity index between two clustering solutions $\lambda^{(a)}$ and $\lambda^{(b)}$. One can express the average normalized similarity measure between a set of r labelings, Λ, and a single

[2]This definition is also valid for overlapping clustering.

consensus labeling $\hat{\lambda}$, by

$$\phi(\Lambda,\hat{\lambda}) = \frac{1}{r}\sum_{q=1}^{r}\phi(\lambda^{(q)},\hat{\lambda}) \tag{22.2}$$

This serves as the objective function in certain cluster ensemble formulations, where the goal is to find the combined clustering $\hat{\lambda}$ with $\hat{k}$ clusters such that $\phi(\Lambda,\hat{\lambda})$ is maximized. It turns out though that this objective is intractable, so heuristic approaches have to be resorted to.

22.3 Measuring Similarity Between Clustering Solutions

Since no ground truth is available for clustering problems, cluster ensemble algorithms instead aim to maximize some similarity measure between the consensus clustering and each of the base clustering solutions. Two of the most desirable properties of such similarity measures are

- The index should be normalized for easy interpretation and comparison across solutions with varying number of clusters.
- The expected value of the index between pairs of independent clusterings should be constant (and preferably zero indicating no similarity). The utility of such a property will be clarified later in this section.

Below, we present different indices and show how these indices are modified to achieve zero expected value of the same. The general rule for any "adjustment for chance" is as follows:

$$\text{adjusted index} = \frac{\text{index} - \text{Expected-index}}{\text{Max-index} - \text{Expected-index}} \tag{22.3}$$

where Expected-index is calculated according to a permutation model [40]. In such a model, cluster labels are assumed to be generated randomly subject to the constraints of a fixed number of clusters and a fixed number of points in each cluster. Max-index is the maximum value the index can take.

Now let us discuss two key approaches for measuring the similarity of two clustering solutions [50]. The first approach is based on counting the number of pairs in agreement or in disagreement in two clustering solutions. The second approach uses concepts from information theory.

1. **Pair Counting-Based Measures:** Measures of this type are built on counting the number of pairs of points on which two candidate clustering solutions agree or disagree. More formally, suppose we have two candidate clusterings $\lambda^{(a)} = \{C_h^{(a)}|h = 1,2,...,k^{(a)}\}$ and $\lambda^{(b)} = \{C_\ell^{(b)}|\ell = 1,2,...,k^{(b)}\}$. Let $n_h^{(a)}$ be the number of objects in cluster $C_h^{(a)}$ and $n_l^{(b)}$ be the number of objects in cluster $C_l^{(b)}$. Table 22.1 is a contingency table that shows the overlap between different clusters of these clusterings, where $n_{h\ell} = |C_h^{(a)} \cap C_\ell^{(b)}|$. The most well-known index of this class is the Rand Index (**RI**) which is a normalized measure of number of agreements between two candidate solutions and is defined as

$$\phi^{(RI)}(\lambda^{(a)},\lambda^{(b)}) = \sum_{h\ell}\binom{n_{h\ell}^{(a)}}{2} / \frac{1}{2}(S_a + S_b) \tag{22.4}$$

Expected value of **RI**, however, is not constant. This led Hubert and Arabie [32] to propose

TABLE 22.1: Contingency Table Explaining Similarity Measurement of Clustering Solutions

	$C_1^{(b)}$	$C_2^{(b)}$	$\cdots$	$C_{k^{(b)}}^{(b)}$	sum
$C_1^{(a)}$	n_{11}	n_{12}	$\cdots$	$n_{1k^{(b)}}$	$n_1^{(a)}$
$C_2^{(a)}$	n_{21}	n_{22}	$\cdots$	$n_{2k^{(b)}}$	$n_2^{(a)}$
$\cdots$	$\cdots$	$\cdots$	$\cdots$	$\cdots$	$\cdots$
$\cdots$	$\cdots$	$\cdots$	$\cdots$	$\cdots$	$\cdots$
$C_{k^{(a)}}^{(a)}$	$n_{k^{(a)}1}$	$n_{k^{(a)}2}$	$\cdots$	$n_{k^{(a)}k^{(b)}}$	$n_{k^{(a)}}^{(a)}$
sum	$n_1^{(b)}$	$n_2^{(b)}$	$\cdots$	$n_{k^{(b)}}^{(b)}$	n

the Adjusted Rand Index (**ARI**) to correct for a zero baseline. **ARI**, according to the general strategy of correcting for chance, is defined as follows:

$$\phi^{(ARI)}(\lambda^{(a)},\lambda^{(b)}) = \frac{\sum_{h\ell}\binom{n_{h\ell}}{2} - S_aS_b/\binom{n}{2}}{\frac{1}{2}(S_a+S_b) - S_aS_b/\binom{n}{2}} \tag{22.5}$$

where $S_a = \sum_h \binom{n_h^{(a)}}{2}$ and $S_b = \sum_\ell \binom{n_\ell^{(b)}}{2}$. The second term in both numerator and denominator adjusts for the expected number of overlaps that will occur "by chance." Values of **RI** lie between 0 and 1. However, values of **ARI** can be negative which is of no practical interest. **ARI** is 1 when the two candidate clustering solutions match exactly and is 0 when the index value equals its expected value.

In [4], the authors mention as many as 22 different indices of this class. Subsequent work [66] shows that after correction for chance, some of these indices become equivalent. However, **ARI** remains the most popular index of this class. Very recently, a probabilistic version of Rand Index (**PRI**) [11] has been proposed in which the agreements and disagreements of the pairs of data points are weighted according to their occurrence by chance.

2. **Information Theoretic-Based Measures:** There are numerous information theory-based measures in the literature [50]. We discuss only two of them here and show how they can be corrected for occurrence by chance.

   (a) **Normalized Mutual Information (NMI)**
   Strehl and Ghosh [56] propose **NMI** to measure the similarity between two candidate clusterings. The entropy associated with clustering $\lambda^{(a)}$ is $H(\lambda^{(a)}) = -\sum_h \frac{n_h^{(a)}}{n} log(\frac{n_h^{(a)}}{n})$ and that with clustering $\lambda^{(b)}$ is $H(\lambda^{(b)}) = -\sum_\ell \frac{n_\ell^{(b)}}{n} log(\frac{n_\ell^{(b)}}{n})$. Similarly, the joint entropy of $\lambda^{(a)}$ and $\lambda^{(b)}$ is defined as $H(\lambda^{(a)},\lambda^{(b)}) = -\sum_{h,\ell} \frac{n_{h\ell}}{n} log(\frac{n_{h\ell}}{n})$. Now, the **NMI** between $\lambda^{(a)}$and $\lambda^{(b)}$ is defined as

$$\phi^{(NMI)}(\lambda^{(a)},\lambda^{(b)}) = \frac{I(\lambda^{(a)},\lambda^{(b)})}{\sqrt{H(\lambda^{(a)})H(\lambda^{(b)})}} \tag{22.6}$$

   Here, $I(\lambda^{(a)},\lambda^{(b)}) = H(\lambda^{(a)}) + H(\lambda^{(b)}) - H(\lambda^{(a)},\lambda^{(b)})$ is the mutual information between two clusterings $\lambda^{(a)}$ and $\lambda^{(b)}$ [7], which is normalized by the geometric mean of $H(\lambda^{(a)})$ and $H(\lambda^{(b)})$ to compute the **NMI**. It should be noted that $I(\lambda^{(a)},\lambda^{(b)})$ is non-negative and has no upper bound. $\phi^{(NMI)}(\lambda^{(a)},\lambda^{(b)})$, on the other hand, lies between 0 and 1 and is suitable for easier interpretation and comparisons.

(b) **Variation of Information (VI)**

VI is another information theoretic *distance* measure proposed for cluster validation [44, 45] and is defined as follows:

$$\phi^{(VI)}(\lambda^{(a)},\lambda^{(b)}) = H(\lambda^{(a)}) + H(\lambda^{(b)}) - 2I(\lambda^{(a)},\lambda^{(b)}) \qquad (22.7)$$

It turns out that **VI** is a metric. But its original definition is not consistent if data sets of different sizes and clusterings with different number of clusters are considered. Therefore, several normalized versions of **VI** have been proposed. The one proposed by [37] takes the following form:

$$\phi^{(\mathrm{NVI}_1)}(\lambda^{(a)},\lambda^{(b)}) = 1 - \frac{I(\lambda^{(a)},\lambda^{(b)})}{\max\{H(\lambda^{(a)}),H(\lambda^{(b)})\}} \qquad (22.8)$$

which again is a metric. However, the one proposed by Wu et. al. [68] is not a metric and takes the following form:

$$\phi^{(\mathrm{NVI}_2)}(\lambda^{(a)},\lambda^{(b)}) = 1 - \frac{2I(\lambda^{(a)},\lambda^{(b)})}{H(\lambda^{(a)}) + H(\lambda^{(b)})} \qquad (22.9)$$

Both of these measures lie between 0 and 1. Note that one could also define a distance measure from **NMI** just by taking its 1-complement. However, such distance measure is not a metric.

Vinh et. al. [50] empirically show that like **RI**, the information theoretic-based indices do not attain any constant baseline. The problem is more severe when the number of clusters is comparable to the number of data points. With increase in the number of clusters, the expected values of the indices either increase (for **NMI**) or decrease (for **VI**). Therefore, if one needs to make a decision based on the observed values of the indices, on average, a clustering solution with more (or fewer for **VI**) clusters will unjustifiably be preferred.

Therefore, following the generalized suggestion for correction by chance given in Equation 22.3, Vinh et. al. [50] proposed to correct these measures with the expected value of the mutual information of the solutions and empirically showed the invariance of the modified measures w.r.t the number of clusters. The expression for the expected value of the mutual information under the permutation model is given by

$$\begin{aligned}\mathbb{E}[I(S_a,S_b)] \;=\; & \sum_{h\ell} \sum_{n_{h\ell}=\max\{n_h^{(a)}+n_\ell^{(b)}-n,0\}}^{\min\{n_h^{(a)}+n_\ell^{(b)}\}} \frac{n_{h\ell}}{n}\log\left(\frac{n_{h\ell}n}{n_h^{(a)}n_\ell^{(b)}}\right) \\ & \cdot \frac{n_h^{(a)}!\,n_\ell^{(b)}!\,(n-n_h^{(a)})!\,(n-n_\ell^{(b)})!}{n_{h\ell}!\,n!\,(n_h^{(a)}-n_{h\ell})!\,(n_\ell^{(b)}-n_{h\ell})!\,(n-n_h^{(a)}-n_\ell^{(b)}+n_{h\ell})!}\end{aligned} \qquad (22.10)$$

However, if the ratio of the number of data points and the number of clusters is more than 100 for both solutions, empirical studies show that $\mathbb{E}[I(S_a,S_b)]$ is fairly close to zero, and hence, no adjustment for chance is necessary. Unfortunately, with such adjustment for chance, none of the distances measures is a metric anymore. Therefore, depending on the applications, one needs to make a choice between the metric property and the zero baseline property.

22.4 Cluster Ensemble Algorithms

Cluster ensemble methods are now presented under three categories: (i) probabilistic approaches, (ii) approaches based on co-association, and (iii) direct and other heuristic methods.

22.4.1 Probabilistic Approaches to Cluster Ensembles

The two basic probabilistic models for solving cluster ensembles are described in this subsection.

22.4.1.1 A Mixture Model for Cluster Ensembles (MMCE)

In a typical mixture model [10] approach to clustering, such as fitting the data using a mixture of Gaussians, there are $\hat{k}$ mixture components, one for each cluster. A component-specific parametric distribution is used to model the distribution of data attributed to a specific component. Such an approach can be applied to form the consensus decision if the number of consensus clusters is specified. This immediately yields the pioneering approach taken in [60]. We describe it in a bit more detail as this work is essential to build an understanding of later works [64, 65].

In the basic mixture model of cluster ensembles [60], each object $\mathbf{x}_i$ is represented by $\mathbf{y}_i = \Lambda(\mathbf{x}_i)$, i.e., the labels provided by the base clusterings. We assume that there are $\hat{k}$ consensus clusters each of which is indexed by $\hat{\ell}$. Corresponding to each consensus cluster $\hat{\ell}$ and each base clustering q, we have a multinomial distribution $\beta_{\hat{\ell}}^{(q)}$ of dimension $k^{(q)}$. Therefore, a sample from this distribution is a cluster label corresponding to the qth base clustering. The underlying generative process is assumed as follows:

For ith data point $\mathbf{x}_i$,

1. Choose $\mathbf{z}_i = \mathbf{I}_{\hat{\ell}}$ such that $\hat{\ell} \sim$ multinomial(θ). Here $\mathbf{I}_{\hat{\ell}}$ is a probability vector of dimension $k^{(q)}$ with only the $\hat{\ell}^{\text{th}}$ component being 1, and θ is a multinomial distribution of dimension $\hat{k}$.
2. For the qth base clustering of the ith data point, choose the base clustering result $y_{iq} = \ell \sim$ multinomial$(\beta_{\hat{\ell}}^{(q)})$.

These probabilistic assumptions give rise to a simple maximum log-likelihood problem that can be solved using the Expectation Maximization (EM) algorithm. This model also takes care of the missing labels in a natural way.

22.4.1.2 Bayesian Cluster Ensembles (BCE)

A Bayesian version of the multinomial mixture model described above was subsequently proposed by Wang et al. [64]. As in the simple mixture model, we assume $\hat{k}$ consensus clusters with $\beta_{\hat{\ell}}^{(q)}$ being the multinomial distribution corresponding to each consensus cluster $\hat{\ell}$ and each base clustering q. The complete generative process for this model is as follows:

For ith data point $\mathbf{x}_i$,

1. Choose $\theta_i \sim$ Dirichlet(α) where θ_i is a multinomial distribution with dimension $\hat{k}$.
2. For the qth base clustering:
   (a) Choose $\mathbf{z}_{iq} = \mathbf{I}_{\hat{\ell}}$ such that $\hat{\ell} \sim$ multinomial(θ_i). $\mathbf{I}_{\hat{\ell}}$ is a probability vector of dimension $\hat{k}$ with only the $\hat{\ell}^{\text{th}}$ component being 1.
   (b) Choose the base clustering result $y_{iq} = \ell \sim$ multinomial$(\beta_{\hat{\ell}}^{(q)})$.

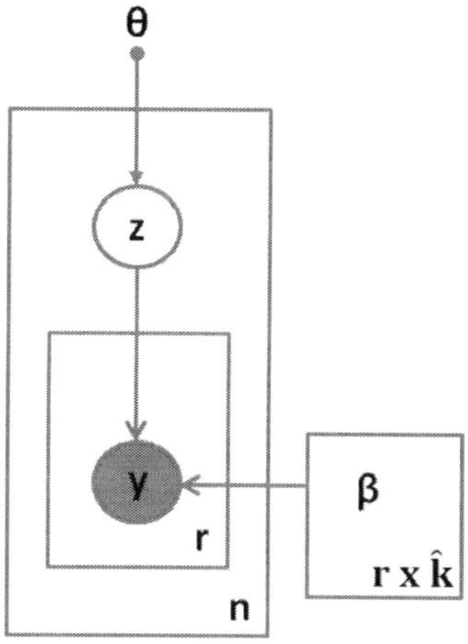

FIGURE 22.1: Graphical model for **MMCE**.

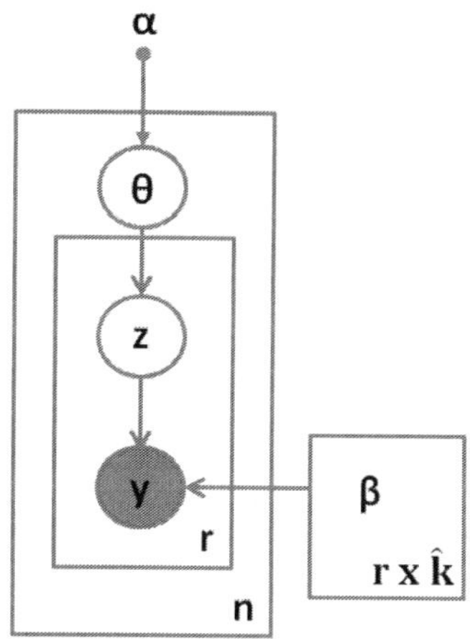

FIGURE 22.2: Graphical model for **BCE**.

So, given the model parameters $(\alpha, \beta = \{\beta_{\hat{\ell}}^{(q)}\})$, the joint distribution of latent and observed variables $\{\mathbf{y}_i, \mathbf{z}_i, \theta_i\}$ is given by

$$p(\mathbf{y}_i, \mathbf{z}_i, \theta_i | \alpha, \beta) = p(\theta_i | \alpha) \prod_{q=1, \exists y_{iq}}^{r} p(\mathbf{z}_{iq} = \mathbf{I}_{\hat{\ell}} | \theta_i) p(y_{iq} | \beta_{\hat{\ell}}^{(q)}) \tag{22.11}$$

where $\exists y_{iq}$ implies that there exists a qth base clustering result for $\mathbf{y}_i$. The marginals $p(\mathbf{y}_i|\alpha,\beta)$ can further be calculated by integrating over the hidden variables $\{\mathbf{z}_i, \theta_i\}$. The authors used variational EM and Gibb's sampling for inference and parameter estimation. The graphical model corresponding to this Bayesian version is given in Figure 22.2. To highlight the difference between Bayesian cluster ensembles and the mixture model for cluster ensembles, the graphical model corresponding to the latter is also shown alongside in Figure 22.1.

22.4.1.3 Nonparametric Bayesian Cluster Ensembles (NPBCE)

Recently, a nonparametric version of Bayesian cluster ensemble (**NPBCE**) has been proposed in [65] which allows the number of consensus clusters to adapt with data. The stick-breaking construction of the generative process of this model is described below. The authors use a truncated stick-breaking construction of the Dirichlet process with truncation enforced at $\hat{k}$. If $\hat{k}$ is made sufficiently large, the resulting Dirichlet Process (truncated) closely approximates a Dirichlet Process.

1. Generate $v_{\hat{\ell}} \sim \beta(1,\alpha)\ \forall \hat{\ell} \in \{1,2,\cdots,\hat{k}\}$. Let $\theta_{\hat{\ell}} = v_{\hat{\ell}} \prod_{j=1}^{\hat{\ell}-1}(1-v_j)$.

2. For each base clustering (indexed by q), generate $\beta_{\hat{\ell}}^{(q)} \sim G_0^{(q)}\ \forall \hat{\ell} \in \{1,2,\cdots,\hat{k}\}$ where $G_0^{(q)}$ is a symmetric Dirichlet distribution of dimension $k^{(q)}$.

3. For the ith data point $\mathbf{x}_i$, generate $\mathbf{z}_i \sim$ multinomial(θ). $\mathbf{z}_i$ is an indicator vector of dimension $\hat{k}$ with only one component being unity and others being zero.

4. For the qth base clustering of the ith data point, generate the base clustering result $y_{iq} = \ell \sim$ multinomial$(\beta_{\hat{\ell}}^{(q)})$.

One should note that **NPBCE** does not allow multiple base clustering solutions of a given data point to be generated from more than one consensus cluster. Therefore, the model is more restrictive

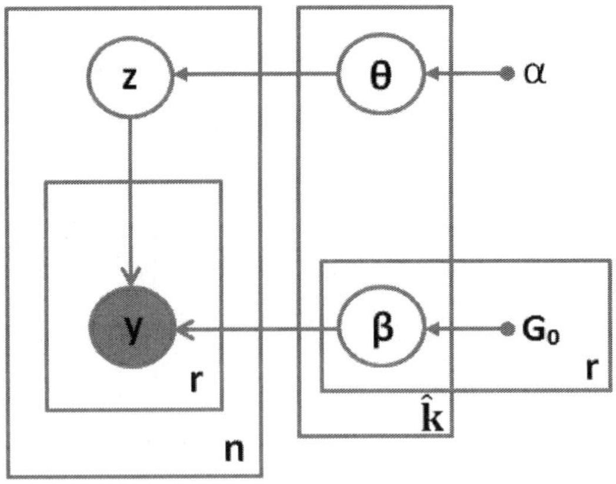

FIGURE 22.3: Graphical model for **NPBCE**.

compared to **BCE** and is really a nonparametric version of **MMCE**. The graphical model shown in Figure 22.3 illustrates this difference more clearly. It should be noted that although all of the generative models presented above were used only with hard partitional clustering, they could be used for overlapping clustering as well.

22.4.2 Pairwise Similarity-Based Approaches

In pairwise similarity-based approaches, one takes the weighted average of all r co-association matrices to form an *ensemble co-association matrix* S which is given as follows:

$$S = \frac{1}{r}\sum_{q=1}^{r} w_q S^{(q)} \tag{22.12}$$

Here w_q specifies the weight assigned to the qth base clustering. This ensemble co-association matrix captures the fraction of times a pair of data points is placed in the same cluster across the r base clusterings. The matrix can now be viewed as a similarity matrix (with a corresponding similarity graph) to be used by the consensus mechanism for creating the consensus clusters. This matrix is different from the similarity matrix $\hat{S}$ that we obtain from the consensus solution $\hat{\lambda}$. We will explain the difference in detail in Section 22.4.2.1.

Note that the co-association matrix size is itself quadratic in n, which thus forms a lower bound on computational complexity as well as memory requirements, inherently handicapping such a technique for applications to very large datasets. However, it is independent of the dimensionality of the data.

22.4.2.1 Methods Based on Ensemble Co-Association Matrix

The Cluster-based Similarity Partitioning Algorithm (CSPA) [56] used METIS [34] to partition the induced consensus similarity graph. METIS was chosen for its scalability and because it tries to enforce comparable sized clusters. This added constraint is desirable in several application domains [57]; however, if the data is actually labeled with imbalanced classes, then it can lower the match between cluster and class labels. Assuming quasi-linear graph clustering, the worst case complexity for this algorithm is $O(n^2kr)$. Punera and Ghosh [52] later proposed a soft version of CSPA, i.e., one that works on soft base clusterings. Al-Razgan and Domeniconi [3] proposed an alternative way of obtaining nonbinary co-association matrices when given access to the raw data.

The Evidence Accumulation approach [21] obtains individual co-association matrices by random initializations of the k-means algorithm, causing some variation in the base cluster solutions. This algorithm is used with a much higher value of k than the range finally desired. The ensemble co-association matrix is then formed, each entry of which signifies the relative co-occurrence of two data points in the same cluster. A minimum spanning tree (MST) algorithm (also called the single-link or nearest neighbor hierarchical clustering algorithm) is then applied on the ensemble co-association matrix. This allows one to obtain nonconvex shaped clusters. Essentially, this approach assumes the designer has access to the raw data, and the consensus mechanism is used to get a more robust solution than what can be achieved by directly applying MST to the raw data.

A related approach is taken by Monti et al [48], where the perturbations in the base clustering are achieved by resampling. Any of bootstrapping, data subsampling, or feature subsampling can be used as a resampling scheme. If either of the first two options is selected, then it is possible that certain objects will be missing in a given base clustering. Hence, when collating the r base co-association matrices, the (i, j)th entry needs to be divided by the number of solutions that included both objects rather than by a fixed r. This work also incorporates a model selection procedure as follows: The consensus co-association matrix is formed multiple times. The number of clusters is kept at k_i for each base clustering during the ith experiment, but this number is changed from one experiment to another. A measurement termed as *consensus distribution* describes how the elements of a consensus matrix are distributed within the 0–1 range. The extent to which the consensus matrix is skewed toward a binary matrix denotes how good the base clusterings match one another. This enables one to choose the most appropriate number of consensus clusters $\hat{k}$. Once $\hat{k}$ is chosen, the corresponding ensemble co-association matrix is fed to a hierarchical clustering algorithm with average linkage. Agglomeration of clusters is stopped when $\hat{k}$ branches are left.

The Iterative Pairwise Consensus (IPC) Algorithm [49] essentially applies model-based k-means [72] to the ensemble co-association matrix S. The consensus clustering solution $\hat{\lambda} = \{C_\ell\}_{\ell=1}^{\hat{k}}$ is initialized to some solution, after which a reassignment of points is carried out based on the current configuration of $\hat{\lambda}$. The point $\mathbf{x}_i$ gets assigned to cluster C_ℓ, if $\mathbf{x}_i$ has maximum average similarity with the points belonging to cluster C_ℓ. Then the consensus solution is updated, and the cycle starts again.

However, both Mirkin [47] and Li et al. [43] show that the problem of consensus clustering can be framed in a different way than what has been discussed so far. In these works, the distance $d(\lambda^{(q_1)}, \lambda^{(q_2)})$ between two clusterings $\lambda^{(q_1)}$ and $\lambda^{(q_2)}$ is defined as the number of pairs of objects that are placed in the same cluster in one of $\lambda^{(q_1)}$ or $\lambda^{(q_2)}$ and in a different cluster in the other, essentially considering the (unadjusted) Rand Index. Using this definition, the consensus clustering problem is formulated as

$$\begin{aligned} \arg\min_{\hat{\lambda}} J &= \arg\min_{\hat{\lambda}} \frac{1}{r} \sum_{q=1}^{r} d(\lambda^{(q)}, \hat{\lambda}) \\ &= \arg\min_{\hat{S}} \frac{1}{r} \sum_{q=1}^{r} w_q \sum_{i<j} [S_{ij}^{(q)} - \hat{S}_{ij}]^2 \end{aligned} \tag{22.13}$$

Mirkin [47, section 5.3.4, p. 260] further proved that the consensus clustering according to criterion (22.13) is equivalent to clustering over the ensemble co-association matrix by subtracting a "soft" and "uniform" threshold from each of the different consensus clusters. This soft threshold, in fact, serves as a tool to balance cluster sizes in the final clustering. The subtracted threshold has also been used in [59] for consensus clustering of gene-expression data.

In [63], a consensus clustering result is obtained by minimizing a weighted sum of the Bregman divergence [8] between the consensus partition and the input partitions w.r.t their co-association matrices. In addition, the authors also show how to generalize their framework in order to incorporate must-link and cannot-link constraints between objects.

Note that the optimization problem in (22.13) is over the domain of $\hat{S}$. The difference between the matrices S and $\hat{S}$ lies in the way the optimization problem is posed. If optimization is performed with cluster labels only (as illustrated in Section 22.4.3), there is no guarantee of achieving the optimum value $\hat{S} = S$. However, if we are optimizing over the domain of the co-association matrix, we can achieve this optimum value in theory.

22.4.2.2 Relating Consensus Clustering to Other Optimization Formulations

The co-association representation of clustering has been used to relate consensus clustering with two other well-known problems.

1. **Consensus Clustering as Nonnegative Matrix Factorization (NNMF)**

   Li et al [43, 42], using the same objective function as mentioned in (22.13), show that the problem of consensus clustering can be reduced to an NNMF problem. Assuming $U_{ij} = \hat{S}_{ij}$ to be a solution to this optimization problem, we can rewrite (22.13) as:

   $$\arg\min_{U} \sum_{i,j=1}^{n} (S_{ij} - U_{ij})^2 = \arg\min_{U} \|S - U\|_F^2 \tag{22.14}$$

   where the matrix norm is the Frobenius norm. This problem formulation is similar to the NNMF formulation [41] and can be solved using an iterative update procedure. In [27], the cost function J used in Equation (22.13) was further modified via normalization to make it consistent with data sets with different numbers of data points (n) and different numbers of base clusterings (r).

2. **Consensus Clustering as Correlation Clustering**

   Gionis et al. [26] show that a certain formulation of consensus clustering is a special case of correlation clustering. Suppose we have a data set $\mathcal{X}$ and some kind of dissimilarity measurement (distance) between every pair of points in $\mathcal{X}$. This dissimilarity measure is denoted by $d_{ij} \in [0,1] \forall \mathbf{x}_i, \mathbf{x}_j \in \mathcal{X}$. The objective of correlation clustering [9] is to find a partition $\hat{\lambda}$ such that

   $$\hat{\lambda} = \arg\min_{\lambda} d(\lambda)$$
   $$= \arg\min_{\lambda} \left[\sum_{(i,j):\lambda(\mathbf{x}_i)=\lambda(\mathbf{x}_j)} d_{ij} + \sum_{(i,j):\lambda(\mathbf{x}_i)\neq\lambda(\mathbf{x}_j)} (1-d_{ij}) \right] \tag{22.15}$$

   In the above equation, $\lambda(\mathbf{x}_i)$ is the cluster label imposed by λ on $\mathbf{x}_i$. The co-association view of the cluster ensemble problem reduces to correlation clustering if the distance d_{ij} is defined as $d_{ij} = \frac{1}{r}|\{\lambda^{(q)} : \lambda^{(q)}(\mathbf{x}_i) \neq \lambda^{(q)}(\mathbf{x}_j)\}| \; \forall\, i,j$.

22.4.3 Direct Approaches Using Cluster Labels

Several consensus mechanisms take only the cluster labels provided by the base clusterings as input and try to optimize an objective function such as (22.2), without computing the co-association matrix.

22.4.3.1 Graph Partitioning

In addition to CSPA, Strehl and Ghosh [56] propose two direct approaches to cluster ensembles: Hyper Graph Partitioning Algorithm (HGPA) which clusters the objects based on their cluster memberships, and Meta Clustering Algorithm (MCLA), which groups the clusters based on which objects are contained in them. HGPA considers a graph with each object being a vertex. A cluster in any base clustering is represented by a hyper-edge connecting the member vertices. The hypergraph clustering package HMETIS (Karypis et al. [35]) was used as it gives quality clusterings and is very scalable. As with CSPA, employing a graph clustering algorithm adds a constraint that favors clusterings of comparable size. Though HGPA is fast with a worst case complexity of $O(nkr)$, it suffers from an additional problem: if all members of a base cluster are not assigned the same cluster in the consensus solution, the corresponding hyper-edge is broken and incurs a constant penalty; however it cannot distinguish between a situation where only one object was clustered differently and one where several objects were allocated to other groups. Due to this issue, HGPA is often not competitive in terms of cluster quality.

MCLA first forms a meta-graph with a vertex for each base cluster. The edge weights of this graph are proportional to the similarity between vertices, computed using the binary Jaccard measure (number of elements in common divided by the total number of distinct elements). Since the base clusterings are partitional, this results in an r-partite graph. The meta-graph is then partitioned into k balanced meta-clusters. Each meta-cluster, therefore, contains approximately r vertices. Finally, each object is assigned to its most closely associated meta-cluster. Ties are broken randomly. The worst case complexity is $O(nk^2r^2)$.

Noting that CSPA and MCLA consider either the similarity of objects or the similarity of clusters only, a Hybrid Bipartite Graph Formulation (HBGF) was proposed in [16]. A bipartite graph models both data points and clusters as vertices, wherein an edge exists only between a cluster vertex and a object vertex if the latter is a member of the former. Either METIS or other multi-way spectral clustering methods are used to partition this bipartite graph. The corresponding soft versions of CSPA, MCLA and HBGF have also been developed by Punera and Ghosh [52]. It should be noted that all of CSPA, MCLA, and HGPA were compared with one other using the NMI measure in [56].

22.4.3.2 Cumulative Voting

The concept of cumulative voting was first introduced in [15] where the authors used bagging to improve the accuracy of clustering procedure. Once clustering is done on a bootstrapped sample, the cluster correspondence problem is solved using iterative relabeling via Hungarian algorithm. Clustering on each bootstrapped sample gives some votes corresponding to each data point and cluster label pair which, in aggregate, decide the final cluster assignment.

A similar approach was adopted in [6]. Each base clustering in this contribution is thought of as providing a soft or probabilistic vote on to which clusters in the consensus solution its data points should belong. These votes are then gathered across the base solutions and thresholded to determine the membership of each object to the consensus clusters. Again, this requires a mapping function from the base clusterings to a stochastic one. An information-theoretic criterion based on the information bottleneck principle was used in [6] for this purpose. The mean of all the stochastic clusterings then yields the consensus partition. This approach is able to cater to a range of k in the base clusterings, is fast as it avoids the quadratic time/space complexity of forming a co-association matrix, and has shown good empirical results as well. Noting that the information bottleneck solutions can be obtained as a special case of Bregman clustering [8], it should be possible to recast this approach as a probabilistic one.

A variety of heuristic search procedures has also been suggested to hunt for a suitable consensus solution. These include a genetic algorithm formulation [71] and one using a multi-ant colony [69]. These approaches tend to be computationally expensive and the lack of extensive comparisons with the methods covered in this article currently make it difficult to assess their quality. Also, one can use several heuristics suggested in [18] to select only a few clustering solutions from a large ensemble.

22.5 Applications of Consensus Clustering

The motivation for consensus clustering has already been introduced in Section 22.1. Since cluster ensemble improves the quality of clustering solution, it can be used for any cluster analysis problem, e.g., image segmentation [67], bioinformatics, document retrieval [28], and automatic malware categorization [70], just to name a few. Gionis et al. [26] show how clustering ensemble algorithms can be used to improve the robustness of clustering solution, for clustering categorical data [29] and heterogeneous data, for identifying the correct number of clusters, and for detecting outliers. Fischer and Buhmann [20] showed how resampling the data and subsequent aggregation of the clustering solutions from the sampled sets can improve the quality of clustering solution. Sawtooth Software (`http://www.sawtoothsoftware.com/`) has commercialized some of the algorithms in [56] for applications in marketing. A package consisting of implementations of all the algorithms in [56] is also available on `http://strehl.com/soft.html`. In this section, we briefly discuss two major application domains.

22.5.1 Gene Expression Data Analysis

Consensus clustering has been applied to microarray data to improve the quality and robustness of the resulting clusters. A resampling based approach is used by Monti et al. [48], in which the agreement across the results obtained by executing a base clustering algorithm on several perturbations of the original dataset is used to obtain the final clustering. Swift et al. [58] use a variety of clustering algorithms on the same dataset to generate different base clustering results and try to find clusters that are consistent across all the base results using simulated annealing. In [19] the consensus clustering problem is treated as a median partition problem, where the aim is to find a partitioning of the data points that minimizes the distance to all the other partitionings. The authors propose greedy heuristic solutions to find a local optimum. Additionally, Deodhar and Ghosh [14] use consensus clustering to find overlapping clusters in microarray data. In this work, two different techniques are used to generate the consensus clustering solution from the candidate solutions. The first one is MCLA with some adjustable threshold, and the second one is soft kernelized k-means that works on an ensemble co-association matrix. In [12], gene expression time-series data is clustered at different time intervals and the solutions from different timestamps are merged into a single solution using graph partitioning of the ensemble co-association matrix.

22.5.2 Image Segmentation

Though there exist several image segmentation algorithms, depending on the application data, some perform better than the others and it is almost impossible to know beforehand which one should be used. The authors in [55] used the cluster ensemble formulation to aggregate the results of multiple segmentation algorithms such as (a) Normalized Cuts, (b) Energy Minimization by Graph Cuts, and (c) Curve Evolution to generate image segmentation. However, they found that the ensemble segmentation outperforms any individual segmentation algorithm. One of such results is shown in Figure 22.4. The first result corresponds to Normalized Cuts, the second one is from Graph

FIGURE 22.4: Segmentation result 1 from [55]

Cuts, the third and fourth ones are from Curve Evolution, and the last one is due to the ensemble segmentation.

In another similar application, the same authors show the utility of ensemble methods for better visualization and interpretation of images obtained from Diffusion Tensor Imaging (DTI) technique (Figure 22.5). DTI images have become popular within neuroimaging because they are useful to infer the underlying structure and organizational pattern in the body (e.g., neural pathways in the brain). To simplify the processing of such images, a number of different measures (or channels) are calculated from the diffusion tensor image. Some of these channels are Apparent Diffusion

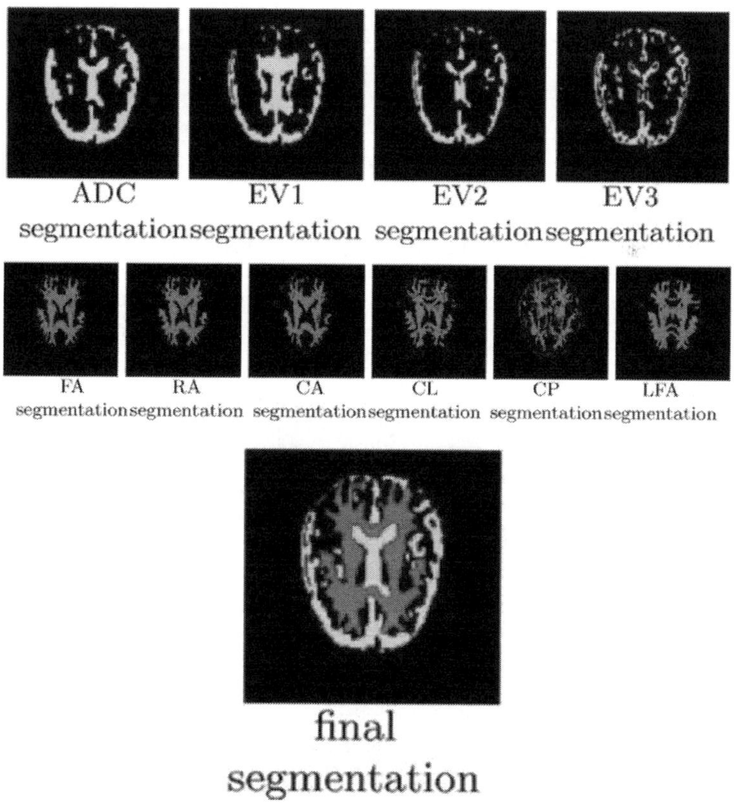

FIGURE 22.5: Segmentation result 2 from [55]

Coefficient (ADC), Fractional Anisotropy (FA), Mean Diffusivity (MD), Planar Anisotropy (PA).[3] Segmentations obtained from 10 of such channels of a brain are given in the first two rows of Figure 22.5. The last row shows the segmentation obtained using the ensemble strategy.

22.6 Concluding Remarks

This article first showed that cluster ensembles are beneficial in a wide variety of scenarios. It then provided a framework for understanding many of the approaches taken so far to design such ensembles. Even though there seems to be many different algorithms for this problem, we showed that there are several commonalities among these approaches. The design domain, however, is still quite rich leaving space for more efficient heuristics as well as formulations that place additional domain constraints to yield consensus solutions that are useful and actionable in diverse applications.

Bibliography

[1] A. Acharya, E. R. Hruschka, J. Ghosh, and S. Acharyya. C^3E: A framework for combining ensembles of classifiers and clusterers. In *10th International Workshop on MCS*, pages 269–278, 2011.

[2] A. Acharya, E.R. Hruschka, J. Ghosh, and S. Acharyya. Transfer learning with cluster ensembles. *JMLR Workshop and Conference Proceedings*, 27:123–132, 2012.

[3] M. Al-Razgan and C. Domeniconi. Weighted cluster ensemble. In *Proceedings of SIAM International Conference on Data Mining*, pages 258–269, 2006.

[4] Ahmed N. Albatineh, Magdalena Niewiadomska-Bugaj, and Daniel Mihalko. On similarity indices and correction for chance agreement. *Journal of Classification*, 23(2):301–313, September 2006.

[5] Sitaram Asur, Srinivasan Parthasarathy, and Duygu Ucar. An ensemble framework for clustering protein-protein interaction networks. In *Proceedings of the 15th Annual International Conference on Intelligent Systems for Molecular Biology (ISMB)*, 2007.

[6] Hanan G. Ayad and Mohamed S. Kamel. Cumulative voting consensus method for partitions with variable number of clusters. *IEEE Transactions on Pattern Analysis and Machine Intelligence*, 30(1):160–173, 2008.

[7] A. Banerjee, I. Dhillon, J. Ghosh, and S. Sra. Clustering on the unit hypersphere using Von Mises-Fisher distributions. *Journal of Machine Learning Research*, 6:1345–1382, 2005.

[8] A. Banerjee, S. Merugu, I. Dhillon, and J. Ghosh. Clustering with Bregman divergences. *Journal of Machine Learning Research*, 6:1705–1749, October 2005.

[9] N. Bansal, A.L. Blum, and S. Chawla. Correlation clustering. In *Proceedings of Foundations of Computer Science*, page 238–247, 2002.

[3] Please see [55] for more details about all of the 10 channels.

[10] C. M. Bishop. *Pattern Recognition and Machine Learning*. Springer, 2006.

[11] Claudio Carpineto and Giovanni Romano. Consensus clustering based on a new probabilistic rand index with application to subtopic retrieval. *IEEE Transactions on Pattern Analysis and Machine Intelligence*, 34(12):2315–2326, 2012.

[12] Tai-Yu Chiu, Ting-Chieh Hsu, and Jia-Shung Wang. AP-Based consensus clustering for gene expression time series. *International Conference on Pattern Recognition*, pages 2512–2515, 2010.

[13] W.H.E. Day. Foreword: Comparison and consensus of classifications. *J. Classification*, 3:183–185, 1986.

[14] Meghana Deodhar and Joydeep Ghosh. Consensus clustering for detection of overlapping clusters in microarray data. In *ICDMW '06: Proceedings of the Sixth IEEE International Conference on Data Mining—Workshops*, pages 104–108, Washington, DC, USA, 2006. IEEE Computer Society.

[15] Sandrine Dudoit and Jane Fridlyand. Bagging to improve the accuracy of a clustering procedure. *Bioinformatics*, 19(9):1090–1099, 2003.

[16] X. Fern and C. Brodley. Solving cluster ensemble problems by bipartite graph partitioning. In *Proceedings of International Conference on Machine Learning*, pages 281–288, 2004.

[17] Xiaoli Z. Fern, and Carla E. Brodley. Random projection for high dimensional data clustering: A cluster ensemble approach. In *Proceedings of the 20th International Conference on Machine Learning*, pages 186–199, Washington, DC, USA, August 2003.

[18] Xiaoli Z. Fern and Wei Lin. Cluster ensemble selection. *Statistical Analysis and Data Mining*, 1(3):128–141, November 2008.

[19] V. Filkov and S. Skiena. Integrating microarray data by consensus clustering. *International Journal on Artificial Intelligence Tools (IJAIT)*, 4:863–880, 2004.

[20] Bernd Fischer and Joachim M. Buhmann. Bagging for path-based clustering. *IEEE Transactions on Pattern Analysis and Machine Intelligence*, 25(11):1411–1415, November 2003.

[21] A. Fred and A. K. Jain. Combining multiple clusterings using evidence accumulation. *IEEE Transactions on Pattern Analysis and Machine Intelligence*, 27(6):835–850, 2005.

[22] J. Gao, W. Fan, J. Jiang, and J. Han. Knowledge transfer via multiple model local structure mapping. In *Proceedings of Knowledge Discovery and Data Mining*, pages 283–291, 2008.

[23] Jing Gao, Feng Liang, Wei Fan, Yizhou Sun, and Jiawei Han. A graph-based consensus maximization approach for combining multiple supervised and unsupervised models. *IEEE Transactions on Knowledge and Data Engineering*, pages 15–28, 2011.

[24] J. Ghosh. Multiclassifier systems: Back to the future (invited paper). In F. Roli and J. Kittler, editors, *Multiple Classifier Systems*, LNCS Vol. 2364, pages 1–15, Springer, 2002.

[25] J. Ghosh, A. Strehl, and S. Merugu. A consensus framework for integrating distributed clusterings under limited knowledge sharing. In *Proceedings of the NSF Workshop on Next Generation Data Mining, Baltimore*, pages 99–108, November 2002.

[26] A. Gionis, H. Mannila, and P. Tsaparas. Clustering aggregation. *ACM Transactions on Knowledge Discovery from Data*, 1(4):109–117, March 2007.

[27] A. Goder and V. Filkov. Consensus clustering algorithms: Comparison and refinement. In *Proceedings of the Tenth Workshop on Algorithm Engineering and Experiments*, pages 109–117, 2008.

[28] Edgar Gonzàlez and Jordi Turmo. Comparing non-parametric ensemble methods for document clustering. In *Proceedings of the 13th International Conference on Natural Language and Information Systems: Applications of Natural Language to Information Systems*, NLDB '08, pages 245–256, Berlin, Heidelberg, Springer-Verlag, 2008.

[29] Zengyou He, Xiaofei Xu, and Shengchun Deng. A cluster ensemble method for clustering categorical data. *Information Fusion*, 6(2):143–151, 2005.

[30] P. Hore, Lawrence O. Hall, and Dmitry B. Goldgof. A scalable framework for cluster ensembles. *Pattern Recognition*, 42(5):676–688, 2009.

[31] Xiaohua Hu and Illhoi Yoo. Cluster ensemble and its applications in gene expression analysis. In *APBC '04: Proceedings of the Second Conference on Asia-Pacific Bioinformatics*, pages 297–302, Darlinghurst, Australia, 2004. Computer Society, Inc.

[32] L. Hubert and P. Arabie. Comparing partitions. *Journal of Classification*, 2:193–218, 1985.

[33] G. Karypis, E.-H. Han, and V. Kumar. Chameleon: Hierarchical clustering using dynamic modeling. *IEEE Computer*, 32(8):68–75, August 1999.

[34] G. Karypis and V. Kumar. A fast and high quality multilevel scheme for partitioning irregular graphs. *SIAM Journal on Scientific Computing*, 20(1):359–392, 1998.

[35] George Karypis, Rajat Aggarwal, Vipin Kumar, and Shashi Shekhar. Multilevel hypergraph partitioning: Applications in VLSI domain. In *Proceedings of the Design and Automation Conference*, pages 526–529, 1997.

[36] J. Kittler and F. Roli, editors. *Multiple Classifier Systems.* LNCS Vol. 2634, Springer, 2002.

[37] A. Kraskov, H. Stögbauer, R. G. Andrzejak, and P. Grassberger. Hierarchical clustering using mutual information. *EPL (Europhysics Letters)*, 70(2):278, 2005.

[38] L. I. Kuncheva. *Combining Pattern Classifiers: Methods and Algorithms.* Wiley, Hoboken, NJ, 2004.

[39] L. I. Kuncheva and S. T. Hadjitodorov. Using diversity in cluster ensemble. In *IEEE International Conference on Systems, Man and Cybernetics*, pages 1214–1219, 2004.

[40] H. O. Lancaster. *The Chi-Squared Distribution*. 1969.

[41] Daniel D. Lee and H. Sebastian Seung. Algorithms for non-negative matrix factorization. In *Neural Information Processing Systems*, pages 556–562. MIT Press, 2000.

[42] T. Li and C. Ding. Weighted consensus clustering. In *Proceedings of Eighth SIAM International Conference on Data Mining*, pages 798–809, 2008.

[43] T. Li, C. Ding, and M. Jordan. Solving consensus and semi-supervised clustering problems using non-negative matrix factorization. In *Proceedings of Eighth IEEE International Conference on Data Mining*, pages 577–582, 2007.

[44] M. Meila. Comparing clusterings by the variation of information. In *Proceedings of Conference on Learning Theory*, pages 173–187, 2003.

[45] Marina Meila. Comparing clusterings—An information based distance. *J. Multivariate Analysis*, 98(5):873–895, May 2007.

[46] S. Merugu and J. Ghosh. A distributed learning framework for heterogeneous data sources. In *Proceedings of the Eleventh ACM SIGKDD International Conference on Knowledge Discovery and Data Mining*, pages 208–217, 2005.

[47] B. Mirkin. *Mathematical Classification and Clustering*. Kluwer, 1996.

[48] S. Monti, P. Tamayo, J. Mesirov, and T. Golub. Consensus clustering—A resampling-based method for class discovery and visualization of gene expression microarray data. In *Journal of Machine Learning*, 52: 91–118, 2003.

[49] N. Nguyen and R. Caruana. Consensus clusterings. In *Proceedings of International Conference on Data Mining*, pages 607–612, 2007.

[50] Xuan Vinh Nguyen, Julien Epps, and James Bailey. Information theoretic measures for clusterings comparison: Variants, properties, normalization and correction for chance. *Journal of Machine Learning Research*, 11:2837–2854, 2010.

[51] N. C. Oza and K. Tumer. Classifier ensembles: Select real-world applications. *Information Fusion*, 9(1):4–20, January 2008.

[52] K. Punera and J. Ghosh. Consensus based ensembles of soft clusterings. In *Proc. MLMTA'07—International Conference on Machine Learning: Models, Technologies & Applications*, 2007.

[53] Xavier Sevillano, Germán Cobo, Francesc Alías, and Joan Claudi Socoró. Feature diversity in cluster ensembles for robust document clustering. In *SIGIR '06: Proceedings of the 29th Annual International ACM SIGIR Conference on Research and Development in Information Retrieval*, pages 697–698, New York, USA, 2006. ACM.

[54] A. Sharkey. *Combining Artificial Neural Nets*. Springer-Verlag, 1999.

[55] Vikas Singh, Lopamudra Mukherjee, Jiming Peng, and Jinhui Xu. Ensemble clustering using semidefinite programming with applications. *Machine Learning*, 79(1–2):177–200, May 2010.

[56] A. Strehl and J. Ghosh. Cluster ensembles—A knowledge reuse framework for combining multiple partitions. *Journal of Machine Learning Research*, 3(Dec):583–617, 2002.

[57] Alexander Strehl and Joydeep Ghosh. A scalable approach to balanced, high-dimensional clustering of market-baskets. In *Proceedings of HiPC 2000, Bangalore*, volume 1970 of *LNCS*, pages 525–536. Springer, December 2000.

[58] S. Swift, A. Tucker, V. Vinciotti, and N. Martin. Consensus clustering and functional interpretation of gene-expression data. *Genome Biology*, 5:R94, 2004.

[59] Stephen Swift, Allan Tucker, Veronica Vinciotti, Nigel Martin, Christine Orengo, Xiaohui Liu, and Paul Kellam. Consensus clustering and functional interpretation of gene-expression data. *Genome Biology*, 5(11):R94, 2004.

[60] A. Topchy, A. Jain, and W. Punch. A mixture model for clustering ensembles. In *Proceedings of SIAM International Conference on Data Mining*, pages 379–390, 2004.

[61] K. Tumer and J. Ghosh. Analysis of decision boundaries in linearly combined neural classifiers. *Pattern Recognition*, 29(2):341–348, 1996.

[62] K. Tumer and J. Ghosh. Robust order statistics based ensembles for distributed data mining. In Hillol Kargupta and Philip Chan, editors, *Advances in Distributed and Parallel Knowledge Discovery*, pages 85–110. AAAI Press, 2000.

[63] Fei Wang, Xin Wang, and Tao Li. Generalized cluster aggregation. In *Proceedings of IJ-CAI'09*, pages 1279–1284, San Francisco, CA, USA, 2009. Morgan Kaufmann Publishers Inc.

[64] H. Wang, H. Shan, and A. Banerjee. Bayesian cluster ensembles. In *Proceedings of the Ninth SIAM International Conference on Data Mining*, pages 211–222, 2009.

[65] Pu Wang, Carlotta Domeniconi, and Kathryn Laskey. Nonparametric Bayesian clustering ensembles. In *Machine Learning and Knowledge Discovery in Databases*, volume 6323 of *Lecture Notes in Computer Science*, pages 435–450. Springer, Berlin/Heidelberg, Berlin, Heidelberg, 2010.

[66] Matthijs Warrens. On similarity coefficients for 2x2 tables and correction for chance. *Psychometrika*, 73(3):487–502, September 2008.

[67] Pakaket Wattuya, Kai Rothaus, J.-S. Prassni, and Xiaoyi Jiang. A random walker based approach to combining multiple segmentations. In *ICPR'08*, pages 1–4, 2008.

[68] Junjie Wu, Jian Chen, Hui Xiong, and Ming Xie. External validation measures for k-means clustering: A data distribution perspective. *Expert Systems with Applications*, 36(3):6050–6061, 2009.

[69] Y. Yang and M.S. Kamel. An aggregated clustering approach using multi-ant colonies algorithms. *Journal of Pattern Recognition*, 39(7):1278–1289, July 2006.

[70] Yanfang Ye, Tao Li, Yong Chen, and Qingshan Jiang. Automatic malware categorization using cluster ensemble. In *Knowledge Discovery and Data Mining '10: Proceedings of the 16th ACM SIGKnowledge Discovery and Data Mining International Conference on Knowledge Discovery and Data Mining*, pages 95–104, New York, USA, 2010. ACM.

[71] H.S. Yoon, S.Y. Ahn, S.H. Lee, S.B. Cho, and J.H. Kim. Heterogeneous clustering ensemble method for combining different cluster results. In *Proceedings of BioDM 2006, Lecture Notes in Computer Science*, volume 3916, pages 82–92, 2006.

[72] S. Zhong and J. Ghosh. A unified framework for model-based clustering. *Journal of Machine Learning Research*, 4:1001–1037, 2003.

Chapter 23

Clustering Validation Measures

Hui Xiong
Rutgers, The State University of New Jersey
Newark, NJ 07102
`hxiong@rutgers.edu`

Zhongmou Li
Rutgers, The State University of New Jersey
Newark, NJ 07102
`mosesli@pegasus.rutgers.edu`

23.1 Introduction

Clustering, one of the most important unsupervised learning problems, is the task of dividing a set of objects into clusters such that objects within the same cluster are similar while objects in different clusters are distinct. Clustering is widely used in many fields, such as text mining, image analysis, and bioinformatics [16, 69, 17]. As an unsupervised learning task, it is necessary to find a way to validate the goodness of partitions after clustering. Otherwise, it would be difficult to make use of different clustering results.

Clustering validation, which evaluates the goodness of clustering results [41], has long been recognized as one of the vital issues essential to the success of clustering applications [26]. Despite the vast amount of expert endeavor spent on this problem [18, 19, 5], there is no consistent and conclusive solution to cluster validation. The best suitable measures to use in practice remain unknown. Indeed, there are many challenging validation issues which have not been fully addressed in the clustering literature. For instance, the importance of normalizing validation measures has not been fully established. Also, the relationship between different validation measures is not clear. Moreover, there are important properties associated with validation measures which are important to the selection of the use of these measures but have not been well characterized. Finally, given the fact that different validation measures may be appropriate for different clustering algorithms, it is necessary to provide a focused study of cluster validation measures on specified clustering algorithms.

Clustering validation measures can be categorized into two main types: external clustering validation and internal clustering validation. The main difference is whether or not external information is used for clustering validation. External validation measures use external information not present in the data to evaluate the extent to which the clustering structure discovered by a clustering algorithm matches some external structure, e.g., the one specified by the given class labels. One example of external validation measure is entropy, which evaluates the "purity" of clusters based on the given class labels [54]. On the other hand, internal measures evaluate the goodness of a clustering structure without respect to external information [57, 6, 53, 34]. For example, the Silhouette index [49] validates the clustering performance based only on the pairwise difference of between- and within-cluster distances of all data points.

Both external and internal validation measures are crucial for many application scenarios. Since external validation measures know the "true" cluster number in advance, they can be used for choosing an optimal clustering algorithm on a specific data set. For instance, if external validation measures show that a document clustering algorithm can lead to the clustering results which can match the categorization performance by human experts, there is a good reason to believe this clustering algorithm has a practical impact on document clustering. On the other hand, internal validation measures can be used to choose the best clustering algorithm as well as the optimal cluster number without any additional information. In practice, external information such as class labels is not available in some real world applications. In that case, internal validation measures are the only option for cluster validation when there is no external information available.

However, there are still scenarios that clustering validations measures have limitations in evaluating the goodness of the clustering results. For example, in the case when external criteria are not available and internal validation measures are not very robust, subjective evaluations such as case studies are often used in many contexts, which is particularly common in network clustering algorithms [56, 70, 66]. Another example would be the case that class labels may not reflect cluster structure well, when the class labels do not necessarily correspond to locality. Some clusters may contain a mixture of objects from different classes, thus objects in widely separated clusters may belong to the same class. In this circumstance, using clustering techniques to reveal the data characteristics may not even be a good idea.

In literature, there is a list of clustering validation measures for soft (fuzzy) clustering algorithms. A fuzzy clustering algorithm generates a fuzzy partition to provide a degree of membership of each object to a given cluster. A fuzzy clustering approach is less prone to local minimum than crisp clustering algorithms since it makes soft decisions in each iteration through the use of membership functions [4]. Kim et al. propose a cluster validation measure for fuzzy partitions obtained from fuzzy C-Means algorithm [31]. The proposed validity index exploits an intercluster proximity between fuzzy clusters, which is used to measure the degree of overlap between clusters. The best fuzzy c-partition is obtained by minimizing the intercluster proximity with respect to c. Smyth proposes a cross-validated likelihood measure to determine the appropriate number of clusters in the context of model-based probabilistic clustering [52]. The Xie-Beni index (XB) [63] defines a fuzzy clustering validation function to measure the overall average compactness and separation of a fuzzy c-partition. The intercluster separation is the minimum square distance between cluster centers, and the intracluster compactness is the mean square distance between each data object and its cluster center. The optimal cluster number is reached when the minimum of XB is found. Gath and Geva also proposed a fuzzy validation index which is based on the concepts of hypervolume and density [14].

Another category of related works is validation measures for subspace clustering algorithms. Subspace clustering is an extension of traditional clustering that seeks to find clusters in different subspaces within a data set [45]. Often in high-dimensional data, many dimensions are irrelevant and can mask existing clusters in noisy data. Subspace clustering algorithms localize the search for relevant dimensions allowing them to find clusters that exist in multiple, possibly overlapping, subspaces. Many of the existing validation measures for traditional clustering approaches, such as entropy [1], F-measure [2], and classification error [46], are also used as validation measures for subspace clustering. Muller et al. [44] provide a systematic and thorough evaluation of subspace clustering paradigms.

This chapter focuses on providing a comprehensive study on various aspects of both the external and internal clustering validation measures for crisp clustering algorithms. The sections of this chapter are organized as follows. In Section 23.2, an organized study on 16 external validation measures for K-means clustering is presented. The importance of measure normalization in cluster evaluation on data with imbalanced class distributions is demonstrated, and the normalization solutions for several measures are also provided. Major properties of the external measures, as well as their interrelationships, are presented [62]. Section 23.3 presents a detailed study on 12 widely used internal validation measures for crisp clustering [37, 38]. Properties of these internal measures in different aspects, such as the impact of data with noise, subclusters, and arbitrary shapes, are well investigated. We conclude this chapter with a summary in Section 23.4.

23.2 External Clustering Validation Measures

In literature, a number of external validation measures for crisp clustering have been proposed. In this section, an organized study on a suite of 16 widely used external clustering validation measures as shown in Table 23.1 is presented for the K-means clustering algorithm. These measures represent a good coverage of the external validation measures available in different fields such as data mining, information retrieval, machine learning, and statistics. A common ground of these measures is that they can be computed by the contingency matrix as follows.

The Contingency Matrix. Given a data set D with n objects, assume that there is a partition $P = \{P_1, \cdots, P_K\}$ of D, where $\bigcup_{i=1}^{K} P_i = D$ and $P_i \bigcap P_j = \phi$ for $1 \leq i \neq j \leq K$, and K is the number of clusters. If the "true" class labels for the data are given, another partition can be generated on D:

TABLE 23.1: External Cluster Validation Measures

	Measure	Definition	Range
1	Entropy (E)	$-\sum_i p_i (\sum_j \frac{p_{ij}}{p_i} \log \frac{p_{ij}}{p_i})$	$[0, \log K']$
2	Purity (P)	$\sum_i p_i (\max_j \frac{p_{ij}}{p_i})$	(0,1]
3	F-measure (F)	$\sum_j p_j \max_i [2 \frac{p_{ij}}{p_i} \frac{p_{ij}}{p_j} / (\frac{p_{ij}}{p_i} + \frac{p_{ij}}{p_j})]$	(0,1]
4	Variation of Information (VI)	$-\sum_i p_i \log p_i - \sum_j p_j \log p_j - 2 \sum_i \sum_j p_{ij} \log \frac{p_{ij}}{p_i p_j}$	$[0, 2\log \max(K, K')]$
5	Mutual Information (MI)	$\sum_i \sum_j p_{ij} \log \frac{p_{ij}}{p_i p_j}$	$(0, \log K']$
6	Rand statistic (R)	$[\binom{n}{2} - \sum_i \binom{n_{i\cdot}}{2} - \sum_j \binom{n_{\cdot j}}{2} + 2\sum_{ij} \binom{n_{ij}}{2}] / \binom{n}{2}$	(0,1]
7	Jaccard coefficient (J)	$\sum_{ij} \binom{n_{ij}}{2} / [\sum_i \binom{n_{i\cdot}}{2} + \sum_j \binom{n_{\cdot j}}{2} - \sum_{ij} \binom{n_{ij}}{2}]$	[0,1]
8	Fowlkes & Mallows index (FM)	$\sum_{ij} \binom{n_{ij}}{2} / \sqrt{\sum_i \binom{n_{i\cdot}}{2} \sum_j \binom{n_{\cdot j}}{2}}$	[0,1]
9	Hubert Γ statistic I (Γ)	$\frac{\binom{n}{2} \sum_{ij} \binom{n_{ij}}{2} - \sum_i \binom{n_{i\cdot}}{2} \sum_j \binom{n_{\cdot j}}{2}}{\sqrt{\sum_i \binom{n_{i\cdot}}{2} \sum_j \binom{n_{\cdot j}}{2} [\binom{n}{2} - \sum_i \binom{n_{i\cdot}}{2}][\binom{n}{2} - \sum_j \binom{n_{\cdot j}}{2}]}}$	(-1,1]
10	Hubert Γ statistic II (Γ')	$[\binom{n}{2} - 2\sum_i \binom{n_{i\cdot}}{2} - 2\sum_j \binom{n_{\cdot j}}{2} + 4\sum_{ij} \binom{n_{ij}}{2}] / \binom{n}{2}$	[0,1]
11	Minkowski score (MS)	$\sqrt{\sum_i \binom{n_{i\cdot}}{2} + \sum_j \binom{n_{\cdot j}}{2} - 2\sum_{ij} \binom{n_{ij}}{2}} / \sqrt{\sum_j \binom{n_{\cdot j}}{2}}$	$[0, +\infty)$
12	classification error (ε)	$1 - \frac{1}{n} \max_\sigma \sum_j n_{\sigma(j),j}$	[0,1)
13	van Dongen criterion (VD)	$(2n - \sum_i \max_j n_{ij} - \sum_j \max_i n_{ij}) / 2n$	[0,1)
14	micro-average precision (MAP)	$\sum_i p_i (\max_j \frac{p_{ij}}{p_i})$	(0,1]
15	Goodman-Kruskal coeff (GK)	$\sum_i p_i (1 - \max_j \frac{p_{ij}}{p_i})$	[0,1)
16	Mirkin metric (M)	$\sum_i n_{i\cdot}^2 + \sum_j n_{\cdot j}^2 - 2\sum_i \sum_j n_{ij}^2$	$[0, 2\binom{n}{2})$

Note: $p_{ij} = n_{ij}/n$, $p_i = n_{i\cdot}/n$, $p_j = n_{\cdot j}/n$.

TABLE 23.2: The Contingency Matrix.

Partition C

		C_1	C_2	$\cdots$	$C_{K'}$	Σ
	P_1	n_{11}	n_{12}	$\cdots$	$n_{1K'}$	$n_{1\cdot}$
Partition P	P_2	n_{21}	n_{22}	$\cdots$	$n_{2K'}$	$n_{2\cdot}$
	$\cdot$	$\cdot$	$\cdot$	$\cdots$	$\cdot$	$\cdot$
	P_K	n_{K1}	n_{K2}	$\cdots$	$n_{KK'}$	$n_{K\cdot}$
	Σ	$n_{\cdot 1}$	$n_{\cdot 2}$	$\cdots$	$n_{\cdot K'}$	n

$C = \{C_1, \cdots, C_{K'}\}$, where $\bigcup_{i=1}^{K'} C_i = D$ and $C_i \bigcap C_j = \phi$ for $1 \leq i \neq j \leq K'$, where K' is the number of classes. Let n_{ij} denote the number of objects in cluster P_i from class C_j, then the information on the overlap between the two partitions can be written in the form of a contingency matrix, as shown in Table 23.2. Consistent notations in this contingency matrix will be used throughout this section.

23.2.1 An Overview of External Clustering Validation Measures

Table 23.1 shows the list of measures to be studied. The "Definition" column gives the computation forms of the measures by using the notations in the contingency matrix. The 16 measures are briefly introduced as follows.

Entropy and purity are frequently used external measures for K-means [54, 67]. They measure the "purity" of the clusters with respect to the given class labels.

F-measure was originally designed for the evaluation of hierarchical clustering [48, 36], but has also been employed for partitional clustering. It combines the precision and recall concepts from the information retrieval community.

The Mutual Information (MI) and Variation of Information (VI) were developed in the field of information theory [8]. MI measures how much information one random variable can tell about another one [55]. VI measures the amount of information that is lost or gained in changing from the class set to the cluster set [42].

The Rand statistic [47], Jaccard coefficient, Fowlkes and Mallows index [13], and Hubert's two statistics [23, 24] evaluate the clustering quality by the agreements and/or disagreements of the pairs of data objects in different partitions.

The Minkowski score [3] measures the difference between the clustering results and a reference clustering (true clusters). And the difference is computed by counting the disagreements of the pairs of data objects in two partitions.

The classification error takes a classification view on clustering [6]. It tries to map each class to a different cluster so as to minimize the total misclassification rate. The "σ" in Table 23.1 is the mapping of class j to cluster $\sigma(j)$.

The van Dongen criterion [60] was originally proposed for evaluating graph clustering. It measures the representativeness of the majority objects in each class and each cluster.

Finally, the micro-average precision, Goodman-Kruskal coefficient [15], and Mirkin metric [43] are also popular measures. However, the former two are equivalent to the purity measure and the Mirkin metric is equivalent to the Rand statistic ($M/2\binom{n}{2}+R=1$). Some discussion on GoodMan–Kruskal and Mirkin can be found in Section 23.2.4.3.

In summary, there are 13 (out of 16) candidate measures. Among them, P, F, MI, R, J, FM, Γ, and Γ' are positive measures—a higher value indicates a better clustering performance. The remainder, however, consists of measures based on the distance notion. The acronyms of these measures will be used throughout this section.

23.2.2 Defective Validation Measures

In this section, some validation measures which produce misleading validation results for K-means on data with skewed class distributions are presented.

23.2.2.1 K-Means: The Uniform Effect

One of the unique characteristic of K-means clustering is the so-called uniform effect; that is, K-means tends to produce clusters with relatively uniform sizes [64]. The coefficient of variation (CV) [10], a statistic which measures the dispersion degree of a random distribution, is used to quantify the uniform effect. CV is defined as the ratio of the standard deviation to the mean. Given a sample of data objects $X=\{x_1,x_2,\ldots,x_n\}$, $CV = s/\bar{x}$, where $\bar{x} = \sum_{i=1}^{n} x_i/n$ and $s = \sqrt{\sum_{i=1}^{n}(x_i-\bar{x})^2/(n-1)}$. CV is a dimensionless number that allows the comparison of the variations of populations that have significantly different mean values. In general, the larger the CV value is, the greater the variability in the data.

Example. Let CV_0 denote the CV value of the "true" class sizes and CV_1 denote the CV value of the resultant cluster sizes. The `sports` data set [59] is used to illustrate the uniform effect by K-means. The "true" class sizes of `sports` have $CV_0=1.02$. Then the CLUTO implementation of K-means [27] with default settings is employed to cluster `sports` into seven clusters, and the CV value of the resultant cluster sizes is 0.42. Therefore, the CV difference is $DCV=CV_1-CV_0=-0.6$, which indicates a significant uniform effect in the clustering result.

Indeed, it has been empirically validated that the 95% confidence interval of CV_1 values produced by K-means is in [0.09, 0.85] [61]. In other words, for data sets with CV_0 values greater than 0.85, the uniform effect of K-means can distort the cluster distribution significantly.

Now the question is: Can these widely used validation measures capture the negative uniform effect by K-means clustering? Next, a necessary but not sufficient criterion is provided to testify whether a validation measure can be effectively used to evaluate K-means clustering.

TABLE 23.3: Two Clustering Results

I	C_1	C_2	C_3	C_4	C_5
P_1	10	0	0	0	0
P_2	10	0	0	0	0
P_3	10	0	0	0	0
P_4	0	0	0	10	0
P_5	0	2	6	0	2

II	C_1	C_2	C_3	C_4	C_5
P_1	27	0	0	2	0
P_2	0	2	0	0	0
P_3	0	0	6	0	0
P_4	3	0	0	8	0
P_5	0	0	0	0	2

23.2.2.2 A Necessary Selection Criterion

Assume that there is a sample document data containing 50 documents from 5 classes. The class sizes are 30, 2, 6, 10 and 2. Thus, $CV_0 = 1.166$, which implies a skewed class distribution.

For this sample data set, assume that there are two clustering results as shown in Table 23.3. In the table, the first result consists of five clusters with extremely balanced sizes. This is also indicated by $CV_1 = 0$. In contrast, for the second result, the five clusters have varied cluster sizes with $CV_1 = 1.125$, much closer to the CV value of the "true" class sizes. Therefore, from a data distribution point of view, the second result should be better than the first one.

Indeed, by taking a closer look on contingency Matrix I in Table 23.3, one can find that the first clustering partitions the objects of the largest class C_1 into three balanced subclusters. Meanwhile, the two small classes C_2 and C_5 have totally "disappeared"—they are overwhelmed in cluster P_5 by the objects from class C_3. In contrast, it is easy to identify all the classes in the second clustering result, since they have the majority of objects in the corresponding clusters. Therefore, it is the conclusion that the first clustering is indeed much worse than the second one.

As shown in Section 23.2.2.1, K-means tends to produce clusters with relatively uniform sizes. Thus the first clustering in Table 23.3 can be regarded as the negative result of the uniform effect. So the first necessary but not sufficient criterion for selecting the measures for K-means is as follows.

Criterion 1 *If an external validation measure cannot capture the uniform effect by K-means on data with skewed class distributions, this measure is not suitable for validating the results of K-means clustering.*

The performance of existing external cluster validation measures for this criterion is presented in next section.

23.2.2.3 The Cluster Validation Results

Table 23.4 shows the validation results for the two clusterings in Table 23.3 by all 13 external validation measures. The better evaluation of each validation measure is highlighted.

As shown in Table 23.4, only three measures, F, P, and MI, cannot capture the uniform effect by K-means and their validation results can be misleading. In other words, these measures are not suitable for evaluating the K-means clustering. These three measures are defective validation measures.

TABLE 23.4: The Cluster Validation Results

	E	P	F	MI	VI	R	J	FM	Γ	Γ'	MS	ε	VD
I	**0.274**	**0.920**	0.617	**1.371**	1.225	0.732	0.375	0.589	0.454	0.464	0.812	0.480	0.240
II	0.396	0.9	**0.902**	1.249	**0.822**	**0.857**	**0.696**	**0.821**	**0.702**	**0.714**	**0.593**	**0.100**	**0.100**

23.2.2.4 The Issues with the Defective Measures

First, the problem of the **entropy** measure lies in the fact that it cannot evaluate the integrity of the classes since $E = -\sum_i p_i \sum_j \frac{p_{ij}}{p_i} \log \frac{p_{ij}}{p_i}$. If a random variable view on cluster P and class C is taken, then $p_{ij} = n_{ij}/n$ is the joint probability of the event: $\{P = P_i \bigwedge C = C_j\}$, and $p_i = n_{i\cdot}/n$ is the marginal probability. Therefore, $E = \sum_i p_i \sum_j -p(C_j|P_i) \log p(C_j|P_i) = \sum_i p_i H(C|P_i) = H(C|P)$, where $H(\cdot)$ is the Shannon entropy [8]. The above implies that the entropy measure is nothing but the conditional entropy of C on P. In other words, if the objects in each large partition are mostly from the same class, the entropy value tends to be small (indicating a better clustering quality). This is usually the case for K-means clustering on highly imbalanced data sets, since K-means tends to partition a large class into several pure subclusters. This leads to the problem that the integrity of the objects from the same class has been damaged. The entropy measure cannot capture this information and penalize it.

The **mutual information** is strongly related to the entropy measure, which is illustrated by the following lemma.

Lemma 23.2.1 *The mutual information measure is equivalent to the entropy measure for cluster validation.*
PROOF. By information theory, $MI = \sum_i \sum_j p_{ij} \log \frac{p_{ij}}{p_i p_j} = H(C) - H(C|P) = H(C) - E$. Since $H(C)$ is a constant for any given data set, MI is essentially equivalent to E. □

The **purity** measure works in a similar fashion as the entropy measure. That is, it measures the "purity" of each cluster by the ratio of the objects from the majority class. Thus, it has the same problem as the entropy measure for evaluating K-means clustering.

In summary, entropy, purity, and mutual information are defective measures for validating K-means clustering.

23.2.2.5 Improving the Defective Measures

In this section, the improved versions of the above three defective measures entropy, mutual information, and purity are provided.

Lemma 23.2.2 *The Variation of Information measure is an improved version of the entropy measure.*
PROOF. Cluster P and class C can be viewed as two random variables, and it has been shown that $VI = H(C) + H(P) - 2MI = H(C|P) + H(P|C)$ [42]. The component $H(C|P)$ is nothing but the entropy measure, and the component $H(P|C)$ is a valuable supplement to $H(C|P)$. That is, $H(P|C)$ evaluates the integrity of each class along different clusters. □

Since MI is equivalent to E according to Lemma 23.2.1, therefore, VI is also an improved version of MI.

Lemma 23.2.3 *The van Dongen criterion is an improved version of the purity measure.*
PROOF. $VD = \frac{2n - \sum_i \max_j n_{ij} - \sum_j \max_i n_{ij}}{2n} = 1 - \frac{1}{2}P - \frac{\sum_j \max_i n_{ij}}{2n}$. Apparently, $\sum_j \max_i n_{ij}/n$ reflects the integrity of the classes and is a supplement to the purity measure. □

23.2.3 Measure Normalization

In this section, discussions on the importance of measure normalization and the normalization solutions of different measures are presented.

23.2.3.1 Normalizing the Measures

Generally speaking, normalizing techniques can be divided into two categories. One is based on a statistical view, which formulates a baseline distribution to correct the measure for randomness. A clustering can then be termed "valid" if it has an unusually high or low value, as measured with respect to the baseline distribution. The other technique uses the minimum and maximum values to normalize the measure into the [0,1] range. From a statistical view, it is equivalent to view this technique with the assumption that each measure takes a uniform distribution over the value interval.

***The Normalizations of* R, FM, Γ, Γ′, J, *and* MS**. The normalization scheme can take the form:

$$S_n = \frac{S - E(S)}{\max(S) - E(S)} \tag{23.1}$$

where $\max(S)$ is the maximum value of the measure S and $E(S)$ is the expected value of S based on the baseline distribution. Some measures derived from the statistics community, such as R, FM, Γ, and Γ', usually take this scheme.

Specifically, Hubert and Arabie (1985) [24] suggested using the multivariate hypergeometric distribution as the baseline distribution in which the row and column sums are fixed in Table 23.2, but the partitions are randomly selected. This determines the expected value as follows:

$$E\left(\sum_i\sum_j \binom{n_{ij}}{2}\right) = \frac{\sum_i \binom{n_{i\cdot}}{2}\sum_j \binom{n_{\cdot j}}{2}}{\binom{n}{2}} \tag{23.2}$$

Based on this value, it is easy to compute the expected values of R, FM, Γ, and Γ' since they are the linear functions of $\sum_i\sum_j\binom{n_{ij}}{2}$ under the hypergeometric distribution assumption. Furthermore, although the exact maximum values of the measures are computationally prohibited under the hypergeometric distribution assumption, it is still reasonable to approximate them by 1. Then, according to Equations (23.1) and (23.2), the normalized R, FM, Γ, and Γ' measures can be calculated as shown in Table 23.5.

The normalization of J and MS is a little bit complex, since they are not linear to $\sum_i\sum_j\binom{n_{ij}}{2}$. Nevertheless, one can still normalize the equivalent measures converted from them. Let $J' = \frac{1-J}{1+J} = \frac{2}{1+J} - 1$ and $MS' = MS^2$.

It is easy to show $J' \Leftrightarrow J$ and $MS' \Leftrightarrow MS$. Then, based on the hypergeometric distribution assumption, the normalized J' and MS' can be calculated as shown in Table 23.5. Since J' and MS' are

TABLE 23.5: The Normalized Measures

	Measure	Normalization
1	R_n	$(m - m_1m_2/M)/(m_1/2 + m_2/2 - m_1m_2/M)$
2	FM_n	$(m - m_1m_2/M)/(\sqrt{m_1m_2} - m_1m_2/M)$
3	Γ_n	$(mM - m_1m_2)/\sqrt{m_1m_2(M-m_1)(M-m_2)}$
4	Γ'_n	$(m - m_1m_2/M)/(m_1/2 + m_2/2 - m_1m_2/M)$
5	J'_n	$(m_1 + m_2 - 2m)/(m_1 + m_2 - 2m_1m_2/M)$
6	MS'_n	$(m_1 + m_2 - 2m)/(m_1 + m_2 - 2m_1m_2/M)$
7	VI_n	$1 + 2\frac{\sum_i\sum_j p_{ij}\log(p_{ij}/p_ip_j)}{(\sum_i p_i \log p_i + \sum_j p_j \log p_j)}$
8	VD_n	$\frac{(2n - \sum_i \max_j n_{ij} - \sum_j \max_i n_{ij})}{(2n - \max_i n_{i\cdot} - \max_j n_{\cdot j})}$
9	F_n	$(F - F_-)/(1 - F_-)$
10	ε_n	$(1 - \frac{1}{n}\max_\sigma \sum_j n_{\sigma(j),j})/(1 - 1/\max(K, K'))$

Note: (1) $m = \sum_{i,j}\binom{n_{ij}}{2}$, $m_1 = \sum_i\binom{n_{i\cdot}}{2}$, $m_2 = \sum_j\binom{n_{\cdot j}}{2}$, $M = \binom{n}{2}$.
(2) $p_i = n_{i\cdot}/n$, $p_j = n_{\cdot j}/n$, $p_{ij} = n_{ij}/n$.
(3) Refer to Table 23.1 for F, and Procedure 1 for F_-.

negative measures—a lower value implies a better clustering—they are normalized by modifying Equation (23.1) as $S_n = (S - \min(S))/(E(S) - \min(S))$.

Finally, there are some interrelationships between these measures as follows.

Proposition 23.2.1

$$\begin{array}{ll} (1) & (R_n \equiv \Gamma'_n) \Leftrightarrow (J'_n \equiv MS'_n) \\ (2) & \Gamma_n \equiv \Gamma \end{array}$$

The above proposition indicates that the normalized Hubert Γ statistic I (Γ_n) is the same as Γ. Also, the normalized Rand statistic (R_n) is the same as the normalized Hubert Γ statistic II (Γ'_n). In addition, the normalized Rand statistic (R_n) is equivalent to J'_n, which is the same as MS'_n. Therefore, there are only three independent normalized measures, R_n, FM_n, and Γ_n, needed for further study. The proposition can be easily proved by mathematical transformation, and due to the space limitation, the proof is omitted.

The Normalizations of VI ***and*** VD. Another normalization scheme is formalized as

$$S_n = \frac{S - \min(S)}{\max(S) - \min(S)} \tag{23.3}$$

Some measures, such as VI and VD, often take this scheme. However, to know the exact maximum and minimum values is often impossible. So it usually turns to a reasonable approximation, e.g., the upper bound for the maximum or the lower bound for the minimum.

When the cluster structure matches the class structure perfectly, $VI = 0$. So, $\min(VI) = 0$. However, finding the exact value of $\max(VI)$ is computationally infeasible. Meila [42] suggested using $2\log\max(K, K')$ to approximate $\max(VI)$, so the normalized VI is $\frac{VI}{2\log\max(K,K')}$.

The VD in Table 23.1 can be regarded as a normalized measure. In this measure, $2n$ has been taken as the upper bound [60], and $\min(VD) = 0$.

However, the above normalized VI and VD cannot well capture the uniform effect of K-means, because the proposed upper bound for VI or VD is not tight enough. Two new tighter upper bounds are introduced as follows.

Lemma 23.2.4 *Let random variables C and P denote the class and cluster sizes, respectively, and* $H(\cdot)$ *be the entropy function; then* $VI \leq H(C) + H(P) \leq 2\log\max(K', K)$.

Lemma 23.2.4 gives a tighter upper bound $H(C) + H(P)$ than $2\log\max(K', K)$ which was provided by Meila [42]. With this new upper bound, the normalized VI_n can be calculated as shown in Table 23.5. In addition, if $H(P)/2 + H(C)/2$ is used as the upper bound to normalize mutual information, the VI_n can be equivalent to the normalized mutual information MI_n ($VI_n + MI_n = 1$).

Lemma 23.2.5 *Let* $n_{i\cdot}$, $n_{\cdot j}$, *and* n *be the values in Table 23.2, then* $VD \leq (2n - \max_i n_{i\cdot} - \max_j n_{\cdot j})/2n \leq 1$.

Due to space limitations, the proofs are omitted. The above two lemmas imply that the tighter upper bounds of VI and VD are the functions of the class and cluster sizes. Using these two new upper bounds, the normalized VI_n and VD_n can be derived as in Table 23.5.

The Normalization of F ***and*** ε have seldom been discussed in the literature. Since $\max(F) = 1$, now the goal is to find a tight lower bound for F, which can be found by Procedure 1.

With Procedure 1, the following lemma can find a lower bound for F.

Lemma 23.2.6 *Given* F_- *computed by Procedure 1,* $F \geq F_-$.

Procedure 1: The computation of F_{-}.

Let $n^* = \max_i n_{i\cdot}$.
Sort the class sizes so that $n_{\cdot[1]} \leq n_{\cdot[2]} \leq \cdots \leq n_{\cdot[K']}$.
Let $a_j = 0$, for $j = 1, 2, \cdots, K'$.
for $j = 1 : K'$
  if $n^* \leq n_{\cdot[j]}$, $a_j = n^*$, **break**.
  else $a_j = n_{\cdot[j]}$, $n^* \leftarrow n^* - n_{\cdot[j]}$.
$F_{-} = (2/n) \sum_{j=1}^{K'} a_j / (1 + \max_i n_{i\cdot}/n_{\cdot[j]})$.

PROOF. It is easy to show

$$F = \sum_j \frac{n_{\cdot j}}{n} \max_i \frac{2n_{ij}}{n_{i\cdot} + n_{\cdot j}} \geq \frac{2}{n} \max_i \sum_j \frac{n_{ij}}{n_{i\cdot}/n_{\cdot j} + 1} \tag{23.4}$$

Consider an optimization problem as follows:

$$\min_{x_{ij}} \sum_j \frac{x_{ij}}{n_{i\cdot}/n_{\cdot j} + 1}$$

$$s.t. \sum_j x_{ij} = n_{i\cdot};\ \forall j,\ x_{ij} \leq n_{\cdot j};\ \forall j,\ x_{ij} \in \mathbb{Z}_+$$

For this optimization problem, to have the minimum objective value, as many objects as possible need to be assigned to the cluster with highest $n_{i\cdot}/n_{\cdot j} + 1$, or equivalently, with smallest $n_{\cdot j}$. Let $n_{\cdot[0]} \leq n_{\cdot[1]} \leq \cdots \leq n_{\cdot[K']}$ where the virtual $n_{\cdot[0]} = 0$, and assume $\sum_{j=0}^{l} n_{\cdot[j]} < n_{i\cdot} \leq \sum_{j=0}^{l+1} n_{\cdot[j]}$, $l \in \{0, 1, \cdots, K'-1\}$. The optimal solution is given as

$$x_{i[j]} = \begin{cases} n_{\cdot[j]},\ 1 \leq j \leq l \\ n_{i\cdot} - \sum_{k=1}^{l} n_{\cdot[k]},\ j = l+1 \\ 0,\ l+1 < j \leq K' \end{cases}$$

Therefore, according to (23.4), $F \geq \frac{2}{n} \max_i \sum_{j=1}^{K'} \frac{x_{i[j]}}{n_{i\cdot}/n_{\cdot[j]} + 1}$.

Let $F_i = \frac{2}{n} \sum_{j=1}^{K'} \frac{x_{i[j]}}{n_{i\cdot}/n_{\cdot[j]} + 1} = \frac{2}{n} \sum_{j=1}^{K'} \frac{x_{i[j]}/n_{i\cdot}}{1/n_{\cdot[j]} + 1/n_{i\cdot}}$. Denote "$x_{i[j]}/n_{i\cdot}$" by "$y_{i[j]}$", and "$\frac{1}{1/n_{\cdot[j]} + 1/n_{i\cdot}}$" by "$p_{i[j]}$", then $F_i = \frac{2}{n} \sum_{j=1}^{K'} p_{i[j]} y_{i[j]}$. Next, it remains to show

$$\arg\max_i F_i = \arg\max_i n_{i\cdot}$$

Assume $n_{i\cdot} < n_{i'\cdot}$, and for some l, $\sum_{j=0}^{l} n_{\cdot[j]} < n_{i\cdot} \leq \sum_{j=0}^{l+1} n_{\cdot[j]}$, $l \in \{0, 1, \cdots, K'-1\}$. This implies that

$$y_{i[j]} \begin{cases} \geq y_{i'[j]},\ 1 \leq j \leq l \\ \leq y_{i'[j]},\ l+1 < j \leq K' \end{cases}$$

Since $\sum_{j=1}^{K'} y_{i[j]} = \sum_{j=1}^{K'} y_{i'[j]} = 1$ and $j \uparrow\ \Rightarrow\ p_{i[j]} \uparrow$, thus $\sum_{j=1}^{K'} p_{i[j]} y_{i[j]} \leq \sum_{j=1}^{K'} p_{i[j]} y_{i'[j]}$. Furthermore, according to the definition of $p_{i[j]}$, $p_{i[j]} \leq p_{i'[j]}$, $\forall\ j \in \{1, \cdots, K'\}$. Therefore,

$$F_i = \frac{2}{n} \sum_{j=1}^{K'} p_{i[j]} y_{i[j]} \leq \frac{2}{n} \sum_{j=1}^{K'} p_{i[j]} y_{i'[j]} \leq \frac{2}{n} \sum_{j=1}^{K'} p_{i'[j]} y_{i'[j]} = F_i'$$

which implies that $n_{i\cdot} \leq n_{i'\cdot}$ is the sufficient condition for $F_i \leq F_i'$. Therefore, by Procedure 1, $F_{-} = \max_i F_i$, which finally leads to $F \geq F_{-}$. □

Therefore, $F_n = (F - F_-)/(1 - F_-)$, as listed in Table 23.5. Finally, the following lemma provides an upper bound of ε.

Lemma 23.2.7 *Given* $K' \leq K$, $\varepsilon \leq 1 - 1/K$.
PROOF. Assume $\sigma_1 : \{1, \cdots, K'\} \rightarrow \{1, \cdots, K\}$ is the optimal mapping of the classes to different clusters, i.e.,

$$\varepsilon = 1 - \frac{\sum_{j=1}^{K'} n_{\sigma_1(j),j}}{n}$$

Then construct a series of mappings $\sigma_s : \{1, \cdots, K'\} \mapsto \{1, \cdots, K\}$ $(s = 2, \cdots, K)$ which satisfy

$$\sigma_{s+1}(j) = \text{mod}(\sigma_s(j), K) + 1, \ \forall j \in \{1, \cdots, K'\}$$

where "mod(x,y)" returns the remainder of positive integer x divided by positive integer y. By definition, σ_s $(s = 2, \cdots, K)$ can also map $\{1, \cdots, K'\}$ to K' different indices in $\{1, \cdots, K\}$ as σ_1. More importantly, $\sum_{j=1}^{K'} n_{\sigma_1(j),j} \geq \sum_{j=1}^{K'} n_{\sigma_s(j),j}$, $\forall s = 2, \cdots, K$, and $\sum_{s=1}^{K} \sum_{j=1}^{K'} n_{\sigma_s(j),j} = n$. Therefore, $\sum_{j=1}^{K'} n_{\sigma_1(j),j} \geq \frac{n}{K}$, which implies $\varepsilon \leq 1 - 1/K$. The proof is completed. □

Therefore, $1 - 1/K$ can be used as the upper bound of ε, and the normalized ε_n is shown in Table 23.5.

23.2.3.2 The *DCV* Criterion

In this section, some experiments are presented to show the importance of *DCV* $(CV_1 - CV_0)$ for selecting validation measures.

Experimental Data Sets. Some synthetic data sets were generated as follows. Assume there is a two-dimensional mixture of two Gaussian distributions. The means of the two distributions are [-2,0] and [2,0]. And their covariance matrices are exactly the same as $[\sigma^2\ 0;\ 0\ \sigma^2]$.

Therefore, given any specific value of σ^2, one can generate a simulated data set with 6000 instances, n_1 instances from the first distribution, and n_2 instances from the second one, where $n_1 + n_2 = 6000$. To produce simulated data sets with imbalanced class sizes, set a series of n_1 values: {3000, 2600, 2200, 1800, 1400, 1000, 600, 200}. If $n_1 = 200$, $n_2 = 5800$, the data set is highly imbalanced with $CV_0 = 1.320$. For each mixture model, 8 simulated data sets were generated with CV_0 ranging from 0 to 1.320. Further, to produce data sets with different clustering tendencies, set a series of σ^2 values: {0.5, 1, 1.5, 2, 2.5, 3, 3.5, 4, 4.5, 5}. As σ^2 increases, the mixture model tends to be more unidentifiable. Finally, for each pair of σ^2 and n_1, the sampling was repeated 10 times to have the average performance evaluation. In summary, $8 \times 10 \times 10 = 800$ data sets were produced. Figure 23.1 shows a sample data set with $n_1 = 1000$ and $\sigma^2 = 2.5$.

A sampling on a real-world data set `hitech` was also conducted to get some sample data sets with imbalanced class distributions. This data set was derived from the San Jose Mercury newspaper articles [59], which contains 2301 documents about computers, electronics, health, medical, research, and technology. Each document is characterized by 126373 terms, and the class sizes are 485, 116, 429, 603, 481, and 187. Carefully setting the sampling ratio for each class, 8 sample data sets were extracted with the class-size distributions (CV_0) ranging from 0.490 to 1.862, as shown in Table 23.6. For each data set, the sampling was repeated 10 times to observe the averaged clustering performance.

Experimental Tools. The MATLAB 7.1 [40] and CLUTO 2.1.2 [27] implementations of K-means were employed for the experiment. The MATLAB version with the squared Euclidean distance is suitable for low-dimensional and dense data sets, while CLUTO with the cosine similarity is used to handle high-dimensional and sparse data sets. Note that the number of clusters, i.e., K, was set to match the number of "true" classes.

The Application of Criterion 1. Here, how Criterion 1 can be applied for selecting measures is

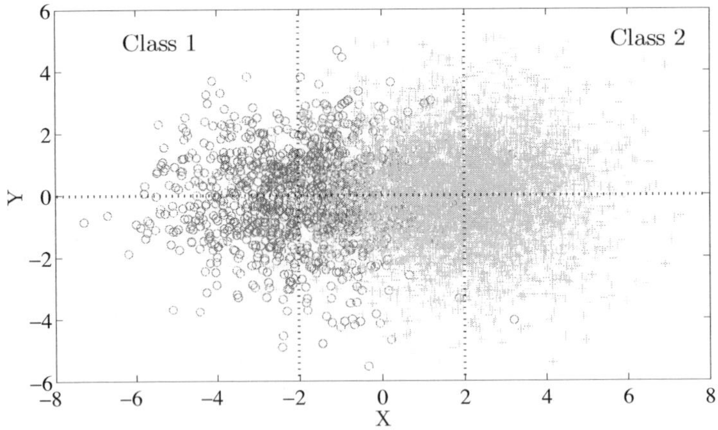

FIGURE 23.1 (See color insert): A simulated data set ($n_1 = 1000$, $\sigma^2 = 2.5$).

presented. As pointed out in Section 23.2.2.1, K-means tends to have the uniform effect on imbalanced data sets. This implies that for data sets with skewed class distributions, the clustering results by K-means tend to be away from "true" class distributions.

To further illustrate this, take a look at Figure 23.2(a) of the simulated data sets. As can be seen, for the extreme case of $\sigma^2 = 5$, the DCV values decrease as the CV_0 values increase. Note that DCV values are usually negative since K-means tends to produce clustering results with relative uniform cluster sizes ($CV_1 < CV_0$). This means that when data become more skewed, the clustering results by K-means tend to be worse. Therefore, the selection of measures can be done by observing the relationship between the measures and the DCV values. As the DCV values go down, the good measures are expected to show worse clustering performances. In this experiment, the MATLAB version of K-means was applied.

A similar trend can be found in Figure 23.2(b) of the sampled data sets. That is, as the CV_0 values go up, the DCV values decrease, which implies worse clustering performances. Indeed, DCV is a good indicator for finding the measures which cannot capture the uniform effect by K-means clustering. In this experiment, the CLUTO version of K-means clustering was applied.

In Section 23.2.4, the Kendall's rank correlation is used (κ) [30] to measure the relationships between external validation measures and DCV. Note that, $\kappa \in [-1, 1]$. $\kappa = 1$ indicates a perfect positive rank correlation, whereas $\kappa = -1$ indicates an extremely negative rank correlation.

TABLE 23.6: The Sizes of the Sampled Data Sets

Data Set	1	2	3	4	5	6	7	8
Class 1	100	90	80	70	60	50	40	30
Class 2	100	90	80	70	60	50	40	30
Class 3	100	90	80	70	60	50	40	30
Class 4	250	300	350	400	450	500	550	600
Class 5	100	90	80	70	60	50	40	30
Class 6	100	90	80	70	60	50	40	30
CV_0	0.49	0.686	0.88	1.078	1.27	1.47	1.666	1.862

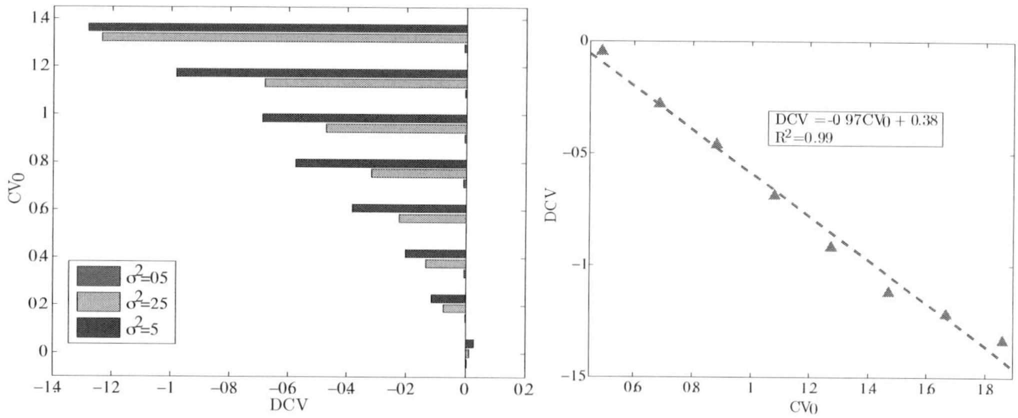

(a) Simulated Data Sets.

(b) Sampled Data Sets.

FIGURE 23.2: Relationship of CV_0 and DCV.

23.2.3.3 The Effect of Normalization

The importance of measure normalization is presented in this section. Along this line, K-means clustering is first applied on the simulated data sets with $\sigma^2 = 5$ and the sampled data sets from `hitech`. Then, both unnormalized and normalized measures are used for cluster validation. Finally, the rank correlation between *DCV* and the measures are computed and the results are shown in Table 23.7.

As can be seen in the table, if the unnormalized measures are used to do cluster validation, only three measures, namely R, Γ, Γ', have strong consistency with *DCV* on both groups of data sets. VI, VD and MS even show strong conflict with *DCV* on the sampled data sets, since their κ values are all close to -1 on sampled data. In addition, notice that F, ε, J, and FM show weak correlation with DCV.

Table 23.7 shows the rank correlations between *DCV* and the normalized measures. As can be seen, all the normalized measures show perfect consistency with *DCV* except for F_n and ε_n. This indicates that the normalization is crucial for evaluating K-means clustering. The proposed bounds for the measures are tight enough to capture the uniform effect in the clustering results.

In Table 23.7, it can be observed that both F_n and ε_n are not consistent with *DCV*. This indicates that normalization does not help F and ε too much. The reason is that the proposed lower bound for F and upper bound for ε are not very tight. Indeed, the normalizations of F and ε are very challenging due to the fact that they both exploit relatively complex optimization schemes in

TABLE 23.7: The Correlation between *DCV* and the Validation Measures

κ	VI	VD	MS	ε	F	R	J	FM	Γ	Γ'
Simulated Data	***-0.71***	***0.79***	***-0.79***	1.00	1.00	1.00	0.91	***0.71***	1.00	1.00
Sampled Data	***-0.93***	***-1.00***	***-1.00***	***0.50***	***0.21***	1.00	***0.50***	***-0.43***	0.93	1.00
κ	VI_n	VD_n	MS'_n	ε_n	F_n	R_n	J'_n	FM_n	Γ_n	Γ'_n
Simulated Data	1.00	1.00	1.00	1.00	1.00	1.00	1.00	1.00	1.00	1.00
Sampled Data	1.00	1.00	1.00	***0.50***	***0.79***	1.00	1.00	1.00	0.93	1.00

Note: Poor or even negative correlations have been highlighted by the bold and italic fonts.

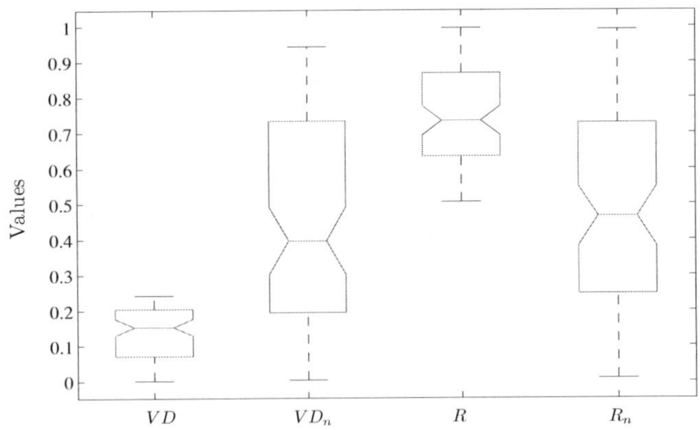

FIGURE 23.3: Unnormalized and normalized measures.

the computations. As a result, it is not easy to compute the expected values from a multivariate hypergeometric distribution perspective, and it is also difficult to find tighter bounds.

Nevertheless, the above experiments show that the normalization is very valuable. In addition, Figure 23.3 shows the cluster validation results of the measures on all the simulated data sets with σ^2 ranging from 0.5 to 5. It is clear that the normalized measures have much wider value range than the unnormalized ones along [0, 1]. This indicates that the values of normalized measures are more spread in [0, 1].

In summary, to compare cluster validation results across different data sets, normalized measures should be used.

23.2.4 Measure Properties

Measure properties, which can serve as the guidance for the selection of measures, are investigated in this section.

23.2.4.1 The Consistency Between Measures

Here, the consistency between a pair of measures is defined as the similarity between their rankings on a series of clustering results. The similarity is measured by the Kendall's rank correlation. And the clustering results are produced by the CLUTO version of K-means clustering on 29 benchmark real-world data sets listed in Table 23.8. In the experiment, for each data set, the cluster number is set to be the same as the "true" class number.

Figures 23.4(a) and 23.4(b) show the correlations between the unnormalized and normalized measures, respectively. One interesting observation is that the normalized measures have a much stronger consistency than the unnormalized measures. For instance, the correlation between VI and R is merely -0.21, but it reaches 0.74 for the corresponding normalized measures. This observation indeed implies that the normalized measures tend to give more robust validation results, which also agrees with previous analysis.

According to the colors in Figure 23.4(b) on the normalized measures, it can be roughly found that R_n, Γ'_n, J'_n, MS'_n, FM_n, and Γ_n are more similar to one another, while VD_n, F_n, VI_n, and ε_n show inconsistency with others in varying degrees. To gain the precise understanding, a hierarchical

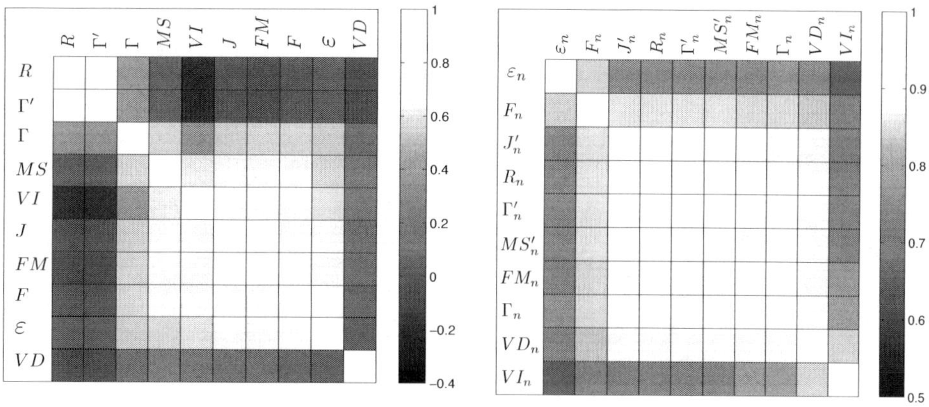

(a) Unnormalized Measures.

(b) Normalized Measures.

FIGURE 23.4 (See color insert): Correlations of the measures.

clustering on the measures is performed by using their correlation matrix. The resultant hierarchy can be found in Figure 23.5 ("s" means the similarity). As mentioned before, R_n, Γ'_n, J'_n, and MS'_n are equivalent, so they have perfect correlation with one another and form the first group. The second group contains FM_n and Γ_n. These two measures behave similarly, and have just slightly weaker consistency with the measures in the first group. Finally, VD_n, F_n, ε_n, and VI_n have obviously weaker consistency with other measures in a descending order.

Furthermore, the data sets in Table 23.8 are divided into two repositories to explore the source of the inconsistency among the measures, where $\Re_1$ contains data sets with $CV_0 < 0.8$, and $\Re_2$ contains the rest. After computing the correlation matrices of the measures on the two repositories (denoted by $M(\Re_1)$ and $M(\Re_2)$), their difference ($M(\Re_1) - M(\Re_2)$) can be calculated as shown in Table 23.9. As can be seen, roughly speaking, all the measures except VI_n show weaker consistency with one another on data sets in $\Re_2$. In other words, while VI_n acts in the opposite way, most measures tend to disagree with one another on data sets with highly imbalanced classes.

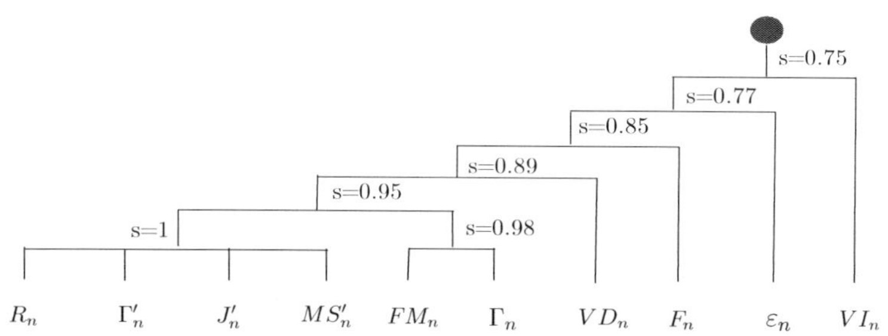

FIGURE 23.5: The measure similarity hierarchy.

TABLE 23.8: The Benchmark Data Sets

Data Set	Source	#Class	#Case	#Feature	CV_0
cacmcisi	CA/CI	2	4663	41681	0.53
classic	CA/CI	4	7094	41681	0.55
cranmed	CR/ME	2	2431	41681	0.21
fbis	TREC	17	2463	2000	0.96
hitech	TREC	6	2301	126373	0.50
k1a	WebACE	20	2340	21839	1.00
k1b	WebACE	6	2340	21839	1.32
la1	TREC	6	3204	31472	0.49
la2	TREC	6	3075	31472	0.52
la12	TREC	6	6279	31472	0.50
mm	TREC	2	2521	126373	0.14
ohscal	OHSUMED	10	11162	11465	0.27
re0	Reuters	13	1504	2886	1.50
re1	Reuters	25	1657	3758	1.39
sports	TREC	7	8580	126373	1.02
tr11	TREC	9	414	6429	0.88
tr12	TREC	8	313	5804	0.64
tr23	TREC	6	204	5832	0.93
tr31	TREC	7	927	10128	0.94
tr41	TREC	10	878	7454	0.91
tr45	TREC	10	690	8261	0.67
wap	WebACE	20	1560	8460	1.04
DLBCL	KRBDSR	3	77	7129	0.25
Leukemia	KRBDSR	7	325	12558	0.58
LungCancer	KRBDSR	5	203	12600	1.36
ecoli	UCI	8	336	7	1.16
pageblocks	UCI	5	5473	10	1.95
letter	UCI	26	20000	16	0.03
pendigits	UCI	10	10992	16	0.04
MIN	-	2	77	7	0.03
MAX	-	26	20000	126373	1.95

Note: CA-CACM, CI-CISI, CR-CRANFIELD, ME-MEDLINE.

TABLE 23.9: $M(\Re_1) - M(\Re_2)$

	R_n	FM_n	Γ_n	VD_n	F_n	ε_n	VI_n
R_n	0.00	0.09	0.13	0.08	0.10	0.26	-0.01
FM_n	0.09	0.00	0.04	0.00	0.10	0.22	-0.10
Γ_n	0.13	0.04	0.00	0.04	0.14	0.22	-0.06
VD_n	0.08	0.00	0.04	0.00	0.05	0.20	-0.18
F_n	0.10	0.10	0.14	0.05	0.00	0.08	-0.08
ε_n	0.26	0.22	0.22	0.20	0.08	0.00	0.04
VI_n	-0.01	-0.10	-0.06	-0.18	-0.08	0.04	0.00

23.2.4.2 Properties of Measures

In this section, some key properties of external clustering validation measures are discussed.

The Sensitivity. The measures have different sensitivity to the clustering results. It can be illustrated by an example. For two clustering results in Table 23.10, the differences between them are

TABLE 23.10: Two Clustering Results

I	C_1	C_2	C_3	Σ
P_1	**3**	**4**	12	19
P_2	**8**	**3**	12	23
P_3	12	12	0	24
Σ	23	19	24	66

II	C_1	C_2	C_3	Σ
P_1	**0**	**7**	12	19
P_2	**11**	**0**	12	23
P_3	12	12	0	24
Σ	23	19	24	66

TABLE 23.11: The Cluster Validation Results

	R_n	FM_n	Γ_n	VD_n	F_n	ε_n	VI_n
I	0.16	0.16	0.16	0.71	0.32	0.77	0.78
II	0.24	0.24	0.24	0.71	0.32	0.70	0.62

the numbers in bold. Validation results of the measures on these two clusterings are shown in Table 23.11. As can be seen, all the measures show different validation results for the two clusterings except for VD_n and F_n. This implies that VD_n and F_n are less sensitive than other measures. This is due to the fact that both VD_n and F_n use maximum functions, which may lose some information in the contingency matrix. Furthermore, VI_n is the most sensitive measure, since the difference of VI_n values for the two clusterings is the largest.

Impact of the Number of Clusters. The impact of the number of clusters on the validation measures is evaluated on data set `la2` in Table 23.8. Here, the cluster number ranges from 2 to 15. As shown in Figure 23.6, the measurement values for all the measures will change as the cluster numbers increase. However, the normalized measures including VI_n, VD_n, and R_n can capture the same optimal cluster number 5. Similar results can also be observed for other normalized measures, such as F_n, FM_n, and Γ_n.

Property 23.2.1 (n-Invariance) *For a contingence matrix M and a positive integer λ, a measure O is n-invariant, if $O(\lambda M) = O(M)$, where n is the number of objects.*

A Summary of Math Properties. Five math properties of measures are listed as follows (see Table 23.12). Due to space limitation, the proofs are omitted here.

Property 23.2.2 (Symmetry) *A measure O is symmetric, if $O(M^T) = O(M)$ for any contingence matrix M.*

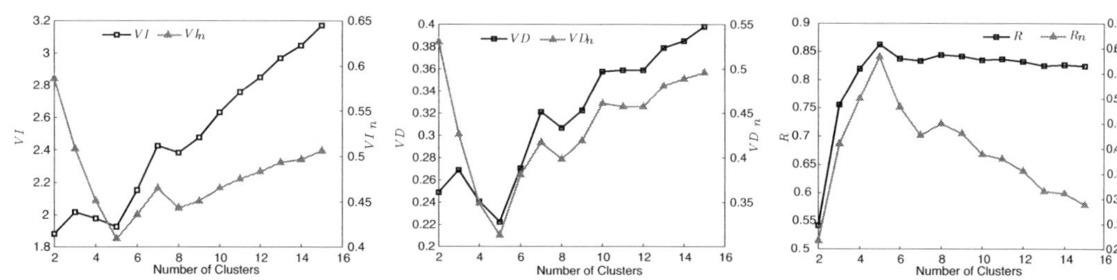

(a) (b) (c)

FIGURE 23.6: Impact of the number of clusters.

TABLE 23.12: Math Properties of Measures

	F_n	VI_n	VD_n	ε_n	R_n	FM_n	Γ_n
P1	No	Yes	Yes	Yes**	Yes	Yes	Yes
P2	Yes	Yes	Yes	Yes	No	No	No
P3	Yes*	Yes*	Yes*	Yes*	No	No	No
P4	No	Yes	Yes	No	No	No	No
P5	Yes	Yes	Yes	Yes	Yes	Yes	Yes

Note: Yes* — Yes for the unnormalized measures.
Yes** — Yes for $K = K'$.

The *symmetry* property treats the predefined class structure as one of the partitions. Therefore, the task of cluster validation is the same as the comparison of partitions. This means transposing two partitions in the contingency matrix should not bring any difference to the measure value. This property is not true for F_n which is a typical measure in asymmetry. Also, ε_n is symmetric if and only if $K = K'$.

Intuitively, a mathematically sound validation measure should satisfy the *n-invariance* property. However, three measures, namely, R_n, FM_n, and Γ_n cannot fulfill this requirement. Nevertheless, they can still be treated as the asymptotically n-invariant measures, since they tend to be n-invariant with the increase of n.

Property 23.2.3 (Convex additivity) *Let* $P = \{P_1, \cdots, P_K\}$ *be a clustering,* P' *be a refinement of* P[1]*, and* P'_l *be the partitioning induced by* P' *on* P_l*. Then a measure* O *is convex additive, if* $O(M(P,P')) = \sum_{l=1}^{K} \frac{n_l}{n} O(M(I_{P_l}, P'_l))$*, where* n_l *is the number of data points in* P_l*,* I_{P_l} *represents the partitioning on* P_l *into one cluster, and* $M(X,Y)$ *is the contingency matrix of* X *and* Y*.*

The *convex additivity* property was introduced by Meila [42]. It requires the measures to show additivity along the lattice of partitions. Unnormalized measures including F, VD, VI, and ε hold this property. However, none of the normalized measures studied in this chapter holds this property.

Property 23.2.4 (Left-domain-completeness) *A measure* O *is left-domain-complete, if for any contingence matrix* M *with statistically independent rows and columns,*

$$O(M) = \begin{cases} 0, & O \textit{ is a positive measure} \\ 1, & O \textit{ is a negative measure} \end{cases}$$

When the rows and columns in the contingency matrix are statistically independent, the poorest values of the measures are expected to be seen, i.e., 0 for positive measures and 1 for negative measures. Among all the measures, however, only VI_n and VD_n can meet this requirement.

Property 23.2.5 (Right-domain-completeness) *A measure* O *is right-domain-complete, if for any contingence matrix* M *with perfectly matched rows and columns,*

$$O(M) = \begin{cases} 1, & O \textit{ is a positive measure} \\ 0, & O \textit{ is a negative measure} \end{cases}$$

This property requires measures to show optimal values when the class structure matches the cluster structure perfectly. The above normalized measures hold this property.

[1] "P' be a refinement of P" means P' is the descendant node of node P in the lattice of partitions. See [42] for details.

23.2.4.3 Discussions

In a nutshell, among 16 external validation measures shown in Table 23.1, it is first known that Mirkin metric (M) is equivalent to Rand statistic (R), and micro-average precision (MAP) and Goodman–Kruskal coefficient (GK) are equivalent to the purity measure (P) by observing their computational forms. Therefore, the scope of the study reduces from 16 measures to 13 measures. In Section 23.2.2, analysis shows that P, mutual information (MI), and entropy (E) are defective measures for evaluating K-means clustering. Also, it is proved that variation of information (VI) is an improved version of MI and E, and van Dongen criterion (VD) is an improved version of P. As a result, the selection pool is further reduced to 10 measures.

In addition, as shown in Section 23.2.3, it is necessary to use the normalized measures for evaluating K-means clustering, since the normalized measures can capture the uniform effect by K-means and allow the evaluation of different clustering results on different data sets. Proposition 23.2.1 on page 579 shows that the normalized Rand statistic (R_n) is the same as the normalized Hubert Γ statistic II (Γ'_n). Also, the normalized Rand statistic is equivalent to J'_n, which is the same as MS'_n. Therefore, only R_n needs further consideration and J'_n, Γ'_n as well as MS'_n can be excluded. The results in Section 23.2.3 show that the normalized F-measure (F_n) and classification error (ε_n) cannot well capture the uniform effect by K-means. Also, these two measures do not satisfy some math properties in Table 23.12. As a result, they are excluded as well. Now, there are only five normalized measures left: VI_n, VD_n, R_n, FM_n, and Γ_n. Figure 23.5 shows that the validation performances of R_n, FM_n, and Γ_n are very similar to each other. Therefore, only R_n needs to be considered.

Based on the above study, it is most suitable to use the normalized van Dongen criterion (VD_n) in most of the general cases, since VD_n has a simple computation form, satisfies all mathematically sound properties as shown in Table 23.12, and can measure well on the data with imbalanced class distributions. However, for the case that the clustering performances are hard to distinguish, one may want to use the normalized variation of information (VI_n) instead,[2] since VI_n has high sensitivity on detecting the clustering changes. Finally, R_n can also be used as a complementary to the above two measures.

23.3 Internal Clustering Validation Measures

In the literature, a number of internal validation measures for crisp clustering have been proposed. In this section, an organized study on a suite of 12 widely used internal clustering validation measures as shown in Table 23.13 are provided for different clustering algorithms. These measures represent a good coverage of the internal validation measures available in different fields such as data mining, information retrieval, machine learning, and statistics. The properties of these validation measures are investigated in six different aspects: monotonicity, noise, density, subclusters, skewed distribution, and arbitrary shapes data. For each aspect, a synthetic data set which best represents the property is generated for studies. Results and discussions will be presented at the end of this section. First, some basic concepts of internal clustering validation measures, as well as the suite of 12 widely used internal clustering validation measures, are introduced.

23.3.1 An Overview of Internal Clustering Validation Measures

As the goal of clustering is to make objects within the same cluster similar and objects in different clusters distinct, internal validation measures are often based on the following two criteria [57, 68, 33].

---

[2] Note that the normalized variation of information is equivalent to the normalized mutual information.

TABLE 23.13: Internal Clustering Validation Measures.

	Measure	Definition
1	$RMSSTD$[1]	$\{\sum_i \sum_{x\in C_i} \parallel x-c_i \parallel^2/[P\sum_i(n_i-1)]\}^{\frac{1}{2}}$
2	R-squared (RS)	$(\sum_{x\in D} \parallel x-c \parallel^2 - \sum_i \sum_{x\in C_i} \parallel x-c_i \parallel^2)/\sum_{x\in D} \parallel x-c \parallel^2$
3	Modified Hubert Γ statistic (Γ)	$\frac{2}{n(n-1)} \sum_{x\in D}\sum_{y\in D} d(x,y) d_{x\in C_i, y\in C_j}(c_i,c_j)$
4	Calinski-Harabasz index (CH)	$\frac{\sum_i n_i d^2(c_i,c)/(NC-1)}{\sum_i \sum_{x\in C_i} d^2(x,c_i)/(n-NC)}$
5	I index (I)	$(\frac{1}{NC}\cdot\frac{\sum_{x\in D} d(x,c)}{\sum_i\sum_{x\in C_i} d(x,c_i)}\cdot \max_{i,j} d(c_i,c_j))^P$
6	Dunn's indices (D)	$\min_i\{\min_j(\frac{\min_{x\in C_i, y\in C_j} d(x,y)}{\max_k\{\max_{x,y\in C_k} d(x,y)\}})\}$
7	Silhouette index (S)	$\frac{1}{NC}\sum_i\{\frac{1}{n_i}\sum_{x\in C_i}\frac{b(x)-a(x)}{\max[b(x),a(x)]}\}$ $a(x) = \frac{1}{n_i-1}\sum_{y\in C_i, y\neq x} d(x,y), b(x) = \min_{j,j\neq i}[\frac{1}{n_j}\sum_{y\in C_j} d(x,y)]$
8	Davies-Bouldin index (DB)	$\frac{1}{NC}\sum_i \max_{j,j\neq i}\{[\frac{1}{n_i}\sum_{x\in C_i} d(x,c_i)+\frac{1}{n_j}\sum_{x\in C_j} d(x,c_j)]/d(c_i,c_j)\}$
9	Xie-Beni index (XB)	$[\sum_i\sum_{x\in C_i} d^2(x,c_i)]/[n\cdot min_{i,j\neq i} d^2(c_i,c_j)]$
10	SD validity index (SD)	$Dis(NC_{max})Scat(NC)+Dis(NC)$ $Scat(NC) = \frac{1}{NC}\sum_i \parallel \sigma(C_i) \parallel / \parallel \sigma(D) \parallel$ $Dis(NC) = \frac{max_{i,j} d(c_i,c_j)}{min_{i,j} d(c_i,c_j)}\sum_i(\sum_j d(c_i,c_j))^{-1}$
11	S_Dbw validity index (S_Dbw)	$Scat(NC)+Dens\_bw(NC)$ $Dens\_bw(NC) = \frac{1}{NC(NC-1)}\sum_i[\sum_{j,j\neq i}\frac{\sum_{x\in C_i\cup C_j} f(x,u_{ij})}{max\{\sum_{x\in C_i} f(x,c_i),\sum_{x\in C_j} f(x,c_j)\}}]$
12	$CVNN$[2] index	$Sep(NC,k)/\max_{NC} Sep(NC,k)+Com(NC)/\max_{NC} Com(NC)$ $Com(NC) = \sum_i[\frac{2}{n_i\cdot(n_i-1)}\sum_{x,y\in C_i} d(x,y)]$ $Sep(NC,k) = \max_i(\frac{1}{n_i}\sum_{j=1,2,\ldots,n_i}\frac{q_j}{k})$

Note: D: data set; n: number of objects in D; c: center of D; P: attributes number of D;
NC: number of clusters; C_i: the ith cluster; n_i: number of objects in C_i; c_i: center of C_i;
k: number of nearest neighbors; q_j: number of C_i's jth object's nearest neighbors which are not in cluster C_i;
$\sigma(C_i)$: variance vector of C_i; $d(x,y)$: distance between x and y; $\parallel X_i \parallel = (X_i^T \cdot X_i)^{1/2}$.
[1] $RMSSTD$: Root-mean-square standard deviation.
[2] $CVNN$: Clustering Validation index based on Nearest Neighbors.

I. Compactness. This measures how closely related the objects in a cluster are. A group of measures evaluates cluster compactness based on variance. Lower variance indicates better compactness. In addition, numerous measures estimate the cluster compactness based on distance, such as maximum or average pairwise distance, and maximum or average center-based distance.

II. Separation. This measures how distinct or well-separated a cluster is from other clusters. For example, the pairwise distances between cluster centers and the pairwise minimum distances between objects in different clusters are widely used as measures of separation. Also, measures based on density are used in some indices.

The general procedure to determine the best partition and optimal cluster number of a set of objects by using internal validation measures is as follows.

Step 1: Initialize a list of clustering algorithms which will be applied to the data set.

Step 2: For each clustering algorithm, use different combinations of parameters to get different clustering results.

Step 3: Compute the corresponding internal validation index of each partition which was obtained in Step 2.

Step 4: Choose the best partition and the optimal cluster number according to the criteria.

Table 23.13 lists the measures to be studied in this section. The "Definition" column gives the computation forms of the measures. While most indices, such as *DB*, *XB*, and *S_Dbw*, consider both of the evaluation criteria (compactness and separation) in the way of ratio or summation some, such as *RMSSTD*, *RS*, and Γ consider only one aspect. The 12 measures are briefly introduced as follows.

The root-mean-square standard deviation (*RMSSTD*) is the square root of the pooled sample variance of all the attributes [51]. It measures the homogeneity of the formed clusters. R-squared (*RS*) is the ratio of sum of squares between clusters to the total sum of squares of the whole data set. It measures the degree of difference between clusters [51, 19]. The Modified Hubert Γ statistic (Γ) [25] evaluates the difference between clusters by counting the disagreements of pairs of data objects in two partitions.

The Calinski–Harabasz index (*CH*) [7] evaluates the cluster validity based on the average between- and within-cluster sum of squares. Index *I* (*I*) [41] measures separation based on the maximum distance between cluster centers, and measures compactness based on the sum of distances between objects and their cluster center. Dunn's index (*D*) [11] uses the minimum pairwise distance between objects in different clusters as the intercluster separation and the maximum diameter among all clusters as the intracluster compactness. These three indices take a form of $Index = (a \cdot Separation)/(b \cdot Compactness)$, where a and b are weights. The optimal cluster number is determined by maximizing the value of these indices.

The Silhouette index (*S*) [49] validates the clustering performance based on the pairwise difference of between- and within-cluster distances. In addition, the optimal cluster number is determined by maximizing the value of this index.

The Davies–Bouldin index (*DB*) [9] is calculated as follows. For each cluster C, the similarities between C and all other clusters are computed, and the highest value is assigned to C as its cluster similarity. Then the *DB* index can be obtained by averaging all the cluster similarities. The smaller the index is, the better the clustering result is. By minimizing this index, clusters are the most distinct from each other and, therefore, achieve the best partition. The Xie-Beni index (*XB*) [63] defines the intercluster separation as the minimum square distance between cluster centers, and the intracluster compactness as the mean square distance between each data object and its cluster center. The optimal cluster number is reached when the minimum of *XB* is found. Kim and Ramakrishna [32] proposed indices DB^{**} and XB^{**} in 2005 as the improvements of *DB* and *XB*. The two improved measures will be used in this study.

The idea of SD index (*SD*) [22] is based on the concepts of the average scattering and the total separation of clusters. The first term evaluates compactness based on variances of cluster objects, and the second term evaluates separation difference based on distances between cluster centers. The *SD* index is the summation of these two terms, and the optimal number of clusters can be obtained by minimizing the value of *SD*.

The S_Dbw index (*S_Dbw*) [20] takes density into account to measure the intercluster separation. The basic idea is that for each pair of cluster centers, at least one of their densities should be larger than the density of their midpoint. The intracluster compactness is the same as it is in *SD*. Similarly, the index is the summation of these two terms and the minimum value of *S_Dbw* indicates the optimal cluster number.

Different from the existing measures, the Clustering Validation index based on Nearest Neighbors (*CVNN*) [38] evaluates the intercluster separation based on objects that carry the geometrical information of each cluster. Sharing the same idea with kNN consistency, *CVNN* uses dynamic multiple objects as representatives for different clusters in different situations when measuring the intercluster separation. If an object is located in the center of a cluster and is surrounded by objects in the same cluster, it is well separated from other clusters and thus contributes little to the intercluster separation. If an object is located at the edge of a cluster and is surrounded mostly by objects in other clusters, it connects to other clusters tightly and thus contributes a lot to the intercluster separation. *CVNN* also employs the average pairwise distance between objects in the same cluster as the measurement of intracluster compactness. Finally, the *CVNN* index takes a form of the summation of the intercluster separation and the intracluster compactness after the normalization for both of them.

There are some other internal validation measures in the literature [50, 21, 58, 35]. However, some have poor performance while some are designed for data sets with specific structures. Take Composed Density between and within clusters index (*CDbw*) and Symmetry distance-based index (*Sym–index*) for examples. It is hard for *CDbw* to find the representatives for each cluster, which makes the result of *CDbw* unstable. On the other hand, *Sym–index* can handle only data sets which are internally symmetrical. A focused study on the above mentioned 12 internal validation measures will be presented in the following sections, and acronyms for these measures will be used.

23.3.2 Understanding of Internal Clustering Validation Measures

In this section, a study of the 12 internal validation measures mentioned in Section 23.3.1 is presented to investigate the validation properties of different internal validation measures in different aspects, which can be helpful for the index selection. If not mentioned, *K*-means is used [39] (implemented by CLUTO) [27] as the clustering algorithm, and the parameter k is set to be 10 for *CVNN*.

23.3.2.1 The Impact of Monotonicity

The monotonicity of different internal validation indices is studied in this subsection. *K*-means algorithm is applied on the data set *Wellseparated* to get the clustering results for different numbers of clusters. As shown in Figure 23.7, *Wellseparated* is a synthetic data set composed of 1000 data objects, which are well separated into five clusters.

As the results shown in Table 23.14, the first three indices monotonically increase or decrease as the cluster number *NC* increases. On the other hand, the remaining nine indices reach their maximum or minimum value as *NC* equals the true cluster number. There are certain reasons for the monotonicity of the first three indices.

$RMSSTD = \sqrt{SSE/P(n-NC)}$, and *SSE* (Sum of Square Error) decreases as *NC* increases. In practice $NC \ll n$; thus, $n - NC$ can be viewed as a constant number. Therefore, *RMSSTD* decreases

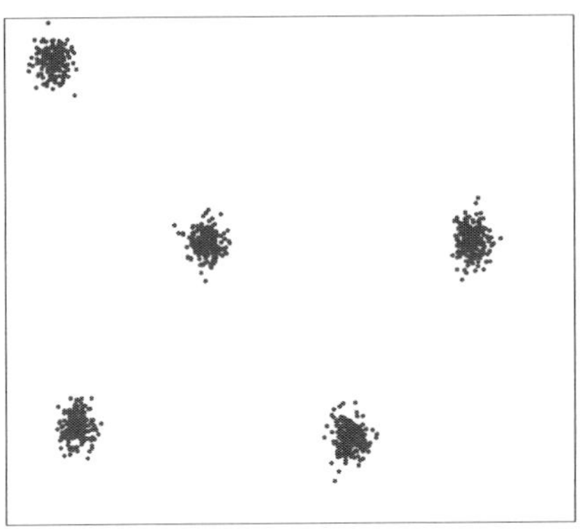

FIGURE 23.7: The data set *Wellseparated*.

TABLE 23.14: Results of the Impact of Monotonicity, True $NC = 5$

	RMSSTD	*RS*	Γ	*CH*	*I*	*D*	*S*	DB^{**}	*SD*	*S_Dbw*	XB^{**}	*CVNN*
2	28.50	0.63	2973	1683	3384	0.49	0.61	0.72	0.22	61.84	0.27	1.00
3	20.80	0.80	3678	2016	5759	0.55	0.71	0.68	0.12	0.15	0.37	0.64
4	14.83	0.90	4007	2968	11230	0.58	0.83	0.52	0.08	0.06	0.50	0.38
5	**3.20**	**0.99**	**4342**	**52863**	**106163**	**2.23**	**0.91**	**0.12**	**0.05**	**0.004**	**0.25**	**0.12**
6	3.08	1.00	4343	45641	82239	0.03	0.72	0.52	0.50	0.07	35.10	0.74
7	2.96	1.00	4344	41291	68894	0.02	0.58	0.80	0.49	0.10	35.10	1.05
8	2.83	1.00	4346	38580	58420	0.01	0.48	1.02	0.54	0.08	36.51	1.11
9	2.72	1.00	4347	36788	50259	0.01	0.39	1.17	0.55	0.11	38.01	1.10

as *NC* increases. Also $RS = (TSS - SSE)/TSS$ (TSS-Total Sum of Squares), and $TSS = SSE + SSB$ (SSB-Between group Sum of Squares) which is a constant number for a certain data set. Thus, *RS* increases as *NC* increases.

From the definition of Γ, only data objects in different clusters will be counted in the equation. As a result, if the data set is divided into two equal clusters, each cluster will have $n/2$ objects, and $n^2/4$ pairs of distances will be actually counted. If the data set is divided into three equal clusters, each cluster will have $n/3$ objects, and $n^2/3$ pairs of distances will be counted. Therefore, with the increasing of the cluster number *NC*, more pairs of distances are counted, which makes Γ increase.

Looking further into these three indices, one can discover that they only take either separation or compactness into account. (*RS* and Γ consider only separation, and *RMSSTD* considers only compactness). As the property of monotonicity, the curves of *RMSSTD*, *RS*, and Γ will be either upward or downward. It is claimed that the optimal cluster number is reached at the shift point of the curves, which is also known as "the elbow" [19]. However, since the judgment of the shift point is very subjective and hard to determine, these three measures are excluded from future studies. The focus will be on the remaining 9 measures.

23.3.2.2 The Impact of Noise

The following study on the data set *Wellseparated.noise* evaluates the influence of noise on internal validation indices. As shown in Figure 23.8, *Wellseparated.noise* is a synthetic data set formulated by adding 5% noise to the data set *Wellseparated*. The cluster numbers selected by indices are shown in Table 23.15. Results show that *D* and *CH* choose the wrong cluster number. There are certain reasons that *D* and *CH* are significantly affected by noise.

D uses the minimum pairwise distance between objects in different clusters ($min_{x \in C_i, y \in C_j} d(x,y)$) as the intercluster separation, and the maximum diameter among all clusters

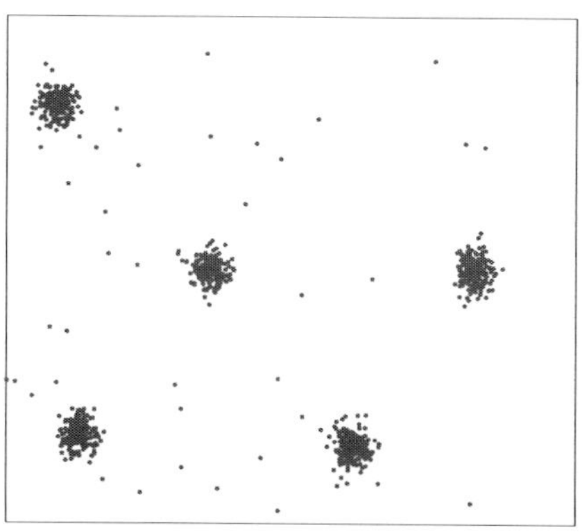

FIGURE 23.8: The data set *Wellseparated.noise*.

TABLE 23.15: Results of the Impact of Noise, True $NC = 5$

	CH	I	D	S	DB^{**}	SD	S_Dbw	XB^{**}	$CVNN$
2	1626	3213	0.0493	0.590	0.739	0.069	20.368	0.264	1.01
3	1846	5073	0.0574	0.670	0.721	0.061	0.523	0.380	0.69
4	2554	9005	**0.0844**	0.783	0.560	0.050	0.087	0.444	0.44
5	10174	**51530**	0.0532	**0.870**	**0.183**	**0.045**	**0.025**	**0.251**	**0.19**
6	**14677**	48682	0.0774	0.802	0.508	0.046	0.044	0.445	0.51
7	12429	37568	0.0682	0.653	0.710	0.055	0.070	0.647	0.92
8	11593	29693	0.0692	0.626	0.863	0.109	0.052	2.404	1.15
9	11088	25191	0.0788	0.596	0.993	0.121	0.056	3.706	0.98

($\max_k\{\max_{x,y \in C_k} d(x,y)\}$) as the intracluster compactness. And the optimal number of clusters can be obtained by maximizing the value of D. When noise is introduced, the intercluster separation can decrease sharply since it uses only the minimum pairwise distance, rather than the average pairwise distance, between objects in different clusters. Thus, the value of D may change dramatically and the corresponding optimal cluster number will be influenced by the noise.

Since $CH = (SSB/SSE) \cdot ((n - NC)/(NC - 1))$ and $((n - NC)/(NC - 1))$ is constant for the same NC, only (SSB/SSE) needs to be considered. By introducing noise, SSE increases in a more significant way compared with SSB. Therefore, for the same NC, CH will decrease by the influence of noise, which makes the value of CH instable. Finally, the optimal cluster number will be affected by noise.

Moreover, the indices other than CH and D will also be influenced by noise in a less sensitive way. Comparing Table 23.15 with Table 23.14, it is clear that the values of other indices change to some degree. If adding 20% noise to the data set *Wellseparated*, the optimal cluster number suggested by I will also be incorrect. Thus, in order to minimize the adverse effect of noise, in practice it is always good to remove noise before clustering.

23.3.2.3 The Impact of Density

A data set with various densities is challenging for many clustering algorithms. Therefore, it is a very interesting topic whether data with different densities also affect the performance of the internal validation measures. A study is conducted on a synthetic data set with different density

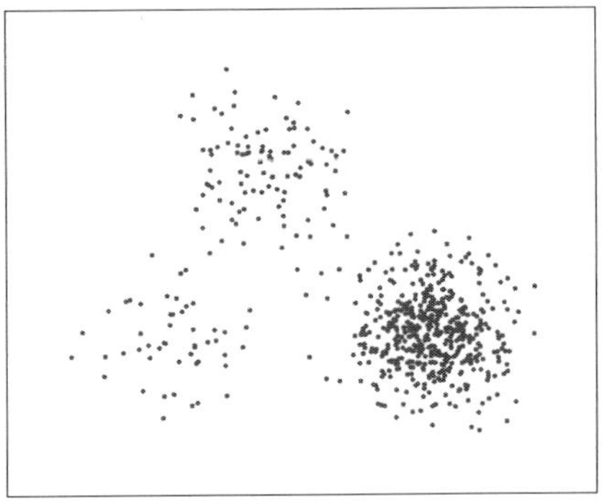

FIGURE 23.9: The data set *Differentdensity*.

TABLE 23.16: Results of the Impact of Density, True $NC = 3$

	CH	I	D	S	DB^{**}	SD	S_Dbw	XB^{**}	$CVNN$
2	1172	**120.1**	0.0493	0.587	0.658	0.705	0.603	0.408	1.03
3	**1197**	104.3	**0.0764**	**0.646**	**0.498**	**0.371**	**0.275**	**0.313**	**0.84**
4	1122	93.5	0.0048	0.463	1.001	0.672	0.401	3.188	0.92
5	932	78.6	0.0049	0.372	1.186	0.692	0.367	3.078	1.22
6	811	59.9	0.0049	0.312	1.457	0.952	0.312	6.192	1.32
7	734	56.1	0.0026	0.278	1.688	1.192	0.298	9.082	1.28
8	657	44.8	0.0026	0.244	1.654	1.103	0.291	8.897	1.26
9	591	45.5	0.0026	0.236	1.696	1.142	0.287	8.897	1.59

named *Differentdensity*. *Differentdensity* has 650 data objects and the details are shown in Figure 23.9. The results listed in Table 23.16 show that only I selects the wrong optimal cluster number.

The reason I does not choose the right cluster number is not easy to explain. One can observe that I keeps decreasing as cluster number NC increases. One possible reason is the uniform effect of the K-means algorithm, which tends to divide objects into relatively equal sizes [64]. I measures compactness based on the sum of distances between objects and their cluster center. When NC is small, objects with high density are likely in the same cluster, which makes the sum of distances remain almost the same. Since most of the objects are in one cluster, the total sum will not change too much. Therefore, as NC increases, I will decrease since NC is in the denominator.

23.3.2.4 The Impact of Subclusters

Subclusters are clusters that are close to each other. Figure 23.10 shows a synthetic data set *Subcluster* which contains five clusters, and four of them are subclusters since they can form two pairs of clusters, respectively. The total number of data objects in *Subcluster* is 1000.

Results presented in Table 23.17 evaluate whether the internal validation measures can handle data set with subclusters. For this data set, D, S, DB^{**}, SD, and XB^{**} get the wrong optimal cluster numbers, while I, CH, S_Dbw, and $CVNN$ have the correct ones. Intercluster separation is supposed to have a sharp decrease when cluster number changes from $NC_{optimal}$ to $NC_{optimal+1}$ [32]. However, for D, S, DB^{**}, SD, and XB^{**}, sharper decreases can be observed at $NC < NC_{optimal}$. The reasons are as follows.

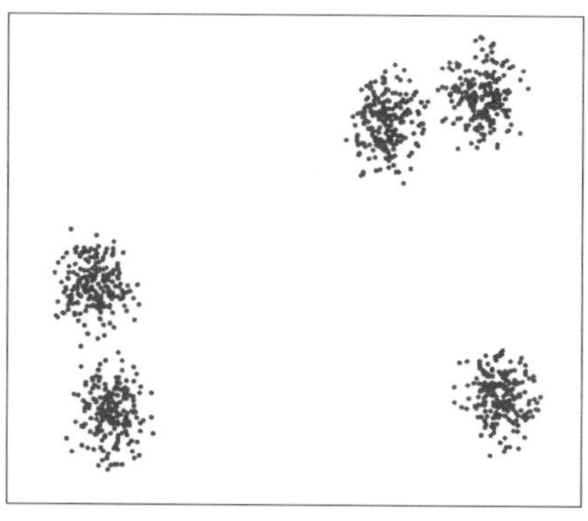

FIGURE 23.10: The data set *Subcluster*.

TABLE 23.17: Results of the Impact of Subclusters, True $NC = 5$

	CH	I	D	S	DB^{**}	SD	S_Dbw	XB^{**}	$CVNN$
2	3474	2616	0.7410	0.736	0.445	0.156	0.207	0.378	1.00
3	7851	5008	**0.7864**	**0.803**	**0.353**	**0.096**	0.056	**0.264**	0.54
4	8670	5594	0.0818	0.737	0.540	0.164	0.039	1.420	0.47
5	**16630**	**9242**	0.0243	0.709	0.414	0.165	**0.026**	1.215	**0.43**
6	14310	7021	0.0243	0.587	0.723	0.522	0.063	12.538	0.79
7	12900	5745	0.0167	0.490	0.953	0.526	0.101	12.978	1.26
8	11948	4803	0.0167	0.402	1.159	0.535	0.105	14.037	1.25
9	11354	4248	0.0107	0.350	1.301	0.545	0.108	14.858	1.06

S uses the average minimum distance between clusters as the intercluster separation. For a data set with subclusters, the intercluster separation will achieve its maximum value when subclusters close to each other are considered as one big cluster. Therefore, the wrong optimal cluster number will be chosen due to subclusters. XB^{**} uses the minimum pairwise distance between cluster centers as the evaluation of separation. For a data set with subclusters, the measure of separation will achieve its maximum value when subclusters close to each other are considered as a big cluster. As a result, the correct cluster number will not be found by using XB^{**}. The reasons for D, SD, and DB^{**} are very similar to the reason of XB^{**}, which will not be elaborated here due to the limit of space.

23.3.2.5 The Impact of Skewed Distributions

It is common that clusters in a data set have unequal sizes. Figure 23.11 shows a synthetic data set *Skewdistribution* with skewed distributions, which contains 1500 data objects. It consists of one large cluster and two small ones. Since K-means has the uniform effect of tending to divide objects into relatively equal sizes, it does not have a good performance when dealing with skewed distributed data sets [65]. In order to demonstrate this statement, four widely used algorithms are employed from four different categories: K-means (prototype-based), DBSCAN (density-based) [12], Agglo (based on average-link, hierarchical) [26], and Chameleon (graph-based) [29]. Each of them are applied on *Skewdistribution* to divide the data set into three clusters, which is the true cluster number. As shown in Figure 23.12, K-means performs the worst while Chameleon is the best.

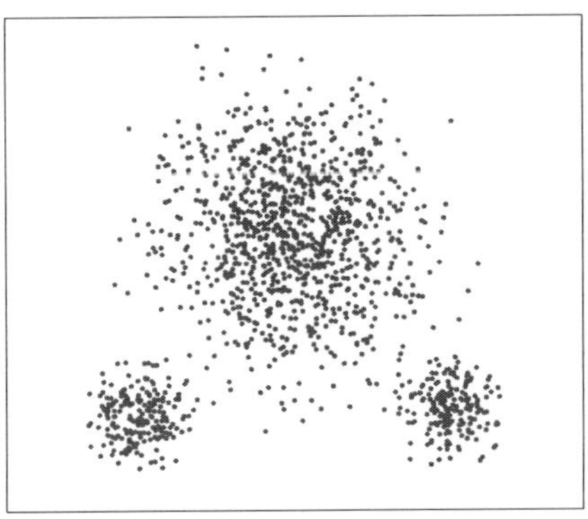

FIGURE 23.11: The data set *Skewdistribution*.

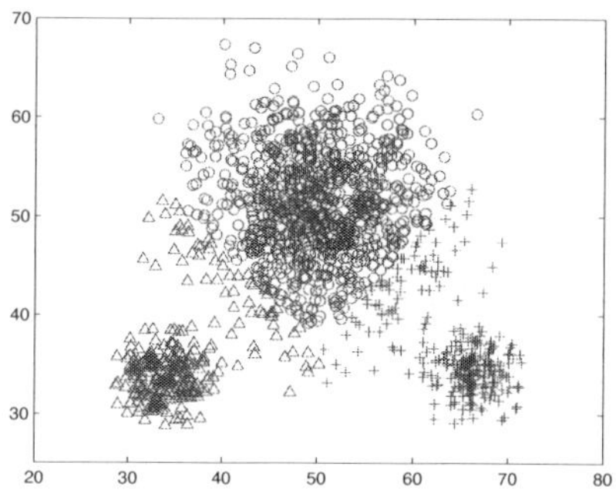

(a) Clustering by K-means

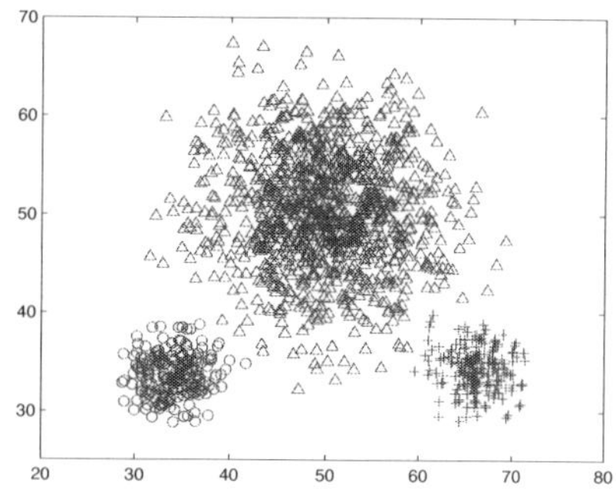

(b) Clustering by Agglo

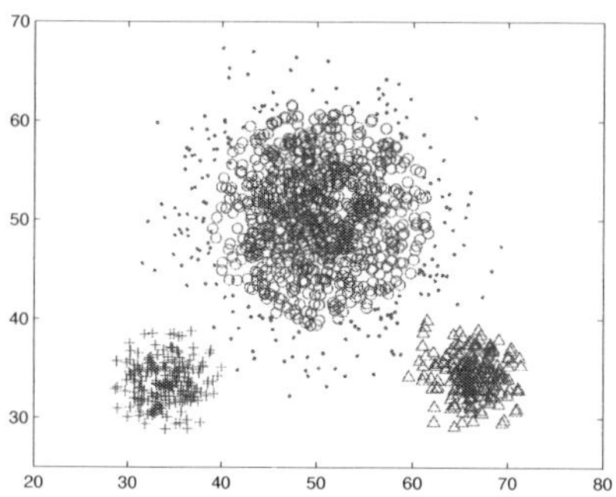

(c) Clustering by DBSCAN

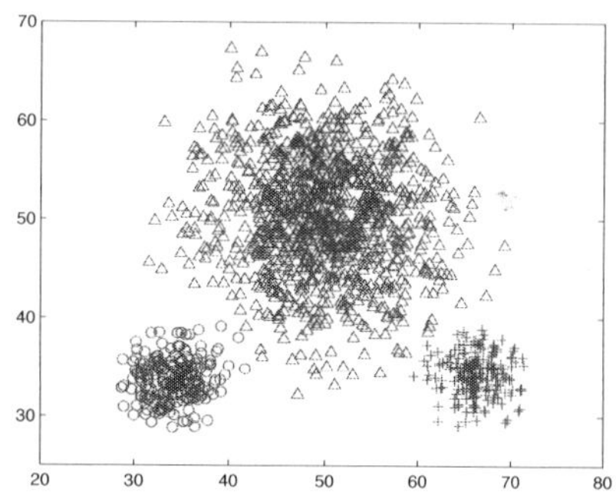

(d) Clustering by Chameleon

FIGURE 23.12: Clustering results on data set *Skewdistribution* by different algorithms where $NC = 3$.

TABLE 23.18: Results of the Impact of Skewed Distributions, True $NC = 3$

	CH	I	D	S	DB^{**}	SD	S_Dbw	XB^{**}	$CVNN$
2	788	232.3	0.0286	0.621	0.571	0.327	0.651	0.369	1.04
3	1590	**417.9**	**0.0342**	**0.691**	**0.466**	**0.187**	**0.309**	**0.264**	**0.73**
4	1714	334.5	0.0055	0.538	0.844	0.294	0.379	1.102	0.85
5	**1905**	282.9	0.0069	0.486	0.807	0.274	0.445	0.865	0.87
6	1886	226.7	0.0075	0.457	0.851	0.308	0.547	1.305	0.96
7	1680	187.1	0.0071	0.371	1.181	0.478	0.378	3.249	1.10
8	1745	172.9	0.0075	0.370	1.212	0.474	0.409	3.463	1.03
9	1317	125.5	0.0061	0.301	1.875	0.681	0.398	7.716	1.41

A study was conducted on the data set *Skewdistribution* to evaluate the performance of different indices on a data set with skewed distributions. Chameleon is applied as the clustering algorithm. Results listed in Table 23.18 show that only CH cannot give the right optimal cluster number. $CH = (TSS/SSE - 1) \cdot ((n - NC)/(NC - 1))$ and TSS is a constant number of a certain data set. Thus, CH

FIGURE 23.13: The data set *T4.8k.modified*.

is essentially based on *SSE*, which shares the same basis with *K*-means algorithm. As mentioned above, *K*-means cannot handle skewed distributed data sets. Therefore, the similar conclusion can be applied to *CH*.

23.3.2.6 The Impact of Arbitrary Shapes

A data set with arbitrary shapes is always hard to handle. Figure 23.13 shows a synthetic data set *T4.8k.modified* which consists of six irregular shape of clusters. It is generated by removing 10% noise from the original data set *T4.8k* which contains 8000 objects[28]. As in the last subsection, the same four algorithms are employed to run on *T4.8k.modified* to divide the data set into six clusters, which is the true cluster number. As shown in Figure 23.14, Chameleon performs the best among these four clustering algorithms.

A study on the data set *T4.8k.modified* was performed to evaluate whether the nine internal validation indices could handle a data set with arbitrary shapes. Chameleon is applied as the clustering algorithm. Results listed in Table 23.19 show that only *CVNN* can deal with data set with arbitrary structures. The reasons are as follows.

D uses the minimum pairwise distance between objects in different clusters to measure the intercluster separation. When dealing with arbitrary shaped data sets, this can be misleading. For example, consider cluster A and cluster B′ shown in Figure 23.15. The minimum pairwise distance between these two clusters is almost zero while they are still separable.

For *CH*, *I*, DB^{**}, *SD*, *S_Dbw*, and XB^{**}, these six indices use the cluster center of each cluster as

TABLE 23.19: Results of the Impact of Arbitrary Shapes, True $NC = 6$

	CH	*I*	*D*	*S*	DB^{**}	*SD*	*S_Dbw*	XB^{**}	*CVNN*
2	301	1808	0.0110	0.231	2.927	0.0442	1.790	7.824	1.03
3	5484	32080	0.0117	0.401	0.984	0.0219	0.579	1.271	0.75
4	**8213**	**34532**	**0.0183**	**0.438**	**0.769**	**0.0197**	0.680	**1.143**	0.65
5	6838	24902	0.0142	0.384	0.828	0.0299	0.509	3.032	0.62
6	7560	24721	0.0074	0.333	1.038	0.0286	∞	2.685	**0.58**
7	7151	20753	0.0080	0.343	0.984	0.0290	0.426	2.674	0.89
8	6445	16922	0.0072	0.367	0.896	0.0293	0.416	2.892	1.39
9	6636	22365	0.0067	0.376	0.865	0.0312	**0.281**	2.755	1.36

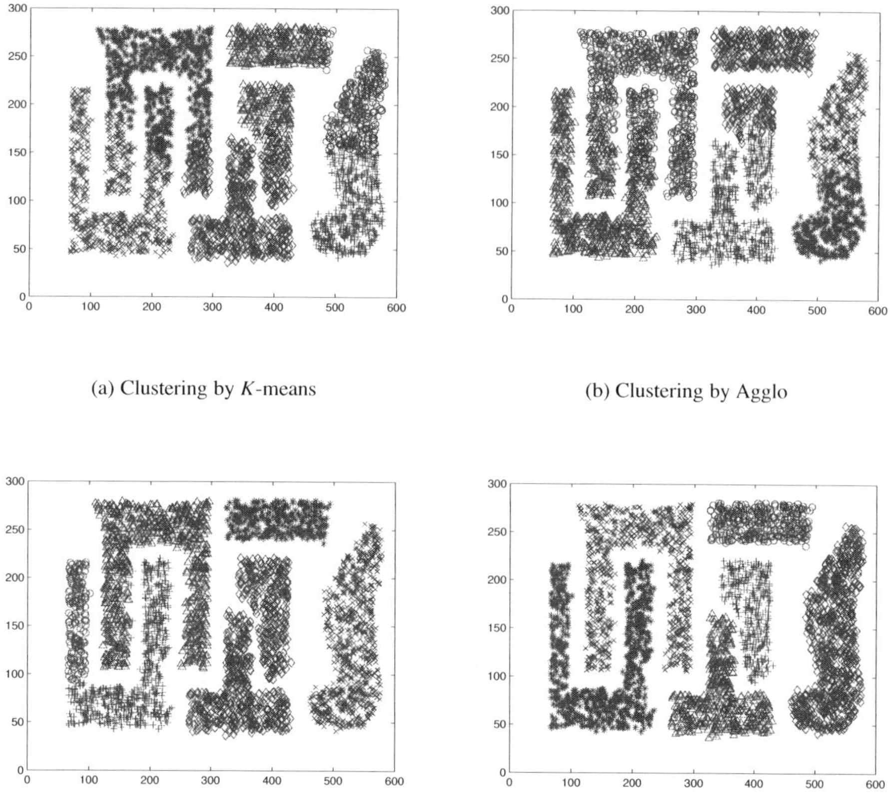

(a) Clustering by *K*-means

(b) Clustering by Agglo

(c) Clustering by DBSCAN

(d) Clustering by Chameleon

FIGURE 23.14: Clustering results on data set *T4.8k.modified* by different algorithms where $NC = 6$.

the representative for that cluster when evaluating the intercluster separation. In addition, S uses the average minimum pairwise distance between objects in each cluster as the separation measurement, which can be viewed as equivalent to the minimum pairwise distance between cluster centers in a sense. Since it is meaningful to use the center to represent the entire cluster only for the sphere-shaped cluster, this implies that these indices can work only in the hypersphere condition. Figure 23.15 gives an illustration for this argument. In this figure, both clusters A and B have an arcuate structure, and the cluster centers are not even in the clusters. If one moves cluster B from the real-line place to the dash-line place B', A and B are getting closer while the distance between their centers becomes larger. In this case, it is meaningless and incorrect to make cluster center representative for the entire cluster.

The above 8 measures use either the average (minimum) pairwise distance between objects in different clusters or one single object (the cluster center) as the representative of the entire cluster when calculating the intercluster separation. These measures consider only the positions of the objects in clusters and fail to take into account the object distributions which form the geometrical information of the cluster. On the other hand, $CVNN$ evaluates the intercluster separation based on objects that carry the geometrical information of each cluster. Sharing the same idea with kNN consistency, $CVNN$ uses dynamic multiple objects as representatives for different clusters in different

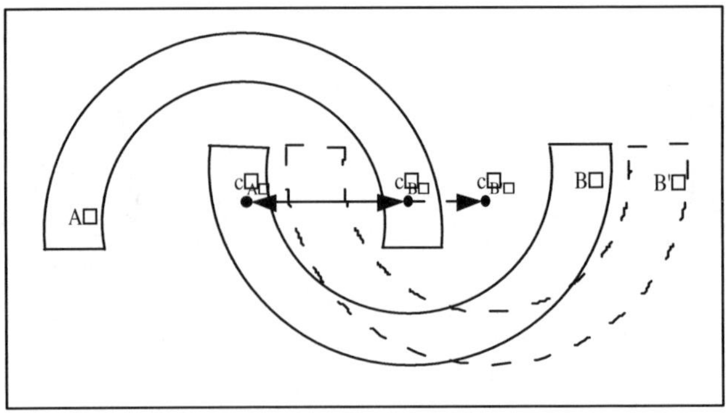

FIGURE 23.15: Arcuate shape intercluster separation.

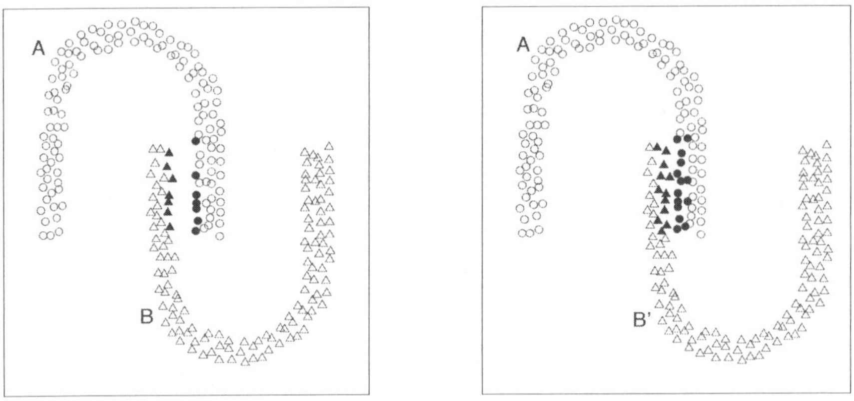

(a) Before movement

(b) After movement

FIGURE 23.16: An illustration of the dynamic effect of cluster representatives of *CVNN*.

situations when measuring the intercluster separation. Figure 23.16 illustrates the dynamic effect of how representatives of clusters evolve in different situations. In this example, both clusters A and B have an arcuate structure and solid objects are the representatives selected by the *CVNN* measure. Comparing subfigure (a) with (b), one can see that A and B' are closer than A and B, which indicates that the intercluster separation is getting worse. Meanwhile, the numbers of representatives for both clusters are growing as well as the intercluster separation measure *Sep* within *CVNN*, which agree with the indication that clusters are becoming more separated. This example illustrates the dynamic effect of *CVNN*'s intercluster separation measure, since the representatives for the same two clusters in different situations are different.

23.3.3 Properties of Measures

Table 23.20 summarizes the properties of different internal validation measures in different aspects, which can be helpful for the index selection. "–" indicates property not tested, and "×"

TABLE 23.20: Overall Performance of Different Measures

Measure	Monotonicity	Noise	Density	Subcluster	Skew Distr.	Arbit. Shape
RMSSTD	×	–	–	–	–	–
RS	×	–	–	–	–	–
Γ	×	–	–	–	–	–
CH		×			×	×
I			×			×
D		×		×		×
S				×		×
DB^{**}				×		×
SD				×		×
S_Dbw						×
XB^{**}				×		×
CVNN						

denotes situation cannot be handled. It suggests that, while the other 11 measures have certain limitations in different scenarios, especially in the aspect of handling data set with arbitrary structures, *CVNN* performs well in all six aspects. *CVNN* exploits the notion of nearest neighbors and uses dynamic multiple objects as representatives for different clusters in different situations, which makes it particularly useful when the data set includes clusters with arbitrary shapes. Thus, *CVNN* can play an important role as a valuable complementary measure in the suite of internal clustering validation measures.

23.4 Summary

This chapter presents detailed studies on 16 external validation measures and 12 internal validation measures in a comprehensive way on various aspects of these validation measure. For the external validation part, different measures are compared and contrasted for K-means clustering. As results revealed, it is necessary to normalize validation measures before they can be employed for clustering validation, since unnormalized measures may lead to inconsistent or even misleading results. This is particularly true for data with imbalanced class distributions. Normalization solutions are also provided for some validation measures. Furthermore, lemmas are provided to show that some validation measures are mathematically equivalent and some measures have very similar validation performances. Finally, key properties of the 16 measures are summarized. These properties should be considered before deciding what is the right measure to use in practice.

For the internal validation part, the validation properties of a suite of 12 existing internal clustering validation measures for crisp clustering are studied in six different aspects: monotonicity, noise, density, subclusters, skewed distribution, and arbitrary shapes data. Six synthetic data sets which best represent the above six aspects are used to evaluate the performance of the 12 validation measures. The results of the studies demonstrate that all measures except for the *CVNN* index have certain limitations in different application scenarios, especially showing difficulties in handling data set with arbitrary structures. On the other hand, *CVNN* exploits the notion of nearest neighbors and uses dynamic multiple objects as representatives for different clusters in different situations, which makes it particularly useful when the data set includes clusters with arbitrary shapes. The summarized validation properties of the 12 internal validation measures may serve as a guide for index selection in practice.

Bibliography

[1] I. Assent, R. Krieger, E. Muller, and T. Seidl. DUSC: Dimensionality unbiased subspace clustering. In *Seventh IEEE International Conference on Data Mining, 2007, ICDM 2007*, pages 409–414. IEEE, 2007.

[2] I. Assent, R. Krieger, E. Muller, and T. Seidl. InSCY: Indexing subspace clusters with in-process-removal of redundancy. In *Eighth IEEE International Conference on Data Mining, 2008. ICDM'08*, pages 719–724. IEEE, 2008.

[3] A. Ben-Hur and I. Guyon. Detecting stable clusters using principal component analysis. In M.J. Brownstein and A.Kohodursky (Eds.) *Methods in Molecular Biology*, pages 159–182, Humana Press, 2003.

[4] J.C. Bezdek. *Pattern Recognition with Fuzzy Objective Function Algorithms*. Kluwer Academic Publishers, 1981.

[5] J.C. Bezdek and N.R. Pal. Some new indexes of cluster validity. *Systems, Man, and Cybernetics, Part B: Cybernetics, IEEE Transactions on*, 28(3):301–315, 1998.

[6] Marcel Brun, Chao Sima, Jianping Hua, James Lowey, Brent Carroll, Edward Suh, and Edward R. Dougherty. Model-based evaluation of clustering validation measures. *Pattern Recognition*, 40:807–824, March 2007.

[7] T. Calinski and J. Harabasz. A dendrite method for cluster analysis. *Communications in Statistics*, 3(1):1–27, 1974.

[8] T.M. Cover and J.A. Thomas. *Elements of Information Theory* (2nd Edition). Wiley-Interscience, 2006.

[9] D.L. Davies and D.W. Bouldin. A cluster separation measure. *IEEE Transactions on Pattern Analysis and Machine Intelligence*, 1(2):224–227, 1979.

[10] M. DeGroot and M. Schervish. *Probability and Statistics* (3rd Edition). Addison Wesley, 2001.

[11] J.C. Dunn. Well separated clusters and optimal fuzzy partitions. *Cybernetics and Systems*, 4(1):95–104, 1974.

[12] Martin Ester, Hans-Peter Kriegel, Jorg Sander, and Xiaowei Xu. A density-based algorithm for discovering clusters in large spatial databases with noise. In *KDD*, pages 226–231, 1996.

[13] E.B. Fowlkes and C.L. Mallows. A method for comparing two hierarchical clusterings. *Journal of the American Statistical Association*, 78:553–569, 1983.

[14] I. Gath and A.B. Geva. Unsupervised optimal fuzzy clustering. *IEEE Transactions on Pattern Analysis and Machine Intelligence*, 11(7):773–780, 1989.

[15] L.A. Goodman and W.H. Kruskal. Measures of association for cross classification. *Journal of the American Statistical Association*, 49:732–764, 1954.

[16] Naiyang Guan, Dacheng Tao, Zhigang Luo, and Bo Yuan. NeNMF: An optimal gradient method for nonnegative matrix factorization. *IEEE Transactions on Signal Processing*, 60:2882–2898, 2012.

[17] Naiyang Guan, Dacheng Tao, Zhigang Luo, and Bo Yuan. Online non-negative matrix factorization with robust stochastic approximation. *IEEE Transactions on Neural Networks and Learning Systems*, 23:1087–1099, 2012.

[18] M. Halkidi, Y. Batistakis, and M. Vazirgiannis. Cluster validity methods: Part I. *SIGMOD Record*, 31(2):40–45, 2002.

[19] Maria Halkidi, Yannis Batistakis, and Michalis Vazirgiannis. On clustering validation techniques. *Journal of Intelligent Information Systems*, 17(2–3):107–145, 2001.

[20] Maria Halkidi and Michalis Vazirgiannis. Clustering validity assessment: Finding the optimal partitioning of a data set. In *Proceedings of the 2001 IEEE International Conference on Data Mining*, pages 187–194, 2001.

[21] Maria Halkidi and Michalis Vazirgiannis. Clustering validity assessment using multi-representatives. In *Proceedings of the SETN*, pages 237–248, 2002.

[22] Maria Halkidi, Michalis Vazirgiannis, and Yannis Batistakis. Quality scheme assessment in the clustering process. In *PKDD '00*, pages 265–276, London, UK, 2000. Springer-Verlag.

[23] L. Hubert. Nominal scale response agreement as a generalized correlation. *British Journal of Mathematical and Statistical Psychology*, 30:98–103, 1977.

[24] L. Hubert and P. Arabie. Comparing partitions. *Journal of Classification*, 2:193–218, 1985.

[25] Lawrence Hubert and Phipps Arabie. Comparing partitions. *Journal of Classification*, 2(1):193–218, 1985.

[26] Anil K. Jain and Richard C. Dubes. *Algorithms for Clustering Data*. Prentice-Hall, Inc., Upper Saddle River, NJ, USA, 1988.

[27] George Karypis. *Cluto—Software for clustering high-dimentional datasets*. version 2.1.2, 2006.

[28] George Karypis. *Karypis Lab*. http://glaros.dtc.umn.edu/gkhome/.

[29] George Karypis, Eui-Hong (Sam) Han, and Vipin Kumar. Chameleon: Hierarchical clustering using dynamic modeling. *Computer*, 32(8):68–75, 1999.

[30] M.G. Kendall. *Rank Correlation Methods*. New York: Hafner Publishing Co., 1955.

[31] D.W. Kim, K.H. Lee, and D. Lee. Fuzzy cluster validation index based on inter-cluster proximity. *Pattern Recognition Letters*, 24(15):2561–2574, 2003.

[32] Minho Kim and R. S. Ramakrishna. New indices for cluster validity assessment. *Pattern Recognition Letters*, 26(15):2353–2363, 2005.

[33] Johann M. Kraus, Christoph Mssel, Gnther Palm, and Hans A. Kestler. Multi-objective selection for collecting cluster alternatives. *Computational Statistics*, 26:341–353, 2011.

[34] Hardy Kremer, Philipp Kranen, Timm Jansen, Thomas Seidl, Albert Bifet, Geoff Holmes, and Bernhard Pfahringer. An effective evaluation measure for clustering on evolving data streams. In *ACM SIGKDD*, pages 868–876, 2011.

[35] Benson S. Y. Lam and Hong Yan. A new cluster validity index for data with merged clusters and different densities. In *IEEE ICSMC*, pages 798–803, 2005.

[36] B. Larsen and C. Aone. Fast and effective text mining using linear-time document clustering. In *KDD*, pages 16–22, 1999.

[37] Yanchi Liu, Zhongmou Li, Hui Xiong, Xuedong Gao, and Junjie Wu. Understanding of internal clustering validation measures. In *Proceedings of the 2010 IEEE International Conference on Data Mining*, pages 911–916, 2010.

[38] Yanchi Liu, Zhongmou Li, Hui Xiong, Xuedong Gao, Junjie Wu, and Sen Wu. Understanding and enhancement of internal clustering validation measures. *IEEE Transactions on Systems, Man, and Cybernetics, Part B*, pages 982–994, 2013.

[39] J. MacQueen. Some methods for classification and analysis of multivariate observations. In *Proceedings of BSMSP*, pages 281–297. University of California Press, 1967.

[40] MathWorks. K-means clustering in statistics toolbox.

[41] Ujjwal Maulik and Sanghamitra Bandyopadhyay. Performance evaluation of some clustering algorithms and validity indices. *IEEE Transactions on Pattern Analysis and Machine Intelligence*, 24:1650–1654, 2002.

[42] M. Meila. Comparing clusterings—An axiomatic view. In *ICML*, pages 577–584, 2005.

[43] B. Mirkin. *Mathematical Classification and Clustering*. Kluwer Academic Press, 1996.

[44] E. Müller, S. Günnemann, I. Assent, and T. Seidl. Evaluating clustering in subspace projections of high dimensional data. *Proceedings of the VLDB Endowment*, 2(1):1270–1281, 2009.

[45] L. Parsons, E. Haque, and H. Liu. Subspace clustering for high dimensional data: A review. *ACM SIGKDD Explorations Newsletter*, 6(1):90–105, 2004.

[46] A. Patrikainen and M. Meila. Comparing subspace clusterings. *IEEE Transactions on Knowledge and Data Engineering*, 18(7):902–916, 2006.

[47] W.M. Rand. Objective criteria for the evaluation of clustering methods. *Journal of the American Statistical Association*, 66:846–850, 1971.

[48] C.J.V. Rijsbergen. *Information Retrieval* (2nd Edition). Butterworths, London, 1979.

[49] Peter Rousseeuw. Silhouettes: a graphical aid to the interpretation and validation of cluster analysis. *Journal of Computation and Applied Mathematics*, 20(1):53–65, 1987.

[50] Sriparna Saha and Sanghamitra Bandyopadhyay. Application of a new symmetry-based cluster validity index for satellite image segmentation. *IEEE Geoscience and Remote Sensing Letters*, pages 166–170, 2002.

[51] Subhash Sharma. *Applied Multivariate Techniques*. John Wiley & Sons, Inc., New York, USA, 1996.

[52] P. Smyth. Model selection for probabilistic clustering using cross-validated likelihood. *Statistics and Computing*, 10(1):63–72, 2000.

[53] Mingzhou (Joe) Song and Lin Zhang. Comparison of cluster representations from partial second- to full fourth-order cross moments for data stream clustering. In *IEEE ICDM*, pages 560–569, 2008.

[54] Michael Steinbach, George Karypis, and Vipin Kumar. A comparison of document clustering techniques. In *Workshop on Text Mining, KDD*, 2000.

[55] A. Strehl, J. Ghosh, and R.J. Mooney. Impact of similarity measures on web-page clustering. In *Workshop on Artificial Intelligence for Web Search, AAAI*, pages 58–64, 2000.

[56] Y. Sun, Y. Yu, and J. Han. Ranking-based clustering of heterogeneous information networks with star network schema. In *Proceedings of the 15th ACM SIGKDD International Conference on Knowledge Discovery and Data Mining*, pages 797–806. ACM, 2009.

[57] Pang-Ning Tan, Michael Steinbach, and Vipin Kumar. *Introduction to Data Mining*, (First Edition). Addison-Wesley Longman Publishing Co., Inc., Boston, MA, USA, 2005.

[58] Robert Tibshirani, Guenther Walther, and Trevor Hastie. Estimating the number of clusters in a data set via the gap statistic. *Journal of the Royal Statistical Society: Series B*, 63(2):411–423, 2001.

[59] TREC. Text retrieval conference. October 2007.

[60] S. van Dongen. Performance criteria for graph clustering and Markov cluster experiments. *TRINS= R0012*, Centrum voor Wiskunde en Informatica. 2000.

[61] J. Wu, H. Xiong, J. Chen, and W. Zhou. A generalization of proximity functions for k-means. In *ICDM*, pages 361–370, 2007.

[62] Junjie Wu, Hui Xiong, and Jian Chen. Adapting the right measures for k-means clustering. In *ACM SIGKDD*, pages 877–886, 2009.

[63] Xuanli Lisa Xie and Gerardo Beni. A validity measure for fuzzy clustering. *IEEE Transactions on Pattern Analysis and Machine Intelligence*, 13(8):841–847, 1991.

[64] H. Xiong, J. Wu, and J. Chen. K-means clustering versus validation measures: A data distribution perspective. In *KDD*, pages 318–331, 2006.

[65] Hui Xiong, Junjie Wu, and Jian Chen. K-means clustering versus validation measures: A data distribution perspective. *IEEE Transactions on Systems, Man, and Cybernetics Part B*, 39(2):318–331, 2009.

[66] X. Xu, N. Yuruk, Z. Feng, and T.A.J. Schweiger. SCAN: A structural clustering algorithm for networks. In *Proceedings of the 13th ACM SIGKDD International Conference on Knowledge Discovery and Data Mining*, pages 824–833. ACM, 2007.

[67] Y. Zhao and G. Karypis. Criterion functions for document clustering: Experiments and analysis. *Machine Learning*, 55(3):311–331, 2004.

[68] Ying Zhao and George Karypis. Evaluation of hierarchical clustering algorithms for document datasets. In *CIKM '02*, pages 515–524, New York, USA, 2002. ACM.

[69] Tianyi Zhou, Dacheng Tao, and Xindong Wu. Manifold elastic net: A unified framework for sparse dimension reduction. *Data Mining and Knowledge Discovery*, 20:340–371, 2010.

[70] Y. Zhou, H. Cheng, and J.X. Yu. Graph clustering based on structural/attribute similarities. *Proceedings of the VLDB Endowment*, 2(1):718–729, 2009.

Chapter 24

Educational and Software Resources for Data Clustering

Charu C. Aggarwal
IBM T. J. Watson Research Center
Yorktown Heights, NY
charu@us.ibm.com

Chandan K. Reddy
Wayne State University
Detroit, MI
reddy@cs.wayne.edu

24.1 Introduction

Since data clustering is a vast area, it is impossible to cover all the material on the topic in a single book. Therefore, this chapter will summarize the key resources in this area. In general, the resources on data clustering can be divided into the categories of (i) books, (ii) survey articles, and (iii) software.

The books on data clustering are mostly generic and are not specific to any particular area or topic. On the other hand, since surveys are generally more focussed, many more surveys have been written, which are specific to particular topics, such as high-dimensional data, nonnegative matrix factorization, spectral clustering, or text data. Finally, since clustering is used extensively in the industry, a significant number of software packages are available for data clustering.

The last of the aforementioned categories is an ever changing landscape, since advancements in data clustering make older software packages obsolete. Therefore, it is possible that the software resources listed in this chapter will eventually become obsolete. Furthermore, it is impossible to fully list all the possible resources available in terms of the different kinds of software. Nevertheless, some of the key commercial and noncommercial sources from which data clustering software can be used will be listed. Commercial software packages are generally based on the classical clustering algorithms, whereas open-source packages are often much more advanced and sometimes implement the latest algorithms available in the literature.

This chapter is organized as follows. Section 24.2 presents educational resources on data clustering. This section is itself divided into two subsections. The key books are discussed in section 24.2.1, whereas the survey papers on data clustering are discussed in section 24.2.2. Finally, the software on data clustering will be discussed in section 24.3. Section 24.4 presents the conclusions and summary.

24.2 Educational Resources

Educational resources are either generic in the form of books or more focussed on the research communities in the form of surveys. Books are generally more useful from a practical perspective, whereas survey articles are more useful for an academic perspective.

24.2.1 Books on Data Clustering

Some of the earlier books on data clustering are by Anderberg [3], Duran [9], and Hartigan [15]. Much of the developments in data clustering occurred *after* the writing of these books. Furthermore, these books were written before the advent of the modern computer age. Therefore, the perspective in these books is not necessarily optimized toward the computational design of clustering algorithms. Many of the classical clustering algorithms were developed *after* the writing of these algorithms, primarily because of the ease in implementation which arose with the use of modern computers.

Two of the earliest and most well-known books after the advent of the modern computer age were written by Jain and Dubes [18] and by Kaufman and Rousseauw [21]. These books discuss most of the classical clustering literature, in a way which is easy to understand and comprehend. In particular, the book by Kaufman and Rousseauw, written in 2005, was updated to the most recent developments in the field at that time. Some of the recent books on data clustering, for example, Xu and Wunsch [33], are focused on the classical methods on data clustering. Mirkin [26] addresses the topic from the perspective of data recovery.

With the advent of database technology and the large amounts of data in the nineties, the issue of scalability has become increasingly important. Therefore, a number of algorithms have been proposed over the last 15 years, which have been designed in order to improve the scalability of the approach. The work in [14] is one of the recent books, which addresses many of the algorithms developed by the database community. Nevertheless, the work over the last 15 years is much broader and addresses many different domains of data, along with scalability issues. To the best of our knowledge, there is no single book addressing all these issues. This book is therefore an attempt to fill the void in the field.

24.2.2 Popular Survey Papers on Data Clustering

There are numerous surveys on the topic of data clustering. The area of data clustering is so vast, that many surveys have been written, which address a *specific area* of data clustering. The surveys broadly fall into three categories:

- *General Data Clustering Survey:* In these surveys, the problems of data clustering is addressed in a very general way and is not specific to a particular technique or data type. However, the area of clustering is so vast, that most of these surveys tend to be overview articles.
- *Technique-Centered Survey:* In these surveys, a specific *technique* such as spectral clustering or k-means clustering is addressed in detail.
- *Data Type-Centered Survey:* In such surveys, a specific kind of data type such as time-series data, high-dimensional data, or text data is addressed by the survey. Different data types provide different challenges to clustering algorithms. Therefore, the clustering methods for a specific data type are often closely related and addressed in a specific survey.

An excellent review article on the topic of data clustering is available in [19]. This article provides an overview of the main issues in data clustering. Two other excellent survey articles are available in [32, 6]. These provide a good overview of the landscape, though not in much detail. This is to be expected, since the area of clustering is too vast to be easily covered by a single publication on the topic.

Many articles have also been written, which cover specific *techniques* of data clustering. An excellent overview article on the k-means algorithm, which also provides a historical perspective, is presented in [17]. The k-means method is a classical technique, which is closely related to hierarchical clustering algorithms. This is covered extensively in [28]. Hierarchical clustering algorithms are also used quite frequently for document data for organization of large corpora. A survey of such algorithms for document data is presented in [31, 34]. Spectral clustering techniques are an important class of methods which have found significant popularity in the clustering literature. Two surveys on spectral clustering may be found in [24, 11]. The survey in [24] is considered classical. While evolutionary clustering algorithms have not found much popularity in the research literature, they have often turned out be quite useful for practitioners. A survey on evolutionary clustering algorithms may be found in [16].

Probabilistic clustering algorithms are very popular in the data mining literature because of their natural interpretability. A closely related class of algorithms is the *fuzzy clustering method.* Techniques for fuzzy clustering are covered extensively in [4, 5].

Numerous articles have also been written on clustering in specific data types or domains. The problem of clustering high dimensional data was first covered in [29] and more recently in [22]. Since high-dimensional clustering is a recent topic, and much research on the area has been performed recently, the latter survey [22] is much more extensive. Document data is a particular case of sparse high-dimensional data, for which clustering methods provide a tremendous challenge. For example, hierarchical methods for document clustering have been studied in [31, 34], and a more general survey on text clustering is provided in [2].

Many data types are *contextual* in which dependencies exist between the data items. Examples of such data sets include (continuous) time-series data, discrete sequential data (which could be either temporal or biological), and graph data. Contextual domains provide a special challenge because the dependencies between the data items need to be accounted for during the clustering process. Numerous surveys have been written to address such contextual domains. An excellent survey on time-series data clustering may be found in [23], while the biological data domain is addressed in [20, 25]. Graph clustering is a recent area which has been popularized by the advent of social and information networks. Graph data clustering is addressed in [12, 30]. The review in [12] is recent and particularly extensive.

24.3 Software for Data Clustering

A significant amount of software is available for data clustering. Interestingly, most of the sophisticated software on data clustering is open-source software, which is freely available at different web sites. On the other hand, most of the commercial software comprises implementations of simpler and more classical algorithms such as k-means or agglomerative clustering.

24.3.1 Free and Open-Source Software

In this section, we will first address the widely used general purpose clustering software. Later, we will provide some additional specialized software that are used for specific data clustering applications.

24.3.1.1 General Clustering Software

The most well-known *general purpose* site offering open source software is the *WEKA* machine learning repository [39]. This is a general purpose repository which contains software not just for clustering, but also for other data mining related tasks such as data preprocessing, classification, and visualization. However, much of the software can also be used for data clustering.

Spider is another widely used data mining software which contains the implementations of several popular clustering algorithms [54]. It is an object-oriented environment for machine learning in MATLAB. Its recent version also provides WEKA interfaces to the machine learning libraries built in Spider.

Cluster is another widely used open-source clustering software that contains several clustering and visualization algorithms [45]. It is very popular in the bioinformatics community. In addition to the standard clustering algorithms, it also provides an excellent graphical environment for analyzing complex datasets. The other advantage of this software is that it has interfaces to allows users to use the clustering algorithms in other programming environments such as Python, MATLAB, and R. In addition to the above mentioned software, programming platforms such as MATLAB and R already have their own implementations of some of the commonly used data clustering algorithms.

Another comprehensive set of data clustering packages is provided in the *ELKI* [1] suite of algorithms that include many classical partitioning algorithms, EM-based probabilistic algorithms, density-based algorithms, and subspace clustering algorithms. The KDnuggets web site [42] also provides access to a significant number of open-source software sites for clustering and segmentation. This can be considered a meta-repository, in that it provides pointers to other relevant sites.

24.3.1.2 Specialized Clustering Software

OpenSubspace is an open-source software that contains the implementations of several *subspace clustering algorithms* [41]. A more detailed description is provided in [27]. It integrates state-of-the-art performance measures and visualization techniques for analyzing subspace clusters in the WEKA environment. The MOA framework also implements a number of stream clustering algorithms and provides tools for their evaluation [7]. Related to *Weka*, it supports the extension with new clustering algorithms, new stream generators, and new evaluation measures. The *Weka* implementation of the framework may be found in [86].

For *text data clustering*, the *cross-bow* method is available for download from [43]. This technique uses EM-clustering of text data, with the use of hierarchical partitioning of text documents. There are also other specialized software such as MALLET [35] which performs clustering along with some statistical natural language processing and topic modeling. CLUTO [46] is a software package for clustering low- and high-dimensional data sets and for analyzing the characteristics of the various clusters. One advantage of this software is that it is rather general purpose. It can be used for text data, multi-dimensional data, or even transaction data. There are also some other packages that perform traditional clustering on text documents [44].

A significant amount of software is also available for contextual data types. For *time-series data clustering*, an open-source software *Gait-CAD* is publicly available under the GNU public license. This is downloadable from sourceforge [49] and is a MATLAB toolbox. For *gene expression data clustering*, one of the most popular tools is *Cluster*, which has an associated *TreeView* software for better visualization [10]. A more detailed discussion of the available software for analyzing gene expression and other forms of biological data can be found in the biological data clustering chapter of this book (Chapter 16).

Numerous clustering tools are available for graph-partitioning and clustering. The network analysis tools (NeAT) [8] provides a number of tools for the analysis of network data, though the focus is primarily on biological networks. In addition, many individual graph clustering tools are available for download at different sites. One of the most famous family of graph partitioning algorithms is METIS [48]. In spite of being one of the earlier methods, it seems to perform competitively with most of the graph clustering algorithms, in terms of both effectiveness and efficiency. CFinder is a free software for finding and visualizing overlapping dense groups of nodes in networks, based on the Clique Percolation Method (CPM) [53]. Another software for Markov Clustering of networks is available at [55]. This approach provides high scalability for the clustering process. For semisupervised clustering, many implementations of different algorithms that perform *constrained clustering* are available in [57].

24.3.2 Commercial Packages

Many mathematical tools such as MATLAB [52] come with built-in methods for data clustering. However, typically, such methods use classical techniques such as the k-means algorithm. This is because these are very general purpose tools, and the clustering capability is not the main purpose of such software. However, numerous other commercial tools have been constructed, with a specific goal of clustering different kinds of data.

The KDnuggets site [42] provides a link to some of the more popular forms of software in this domain. This site provides pointers to software developed by other vendors such as *IBM* and *SAS* rather than only its own dedicated software. One of the most well-known ones is the *IBM SPSS data mining workbench* [50]. This includes two step, k-means, and Kohonen clustering algorithms. Another well-known package is the *SAS Enterprise Modeler* [51]. In this case, the clustering tool is available with other forms of visual and decision support. SAS contains built-in procedures that can perform clustering of even large-scale datasets.

Providing full graphics and visualization capabilities is often important for getting an intuitive *understanding* of the clustering process. *Clustan* [56] offers such capabilities with strong graphics support. The *NeuroXL Clusterizer* [58] is a commercial tool for clustering with neural networks. The advantage of this tool is that it naturally integrates with *Microsoft Excel* and can, therefore, be easily used for spreadsheet data.

24.3.3 Data Benchmarks for Software and Research

Numerous data sets are available for testing clustering software and research. Some of the data sets and web sites are general purpose, whereas other web sites are tailored to specific kinds of data. The most well-known among the general-purpose web sites is the UCI machine learning repository [13]. This resource is generally intended for classification, though many of the data sets can also be used for clustering. The KDnuggets web site [38] also provides access to many general-purpose data sets. This can be considered a meta-repository, in that it provides pointers to other sites containing data sets. Similar to this, the other popularly known meta-repositories that give a wide range of links to other data repositories are STATOO [71] and David Dowe's data links [72]. The latter also provides several online resources on data mining case studies and competitions.

More large-scale datasets are available from the KDD Cup dataset repository [73] and other data mining competitions [74]. These datasets are not only large scale but are also directly collected from complex real-world problems and hence provide several challenges. Clustering is often used in solving such real-world challenges since many of these issues typically fall into the unsupervised category. For the statistics community, Statlib dataset archive [75] is a widely used data collection.

In addition to the above mentioned real-world datasets, there is also a Fundamental Clustering Problem Suite [47] that provides a simple collection of different synthetically generated 2-D and 3-D datasets. This provides some test datasets with different sizes, shapes, and densities.

For specific data types, numerous resources are available. For *time-series clustering*, the UCR time-series web site [36] provides access to very long continuous series of data for clustering. Another time-series data library is available at [37]. For the case of *network data*, the SNAP repository [40] hosted by Jure Leskovec at Stanford University provides access to a large number of network data sets.

For testing the performance of *text document clustering* algorithm, there are several publicly available text document repositories available. Some of the most popular text document collections include Reuters [76], 20NewsGroups [77], TREC [78], and Cora [79].

To evaluate the *biological data clustering*, there is a plethora of websites that host gene expression datasets which can be used to evaluate the clustering algorithms. Gene Expression Omnibus (GEO) repository [68] contains a comprehensive collection of gene expression datasets along with raw source files used to generate this data. Gene Expression Model Selector provides a simple repository of the most widely studied gene expression datasets [69] and several other cancer-specific gene expression datasets are available at [70]. Similarly, to test the performance of *biological network clustering* algorithms, there are a plenty of databases that contain protein-protein interactions. The most popular ones are DIP (Database of Interacting Proteins) [66], BioGRID [67], STRING (Search Tool for the Retrieval of Interacting Genes/Proteins) [65] and MIPS (Mammalian Protein-Protein Interaction Database) [64]. The last resource contains links to several other interaction network databases.

To evaluate the *sequence data clustering* algorithms, several biological sequence database repositories are available at [63]. The most widely used repositories are GenBank [61] and EMBL [62] for nucleic acid sequences and Protein Information Resources (PIR) [60] and UniProt [59] for protein sequences.

In the context of *image applications*, researchers in machine learning and computer vision communities have used clustering for solving automatic grouping of images and image segmentation problems. Clustering can help in efficient organization and retrieval of image databases. ImageCLEF [82] and ImageNet [83] are two widely used image data repositories which are used to demonstrate the performance of image data retrieval tasks. Data clustering can also be used in the context of image segmentation. Vision and Autonomous Systems Center's Image Database [80] from Carnegie Mellon University and the Berkeley Segmentation dataset [81] can be used to test the performance of clustering for image segementation problems. An extensive list of websites that provide image databases is given in [84] and [85].

24.4 Summary

This chapter presents a summary of the key resources for data clustering in terms of books, surveys, and commercial and noncommercial software packages. While many of the books and surveys address different aspects of data clustering in terms of either techniques or data types, there seems to be a gap in creating a single integrated book, which covers the modern literature on this topic. This book is intended to fill that gap.

Bibliography

[1] E. Achtert, S. Goldhofer, H.-P. Kriegel, E. Schubert, and A. Zimek. Evaluation of clusterings – metrics and visual support. In *ICDE Conference*, pages 1285–1288, 2012. `http://elki.dbs.ifi.lmu.de/wiki/Algorithms#`

[2] C. Aggarwal, and C. Zhai. A survey of text clustering algorithms, In *Mining Text Data*, pages 77–128, Springer, 2012.

[3] M. Anderberg. *Cluster Analysis for Applications.* Academic Press, 1973.

[4] A. Baraldi and P. Blonda. A survey of fuzzy clustering algorithms for pattern recognition. I. *IEEE Transactions on Systems, Man, and Cybernetics, Part B*, 29(6):778–785, 1999.

[5] A. Baraldi and P. Blonda. A survey of fuzzy clustering algorithms for pattern recognition. II. *IEEE Transactions on Systems, Man, and Cybernetics, Part B: Cybernetics*, 29(6):786–801, 1999.

[6] P. Berkhin. A survey of clustering data mining techniques. *Grouping Multidimensional Data*, pages 25–71, 2006.

[7] A. Bifet, G. Holmes, B. Pfahringer, P. Kranen, H. Kremer, T. Jansen, and T. Seidl. MOA: Massive online analysis, a framework for stream classification and clustering, *Journal of Machine Learning Research-Proceedings Track*, 11:44–50, 2010. `http://moa.cms.waikato.ac.nz/publications/`

[8] S. Brohee, K. Faust, G. Lima-Mendez, O. Sand, R. Janky, G. Vanderstocken, Y. Deville, and J. van Helden. NeAT: A toolbox for the analysis of biological networks, clusters, classes and pathways. *Nucleic Acids Research*, 36(Web Server issue), July 2008.

[9] B. Duran. *Cluster Analysis: A Survey.* Springer-Verlag, 1974.

[10] M. B. Eisen, P. T. Spellman, P. O. Brown, and D. Botstein. Cluster analysis and display of genome-wide expression patterns. *Proceedings of the National Academy of Sciences*, 95(25):14863–14868, 1998. `http://rana.lbl.gov/EisenSoftware.htm`

[11] M. Filippone, F. Camastra, F. Masulli, and S. Rovetta. A survey of kernel and spectral methods for clustering. *Pattern Recognition*, 41(1):176–190, 2008.

[12] S. Fortunato. Community detection in graphs, *Physics Reports*, 486(3–5):75–174, February 2010.

[13] A. Frank, and A. Asuncion. UCI Machine Learning Repository, Irvine, CA: University of California, School of Information and Computer Science, 2010. `http://archive.ics.uci.edu/ml`

[14] G. Gan, C. Ma, and J. Wu. *Data Clustering: Theory, Algorithms, and Applications.* SIAM, Society for Industrial and Applied Mathematics, 2007.

[15] J. A. Hartigan. *Clustering Algorithms.* John Wiley and Sons, 1975.

[16] E. Hruschka, R.. Campello, A. Freitas, and André C. Ponce Leon F. De Carvalho. A survey of evolutionary algorithms for clustering. *IEEE Transactions on Systems, Man, and Cybernetics, Part C*, 39(2):133–155, 2009.

[17] A. Jain. Data clustering: 50 years beyond k-means. *Pattern Recognition Letters*, 31(8):651–666, 2010.

[18] A. Jain and R. Dubes. *Algorithms for Clustering Data.* Prentice Hall, 1988.

[19] A. Jain, M. Murty, and P. Flynn. Data clustering: A review. *ACM Computing Surveys (CSUR)*, 31(3):264–323, 1999.

[20] D. Jiang, C. Tang, and A. Zhang. Cluster analysis for gene expression data: A survey. *IEEE Transactions on Knowledge and Data Engineering*, 16(11):1370–1386, 2004.

[21] L. Kaufman and P. Rousseeuw. *Finding Groups in Data: An Introduction to Cluster Analysis.* Wiley-Interscience, 2005.

[22] H.-P. Kriegel, P. Kröger, and A. Zimek. Clustering high-dimensional data: A survey on subspace clustering, pattern-based clustering, and correlation clustering. *ACM Transactions on Knowledge Discovery from Data (TKDD)*, 3(1):1–58, 2009.

[23] T. Liao. Clustering of time series data—A survey. *Pattern Recognition*, 38(11):1857–1874, 2005.

[24] U. von Luxberg. A tutorial on spectral clustering, *Statistics and Computing*, 17(4):395–416, 2007.

[25] S. Madeira and A. Oliveira. Biclustering algorithms for biological data analysis: A survey. *IEEE/ACM Transactions on Computational Biology and Bioinformatics*, 1(1):24–45, 2004.

[26] B. Mirkin. *Clustering for Data Mining: A Data Recovery Approach.* Chapman and Hall/CRC, Boca raton, FL, 2005.

[27] E. Muller, S. Gnnemann, I. Assent, and T. Seidl. Evaluating clustering in subspace projections of high dimensional data. *PVLDB*, 2(1): 1270-1281, 2009.

[28] F. Murtagh. A survey of recent advances in hierarchical clustering algorithms. *The Computer Journal*, 26(4):354–359, 1983.

[29] L. Parsons, E. Haque, and H. Liu. Subspace clustering for high dimensional data: A review. *ACM SIGKDD Explorations*, 6(1):90–105, 2004.

[30] S. Schaeffer. Graph clustering. *Computer Science Review*, 1(1):27–64, 2007.

[31] P. Willett. Recent trends in hierarchical document clustering: A critical review. *Information Processing & Management*, 24(5):577–597, 1988.

[32] R. Xu and D. Wunsch. Survey of clustering algorithms. *IEEE Transactions on Neural Networks*, 16(3):645–678, 2005.

[33] R. Xu and D. Wunsch. *Clustering.* Wiley-IEEE Press, 2008.

[34] Y. Zhao, G. Karypis, and U. Fayyad. Hierarchical clustering algorithms for document datasets. *Data Mining and Knowledge Discovery*, 10(2):141–168, 2005.

[35] http://mallet.cs.umass.edu/

[36] http://www.cs.ucr.edu/~eamonn/time_series_data/

[37] http://datamarket.com/data/list/?q=provider:tsdl

[38] http://www.kdnuggets.com/datasets/

[39] http://www.cs.waikato.ac.nz/ml/weka/

[40] http://snap.stanford.edu/data/

[41] http://dme.rwth-aachen.de/en/OpenSubspace

[42] http://www.kdnuggets.com/software/clustering.html

[43] http://www.cs.cmu.edu/~mccallum/bow/

[44] http://rubyforge.org/projects/clusterer/

[45] http://bonsai.hgc.jp/~mdehoon/software/cluster/software.htm

[46] http://glaros.dtc.umn.edu/gkhome/views/cluto

[47] http://www.uni-marburg.de/fb12/datenbionik/data?language_sync=1

[48] http://glaros.dtc.umn.edu/gkhome/views/metis

[49] http://sourceforge.net/projects/gait-cad/

[50] http://www-01.ibm.com/software/analytics/spss/products/modeler/

[51] http://www.sas.com/technologies/analytics/datamining/miner/index.html

[52] http://www.mathworks.com/

[53] http://www.cfinder.org

[54] http://people.kyb.tuebingen.mpg.de/spider/

[55] http://www.micans.org/mcl/

[56] http://www.clustan.com/

[57] http://www.constrained-clustering.org/

[58] http://neuroxl.com/products/excel-cluster-analysis-software/neuroxl-clusterizer.htm

[59] http://www.ebi.ac.uk/uniprot/

[60] http://www-nbrf.georgetown.edu/pirwww/

[61] http://www.ncbi.nlm.nih.gov/genbank/

[62] http://www.ebi.ac.uk/embl/

[63] http://www.ebi.ac.uk/Databases/

[64] http://mips.helmholtz-muenchen.de/proj/ppi/

[65] http://string.embl.de/

[66] http://dip.doe-mbi.ucla.edu/dip/Main.cgi

[67] http://thebiogrid.org/

[68] http://www.ncbi.nlm.nih.gov/geo/

[69] http://www.gems-system.org/

[70] http://www.broadinstitute.org/cgi-bin/cancer/datasets.cgi

[71] http://www.statoo.com/en/resources/anthill/Datamining/Data/

[72] http://www.csse.monash.edu.au/~dld/datalinks.html

[73] http://www.sigkdd.org/kddcup/

[74] http://www.kdnuggets.com/competitions/index.html

[75] http://lib.stat.cmu.edu/datasets/

[76] http://www.daviddlewis.com/resources/testcollections/reuters21578/

[77] http://qwone.com/~jason/20Newsgroups/

[78] http://trec.nist.gov/data.html

[79] http://people.cs.umass.edu/~mccallum/data.html

[80] http://vasc.ri.cmu.edu/idb/

[81] http://www.eecs.berkeley.edu/Research/Projects/CS/vision/bsds/

[82] http://www.imageclef.org/

[83] http://www.image-net.org/

[84] http://www.imageprocessingplace.com/root_files_V3/image_databases.htm

[85] http://www.cs.cmu.edu/~cil/v-images.html

[86] http://moa.cms.waikato.ac.nz

Index